D0152471

Also from Wiley...

Circuit Solutions powered by JustAsk!

A WEBSITE WITH ANSWERS!

Circuit Solutions invites you to be a part of the solution as it walks you step-by-step through a large number of end-of-chapter problems and interactive applications from the text. This powerful online problem-solving tool takes select end-of-chapter problems and provides you with more than just the answers. Circuit Solutions is your passport to greater understanding of key concepts. Save time in your studies while building valuable analysis skills.

Wherever you see **CS** next to an end-of-chapter problem, you know that the Circuit Solution will provide you with your choice of:

➤ Step-by-step, detailed solutions and aswers

➤ Problem-Solving Videos **PSV**, step-by-step solutions to learning extensions and supplemental end-of-chapter problems

➤ Convenient pop-up windows that highlight relevant concepts, background theory methods, and laws that should be applied when solving each problem

➤ Solution guidelines illustrate the steps you need to take in order to solve the problem, while allowing you to solve the problem on your own

➤ Answers to Selected Problems from the end-of-chapter problems

➤ A complete glossary of key terms and definitions

➤ All relevant concepts, theories, and rules databased and linked to problems and solutions

Use the registration code on the opposite page to find out more about Circuit Solutions.

www.wiley.com/college/irwin
Circuits like you have never seen them!

The Range of Current

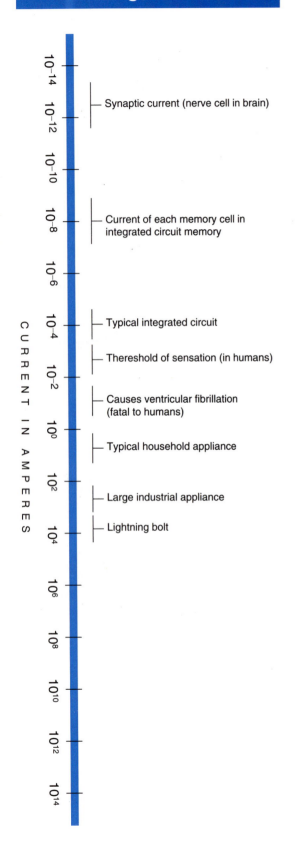

CURRENT IN AMPERES

- 10^{-14}
- 10^{-12} — Synaptic current (nerve cell in brain)
- 10^{-10}
- 10^{-8} — Current of each memory cell in integrated circuit memory
- 10^{-6}
- 10^{-4} — Typical integrated circuit
- 10^{-2} — Thereshold of sensation (in humans)
- 10^{0} — Causes ventricular fibrillation (fatal to humans)
- — Typical household appliance
- 10^{2} — Large industrial appliance
- 10^{4} — Lightning bolt
- 10^{6}
- 10^{8}
- 10^{10}
- 10^{12}
- 10^{14}

The Range of Voltage

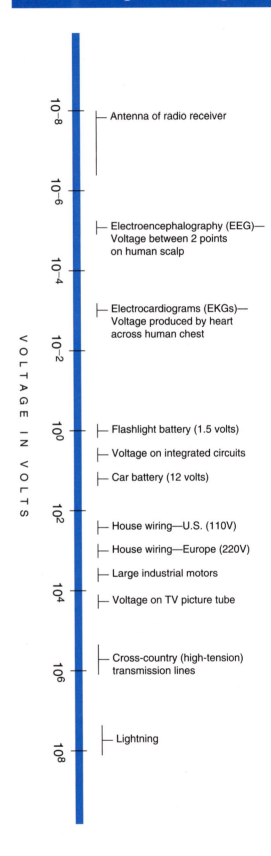

VOLTAGE IN VOLTS

- 10^{-8} — Antenna of radio receiver
- 10^{-6}
- 10^{-4} — Electroencephalography (EEG)— Voltage between 2 points on human scalp
- — Electrocardiograms (EKGs)— Voltage produced by heart across human chest
- 10^{-2}
- 10^{0} — Flashlight battery (1.5 volts)
- — Voltage on integrated circuits
- — Car battery (12 volts)
- 10^{2} — House wiring—U.S. (110V)
- — House wiring—Europe (220V)
- — Large industrial motors
- 10^{4} — Voltage on TV picture tube
- 10^{6} — Cross-country (high-tension) transmission lines
- 10^{8} — Lightning

eGrade Plus

www.wiley.com/college/irwin

Based on the Activities You Do Every Day

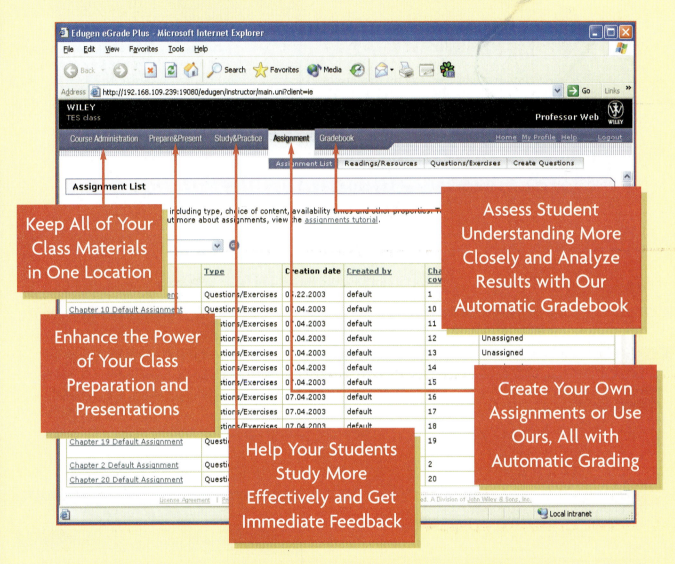

Keep All of Your Class Materials in One Location

Enhance the Power of Your Class Preparation and Presentations

Help Your Students Study More Effectively and Get Immediate Feedback

Assess Student Understanding More Closely and Analyze Results with Our Automatic Gradebook

Create Your Own Assignments or Use Ours, All with Automatic Grading

All the content and tools you need, all in one location, in an easy-to-use browser format.

Choose the resources you need, or rely on the arrangement supplied by us.

Now, many of Wiley's textbooks are available with eGrade Plus, a powerful online tool that provides a completely integrated suite of teaching and learning resources in one easy-to-use website. eGrade Plus integrates Wiley's world-renowned content with media, including a multimedia version of the text, PowerPoint slides, and more. Upon adoption of eGrade Plus, you can begin to customize your course with the resources shown here.

See for yourself!

Go to www.wiley.com/college/egradeplus for an online demonstration of this powerful new software.

Students,
eGrade Plus Allows You to:

Study More Effectively

Get Immediate Feedback When You Practice on Your Own

eGrade Plus problems link directly to relevant sections of the **electronic book content,** so that you can review the text while you study and complete homework online. Additional resources include **problem solving videos, seleced problem solutions,** and **hyperlinks to Circuit Solutions, powered by JustAsk!**

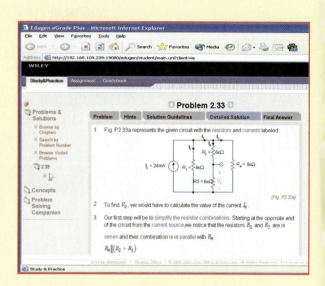

Complete Assignments / Get Context-Sensitive Help

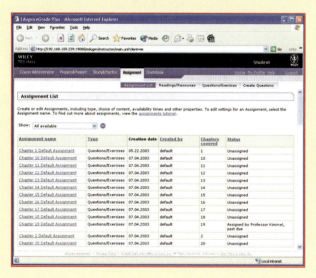

An **Assignment** area keeps all your assigned work in one location, making it easy for you to stay "on task." In addition, many homework problems contain a **link** to the relevant section of the **multimedia book,** providing you with a text explanation to help conquer problem-solving obstacles as they arise.

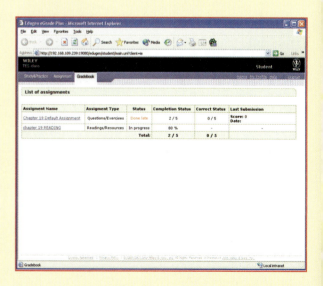

Keep Track of How You're Doing

A **Personal Gradebook** allows you to view your results from past assignments at any time.

Basic Engineering Circuit Analysis

Basic Engineering Circuit Analysis

Eighth Edition

J. David Irwin

Auburn University

R. Mark Nelms

Auburn University

WILEY

John Wiley & Sons, Inc.

EXECUTIVE EDITOR: Bill Zobrist
ASSISTANT EDITOR: Catherine Mergen, Kelly Boyle
EDITORIAL ASSISTANT: Bridget Morrissey
PRODUCT MANAGER: Catherine Shultz
SENIOR PRODUCTION EDITOR: William A. Murray; Production Management Services provided
by Christine Cervoni, Camelot Editorial Services
SENIOR DESIGNER: Karin Kincheloe
TEXT DESIGN: Nancy Field
ILLUSTRATION COORDINATOR: Anna Melhorn; Christine Cervoni, Camelot Editorial Services
SENIOR PHOTO EDITOR: Lisa Gee
PHOTO RESEARCHER: Ramon Rivera Moret
COVER PHOTO: Photodisc Green/Getty Images

This book was set in 10/12 Times Roman by Preparé Inc., and printed and bound by
VonHoffmann Press, Inc. The cover was printed by VonHoffmann Press, Inc.

This book is printed on acid free paper.

Copyright © 2005 John Wiley & Sons, Inc. All rights reserved.

No part of this publication may be reproduced, stored in a retrieval system or transmitted
in any form or by any means, electronic, mechanical, photocopying, recording, scanning
or otherwise, except as permitted under Sections 107 or 108 of the 1976 United States
Copyright Act, without either the prior written permission of the Publisher, or
authorization through payment of the appropriate per-copy fee to the Copyright
Clearance Center, 222 Rosewood Drive, Danvers, MA 01923, (978) 750-8400, fax
(978) 646-8600. Requests to the Publisher for permission should be addressed to the
Permissions Department, John Wiley & Sons, Inc., 111 River Street, Hoboken, NJ 07030,
(201) 748-6011, fax (201) 748-6008.

To order books or for customer service please, call 1(800)-CALL-WILEY (225-5945).

Library of Congress Cataloging-in-Publication Data

Irwin, J. David
    Basic engineering circuit analysis / J. David Irwin.—8th ed.
        p. cm.
    Includes index.
    ISBN 0-471-48728-7 (cloth)
        1. Electric circuit analysis. I. Title.

TK454.I78 2004
621.319'2—dc22

2004042293

Printed in the United States of America

10 9 8 7 6 5 4 3

To my loving family:
Edie
Geri, Bruno, Andrew and Ryan
John, Julie, John David and Abi
Laura

To my parents:
Robert and Elizabeth Nelms

Brief
Contents

Contents

Preface

To The Student

Circuit analysis is not only fundamental to the areas of electrical and computer engineering—the concepts studied here also have tentacles that extend far beyond those boundaries. It remains the starting point for many future engineers who wish to work in this field. The text and all the supplementary materials associated with it are intended to aid you in that quest. We highly recommend you study the Preface closely in order to gauge the full breadth of what resources are available to you as a learner and take every opportunity to work one more problem or study one more hour than you planned. In the end, you'll be thankful you did.

To The Instructor

Although the content of this text remains essentially the same, the Eighth Edition marks an exciting new beginning for this text and its authors. Mark Nelms has joined the team as a co-author and is responsible for some of the key pedagogical innovations of this new edition. Other important changes we have undertaken are the result of a careful study of feedback received from instructors and students and should appeal to a wide variety of instructors. We recognize the abundant changes in the way this material is being taught and learned. Consequently, the authors and Wiley have created a formidable offering of traditional and nontraditional content sources that will meet the needs of a modern circuit analysis course.

New To The Eighth Edition

- A new four-color design is used to provide an enhanced presentation of text and figures. This aids in the pedagogical presentation, particularly in complex illustrations. It is stunning.

- A new Chapter 4 has been created on op-amps. A wide variety of standard and important op-amp configurations are presented.

- The end-of-chapter homework problems have been substantially revised and augmented. The problems in previous editions often lacked sufficient variation and depth. This should no longer be a concern, and every professor will find problems of varying degrees of difficulty. There are over 1200 problems in the Eighth Edition!

- Practical applications have been added for nearly every topic in the text. Since they are items students will naturally encounter on a regular basis, they serve to answer questions such as, "Why is this important?" or "How am I going to use what I learn from this course?"

- Problem-Solving Strategies are a feature that never received the attention it deserved until the Eighth Edition. This feature has been greatly enhanced and augmented. (See below under Pedagogy.)

- Problem-Solving Videos have been created that show students step-by-step how to solve learning extensions and supplemental end-of-chapter problems. Look for the PSV (Problem-Solving Video) icon for the application of this unique feature. **PSV**

- **eGrade Plus**: This innovative e-learning environment offers instructors and students a powerful combination of course management tools and integrated online resources. If your instructor has adopted eGrade Plus for use in the course, students have the opportunity to purchase access to the content of this book in an online environment, at a greatly reduced cost compared to the regular text. For further information, visit the Wiley College website at: *www.wiley.com/college/egradeplus*.

Making The Most of The Pedagogy of This Text

The pedagogy of this text is rich and varied. It includes print and media, and much thought has been put into trying to integrate its use. In order to make the most of the pedagogy that appears in most chapters, please review the following list of the elements you will commonly find.

Learning Goals are provided at the outset of each chapter. This simple list tells the student what they should expect to get out of studying this chapter.

Examples are a mainstay of any circuit analysis text, and this text has always had many examples as its trademark. Significant work has gone into revising these examples in order to provide a more graduated level of presentation with simple, medium, and challenging examples. Besides regular examples, numerous **Design Examples** and **Application Examples** can be found throughout the text.

Hints can often be found in the margins. They augment understanding and serve as a reminder of key issues.

Learning Extensions are a critical learning tool in this text. These exercises are meant to test the cumulative concepts to that point in a given section or sections. Both a problem and the answer are provided. The student who masters these is ready to move forward. These make excellent quiz problems.

Problem-Solving Strategies are step-by-step problem-solving techniques that students find particularly useful. They can solve the age-old question, "Where do I start?" Nearly every chapter has one or more of these strategies, which are a kind of summation of problem solving for concepts presented.

The **Problems** have been greatly revised for the Eighth Edition. In the previous edition there were too many simple problems or problems that were alike. The new edition has hundreds of new problems of varying depth and level. Any instructor will find numerous problems appropriate for any level class. Included with the over 1200 Problems are **FE Exam Problems** for each chapter. These problems closely correlate to problems you would typically find on the FE Exam.

The use of **PSPICE** and **MATLAB** has been integrated throughout the text and has been lauded as the finest of its kind by users. These examples are color coded with an icon for easy reference.

Supplements

The supplements list is large and provides instructors and students with a wealth of traditional and modern resources from which to choose to learn.

The Instructor's or Student's Companion Sites at Wiley—The web pages are an important focal point for students and instructors. You will find numerous files, supplementary documents, and possibly errata at these sites. We recommend you visit early and often. Go to Wiley's book homepage by going to *www.wiley.com/college/irwin* and finding the Eighth Edition. Once you go to the Eighth Edition homepage, you'll see these sites listed somewhere in the navigation bar. Instructors will have to register for access, whereas most files on the student site will likely be freely available.

The Student Study Guide—This supplement contains detailed examples that track the chapter presentation to aid and check the student's understanding of the problem-solving process. Many of these examples involve use of software programs such as EXCEL, MATLAB, and PSPICE.

The Problem-Solving Companion is freely available for download as a PDF file off the Student Companion Site at Wiley. This companion contains over 70 additional problems with extremely detailed worked-out solutions. The problem-solving process is walked through in excruciating detail.

Problem-Solving Videos are an innovation of the Eighth Edition. Throughout the text are icons **PSV** that inform you when a video can be viewed. The videos provide step-by-step solutions to learning extensions and supplemental end-of-chapter problems. Videos for learning extensions will follow directly after a chapter feature called *Problem-Solving Strategy*. Icons with end-of-chapter problems **PSV** indicate that a video solution is available for a similar problem —not the actual end-of-chapter problem. Students who use these videos have found them to be very helpful. The Problem-Solving Videos can be viewed within the Circuit Solutions website. Please go to *http://www.justask4u.com/irwin* and use the Circuit Solutions powered by JustAsk registration code included on the card adhered to the inside front cover of this text.

The Solutions Manual for the Eighth Edition has been completely redone, checked, and double-checked for accuracy. Although it is handwritten, we believe it is the most accurate solutions manual ever created for this text. Qualified instructors can download it off Wiley's Instructor's Companion Site.

PowerPoint Lecture Slides are an especially valuable supplementary aid. Whereas most publishers make only figures available, these slides are true lecture tools that summarize the key learning points for each chapter and can be edited easily in PowerPoint. These are available for download off Wiley's Instructor's Companion Site for qualified adopters.

Answers to Selected Problems from the end-of-chapter problems in the text are freely available for download as a PDF file off the Student Companion Site at Wiley and are located at: *http://www.justask4u.com/irwin*.

Circuit Solutions powered by JustAsk!—This is one of the most innovative uses of the web in circuits today. Free access is included with the purchase of every new text. This website contains detailed solutions to end-of-chapter problems marked with a **CS**. The beauty of these detailed solutions is that as a concept or theorem is used to solve a step in the problem, you are also provided with a hypertext link for that concept or theorem. Included throughout are page references giving the student a deeply integrated study tool that has never before been available in this circuits course. Throughout these solutions are numerous visualizations that enable learning in a mode that a printed text cannot equal. Another major section of Circuit Solutions is an Interactive Applications section where the user is free to play and experiment with various circuit analysis examples in a highly visual mode. This is truly one of the most interesting uses of the web for this course we've ever seen, and it is highly recommended you visit the site. In the front of this text, find the Circuit Solutions registration code card and the instructions on it to find the website and get access. A demonstration can be found by going to: *http://www.justask4u.com/irwin* and by clicking on DEMO.

eGrade Plus Helping Teachers Teach and Students Learn
www.wiley.com/college/irwin

This title is available with eGrade Plus, a powerful online tool that provides instructors and students with an integrated suite of teaching and learning resources in one easy-to-use website. eGrade Plus is organized around the essential activities you and your students perform in class:

FOR INSTRUCTORS

- **Prepare and Present:** Create class presentations using a wealth of Wiley-provided resources—such as an online version of the textbook, PowerPoint slides, and interactive simulations—making your preparation time more efficient. You may easily adapt, customize, and add to this content to meet the needs of your course.

- **Create Assignments:** Automate the assigning and grading of homework or quizzes by using Wiley-provided question banks or by writing your own. Student results will be automatically graded and recorded in your gradebook. eGrade Plus can link homework problems to the relevant section of the online text, providing students with context-sensitive help.

- **Track Student Progress:** Keep track of your students' progress via an instructor's gradebook, which allows you to analyze individual and overall class results to determine their progress and level of understanding.

- **Administer Your Course:** eGrade Plus can easily be integrated with another course management system, gradebook, or other resources you are using in your class, providing you with the flexibility to build your course, your way.

FOR STUDENTS

Wiley's eGrade Plus provides immediate feedback on student assignments and a wealth of support materials. This powerful study tool will help your students develop their conceptual understanding of the class material and increase their ability to solve problems.

- **A "Study and Practice"** area links directly to text content, allowing students to review the text while they study and complete homework assignments. Additional resources can include interactive simulations, study guide and solutions manual material, and other problem-solving resources.

- **An "Assignment"** area keeps all the work you want your students to complete in one location, making it easy for them to stay "on task." Students will have access to a variety of interactive problem-solving tools, as well as other resources for building their confidence and understanding. In addition, many homework problems contain a link to the relevant section of the multimedia book, providing students with context-sensitive help that allows them to conquer problem-solving obstacles as they arise.

- **A Personal Gradebook** for each student will allow students to view their results from past assignments at any time.

Please view our online demo at *www.wiley.com/college/egradeplus*. Here you will find additional information about the features and benefits of eGrade Plus, how to request a "test drive" of eGrade Plus for this title, and how to adopt it for class use.

Acknowledgments

Over the more than 20-year period that this text has been employed, it is estimated that more than one thousand instructors have used my book to teach circuit analysis to hundreds of thousands of students. I am most grateful for the confidence that has been demonstrated in the educational soundness of the text. In addition, I have received numerous evaluations and suggestions from professors and their students over the years, and their feedback has helped me continuously improve the presentation. For this Eighth Edition, I owe Bill Dillard at Auburn University a special debt of gratitude for the numerous contributions that he made to the development of new material in the text, as well as some of the supporting material. I am also most appreciative of the suggestions for improvement that were made by Professor John Choma of the University of Southern California and Professor Mark Nelms of Auburn University, who now joins this edition as a co-author.

We were fortunate to have an outstanding group of reviewers for the book, notably:

David Conner, University of Alabama at Birmingham
Keith Holbert, Arizona State University
Kevin Donahue, University of Kentucky
Jim Rowland, University of Kansas
Armando Rodriguez, Arizona State University
Carl Wells, Washington State University
Walter Green, University of Tennessee
Aileen Honka, The MOSIS Service-USC Inf. Sciences Institute
Mark Rabalais, Louisiana State University
Yasser Hegazy, University of Waterloo
Gene Stuffle, Idaho State University
Thomas Thomas, University of South Alabama

The preparation of this book and the materials that support it have been handled with both enthusiasm and great care. The combined wisdom and leadership of Bill Zobrist, our executive editor, has resulted in a tremendous team effort that has addressed every aspect of the presentation. This team included the following individuals:

Marketing Manager, Jenny Powers
Senior Production Editor, William Murray
Senior Designer, Karin Kincheloe
Product Manager, Catherine F. Shultz
Illustration Coordinators, Gene Aiello and Anna Melhorn
Assistant Editor, Kelly Boyle
Senior Marketing Research Analyst, Carl Kulo
Editorial Assistant, Bridget Morrisey

Each member of this team played a vital role in preparing the package that is the Eighth Edition of *Basic Engineering Circuit Analysis*, and we are most appreciative of their many contributions.

As in the past, we are most pleased to acknowledge the support that has been provided by numerous individuals to earlier editions of this book. The Auburn personnel who have helped are:

Paulo R. Marino	M. S. Morse
Thomas A. Baginski	Sung-Won Park
Charles A. Gross	C. L. Rogers
James L. Lowry	Travis Blalock
Zhi Ding	Kevin Driscoll
David C. Hill	Keith Jones
Henry Cobb	George Lindsey
Les Simonton	David Mack
Betty Kelley	John Parr
E. R. Graf	Monty Rickles
L. L. Grigsby	James Trivltayakhum
M. A. Honnell	Susan Williamson
R. C. Jaeger	Jacinda Woodward
Jo Ann Loden	Tom Shumpert

Many of our colleagues throughout the United States, some of whom are now retired, have also made numerous suggestions for improving the book:

M. E. Shafeei, Penn State University at Harrisbury

Ian McCausland, University of Toronto

Leonard J. Tung, Florida A&M University/Florida State University

Arthur C. Moeller, Marquette University

Darrell Vines, Texas Tech University

M. Paul Murray, Mississippi State University

David Anderson, University of Iowa

Burks Oakley II, University of Illinois at Champaign-Urbana

Richard Baker, UCLA

John O'Malley, University of Florida

James L. Dodd, Mississippi State University

William R. Parkhurst, Wichita State University

Earl D. Eyman, University of Iowa

George Prans, Manhattan College

Arvin Grabel, Northeastern University

James Rowland, University of Kansas

Paul Gray, University of Wisconsin-Platteville

Robert N. Sackett, Normandale Community College

Mohammad Habli, University of New Orleans

Richard Sanford, Clarkson University

John Hadjilogiou, Florida Institute of Technology

Ronald Schulz, Cleveland State University

Ralph Kinney, LSU

Karen M. St. Germaine, University of Nebraska

K. S. P. Kumar, University of Minnesota

Janusz Strazyk, Ohio University

James Luster, Snow College

Saad Tabet, Florida State University

Robert Krueger, University of Wisconsin

Seth Wolpert, University of Maine

Ashok Goel, Michigan Technological University

Clifford Pollock, Cornell University

Peyton Peebles, University of Florida

Marty Kaliski, Cal Poly, San Luis Obispo

Darryl Morrell, Arizona State University

Peddapullaiah Sannuti, Rutgers University

Paul Greiling, UCLA

Jorge Aravena, Louisiana State University

Scott F. Smith, Boise State University

Val Tareski, North Dakota State University

John Durkin, University of Akron

Jung Young Lee, UC Berkely student

Finally, Dave Irwin wishes to express his deep appreciation to his wife, Edie, who has been most supportive of our efforts in this book. Mark Nelms would like to thank his parents, Robert and Elizabeth, for their support and encouragement.

J. David Irwin and R. Mark Nelms

Basic Concepts

Today we live in a predominantly electrical world. Although this statement may sound strange at first, a moment's reflection will indicate its inherent truth. The two primary areas of electrotechnology that permeate essentially every aspect of our lives are power and information. Without them, life as we know it would undergo stupendous changes. We have learned to generate, convert, transmit, and utilize these technologies for the enhancement of the whole human race.

Electrotechnology is a driving force in the changes that are occurring in every engineering discipline. For example, surveying is now done with lasers and electronic range finders, and automobiles employ electronic dashboards and electronic ignition systems. Industrial processes that range from chemical refineries and metal foundries to wastewater treatment plants use (1) electronic sensors to obtain information about the process, (2) instrumentation systems to gather the information, and (3) computer control systems to process the information and generate electronic commands to actuators, which correct and control the process.

Fundamental to electrotechnology is the area of circuit analysis. A thorough knowledge of this subject provides an understanding of such things as cause and effect, amplification and attenuation, feedback and control, and stability and oscillation. Of critical importance is the fact that the same principles applied to engineering systems can also be applied to economic and social systems. Thus, the ramifications of circuit analysis are immense, and a solid understanding of this subject is well worth the effort expended to obtain it.

In this chapter we will introduce some of the basic quantities that will be used throughout the text. Specifically, we will define electric current, voltage, power and energy, as well as the difference between direct current and alternating current. In addition, we will classify electric elements as either passive or active, the latter of which can be further subdivided into both independent and dependent. This basic introduction will lay the groundwork for our further study of a wide variety of electric circuits. ●

Access Problem-Solving Videos **PSV**
and Circuit Solutions **CS** ***at:***
http://www.justask4u.com/irwin
using the registration code on the inside cover and see a website with answers and more!

1.1   System of Units

The system of units we employ is the international system of units, the Système International des Unités, which is normally referred to as the SI standard system. This system, which is composed of the basic units meter (m), kilogram (kg), second (s), ampere (A), degree kelvin (K), and candela (cd), is defined in all modern physics texts and therefore will not be defined here. However, we will discuss the units in some detail as we encounter them in our subsequent analyses.

The standard prefixes that are employed in SI are shown in Fig. 1.1. Note the decimal relationship between these prefixes. These standard prefixes are employed throughout our study of electric circuits.

Circuit technology has changed drastically over the years. For example, in the early 1960s the space on a circuit board occupied by the base of a single vacuum tube was about the size of a quarter (25-cent coin). Today that same space could be occupied by an Intel Pentium integrated circuit chip containing 50 million transistors. These chips are the engine for a host of electronic equipment.

Figure 1.1
Standard SI prefixes.

1.2   Basic Quantities

Before we begin our analysis of electric circuits, we must define terms that we will employ. However, in this chapter and throughout the book our definitions and explanations will be as simple as possible to foster an understanding of the use of the material. No attempt will be made to give complete definitions of many of the quantities because such definitions are not only unnecessary at this level but are often confusing. Although most of us have an intuitive concept of what is meant by a circuit, we will simply refer to an *electric circuit* as an interconnection of electrical components, each of which we will describe with a mathematical model.

The most elementary quantity in an analysis of electric circuits is the electric *charge*. Our interest in electric charge is centered around its motion, since charge in motion results in an energy transfer. Of particular interest to us are those situations in which the motion is confined to a definite closed path.

An electric circuit is essentially a pipeline that facilitates the transfer of charge from one point to another. The time rate of change of charge constitutes an electric *current*. Mathematically, the relationship is expressed as

$$i(t) = \frac{dq(t)}{dt} \qquad \text{or} \qquad q(t) = \int_{-\infty}^{t} i(x)\, dx \qquad\qquad \textbf{1.1}$$

where i and q represent current and charge, respectively (lowercase letters represent time dependency, and capital letters are reserved for constant quantities). The basic unit of current is the ampere (A), and 1 ampere is 1 coulomb per second.

Although we know that current flow in metallic conductors results from electron motion, the conventional current flow, which is universally adopted, represents the movement of positive charges. It is important that the reader think of current flow as the movement of positive charge regardless of the physical phenomena that take place. The symbolism that will be used to represent current flow is shown in Fig. 1.2. $I_1 = 2$ A in Fig. 1.2a indicates that at any point in the wire shown, 2 C of charge pass from left to right each second. $I_2 = -3$ A in Fig. 1.2b indicates that at any point in the wire shown, 3 C of charge pass from right to left each second. Therefore, it is important to specify not only the magnitude of the variable representing the current, but also its direction.

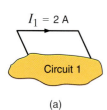

(a)

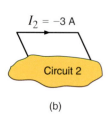

(b)

Figure 1.2
Conventional current flow:
(a) positive current flow;
(b) negative current flow.

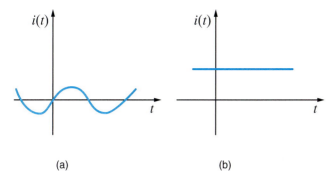

Figure 1.3
Two common types of current: (a) alternating current (ac); (b) direct current (dc).

(a)   (b)

There are two types of current that we encounter often in our daily lives, alternating current (ac) and direct current (dc), which are shown as a function of time in Fig. 1.3. *Alternating current* is the common current found in every household and is used to run the refrigerator, stove, washing machine, and so on. Batteries, which are used in automobiles or flashlights, are one source of *direct current*. In addition to these two types of currents, which have a wide variety of uses, we can generate many other types of currents. We will examine some of these other types later in the book. In the meantime, it is interesting to note that the magnitude of currents in elements familiar to us ranges from soup to nuts, as shown in Fig. 1.4.

We have indicated that charges in motion yield an energy transfer. Now we define the *voltage* (also called the *electromotive force* or *potential*) between two points in a circuit as the difference in energy level of a unit charge located at each of the two points. Voltage is very similar to a gravitational force. Think about a bowling ball being dropped from a ladder into a tank of water. As soon as the ball is released, the force of gravity pulls it toward the bottom of the tank. The potential energy of the bowling ball decreases as it approaches the bottom. The gravitational force is pushing the bowling ball through the water. Think of the bowling ball as a charge and the voltage as the force pushing the charge through a circuit. Charges in motion represent a current, so the motion of the bowling ball could be thought of as a current. The water in the tank will resist the motion of the bowling ball. The motion of charges in an electric circuit will be impeded or resisted as well. We will introduce the concept of resistance in Chapter 2 to describe this effect.

Work or energy, $w(t)$ or W, is measured in joules (J); 1 joule is 1 newton meter (N·m). Hence, voltage [$v(t)$ or V] is measured in volts (V) and 1 volt is 1 joule per coulomb; that is, 1 volt = 1 joule per coulomb = 1 newton meter per coulomb. If a unit positive charge is moved between two points, the energy required to move it is the difference in energy level between the two points and is the defined voltage. It is extremely important that the variables used to represent voltage between two points be defined in such a way that the solution will let us interpret which point is at the higher potential with respect to the other.

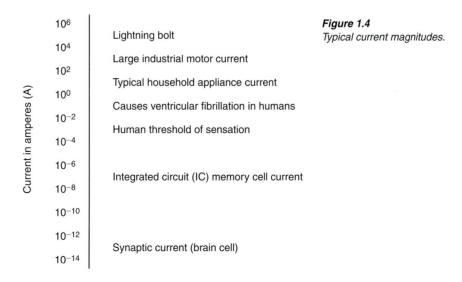

Figure 1.4
Typical current magnitudes.

Figure 1.5
Voltage representations.

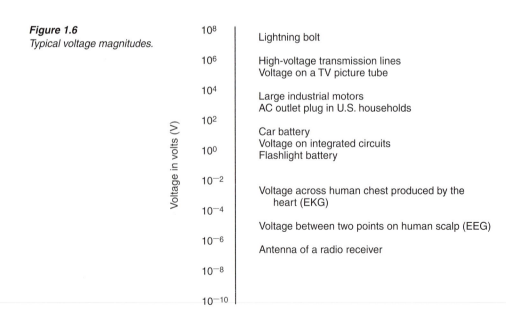

In Fig. 1.5a the variable that represents the voltage between points A and B has been defined as V_1, and it is assumed that point A is at a higher potential than point B, as indicated by the $+$ and $-$ signs associated with the variable and defined in the figure. The $+$ and $-$ signs define a reference direction for V_1. If $V_1 = 2$ V, then the difference in potential of points A and B is 2 V and point A is at the higher potential. If a unit positive charge is moved from point A through the circuit to point B, it will give up energy to the circuit and have 2 J less energy when it reaches point B. If a unit positive charge is moved from point B to point A, extra energy must be added to the charge by the circuit, and hence the charge will end up with 2 J more energy at point A than it started with at point B.

For the circuit in Fig. 1.5b, $V_2 = -5$ V means that the potential between points A and B is 5 V and point B is at the higher potential. The voltage in Fig. 1.5b can be expressed as shown in Fig. 1.5c. In this equivalent case, the difference in potential between points A and B is $V_2 = 5$ V, and point B is at the higher potential.

Note that it is important to define a variable with a reference direction so that the answer can be interpreted to give the physical condition in the circuit. We will find that it is not possible in many cases to define the variable so that the answer is positive, and we will also find that it is not necessary to do so.

As demonstrated in Figs. 1.5b and c, a negative number for a given variable, for example, V_2 in Fig. 1.5b, gives exactly the same information as a positive number, that is, V_2 in Fig. 1.5c, except that it has an opposite reference direction. Hence, when we define either current or voltage, it is absolutely necessary that we specify both magnitude and direction. Therefore, it is incomplete to say that the voltage between two points is 10 V or the current in a line is 2 A, since only the magnitude and not the direction for the variables has been defined.

The range of magnitudes for voltage, equivalent to that for currents in Fig. 1.4, is shown in Fig. 1.6. Once again, note that this range spans many orders of magnitude.

Figure 1.6
Typical voltage magnitudes.

Voltage in volts (V)

10^8	Lightning bolt
10^6	High-voltage transmission lines Voltage on a TV picture tube
10^4	Large industrial motors AC outlet plug in U.S. households
10^2	Car battery
10^0	Voltage on integrated circuits Flashlight battery
10^{-2}	Voltage across human chest produced by the heart (EKG)
10^{-4}	Voltage between two points on human scalp (EEG)
10^{-6}	Antenna of a radio receiver
10^{-8}	
10^{-10}	

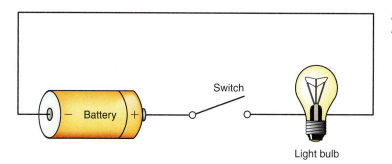

Figure 1.7
Flashlight circuit.

At this point we have presented the conventions that we employ in our discussions of current and voltage. *Energy* is yet another important term of basic significance. Let's investigate the voltage–current relationships for energy transfer using the flashlight shown in Fig. 1.7. The basic elements of a flashlight are a battery, a switch, a light bulb, and connecting wires. Assuming a good battery, we all know that the light bulb will glow when the switch is closed. A current now flows in this closed circuit as charges flow out of the positive terminal of the battery through the switch and light bulb and back into the negative terminal of the battery. The current heats up the filament in the bulb, causing it to glow and emit light. The light bulb converts electrical energy to thermal energy; as a result, charges passing through the bulb lose energy. These charges acquire energy as they pass through the battery as chemical energy is converted to electrical energy. An energy conversion process is occurring in the flashlight as the chemical energy in the battery is converted to electrical energy, which is then converted to thermal energy in the light bulb.

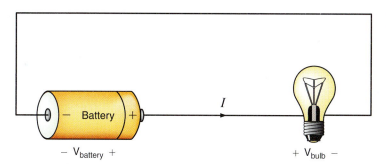

Figure 1.8
Flashlight circuit with voltages and current.

Let's redraw the flashlight as shown in Fig. 1.8. There is a current *I* flowing in this diagram. Since we know that the light bulb uses energy, the charges coming out of the bulb have less energy than those entering the light bulb. In other words, the charges expend energy as they move through the bulb. This is indicated by the voltage shown across the bulb. The charges gain energy as they pass through the battery, which is indicated by the voltage across the battery. Note the voltage–current relationships for the battery and bulb. We know that the bulb is absorbing energy; the current is entering the positive terminal of the voltage. For the battery, the current is leaving the positive terminal, which indicates that energy is being supplied.

This is further illustrated in Fig. 1.9 where a circuit element has been extracted from a larger circuit for examination. In Fig. 1.9a, energy is being supplied *to* the element by whatever is attached to the terminals. Note that 2 A, that is, 2 C of charge are moving from point *A* to point *B* through the element each second. Each coulomb loses 3 J of energy as it passes through the element from point *A* to point *B*. Therefore, the element is absorbing 6 J of energy per second. Note that when the element is *absorbing* energy, a positive current enters the positive terminal. In Fig. 1.9b energy is being supplied *by* the element to whatever is connected to terminals *A-B*. In this case, note that when the element is *supplying* energy, a positive current enters the negative terminal and leaves via the positive terminal. In this convention, a negative current in one direction is equivalent to a positive current in the opposite direction, and vice versa. Similarly, a negative voltage in one direction is equivalent to a positive voltage in the opposite direction.

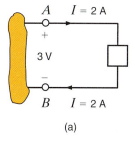

(a)

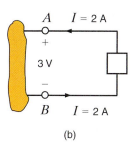

(b)

Figure 1.9
Voltage–current relationships for (a) energy absorbed and (b) energy supplied.

Example 1.1

Suppose that your car will not start. To determine whether the battery is faulty, you turn on the light switch and find that the lights are very dim, indicating a weak battery. You borrow a friend's car and a set of jumper cables. However, how do you connect his car's battery to yours? What do you want his battery to do?

SOLUTION   Essentially, his car's battery must supply energy to yours, and therefore it should be connected in the manner shown in Fig. 1.10. Note that the positive current leaves the positive terminal of the good battery (supplying energy) and enters the positive terminal of the weak battery (absorbing energy). Note that the same connections are used when charging a battery.

Figure 1.10
Diagram for Example 1.1.

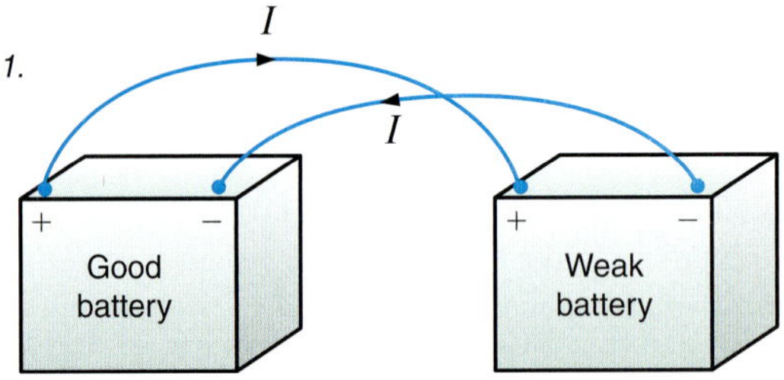

In practical applications there are often considerations other than simply the electrical relations (e.g., safety). Such is the case with jump-starting an automobile. Automobile batteries produce explosive gases that can be ignited accidentally, causing severe physical injury. Be safe—follow the procedure described in your auto owner's manual.

We have defined voltage in joules per coulomb as the energy required to move a positive charge of 1 C through an element. If we assume that we are dealing with a differential amount of charge and energy, then

$$v = \frac{dw}{dq} \qquad\qquad \textbf{1.2}$$

Multiplying this quantity by the current in the element yields

$$vi = \frac{dw}{dq}\left(\frac{dq}{dt}\right) = \frac{dw}{dt} = p \qquad\qquad \textbf{1.3}$$

which is the time rate of change of energy or power measured in joules per second, or watts (W). Since, in general, both v and i are functions of time, p is also a time-varying quantity. Therefore, the change in energy from time t_1 to time t_2 can be found by integrating Eq. (1.3); that is,

$$\Delta w = \int_{t_1}^{t_2} p \, dt = \int_{t_1}^{t_2} vi \, dt \qquad\qquad \textbf{1.4}$$

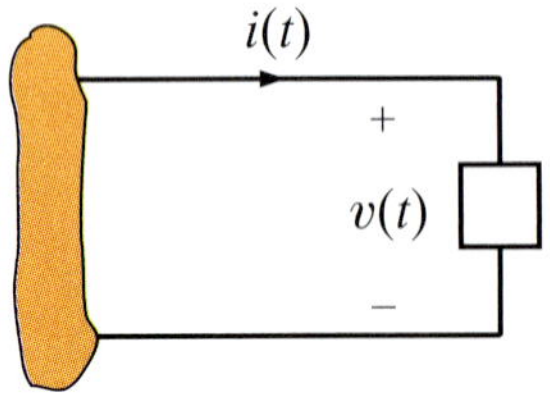

Figure 1.11
Sign convention for power.

HINT

The passive sign convention is used to determine whether power is being absorbed or supplied.

At this point, let us summarize our sign convention for power. To determine the sign of any of the quantities involved, the variables for the current and voltage should be arranged as shown in Fig. 1.11. The variable for the voltage $v(t)$ is defined as the voltage across the element with the positive reference at the same terminal that the current variable $i(t)$ is entering. This convention is called the *passive sign convention* and will be so noted in the remainder of this book. The product of v and i, with their attendant signs, will determine the magnitude and sign of the power. If the sign of the power is positive, power is being absorbed by the element; if the sign is negative, power is being supplied by the element.

Example 1.2

Given the two diagrams shown in Fig. 1.12, determine whether the element is absorbing or supplying power and how much.

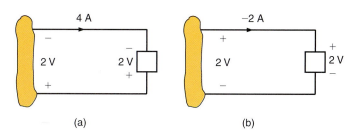

(a)   (b)

Figure 1.12
Elements for Example 1.2.

SOLUTION   In Fig. 1.12a the power is $P = (2\text{ V})(-4\text{ A}) = -8$ W. Therefore, the element is supplying power. In Fig. 1.12b, the power is $P = (2\text{ V})(-2\text{ A}) = -4$ W. Therefore, the element is absorbing power.

LEARNING EXTENSION

E1.1 Determine the amount of power absorbed or supplied by the elements in Fig. E1.1.

ANSWER:
(a) $P = -48$ W;
(b) $P = 8$ W.

Figure E1.1

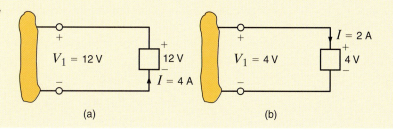

(a)   (b)

Example 1.3

We wish to determine the unknown voltage or current in Fig. 1.13.

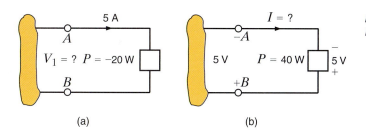

(a)   (b)

Figure 1.13
Elements for Example 1.3.

SOLUTION   In Fig. 1.13a, a power of -20 W indicates that the element is delivering power. Therefore, the current enters the negative terminal (terminal A), and from Eq. (1.3) the voltage is 4 V. Thus, B is the positive terminal, A is the negative terminal, and the voltage between them is 4 V.

In Fig 1.13b, a power of $+40$ W indicates that the element is absorbing power and, therefore, the current should enter the positive terminal B. The current thus has a value of -8 A, as shown in the figure.

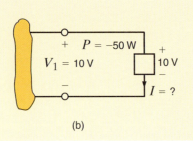

LEARNING EXTENSION

E1.2 Determine the unknown variables in Fig. E1.2.

Figure E1.2

(a)

(b)

ANSWER:
(a) $V_1 = -20$ V;
(b) $I = -5$ A.

Finally, it is important to note that these electrical networks satisfy the principle of conservation of energy. For our present purposes this means that the power supplied in a network is exactly equal to the power absorbed.

1.3 Circuit Elements

Thus far we have defined voltage, current, and power. In the remainder of this chapter we will define both independent and dependent current and voltage sources. Although we will assume ideal elements, we will try to indicate the shortcomings of these assumptions as we proceed with the discussion.

In general, the elements we will define are terminal devices that are completely characterized by the current through the element and/or the voltage across it. These elements, which we will employ in constructing electric circuits, will be broadly classified as being either active or passive. The distinction between these two classifications depends essentially on one thing—whether they supply or absorb energy. As the words themselves imply, an *active* element is capable of generating energy and a *passive* element cannot generate energy.

However, later we will show that some passive elements are capable of storing energy. Typical active elements are batteries and generators. The three common passive elements are resistors, capacitors, and inductors.

In Chapter 2 we will launch an examination of passive elements by discussing the resistor in detail. However, before proceeding with that element, we first present some very important active elements.

1. Independent voltage source

2. Independent current source

3. Two dependent voltage sources

4. Two dependent current sources

INDEPENDENT SOURCES An *independent voltage source* is a two-terminal element that maintains a specified voltage between its terminals *regardless of the current through it* as shown by the *v-i* plot in Fig. 1.14a. The general symbol for an independent source, a circle, is also shown in Fig. 1.14a. As the figure indicates, terminal *A* is $v(t)$ volts positive with respect to terminal *B*.

In contrast to the independent voltage source, the *independent current source* is a two-terminal element that maintains a specified current *regardless of the voltage across its terminals*, as illustrated by the *v-i* plot in Fig. 1.14b. The general symbol for an independent current source is also shown in Fig. 1.14b, where $i(t)$ is the specified current and the arrow indicates the positive direction of current flow.

In their normal mode of operation, independent sources supply power to the remainder of the circuit. However, they may also be connected into a circuit in such a way that they absorb power. A simple example of this latter case is a battery-charging circuit such as that shown in Example 1.1.

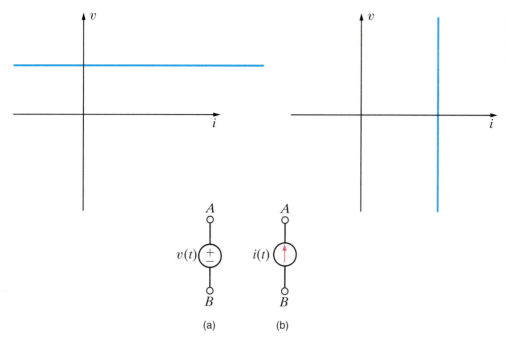

Figure 1.14
Symbols for (a) independent
voltage source, (b) independent
current source.

It is important that we pause here to interject a comment concerning a shortcoming of the models. In general, mathematical models approximate actual physical systems only under a certain range of conditions. Rarely does a model accurately represent a physical system under every set of conditions. To illustrate this point, consider the model for the voltage source in Fig. 1.14a. We assume that the voltage source delivers v volts regardless of what is connected to its terminals. Theoretically, we could adjust the external circuit so that an infinite amount of current would flow, and therefore the voltage source would deliver an infinite amount of power. This is, of course, physically impossible. A similar argument could be made for the independent current source. Hence, the reader is cautioned to keep in mind that models have limitations and thus are valid representations of physical systems only under certain conditions.

For example, can the independent voltage source be utilized to model the battery in an automobile under all operating conditions? With the headlights on, turn on the radio? Do the headlights dim with the radio on? They probably won't if the sound system in your automobile was installed at the factory. If you try to crank your car with the headlights on, you will notice that the lights dim. The starter in your car draws considerable current, thus causing the voltage at the battery terminals to drop and dimming the headlights. The independent voltage source is a good model for the battery with the radio turned on; however, an improved model is needed for your battery to predict its performance under cranking conditions.

Example 1.4

Determine the power absorbed or supplied by the elements in the network in Fig. 1.15.

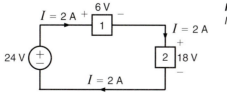

Figure 1.15
Network for Example 1.4.

SOLUTION   The current flow is out of the positive terminal of the 24-V source, and therefore this element is supplying $(2)(24) = 48$ W of power. The current is into the positive terminals of elements 1 and 2, and therefore elements 1 and 2 are absorbing $(2)(6) = 12$ W and $(2)(18) = 36$ W, respectively. Note that the power supplied is equal to the power absorbed.

HINT

Elements that are connected in series have the same current.

E1.3 Find the power that is absorbed or supplied by the elements in Fig. E1.3.

Figure E1.3

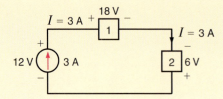

ANSWER: Current source supplies 36 W, element 1 absorbs 54 W, and element 2 supplies 18 W.

DEPENDENT SOURCES   In contrast to the independent sources, which produce a particular voltage or current completely unaffected by what is happening in the remainder of the circuit, dependent sources generate a voltage or current that is determined by a voltage or current at a specified location in the circuit. These sources are very important because they are an integral part of the mathematical models used to describe the behavior of many electronic circuit elements.

For example, metal-oxide-semiconductor field-effect transistors (MOSFETs) and bipolar transistors, both of which are commonly found in a host of electronic equipment, are modeled with dependent sources, and therefore the analysis of electronic circuits involves the use of these controlled elements.

In contrast to the circle used to represent independent sources, a diamond is used to represent a dependent or controlled source. Figure 1.16 illustrates the four types of dependent sources. The input terminals on the left represent the voltage or current that controls the dependent source, and the output terminals on the right represent the output current or voltage of the controlled source. Note that in Figs. 1.16a and d the quantities μ and β are dimensionless constants because we are transforming voltage to voltage and current to current. This is not the case in Figs. 1.16b and c; hence, when we employ these elements a short time later, we must describe the units of the factors r and g.

Figure 1.16
Four different types of dependent sources.

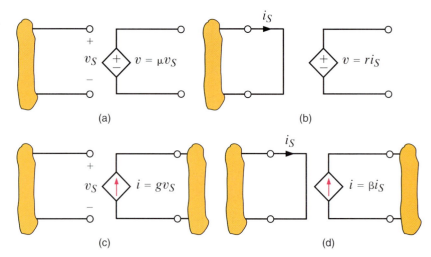

(a)

(b)

(c)

(d)

Example 1.5

Given the two networks shown in Fig. 1.17, we wish to determine the outputs.

SOLUTION   In Fig. 1.17a the output voltage is $V_o = \mu V_S$ or $V_o = 20\,V_S = (20)(2\text{ V}) = 40\text{ V}$. Note that the output voltage has been amplified from 2 V at the input terminals to 40 V at the output terminals; that is, the circuit is a voltage amplifier with an amplification factor of 20.

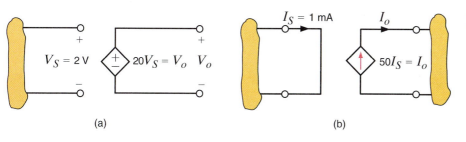

Figure 1.17
Circuits for Example 1.5.

In Fig. 1.17b, the output current is $I_o = \beta I_S = (50)(1\text{ mA}) = 50\text{ mA}$; that is, the circuit has a current gain of 50, meaning that the output current is 50 times greater than the input current.

LEARNING EXTENSION

E1.4 Determine the power supplied by the dependent sources in Fig. E1.4.

ANSWER:
(a) Power supplied = 80 W;
(b) power supplied = 160 W.

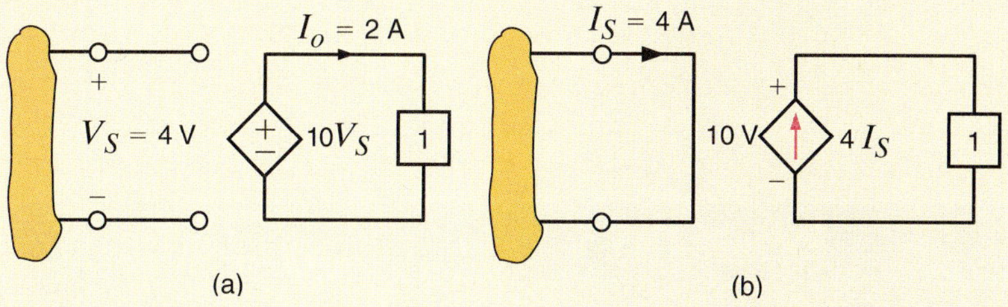

Figure E1.4

Example 1.6

Let us find the current I_o in the network in Fig. 1.18.

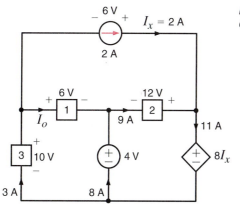

Figure 1.18
Circuit used in Example 1.6.

SOLUTION   First, we must determine the power absorbed or supplied by each element in the network. Using the sign convention for power, we find

$$P_{2\,A} = (6)(-2) = -12\text{ W}$$
$$P_1 = (6)(I_o) = 6I_o\text{ W}$$
$$P_2 = (12)(-9) = -108\text{ W}$$
$$P_3 = (10)(-3) = -30\text{ W}$$
$$P_{4\,V} = (4)(-8) = -32\text{ W}$$
$$P_{DS} = (8I_x)(11) = (16)(11) = 176\text{ W}$$

Since energy must be conserved,

$$-12 + 6I_o - 108 - 30 - 32 + 176 = 0$$

or

$$6I_o + 176 = 12 + 108 + 30 + 32$$

Hence,

$$I_o = 1A$$

LEARNING EXTENSION

E1.5 Find the power that is absorbed or supplied by the circuit elements in the network in Fig. E1.5.

Figure E1.5

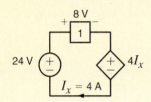

ANSWER:
$P_{24\,V} = 96$ W supplied;
$P_1 = 32$ W absorbed;
$P_{4I_x} = 64$ W absorbed.

SUMMARY

- **The standard prefixes employed**

$p = 10^{-12}$	$k = 10^3$
$n = 10^{-9}$	$M = 10^6$
$\mu = 10^{-6}$	$G = 10^9$
$m = 10^{-3}$	$T = 10^{12}$

- **The relationships between current and charge**

$$i(t) = \frac{dq(t)}{dt} \quad \text{or} \quad q(t) = \int_{-\infty}^{t} i(x)\,dx$$

- **The relationships among power, energy, current, and voltage**

$$p = \frac{dw}{dt} = vi$$

$$\Delta w = \int_{t_1}^{t_2} p\,dt = \int_{t_1}^{t_2} vi\,dt$$

- **The passive sign convention** The passive sign convention states that if the voltage and current associated with an element are as shown in Fig. 1.11, the product of v and i, with their attendant signs, determines the magnitude and sign of the power. If the sign is positive, power is being absorbed by the element, and if the sign is negative, the element is supplying power.

- **Independent and dependent sources** An ideal independent voltage (current) source is a two-terminal element that maintains a specified voltage (current) between its terminals regardless of the current (voltage) through (across) the element. Dependent or controlled sources generate a voltage or current that is determined by a voltage or current at a specified location in the circuit.

- **Conservation of energy** The electric circuits under investigation satisfy the conservation of energy.

PROBLEMS

PSV **CS** both available on the web at: http://www.justask4u.com/irwin

SECTION 1.2

1.1 If 60 C of charge pass through an electric conductor in 30 seconds, determine the current in the conductor. **CS**

1.2 In an electric conductor, a charge of 300 C passes any point in a 5-s interval. Determine the current in the conductor.

1.3 The current in a conductor is 1.5 A. How many coulombs of charge pass any point in a time interval of 1.5 min?

1.4 Determine the number of coulombs of charge produced by a 12-A battery charger in an hour.

1.5 A lightning bolt carrying 20,000 A lasts for 70 μs. If the lightning strikes a tractor, determine the charge deposited on the tractor if the tires are assumed to be perfect insulators.

1.6 If a 12-V battery supplies 10 A, find the amount of energy delivered in 1 hour.

1.7 Determine the energy required to move 240 C through 6 V. **cs**

1.8 Five coulombs of charge pass through the element in Fig. P1.8 from point A to point B. If the energy absorbed by the element is 120 J, determine the voltage across the element.

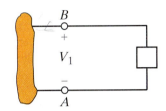

Figure P1.8

1.9 The charge entering an element is shown in Fig. P1.9. Find the current in the element in the time interval $0 \leq t \leq 0.5$ s. [*Hint:* The equation for $q(t)$ is $q(t) = 1 + (1/0.5)t, t \geq 0$.]

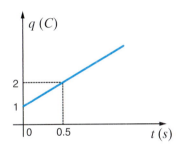

Figure P1.9

1.10 Determine the amount of power absorbed or supplied by the element in Fig. P1.10 if

(a) $V_1 = 9$ V and $I = 2$ A.

(b) $V_1 = 9$ V and $I = -3$ A.

(c) $V_1 = -12$ V and $I = 2$ A.

(d) $V_1 = -12$ V and $I = -3$ A.

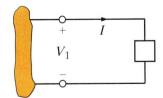

Figure P1.10

1.11 Determine the magnitude and direction of the voltage across the elements in Fig. P1.11.

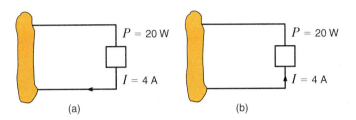

Figure P1.11

1.12 Determine the missing quantity in the circuits in Fig. P1.12.

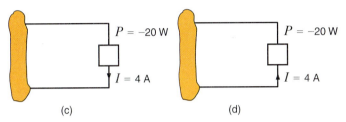

Figure P1.12

1.13 Determine the missing quantity in the circuits in Fig. P1.13.

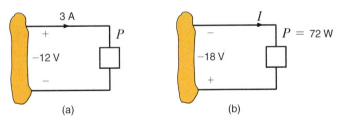

Figure P1.13

1.14 Determine the missing quantity in the circuits in Fig. P1.14. **cs**

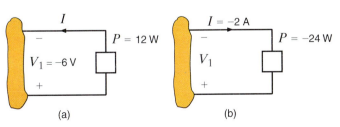

Figure P1.14

1.15 Two elements are connected in series, as shown in Fig. P1.15. Element 1 supplies 24 W of power. Is element 2 absorbing or supplying power, and how much? **CS**

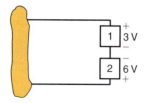

Figure P1.15

1.16 Determine the power supplied to the elements in Fig. P1.16.

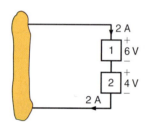

Figure P1.16

1.17 Determine the power supplied to the elements in Fig. P1.17.

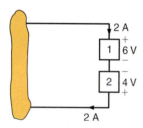

Figure P1.17

1.18 Determine the power supplied to the elements in Fig. P1.18.

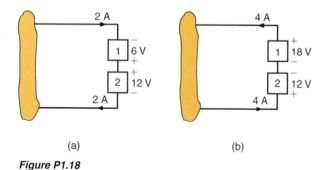

(a)   (b)

Figure P1.18

1.19 (a) In Fig. P1.19(a), $P_1 = 36$ W. Is element 2 absorbing or supplying power, and how much?

(b) In Fig. P1.19(b), $P_2 = -48$ W. Is element 1 absorbing or supplying power, and how much?

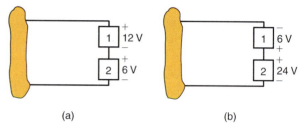

(a)   (b)

Figure P1.19

1.20 Two elements are connected in series, as shown in Fig. P1.20. Element 1 supplies 24 W of power. Is element 2 absorbing or supplying power, and how much?

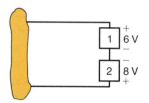

Figure P1.20

1.21 Two elements are connected in series, as shown in Fig. P1.21. Element 1 supplies 24 W of power. Is element 2 absorbing or supplying power, and how much? **CS**

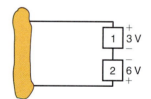

Figure P1.21

1.22 Two elements are connected in series, as shown in Fig. P1.22. Element 1 absorbs 36 W of power. Is element 2 absorbing or supplying power, and how much?

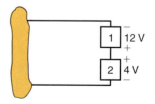

Figure P1.22

SECTION 1.3

Important note: The values used in the problems of Section 1.3 are not arbitrary. They have been selected to satisfy the basic laws of circuit analysis that will be studied in the following chapters.

1.23 Determine the power that is absorbed or supplied by the circuit elements in Fig. P1.23.

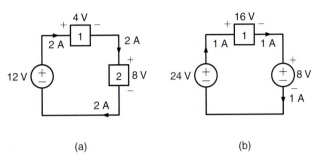

(a)   (b)

Figure P1.23

1.24 Find the power that is absorbed or supplied by the network elements in Fig. P1.24. **PSV**

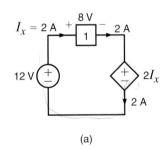

(a)

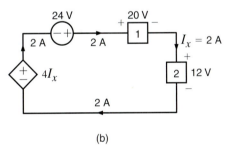

(b)

Figure P1.24

1.25 Is the source V_S in the network in Fig. P1.25 absorbing or supplying power, and how much?

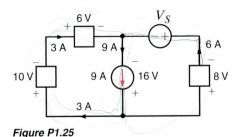

Figure P1.25

1.26 Find V_x in the network in Fig. P1.26.

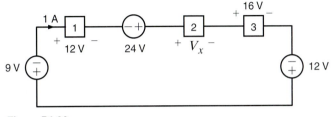

Figure P1.26

1.27 Find V_x in the network in Fig. P1.27. **PSV**

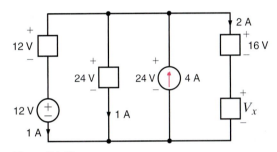

Figure P1.27

1.28 Compute the power that is absorbed or supplied by the elements in the network in Fig. P1.28. **CS**

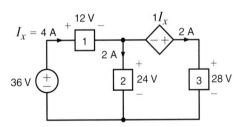

Figure P1.28

1.29 Find I_o in the network in Fig. P1.29. **CS**

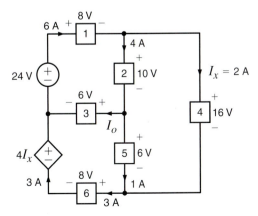

Figure P1.29

1.30 Find I_x in the circuit in Fig. P1.30.

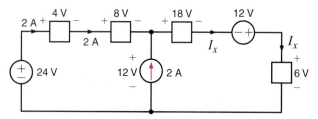

Figure P1.30

1.31 Find V_x in the network in Fig. P1.31.

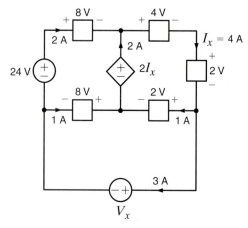

Figure P1.31

1.32 Find I_s such that the power absorbed by the two elements in Fig. P1.32 is 24 W.

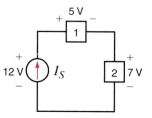

Figure P1.32

Resistive Circuits 2

In this chapter we introduce the basic concepts and laws that are fundamental to circuit analysis. These laws are Ohm's law, Kirchhoff's current law (KCL), and Kirchhoff's voltage law (KVL). We cannot overemphasize the importance of these three laws because they will be used extensively throughout our entire study of circuit analysis. The reader who masters their use quickly will not only find the material in this text easy to learn, but will be well positioned to grasp subsequent topics in the field of electrical engineering.

As a general rule, most of our activities will be confined to analysis— that is, to the determination of a specific voltage, current, or power somewhere in a network. The techniques we introduce have wide application in circuit analysis, even though we will discuss them within the framework of simple networks.

Our approach here is to begin with the simplest passive element, the resistor, and the mathematical relationship that exists between the voltage across it and the current through it, as specified by Ohm's law. As we build our confidence and proficiency by successfully analyzing some elementary circuits, we will introduce other techniques, such as voltage division and current division, that will accelerate our work.

In this chapter we introduce circuits containing dependent sources, which are used to model active devices such as transistors. Thus, our study of circuit analysis provides a natural introduction to many topics in the area of electronics.

Finally, we present a real-world application to indicate the usefulness of circuit analysis, and then we briefly introduce the topic of circuit design in an elementary fashion. In future chapters these topics will be revisited often to present some fascinating examples that describe problems we encounter in our everyday lives. ●

LEARNING GOALS

Access Problem-Solving Videos **PSV** *and Circuit Solutions* **CS** *at:*
http://www.justask4u.com/irwin
using the registration code on the inside cover and see a website with answers and more!

2.1 Ohm's Law

HINT

The passive sign convention will be employed in conjunction with Ohm's law.

Ohm's law is named for the German physicist Georg Simon Ohm, who is credited with establishing the voltage–current relationship for resistance. As a result of his pioneering work, the unit of resistance bears his name.

Ohm's law states that the voltage across a resistance is directly proportional to the current flowing through it. The resistance, measured in ohms, is the constant of proportionality between the voltage and current.

A circuit element whose electrical characteristic is primarily resistive is called a resistor and is represented by the symbol shown in Fig. 2.1a. A resistor is a physical device that can be purchased in certain standard values in an electronic parts store. These resistors, which find use in a variety of electrical applications, are normally carbon composition or wirewound. In addition, resistors can be fabricated using thick oxide or thin metal films for use in hybrid circuits, or they can be diffused in semiconductor integrated circuits. Some typical discrete resistors are shown in Fig. 2.1b.

The mathematical relationship of Ohm's law is illustrated by the equation

$$v(t) = R \times i(t), \text{ where } R \geqq 0 \qquad \textbf{2.1}$$

or equivalently, by the voltage–current characteristic shown in Fig. 2.2a. Note carefully the relationship between the polarity of the voltage and the direction of the current. In addition, note that we have tacitly assumed that the resistor has a constant value and therefore that the voltage–current characteristic is linear.

The symbol Ω is used to represent ohms, and therefore,

$$1 \ \Omega = 1 \ \text{V/A}$$

Figure 2.1
*(a) Symbol for a resistor;
(b) some practical devices.
(1), (2), and (3) are high-power resistors. (4) and (5) are high-wattage fixed resistors.
(6) is a high-precision resistor.
(7)–(12) are fixed resistors with different power ratings.*

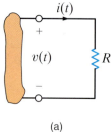

(a)

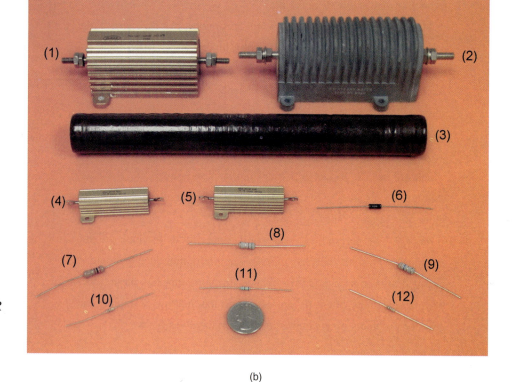

(b)

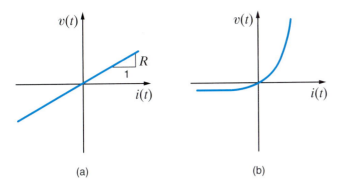

Figure 2.2
Graphical representation of the voltage–current relationship for (a) a linear resistor and (b) a light bulb.

(a)   (b)

Although in our analysis we will always assume that the resistors are *linear* and are thus described by a straight-line characteristic that passes through the origin, it is important that readers realize that some very useful and practical elements do exist that exhibit a *nonlinear* resistance characteristic; that is, the voltage–current relationship is not a straight line.

The light bulb from the flashlight in Chapter 1 is an example of an element that exhibits a nonlinear characteristic. A typical characteristic for a light bulb is shown in Fig. 2.2b.

Since a resistor is a passive element, the proper current–voltage relationship is illustrated in Fig. 2.1a. The power supplied to the terminals is absorbed by the resistor. Note that the charge moves from the higher to the lower potential as it passes through the resistor and the energy absorbed is dissipated by the resistor in the form of heat. As indicated in Chapter 1, the rate of energy dissipation is the instantaneous power, and therefore

$$p(t) = v(t)i(t) \qquad \textbf{2.2}$$

which, using Eq. (2.1), can be written as

$$p(t) = Ri^2(t) = \frac{v^2(t)}{R} \qquad \textbf{2.3}$$

This equation illustrates that the power is a nonlinear function of either current or voltage and that it is always a positive quantity.

Conductance, represented by the symbol G, is another quantity with wide application in circuit analysis. By definition, conductance is the reciprocal of resistance; that is,

$$G = \frac{1}{R} \qquad \textbf{2.4}$$

The unit of conductance is the siemens, and the relationship between units is

$$1\,\text{S} = 1\,\text{A/V}$$

Using Eq. (2.4), we can write two additional expressions,

$$i(t) = Gv(t) \qquad \textbf{2.5}$$

and

$$p(t) = \frac{i^2(t)}{G} = Gv^2(t) \qquad \textbf{2.6}$$

Equation (2.5) is another expression of Ohm's law.

Two specific values of resistance, and therefore conductance, are very important: $R = 0$ and $R = \infty$.

Figure 2.3
Short-circuit and open-circuit descriptions.

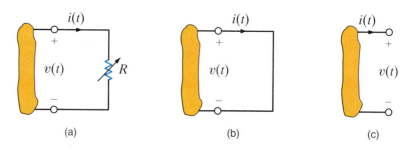

(a)   (b)   (c)

In examining the two cases, consider the network in Fig. 2.3a. The variable resistance symbol is used to describe a resistor such as the volume control on a radio or television set. As the resistance is decreased and becomes smaller and smaller, we finally reach a point where the resistance is zero and the circuit is reduced to that shown in Fig. 2.3b; that is, the resistance can be replaced by a short circuit. On the other hand, if the resistance is increased and becomes larger and larger, we finally reach a point where it is essentially infinite and the resistance can be replaced by an open circuit, as shown in Fig. 2.3c. Note that in the case of a short circuit where $R = 0$,

$$v(t) = Ri(t)$$
$$= 0$$

Therefore, $v(t) = 0$, although the current could theoretically be any value. In the open-circuit case where $R = \infty$,

$$i(t) = v(t)/R$$
$$= 0$$

Therefore, the current is zero regardless of the value of the voltage across the open terminals.

Example 2.1

In the circuit in Fig. 2.4a, determine the current and the power absorbed by the resistor.

SOLUTION Using Eq. (2.1), we find the current to be

$$I = V/R = 12/2k = 6 \text{ mA}$$

Note that because many of the resistors employed in our analysis are in $k\Omega$, we will use k in the equations in place of 1000. The power absorbed by the resistor is given by Eq. (2.2) or (2.3) as

$$P = VI = (12)(6 \times 10^{-3}) = 0.072 \text{ W}$$
$$= I^2R = (6 \times 10^{-3})^2(2k) = 0.072 \text{ W}$$
$$= V^2/R = (12)^2/2k = 0.072 \text{ W}$$

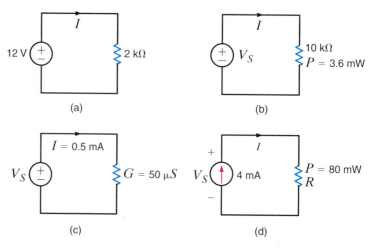

(a)   (b)

(c)   (d)

Figure 2.4 *Circuits for Examples 2.1 to 2.4.*

Example 2.2

The power absorbed by the 10-kΩ resistor in Fig. 2.4b is 3.6 mW. Determine the voltage and the current in the circuit.

SOLUTION   Using the power relationship, we can determine either of the unknowns.

$$V_S^2/R = P$$
$$V_S^2 = (3.6 \times 10^{-3})(10\text{k})$$
$$V_S = 6 \text{ V}$$

and

$$I^2R = P$$
$$I^2 = (3.6 \times 10^{-3})/10\text{k}$$
$$I = 0.6 \text{ mA}$$

Furthermore, once V_S is determined, I could be obtained by Ohm's law, and likewise once I is known, then Ohm's law could be used to derive the value of V_S. Note carefully that the equations for power involve the terms I^2 and V_S^2. Therefore, $I = -0.6$ mA and $V_S = -6$ V also satisfy the mathematical equations and, in this case, the direction of *both* the voltage and current is reversed.

Example 2.3

Given the circuit in Fig. 2.4c, we wish to find the value of the voltage source and the power absorbed by the resistance.

SOLUTION   The voltage is

$$V_S = I/G = (0.5 \times 10^{-3})/(50 \times 10^{-6}) = 10 \text{ V}$$

The power absorbed is then

$$P = I^2/G = (0.5 \times 10^{-3})^2/(50 \times 10^{-6}) = 5 \text{ mW}$$

Or we could simply note that

$$R = 1/G = 20 \text{ k}\Omega$$

and therefore

$$V_S = IR = (0.5 \times 10^{-3})(20\text{k}) = 10 \text{ V}$$

and the power could be determined using $P = I^2R = V_S^2/R = V_SI$.

Example 2.4

Given the network in Fig. 2.4d, we wish to find R and V_S.

SOLUTION   Using the power relationship, we find that

$$R = P/I^2 = (80 \times 10^{-3})/(4 \times 10^{-3})^2 = 5 \text{ k}\Omega$$

The voltage can now be derived using Ohm's law as

$$V_S = IR = (4 \times 10^{-3})(5\text{k}) = 20 \text{ V}$$

The voltage could also be obtained from the remaining power relationships in Eqs. (2.2) and (2.3).

Before leaving this initial discussion of circuits containing sources and a single resistor, it is important to note a phenomenon that we will find to be true in circuits containing many sources and resistors. The presence of a voltage source between a pair of terminals tells us precisely what the voltage is between the two terminals regardless of what is happening in the balance of the network. What we do not know is the current in the voltage source. We must apply circuit analysis to the entire network to determine this current. Likewise, the presence of a current source connected between two terminals specifies the exact value of the current through the source between the terminals. What we do not know is the value of the voltage across the current source. This value must be calculated by applying circuit analysis to the entire network. Furthermore, it is worth emphasizing that when applying Ohm's law, the relationship $V = IR$ specifies a relationship between the voltage *directly across* a resistor R and the current that is *present* in this resistor. Ohm's law does not apply when the voltage is present in one part of the network and the current exists in another. This is a common mistake made by students who try to apply $V = IR$ to a resistor R in the middle of the network while using a V at some other location in the network.

LEARNING EXTENSIONS

E2.1 Given the circuits in Fig. E2.1, find (a) the current I and the power absorbed by the resistor in Fig. E2.1a, and (b) the voltage across the current source and the power supplied by the source in Fig. E2.1b.

ANSWER: (a) $I = 0.3$ mA, $P = 3.6$ mW; (b) $V_S = 3.6$ V, $P = 2.16$ mW.

Figure E2.1

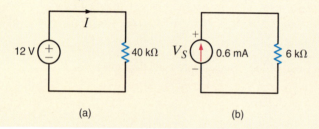

(a)   (b)

E2.2 Given the circuits in Fig. E2.2, find (a) R and V_S in the circuit in Fig. E2.2a, and (b) find I and R in the circuit in Fig. E2.2b.

ANSWER: (a) $R = 10$ kΩ, $V_S = 4$ V; (b) $I = 20.8$ mA, $R = 576$ Ω.

Figure E2.2

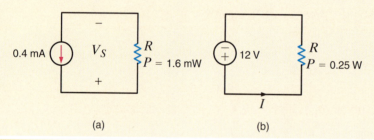

(a)   (b)

2.2 Kirchhoff's Laws

The previous circuits that we have considered have all contained a single resistor and were analyzed using Ohm's law. At this point we begin to expand our capabilities to handle more complicated networks that result from an interconnection of two or more of these simple elements. We will assume that the interconnection is performed by electrical conductors (wires) that have zero resistance—that is, perfect conductors. Because the wires have zero resistance, the energy in the circuit is in essence lumped in each element, and we employ the term *lumped-parameter circuit* to describe the network.

To aid us in our discussion, we will define a number of terms that will be employed throughout our analysis. As will be our approach throughout this text, we will use examples to illustrate the concepts and define the appropriate terms. For example, the circuit shown

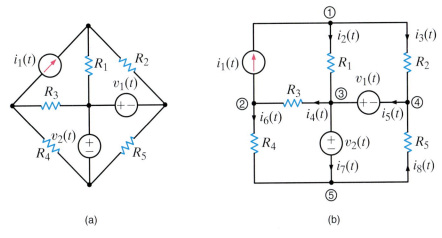

Figure 2.5
Circuit used to illustrate KCL.

(a)　　　　　　　　　　(b)

in Fig. 2.5a will be used to describe the terms *node, loop,* and *branch.* A node is simply a point of connection of two or more circuit elements. The reader is cautioned to note that, although one node can be spread out with perfect conductors, it is still only one node. This is illustrated in Fig. 2.5b where the circuit has been redrawn. Node 5 consists of the entire bottom connector of the circuit.

If we start at some point in the circuit and move along perfect conductors in any direction until we encounter a circuit element, the total path we cover represents a single node. Therefore, we can assume that a node is one end of a circuit element together with all the perfect conductors that are attached to it. Examining the circuit, we note that there are numerous paths through it. A *loop* is simply any *closed path* through the circuit in which no node is encountered more than once. For example, starting from node 1, one loop would contain the elements R_1, v_2, R_4, and i_1; another loop would contain R_2, v_1, v_2, R_4, and i_1; and so on. However, the path R_1, v_1, R_5, v_2, R_3, and i_1 is not a loop because we have encountered node 3 twice. Finally, a *branch* is a portion of a circuit containing only a single element and the nodes at each end of the element. The circuit in Fig. 2.5 contains eight branches.

Given the previous definitions, we are now in a position to consider Kirchhoff's laws, named after German scientist Gustav Robert Kirchhoff. These two laws are quite simple but extremely important. We will not attempt to prove them because the proofs are beyond our current level of understanding. However, we will demonstrate their usefulness and attempt to make the reader proficient in their use. The first law is *Kirchhoff's current law* (KCL), which states that *the algebraic sum of the currents entering any node is zero.* In mathematical form the law appears as

HINT

KCL is an extremely important and useful law.

$$\sum_{j=1}^{N} i_j(t) = 0 \qquad\qquad \textbf{2.7}$$

where $i_j(t)$ is the *j*th current entering the node through branch *j* and *N* is the number of branches connected to the node. To understand the use of this law, consider node 3 shown in Fig. 2.5. Applying Kirchhoff's current law to this node yields

$$i_2(t) - i_4(t) + i_5(t) - i_7(t) = 0$$

We have assumed that the algebraic signs of the currents entering the node are positive and, therefore, that the signs of the currents leaving the node are negative.

If we multiply the foregoing equation by −1, we obtain the expression

$$-i_2(t) + i_4(t) - i_5(t) + i_7(t) = 0$$

which simply states that *the algebraic sum of the currents leaving a node is zero.* Alternatively, we can write the equation as

$$i_2(t) + i_5(t) = i_4(t) + i_7(t)$$

which states that *the sum of the currents entering a node is equal to the sum of the currents leaving the node.* Both of these italicized expressions are alternative forms of Kirchhoff's current law.

Once again it must be emphasized that the latter statement means that the sum of the *variables* that have been defined entering the node is equal to the sum of the *variables* that have been defined leaving the node, not the actual currents. For example, $i_j(t)$ may be defined entering the node, but if its actual value is negative, there will be positive charge leaving the node.

Note carefully that Kirchhoff's current law states that the *algebraic* sum of the currents either entering or leaving a node must be zero. We now begin to see why we stated in Chapter 1 that it is critically important to specify both the magnitude and the direction of a current. Recall that current is charge in motion. Based on our background in physics, charges cannot be stored at a node. In other words, if we have a number of charges entering a node, then an equal number must be leaving that same node. Kirchhoff's current law is based on this principle of conservation of charge.

Example 2.5

Let us write KCL for every node in the network in Fig. 2.5 assuming that the currents leaving the node are positive.

SOLUTION The KCL equations for nodes 1 through 5 are

$$-i_1(t) + i_2(t) + i_3(t) = 0$$
$$i_1(t) - i_4(t) + i_6(t) = 0$$
$$-i_2(t) + i_4(t) - i_5(t) + i_7(t) = 0$$
$$-i_3(t) + i_5(t) - i_8(t) = 0$$
$$-i_6(t) - i_7(t) + i_8(t) = 0$$

Note carefully that if we add the first four equations, we obtain the fifth equation. What does this tell us? Recall that this means that this set of equations is not linearly independent. We can show that the first four equations are, however, linearly independent. Store this idea in memory because it will become very important when we learn how to write the equations necessary to solve for all the currents and voltages in a network in the following chapter.

Example 2.6

The network in Fig. 2.5 is represented by the topological diagram shown in Fig. 2.6. We wish to find the unknown currents in the network.

Figure 2.6
Topological diagram for the circuit in Fig. 2.5.

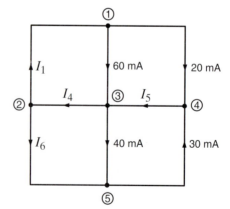

SOLUTION Assuming the currents leaving the node are positive, the KCL equations for nodes 1 through 4 are

$$-I_1 + 0.06 + 0.02 = 0$$
$$I_1 - I_4 + I_6 = 0$$
$$-0.06 + I_4 - I_5 + 0.04 = 0$$
$$-0.02 + I_5 - 0.03 = 0$$

The first equation yields I_1 and the last equation yields I_5. Knowing I_5 we can immediately obtain I_4 from the third equation. Then the values of I_1 and I_4 yield the value of I_6 from the second equation. The results are $I_1 = 80$ mA, $I_4 = 70$ mA, $I_5 = 50$ mA, and $I_6 = -10$ mA.

As indicated earlier, dependent or controlled sources are very important because we encounter them when analyzing circuits containing active elements such as transistors. The following example presents a circuit containing a current-controlled current source.

Example 2.7

Let us write the KCL equations for the circuit shown in Fig. 2.7.

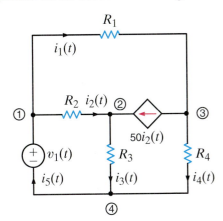

Figure 2.7
Circuit containing a dependent current source.

SOLUTION   The KCL equations for nodes 1 through 4 follow.

$$i_1(t) + i_2(t) - i_5(t) = 0$$
$$-i_2(t) + i_3(t) - 50i_2(t) = 0$$
$$-i_1(t) + 50i_2(t) + i_4(t) = 0$$
$$i_5(t) - i_3(t) - i_4(t) = 0$$

If we added the first three equations, we would obtain the negative of the fourth. What does this tell us about the set of equations?

Finally, it is possible to generalize Kirchhoff's current law to include a closed surface. By a closed surface we mean some set of elements completely contained within the surface that are interconnected. Since the current entering each element within the surface is equal to that leaving the element (i.e., the element stores no net charge), it follows that the current entering an interconnection of elements is equal to that leaving the interconnection. Therefore, Kirchhoff's current law can also be stated as follows: *The algebraic sum of the currents entering any closed surface is zero.*

Example 2.8

Let us find I_4 and I_1 in the network represented by the topological diagram in Fig. 2.6.

SOLUTION   This diagram is redrawn in Fig. 2.8; node 1 is enclosed in surface 1, and nodes 3 and 4 are enclosed in surface 2. A quick review of the previous example indicates that we derived a value for I_4 from the value of I_5. However, I_5 is now completely enclosed in surface 2. If we apply KCL to surface 2, assuming the currents out of the surface are positive, we obtain

$$I_4 - 0.06 - 0.02 - 0.03 + 0.04 = 0$$

or

$$I_4 = 70 \text{ mA}$$

which we obtained without any knowledge of I_5. Likewise for surface 1, what goes in must come out and, therefore, $I_1 = 80$ mA. The reader is encouraged to cut the network in Fig. 2.6 into two pieces in any fashion and show that KCL is always satisfied at the boundaries.

Figure 2.8
Diagram used to
demonstrate KCL for a
surface.

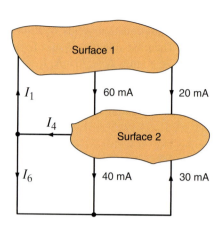

LEARNING EXTENSIONS

E2.3 Given the networks in Fig. E2.3, find (a) I_1 in Fig. E2.3a and (b) I_T in Fig. E2.3b.

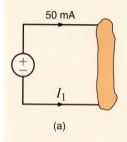

(a)

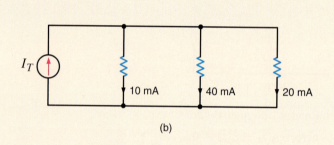

(b)

Figure E2.3

ANSWER:
(a) $I_1 = -50$ mA;
(b) $I_T = 70$ mA.

E2.4 Find (a) I_1 in the network in Fig. E2.4a and (b) I_1 and I_2 in the circuit in Fig. E2.4b.

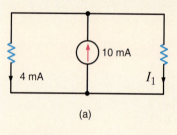

(a)

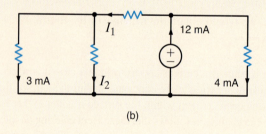

(b)

Figure E2.4

ANSWER: (a) $I_1 = 6$ mA;
(b) $I_1 = 8$ mA and
$I_2 = 5$ mA.

E2.5 Find the current i_x in the circuits in Fig. E2.5.

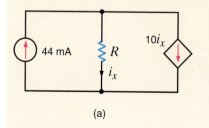

(a)

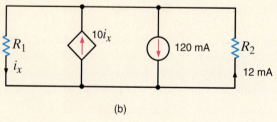

(b)

Figure E2.5

ANSWER: (a) $i_x = 4$ mA;
(b) $i_x = 12$ mA.

Kirchhoff's second law, called *Kirchhoff's voltage law* (KVL), states that the *algebraic sum of the voltages around any loop is zero*. As was the case with Kirchhoff's current law, we will defer the proof of this law and concentrate on understanding how to apply it. Once again the reader is cautioned to remember that we are dealing only with lumped-parameter circuits. These circuits are conservative, meaning that the work required to move a unit charge around any loop is zero.

In Chapter 1, we related voltage to the difference in energy levels within a circuit and talked about the energy conversion process in a flashlight. Because of this relationship between voltage and energy, Kirchhoff's voltage law is based on the conservation of energy.

Recall that in Kirchhoff's current law, the algebraic sign was required to keep track of whether the currents were entering or leaving a node. In Kirchhoff's voltage law, the algebraic sign is used to keep track of the voltage polarity. In other words, as we traverse the circuit, it is necessary to sum to zero the increases and decreases in energy level. Therefore, it is important we keep track of whether the energy level is increasing or decreasing as we go through each element.

In applying KVL, we must traverse any loop in the circuit and sum to zero the increases and decreases in energy level. At this point, we have a decision to make. Do we want to consider a decrease in energy level as positive or negative? We will adopt a policy of considering a decrease in energy level as positive and an increase in energy level as negative. As we move around a loop, we encounter the plus sign first for a decrease in energy level and a negative sign first for an increase in energy level.

Example 2.9

Consider the circuit shown in Fig. 2.9. If V_{R_1} and V_{R_2} are known quantities, let us find V_{R_3}.

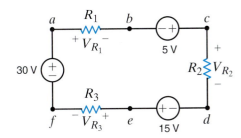

Figure 2.9
Circuit used to illustrate KVL.

SOLUTION Starting at point a in the network and traversing it in a clockwise direction, we obtain the equation

$$+V_{R_1} - 5 + V_{R_2} - 15 + V_{R_3} - 30 = 0$$

which can be written as

$$+V_{R_1} + V_{R_2} + V_{R_3} = 5 + 15 + 30$$
$$= 50$$

Now suppose that V_{R_1} and V_{R_2} are known to be 18 V and 12 V, respectively. Then $V_{R_3} = 20$ V.

Example 2.10

Consider the network in Fig. 2.10.

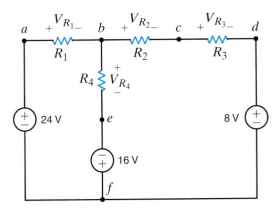

Figure 2.10
Circuit used to explain KVL.

Let us demonstrate that only two of the three possible loop equations are linearly independent.

SOLUTION Note that this network has three closed paths: the left loop, right loop, and outer loop. Applying our policy for writing KVL equations and traversing the left loop starting at point a, we obtain

$$V_{R_1} + V_{R_4} - 16 - 24 = 0$$

The corresponding equation for the right loop starting at point b is

$$V_{R_2} + V_{R_3} + 8 + 16 - V_{R_4} = 0$$

The equation for the outer loop starting at point a is

$$V_{R_1} + V_{R_2} + V_{R_3} + 8 - 24 = 0$$

Note that if we add the first two equations, we obtain the third equation. Therefore, as we indicated in Example 2.5, the three equations are not linearly independent. Once again, we will address this issue in the next chapter and demonstrate that we need only the first two equations to solve for the voltages in the circuit.

Finally, we employ the convention V_{ab} to indicate the voltage of point a with respect to point b: that is, the variable for the voltage between point a and point b, with point a considered positive relative to point b. Since the potential is measured between two points, it is convenient to use an arrow between the two points, with the head of the arrow located at the positive node. Note that the double-subscript notation, the + and − notation, and the single-headed arrow notation are all the same if the head of the arrow is pointing toward the positive terminal and the first subscript in the double-subscript notation. All of these equivalent forms for labeling voltages are shown in Fig. 2.11. The usefulness of the arrow notation stems from the fact that we may want to label the voltage between two points that are far apart in a network. In this case, the other notations are often confusing.

Figure 2.11

Equivalent forms for labeling voltage.

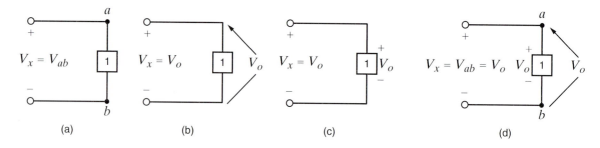

(a)   (b)   (c)   (d)

Example 2.11

Consider the network in Fig. 2.12a. Let us apply KVL to determine the voltage between two points. Specifically, in terms of the double-subscript notation, let us find V_{ae} and V_{ec}.

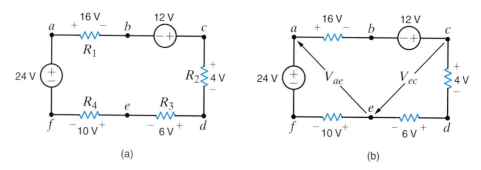

(a)   (b)

Figure 2.12 Network used in Example 2.11.

SOLUTION The circuit is redrawn in Fig. 2.12b. Since points a and e as well as e and c are not physically close, the arrow notation is very useful. Our approach to determining the unknown voltage is to apply KVL with the unknown voltage in the closed path. Therefore, to determine V_{ae} we can use the path *aefa* or *abcdea*. The equations for the two paths in which V_{ae} is the only unknown are

$$V_{ae} + 10 - 24 = 0$$

and

$$16 - 12 + 4 + 6 - V_{ae} = 0$$

Note that both equations yield $V_{ae} = 14$ V. Even before calculating V_{ae}, we could calculate V_{ec} using the path *cdec* or *cefabc*. However, since V_{ae} is now known, we can also use the path *ceabc*. KVL for each of these paths is

$$4 + 6 + V_{ec} = 0$$
$$-V_{ec} + 10 - 24 + 16 - 12 = 0$$

and

$$-V_{ec} - V_{ae} + 16 - 12 = 0$$

Each of these equations yields $V_{ec} = -10$ V.

In general, the mathematical representation of Kirchhoff's voltage law is

$$\sum_{j=1}^{N} v_j(t) = 0 \qquad \textbf{2.8}$$

HINT

KVL is an extremely important and useful law.

where $v_j(t)$ is the voltage across the *j*th branch (with the proper reference direction) in a loop containing N voltages. This expression is analogous to Eq. (2.7) for Kirchhoff's current law.

Example 2.12

Given the network in Fig. 2.13 containing a dependent source, let us write the KVL equations for the two closed paths *abda* and *bcdb*.

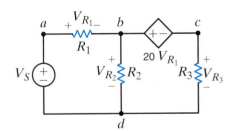

Figure 2.13
Network containing a dependent source.

SOLUTION   The two KVL equations are

$$V_{R_1} + V_{R_2} - V_S = 0$$
$$20V_{R_1} + V_{R_3} - V_{R_2} = 0$$

LEARNING EXTENSIONS

E2.6 Find V_{ad} and V_{eb} in the network in Fig. E2.6.

ANSWER: $V_{ad} = 26$ V, $V_{eb} = 10$ V.

Figure E2.6

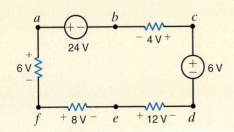

(continues on the next page)

E2.7 Find V_{bd} in the circuit in Fig. E2.7.

ANSWER: $V_{bd} = 11$ V.

Figure E2.7

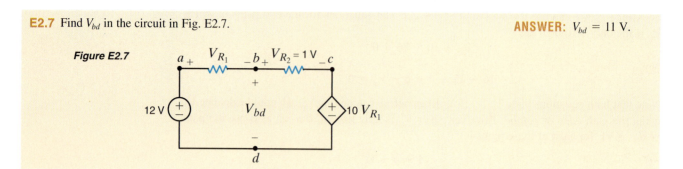

HINT

The subtleties associated with Ohm's law, as described here, are important and must be adhered to in order to ensure that the variables have the proper sign.

Before proceeding with the analysis of simple circuits, it is extremely important that we emphasize a subtle but very critical point. Ohm's law as defined by the equation $V = IR$ refers to the relationship between the voltage and current as defined in Fig. 2.14a. If the direction of either the current or the voltage, but not both, is reversed, the relationship between the current and the voltage would be $V = -IR$. In a similar manner, given the circuit in Fig. 2.14b, if the polarity of the voltage between the terminals A and B is specified as shown, then the direction of the current I is from point B through R to point A. Likewise, in Fig. 2.14c, if the direction of the current is specified as shown, then the polarity of the voltage must be such that point D is at a higher potential than point C and, therefore, the arrow representing the voltage V is from point C to point D.

Figure 2.14
Circuits used to explain Ohm's law.

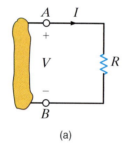 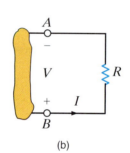 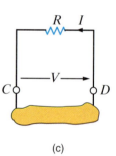

(a)　　　　　　(b)　　　　　　(c)

2.3 Single-Loop Circuits

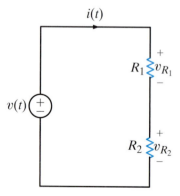

Figure 2.15
Single-loop circuit.

VOLTAGE DIVISION At this point we can begin to apply the laws we have presented earlier to the analysis of simple circuits. To begin, we examine what is perhaps the simplest circuit—a single closed path, or loop, of elements.

Applying KCL to every node in a single-loop circuit reveals that the same current flows through all elements. We say that these elements are connected in series because they carry the same current. We will apply Kirchhoff's voltage law and Ohm's law to the circuit to determine various quantities in the circuit.

Our approach will be to begin with a simple circuit and then generalize the analysis to more complicated ones. The circuit shown in Fig. 2.15 will serve as a basis for discussion. This circuit consists of an independent voltage source that is in series with two resistors. We have assumed that the current flows in a clockwise direction. If this assumption is correct, the solution of the equations that yields the current will produce a positive value. If the current is actually flowing in the opposite direction, the value of the current variable will simply be negative, indicating that the current is flowing in a direction opposite to that assumed. We have also made voltage polarity assignments for v_{R_1} and v_{R_2}. These assignments have been made using the convention employed in our discussion of Ohm's law and our choice for the direction of $i(t)$—that is, the convention shown in Fig. 2.14a.

Applying Kirchhoff's voltage law to this circuit yields

$$-v(t) + v_{R_1} + v_{R_2} = 0$$

or

$$v(t) = v_{R_1} + v_{R_2}$$

However, from Ohm's law we know that

$$v_{R_1} = R_1 i(t)$$

$$v_{R_2} = R_2 i(t)$$

Therefore,

$$v(t) = R_1 i(t) + R_2 i(t)$$

Solving the equation for $i(t)$ yields

$$i(t) = \frac{v(t)}{R_1 + R_2} \qquad\qquad \textbf{2.9}$$

Knowing the current, we can now apply Ohm's law to determine the voltage across each resistor:

$$v_{R_1} = R_1 i(t)$$

$$= R_1 \left[\frac{v(t)}{R_1 + R_2} \right] \qquad\qquad \textbf{2.10}$$

$$= \frac{R_1}{R_1 + R_2} v(t)$$

Similarly,

$$v_{R_2} = \frac{R_2}{R_1 + R_2} v(t) \qquad\qquad \textbf{2.11}$$

HINT

The manner in which voltage divides between two series resistors

Though simple, Eqs. (2.10) and (2.11) are very important because they describe the operation of what is called a *voltage divider*. In other words, the source voltage $v(t)$ *is divided between the resistors R_1 and R_2 in direct proportion to their resistances.*

In essence, if we are interested in the voltage across the resistor R_1, we bypass the calculation of the current $i(t)$ and simply multiply the input voltage $v(t)$ by the ratio

$$\frac{R_1}{R_1 + R_2}$$

As illustrated in Eq. (2.10), we are using the current in the calculation, but not explicitly.

Note that the equations satisfy Kirchhoff's voltage law, since

$$-v(t) + \frac{R_1}{R_1 + R_2} v(t) + \frac{R_2}{R_1 + R_2} v(t) = 0$$

Example 2.13

Consider the circuit shown in Fig. 2.16. The circuit is identical to Fig. 2.15 except that R_1 is a variable resistor such as the volume control for a radio or television set. Suppose that $V_S = 9\,\text{V}$, $R_1 = 90\,\text{k}\Omega$, and $R_2 = 30\,\text{k}\Omega$.

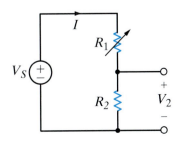

Figure 2.16
Voltage-divider circuit.

Let us examine the change in both the voltage across R_2 and the power absorbed in this resistor as R_1 is changed from 90 kΩ to 15 kΩ.

SOLUTION Since this is a voltage-divider circuit, the voltage V_2 can be obtained directly as

$$V_2 = \left[\frac{R_2}{R_1 + R_2}\right]V_S$$
$$= \left[\frac{30k}{90k + 30k}\right](9)$$
$$= 2.25 \text{ V}$$

Now suppose that the variable resistor is changed from 90 kΩ to 15 kΩ. Then

$$V_2 = \left[\frac{30k}{30k + 15k}\right]9$$
$$= 6 \text{ V}$$

The direct voltage-divider calculation is equivalent to determining the current I and then using Ohm's law to find V_2. Note that the larger voltage is across the larger resistance. This voltage-divider concept and the simple circuit we have employed to describe it are very useful because, as will be shown later, more complicated circuits can be reduced to this form.

Finally, let us determine the instantaneous power absorbed by the resistor R_2 under the two conditions $R_1 = 90$ kΩ and $R_1 = 15$ kΩ. For the case $R_1 = 90$ kΩ, the power absorbed by R_2 is

$$P_2 = I^2 R_2 = \left(\frac{9}{120k}\right)^2 (30k)$$
$$= 0.169 \text{ mW}$$

In the second case

$$P_2 = \left(\frac{9}{45k}\right)^2 (30k)$$
$$= 1.2 \text{ mW}$$

The current in the first case is 75 μA, and in the second case it is 200 μA. Since the power absorbed is a function of the square of the current, the power absorbed in the two cases is quite different.

Let us now demonstrate the practical utility of this simple voltage-divider network.

Example 2.14

Consider the circuit in Fig. 2.17a, which is an approximation of a high-voltage dc transmission facility. We have assumed that the bottom portion of the transmission line is a perfect conductor and will justify this assumption in the next chapter. The load can be represented by a resistor of value 183.5 Ω. Therefore, the equivalent circuit of this network is shown in Fig. 2.17b.

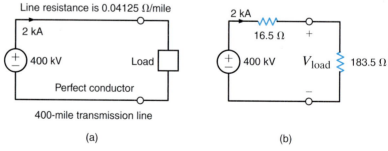

Figure 2.17 *A high-voltage dc transmission facility.*

Let us determine both the power delivered to the load and the power losses in the line.

SOLUTION Using voltage division, the load voltage is

$$V_{load} = \left[\frac{183.5}{183.5 + 16.5}\right] 400k$$

$$= 367 \text{ kV}$$

The input power is 800 MW and the power transmitted to the load is

$$P_{load} = I^2 R_{load}$$

$$= 734 \text{ MW}$$

Therefore, the power loss in the transmission line is

$$P_{line} = P_{in} - P_{load} = I^2 R_{line}$$

$$= 66 \text{ MW}$$

Since $P = VI$, suppose now that the utility company supplied power at 200 kV and 4 kA. What effect would this have on our transmission network? Without making a single calculation, we know that because power is proportional to the square of the current, there would be a large increase in the power loss in the line and, therefore, the efficiency of the facility would decrease substantially. That is why, in general, we transmit power at high voltage and low current.

MULTIPLE SOURCE/RESISTOR NETWORKS At this point we wish to extend our analysis to include a multiplicity of voltage sources and resistors. For example, consider the circuit shown in Fig. 2.18a. Here we have assumed that the current flows in a clockwise direction, and we have defined the variable $i(t)$ accordingly. This may or may not be the case, depending on the value of the various voltage sources. Kirchhoff's voltage law for this circuit is

$$+v_{R_1} + v_2(t) - v_3(t) + v_{R_2} + v_4(t) + v_5(t) - v_1(t) = 0$$

or, using Ohm's law,

$$\left(R_1 + R_2\right)i(t) = v_1(t) - v_2(t) + v_3(t) - v_4(t) - v_5(t)$$

which can be written as

$$\left(R_1 + R_2\right)i(t) = v(t)$$

where

$$v(t) = v_1(t) + v_3(t) - \left[v_2(t) + v_4(t) + v_5(t)\right]$$

so that under the preceding definitions, Fig. 2.18a is equivalent to Fig. 2.18b. In other words, the sum of several voltage sources in series can be replaced by one source whose value is the algebraic sum of the individual sources. This analysis can, of course, be generalized to a circuit with N series sources.

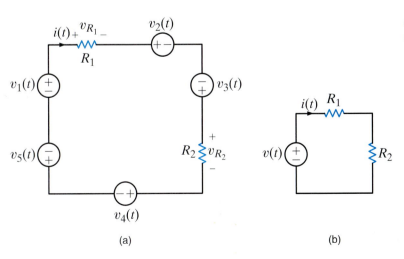

Figure 2.18
Equivalent circuits with multiple sources.

(a)   (b)

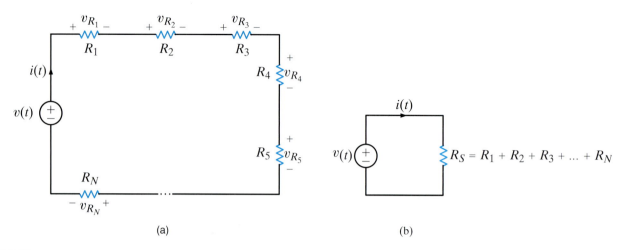

Figure 2.19
Equivalent circuits.

Now consider the circuit with N resistors in series, as shown in Fig. 2.19a. Applying Kirchhoff's voltage law to this circuit yields

$$v(t) = v_{R_1} + v_{R_2} + \cdots + v_{R_N}$$
$$= R_1 i(t) + R_2 i(t) + \cdots + R_N i(t)$$

and therefore,

$$v(t) = R_S i(t) \qquad \textbf{2.12}$$

where

$$R_S = R_1 + R_2 + \cdots + R_N \qquad \textbf{2.13}$$

and hence,

$$i(t) = \frac{v(t)}{R_S} \qquad \textbf{2.14}$$

Note also that for any resistor R_i in the circuit, the voltage across R_i is given by the expression

$$v_{R_i} = \frac{R_i}{R_S} v(t) \qquad \textbf{2.15}$$

which is the voltage-division property for multiple resistors in series.

Equation (2.13) illustrates that *the equivalent resistance of N resistors in series is simply the sum of the individual resistances.* Thus, using Eq. (2.13), we can draw the circuit in Fig. 2.19b as an equivalent circuit for the one in Fig. 2.19a.

Example 2.15

Given the circuit in Fig. 2.20a, let us find I, V_{bd}, and the power absorbed by the 30-kΩ resistor. Finally, let us use voltage division to find V_{bc}.

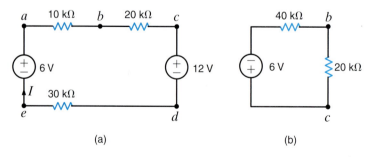

Figure 2.20 *Circuit used in Example 2.15.*

SOLUTION   KVL for the network yields the equation

$$10kI + 20kI + 12 + 30kI - 6 = 0$$
$$60kI = -6$$
$$I = -0.1 \text{ mA}$$

Therefore, the magnitude of the current is 0.1 mA, but its direction is opposite to that assumed.

The voltage V_{bd} can be calculated using either of the closed paths *abdea* or *bcdb*. The equations for both cases are

$$10kI + V_{bd} + 30kI - 6 = 0$$

and

$$20kI + 12 - V_{bd} = 0$$

Using $I = -0.1$ mA in either equation yields $V_{bd} = 10$ V. Finally, the power absorbed by the 30-kΩ resistor is

$$P = I^2R = 0.3 \text{ mW}$$

Now from the standpoint of determining the voltage V_{bc}, we can simply add the sources since they are in series, add the remaining resistors since they are in series, and reduce the network to that shown in Fig. 2.20b. Then

$$V_{bc} = \frac{20k}{20k + 40k}(-6)$$
$$= -2 \text{ V}$$

Example 2.16

A dc transmission facility is modeled by the approximate circuit shown in Fig. 2.21. If the load voltage is known to be $V_{load} = 458.3$ kV, we wish to find the voltage at the sending end of the line and the power loss in the line.

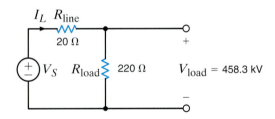

Figure 2.21
Circuit used in Example 2.16.

SOLUTION   Knowing the load voltage and load resistance, we can obtain the line current using Ohm's law:

$$I_L = 458.3k/220$$
$$= 2.083 \text{ kA}$$

The voltage drop across the line is

$$V_{line} = (I_L)(R_{line})$$
$$= 41.66 \text{ kV}$$

Now, using KVL,

$$V_S = V_{line} + V_{load}$$
$$= 500 \text{ kV}$$

Note that since the network is simply a voltage-divider circuit, we could obtain V_S immediately from our knowledge of R_{line}, R_{load}, and V_{load}. That is,

$$V_{load} = \left[\frac{R_{load}}{R_{load} + R_{line}} \right] V_S$$

and V_S is the only unknown in this equation.

The power absorbed by the line is

$$P_{\text{line}} = I_L^2 R_{\text{line}}$$

$$= 86.79 \text{ MW}$$

PROBLEM-SOLVING STRATEGY

Single-Loop Circuits

Step 1. Define a current $i(t)$. We know from KCL that there is only one current for a single-loop circuit. This current is assumed to be flowing either clockwise or counterclockwise around the loop.

Step 2. Using Ohm's law, define a voltage across each resistor in terms of the defined current.

Step 3. Apply KVL to the single-loop circuit.

Step 4. Solve the single KVL equation for the current $i(t)$. If $i(t)$ is positive, the current is flowing in the direction assumed; if not, then the current is actually flowing in the opposite direction.

LEARNING EXTENSIONS

E2.8 Find I and V_{bd} in the circuit in Fig. E2.8. **PSV**

ANSWER: $I = -0.05$ mA and $V_{bd} = 10$ V.

Figure E2.8

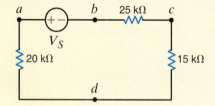

E2.9 In the network in Fig. E2.9, if V_{ad} is 3 V, find V_S.

ANSWER: $V_S = 9$ V.

Figure E2.9

2.4 Single-Node-Pair Circuits

CURRENT DIVISION An important circuit is the single-node-pair circuit. If we apply KVL to every loop in a single-node-pair circuit, we discover that all of the elements have the same voltage across them and, therefore, are said to be connected in parallel. We will, however, apply Kirchhoff's current law and Ohm's law to determine various unknown quantities in the circuit.

Following our approach with the single-loop circuit, we will begin with the simplest case and then generalize our analysis. Consider the circuit shown in Fig. 2.22. Here we have an independent current source in parallel with two resistors.

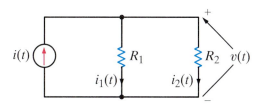

Figure 2.22
Simple parallel circuit.

Since all of the circuit elements are in parallel, the voltage $v(t)$ appears across each of them. Furthermore, an examination of the circuit indicates that the current $i(t)$ is into the upper node of the circuit and the currents $i_1(t)$ and $i_2(t)$ are out of the node. Since KCL essentially states that what goes in must come out, the question we must answer is how $i_1(t)$ and $i_2(t)$ divide the input current $i(t)$.

Applying Kirchhoff's current law to the upper node, we obtain

$$i(t) = i_1(t) + i_2(t)$$

and, employing Ohm's law, we have

$$i(t) = \frac{v(t)}{R_1} + \frac{v(t)}{R_2}$$

$$= \left(\frac{1}{R_1} + \frac{1}{R_2} \right) v(t)$$

$$= \frac{v(t)}{R_p}$$

where

$$\frac{1}{R_p} = \frac{1}{R_1} + \frac{1}{R_2} \qquad \textbf{2.16}$$

> **HINT**
> The parallel resistance equation

$$R_p = \frac{R_1 R_2}{R_1 + R_2} \qquad \textbf{2.17}$$

Therefore, the equivalent resistance of two resistors connected in parallel is equal to the product of their resistances divided by their sum. Note also that this equivalent resistance R_p is always less than either R_1 or R_2. Hence, by connecting resistors in parallel we reduce the overall resistance. In the special case when $R_1 = R_2$, the equivalent resistance is equal to half of the value of the individual resistors.

The manner in which the current $i(t)$ from the source divides between the two branches is called *current division* and can be found from the preceding expressions. For example,

$$v(t) = R_p i(t)$$

$$= \frac{R_1 R_2}{R_1 + R_2} i(t) \qquad \textbf{2.18}$$

and

$$i_1(t) = \frac{v(t)}{R_1}$$

> **HINT**
> The manner in which current divides between two parallel resistors

$$i_1(t) = \frac{R_2}{R_1 + R_2} i(t) \qquad \textbf{2.19}$$

and

$$i_2(t) = \frac{v(t)}{R_2}$$

$$= \frac{R_1}{R_1 + R_2} i(t) \qquad \textbf{2.20}$$

Equations (2.19) and (2.20) are mathematical statements of the current-division rule.

Example 2.17

Given the network in Fig. 2.23a, let us find I_1, I_2, and V_o.

SOLUTION First, it is important to recognize that the current source feeds two parallel paths. To emphasize this point, the circuit is redrawn as shown in Fig. 2.23b. Applying current division, we obtain

$$I_1 = \left[\frac{40k + 80k}{60k + (40k + 80k)}\right](0.9 \times 10^{-3})$$

$$= 0.6 \text{ mA}$$

and

$$I_2 = \left[\frac{60k}{60k + (40k + 80k)}\right](0.9 \times 10^{-3})$$

$$= 0.3 \text{ mA}$$

Note that the larger current flows through the smaller resistor, and vice versa. In addition, note that if the resistances of the two paths are equal, the current will divide equally between them. KCL is satisfied since $I_1 + I_2 = 0.9$ mA.

The voltage V_o can be derived using Ohm's law as

$$V_o = 80kI_2$$

$$= 24 \text{ V}$$

The problem can also be approached in the following manner. The total resistance seen by the current source is 40 kΩ, that is, 60 kΩ in parallel with the series combination of 40 kΩ and 80 kΩ as shown in Fig. 2.23c. The voltage across the current source is then

$$V_1 = (0.9 \times 10^{-3})40k$$

$$= 36 \text{ V}$$

Now that V_1 is known, we can apply voltage division to find V_o.

$$V_o = \left(\frac{80k}{80k + 40k}\right)V_1$$

$$= \left(\frac{80k}{120k}\right)36$$

$$= 24 \text{ V}$$

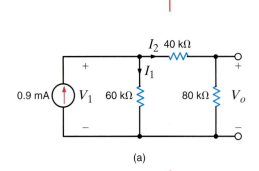

(a)

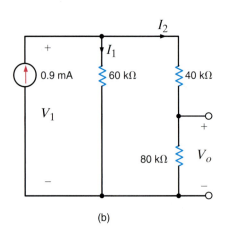

(b)

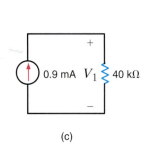

(c)

Figure 2.23 *Circuits used in Example 2.17.*

Example 2.18

A typical car stereo consists of a 2-W audio amplifier and two speakers represented by the diagram shown in Fig. 2.24a. The output circuit of the audio amplifier is in essence a 430-mA current source, and each speaker has a resistance of 4 Ω. Let us determine the power absorbed by the speakers.

SOLUTION   The audio system can be modeled as shown in Fig. 2.24b. Since the speakers are both 4-Ω devices, the current will split evenly between them, and the power absorbed by each speaker is

$$P = I^2R$$
$$= (215 \times 10^{-3})^2(4)$$
$$= 184.9 \text{ mW}$$

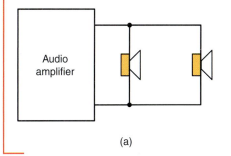

(a)

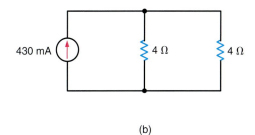

(b)

Figure 2.24
Circuits used in Example 2.18.

LEARNING EXTENSION

E2.10 Find the currents I_1 and I_2 and the power absorbed by the 40-kΩ resistor in the network in Fig. E2.10.

ANSWER: $I_1 = 12$ mA, $I_2 = -4$ mA, and $P_{40 \text{ k}\Omega} = 5.76$ W.

Figure E2.10

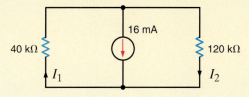

MULTIPLE SOURCE/RESISTOR NETWORKS

Let us now extend our analysis to include a multiplicity of current sources and resistors in parallel. For example, consider the circuit shown in Fig. 2.25a. We have assumed that the upper node is $v(t)$ volts positive with respect to the lower node. Applying Kirchhoff's current law to the upper node yields

$$i_1(t) - i_2(t) - i_3(t) + i_4(t) - i_5(t) - i_6(t) = 0$$

or

$$i_1(t) - i_3(t) + i_4(t) - i_6(t) = i_2(t) + i_5(t)$$

Figure 2.25
Equivalent circuits.

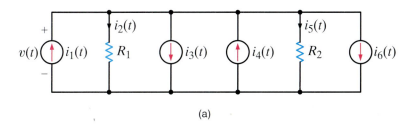

(a)

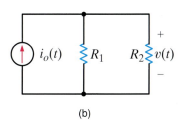

(b)

Figure 2.26
Equivalent circuits.

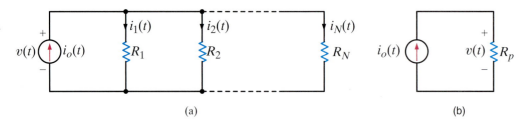

(a)   (b)

The terms on the left side of the equation all represent sources that can be combined algebraically into a single source; that is,

$$i_o(t) = i_1(t) - i_3(t) + i_4(t) - i_6(t)$$

which effectively reduces the circuit in Fig. 2.25a to that in Fig. 2.25b. We could, of course, generalize this analysis to a circuit with N current sources. Using Ohm's law, we can express the currents on the right side of the equation in terms of the voltage and individual resistances so that the KCL equation reduces to

$$i_o(t) = \left(\frac{1}{R_1} + \frac{1}{R_2} \right) v(t)$$

Now consider the circuit with N resistors in parallel, as shown in Fig. 2.26a. Applying Kirchhoff's current law to the upper node yields

$$i_o(t) = i_1(t) + i_2(t) + \cdots + i_N(t)$$

$$= \left(\frac{1}{R_1} + \frac{1}{R_2} + \cdots + \frac{1}{R_N} \right) v(t) \qquad \textbf{2.21}$$

or

$$i_o(t) = \frac{v(t)}{R_p} \qquad \textbf{2.22}$$

where

$$\frac{1}{R_p} = \sum_{i=1}^{N} \frac{1}{R_i} \qquad \textbf{2.23}$$

so that as far as the source is concerned, Fig. 2.26a can be reduced to an equivalent circuit, as shown in Fig. 2.26b.

The current division for any branch can be calculated using Ohm's law and the preceding equations. For example, for the jth branch in the network of Fig. 2.26a,

$$i_j(t) = \frac{v(t)}{R_j}$$

Using Eq. (2.22), we obtain

$$i_j(t) = \frac{R_p}{R_j} i_o(t) \qquad \textbf{2.24}$$

which defines the current-division rule for the general case.

Example 2.19

Given the circuit in Fig. 2.27a, we wish to find the current in the 12-kΩ load resistor.

SOLUTION   To simplify the network in Fig. 2.27a, we add the current sources algebraically and combine the parallel resistors in the following manner:

$$\frac{1}{R_p} = \frac{1}{18\text{k}} + \frac{1}{9\text{k}} + \frac{1}{12\text{k}}$$

$$R_p = 4 \text{ k}\Omega$$

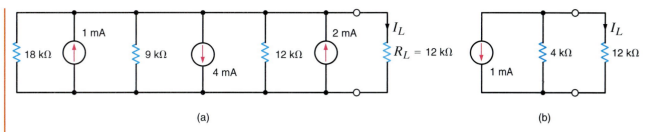

(a)

Figure 2.27
Circuits used in Example 2.19.

(b)

Using these values we can reduce the circuit in Fig. 2.27a to that in Fig. 2.27b. Now, applying current division, we obtain

$$I_L = -\left[\frac{4k}{4k + 12k}\right](1 \times 10^{-3})$$

$$= -0.25 \text{ mA}$$

PROBLEM-SOLVING STRATEGY

Single-Node-Pair Circuits

Step 1. Define a voltage $v(t)$ between the two nodes in this circuit. We know from KVL that there is only one voltage for a single-node-pair circuit. A polarity is assigned to the voltage such that one of the nodes is assumed to be at a higher potential than the other node, which we will call the reference node.

Step 2. Using Ohm's law, define a current flowing through each resistor in terms of the defined voltage.

Step 3. Apply KCL at one of the two nodes in the circuit.

Step 4. Solve the single KCL equation for $v(t)$. If $v(t)$ is positive, then the reference node is actually at a lower potential than the other node; if not, the reference node is actually at a higher potential than the other node.

LEARNING EXTENSION

E2.11 Find the power absorbed by the 6-kΩ resistor in the network in Fig. E2.11.   **PSV**   **ANSWER:** $P = 2.67$ mW.

Figure E2.11

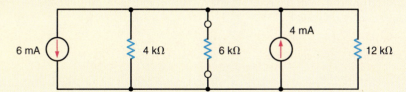

2.5   Series and Parallel Resistor Combinations

We have shown in our earlier developments that the equivalent resistance of N resistors in series is

$$R_S = R_1 + R_2 + \cdots + R_N \qquad \textbf{2.25}$$

and the equivalent resistance of N resistors in parallel is found from

$$\frac{1}{R_p} = \frac{1}{R_1} + \frac{1}{R_2} + \cdots + \frac{1}{R_N} \qquad \textbf{2.26}$$

Let us now examine some combinations of these two cases.

Example 2.20

We wish to determine the resistance at terminals A-B in the network in Fig. 2.28a.

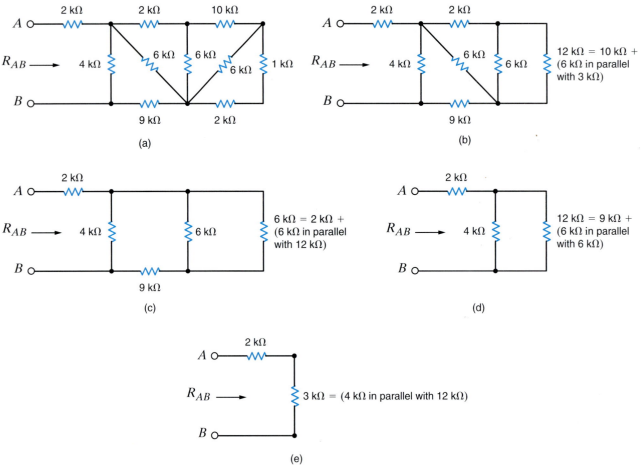

(a)

(b)

(c)

(d)

(e)

Figure 2.28 *Simplification of a resistance network.*

SOLUTION   Starting at the opposite end of the network from the terminals and combining resistors as shown in the sequence of circuits in Fig. 2.28, we find that the equivalent resistance at the terminals is 5 kΩ.

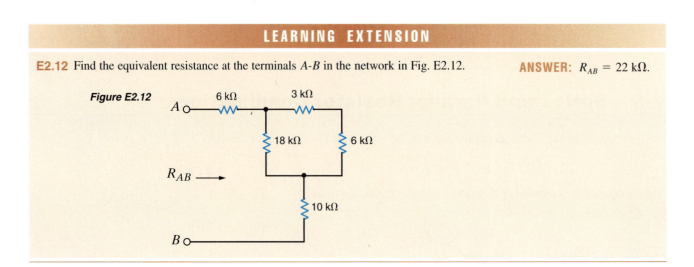

LEARNING EXTENSION

E2.12 Find the equivalent resistance at the terminals A-B in the network in Fig. E2.12.

ANSWER: $R_{AB} = 22$ kΩ.

Figure E2.12

PROBLEM-SOLVING STRATEGY

Simplifying Resistor Combinations

When trying to determine the equivalent resistance at a pair of terminals of a network composed of an interconnection of numerous resistors, it is recommended that the analysis begin at the end of the network opposite the terminals. Two or more resistors are combined to form a single resistor, thus simplifying the network by reducing the number of components as the analysis continues in a steady progression toward the terminals. The simplification involves the following:

Step 1. *Resistors in series.* Resistors R_1 and R_2 are in series if they are connected in tandem and carry exactly the same current. They can then be combined into a single resistor R_S, where $R_S = R_1 + R_2$.

Step 2. *Resistors in parallel.* Resistors R_1 and R_2 are in parallel if they are connected to the same two nodes and have exactly the same voltage across their terminals. They can then be combined into a single resistor R_p, where $R_p = R_1 R_2 / (R_1 + R_2)$.

These two combinations are used repeatedly, as needed, to reduce the network to a single resistor at the pair of terminals.

LEARNING EXTENSION

E2.13 Find the equivalent resistance at the terminals *A-B* in the circuit in Fig. E2.13.   **PSV**   **ANSWER:** $R_{AB} = 3\text{ k}\Omega$.

Figure E2.13

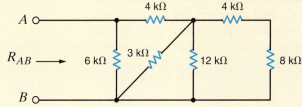

Example 2.21

A standard dc current-limiting power supply shown in Fig. 2.29a provides 0–18 V at 3 A to a load. The voltage drop, V_R, across a resistor, R, is used as a current-sensing device, fed back to the power supply and used to limit the current I. That is, if the load is adjusted so that the current tries to exceed 3 A, the power supply will act to limit the current to that value. The feedback voltage, V_R, should typically not exceed 600 mV.

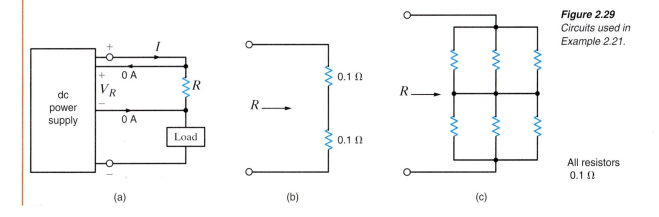

Figure 2.29
Circuits used in Example 2.21.

All resistors
0.1 Ω

(a)   (b)   (c)

If we have a box of standard 0.1-Ω, 5-W resistors, let us determine the configuration of these resistors that will provide $V_R = 600$ mV when the current is 3 A.

SOLUTION   Using Ohm's law, the value of R should be

$$R = \frac{V_R}{I}$$

$$= \frac{0.6}{3}$$

$$= 0.2 \ \Omega$$

Therefore, two 0.1-Ω resistors connected in series, as shown in Fig. 2.29b, will provide the proper feedback voltage. Suppose, however, that the power supply current is to be limited to 9 A. The resistance required in this case to produce $V_R = 600$ mV is

$$R = \frac{0.6}{9}$$

$$= 0.0667 \ \Omega$$

We must now determine how to interconnect the 0.1-Ω resistor to obtain $R = 0.0667 \ \Omega$. Since the desired resistance is less than the components available (i.e., 0.1-Ω), we must connect the resistors in some type of parallel configuration. Since all the resistors are of equal value, note that three of them connected in parallel would provide a resistance of one-third their value, or 0.0333 Ω. Then two such combinations connected in series, as shown in Fig. 2.29c, would produce the proper resistance.

Finally, we must check to ensure that the configurations in Figs. 2.29b and c have not exceeded the power rating of the resistors. In the first case, the current $I = 3$ A is present in each of the two series resistors. Therefore, the power absorbed in each resistor is

$$P = I^2R$$

$$= (3)^2(0.1)$$

$$= 0.9 \ W$$

which is well within the 5-W rating of the resistors.

In the second case, the current $I = 9$ A. The resistor configuration for R in this case is a series combination of two sets of three parallel resistors of equal value. Using current division, we know that the current I will split equally among the three parallel paths and, hence, the current in each resistor will be 3 A. Therefore, once again, the power absorbed by each resistor is within its power rating.

RESISTOR SPECIFICATIONS   Some important parameters that are used to specify resistors are the resistor's value, tolerance, and power rating. The tolerance specifications for resistors are typically 5% and 10%. A listing of standard resistor values with their specified tolerances is shown in Table 2.1.

The power rating for a resistor specifies the maximum power that can be dissipated by the resistor. Some typical power ratings for resistors are 1/4 W, 1/2 W, 1 W, 2 W, and so forth, up to very high values for high-power applications. Thus, in selecting a resistor for some particular application, one important selection criterion is the expected power dissipation.

Table 2.1 Standard resistor values for 5% and 10% tolerances (10% values shown in boldface)

1.0	**10**	**100**	**1.0k**	**10k**	**100k**	**1.0M**	**10M**
1.1	11	110	1.1k	11k	110k	**1.1M**	11M
1.2	**12**	**120**	**1.2k**	**12k**	**120k**	1.2M	**12M**
1.3	13	130	1.3k	13k	130k	1.3M	13M
1.5	**15**	**150**	**1.5k**	**15k**	**150k**	**1.5M**	**15M**
1.6	16	160	1.6k	16k	160k	1.6M	16M
1.8	**18**	**180**	**1.8k**	**18k**	**180k**	**1.8M**	**18M**
2.0	20	200	2.0k	20k	200k	2.0M	20M
2.2	**22**	**220**	**2.2k**	**22k**	**220k**	**2.2M**	**22M**
2.4	24	240	2.4k	24k	240k	2.4M	
2.7	**27**	**270**	**2.7k**	**27k**	**270k**	**2.7M**	
3.0	30	300	3.0k	30k	300k	3.0M	
3.3	**33**	**330**	**3.3k**	**33k**	**330k**	**3.3M**	
3.6	36	360	3.6k	36k	360k	3.6M	
3.9	**39**	**390**	**3.9k**	**39k**	**390k**	**3.9M**	
4.3	43	430	4.3k	43k	430k	4.3M	
4.7	**47**	**470**	**4.7k**	**47k**	**470k**	**4.7M**	
5.1	51	510	5.1k	51k	510k	5.1M	
5.6	**56**	**560**	**5.6k**	**56k**	**560k**	**5.6M**	
6.2	62	620	6.2k	62k	620k	6.2M	
6.8	**68**	**680**	**6.8k**	**68k**	**680k**	**6.8M**	
7.5	75	750	7.5k	75k	750k	7.5M	
8.2	**82**	**820**	**8.2k**	**82k**	**820k**	**8.2M**	
9.1	91	910	9.1k	91k	910k	9.1M	

Example 2.22

Given the network in Fig. 2.30, we wish to find the range for both the current and power dissipation in the resistor if R is a 2.7-kΩ resistor with a tolerance of 10%.

SOLUTION Using the equations $I = V/R = 10/R$ and $P = V^2/R = 100/R$, the minimum and maximum values for the resistor, current, and power are outlined next.

$$\text{Minimum resistor value} = R(1 - 0.1) = 0.9\,R = 2.43\ \text{k}\Omega$$
$$\text{Maximum resistor value} = R(1 + 0.1) = 1.1\,R = 2.97\ \text{k}\Omega$$
$$\text{Minimum current value} = 10/2970 = 3.37\ \text{mA}$$
$$\text{Maximum current value} = 10/2430 = 4.12\ \text{mA}$$
$$\text{Minimum power value} = 100/2970 = 33.7\ \text{mW}$$
$$\text{Maximum power value} = 100/2430 = 41.2\ \text{mW}$$

Thus, the ranges for the current and power are 3.37 mA to 4.12 mA and 33.7 mW to 41.2 mW, respectively.

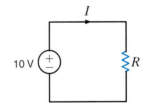

Figure 2.30
Circuit used in Example 2.22.

Example 2.23

Given the network shown in Fig. 2.31: (a) find the required value for the resistor R; (b) use Table 2.1 to select a standard 10% tolerance resistor for R; (c) using the resistor selected in (b), determine the voltage across the 3.9-kΩ resistor; (d) calculate the percent error in the voltage V_1, if the standard resistor selected in (b) is used; and (e) determine the power rating for this standard component.

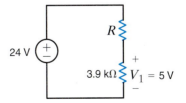

Figure 2.31
Circuit used in Example 2.23.

SOLUTION

a. Using KVL, the voltage across R is 19 V. Then using Ohm's law, the current in the loop is

$$I = 5/3.9k = 1.282 \text{ mA}$$

The required value of R is then

$$R = 19/0.001282 = 14.82 \text{ k}\Omega$$

b. As shown in Table 2.1, the nearest standard 10% tolerance resistor is 15 kΩ.

c. Using the standard 15-kΩ resistor, the actual current in the circuit is

$$I = 24/18.9k = 1.2698 \text{ mA}$$

and the voltage across the 3.9-kΩ resistor is

$$V = IR = (0.0012698)(3.9k) = 4.952 \text{ V}$$

d. The percent error involved in using the standard resistor is

$$\% \text{ Error} = (4.952 - 5)/5 \times 100 = -0.96\%$$

e. The power absorbed by the resistor R is then

$$P = IR = (0.0012698)^2(15k) = 24.2 \text{ mW}$$

Therefore, even a quarter-watt resistor is adequate in this application.

2.6 Circuits with Series-Parallel Combinations of Resistors

At this point we have learned many techniques that are fundamental to circuit analysis. Now we wish to apply them and show how they can be used in concert to analyze circuits. We will illustrate their application through a number of examples that will be treated in some detail.

Example 2.24

We wish to find all the currents and voltages labeled in the ladder network shown in Fig. 2.32a.

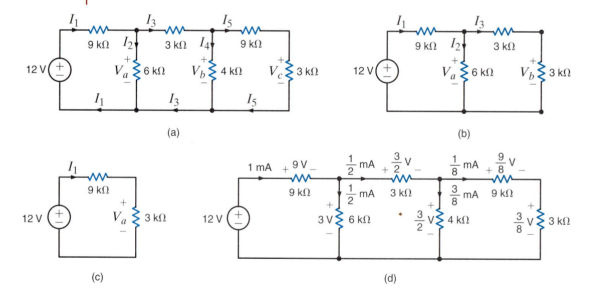

Figure 2.32 *Analysis of a ladder network.*

SOLUTION To begin our analysis of the network, we start at the right end of the circuit and combine the resistors to determine the total resistance seen by the 12-V source. This will allow us to calculate the current I_1. Then employing KVL, KCL, Ohm's law, and/or voltage and current division, we will be able to calculate all currents and voltages in the network.

At the right end of the circuit, the 9-kΩ and 3-kΩ resistors are in series and, thus, can be combined into one equivalent 12-kΩ resistor. This resistor is in parallel with the 4-kΩ resistor, and their combination yields an equivalent 3-kΩ resistor, shown at the right edge of the circuit in Fig. 2.32b. In Fig. 2.32b the two 3-kΩ resistors are in series, and their combination is in parallel with the 6-kΩ resistor. Combining all three resistances yields the circuit shown in Fig. 2.32c.

Applying Kirchhoff's voltage law to the circuit in Fig. 2.32c yields

$$I_1(9k + 3k) = 12$$
$$I_1 = 1 \text{ mA}$$

V_a can be calculated from Ohm's law as

$$V_a = I_1(3k)$$
$$= 3 \text{ V}$$

or, using Kirchhoff's voltage law,

$$V_a = 12 - 9kI_1$$
$$= 12 - 9$$
$$= 3 \text{ V}$$

Knowing I_1 and V_a, we can now determine all currents and voltages in Fig. 2.32b. Since $V_a = 3$ V, the current I_2 can be found using Ohm's law as

$$I_2 = \frac{3}{6k}$$
$$= \frac{1}{2} \text{ mA}$$

Then, using Kirchhoff's current law, we have

$$I_1 = I_2 + I_3$$
$$1 \times 10^{-3} = \frac{1}{2} \times 10^{-3} + I_3$$
$$I_3 = \frac{1}{2} \text{ mA}$$

Note that the I_3 could also be calculated using Ohm's law:

$$V_a = (3k + 3k)I_3$$
$$I_3 = \frac{3}{6k}$$
$$= \frac{1}{2} \text{ mA}$$

Applying Kirchhoff's voltage law to the right-hand loop in Fig. 2.32b yields

$$V_a - V_b = 3kI_3$$
$$3 - V_b = \frac{3}{2}$$
$$V_b = \frac{3}{2} \text{ V}$$

or, since V_b is equal to the voltage drop across the 3-kΩ resistor, we could use Ohm's law as

$$V_b = 3kI_3$$
$$= \frac{3}{2} \text{ V}$$

We are now in a position to calculate the final unknown currents and voltages in Fig. 2.32a. Knowing V_b, we can calculate I_4 using Ohm's law as

$$V_b = 4kI_4$$

$$I_4 = \frac{\frac{3}{2}}{4k}$$

$$= \frac{3}{8}\,\text{mA}$$

Then, from Kirchhoff's current law, we have

$$I_3 = I_4 + I_5$$

$$\frac{1}{2} \times 10^{-3} = \frac{3}{8} \times 10^{-3} + I_5$$

$$I_5 = \frac{1}{8}\,\text{mA}$$

We could also have calculated I_5 using the current-division rule. For example,

$$I_5 = \frac{4k}{4k + (9k + 3k)}\,I_3$$

$$= \frac{1}{8}\,\text{mA}$$

Finally, V_c can be computed as

$$V_c = I_5(3k)$$

$$= \frac{3}{8}\,\text{V}$$

V_c can also be found using voltage division (i.e., the voltage V_b will be divided between the 9-kΩ and 3-kΩ resistors). Therefore,

$$V_c = \left[\frac{3k}{3k + 9k} \right] V_b$$

$$= \frac{3}{8}\,\text{V}$$

Note that Kirchhoff's current law is satisfied at every node and Kirchhoff's voltage law is satisfied around every loop, as shown in Fig. 2.32d.

The following example is, in essence, the reverse of the previous example in that we are given the current in some branch in the network and are asked to find the value of the input source.

Example 2.25

Given the circuit in Fig. 2.33 and $I_4 = 1/2$ mA, let us find the source voltage V_o.

SOLUTION If $I_4 = 1/2$ mA, then from Ohm's law, $V_b = 3$ V. V_b can now be used to calculate $I_3 = 1$ mA. Kirchhoff's current law applied at node y yields

$$I_2 = I_3 + I_4$$

$$= 1.5\,\text{mA}$$

Then, from Ohm's law, we have

$$V_a = \left(1.5 \times 10^{-3}\right)(2k)$$

$$= 3\,\text{V}$$

Since $V_a + V_b$ is now known, I_5 can be obtained:

$$I_5 = \frac{V_a + V_b}{3k + 1k}$$

$$= 1.5 \, mA$$

Applying Kirchhoff's current law at node x yields

$$I_1 = I_2 + I_5$$

$$= 3 \, mA$$

Now KVL applied to any closed path containing V_o will yield the value of this input source. For example, if the path is the outer loop, KVL yields

$$-V_o + 6kI_1 + 3kI_5 + 1kI_5 + 4kI_1 = 0$$

Since $I_1 = 3 \, mA$ and $I_5 = 1.5 \, mA$,

$$V_o = 36 \, V$$

If we had selected the path containing the source and the points x, y, and z, we would obtain

$$-V_o + 6kI_1 + V_a + V_b + 4kI_1 = 0$$

Once again, this equation yields

$$V_o = 36 \, V.$$

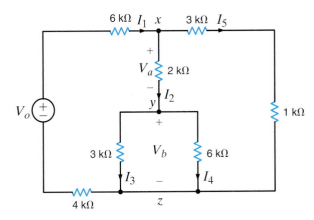

Figure 2.33
Example circuit for analysis.

PROBLEM-SOLVING STRATEGY

Analyzing Circuits Containing a Single Source and a Series-Parallel Interconnection of Resistors

Step 1. Systematically reduce the resistive network so that the resistance seen by the source is represented by a single resistor.

Step 2. Determine the source current for a voltage source or the source voltage if a current source is present.

Step 3. Expand the network, retracing the simplification steps, and apply Ohm's law, KVL, KCL, voltage division, and current division to determine all currents and voltages in the network.

E2.14 Find V_o in the network in Fig. E2.14.   **PSV**

ANSWER: $V_o = 2$ V.

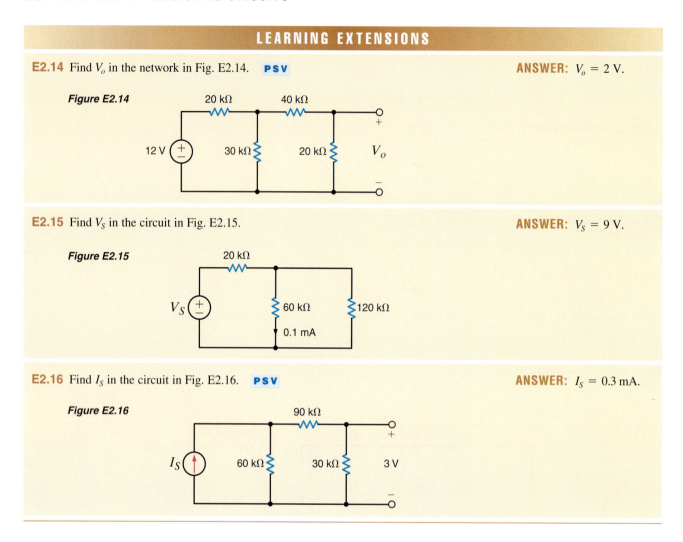

Figure E2.14

E2.15 Find V_S in the circuit in Fig. E2.15.

ANSWER: $V_S = 9$ V.

Figure E2.15

E2.16 Find I_S in the circuit in Fig. E2.16.   **PSV**

ANSWER: $I_S = 0.3$ mA.

Figure E2.16

2.7 Wye ⇌ Delta Transformations

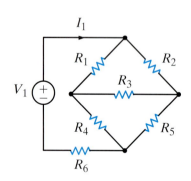

Figure 2.34
Network used to illustrate the need for the wye ⇌ delta transformation.

To provide motivation for this topic, consider the circuit in Fig. 2.34. Note that this network has essentially the same number of elements as contained in our recent examples. However, when we attempt to reduce the circuit to an equivalent network containing the source V_1 and an equivalent resistor R, we find that nowhere is a resistor in series or parallel with another. Therefore, we cannot attack the problem directly using the techniques that we have learned thus far. We can, however, replace one portion of the network with an equivalent circuit, and this conversion will permit us, with ease, to reduce the combination of resistors to a single equivalent resistance. This conversion is called the wye-to-delta or delta-to-wye transformation.

Consider the networks shown in Fig. 2.35. Note that the resistors in Fig. 2.35a form a Δ (delta) and the resistors in Fig. 2.35b form a Y (wye). If both of these configurations are connected at only three terminals *a*, *b*, and *c*, it would be very advantageous if an equivalence could be established between them. It is, in fact, possible to relate the resistances of one network to those of the other such that their terminal characteristics are the same. This relationship between the two network configurations is called the Y-Δ transformation.

The transformation that relates the resistances R_1, R_2, and R_3 to the resistances R_a, R_b, and R_c is derived as follows. For the two networks to be equivalent at each corresponding pair of terminals, it is necessary that the resistance at the corresponding terminals be equal (e.g., the resistance at terminals *a* and *b* with *c* open-circuited must be the same for both networks).

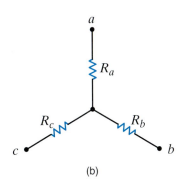

Figure 2.35
Delta and wye resistance networks.

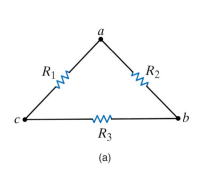

(a)   (b)

Therefore, if we equate the resistances for each corresponding set of terminals, we obtain the following equations:

$$R_{ab} = R_a + R_b = \frac{R_2(R_1 + R_3)}{R_2 + R_1 + R_3}$$

$$R_{bc} = R_b + R_c = \frac{R_3(R_1 + R_2)}{R_3 + R_1 + R_2}$$

$$R_{ca} = R_c + R_a = \frac{R_1(R_2 + R_3)}{R_1 + R_2 + R_3}$$

2.27

Solving this set of equations for R_a, R_b, and R_c yields

$$R_a = \frac{R_1 R_2}{R_1 + R_2 + R_3}$$

$$R_b = \frac{R_2 R_3}{R_1 + R_2 + R_3}$$

$$R_c = \frac{R_1 R_3}{R_1 + R_2 + R_3}$$

2.28

Similarly, if we solve Eq. (2.27) for R_1, R_2, and R_3, we obtain

$$R_1 = \frac{R_a R_b + R_b R_c + R_a R_c}{R_b}$$

$$R_2 = \frac{R_a R_b + R_b R_c + R_a R_c}{R_c}$$

$$R_3 = \frac{R_a R_b + R_b R_c + R_a R_c}{R_a}$$

2.29

Equations (2.28) and (2.29) are general relationships and apply to any set of resistances connected in a Y or Δ. For the balanced case where $R_a = R_b = R_c$ and $R_1 = R_2 = R_3$, the equations above reduce to

$$R_Y = \frac{1}{3} R_\Delta$$

2.30

and

$$R_\Delta = 3 R_Y$$

2.31

It is important to note that it is not necessary to memorize the formulas in Eqs. (2.28) and (2.29). Close inspection of these equations and Fig. 2.35 illustrates a definite pattern to the relationships between the two configurations. For example, the resistance connected to point *a* in the wye (i.e., R_a) is equal to the product of the two resistors in the Δ that are connected to point *a* divided by the sum of all the resistances in the delta. R_b and R_c are determined in a similar manner. Similarly, there are geometrical patterns associated with the equations for calculating the resistors in the delta as a function of those in the wye.

Let us now examine the use of the delta ⇌ wye transformation in the solution of a network problem.

Example 2.26

Given the network in Fig. 2.36a, let us find the source current I_S.

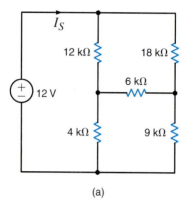
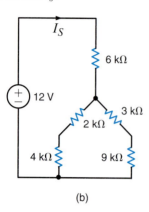

(a) (b)

Figure 2.36 Circuits used in Example 2.26.

SOLUTION   Note that none of the resistors in the circuit are in series or parallel. However, careful examination of the network indicates that the 12k-, 6k-, and 18k-ohm resistors, as well as the 4k-, 6k-, and 9k-ohm resistors each form a delta that can be converted to a wye. Furthermore, the 12k-, 6k-, and 4k-ohm resistors, as well as the 18k-, 6k-, and 9k-ohm resistors, each form a wye that can be converted to a delta. Any one of these conversions will lead to a solution. We will perform a delta-to-wye transformation on the 12k-, 6k-, and 18k-ohm resistors, which leads to the circuit in Fig. 2.36b. The 2k- and 4k-ohm resistors, like the 3k- and 9k-ohm resistors, are in series and their parallel combination yields a 4k-ohm resistor. Thus, the source current is

$$I_S = 12/(6k + 4k)$$
$$= 1.2 \text{ mA}.$$

LEARNING EXTENSIONS

E2.17 Determine the total resistance R_T in the circuit in Fig. E2.17.

ANSWER: $R_T = 34 \text{ k}\Omega$.

Figure E2.17

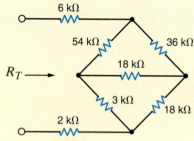

E2.18 Find V_o in the network in Fig. E2.18.

ANSWER: $V_o = 24 \text{ V}$.

Figure E2.18

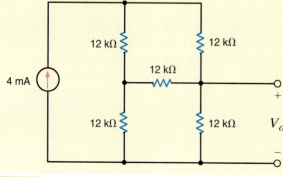

2.8   Circuits with Dependent Sources

In Chapter 1 we outlined the different kinds of dependent sources. These controlled sources are extremely important because they are used to model physical devices such as *npn* and *pnp* bipolar junction transistors (BJTs) and field-effect transistors (FETs) that are either metal-oxide-semiconductor field-effect transistors (MOSFETs) or insulated-gate field-effect transistors (IGFETs). These basic structures are, in turn, used to make analog and digital devices. A typical analog device is an operational amplifier (op-amp). This device is presented in Chapter 4. Typical digital devices are random access memories (RAMs), read-only memories (ROMs), and microprocessors. We will now show how to solve simple one-loop and one-node circuits that contain these dependent sources. Although the following examples are fairly simple, they will serve to illustrate the basic concepts.

PROBLEM-SOLVING STRATEGY

Circuits with Dependent Sources

Step 1. When writing the KVL and/or KCL equations for the network, treat the dependent source as though it were an independent source.

Step 2. Write the equation that specifies the relationship of the dependent source to the controlling parameter.

Step 3. Solve the equations for the unknowns. Be sure that the number of linearly independent equations matches the number of unknowns.

The following four examples will each illustrate one of the four types of dependent sources: current-controlled voltage source, current-controlled current source, voltage-controlled voltage source, and voltage-controlled current source.

Example 2.27

Let us determine the voltage V_o in the circuit in Fig. 2.37.

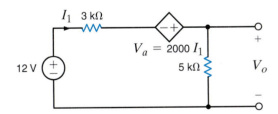

Figure 2.37
Circuit used in Example 2.27.

SOLUTION   Applying KVL, we obtain

$$-12 + 3kI_1 - V_A + 5kI_1 = 0$$

where

$$V_A = 2000I_1$$

and the units of the multiplier, 2000, are ohms. Solving these equations yields

$$I_1 = 2 \text{ mA}$$

Then

$$V_o = (5 \text{ k})I_1$$
$$= 10 \text{ V}$$

Example 2.28

Given the circuit in Fig. 2.38 containing a current-controlled current source, let us find the voltage V_o.

Figure 2.38
Circuit used in
Example 2.28.

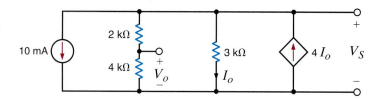

SOLUTION Applying KCL at the top node, we obtain

$$10 \times 10^{-3} + \frac{V_S}{2k + 4k} + \frac{V_S}{3k} - 4I_o = 0$$

where

$$I_o = \frac{V_S}{3k}$$

Substituting this expression for the controlled source into the KCL equation yields

$$10^{-2} + \frac{V_S}{6k} + \frac{V_S}{3k} - \frac{4V_S}{3k} = 0$$

Solving this equation for V_S, we obtain

$$V_S = 12 \text{ V}$$

The voltage V_o can now be obtained using a simple voltage divider; that is,

$$V_o = \left[\frac{4k}{2k + 4k} \right] V_S$$

$$= 8 \text{ V}$$

Example 2.29

The network in Fig. 2.39 contains a voltage-controlled voltage source. We wish to find V_o in this circuit.

Figure 2.39
Circuit used in
Example 2.29.

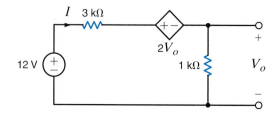

SOLUTION Applying KVL to this network yields

$$-12 + 3kI + 2V_o + 1kI = 0$$

where

$$V_o = 1kI$$

Hence, the KVL equation can be written as

$$-12 + 3kI + 2kI + 1kI = 0$$

or

$$I = 2 \text{ mA}$$

Therefore,

$$V_o = 1kI$$
$$= 2 \text{ V}$$

Example 2.30

An equivalent circuit for a FET common-source amplifier or BJT common-emitter amplifier can be modeled by the circuit shown in Fig. 2.40a. We wish to determine an expression for the gain of the amplifier, which is the ratio of the output voltage to the input voltage.

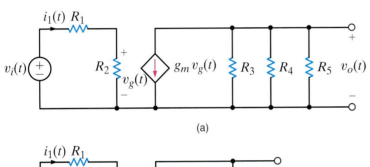

(a)

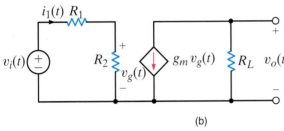

(b)

Figure 2.40
Example circuit containing a voltage-controlled current source.

SOLUTION Note that although this circuit, which contains a voltage-controlled current source, appears to be somewhat complicated, we are actually in a position now to solve it with techniques we have studied up to this point. The loop on the left, or input to the amplifier, is essentially detached from the output portion of the amplifier on the right. The voltage across R_2 is $v_g(t)$, which controls the dependent current source.

To simplify the analysis, let us replace the resistors R_3, R_4, and R_5 with R_L such that

$$\frac{1}{R_L} = \frac{1}{R_3} + \frac{1}{R_4} + \frac{1}{R_5}$$

Then the circuit reduces to that shown in Fig. 2.40b. Applying Kirchhoff's voltage law to the input portion of the amplifier yields

$$v_i(t) = i_1(t)(R_1 + R_2)$$

and

$$v_g(t) = i_1(t)R_2$$

Solving these equations for $v_g(t)$ yields

$$v_g(t) = \frac{R_2}{R_1 + R_2} v_i(t)$$

From the output circuit, note that the voltage $v_o(t)$ is given by the expression

$$v_o(t) = -g_m v_g(t)R_L$$

Combining this equation with the preceding one yields

$$v_o(t) = \frac{-g_m R_L R_2}{R_1 + R_2} v_i(t)$$

Therefore, the amplifier gain, which is the ratio of the output voltage to the input voltage, is given by

$$\frac{v_o(t)}{v_i(t)} = -\frac{g_m R_L R_2}{R_1 + R_2}$$

Reasonable values for the circuit parameters in Fig. 2.40a are $R_1 = 100\ \Omega$, $R_2 = 1\ k\Omega$, $g_m = 0.04\ S$, $R_3 = 50\ k\Omega$, and $R_4 = R_5 = 10\ k\Omega$. Hence, the gain of the amplifier under these conditions is

$$\frac{v_o(t)}{v_i(t)} = \frac{-(0.04)(4.545)(10^3)(1)(10^3)}{(1.1)(10^3)}$$

$$= -165.29$$

Thus, the magnitude of the gain is 165.29.

At this point it is perhaps helpful to point out again that when analyzing circuits with dependent sources, we first treat the dependent source as though it were an independent source when we write a Kirchhoff's current or voltage law equation. Once the equation is written, we then write the controlling equation that specifies the relationship of the dependent source to the unknown variable. For instance, the first equation in Example 2.28 treats the dependent source like an independent source. The second equation in the example specifies the relationship of the dependent source to the voltage, which is the unknown in the first equation.

LEARNING EXTENSIONS

E2.19 Find V_o in the circuit in Fig. E2.19.   **PSV**

ANSWER: $V_o = 12\ V$.

Figure E2.19

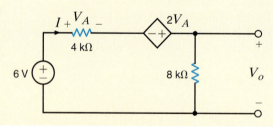

E2.20 Find V_o in the network in Fig. E2.20.

ANSWER: $V_o = 8\ V$.

Figure E2.20

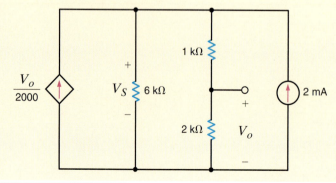

2.9 Application Examples

Throughout this book we endeavor to present a wide variety of examples that demonstrate the usefulness of the material under discussion in a practical environment. To enhance our presentation of the practical aspects of circuit analysis and design, we have dedicated sections, such as this one, in most chapters for the specific purpose of presenting additional application-oriented examples.

Application Example 2.31

The eyes (heating elements) of an electric range are frequently made of resistive nichrome strips. Operation of the eye is quite simple. A current is passed through the heating element causing it to dissipate power in the form of heat. Also, a four-position selector switch, shown in Fig. 2.41, controls the power (heat) output. In this case the eye consists of two nichrome strips modeled by the resistors R_1 and R_2, where $R_1 < R_2$.

1. How should positions A, B, C, and D be labeled with regard to high, medium, low, and off settings?

2. If we desire that high and medium correspond to 2000 W and 1200 W power dissipation, respectively, what are the values of R_1 and R_2?

3. What is the power dissipation at the low setting?

SOLUTION   Position A is the off setting since no current flows to the heater elements. In position B, current flows through R_2 only, while in position C current flows through R_1 only. Since $R_1 < R_2$, more power will be dissipated when the switch is at position C. Thus, position C is the medium setting, B is the low setting, and, by elimination, position D is the high setting.

When the switch is at the medium setting, only R_1 dissipates power, and we can write R_1 as

$$R_1 = \frac{V_S^2}{P_1} = \frac{230^2}{1200}$$

or

$$R_1 = 44.08\ \Omega$$

On the high setting, 2000 W of total power is delivered to R_1 and R_2. Since R_1 dissipates 1200 W, R_2 must dissipate the remaining 800 W. Therefore, R_2 is

$$R_2 = \frac{V_S^2}{P_2} = \frac{230^2}{800}$$

or

$$R_2 = 66.13\ \Omega$$

Finally, at the low setting, only R_2 is connected to the voltage source; thus, the power dissipation at this setting is 800 W.

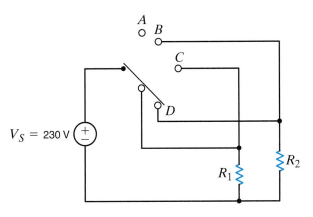

Figure 2.41
Simple resistive heater selector circuit.

58 CHAPTER 2 RESISTIVE CIRCUITS

Application Example 2.32

Have you ever cranked your car with the headlights on? While the starter kicked the engine, you probably saw the headlights dim then return to normal brightness once the engine was running on its own. Can we create a model to predict this phenomenon?

SOLUTION Yes, we can. Consider the conceptual circuit in Fig. 2.42a, and the model circuit in Fig. 2.42b, which isolates just the battery, headlights, and starter. Note the resistor R_{batt}. It is included to model several power loss mechanisms that can occur between the battery and the loads, that is, the headlights and starter. First, there are the chemical processes within the battery itself which are not 100% efficient. Second, there are the electrical connections at both the battery posts and the loads. Third, the wiring itself has some resistance, although this is usually so small that it is negligible. The sum of these losses is modeled by R_{batt}, and we expect the value of R_{batt} to be small. A reasonable value is 25 mΩ.

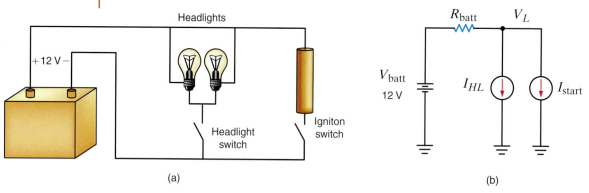

(a)   (b)

Figure 2.42 *A conceptual (a) model and (b) circuit for examining the effect of starter current on headlight intensity.*

Next we address the starter. When energized, a typical automobile starter will draw between 90 and 120 A. We will use 100 A as a typical number. Finally, the headlights will draw much less current—perhaps only 1 A. Now we have values to use in our model circuit.

Assume first that the starter is off. By applying KCL at the node labeled V_L, we find that the voltage applied to the headlights can be written as

$$V_L = V_{batt} - I_{HL}R_{batt}$$

Substituting our model values into this equation yields $V_L = 11.75$ V—very close to 12 V. Now we energize the starter and apply KCL again.

$$V_L = V_{batt} - (I_{HL} + I_{start})R_{batt}$$

Now the voltage across the headlights is only 9.25 V. No wonder the headlights dim! How would corrosion or loose connections on the battery posts change the situation? In this case, we would expect the quality of the connection from battery to load to deteriorate, increasing R_{batt} and compounding the headlight dimming issue.

Application Example 2.33

A Wheatstone Bridge circuit is an accurate device for measuring resistance. The circuit, shown in Fig. 2.43, is used to measure the unknown resistor R_x. The center leg of the circuit contains a galvanometer, which is a very sensitive device that can be used to measure current in the microamp range. When the unknown resistor is connected to the bridge, R_3 is adjusted until the current in the galvanometer is zero, at which point the bridge is balanced. In this balanced condition

$$\frac{R_1}{R_3} = \frac{R_2}{R_x}$$

so that

$$R_x = \left(\frac{R_2}{R_1}\right)R_3$$

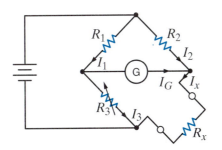

Figure 2.43
The Wheatstone bridge circuit.

Engineers also use this bridge circuit to measure strain in solid material. For example, a system used to determine the weight of a truck is shown in Fig. 2.44a. The platform is supported by cylinders on which strain gauges are mounted. The strain gauges, which measure strain when the cylinder deflects under load, are connected to a Wheatstone bridge as shown in Fig. 2.44b. The strain gauge has a resistance of 120 Ω under no-load conditions and changes value under load. The variable resistor in the bridge is a calibrated precision device.

Weight is determined in the following manner. The ΔR_3 required to balance the bridge represents the Δ strain, which when multiplied by the modulus of elasticity yields the Δ stress. The Δ stress multiplied by the cross-sectional area of the cylinder produces the Δ load, which is used to determine weight.

Let us determine the value of R_3 under no load when the bridge is balanced and its value when the resistance of the strain gauge changes to 120.24 Ω under load.

SOLUTION   Using the balance equation for the bridge, the value of R_3 at no load is

$$R_3 = \left(\frac{R_1}{R_2}\right) R_x$$

$$= \left(\frac{100}{110}\right)(120)$$

$$= 109.0909 \ \Omega$$

Under load, the value of R_3 is

$$R_3 = \left(\frac{100}{110}\right)(120.24)$$

$$= 109.3091 \ \Omega$$

Therefore, the ΔR_3 is

$$\Delta R_3 = 109.3091 - 109.0909$$

$$= 0.2182 \ \Omega$$

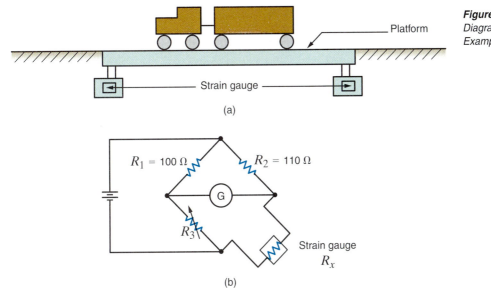

Figure 2.44
Diagrams used in
Example 2.33.

2.10 Design Examples

Most of this text is concerned with circuit analysis; that is, given a circuit in which all the components are specified, analysis involves finding such things as the voltage across some element or the current through another. Furthermore, the solution of an analysis problem is generally unique. In contrast, design involves determining the circuit configuration that will meet certain specifications. In addition, the solution is generally not unique in that there may be many ways to satisfy the circuit/performance specifications. It is also possible that there is no solution that will meet the design criteria.

In addition to meeting certain technical specifications, designs normally must also meet other criteria, such as economic, environmental, and safety constraints. For example, if a circuit design that meets the technical specifications is either too expensive or unsafe, it is not viable regardless of its technical merit.

At this point, the number of elements that we can employ in circuit design is limited primarily to the linear resistor and the active elements we have presented. However, as we progress through the text we will introduce a number of other elements (for example, the op-amp, capacitor, and inductor), which will significantly enhance our design capability.

We begin our discussion of circuit design by considering a couple of simple examples that demonstrate the selection of specific components to meet certain circuit specifications.

Design Example 2.34

An electronics hobbyist who has built his own stereo amplifier wants to add a back-lit display panel to his creation for that professional look. His panel design requires seven light bulbs—two operate at 12 V/15 mA and five at 9 V/5 mA. Luckily, his stereo design already has a quality 12-V dc supply; however, there is no 9-V supply. Rather than building a new dc power supply, let us use the inexpensive circuit shown in Fig. 2.45a to design a 12-V to 9-V converter with the restriction that the variation in V_2 be no more than $\pm 5\%$. In particular, we must determine the necessary values of R_1 and R_2.

SOLUTION First, lamps L_1 and L_2 have no effect on V_2. Second, when lamps L_3–L_7 are on, they each have an equivalent resistance of

$$R_{eq} = \frac{V_2}{I} = \frac{9}{0.005} = 1.8 \text{ k}\Omega$$

As long as V_2 remains fairly constant, the lamp resistance will also be fairly constant. Thus, the requisite model circuit for our design is shown in Fig. 2.45b. The voltage V_2 will be at its maximum value of $9 + 5\% = 9.45$ V when L_3–L_7 are all off. In this case R_1 and R_2 are in series, and V_2 can be expressed by simple voltage division as

$$V_2 = 9.45 = 12\left[\frac{R_2}{R_1 + R_2}\right]$$

Rearranging the equation yields

$$\frac{R_1}{R_2} = 0.27$$

A second expression involving R_1 and R_2 can be developed by considering the case when L_3–L_7 are all on, which causes V_2 to reach its minimum value of $9 - 5\%$, or 8.55 V. Now, the effective resistance of the lamps is five 1.8-kΩ resistors in parallel, or 360 Ω. The corresponding expression for V_2 is

$$V_2 = 8.55 = 12\left[\frac{R_2//360}{R_1 + (R_2//360)}\right]$$

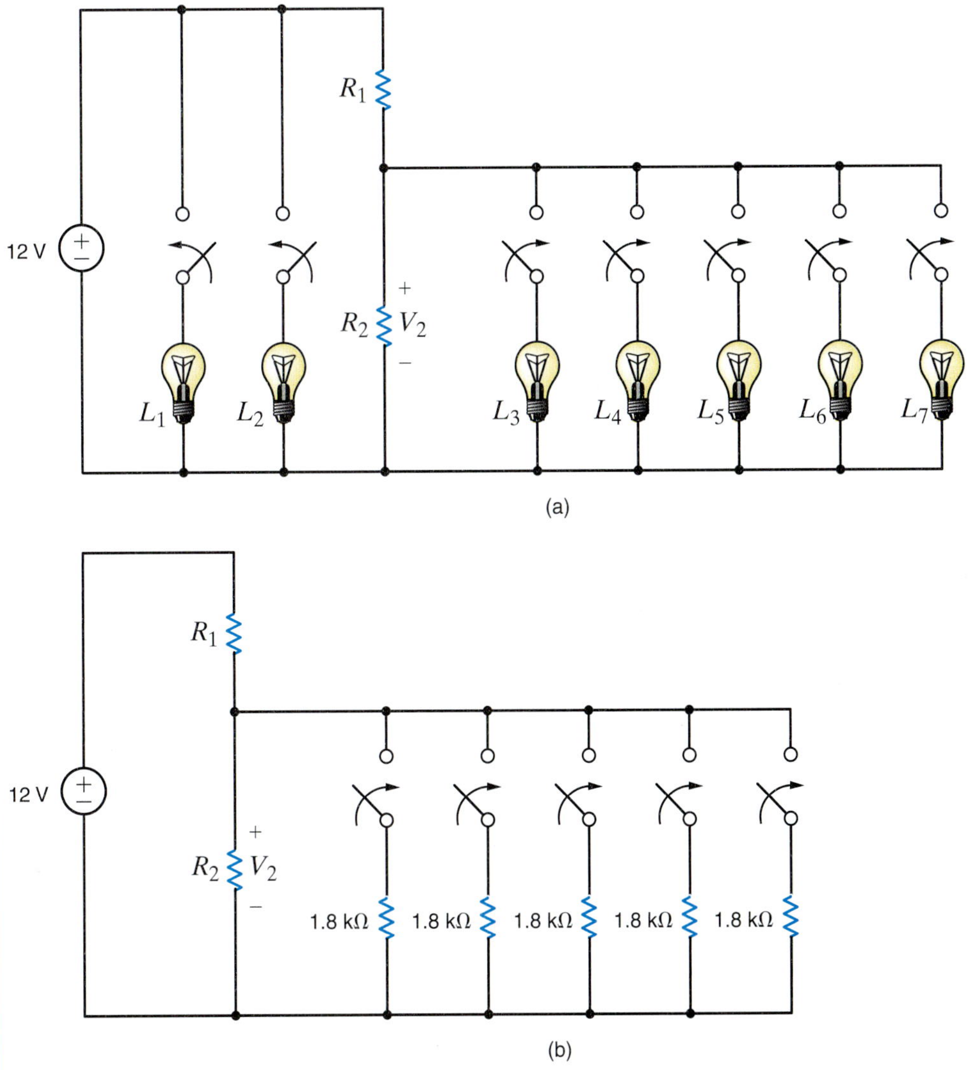

Figure 2.45
*12-V to 9-V converter circuit
for powering panel lighting.*

(a)

(b)

which can be rewritten in the form

$$\frac{\dfrac{360R_1}{R_2} + 360 + R_1}{360} = \frac{12}{8.55} = 1.4$$

Substituting the value determined for R_1/R_2 into the preceding equation yields

$$R_1 = 360[1.4 - 1 - 0.27]$$

or

$$R_1 = 48.1 \ \Omega$$

and so for R_2

$$R_2 = 178.3 \ \Omega$$

Design Example 2.35

Let's design a circuit that produces a 5-V output from a 12-V input. We will arbitrarily fix the power consumed by the circuit at 240 mW. Finally, we will choose the best possible standard resistor values from Table 2.1 and calculate the percent error in the output voltage that results from that choice.

SOLUTION The simple voltage divider, shown in Fig. 2.46, is ideally suited for this application. We know that V_o is given by

$$V_o = V_{in}\left[\frac{R_2}{R_1 + R_2}\right]$$

which can be written as

$$R_1 = R_2\left[\frac{V_{in}}{V_o} - 1\right]$$

Figure 2.46
A simple voltage divider.

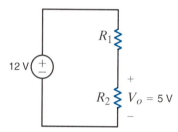

Since all of the circuit's power is supplied by the 12-V source, the total power is given by

$$P = \frac{V_{in}^2}{R_1 + R_2} \leq 0.24$$

Using the second equation to eliminate R_1, we find that R_2 has a lower limit of

$$R_2 \geq \frac{V_o V_{in}}{P} = \frac{(5)(12)}{0.24} = 250 \ \Omega$$

Substituting these results into the second equation yields the lower limit of R_1, that is

$$R_1 = R_2\left[\frac{V_{in}}{V_o} - 1\right] \geq 350 \ \Omega$$

Thus, we find that a significant portion of Table 2.1 is not applicable to this design. However, determining the best pair of resistor values is primarily a trial-and-error operation that can be enhanced by using an Excel spreadsheet. After a number of trials, we find that the best values that satisfy the constraint equations are

$$R_2 = 750 \ \Omega \quad \text{and} \quad R_1 = 1000 \ \Omega$$

and these values yield an output voltage of 5.14 V. The resulting error in the output voltage can be determined from the expression

$$\text{percent error} = \left[\frac{5.14 - 5}{5}\right]100\% = 2.8\%$$

Note that by selecting the best possible resistor values, we have an output error of 2.8% that is less than the 5% resistor tolerance.

It should be noted, however, that these resistor values are nominal, that is, typical values. To find the worst-case error, we must consider that each resistor as purchased may be as much as $\pm 5\%$ off from the nominal value. In this application, since V_o is already greater than the target of 5 V, the worst-case scenario occurs when V_o increases even further, that is, when R_1 is 5% too low (950 Ω) and R_2 is 5% too high (787.5 Ω). The resulting output voltage is 5.44 V, that is, an 8.8% error. Of course, most resistor values are closer to the nominal value than to the guaranteed maximum/minimum values. However, if we intend to build this circuit with a guaranteed tight output error, for example, 5%, we should purchase resistors with lower tolerances.

How much lower should the tolerances be? Our first equation can be altered to yield the worst-case output voltage by adding a tolerance, Δ, to R_2 and subtracting the tolerance from R_1. Let us choose a worst-case output voltage of $V_{o\max} = 5.25$ V, that is, a 5% error.

$$V_{o\max} = 5.25 = V_{in}\left[\frac{R_2(1 + \Delta)}{R_1(1 - \Delta) + R_2(1 + \Delta)}\right] = 12\left[\frac{750(1 + \Delta)}{750(1 + \Delta) + 1000(1 - \Delta)}\right]$$

The resulting value of Δ is 0.018, or 1.8%. Standard resistors are available in tolerances of 10%, 5%, 2%, and 1%. Tighter tolerances are available but very expensive. Thus, based on nominal values of 750 Ω and 1000 Ω, we should choose 1% resistors to assure an output voltage error less than 5%.

Design Example 2.36

In factory instrumentation, process parameters such as pressure and flowrate are measured, converted to electrical signals, and sent some distance to an electronic controller. The controller then decides what actions should be taken. One of the main concerns in these systems is the physical distance between the sensor and the controller. An industry standard format for encoding the measurement value is called the 4–20 mA standard, where the parameter range is linearly distributed from 4 to 20 mA. For example, a 100 psi pressure sensor would output 4 mA if the pressure were 0 psi, 20 mA at 100 psi, and 12 mA at 50 psi. But most instrumentation is based on voltages between 0 and 5 V, not on currents.

Therefore, let us design a current-to-voltage converter that will output 5 V when the current signal is 20 mA.

SOLUTION   The circuit in Fig. 2.47a is a very accurate model of our situation. The wiring from the sensor unit to the controller has some resistance, R_{wire}. If the sensor output were a voltage proportional to pressure, the voltage drop in the line would cause measurement error even if the sensor output were an ideal source of voltage. But, since the data are contained in the current value, R_{wire} does not affect the accuracy at the controller as long as the sensor acts as an ideal current source.

Figure 2.47
*The 4–20 mA control loop
(a) block diagram, (b) with the
current-to-voltage converter.*

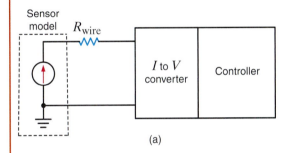

(a)

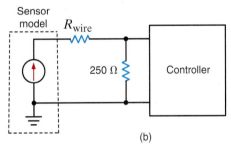

(b)

As for the current-to-voltage converter, it is extremely simple—a resistor. For 5 V at 20 mA, we employ Ohm's law to find

$$R = \frac{5}{0.02} = 250 \ \Omega$$

The resulting converter is added to the system in Fig. 2.47b, where we tacitly assume that the controller does not load the remaining portion of the circuit.

Note that the model indicates that the distance between the sensor and controller could be infinite. Intuitively, this situation would appear to be unreasonable, and it is. Losses that would take place over distance can be accounted for by using a more accurate model of the sensor, as shown in Fig. 2.48. The effect of this new sensor model can be seen from the equations that describe this new network. The model equations are

$$I_S = \frac{V_S}{R_S} + \frac{V_S}{R_{\text{wire}} + 250}$$

and

$$I_{\text{signal}} = \frac{V_S}{R_{\text{wire}} + 250}$$

Combining these equations yields

$$\frac{I_{\text{signal}}}{I_S} = \frac{1}{1 + \dfrac{R_{\text{wire}} + 250}{R_S}}$$

Thus, we see that it is the size of R_S relative to $(R_{\text{wire}} + 250\ \Omega)$ that determines the accuracy of the signal at the controller. Therefore, we want R_S as large as possible. Both the maximum sensor output voltage and output resistance, R_S, are specified by the sensor manufacturer.

We will revisit this current-to-voltage converter in Chapter 4.

Figure 2.48
A more accurate model for the 4–20 mA control loop.

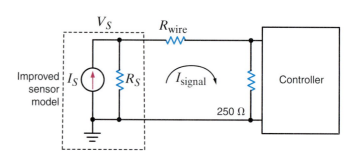

Design Example 2.37

The network in Fig. 2.49 is an equivalent circuit for a transistor amplifier used in a stereo preamplifier. The input circuitry, consisting of a 2-mV source in series with a 500-Ω resistor, models the output of a compact disk player. The dependent source, R_{in}, and R_o model the transistor, which amplifies the signal and then sends it to the power amplifier. The 10-kΩ load resistor models the input to the power amplifier that actually drives the speakers. We must design a transistor amplifier as shown in Fig. 2.49 that will provide an overall gain of -200. In practice we do not actually vary the device parameters to achieve the desired gain; rather, we select a transistor from the manufacturer's data books that will satisfy the required specification. The model parameters for three different transistors are listed as follows:

Manufacturer's transistor parameter values

Part Number	R_{in} (kΩ)	R_o (kΩ)	g_m (mA/V)
1	1.0	50	50
2	2.0	75	30
3	8.0	80	20

Design the amplifier by choosing the transistor that produces the most accurate gain. What is the percent error of your choice?

SOLUTION   The output voltage can be written

$$V_o = -g_m V(R_o // R_L)$$

Using voltage division at the input to find V,

$$V = V_S \left(\frac{R_{\text{in}}}{R_{\text{in}} + R_S} \right)$$

Combining these two expressions, we can solve for the gain:

$$A_V = \frac{V_o}{V_S} = -g_m \left(\frac{R_{\text{in}}}{R_{\text{in}} + R_S} \right)(R_o // R_L)$$

Using the parameter values for the three transistors, we find that the best alternative is transistor number 2, which has a gain error of

$$\text{Percent error} = \left(\frac{211.8 - 200}{200} \right) \times 100\% = 5.9\%$$

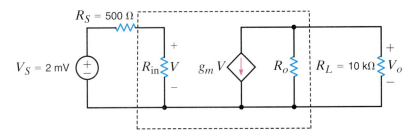

Figure 2.49
Transistor amplifier circuit model.

SUMMARY

- **Ohm's law** $V = IR$
- **The passive sign convention with Ohm's law** The current enters the resistor terminal with the positive voltage reference.
- **Kirchhoff's current law (KCL)** The algebraic sum of the currents leaving (entering) a node is zero.
- **Kirchhoff's voltage law (KVL)** The algebraic sum of the voltages around any closed path is zero.
- **Solving a single-loop circuit** Determine the loop current by applying KVL and Ohm's law.
- **Solving a single-node-pair circuit** Determine the voltage between the pair of nodes by applying KCL and Ohm's law.
- **The voltage-division rule** The voltage is divided between two series resistors in direct proportion to their resistance.

- **The current-division rule** The current is divided between two parallel resistors in reverse proportion to their resistance.
- **The equivalent resistance of a network of resistors** Combine resistors in series by adding their resistances. Combine resistors in parallel by adding their conductances. The wye-to-delta and delta-to-wye transformations are also an aid in reducing the complexity of a network.
- **Short circuit** Zero resistance, zero voltage; the current in the short is determined by the rest of the circuit.
- **Open circuit** Zero conductance, zero current; the voltage across the open terminals is determined by the rest of the circuit.

PROBLEMS

PSV **CS** both available on the web at: **http://www.justask4u.com/irwin**

SECTION 2.1

2.1 Find the current I and the power supplied by the source in the network in Fig. P2.1. **CS**

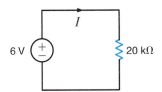

Figure P2.1

2.2 In the network in Fig. P2.2, the power absorbed by R_x is 20 mW. Find R_x. **CS**

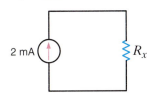

Figure P2.2

2.3 Find the current I and the power supplied by the source in the network in Fig. P2.3.

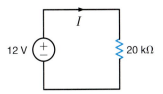

Figure P2.3

2.4 In the circuit in Fig. P2.4, find the voltage across the current source and the power absorbed by the resistor.

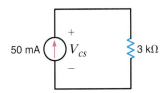

Figure P2.4

2.5 If the 5-kΩ resistor in the network in Fig. P2.5 absorbs 200 mW, find V_S.

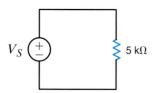

Figure P2.5

2.6 In the network in Fig. P2.6, the power absorbed by G_x is 20 mW. Find G_x. **PSV**

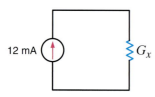

Figure P2.6

2.7 A model for a standard two D-cell flashlight is shown in Fig. P2.7. Find the power dissipated in the lamp.

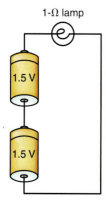

Figure P2.7

2.8 An automobile uses two halogen headlights connected as shown in Fig. P2.8. Determine the power supplied by the battery if each headlight draws 3 A of current.

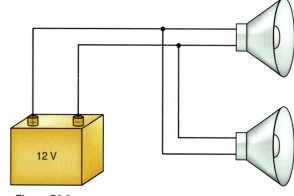

Figure P2.8

2.9 Many years ago a string of Christmas tree lights was manufactured in the form shown in Fig. P2.9a. Today the lights are manufactured as shown in Fig. P2.9b. Is there a good reason for this change?

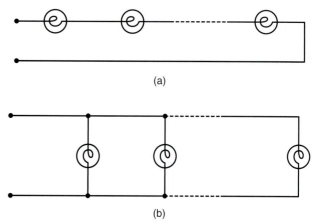

Figure P2.9

SECTION 2.2

2.10 Find I_1 in the network in Fig. P2.10. **CS**

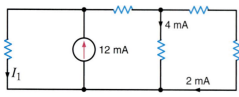

Figure P2.10

2.11 Find I_1 and I_2 in the circuit in Fig. P2.11. **CS**

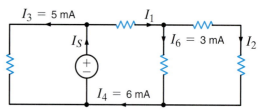

Figure P2.11

2.12 Find I_o and I_1 in the circuit in Fig. P2.12.

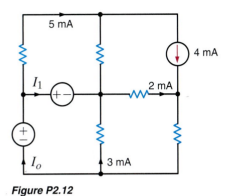

Figure P2.12

2.13 Find I_x in the circuit in Fig. P2.13.

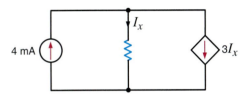

Figure P2.13

2.14 Find I_x in the circuit in Fig. P2.14. **PSV**

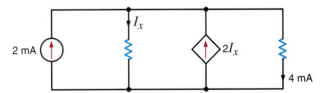

Figure P2.14

2.15 Find I_x in the circuit in Fig. P2.15. **CS**

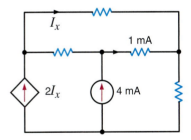

Figure P2.15

2.16 Find V_x in the circuit in Fig. P2.16. **CS**

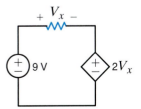

Figure P2.16

2.17 Find V_{fb} and V_{ec} in the circuit in Fig. P2.17.

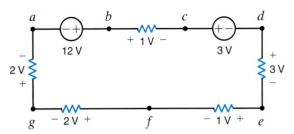

Figure P2.17

2.18 Find V_{ac} in the circuit in Fig. P2.18. **CS**

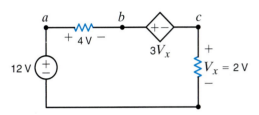

Figure P2.18

2.19 Find V_{da} and V_{be} in the circuit in Fig. P2.19.

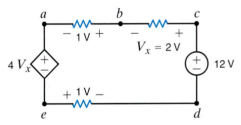

Figure P2.19

2.20 The 10-V source absorbs 2.5 mW of power. Calculate V_{ba} and the power absorbed by the dependent voltage source in Fig. P2.20.

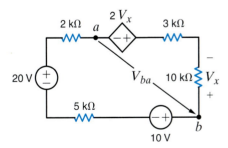

Figure P2.20

2.21 Find V_o in the network in Fig. P2.21.

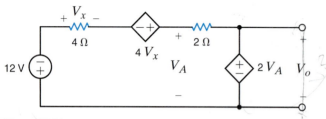

Figure P2.21

2.22 Find V_o in the circuit in Fig. P2.22. **PSV**

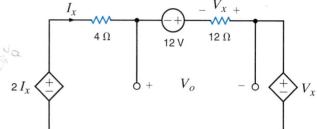

Figure P2.22

SECTION 2.3

2.23 Find V_{ac} in the network in Fig. P2.23.

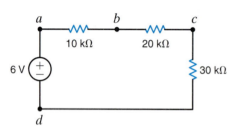

Figure P2.23

2.24 Find both I and V_{bd} in the circuit in Fig. P2.24.

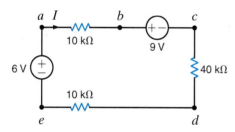

Figure P2.24

2.25 Find V_x in the circuit in Fig. P2.25. **CS**

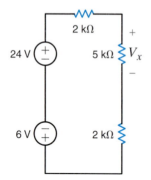

Figure P2.25

2.26 Find V_1 in the network in Fig. P2.26. **PSV**

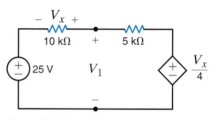

Figure P2.26

2.27 Find the power absorbed by the 30-kΩ resistor in the circuit in Fig. P2.27. **CS**

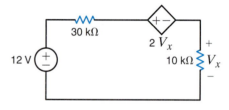

Figure P2.27

2.28 In the network in Fig. P2.28, if $V_x = 12$ V, find V_S.

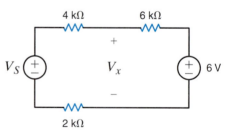

Figure P2.28

2.29 In the circuit in Fig. P2.29, $P_{3k\Omega} = 12$ mW. Find V_S.

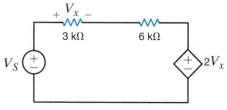

Figure P2.29

2.30 If $V_o = 4$ V in the network in Fig. P2.30, find V_S.

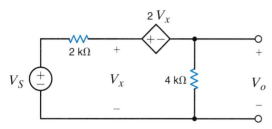

Figure P2.30

2.31 If $V_A = 12$ V in the circuit in Fig. P2.31, find V_S.

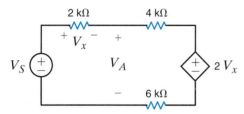

Figure P2.31

2.32 A commercial power supply is modeled by the network shown in Fig. P2.32.

(a) Plot V_o versus R_{load} for $1\ \Omega \le R_{\text{load}} \le \infty$.

(b) What is the maximum value of V_o in (a)?

(c) What is the minimum value of V_o in (a)?

(d) If for some reason the output should become short circuited, that is, $R_{\text{load}} \to 0$, what current is drawn from the supply?

(e) What value of R_{load} corresponds to maximum power consumed?

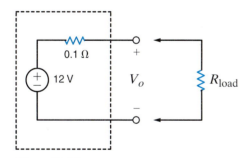

Figure P2.32

2.33 A commercial power supply is guaranteed by the manufacturer to deliver 5 V ±1% across a load range of 0 to 10 A. Using the circuit in Fig. P2.33 to model the supply, determine the appropriate values of R and V.

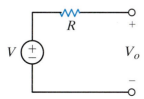

Figure P2.33

2.34 A power supply is specified to provide 48 ±2 V at 0–200 A and is modeled by the circuit in Fig. P2.34.

(a) What are the appropriate values for V and R?

(b) What is the maximum power the supply can deliver? What values of I_{load} and V_o correspond to that level?

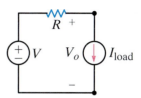

Figure P2.34

2.35 Although power supply loads are often modeled as either resistors or constant current sources, some loads are best modeled as constant power loads, as indicated in Fig. P2.35. Given the model shown in the figure,

(a) Write a V–I expression for a constant power load that always draws P_L watts.

(b) If $P_L = 40$ W, $V_{ps} = 9$ V and $I_o = 5$ A, determine the values of V_o and R_{ps}.

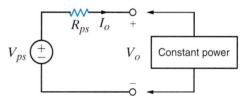

Figure P2.35

2.36 A student needs a 15-V voltage source for research. She has been able to locate two power supplies, a 10-V supply and a 5-V supply. The equivalent circuits for the two supplies are shown in Fig. P2.36.

(a) Draw an equivalent circuit for the effective 15-V supply.

(b) If she can tolerate a 0.5-V deviation from 15 V, what is the maximum current change the combined supply can satisfy?

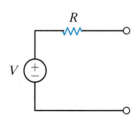

Voltage	5 V	10 V
Resistance	0.25 Ω	0.05 Ω

Figure P2.36

2.37 Given the network in Fig. P2.37, we wish to obtain a voltage of $2\ V \le V_o \le 9\ V$ across the full range of the pot. Determine the values of R_1 and R_2.

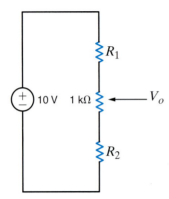

Figure P2.37

SECTION 2.4

2.38 Determine I_L in the circuit in Fig. P2.38.

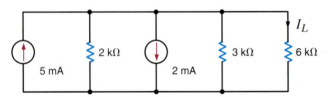

Figure P2.38

2.39 Find V_o in the circuit in Fig. P2.39. **CS**

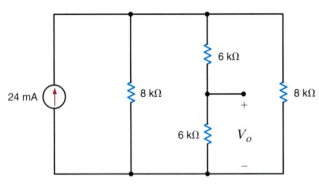

Figure P2.39

2.40 Find I_o in the network in Fig. P2.40. **PSV**

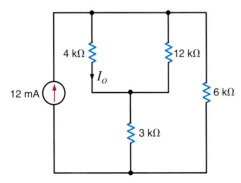

Figure P2.40

2.41 Find V_o in the network in Fig. P2.41.

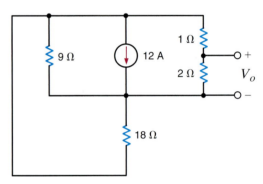

Figure P2.41

2.42 In the network in Fig. P2.42, $P_{6\,k\Omega} = 96$ mW. Find I_S.

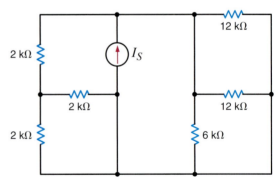

Figure P2.42

2.43 In the circuit in Fig. P2.43, $V_x = 12$ V. Find V_S.

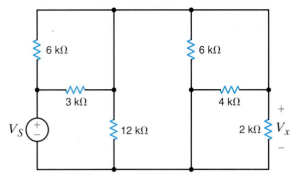

Figure P2.43

2.44 In the circuit in Fig. P2.44, $V_x = 6$ V. Find I_S.

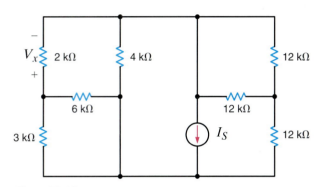

Figure P2.44

2.45 Determine I_L in the circuit in Fig. P2.45.

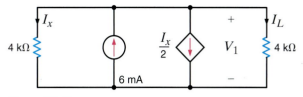

Figure P2.45

2.46 Determine I_L in the circuit in Fig. P2.46.   **PSV**

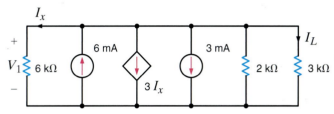

Figure P2.46

SECTION 2.5

2.47 Find R_{AB} in the circuit in Fig. P2.47.   **CS**

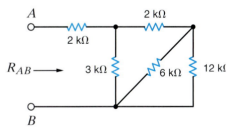

Figure P2.47

2.48 Find R_{AB} in the network in Fig. P2.48.

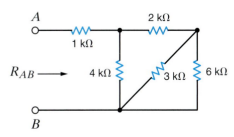

Figure P2.48

2.49 Find R_{AB} in the circuit in Fig. P2.49.

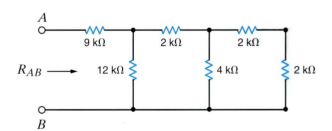

Figure P2.49

2.50 Find R_{AB} in the circuit in Fig. P2.50.   **PSV**

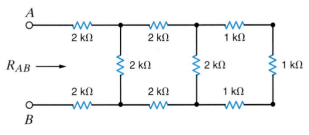

Figure P2.50

2.51 Determine R_{AB} in the circuit in Fig. P2.51.

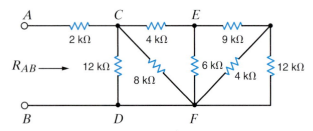

Figure P2.51

2.52 Find R_{AB} in the network in Fig. P2.52.

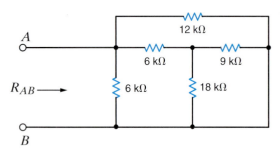

Figure P2.52

2.53 Find R_{AB} in the network in Fig. P2.53. **cs**

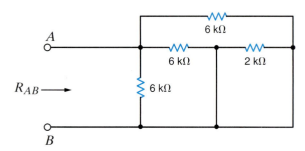

Figure P2.53

2.54 Find the equivalent resistance, R_{eq}, in the circuit in Fig. P2.54.

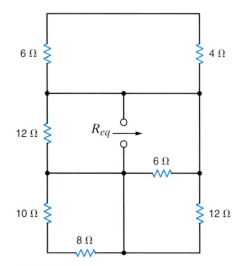

Figure P2.54

2.55 Find the equivalent resistance, R_{eq}, in the network in Fig. P2.55.

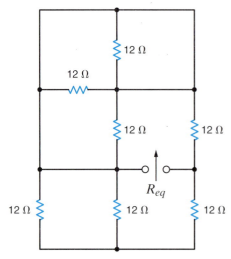

Figure P2.55

2.56 Find the range of resistance for the following resistors.

(a) 1 kΩ with a tolerance of 5%

(b) 470 Ω with a tolerance of 2%

(c) 22 kΩ with a tolerance of 10%

2.57 Given the network in Fig. P2.57, find the possible range of values for the current and power dissipated by the following resistors. **cs**

(a) 390 Ω with a tolerance of 1%

(b) 560 Ω with a tolerance of 2%

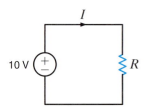

Figure P2.57

2.58 Given the circuit in Fig. P2.58,

(a) find the required value of R.

(b) use Table 2.1 to select a standard 10% tolerance resistor for R.

(c) calculate the actual value of I.

(d) determine the percent error between the actual value of I and that shown in the circuit.

(e) determine the power rating for the resistor R.

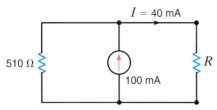

Figure P2.58

2.59 The resistors R_1 and R_2 shown in the circuit in Fig. P2.59 are 1 Ω with a tolerance of 5% and 2 Ω with a tolerance of 10%, respectively.

  (a) What is the nominal value of the equivalent resistance?

  (b) Determine the positive and negative tolerance for the equivalent resistance.

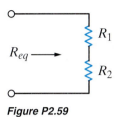

Figure P2.59

SECTION 2.6

2.60 Find V_{ab} and V_{dc} in the circuit in Fig. P2.60.

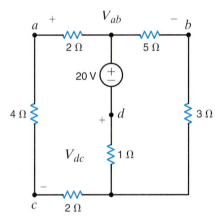

Figure P2.60

2.61 Find I_1 and V_o in the circuit in Fig. P2.61.

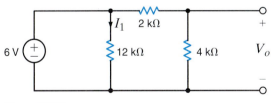

Figure P2.61

2.62 Find I_1 and V_o in the circuit in Fig. P2.62.

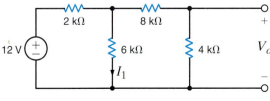

Figure P2.62

2.63 Find I_o in the network in Fig. P2.63.

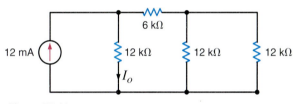

Figure P2.63

2.64 Find I_1 in the circuit in Fig. P2.64.

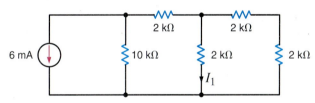

Figure P2.64

2.65 Determine V_o in the network in Fig. P2.65. **PSV**

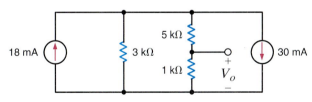

Figure P2.65

2.66 Determine I_o in the circuit in Fig. P2.66.

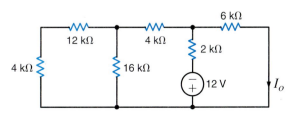

Figure P2.66

2.67 Determine V_o in the network in Fig. P2.67.

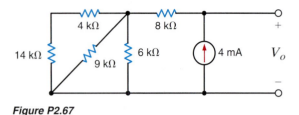

Figure P2.67

2.68 Find I_o in the circuit in Fig. P2.68. **CS**

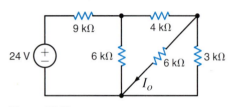

Figure P2.68

2.69 Find the value of V_x in the network in Fig. P2.69 such that the 5-A current source supplies 50 W. **PSV**

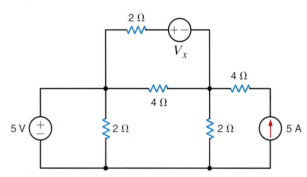

Figure P2.69

2.70 Find the value of V_1 in the network in Fig. P2.70 such that $V_a = 0$.

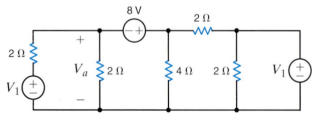

Figure P2.70

2.71 Find the value of V_x in the circuit in Fig. P2.71 such that the power supplied by the 5-A source is 60 W.

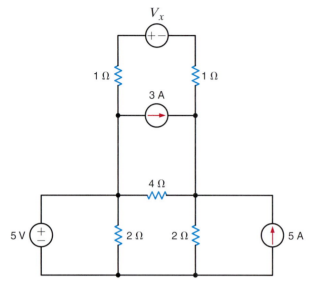

Figure P2.71

2.72 Find the value of V_S in the network in Fig. P2.72 such that the power supplied by the current source is 0.

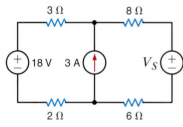

Figure P2.72

2.73 Find V_o in the circuit in Fig. P2.73.

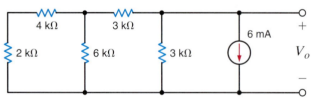

Figure P2.73

2.74 Find I_o in the network in Fig. P2.74.

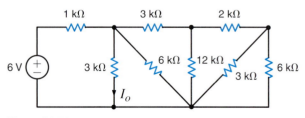

Figure P2.74

2.75 Find I_o in the circuit in Fig. P2.75. **CS**

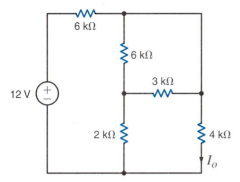

Figure P2.75

2.76 Determine V_o in the circuit in Fig. P2.76. **PSV**

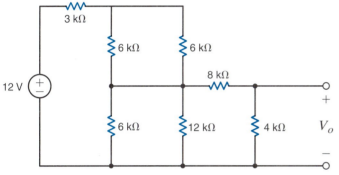

Figure P2.76

2.77 Find V_o in the circuit in Fig. P2.77.

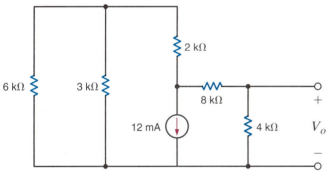

Figure P2.77

2.78 Find V_o in the circuit in Fig. P2.78.

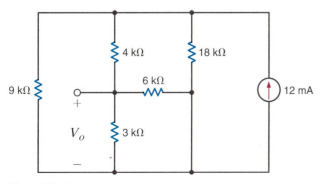

Figure P2.78

2.79 Find I_o in the circuit in Fig. P2.79.

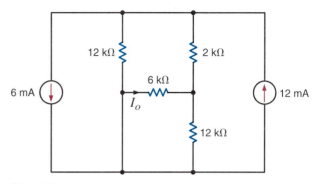

Figure P2.79

2.80 Find I_o in the circuit in Fig. P2.80.

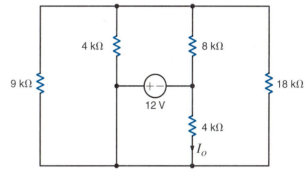

Figure P2.80

2.81 Find I_o in the circuit in Fig. P2.81.

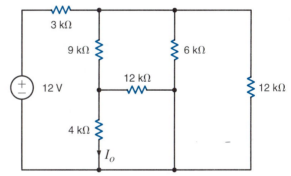

Figure P2.81

2.82 Find V_o in the circuit in Fig. P2.82. **PSV**

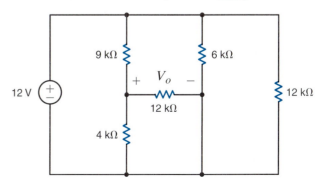

Figure P2.82

2.83 Find I_o in the circuit in Fig. P2.83.

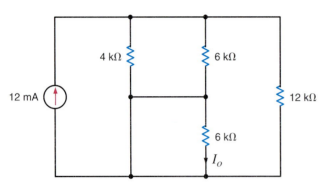

Figure P2.83

2.84 Determine the value of V_o in the circuit in Fig. P2.84.

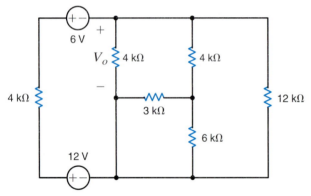

Figure P2.84

2.85 Find $P_{4\Omega}$ in the network in Fig. P2.85.

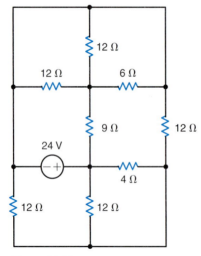

Figure P2.85

2.86 Find I_o in the network in Fig. P2.86.

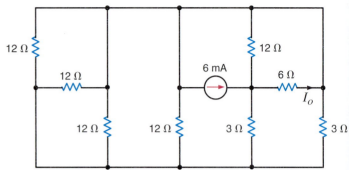

Figure P2.86

2.87 In the network in Fig. P2.87, the power absorbed by the 4-Ω resistor is 100 W. Find V_S. **CS**

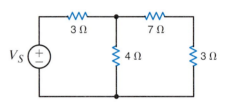

Figure P2.87

2.88 If $V_o = 2$ V in the circuit in Fig. P2.88, find V_S.

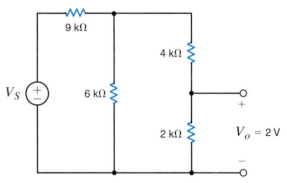

Figure P2.88

2.89 If $V_o = 6$ V in the circuit in Fig. P2.89, find I_S.

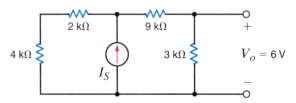

Figure P2.89

2.90 If $I_o = 2$ mA in the circuit in Fig. P2.90, find V_S.

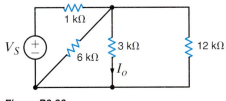

Figure P2.90

2.91 If $V_1 = 5$ V in the circuit in Fig. P2.91, find I_S.

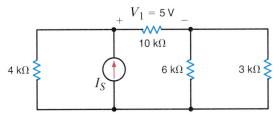

Figure P2.91

2.92 In the network in Fig. P2.92, $V_1 = 12$ V. Find V_S. **cs**

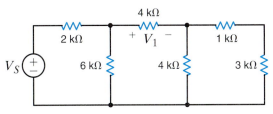

Figure P2.92

2.93 In the circuit in Fig. P2.93, $V_o = 2$ V. Find I_S.

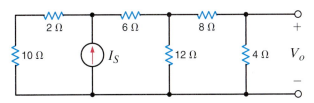

Figure P2.93

2.94 In the network in Fig. P2.94, $V_o = 6$ V. Find I_S.

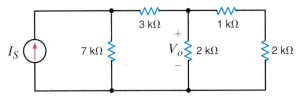

Figure P2.94

2.95 In $I_o = 4$ mA in the circuit in Fig. P2.95, find I_S. **cs**

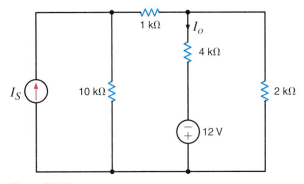

Figure P2.95

2.96 If $V_o = 6$ V in the circuit in Fig. P2.96, find I_S.

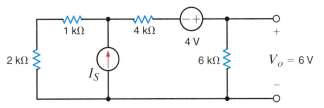

Figure P2.96

2.97 Given that $V_o = 4$ V in the network in Fig. P2.97, find V_S.

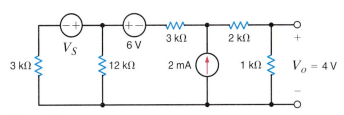

Figure P2.97

2.98 Find I_o in the circuit in Fig. P2.98.

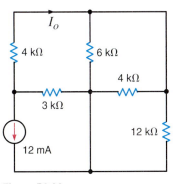

Figure P2.98

2.99 Given V_o in the network in Fig. P2.99, find I_A.

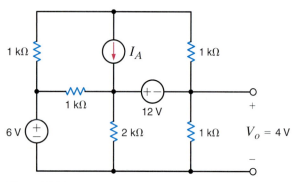

Figure P2.99

2.101 Given $I_o = 2$ mA in the network in Fig. P2.101, find V_A.

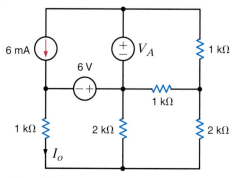

Figure P2.101

2.100 Given $I_o = 2$ mA in the circuit in Fig. P2.100, find I_A.

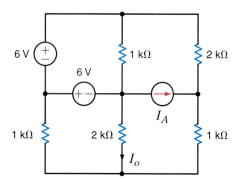

Figure P2.100

SECTION 2.7

2.102 Find the power absorbed by the network in Fig. P2.102.

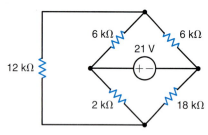

Figure P2.102

2.103 Find the value of g in the network in Fig. P2.103 such that the power supplied by the 3-A source is 20 W.

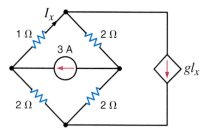

Figure P2.103

2.104 Find the power supplied by the 24-V source in the circuit in Fig. P2.104.

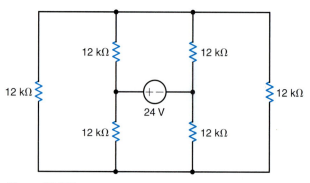

Figure P2.104

2.105 Find I_o in the circuit in Fig. P2.105. **PSV**

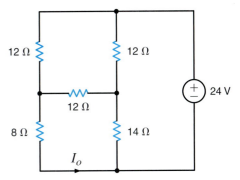

Figure P2.105

2.106 Find I_o in the circuit in Fig. P2.106. **CS**

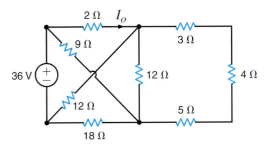

Figure P2.106

2.107 Find V_o in the network in Fig. P2.107.

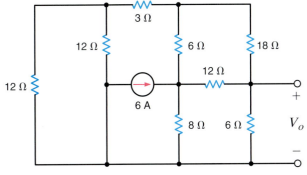

Figure P2.107

2.108 Find I_x in the circuit in Fig. P2.108.

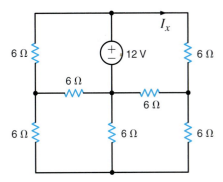

Figure P2.108

2.109 Find I_o in the circuit in Fig. P2.109. **CS**

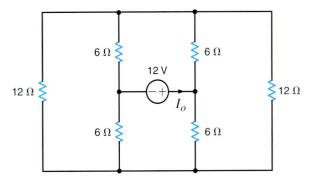

Figure P2.109

SECTION 2.8

2.110 Find V_o in the circuit in Fig. P2.110.

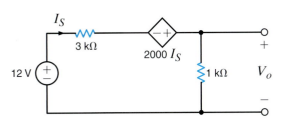

Figure P2.110

2.111 Find V_o in the network in Fig. P2.111.

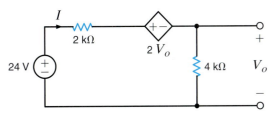

Figure P2.111

2.112 Find V_o in the network in Fig. P2.112. **PSV**

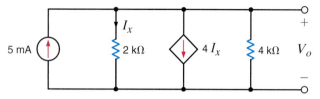

Figure P2.112

2.113 Find I_o in the network in Fig. P2.113. **CS**

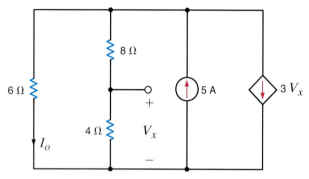

Figure P2.113

2.114 Find the power absorbed by the 10-kΩ resistor in the circuit in Fig. P2.114.

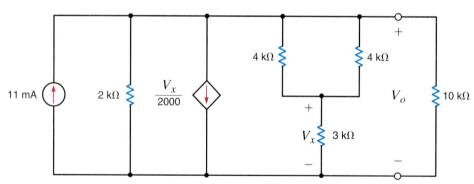

Figure P2.114

2.115 Find the value of k in the network in Fig. P2.115 such that the power supplied by the 6-A source is 108 W.

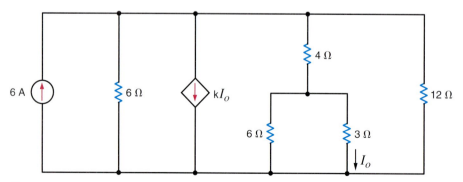

Figure P2.115

2.116 For the network in Fig. P2.116, choose the values of R_{in} and R_o such that is V_o maximized. What is the resulting ratio, V_o/V_S? **CS**

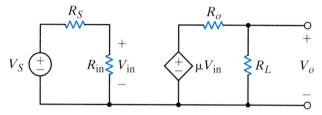

Figure P2.116

2.117 A typical transistor amplifier is shown in Fig. P2.117. Find the amplifier gain G (i.e., the ratio of the output voltage to the input voltage).

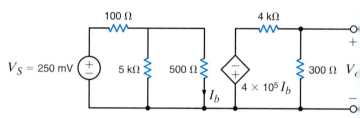

Figure P2.117

2.118 In many amplifier applications we are concerned not only with voltage gain, but also with power gain.

Power gain = A_p (power delivered to the load)/ (power delivered to the input)

Find the power gain for the circuit in Fig. P2.118, where $R_L = 60$ kΩ.

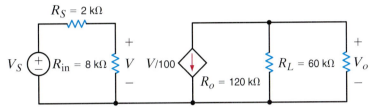

Figure P2.118

TYPICAL PROBLEMS FOUND ON THE FE EXAM

2FE-1 Find the power generated by the source in the network in Fig. 2PFE-1. **CS**

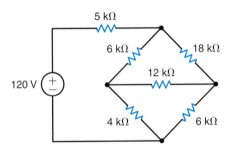

Fig. 2PFE-1

2FE-2 Find the equivalent resistance of the circuit in Fig. 2PFE-2 at the terminals A-B.

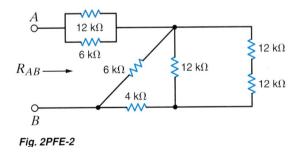

Fig. 2PFE-2

2FE-3 Find the voltage V_o in the network in Fig. 2PFE-3. **CS**

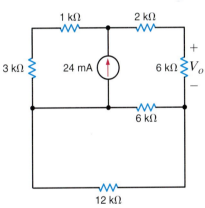

Fig. 2PFE-3

2FE-4 Find the current I_o in the circuit in Fig. 2PFE-4.

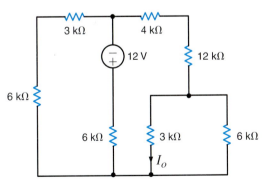

Fig. 2PFE-4

Nodal and Loop Analysis Techniques

Access Problem-Solving Videos **PSV**
and Circuit Solutions **CS** *at:*
http://www.justask4u.com/irwin
using the registration code on the inside cover
and see a website with answers and more!

In Chapter 2 we analyzed the simplest possible circuits, those containing only a single-node pair or a single loop. We found that these circuits can be completely analyzed via a single algebraic equation. In the case of the single-node-pair circuit (i.e., one containing two nodes, one of which is a reference node), once the node voltage is known, we can calculate all the currents. In a single-loop circuit, once the loop current is known, we can calculate all the voltages.

In this chapter we extend our capabilities in a systematic manner so that we can calculate all currents and voltages in circuits that contain multiple nodes and loops. Our analyses are based primarily on two laws with which we are already familiar: Kirchhoff's current law (KCL) and Kirchhoff's voltage law (KVL). In a nodal analysis we employ KCL to determine the node voltages, and in a loop analysis we use KVL to determine the loop currents. ●

3.1   Nodal analysis

In a nodal analysis the variables in the circuit are selected to be the node voltages. The node voltages are defined with respect to a common point in the circuit. One node is selected as the reference node, and all other node voltages are defined with respect to that node. Quite often this node is the one to which the largest number of branches are connected. It is commonly called *ground* because it is said to be at ground-zero potential, and it sometimes represents the chassis or ground line in a practical circuit.

We will select our variables as being positive with respect to the reference node. If one or more of the node voltages are actually negative with respect to the reference node, the analysis will indicate it.

In order to understand the value of knowing all the node voltages in a network, we consider once again the network in Fig. 2.32, which is redrawn in Fig. 3.1. The voltages, V_S, V_a, V_b, and V_c, are all measured with respect to the bottom node, which is selected as the reference and labeled with the ground symbol ⏚. Therefore, the voltage at node 1 is $V_S = 12$ V with respect to the reference node 5; the voltage at node 2 is $V_a = 3$ V with respect to the reference node 5, and so on. Now note carefully that once these node voltages are known, we can immediately calculate any branch current or the power supplied or absorbed by any element, since we know the voltage across every element in the network. For example, the voltage V_1 across the leftmost 9-kΩ resistor is the difference in potential between the two ends of the resistor; that is,

$$V_1 = V_S - V_a$$
$$= 12 - 3$$
$$= 9 \text{ V}$$

This equation is really nothing more than an application of KVL around the leftmost loop; that is,

$$-V_S + V_1 + V_a = 0$$

In a similar manner, we find that

$$V_3 = V_a - V_b$$

and

$$V_5 = V_b - V_c$$

Then the currents in the resistors are

$$I_1 = \frac{V_1}{9k} = \frac{V_S - V_a}{9k}$$

$$I_3 = \frac{V_3}{3k} = \frac{V_a - V_b}{3k}$$

$$I_5 = \frac{V_5}{9k} = \frac{V_b - V_c}{9k}$$

In addition,

$$I_2 = \frac{V_a - 0}{6k}$$

$$I_4 = \frac{V_b - 0}{4k}$$

since the reference node 5 is at zero potential.

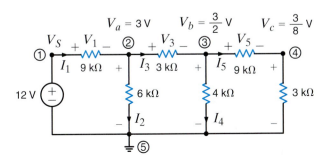

Figure 3.1

Circuit with known node voltages.

Figure 3.2

Circuit used to illustrate Ohm's law in a multiple-node network.

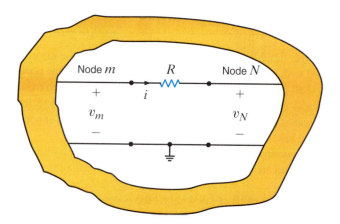

Thus, as a general rule, if we know the node voltages in a circuit, we can calculate the current through any resistive element using Ohm's law; that is,

$$i = \frac{v_m - v_N}{R}$$

3.1

as illustrated in Fig. 3.2.

Now that we have demonstrated the value of knowing all the node voltages in a network, let us determine the manner in which to calculate them. In a nodal analysis, we employ KCL equations in such a way that the variables contained in these equations are the unknown node voltages of the network. As we have indicated, one of the nodes in an N-node circuit is selected as the reference node, and the voltages at all the remaining $N - 1$ nonreference nodes are measured with respect to this reference node. Using network topology, it can be shown that exactly $N - 1$ linearly independent KCL equations are required to determine the $N - 1$ unknown node voltages. Therefore, theoretically once one of the nodes in an N-node circuit has been selected as the reference node, our task is reduced to identifying the remaining $N - 1$ nonreference nodes and writing one KCL equation at each of them.

In a multiple-node circuit, this process results in a set of $N - 1$ linearly independent simultaneous equations in which the variables are the $N - 1$ unknown node voltages. To help solidify this idea, consider once again Example 2.5. Note that in this circuit only four (i.e., any four) of the five KCL equations, one of which is written for each node in this five-node network, are linearly independent. Furthermore, many of the branch currents in this example (those not contained in a source) can be written in terms of the node voltages as illustrated in Fig. 3.2 and expressed in Eq. (3.1). It is in this manner, as we will illustrate in the sections that follow, that the KCL equations contain the unknown node voltages.

It is instructive to treat nodal analysis by examining several different types of circuits and illustrating the salient features of each. We begin with the simplest case. However, as a prelude to our discussion of the details of nodal analysis, experience indicates that it is worthwhile to digress for a moment to ensure that the concept of node voltage is clearly understood.

At the outset it is important to specify a reference. For example, to state that the voltage at node A is 12 V means nothing unless we provide the reference point; that is, the voltage at node A is 12 V with respect to what? The circuit in Fig. 3.3 illustrates a portion of a network containing three nodes, one of which is the reference node.

Figure 3.3

An illustration of node voltages.

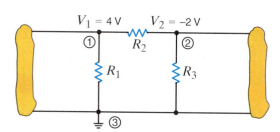

The voltage $V_1 = 4$ V is the voltage at node 1 with respect to the reference node 3. Similarly, the voltage $V_2 = -2$ V is the voltage at node 2 with respect to node 3. In addition, however, the voltage at node 1 with respect to node 2 is +6 V, and the voltage at node 2 with respect to node 1 is −6 V. Furthermore, since the current will flow from the node of higher potential to the node of lower potential, the current in R_1 is from top to bottom, the current in R_2 is from left to right, and the current in R_3 is from bottom to top.

These concepts have important ramifications in our daily lives. If a man were hanging in midair with one hand on one line and one hand on another and the dc line voltage of each line was exactly the same, the voltage across his heart would be zero and he would be safe. If, however, he let go of one line and let his feet touch the ground, the dc line voltage would then exist from his hand to his foot with his heart in the middle. He would probably be dead the instant his foot hit the ground.

In the town where we live, a young man tried to retrieve his parakeet that had escaped its cage and was outside sitting on a power line. He stood on a metal ladder and with a metal pole reached for the parakeet; when the metal pole touched the power line, the man was killed instantly. Electric power is vital to our standard of living, but it is also very dangerous. The material in this book *does not* qualify you to handle it safely. Therefore, always be extremely careful around electric circuits.

Now as we begin our discussion of nodal analysis, our approach will be to begin with simple cases and proceed in a systematic manner to those that are more challenging. Numerous examples will be the vehicle used to demonstrate each facet of this approach. Finally, at the end of this section, we will outline a strategy for attacking any circuit using nodal analysis.

CIRCUITS CONTAINING ONLY INDEPENDENT CURRENT

SOURCES Consider the network shown in Fig. 3.4. Note that this network contains three nodes, and thus we know that exactly $N - 1 = 3 - 1 = 2$ linearly independent KCL equations will be required to determine the $N - 1 = 2$ unknown node voltages. First, we select the bottom node as the reference node, and then the voltage at the two remaining nodes labeled v_1 and v_2 will be measured with respect to this node.

The branch currents are assumed to flow in the directions indicated in the figures. If one or more of the branch currents are actually flowing in a direction opposite to that assumed, the analysis will simply produce a branch current that is negative.

Applying KCL at node 1 yields

$$-i_A + i_1 + i_2 = 0$$

Using Ohm's law ($i = Gv$) and noting that the reference node is at zero potential, we obtain

$$-i_A + G_1(v_1 - 0) + G_2(v_1 - v_2) = 0$$

or

$$(G_1 + G_2)v_1 - G_2 v_2 = i_A$$

KCL at node 2 yields

$$-i_2 + i_B + i_3 = 0$$

or

$$-G_2(v_1 - v_2) + i_B + G_3(v_2 - 0) = 0$$

which can be expressed as

$$-G_2 v_1 + (G_2 + G_3)v_2 = -i_B$$

HINT

Employing the passive sign convention.

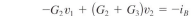

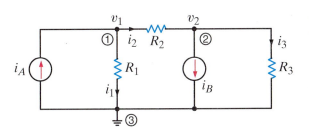

Figure 3.4
A three-node circuit.

Therefore, the two equations for the two unknown node voltages v_1 and v_2 are

$$(G_1 + G_2)v_1 - G_2v_2 = i_A$$

$$-G_2v_1 + (G_2 + G_3)v_2 = -i_B \qquad \textbf{3.2}$$

Note that the analysis has produced two simultaneous equations in the unknowns v_1 and v_2. They can be solved using any convenient technique, and modern calculators and personal computers are very efficient tools for their application.

In what follows, we will demonstrate three techniques for solving linearly independent simultaneous equations: Gaussian elimination, matrix analysis, and the MATLAB mathematical software package. A brief refresher that illustrates the use of both Gaussian elimination and matrix analysis in the solution of these equations is provided in the Problem-Solving Companion for this text. Use of the MATLAB software is straightforward, and we will demonstrate its use as we encounter the application.

The KCL equations at nodes 1 and 2 produced two linearly independent simultaneous equations:

$$-i_A + i_1 + i_2 = 0$$

$$-i_2 + i_B + i_3 = 0$$

The KCL equation for the third node (reference) is

$$+i_A - i_1 - i_B - i_3 = 0$$

Note that if we add the first two equations, we obtain the third. Furthermore, any two of the equations can be used to derive the remaining equation. Therefore, in this $N = 3$ node circuit, only $N - 1 = 2$ of the equations are linearly independent and required to determine the $N - 1 = 2$ unknown node voltages.

Note that a nodal analysis employs KCL in conjunction with Ohm's law. Once the direction of the branch currents has been *assumed*, then Ohm's law, as illustrated by Fig. 3.2 and expressed by Eq. (3.1), is used to express the branch currents in terms of the unknown node voltages. We can assume the currents to be in any direction. However, once we assume a particular direction, we must be very careful to write the currents correctly in terms of the node voltages using Ohm's law.

Example 3.1

Suppose that the network in Fig. 3.4 has the following parameters: $I_A = 1$ mA, $R_1 = 12$ kΩ, $R_2 = 6$ kΩ, $I_B = 4$ mA, and $R_3 = 6$ kΩ. Let us determine all node voltages and branch currents.

SOLUTION For purposes of illustration we will solve this problem using Gaussian elimination, matrix analysis, and MATLAB. Using the parameter values Eq. (3.2) becomes

$$V_1\left[\frac{1}{12k} + \frac{1}{6k}\right] - V_2\left[\frac{1}{6k}\right] = 1 \times 10^{-3}$$

$$-V_1\left[\frac{1}{6k}\right] + V_2\left[\frac{1}{6k} + \frac{1}{6k}\right] = -4 \times 10^{-3}$$

where we employ capital letters because the voltages are constant. The equations can be written as

$$\frac{V_1}{4k} - \frac{V_2}{6k} = 1 \times 10^{-3}$$

$$-\frac{V_1}{6k} + \frac{V_2}{3k} = -4 \times 10^{-3}$$

Using Gaussian elimination, we solve the first equation for V_1 in terms of V_2:

$$V_1 = V_2\left(\frac{2}{3}\right) + 4$$

This value is then substituted into the second equation to yield

$$\frac{-1}{6k}\left(\frac{2}{3}V_2 + 4\right) + \frac{V_2}{3k} = -4 \times 10^{-3}$$

or

$$V_2 = -15 \text{ V}$$

This value for V_2 is now substituted back into the equation for V_1 in terms of V_2, which yields

$$V_1 = \frac{2}{3}V_2 + 4$$

$$= -6 \text{ V}$$

The circuit equations can also be solved using matrix analysis. The general form of the matrix equation is

$$\mathbf{GV} = \mathbf{I}$$

where in this case

$$\mathbf{G} = \begin{bmatrix} \dfrac{1}{4k} & -\dfrac{1}{6k} \\ -\dfrac{1}{6k} & \dfrac{1}{3k} \end{bmatrix}, \mathbf{V} = \begin{bmatrix} V_1 \\ V_2 \end{bmatrix}, \text{ and } \mathbf{I} = \begin{bmatrix} 1 \times 10^{-3} \\ -4 \times 10^{-3} \end{bmatrix}$$

The solution to the matrix equation is

$$\mathbf{V} = \mathbf{G}^{-1}\mathbf{I}$$

and therefore,

$$\begin{bmatrix} V_1 \\ V_2 \end{bmatrix} = \begin{bmatrix} \dfrac{1}{4k} & \dfrac{-1}{6k} \\ \dfrac{-1}{6k} & \dfrac{1}{3k} \end{bmatrix}^{-1} \begin{bmatrix} 1 \times 10^{-3} \\ -4 \times 10^{-3} \end{bmatrix}$$

To calculate the inverse of $\mathbf{G}$, we need the adjoint and the determinant. The adjoint is

$$\text{Adj } \mathbf{G} = \begin{bmatrix} \dfrac{1}{3k} & \dfrac{1}{6k} \\ \dfrac{1}{6k} & \dfrac{1}{4k} \end{bmatrix}$$

and the determinant is

$$|\mathbf{G}| = \left(\frac{1}{3k}\right)\left(\frac{1}{4k}\right) - \left(\frac{-1}{6k}\right)\left(\frac{-1}{6k}\right)$$

$$= \frac{1}{18k^2}$$

Therefore,

$$\begin{bmatrix} V_1 \\ V_2 \end{bmatrix} = 18k^2 \begin{bmatrix} \dfrac{1}{3k} & \dfrac{1}{6k} \\ \dfrac{1}{6k} & \dfrac{1}{4k} \end{bmatrix} \begin{bmatrix} 1 \times 10^{-3} \\ -4 \times 10^{-3} \end{bmatrix}$$

$$= 18k^2 \begin{bmatrix} \dfrac{1}{3k^2} - \dfrac{4}{6k^2} \\ \dfrac{1}{6k^2} - \dfrac{1}{k^2} \end{bmatrix}$$

$$= \begin{bmatrix} -6 \\ -15 \end{bmatrix}$$

The MATLAB solution begins with the set of equations expressed in matrix form as

```
G*V=I
```

where the symbol * denotes the multiplication of the voltage vector **V** by the coefficient matrix **G**. Then once the MATLAB software is loaded into the PC, the coefficient matrix (**G**) and the vector **V** can be expressed in MATLAB notation by typing in the rows of the matrix or vector at the prompt >>. Use semicolons to separate rows and spaces to separate columns. Brackets are used to denote vectors or matrices. When the matrix **G** and the vector **I** have been defined, then the solution equation

```
V=inv(G)*I
```

which is also typed in at the prompt >>, will yield the unknown vector **V**.

The matrix equation for our circuit expressed in decimal notation is

$$\begin{bmatrix} 0.00025 & -0.00016666 \\ -0.00016666 & 0.0003333 \end{bmatrix} \begin{bmatrix} V_1 \\ V_2 \end{bmatrix} = \begin{bmatrix} 0.001 \\ -0.004 \end{bmatrix}$$

If we now input the coefficient matrix **G**, then the vector **I** and finally the equation V = inv(G)*I, the computer screen containing these data and the solution vector **V** appears as follows:

```
>> G = [0.00025 -0.000166666;
 -0.000166666 0.00033333]

   G =

      1.0e-003 *

       0.2500      -0.1667
      -0.1667       0.3333
>> I = [0.001 ; -0.004]

   I =

       0.0010
      -0.0040
>> V = inv(G)*I

   V =

      -6.0001
     -15.0002
```

Knowing the node voltages, we can determine all the currents using Ohm's law:

$$I_1 = \frac{V_1}{R_1} = \frac{-6}{12k} = -\frac{1}{2}\,\text{mA}$$

$$I_2 = \frac{V_1 - V_2}{6k} = \frac{-6 - (-15)}{6k} = \frac{3}{2}\,\text{mA}$$

and

$$I_3 = \frac{V_2}{6k} = \frac{-15}{6k} = -\frac{5}{2}\,\text{mA}$$

Figure 3.5 illustrates the results of all the calculations. Note that KCL is satisfied at every node.

Figure 3.5
Circuit used in
Example 3.1.

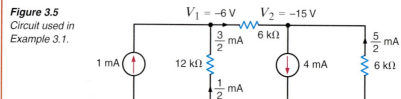

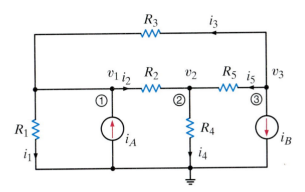

Figure 3.6
A four-node circuit.

Let us now examine the circuit in Fig. 3.6. The current directions are assumed as shown in the figure.

We note that this network has four nodes. The node at the bottom of the circuit is selected as the reference node and labeled with the ground symbol. Since $N = 4$, $N - 1 = 3$ linearly independent KCL equations will be required to determine the three unknown nonreference node voltages labeled v_1, v_2, and v_3.

At node 1, KCL yields

$$i_1 - i_A + i_2 - i_3 = 0$$

or

$$\frac{v_1}{R_1} - i_A + \frac{v_1 - v_2}{R_2} - \frac{v_3 - v_1}{R_3} = 0$$

$$v_1\left(\frac{1}{R_1} + \frac{1}{R_2} + \frac{1}{R_3}\right) - v_2\frac{1}{R_2} - v_3\frac{1}{R_3} = i_A$$

At node 2, KCL yields

$$-i_2 + i_4 - i_5 = 0$$

or

$$-\frac{v_1 - v_2}{R_2} + \frac{v_2}{R_4} - \frac{v_3 - v_2}{R_5} = 0$$

$$-v_1\frac{1}{R_2} + v_2\left(\frac{1}{R_2} + \frac{1}{R_4} + \frac{1}{R_5}\right) - v_3\frac{1}{R_5} = 0$$

At node 3, the equation is

$$i_3 + i_5 + i_B = 0$$

or

$$\frac{v_3 - v_1}{R_3} + \frac{v_3 - v_2}{R_5} + i_B = 0$$

$$-v_1\frac{1}{R_3} - v_2\frac{1}{R_5} + v_3\left(\frac{1}{R_3} + \frac{1}{R_5}\right) = -i_B$$

Grouping the node equations together, we obtain

$$v_1\left(\frac{1}{R_1} + \frac{1}{R_2} + \frac{1}{R_3}\right) - v_2\frac{1}{R_2} - v_3\frac{1}{R_3} = i_A$$

$$-v_1\frac{1}{R_2} + v_2\left(\frac{1}{R_2} + \frac{1}{R_4} + \frac{1}{R_5}\right) - v_3\frac{1}{R_5} = 0 \qquad \textbf{3.3}$$

$$-v_1\frac{1}{R_3} - v_2\frac{1}{R_5} + v_3\left(\frac{1}{R_3} + \frac{1}{R_5}\right) = -i_B$$

Note that our analysis has produced three simultaneous equations in the three unknown node voltages v_1, v_2, and v_3. The equations can also be written in matrix form as

$$
\begin{bmatrix}
\dfrac{1}{R_1} + \dfrac{1}{R_2} + \dfrac{1}{R_3} & -\dfrac{1}{R_2} & -\dfrac{1}{R_3} \\[2mm]
-\dfrac{1}{R_2} & \dfrac{1}{R_2} + \dfrac{1}{R_4} + \dfrac{1}{R_5} & -\dfrac{1}{R_5} \\[2mm]
-\dfrac{1}{R_3} & -\dfrac{1}{R_5} & \dfrac{1}{R_3} + \dfrac{1}{R_5}
\end{bmatrix}
\begin{bmatrix} v_1 \\ v_2 \\ v_3 \end{bmatrix}
=
\begin{bmatrix} i_A \\ 0 \\ -i_B \end{bmatrix}
\qquad \textbf{3.4}
$$

At this point it is important that we note the symmetrical form of the equations that describe the two previous networks. Equations (3.2) and (3.3) exhibit the same type of symmetrical form. The **G** matrix for each network is a symmetrical matrix. This symmetry is not accidental. The node equations for networks containing only resistors and independent current sources can always be written in this symmetrical form. We can take advantage of this fact and learn to write the equations by inspection. Note in the first equation of (3.2) that the coefficient of v_1 is the sum of all the conductances connected to node 1 and the coefficient of v_2 is the negative of the conductances connected between node 1 and node 2. The right-hand side of the equation is the sum of the currents entering node 1 through current sources. This equation is KCL at node 1. In the second equation in (3.2), the coefficient of v_2 is the sum of all the conductances connected to node 2, the coefficient of v_1 is the negative of the conductance connected between node 2 and node 1, and the right-hand side of the equation is the sum of the currents entering node 2 through current sources. This equation is KCL at node 2. Similarly, in the first equation in (3.3) the coefficient of v_1 is the sum of the conductances connected to node 1, the coefficient of v_2 is the negative of the conductance connected between node 1 and node 2, the coefficient of v_3 is the negative of the conductance connected between node 1 and node 3, and the right-hand side of the equation is the sum of the currents entering node 1 through current sources. The other two equations in (3.3) are obtained in a similar manner. In general, if KCL is applied to node j with node voltage v_j, the coefficient of v_j is the sum of all the conductances connected to node j and the coefficients of the other node voltages $(e.g., v_{j-1}, v_{j+1})$ are the negative of the sum of the conductances connected directly between these nodes and node j. The right-hand side of the equation is equal to the sum of the currents entering the node via current sources. Therefore, the left-hand side of the equation represents the sum of the currents leaving node j and the right-hand side of the equation represents the currents entering node j.

Example 3.2

Let us apply what we have just learned to write the equations for the network in Fig. 3.7 by inspection. Then given the following parameters, we will determine the node voltages using MATLAB: $R_1 = R_2 = 2\ \text{k}\Omega$, $R_3 = R_4 = 4\ \text{k}\Omega$, $R_5 = 1\ \text{k}\Omega$, $i_A = 4\ \text{mA}$, and $i_B = 2\ \text{mA}$.

Figure 3.7
Circuit used in Example 3.2.

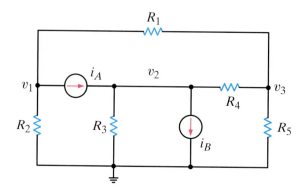

SOLUTION   The equations are

$$v_1\left(\frac{1}{R_1} + \frac{1}{R_2}\right) - v_2(0) - v_3\left(\frac{1}{R_1}\right) = -i_A$$

$$-v_1(0) + v_2\left(\frac{1}{R_3} + \frac{1}{R_4}\right) - v_3\left(\frac{1}{R_4}\right) = i_A - i_B$$

$$-v_1\left(\frac{1}{R_1}\right) - v_2\left(\frac{1}{R_4}\right) + v_3\left(\frac{1}{R_1} + \frac{1}{R_4} + \frac{1}{R_5}\right) = 0$$

which can also be written directly in matrix form as

$$\begin{bmatrix} \frac{1}{R_1} + \frac{1}{R_2} & 0 & -\frac{1}{R_1} \\ 0 & \frac{1}{R_3} + \frac{1}{R_4} & -\frac{1}{R_4} \\ -\frac{1}{R_1} & -\frac{1}{R_4} & \frac{1}{R_1} + \frac{1}{R_4} + \frac{1}{R_5} \end{bmatrix} \begin{bmatrix} v_1 \\ v_2 \\ v_3 \end{bmatrix} = \begin{bmatrix} -i_A \\ i_A - i_B \\ 0 \end{bmatrix}$$

Both the equations and the **G** matrix exhibit the symmetry that will always be present in circuits that contain only resistors and current sources.

If the component values are now used, the matrix equation becomes

$$\begin{bmatrix} \frac{1}{2k} + \frac{1}{2k} & 0 & -\frac{1}{2k} \\ 0 & \frac{1}{4k} + \frac{1}{4k} & -\frac{1}{4k} \\ -\frac{1}{2k} & -\frac{1}{4k} & \frac{1}{2k} + \frac{1}{4k} + \frac{1}{1k} \end{bmatrix} \begin{bmatrix} v_1 \\ v_2 \\ v_3 \end{bmatrix} = \begin{bmatrix} -0.004 \\ 0.002 \\ 0 \end{bmatrix}$$

or

$$\begin{bmatrix} 0.001 & 0 & -0.0005 \\ 0 & 0.0005 & -0.00025 \\ -0.0005 & -0.00025 & 0.00175 \end{bmatrix} \begin{bmatrix} v_1 \\ v_2 \\ v_3 \end{bmatrix} = \begin{bmatrix} -0.004 \\ 0.002 \\ 0 \end{bmatrix}$$

If we now employ these data with the MATLAB software, the computer screen containing the data and the results of the MATLAB analysis is as shown next.

```
>> G = [0.001 0 -0.0005 ; 0 0.0005 -0.00025 ;
-0.0005 -0.00025 0.00175]

   G =

       0.0010          0     -0.0005
            0     0.0005     -0.0003
      -0.0005    -0.0003      0.0018

>> I = [-0.004 ; 0.002 ; 0]

   I =

      -0.0040
       0.0020
            0

>> V = inv(G)*I

   V =

      -4.3636
       3.6364
      -0.7273
```

LEARNING EXTENSIONS

E3.1 Write the node equations for the circuit in Fig. E3.1.

Figure E3.1

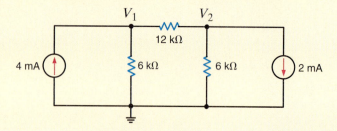

V_1 V_2

12 kΩ

4 mA   6 kΩ   6 kΩ   2 mA

ANSWER:

$$\frac{1}{4k} V_1 - \frac{1}{12k} V_2 = 4 \times 10^{-3},$$

$$\frac{-1}{12k} V_1 + \frac{1}{4k} V_2 = -2 \times 10^{-3}.$$

E3.2 Find all the node voltages in the network in Fig. E3.2 using MATLAB.

Figure E3.2

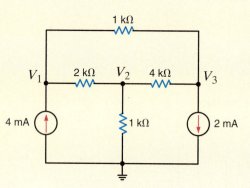

1 kΩ

V_1 2 kΩ V_2 4 kΩ V_3

4 mA   1 kΩ   2 mA

ANSWER: $V_1 = 5.4286$ V, $V_2 = 2.000$ V, $V_3 = 3.1429$ V.

CIRCUITS CONTAINING DEPENDENT CURRENT SOURCES The presence of a dependent source may destroy the symmetrical form of the nodal equations that define the circuit. Consider the circuit shown in Fig. 3.8, which contains a current-controlled current source. The KCL equations for the nonreference nodes are

$$\beta i_o + \frac{v_1}{R_1} + \frac{v_1 - v_2}{R_2} = 0$$

and

$$\frac{v_2 - v_1}{R_2} + i_o - i_A = 0$$

where $i_o = v_2/R_3$. Simplifying the equations, we obtain

$$(G_1 + G_2)v_1 - (G_2 - \beta G_3)v_2 = 0$$

$$-G_2 v_1 + (G_2 + G_3)v_2 = i_A$$

or in matrix form

$$\begin{bmatrix} (G_1 + G_2) & -(G_2 - \beta G_3) \\ -G_2 & (G_2 + G_3) \end{bmatrix} \begin{bmatrix} v_1 \\ v_2 \end{bmatrix} = \begin{bmatrix} 0 \\ i_A \end{bmatrix}$$

Note that the presence of the dependent source has destroyed the symmetrical nature of the node equations.

Figure 3.8
Circuit with a dependent source.

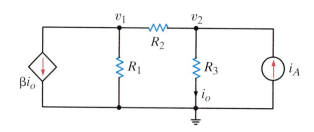

v_1 R_2 v_2

βi_o R_1 R_3 i_A

i_o

Example 3.3

Let us determine the node voltages for the network in Fig. 3.8 given the following parameters:

$$\beta = 2 \qquad R_2 = 6 \text{ k}\Omega \qquad i_A = 2 \text{ mA}$$
$$R_1 = 12 \text{ k}\Omega \qquad R_3 = 3 \text{ k}\Omega$$

SOLUTION   Using these values with the equations for the network yields

$$\frac{1}{4k} V_1 + \frac{1}{2k} V_2 = 0$$

$$-\frac{1}{6k} V_1 + \frac{1}{2k} V_2 = 2 \times 10^{-3}$$

Solving these equations using any convenient method yields $V_1 = -24/5$ V and $V_2 = 12/5$ V. We can check these answers by determining the branch currents in the network and then using that information to test KCL at the nodes. For example, the current from top to bottom through R_3 is

$$I_o = \frac{V_2}{R_3} = \frac{12/5}{3k} = \frac{4}{5k} \text{ A}$$

Similarly, the current from right to left through R_2 is

$$I_2 = \frac{V_2 - V_1}{R_2} = \frac{12/5 - (-24/5)}{6k} = \frac{6}{5k} \text{ A}$$

All the results are shown in Fig. 3.9. Note that KCL is satisfied at every node.

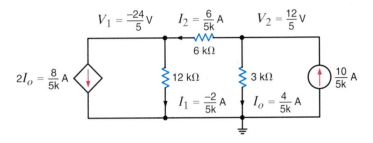

Figure 3.9
Circuit used in Example 3.3.

Example 3.4

Let us determine the set of linearly independent equations that when solved will yield the node voltages in the network in Fig. 3.10. Then given the following component values, we will compute the node voltages using MATLAB: $R_1 = 1 \text{ k}\Omega$, $R_2 = R_3 = 2 \text{ k}\Omega$, $R_4 = 4 \text{ k}\Omega$, $i_A = 2 \text{ mA}$, $i_B = 4 \text{ mA}$, and $\alpha = 2$.

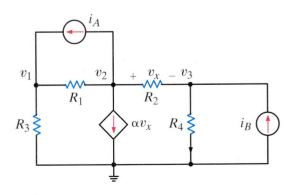

Figure 3.10
Circuit containing a voltage-controlled current source.

SOLUTION Applying KCL at each of the nonreference nodes yields the equations

$$G_3 v_1 + G_1(v_1 - v_2) - i_A = 0$$

$$i_A + G_1(v_2 - v_1) + \alpha v_x + G_2(v_2 - v_3) = 0$$

$$G_2(v_3 - v_2) + G_4 v_3 - i_B = 0$$

where $v_x = v_2 - v_3$. Simplifying these equations, we obtain

$$(G_1 + G_3)v_1 - G_1 v_2 = i_A$$

$$-G_1 v_1 + (G_1 + \alpha + G_2)v_2 - (\alpha + G_2)v_3 = -i_A$$

$$-G_2 v_2 + (G_2 + G_4)v_3 = i_B$$

Given the component values, the equations become

$$\begin{bmatrix} \dfrac{1}{1k} + \dfrac{1}{2k} & -\dfrac{1}{k} & 0 \\[2mm] -\dfrac{1}{k} & \dfrac{1}{k} + 2 + \dfrac{1}{2k} & -\left(2 + \dfrac{1}{2k}\right) \\[2mm] 0 & -\dfrac{1}{2k} & \dfrac{1}{2k} + \dfrac{1}{4k} \end{bmatrix} \begin{bmatrix} v_1 \\ v_2 \\ v_3 \end{bmatrix} = \begin{bmatrix} 0.002 \\ -0.002 \\ 0.004 \end{bmatrix}$$

or

$$\begin{bmatrix} 0.0015 & -0.001 & 0 \\ -0.001 & 2.0015 & -2.0005 \\ 0 & -0.0005 & 0.00075 \end{bmatrix} \begin{bmatrix} v_1 \\ v_2 \\ v_3 \end{bmatrix} = \begin{bmatrix} 0.002 \\ -0.002 \\ 0.004 \end{bmatrix}$$

The MATLAB input and output listings are shown next.

```
>> G = [0.0015 -0.001 0 ; -0.001 2.0015 -2.0005 ;
0 -0.0005 0.00075]

   G =

        0.0015        -0.0010             0
       -0.0010         2.0015        -2.0005
             0        -0.0005         0.0008

>> I = [0.002 ; -0.002 ; 0.004]

   I =

        0.0020
       -0.0020
        0.0040

>> V = inv(G)*I

   V =

       11.9940
       15.9910
       15.9940
```

LEARNING EXTENSIONS

E3.3 Find the node voltages in the circuit in Fig. E3.3.

Figure E3.3

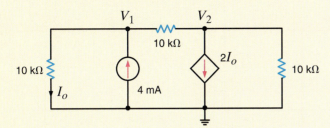

ANSWER: $V_1 = 16$ V, $V_2 = -8$ V.

E3.4 Find the voltage V_o in the network in Fig. E3.4.

Figure E3.4

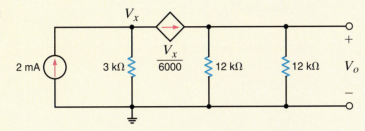

ANSWER: $V_o = 4$ V.

CIRCUITS CONTAINING INDEPENDENT VOLTAGE SOURCES
As is our practice, in our discussion of this topic we will proceed from the simplest case to those cases that are more complicated. The simplest case is that in which an independent voltage source is connected to the reference node. The following example illustrates this case.

Example 3.5

Consider the circuit shown in Fig. 3.11a. Let us determine all node voltages and branch currents.

SOLUTION This network has three nonreference nodes with labeled node voltages V_1, V_2, and V_3. Based on our previous discussions, we would assume that in order to find all the node voltages we would need to write a KCL equation at each of the nonreference nodes. The resulting three linearly independent simultaneous equations would produce the unknown node voltages. However, note that V_1 and V_3 are known quantities because an independent voltage source is connected directly between the nonreference node and each of these nodes. Therefore, $V_1 = 12$ V and $V_3 = -6$ V. Furthermore, note that the current through the 9-kΩ resistor is $[12 - (-6)]/9k = 2$ mA from left to right. We do not know V_2 or the current in the remaining resistors. However, since only one node voltage is unknown, a single-node equation will produce it. Applying KCL to this center node yields

$$\frac{V_2 - V_1}{12k} + \frac{V_2 - 0}{6k} + \frac{V_2 - V_3}{12k} = 0$$

or

$$\frac{V_2 - 12}{12k} + \frac{V_2}{6k} + \frac{V_2 - (-6)}{12k} = 0$$

from which we obtain

$$V_2 = \frac{3}{2} \text{ V}$$

Once all the node voltages are known, Ohm's law can be used to find the branch currents shown in Fig. 3.11b. The diagram illustrates that KCL is satisfied at every node.

Note that the presence of the voltage sources in this example has simplified the analysis, since two of the three linear independent equations are $V_1 = 12$ V and $V_3 = -6$ V. We will find that as a general rule, whenever voltage sources are present between nodes, the node voltage equations that describe the network will be simpler.

HINT

Any time an independent voltage source is connected between the reference node and a nonreference node, the nonreference node voltage is known.

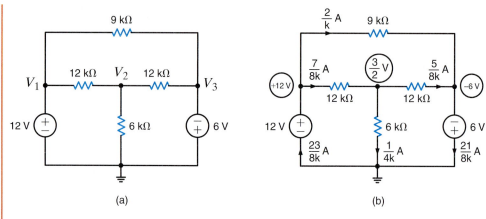

(a)   (b)

Figure 3.11 Circuit used in Example 3.5.

LEARNING EXTENSION

E3.5 Use nodal analysis to find the current I_o in the network in Fig. E3.5.

ANSWER: $I_o = \dfrac{3}{4}$ mA.

Figure E3.5

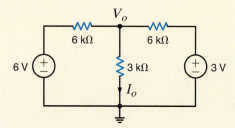

Next let us consider the case in which an independent voltage source is connected between two nonreference nodes.

Example 3.6

Suppose we wish to find the currents in the two resistors in the circuit of Fig. 3.12a.

SOLUTION If we try to attack this problem in a brute force manner, we immediately encounter a problem. Thus far, branch currents were either known source values or could be expressed as the branch voltage divided by the branch resistance. However, the branch current through the 6-V source is certainly not known and cannot be directly expressed using Ohm's law. We can, of course, give this current a name and write the KCL equations at the two nonreference nodes in terms of this current. However, this approach is no panacea because this technique will result in *two* linearly independent simultaneous equations in terms of *three* unknowns—that is, the two node voltages and the current in the voltage source.

To solve this dilemma, we recall that $N - 1$ linearly independent equations are required to determine the $N - 1$ nonreference node voltages in an N-node circuit. Since our network has three nodes, we need two linearly independent equations. Now note that if somehow one of the node voltages is known, we immediately know the other; that is, if V_1 is known, then $V_2 = V_1 - 6$. If V_2 is known, then $V_1 = V_2 + 6$. Therefore, the difference in potential between the two nodes is *constrained* by the voltage source and, hence,

$$V_1 - V_2 = 6$$

This constraint equation is one of the two linearly independent equations needed to determine the node voltages.

Next consider the network in Fig. 3.12b, in which the 6-V source is completely enclosed within the dashed surface. The constraint equation governs this dashed portion of the network. The remaining equation is obtained by applying KCL to this dashed surface, which is commonly called a *supernode*. Recall that in Chapter 2 we demonstrated that KCL must hold for a surface, and this technique eliminates the problem of dealing with a current through a voltage source. KCL for the supernode is

$$-6 \times 10^{-3} + \frac{V_1}{6k} + \frac{V_2}{12k} + 4 \times 10^{-3} = 0$$

Solving these equations yields $V_1 = 10$ V and $V_2 = 4$ V and, hence, $I_1 = 5/3$ mA and $I_2 = 1/3$ mA. A quick check indicates that KCL is satisfied at every node.

Note that applying KCL at the reference node yields the same equation as shown above. The student may feel that the application of KCL at the reference node saves one from having to deal with supernodes. Recall that we do not apply KCL at any node—even the reference node—that contains an independent voltage source. This idea can be illustrated with the circuit in the next example.

Figure 3.12
Circuits used in Example 3.6.

Next consider the network in Fig. 3.12b...

(a)

(b)

Example 3.7

Let us determine the current I_o in the network in Fig. 3.13a.

SOLUTION Examining the network, we note that node voltages V_2 and V_4 are known and the node voltages V_1 and V_3 are constrained by the equation

$$V_1 - V_3 = 12$$

The network is redrawn in Fig. 3.13b.

Figure 3.13
Example circuit with supernodes.

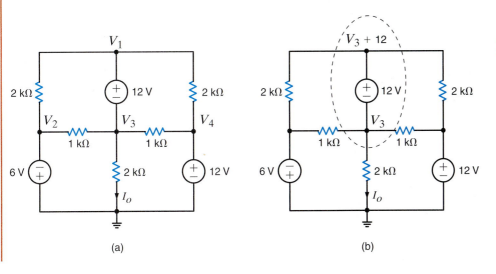

(a)

(b)

Since we want to find the current I_o, V_1 (in the supernode containing V_1 and V_3) is written as $V_3 + 12$. The KCL equation at the supernode is then

$$\frac{V_3 + 12 - (-6)}{2k} + \frac{V_3 + 12 - 12}{2k} + \frac{V_3 - (-6)}{1k} + \frac{V_3 - 12}{1k} + \frac{V_3}{2k} = 0$$

Solving the equation for V_3 yields

$$V_3 = -\frac{6}{7}\ \text{V}$$

I_o can then be computed immediately as

$$I_o = \frac{-\dfrac{6}{7}}{2k} = -\frac{3}{7}\ \text{mA}$$

LEARNING EXTENSION

E3.6 Use nodal analysis to find I_o in the network in Fig. E3.6.

ANSWER: $I_o = 3.8$ mA.

Figure E3.6

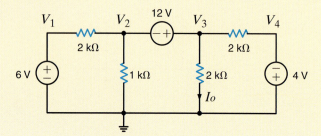

CIRCUITS CONTAINING DEPENDENT VOLTAGE SOURCES As the following examples will indicate, networks containing dependent (controlled) sources are treated in the same manner as described earlier.

Example 3.8

We wish to find I_o in the network in Fig. 3.14.

SOLUTION Since the dependent voltage source is connected between the node labeled V_1 and the reference node,

$$V_1 = 2kI_x$$

KCL at the node labeled V_2 is

$$\frac{V_2 - V_1}{2k} - \frac{4}{k} + \frac{V_2}{1k} = 0$$

where

$$I_x = \frac{V_2}{1k}$$

Solving these equations yields $V_2 = 8$ V and $V_1 = 16$ V. Therefore

$$I_o = \frac{V_1 - V_2}{2k}$$

$$= 4\ \text{mA}$$

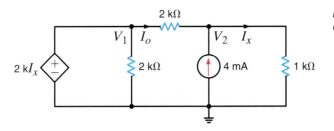

Figure 3.14
Circuit used in Example 3.8.

Example 3.9

Let us find the current I_o in the network in Fig. 3.15.

SOLUTION This circuit contains both an independent voltage source and a voltage-controlled voltage source. Note that $V_3 = 6$ V, $V_2 = V_x$, and a supernode exists between the nodes labeled V_1 and V_2.

Applying KCL to the supernode, we obtain

$$\frac{V_1 - V_3}{6k} + \frac{V_1}{12k} + \frac{V_2}{6k} + \frac{V_2 - V_3}{12k} = 0$$

where the constraint equation for the supernode is

$$V_1 - V_2 = 2V_x$$

The final equation is

$$V_3 = 6$$

Solving these equations, we find that

$$V_1 = \frac{9}{2} \text{ V}$$

and, hence,

$$I_o = \frac{V_1}{12k} = \frac{3}{8} \text{ mA}$$

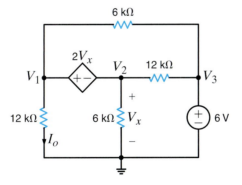

Figure 3.15
Circuit used in Example 3.9.

Finally, let us consider two additional circuits that, for purposes of comparison, we will examine using more than one method.

Example 3.10

Let us find V_o in the network in Fig. 3.16a. Note that the circuit contains two voltage sources, one of which is a controlled source, and two independent current sources. The circuit is redrawn in Fig. 3.16b in order to label the nodes and identify the supernode surrounding the controlled source. Because of the presence of the independent voltage source, the voltage at node 4 is known to be 4 V. We will use this knowledge in writing the node equations for the network.

Since the network has five nodes, four linear independent equations are sufficient to determine all the node voltages. Within the supernode, the defining equation is

$$V_1 - V_2 = 2V_x$$

where

$$V_2 = V_x$$

and thus

$$V_1 = 3V_x$$

Furthermore, we know that one additional equation is

$$V_4 = 4$$

Thus, given these two equations, only two more equations are needed in order to solve for the unknown node voltages. These additional equations result from applying KCL at the supernode and at the node labeled V_3. The equations are

$$-\frac{2}{k} + \frac{V_x}{1k} + \frac{V_x - V_3}{1k} + \frac{3V_x - V_3}{1k} + \frac{3V_x - 4}{1k} = 0$$

$$\frac{V_3 - 3V_x}{1k} + \frac{V_3 - V_x}{1k} = \frac{2}{k}$$

Combining the equations yields the two equations

$$8V_x - 2V_3 = 6$$

$$-4V_x + 2V_3 = 2$$

Solving these equations, we obtain

$$V_x = 2\ \text{V} \quad \text{and} \quad V_3 = 5\ \text{V}$$

$$V_o = 3V_x - V_3 = 1\ \text{V}$$

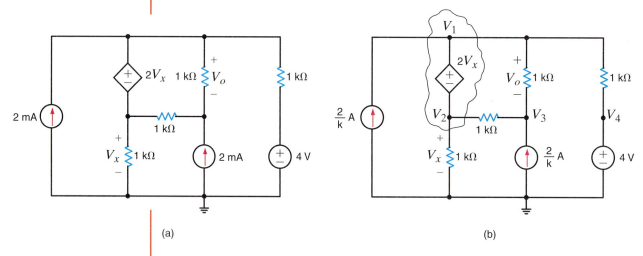

(a)   (b)

Figure 3.16 *Circuit used in Example 3.10.*

Example 3.11

We wish to find I_o in the network in Fig. 3.17a. Note that this circuit contains three voltage sources, one of which is a controlled source and another is a controlled current source. Because two of the voltage sources are connected to the reference node, one node voltage is known directly and one is specified by the dependent source. Furthermore, the difference in voltage between two nodes is defined by the 6-V independent source.

The network is redrawn in Fig. 3.17b in order to label the nodes and identify the supernode. Since the network has six nodes, five linear independent equations are needed to determine the unknown node voltages.

The two equations for the supernode are

$$V_1 - V_4 = -6$$

$$\frac{V_1 - 12}{1k} + \frac{V_1 - V_3}{1k} + 2I_x + \frac{V_4 - V_3}{1k} + \frac{V_4}{1k} + \frac{V_4 - V_5}{1k} = 0$$

The three remaining equations are

$$V_2 = 12$$

$$V_3 = 2V_x$$

$$\frac{V_5 - V_4}{1k} + \frac{V_5}{1k} = 2I_x$$

The equations for the control parameters are

$$V_x = V_1 - 12$$

$$I_x = \frac{V_4}{1k}$$

Combining these equations yields the following set of equations

$$-2V_1 + 5V_4 - V_5 = -36$$

$$V_1 - V_4 = -6$$

$$-3V_4 + 2V_5 = 0$$

Solving these equations by any convenient means yields

$$V_1 = -38 \text{ V}$$

$$V_4 = -32 \text{ V}$$

$$V_5 = -48 \text{ V}$$

Then, since $V_3 = 2V_x$, $V_3 = -100$ V. I_o is -48 mA. The reader is encouraged to verify that KCL is satisfied at every node.

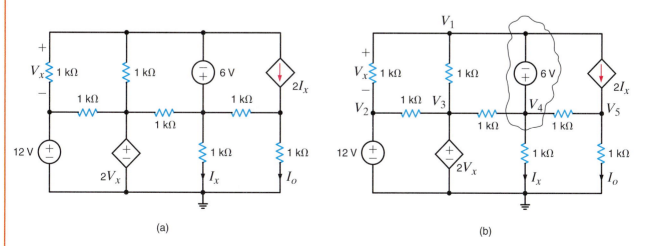

(a)   (b)

Figure 3.17 *Circuit used in Example 3.11.*

PROBLEM-SOLVING STRATEGY

Nodal Analysis

Step 1. Select one node in the N-node circuit as the reference node. Assume that the node voltage is zero and measure all node voltages with respect to this node.

Step 2. If only independent current sources are present in the network, write the KCL equations at the $N - 1$ nonreference nodes. If dependent current sources are present, write the KCL equations as is done for networks with only independent current sources; then write the controlling equations for the dependent sources.

Step 3. If voltage sources are present in the network, they may be connected (1) between the reference node and a nonreference node or (2) between two nonreference nodes. In the former case, if the voltage source is an independent source, then the voltage at one of the nonreference nodes is known. If the source is dependent, it is treated as an independent source when writing the KCL equation, but an additional constraint equation is necessary, as described previously.

In the latter case, if the source is independent, the voltage between the two nodes is constrained by the value of the voltage source, and an equation describing this constraint represents one of the $N - 1$ linearly independent equations required to determine the N-node voltages. The surface of the network described by the constraint equation (i.e., the source and two connecting nodes) is called a supernode. One of the remaining $N - 1$ linearly independent equations is obtained by applying KCL at this supernode. If the voltage source is dependent, it is treated as an independent source when writing the KCL equations, but an additional constraint equation is necessary, as described previously.

LEARNING EXTENSION

E3.7 Use nodal analysis to find I_o in the circuit in Fig. E3.7. **PSV**

ANSWER: $I_o = \dfrac{4}{3}$ mA.

Figure E3.7

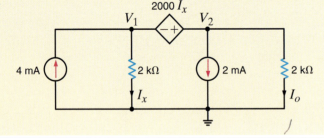

3.2 Loop Analysis

We found that in a nodal analysis the unknown parameters are the node voltages and KCL was employed to determine them. Once these node voltages have been calculated, all the branch currents in the network can easily be determined using Ohm's law. In contrast to this approach, a loop analysis uses KVL to determine a set of loop currents in the circuit. Once these loop currents are known, Ohm's law can be used to calculate any voltages in the network. Via network topology we can show that, in general, there are exactly $B - N + 1$ linearly independent KVL equations for any network, where B is the number of branches in the circuit and N is the number of nodes. For example, if we once again examine the circuit in Fig. 2.5, we find that there are eight branches and five nodes. Thus, the number

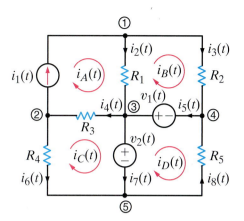

Figure 3.18
Figure 2.5 redrawn with loop currents.

of linearly independent KVL equations necessary to determine all currents in the network is $B - N + 1 = 8 - 5 + 1 = 4$. The network in Fig. 2.5 is redrawn as shown in Fig. 3.18 with 4 loop currents labeled as shown. The branch currents are then determined as

$$i_1(t) = i_A(t)$$
$$i_2(t) = i_A(t) - i_B(t)$$
$$i_3(t) = i_B(t)$$
$$i_4(t) = i_A(t) - i_C(t)$$
$$i_5(t) = i_B(t) - i_D(t)$$
$$i_6(t) = -i_C(t)$$
$$i_7(t) = i_C(t) - i_D(t)$$
$$i_8(t) = -i_D(t)$$

All the circuits we will examine in this text will be *planar*, which simply means that we can draw the circuit on a sheet of paper in such a way that no conductor crosses another conductor. If a circuit is planar, the loops are more easily identified. For example, recall in Chapter 2 that we found that a single equation was sufficient to determine the current in a circuit containing a single loop. If the circuit contains N independent loops, we will show (and the general topological formula $B - N + 1$ can be used for verification), that N independent simultaneous equations will be required to describe the network.

Our approach to loop analysis will mirror the approach used in nodal analysis (i.e., we will begin with simple cases and systematically proceed to those that are more difficult). Then at the end of this section we will outline a general strategy for employing loop analysis.

CIRCUITS CONTAINING ONLY INDEPENDENT VOLTAGE SOURCES

To begin our analysis, consider the circuit shown in Fig. 3.19. We note that this network has seven branches and six nodes, and thus the number of linearly independent KVL equations necessary to determine all currents in the circuit is $B - N + 1 = 7 - 6 + 1 = 2$. Since two linearly independent KVL equations are required, we identify two independent loops, A-B-E-F-A and B-C-D-E-B. We now define a new set of current variables called *loop currents*, which can be used to find the physical currents in the circuit. Let us assume that current i_1 flows in the first loop and that current i_2 flows in the second loop. Then the branch current flowing from B to E through R_3 is $i_1 - i_2$. The directions of the currents have been assumed. As was the case in the nodal analysis, if the actual currents are not in the direction indicated, the values calculated will be negative.

Applying KVL to the first loop yields

$$+v_1 + v_3 + v_2 - v_{S1} = 0$$

KVL applied to loop 2 yields

$$+v_{S2} + v_4 + v_5 - v_3 = 0$$

where $v_1 = i_1 R_1$, $v_2 = i_1 R_2$, $v_3 = \left(i_1 - i_2\right) R_3$, $v_4 = i_2 R_4$, and $v_5 = i_2 R_5$.

HINT

The equations employ the passive sign convention.

Substituting these values into the two KVL equations produces the two simultaneous equations required to determine the two loop currents; that is,

$$i_1(R_1 + R_2 + R_3) - i_2(R_3) = v_{S1}$$
$$-i_1(R_3) + i_2(R_3 + R_4 + R_5) = -v_{S2}$$

or in matrix form

$$\begin{bmatrix} R_1 + R_2 + R_3 & -R_3 \\ -R_3 & R_3 + R_4 + R_5 \end{bmatrix} \begin{bmatrix} i_1 \\ i_2 \end{bmatrix} = \begin{bmatrix} v_{S1} \\ -v_{S2} \end{bmatrix}$$

At this point, it is important to define what is called a *mesh*. A mesh is a special kind of loop that does not contain any loops within it. Therefore, as we traverse the path of a mesh, we do not encircle any circuit elements. For example, the network in Fig. 3.19 contains two meshes defined by the paths *A-B-E-F-A* and *B-C-D-E-B*. The path *A-B-C-D-E-F-A* is a loop, but it is not a mesh. Since the majority of our analysis in this section will involve writing KVL equations for meshes, we will refer to the currents as mesh currents and the analysis as a *mesh analysis*.

Figure 3.19
A two-loop circuit.

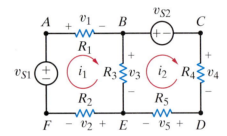

Example 3.12

Consider the network in Fig. 3.20a. We wish to find the current I_o.

SOLUTION   We will begin the analysis by writing mesh equations. Note that there are no + and − signs on the resistors. However, they are not needed, since we will apply Ohm's law to each resistive element as we write the KVL equations. The equation for the first mesh is

$$-12 + 6kI_1 + 6k(I_1 - I_2) = 0$$

The KVL equation for the second mesh is

$$6k(I_2 - I_1) + 3kI_2 + 3 = 0$$

where $I_o = I_1 - I_2$.

Solving the two simultaneous equations yields $I_1 = 5/4$ mA and $I_2 = 1/2$ mA. Therefore, $I_o = 3/4$ mA. All the voltages and currents in the network are shown in Fig. 3.20b. Recall from nodal analysis that once the node voltages were determined, we could check our analysis using KCL at the nodes. In this case, we know the branch currents and can use KVL around any closed path to check our results. For example, applying KVL to the outer loop yields

$$-12 + \frac{15}{2} + \frac{3}{2} + 3 = 0$$
$$0 = 0$$

Since we want to calculate the current I_o, we could use loop analysis, as shown in Fig. 3.20c. Note that the loop current I_1 passes through the center leg of the network and, therefore, $I_1 = I_o$. The two loop equations in this case are

$$-12 + 6k(I_1 + I_2) + 6kI_1 = 0$$

and

$$-12 + 6k(I_1 + I_2) + 3kI_2 + 3 = 0$$

Solving these equations yields $I_1 = 3/4$ mA and $I_2 = 1/2$ mA. Since the current in the 12-V source is $I_1 + I_2 = 5/4$ mA, these results agree with the mesh analysis.

Finally, for purposes of comparison, let us find I_o using nodal analysis. The presence of the two voltage sources would indicate that this is a viable approach. Applying KCL at the top center node, we obtain

$$\frac{V_o - 12}{6k} + \frac{V_o}{6k} + \frac{V_o - 3}{3k} = 0$$

and hence,

$$V_o = \frac{9}{2} \text{ V}$$

and then

$$I_o = \frac{V_o}{6k} = \frac{3}{4} \text{ mA}$$

Note that in this case we had to solve only one equation instead of two.

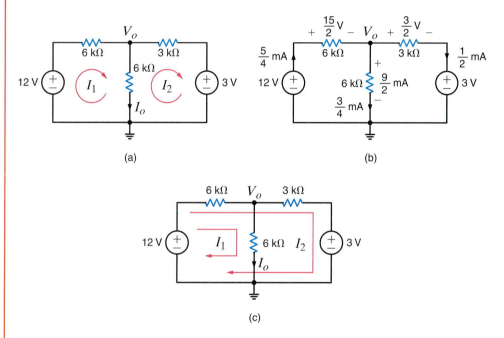

Figure 3.20 Circuits used in Example 3.12.

Once again we are compelled to note the symmetrical form of the mesh equations that describe the circuit in Fig. 3.19. Note that the coefficient matrix for this circuit is symmetrical.

Since this symmetry is generally exhibited by networks containing resistors and independent voltage sources, we can learn to write the mesh equations by inspection. In the first equation, the coefficient of i_1 is the sum of the resistances through which mesh current 1 flows, and the coefficient of i_2 is the negative of the sum of the resistances common to mesh current 1 and mesh current 2. The right-hand side of the equation is the algebraic sum of the voltage sources in mesh 1. The sign of the voltage source is positive if it aids the assumed direction of the current flow and negative if it opposes the assumed flow. The first equation is KVL for mesh 1. In the second equation, the coefficient of i_2 is the sum of all the resistances in mesh 2, the coefficient of i_1 is the negative of the sum of the resistances common to mesh 1 and mesh 2, and the right-hand side of the equation is the algebraic sum of the voltage sources in mesh 2. In general, if we assume all of the mesh currents to be in the same direction (clockwise or counterclockwise), then if KVL is applied to mesh j with mesh current i_j, the coefficient of i_j is the sum of the resistances in mesh j and the coefficients of the other mesh currents $\left(e.g., i_{j-1}, i_{j+1}\right)$ are the negatives of the resistances common to these meshes and mesh j. The right-hand side of the equation is equal to the algebraic sum of the voltage sources in mesh j. These voltage sources have a positive sign if they aid the current flow i_j and a negative sign if they oppose it.

┌─ **Example 3.13**

Let us write the mesh equations by inspection for the network in Fig. 3.21. Then we will use MATLAB to solve for the mesh currents.

SOLUTION The three linearly independent simultaneous equations are

$$(4k + 6k)I_1 - (0)I_2 - (6k)I_3 = -6$$
$$-(0)I_1 + (9k + 3k)I_2 - (3k)I_3 = 6$$
$$-(6k)I_1 - (3k)I_2 + (3k + 6k + 12k)I_3 = 0$$

or in matrix form

$$\begin{bmatrix} 10k & 0 & -6k \\ 0 & 12k & -3k \\ -6k & -3k & 21k \end{bmatrix} \begin{bmatrix} I_1 \\ I_2 \\ I_3 \end{bmatrix} = \begin{bmatrix} -6 \\ 6 \\ 0 \end{bmatrix}$$

Note the symmetrical form of the equations. The general form of the matrix equation is

$$\mathbf{RI} = \mathbf{V}$$

and the solution of this matrix equation is

$$\mathbf{I} = \mathbf{R^{-1}V}$$

The input/output data for a MATLAB solution are as follows:

```
>> R = [10e3 0 -6e3; 0 12e3 -3e3;
-6e3 -3e3 21e3]

    R =

        10000           0       -6000
            0       12000       -3000
        -6000       -3000       21000

>> V = [-6 ; 6 ; 0]

    V =

        -6
         6
         0

>> I = inv(R)*V

    I =

        1.0e-003  *

        -0.6757
         0.4685
        -0.1261
```

Figure 3.21
Circuit used in
Example 3.13.

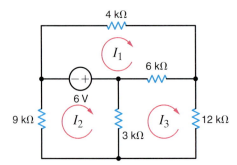

CIRCUITS CONTAINING INDEPENDENT CURRENT SOURCES

Just as the presence of a voltage source in a network simplified the nodal analysis, the presence of a current source simplifies a loop analysis. The following examples illustrate the point.

E3.8 Use mesh equations to find V_o in the circuit in Fig. E3.8.

ANSWER: $V_o = \dfrac{33}{5}$ V.

Figure E3.8

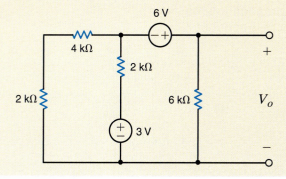

Example 3.14

Let us find both V_o and V_1 in the circuit in Fig. 3.22.

SOLUTION   Although it appears that there are two unknown mesh currents, the current I_1 goes directly through the current source and, therefore, I_1 is constrained to be 2 mA. Hence, only the current I_2 is unknown. KVL for the rightmost mesh is

$$2k(I_2 - I_1) - 2 + 6kI_2 = 0$$

And, of course,

$$I_1 = 2 \times 10^{-3}$$

These equations can be written as

$$-2kI_1 + 8kI_2 = 2$$

$$I_1 = 2/k$$

The input/output data for a MATLAB solution are as follows:

```
>> R = [-2000 8000; 1 0]

   R =

      -2000      8000
          1         0

>> V = [2 ; 0.002]

   V =

       2.0000
       0.0020

>> I = inv(R)*V

   I =

       0.0020
       0.0008
```

```
>> format long

>> I

  I =

       0.00200000000000
       0.00075000000000
```

Note carefully that the first solution for I_2 contains a single digit in the last decimal place. We are naturally led to question whether a number has been rounded off to this value. If we type "format long," MATLAB will provide the answer using 15 digits. Thus, instead of 0.008, the more accurate answer is 0.0075. And hence,

$$V_o = 6kI_2 = \frac{9}{2} \text{ V}$$

To obtain V_1 we apply KVL around any closed path. If we use the outer loop, the KVL equation is

$$-V_1 + 4kI_1 - 2 + 6kI_2 = 0$$

And therefore,

$$V_1 = \frac{21}{2} \text{ V}$$

Note that since the current I_1 is known, the 4-kΩ resistor did not enter the equation in finding V_o. However, it appears in every loop containing the current source and, thus, is used in finding V_1.

Figure 3.22
Circuit used in Example 3.14.

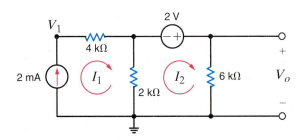

Example 3.15

We wish to find V_o in the network in Fig. 3.23.

SOLUTION Since the currents I_1 and I_2 pass directly through a current source, two of the three required equations are

$$I_1 = 4 \times 10^{-3}$$

$$I_2 = -2 \times 10^{-3}$$

The third equation is KVL for the mesh containing the voltage source; that is,

$$4k(I_3 - I_2) + 2k(I_3 - I_1) + 6kI_3 - 3 = 0$$

These equations yield

$$I_3 = \frac{1}{4} \text{ mA}$$

and hence,

$$V_o = 6kI_3 - 3 = \frac{-3}{2} \text{ V}$$

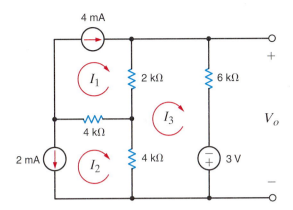

Figure 3.23
Circuit used in Example 3.15.

What we have demonstrated in the previous example is the general approach for dealing with independent current sources when writing KVL equations; that is, use one loop through each current source. The number of "window panes" in the network tells us how many equations we need. Additional KVL equations are written to cover the remaining circuit elements in the network. The following example illustrates this approach.

Example 3.16

Let us find I_o in the network in Fig. 3.24a.

SOLUTION   First, we select two loop currents I_1 and I_2 such that I_1 passes directly through the 2-mA source, and I_2 passes directly through the 4-mA source, as shown in Fig. 3.24b. Therefore, two of our three linearly independent equations are

$$I_1 = 2 \times 10^{-3}$$
$$I_2 = 4 \times 10^{-3}$$

The remaining loop current I_3 must pass through the circuit elements not covered by the two previous equations and cannot, of course, pass through the current sources. The path for this remaining loop current can be obtained by open-circuiting the current sources, as shown in Fig. 3.24c. When all currents are labeled on the original circuit, the KVL equation for this last loop, as shown in Fig. 3.24d, is

$$-6 + 1\text{k}I_3 + 2\text{k}(I_2 + I_3) + 2\text{k}(I_3 + I_2 - I_1) + 1\text{k}(I_3 - I_1) = 0$$

Solving the equations yields

$$I_3 = \frac{-2}{3}\ \text{mA}$$

and therefore,

$$I_o = I_1 - I_2 - I_3 = \frac{-4}{3}\ \text{mA}$$

Next consider the supermesh technique. In this case the three mesh currents are specified as shown in Fig. 3.24e, and since the voltage across the 4-mA current source is unknown, it is assumed to be V_x. The mesh currents constrained by the current sources are

$$I_1 = 2 \times 10^{-3}$$
$$I_2 - I_3 = 4 \times 10^{-3}$$

The KVL equations for meshes 2 and 3, respectively, are

$$2\text{k}I_2 + 2\text{k}(I_2 - I_1) - V_x = 0$$
$$-6 + 1\text{k}I_3 + V_x + 1\text{k}(I_3 - I_1) = 0$$

HINT

In this case the 4-mA current source is located on the boundary between two meshes. Thus, we will demonstrate two techniques for dealing with this type of situation. One is a special loop technique, and the other is known as the supermesh approach.

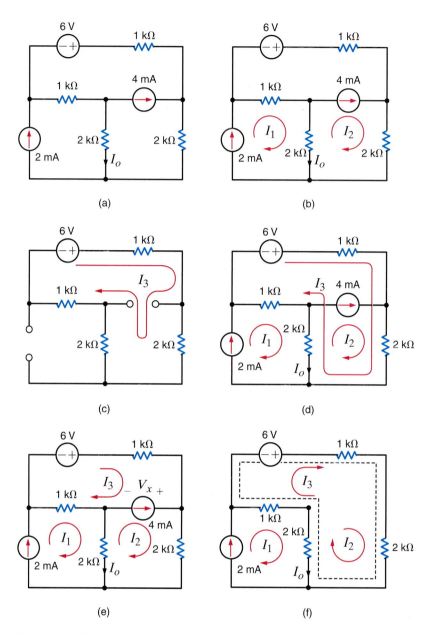

Figure 3.24 *Circuits used in Example 3.16.*

Adding the last two equations yields

$$-6 + 1kI_3 + 2kI_2 + 2k(I_2 - I_1) + 1k(I_3 - I_1) = 0$$

Note that the unknown voltage V_x has been eliminated. The two constraint equations, together with this latter equation, yield the desired result.

The purpose of the supermesh approach is to avoid introducing the unknown voltage V_x. The supermesh is created by mentally removing the 4-mA current source, as shown in Fig. 3.24f. Then writing the KVL equation around the dotted path, which defines the supermesh, using the original mesh currents as shown in Fig. 3.20e, yields

$$-6 + 1kI_3 + 2kI_2 + 2k(I_2 - I_1) + 1k(I_3 - I_1) = 0$$

Note that this supermesh equation is the same as that obtained earlier by introducing the voltage V_x.

E3.9 Find V_o in the network in Fig. E3.9.

ANSWER: $V_o = \dfrac{33}{5}$ V.

Figure E3.9

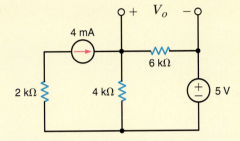

E3.10 Find V_o in the network in Fig. E3.10.

ANSWER: $V_o = \dfrac{32}{5}$ V.

Figure E3.10

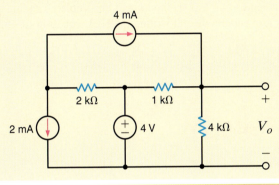

CIRCUITS CONTAINING DEPENDENT SOURCES   We deal with circuits containing dependent sources just as we have in the past. First, we treat the dependent source as though it were an independent source when writing the KVL equations. Then we write the controlling equation for the dependent source. The following examples illustrate the point.

Example 3.17

Let us find V_o in the circuit in Fig. 3.25, which contains a voltage-controlled voltage source.

SOLUTION   The equation for the loop currents shown in the figure are

$$-2V_x + 2k(I_1 + I_2) + 4kI_1 = 0$$
$$-2V_x + 2k(I_1 + I_2) - 3 + 6kI_2 = 0$$

where

$$V_x = 4kI_1$$

These equations can be combined to produce

$$-2kI_1 + 2kI_2 = 0$$
$$-6kI_1 + 8kI_2 = 3$$

The input/output data for a MATLAB solution are

```
>> R = [-2000 2000; -6000 8000]

   R =

      -2000      2000
      -6000      8000
```

```
>> V = [0; 3]

   V =

       0
       3

>> I = inv(R)*V

   I =

       0.00150000000000
       0.00150000000000
```

and therefore,

$$V_o = 6kI_2 = 9 \text{ V}$$

For comparison, we will also solve the problem using nodal analysis. The presence of the voltage sources indicates that this method could be simpler. Treating the 3-V source and its connecting nodes as a supernode and writing the KCL equation for this supernode yields

$$\frac{V_x - 2V_x}{2k} + \frac{V_x}{4k} + \frac{V_x + 3}{6k} = 0$$

where

$$V_o = V_x + 3$$

These equations also yield $V_o = 9$ V.

Figure 3.25
Circuit used in Example 3.17.

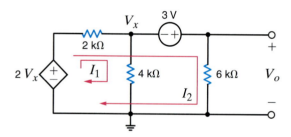

Example 3.18

Let us find V_o in the circuit in Fig. 3.26, which contains a voltage-controlled current source.

SOLUTION The currents I_1 and I_2 are drawn through the current sources. Therefore, two of the equations needed are

$$I_1 = \frac{V_x}{2000}$$

$$I_2 = 2 \times 10^{-3}$$

The KVL equation for the third mesh is

$$-3 + 2k(I_3 - I_1) + 6kI_3 = 0$$

where

$$V_x = 4k\,(I_1 - I_2)$$

Combining these equations yields

$$-I_1 + 2I_2 = 0$$

$$I_2 = 2/k$$

$$-2kI_2 + 8kI_3 = 3$$

The MATLAB solution for these equations is

```
>> R = [-1 2 0; 0 1 0; -2000 0 8000]

   R =

           -1        2          0
            0        1          0
        -2000        0       8000

>> V = [0; 0.002; 3]

   V =

                         0
         0.00200000000000
         3.00000000000000

>> I = inv(R)*V

   I =

         0.00400000000000
         0.00200000000000
         0.00137500000000
```

And hence, $V_o = 8.25$ V

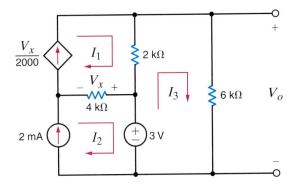

Figure 3.26
Circuit used in Example 3.18.

Example 3.19

The network in Fig. 3.27 contains both a current-controlled voltage source and a voltage-controlled current source. Let us use MATLAB to determine the loop currents.

SOLUTION   The equations for the loop currents shown in the figure are

$$I_1 = \frac{4}{k}$$

$$I_2 = \frac{V_x}{2k}$$

$$-1kI_x + 2k(I_3 - I_1) + 1k(I_3 - I_4) = 0$$

$$1k(I_4 - I_3) + 1k(I_4 - I_2) + 12 = 0$$

where

$$V_x = 2k(I_3 - I_1)$$

$$I_x = I_4 - I_2$$

Combining these equations yields

$$I_1 = \frac{4}{k}$$

$$I_1 + I_2 - I_3 = 0$$

$$1kI_2 + 3kI_3 - 2kI_4 = 8$$

$$1kI_2 + 1kI_3 - 2kI_4 = 12$$

In matrix form the equations are

$$\begin{bmatrix} 1 & 0 & 0 & 0 \\ 1 & 1 & -1 & 0 \\ 0 & 1k & 3k & -2k \\ 0 & 1k & 1k & -2k \end{bmatrix} \begin{bmatrix} I_1 \\ I_2 \\ I_3 \\ I_4 \end{bmatrix} = \begin{bmatrix} \frac{4}{k} \\ 0 \\ 8 \\ 12 \end{bmatrix}$$

The input and output data for the MATLAB solution are as follows:

```
>> R = [1 0 0 0 ; 1 1 -1 0;    0 1000 3000 -2000;
0 1000 1000 -2000]

    R =

        1           0           0           0
        1           1          -1           0
        0        1000        3000       -2000
        0        1000        1000       -2000

>> V = [0.004; 0; 8; 12]

    V =

        0.0040
             0
        8.0000
       12.0000

>> I = inv(R)*V

    I =

        0.0040
       -0.0060
       -0.0020
       -0.0100
```

Figure 3.27
Circuit used in
Example 3.19.

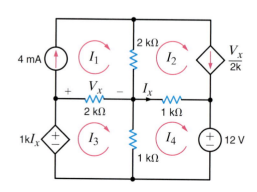

Example 3.20

At this point we will again examine the circuit in Example 3.10 and analyze it using loop equations. Recall that because the network has two voltage sources, the nodal analysis was somewhat simplified. In a similar manner, the presence of the current sources should simplify a loop analysis.

Clearly, the network has four loops, and thus four linearly independent equations are required to determine the loop currents. The network is redrawn in Fig. 3.28 where the loop currents are specified. Note that we have drawn one current through each of the independent current sources. This choice of currents simplifies the analysis since two of the four equations are

$$I_1 = 2/k$$
$$I_3 = -2/k$$

The two remaining KVL equations for loop currents I_2 and I_4 are

$$-2V_x + 1kI_2 + (I_2 - I_3)1k = 0$$
$$(I_4 + I_3 - I_1)1k - 2V_x + 1kI_4 + 4 = 0$$

where

$$V_x = 1k(I_1 - I_3 - I_4)$$

Substituting the equations for I_1 and I_3 into the two KVL equations yields

$$2kI_2 + 2kI_4 = 6$$
$$4kI_4 = 8$$

Solving these equations for I_2 and I_4, we obtain

$$I_4 = 2 \text{ mA}$$
$$I_2 = 1 \text{ mA}$$

and thus

$$V_o = 1 \text{V}$$

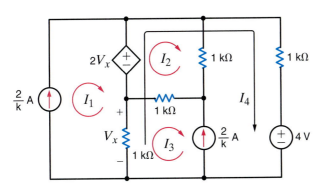

Figure 3.28
Circuit used in Example 3.20.

Example 3.21

Let us once again consider Example 3.11. In this case we will examine the network using loop analysis. Although there are four sources, two of which are dependent, only one of them is a current source. Thus, from the outset we expect that a loop analysis will be more difficult than a nodal analysis. Clearly, the circuit contains six loops. Thus, six linearly independent equations are needed to solve for all the unknown currents.

The network is redrawn in Fig. 3.29 where the loops are specified. The six KVL equations that describe the network are

$$1kI_1 + 1k(I_1 - I_2) + 1k(I_1 - I_4) = 0$$
$$1k(I_2 - I_1) - 6 + 1k(I_2 - I_5) = 0$$
$$I_3 = 2I_x$$

$$-12 + 1k(I_4 - I_1) + 2V_x = 0$$
$$-2V_x + 1k(I_5 - I_2) + 1k(I_5 - I_o) = 0$$
$$1k(I_o - I_5) + 1k(I_o - I_3) + 1kI_o = 0$$

Figure 3.29
Circuit used in Example 3.21.

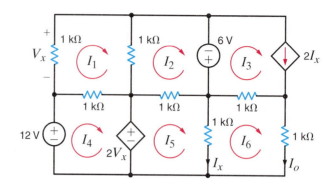

And the control variables for the two dependent sources are

$$V_x = -1kI_1$$
$$I_x = I_5 - I_o$$

Substituting the control parameters into the six KVL equations yields

$$
\begin{array}{cccccccl}
3I_1 & -I_2 & 0 & -I_4 & 0 & 0 & = 0 \\
-I_1 & +2I_2 & 0 & 0 & -I_5 & 0 & = 6/k \\
0 & 0 & I_3 & 0 & -2I_5 & +2I_o & = 0 \\
-3I_1 & 0 & 0 & +I_4 & 0 & 0 & = 12/k \\
2I_1 & -I_2 & 0 & 0 & +2I_5 & -I_o & = 0 \\
0 & 0 & 0 & 0 & -3I_5 & +5I_o & = 0
\end{array}
$$

which can be written in matrix form as

$$
\begin{bmatrix}
3 & -1 & 0 & -1 & 0 & 0 \\
-1 & 2 & 0 & 0 & -1 & 0 \\
0 & 0 & 1 & 0 & -2 & 2 \\
-3 & 0 & 0 & 1 & 0 & 0 \\
2 & -1 & 0 & 0 & 2 & -1 \\
0 & 0 & 0 & 0 & -3 & 5
\end{bmatrix}
\begin{bmatrix}
I_1 \\ I_2 \\ I_3 \\ I_4 \\ I_5 \\ I_o
\end{bmatrix}
=
\begin{bmatrix}
0 \\ 6/k \\ 0 \\ 12/k \\ 0 \\ 0
\end{bmatrix}
$$

Although these six linearly independent simultaneous equations can be solved by any convenient method, we will employ a MATLAB solution. As the results listed below indicate, the current I_o is -48 mA.

```
>> R = [3 -1 0 -1 0 0 ; -1 2 0 0 -1 0 ; 0 0 1 0 -2 2 ;
-3 0 0 1 0 0 ; 2 -1 0 0 2 -1 ; 0 0 0 0 -3 5]

   R =

        3      -1       0      -1       0       0
       -1       2       0       0      -1       0
        0       0       1       0      -2       2
       -3       0       0       1       0       0
        2      -1       0       0       2      -1
        0       0       0       0      -3       5

>> V = [0; 0.006; 0; 0.012; 0; 0]
```

```
    V  =

                  0
             0.0060
                  0
             0.0120
                  0
                  0

>>  I  =  inv(R)*V

    I  =

             0.0500
            -0.0120
            -0.0640
             0.1620
            -0.0800
            -0.0480


>>
```

As a final point, it is very important to examine the circuit carefully before selecting an analysis approach. One method could be much simpler than another, and a little time invested up front may save a lot of time in the long run.

PROBLEM-SOLVING STRATEGY

Loop Analysis

Step 1. One loop current is assigned to each independent loop in a circuit that contains N independent loops.

Step 2. If only independent voltage sources are present in the network, write the N linearly independent KVL equations, one for each loop. If dependent voltage sources are present, write the KVL equation as is done for circuits with only independent voltage sources; then write the controlling equations for the dependent sources.

Step 3. If current sources are present in the network, either of two techniques can be used. In the first case, one loop current is selected to pass through one of the current sources. This is done for each current source in the network. The remaining loop currents (N − the number of current sources) are determined by open-circuiting the current sources in the network and using this modified network to select them. Once all these currents are defined in the original network, the N loop equations can be written. The second approach is similar to the first with the exception that if two mesh currents pass through a particular current source, a supermesh is formed around this source. The two required equations for the meshes containing this source are the constraint equations for the two mesh currents that pass through the source and the supermesh equation. As indicated earlier, if dependent current sources are present, the controlling equations for these sources are also necessary.

E3.11 Use mesh analysis to find V_o in the circuit in Fig. E3.11. **PSV**   **ANSWER:** $V_o = 12$ V.

Figure E3.11

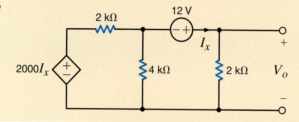

E3.12 Use loop analysis to solve the network in Example 3.5 and compare the time and effort involved in the two solution techniques.

E3.13 Use nodal analysis to solve the circuit in Example 3.15 and compare the time and effort involved in the two solution strategies.

3.3 Application Example

┌─ **Application Example 3.22**

A conceptual circuit for manually setting the speed of a dc electric motor is shown in Figure 3.30a. The resistors R_1 and R_2 are inside a component called a potentiometer, or pot, which is nothing more than an adjustable resistor, for example, a volume control. Turning the knob changes the ratio $\alpha = R_2/(R_1 + R_2)$, but the total resistance, $R_{pot} = R_1 + R_2$, is unchanged. In this way the pot forms a voltage divider that sets the voltage V_{speed}. The power amplifier output, V_M, is four times V_{speed}. Power amplifiers can output the high currents needed to drive the motor. Finally, the dc motor speed is proportional to V_M, that is, the speed in rpm is some constant k times V. Without knowing the details of the power amplifier, can we analyze this system? In particular, can we develop a relationship between rpm and α?

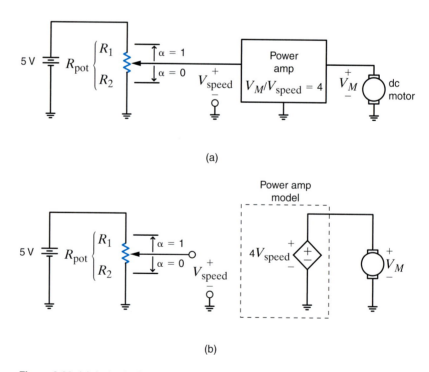

(a)

(b)

Figure 3.30 (a) A simple dc motor driver and (b) the circuit model used to analyze it.

SOLUTION   Since the power amplifier output voltage is proportional to its input, we can model the amplifier as a simple dependent source. The resulting circuit diagram is shown in Fig. 3.30b. Now we can easily develop a relationship between motor speed and the pot position, α. The equations that govern the operation of the motor, power amplifier, and the voltage divider are

$$\text{speed (rpm)} = K_M V_M$$

$$V_M = 4V_{\text{speed}}$$

$$V_{\text{speed}} = 5\frac{R_2}{R_1 + R_2} = 5\left[\frac{R_2}{R_{\text{pot}}}\right] = 5\alpha$$

$$R_2 = \alpha R_{\text{pot}} \qquad R_1 = (1 - \alpha)R_{\text{pot}}$$

Combining these relationships to eliminate V_{speed} yields a relationship between motor speed and α, that is, rpm $= 20\alpha$. If, for example, the motor constant K_M is 50 rpm/V, then

$$\text{rpm} = 1000\alpha$$

This relationship specifies that the motor speed is proportional to the pot knob position. Since the maximum value of α is 1, the motor speed ranges from 0 to 1000 rpm.

Note that in our model, the power amplifier, modeled by the dependent source, can deliver *any* current the motor requires. Of course, this is not possible, but it does demonstrate some of the tradeoffs we experience in modeling. By choosing a simple model, we were able to develop the required relationship quickly. However, other characteristics of an actual power amplifier have been omitted in this model.

3.4   Design Example

Design Example 3.23

An 8-volt source is to be used in conjunction with two standard resistors to design a voltage divider that will output 5 V when connected to a 100-μA load. While keeping the consumed power as low as possible, we wish to minimize the error between the actual output and the required 5 volts.

SOLUTION   The divider can be modeled as shown in Fig. 3.31. Applying KCL at the output node yields the equation

$$\frac{V_S - V_o}{R_1} = \frac{V_o}{R_2} + I_o$$

Using the specified parameters for the input voltage, desired output voltage, and the current source, we obtain

$$R_1 = \frac{3R_2}{5 + (100\mu)R_2}$$

By trial and error, we find that excellent values for the two standard resistors are $R_1 = 10$ kΩ and $R_2 = 27$ kΩ. Large resistor values are used to minimize power consumption. With this selection of resistors the output voltage is 5.11 V, which is a percent error of only 2.15%.

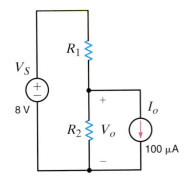

Figure 3.31
A simple voltage-divider circuit with a 100-µA load.

SUMMARY

Nodal analysis for an N-node circuit

- Select one node in the *N*-node circuit as the reference node. Assume that the node voltage is zero and measure all node voltages with respect to this node.

- If only independent current sources are present in the network, write the KCL equations at the *N* − 1 nonreference nodes. If dependent current sources are present, write the KCL equations as is done for networks with only independent current sources; then write the controlling equations for the dependent sources.

- If voltage sources are present in the network, they may be connected (1) between the reference node and a nonreference node or (2) between two nonreference nodes. In the former case, if the voltage source is an independent source, then the voltage at one of the nonreference nodes is known. If the source is dependent, it is treated as an independent source when writing the KCL equations, but an additional constraint equation is necessary.

  In the latter case, if the source is independent, the voltage between the two nodes is constrained by the value of the voltage source and an equation describing this constraint represents one of the *N* − 1 linearly independent equations required to determine the *N*-node voltages. The surface of the network described by the constraint equation (i.e., the source and two connecting nodes) is called a supernode. One of the remaining *N* − 1 linearly independent equations is obtained by applying KCL at this supernode. If the voltage source is dependent, it is treated as an independent source when writing the KCL equations, but an additional constraint equation is necessary.

Loop analysis for an N-loop circuit

- One loop current is assigned to each independent loop in a circuit that contains *N* independent loops.

- If only independent voltage sources are present in the network, write the *N* linearly independent KVL equations, one for each loop. If dependent voltage sources are present, write the KVL equations as is done for circuits with only independent voltage sources; then write the controlling equations for the dependent sources.

- If current sources are present in the network, either of two techniques can be used. In the first case, one loop current is selected to pass through one of the current sources. This is done for each current source in the network. The remaining loop currents (*N* − the number of current sources) are determined by open-circuiting the current sources in the network and using this modified network to select them. Once all these currents are defined in the original network, the *N*-loop equations can be written. The second approach is similar to the first with the exception that if two mesh currents pass through a particular current source, a supermesh is formed around this source. The two required equations for the meshes containing this source are the constraint equations for the two mesh currents that pass through the source and the supermesh equation. If dependent current sources are present, the controlling equations for these sources are also necessary.

PROBLEMS

PSV **CS** both available on the web at: http://www.justask4u.com/irwin

SECTION 3.1

3.1 Find I_o in the circuit in Fig. P3.1 using nodal analysis. **CS**

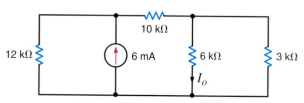

Figure P3.1

3.2 Use nodal analysis to find V_1 in the circuit in Fig. P3.2.

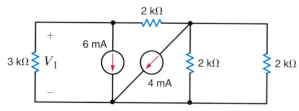

Figure P3.2

3.3 Use nodal analysis to find both V_1 and V_o in the circuit in Fig. P3.3. **PSV**

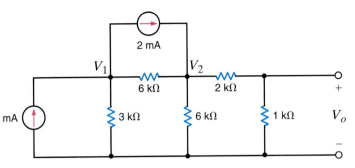

Figure P3.3

3.4 Find V_1 and V_2 in the circuit in Fig. P3.4 using nodal analysis. Then solve the problem using MATLAB and compare your answers.

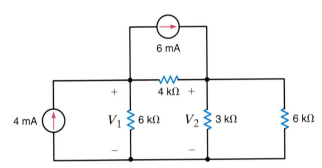

Figure P3.4

3.5 Find I_o in the circuit in Fig. P3.5 using nodal analysis. **CS**

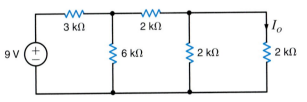

Figure P3.5

3.6 Find I_o in the network in Fig. P3.6 using nodal analysis.

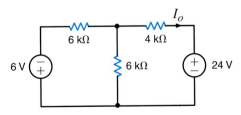

Figure P3.6

3.7 Find V_o in the network in Fig. P3.7 using nodal analysis. **CS**

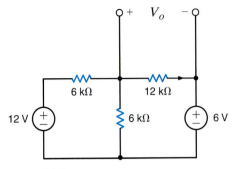

Figure P3.7

3.8 Find V_o in the circuit in Fig. P3.8 using nodal analysis. **PSV**

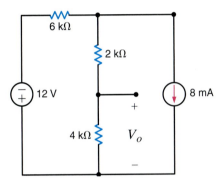

Figure P3.8

3.9 Use nodal analysis to find V_o in the circuit in Fig. P3.9.

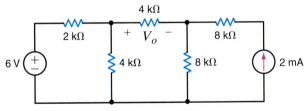

Figure P3.9

3.10 Find I_o in the circuit in Fig. P3.10 using nodal analysis.

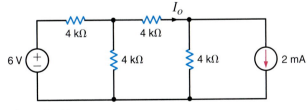

Figure P3.10

3.11 Use nodal analysis to find V_o in the network in Fig. P3.11. Then solve the problem using MATLAB and compare your answers.

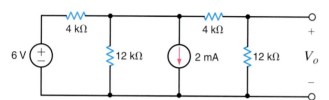

Figure P3.11

3.12 Use nodal analysis to find V_o in the circuit in Fig. P3.12.

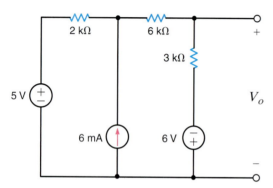

Figure P3.12

3.13 Use nodal analysis to find V_o in the circuit in Fig. P3.13.

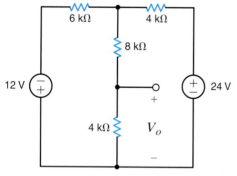

Figure P3.13

3.14 Find I_o in the network in Fig. P3.14. **cs**

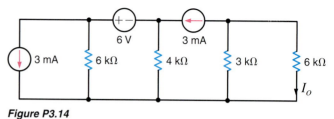

Figure P3.14

3.15 Find I_1 in the network in Fig. P3.15. **cs**

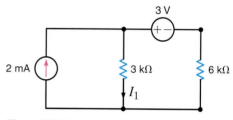

Figure P3.15

3.16 Find I_o in the network in Fig. P3.16.

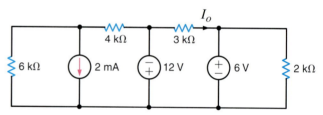

Figure P3.16

3.17 Use nodal analysis to find V_x and V_y in the circuit in Fig. P3.17.

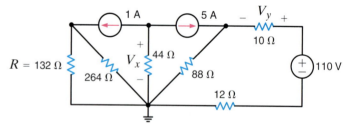

Figure P3.17

3.18 For the network in Fig P3.17, explain why the resistor R plays no role in determining V_x and V_y.

3.19 Use nodal analysis to find V_o in the network in Fig P3.19.

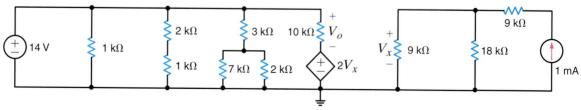

Figure P3.19

3.20 Use nodal analysis to find V_A and V_B in the network in Fig. P3.20. Simplify the analysis by making an insightful choice for the reference node.

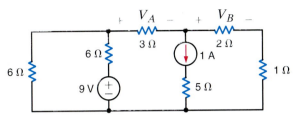

Figure P3.20

3.21 Find I_o in the circuit in Fig. P3.21.

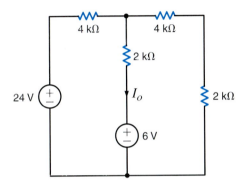

Figure P3.21

3.22 Use nodal analysis to find I_o and I_S in the circuit in Fig. P3.22.

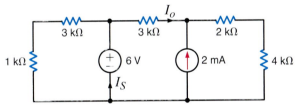

Figure P3.22

3.23 Use nodal analysis to find V_o in the network in Fig. P3.23.

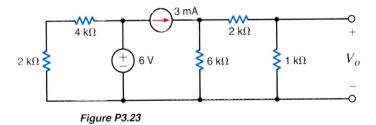

Figure P3.23

3.24 Use nodal analysis to find I_o in the circuit in Fig. P3.24.

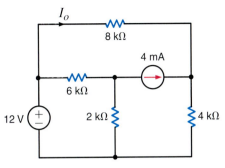

Figure P3.24

3.25 Find I_o in the network in Fig. P3.25 using nodal analysis.

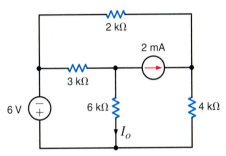

Figure P3.25

3.26 Use nodal analysis to find I_o in the network in Fig. P3.26.

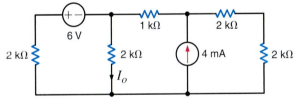

Figure P3.26

3.27 Use nodal analysis to find V_o in the network in Fig. P3.27. Then solve this problem using MATLAB and compare your answers. **CS**

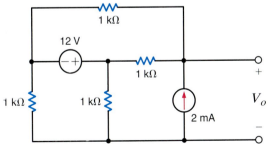

Figure P3.27

3.28 Find V_o in the circuit in Fig. P3.28 using nodal analysis.

PSV

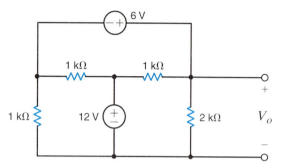

Figure P3.28

3.29 Use nodal analysis to find V_o in the circuit in Fig. P3.29.

CS

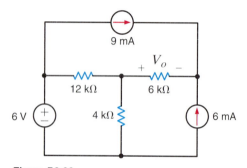

Figure P3.29

3.30 Use nodal analysis to find V_o in the circuit in Fig. P3.30.

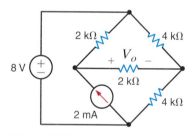

Figure P3.30

3.31 Use nodal analysis to find V_o in the circuit in Fig. P3.31.

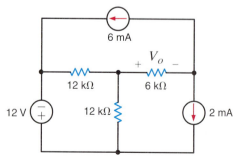

Figure P3.31

3.32 Find V_o in the network in Fig. P3.32 using nodal analysis.

CS

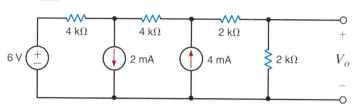

Figure P3.32

3.33 Use nodal analysis to find V_o in the network in Fig. P3.33.

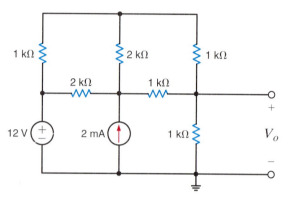

Figure P3.33

3.34 Find V_o in the circuit in Fig. P3.34 using nodal analysis.

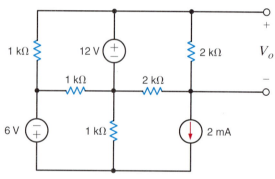

Figure P3.34

3.35 Use nodal analysis to find V_o in the network in Fig. P3.35.

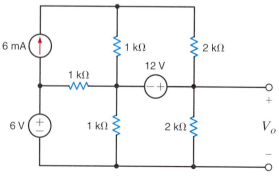

Figure P3.35

3.36 Use MATLAB to find the node voltages in the network in Fig. P3.36. **cs**

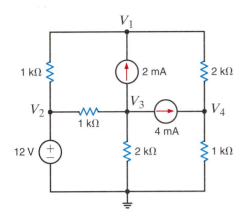

Figure P3.36

3.37 Determine V_o in the network in Fig. P3.37 using nodal analysis.

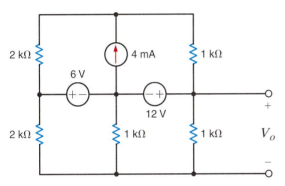

Figure P3.37

3.38 Find V_o in the circuit in Fig. P3.38.

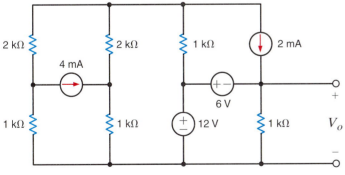

Figure P3.38

3.39 Find V_o in the network in Fig. P3.39.

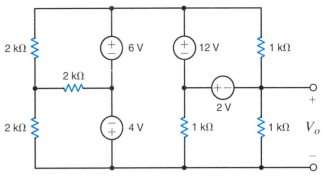

Figure P3.39

3.40 Use nodal analysis to find V_o in the circuit in Fig. P3.40.

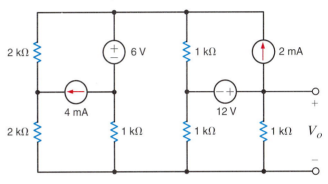

Figure P3.40

3.41 Determine V_o in the network in Fig. P3.41.

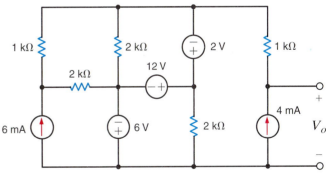

Figure P3.41

3.42 Find V_o in the circuit in Fig. P3.42.

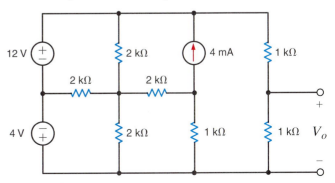

Figure P3.42

3.43 Find I_o in the circuit in Fig. P3.43 using nodal analysis.

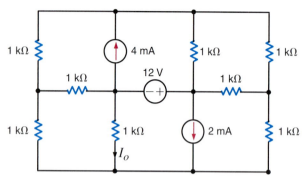

Figure P3.43

3.44 Use nodal analysis to find V_o in Fig. P3.44.

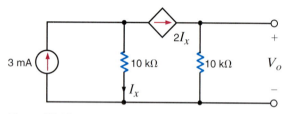

Figure P3.44

3.45 Find V_o in the circuit in Fig. P3.45 using nodal analysis.

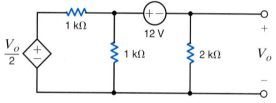

Figure P3.45

3.46 Find V_o in the circuit in Fig. P3.46 using nodal analysis. Then solve the problem using MATLAB and compare your answers. **CS**

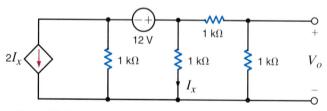

Figure P3.46

3.47 Find I_o in the network in Fig. P3.47. **PSV**

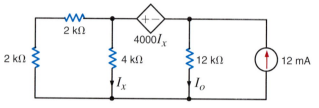

Figure P3.47

3.48 Find I_o in the circuit in Fig. P3.48 using nodal analysis.

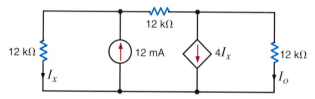

Figure P3.48

3.49 Find V_o in the network in Fig. P3.49 using nodal analysis. **CS**

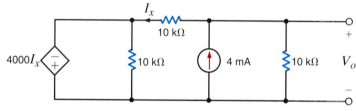

Figure P3.49

3.50 Find V_o in the circuit in Fig. P3.50. **PSV**

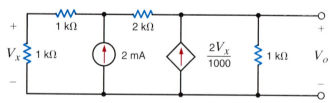

Figure P3.50

3.51 Use nodal analysis to find V_o in the circuit in Fig. P3.51. In addition, find all branch currents and check your answers using KCL at every node.

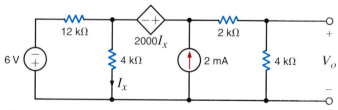

Figure P3.51

3.52 Find the power supplied by the 2-A current source in the network in Fig. P3.52 using nodal analysis.

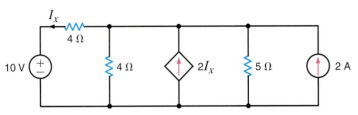

Figure P3.52

3.53 Use nodal equations for the circuit in Fig. P3.53 to determine V_o.

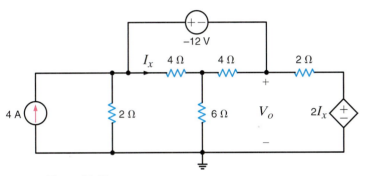

Figure P3.53

3.54 Determine V_o in the network in Fig. P3.54 using nodal analysis.

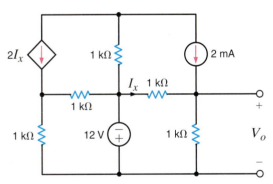

Figure P3.54

3.55 Calculate V_o in the circuit in Fig. P3.55 using nodal analysis.

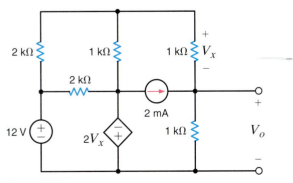

Figure P3.55

3.56 Using nodal analysis, find V_o in the network in Fig. P3.56.

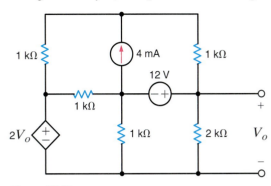

Figure P3.56

3.57 Use nodal analysis to find V_o in the circuit in Fig. P3.57.

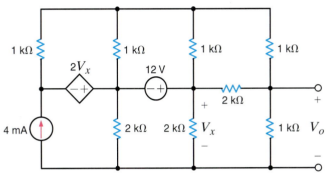

Figure P3.57

3.58 Use nodal analysis to determine I_o in the circuit in Fig. P3.58.

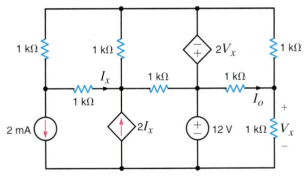

Figure P3.58

3.59 Find I_o in the network in Fig. P3.59 using nodal analysis.

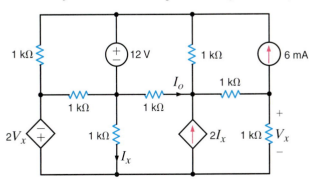

Figure P3.59

3.60 Given the network in Fig. P3.60, we wish to determine the power dissipated in the resistor R_3.

(a) Is mesh or nodal analysis the most efficient approach? Why?

(b) For a nodal analysis, comment on the advantages of selecting node 1 as the reference node. Repeat for nodes 2, 3, and 4.

(c) Based on your results in (b), write the node equations.

3.61 In the circuit in Fig. P3.61, use Gaussian elimination to determine V_o.

(a) Would mesh or nodal analysis be the most efficient approach? Why?

(b) If mesh analysis is used, are any supermeshes required? Write the mesh equations. If nodal analysis is used, are any supernodes required? If so, how many? What is the best location for the reference node and why? Write the node equations.

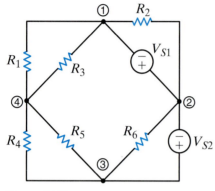

Figure P3.60

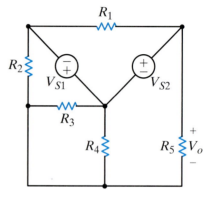

Figure P3.61

SECTION 3.2

3.62 Use mesh equations to find V_o in the circuit in Fig. P3.62.

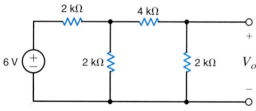

Figure P3.62

3.63 Find V_o in the network in Fig. P3.63 using mesh equations. **PSV**

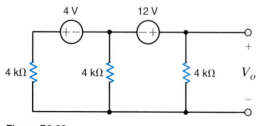

Figure P3.63

3.64 Use mesh analysis to find V_o in the circuit in Fig. P3.64.

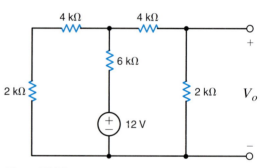

Figure P3.64

3.65 Use mesh analysis to find V_o in the circuit in Fig. P3.65.

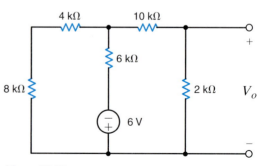

Figure P3.65

3.66 Use mesh analysis to find V_o in the network in Fig. P3.66. **CS**

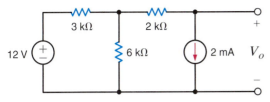

Figure P3.66

3.67 Use loop analysis to find V_o in the circuit in Fig. P3.67.

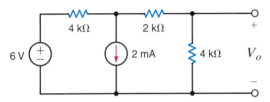

Figure P3.67

3.68 Use loop analysis to find V_o in the network in Fig. P3.68. **PSV**

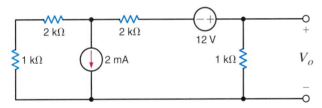

Figure P3.68

3.69 Find I_o in the network in Fig. P3.69 using mesh analysis. **CS**

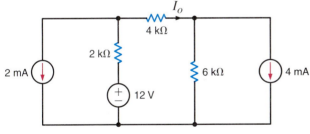

Figure P3.69

3.70 Use both nodal analysis and mesh analysis to find I_o in the circuit in Fig. P3.70.

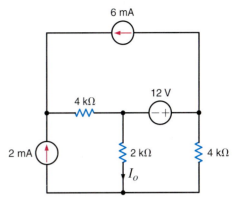

Figure P3.70

3.71 Find I_o in the network in Fig. P3.71 using loop analysis. Then solve the problem using MATLAB and compare your answers. **CS**

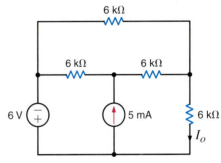

Figure P3.71

3.72 Find V_o in the network in Fig. P3.72 using both mesh and nodal analysis. **PSV**

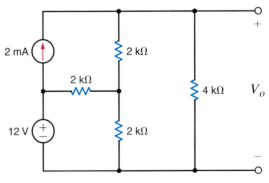

Figure P3.72

3.73 Use loop analysis to find I_o in the network in Fig. P3.73.
CS

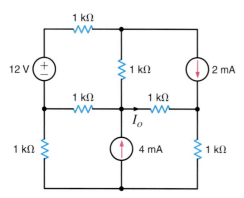

Figure P3.73

3.74 Find I_o in the circuit in Fig. P3.74. **CS**

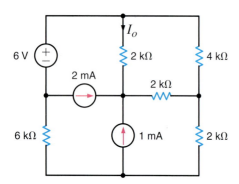

Figure P3.74

3.75 Solve Problem 3.33 using loop analysis.

3.76 Solve Problem 3.34 using loop analysis.

3.77 Solve Problem 3.35 using loop analysis.

3.78 Solve Problem 3.37 using loop analysis.

3.79 Solve Problem 3.40 using loop analysis.

3.80 Solve Problem 3.43 using loop analysis.

3.81 Use MATLAB to find the mesh currents in the network in Fig. P3.81. **CS**

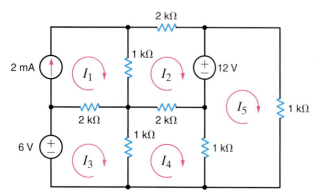

Figure P3.81

3.82 Write mesh equations for the circuit in Fig. P3.82 using the assigned currents.

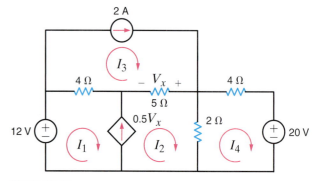

Figure P3.82

3.83 Use mesh analysis to find V_o in the circuit in Fig. P3.83.
PSV

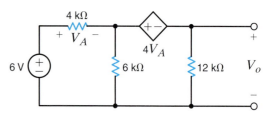

Figure P3.83

3.84 Find V_o in the circuit in Fig. P3.84 using mesh analysis.
CS

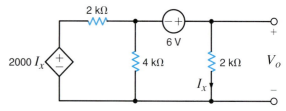

Figure P3.84

3.85 Use loop analysis to find V_o in the network in Fig. P3.85.

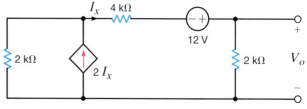

Figure P3.85

3.86 Use loop analysis to find V_o in the circuit in Fig. P3.86.

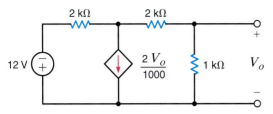

Figure P3.86

3.87 Use both nodal analysis and mesh analysis to find V_o in the circuit in Fig. P3.87.

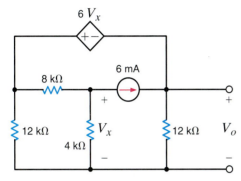

Figure P3.87

3.88 Using mesh analysis, find V_o in the circuit in Fig. P3.88.

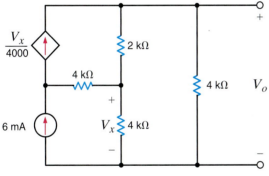

Figure P3.88

3.89 Find V_o in the network in Fig. P3.89. **PSV**

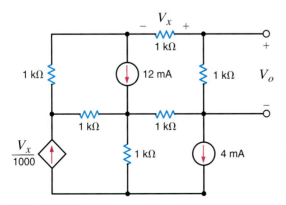

Figure P3.89

3.90 Solve Problem 3.54 using loop analysis.

3.91 Solve Problem 3.55 using loop analysis.

3.92 Solve Problem 3.56 using loop analysis.

3.93 Solve Problem 3.57 using loop analysis.

3.94 Solve Problem 3.58 using loop analysis.

3.95 Solve Problem 3.59 using loop analysis.

3.96 Use mesh analysis to determine the power delivered by the independent 3-V source in the network in Fig. P3.96.

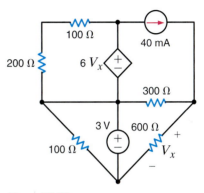

Figure P3.96

3.97 Use mesh analysis to find the power delivered by the current-controlled voltage source in the circuit in Fig. P3.97.

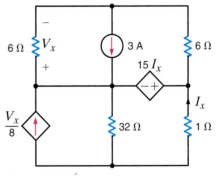

Figure P3.97

TYPICAL PROBLEMS FOUND ON THE FE EXAM

3FE-1 Find V_o in the circuit in Fig. 3PFE-1. **cs**

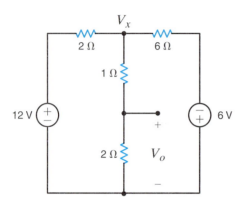

Figure 3PFE-1

3FE-2 Determine the power dissipated in the 6-ohm resistor in the network in Fig. 3PFE-2.

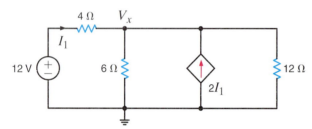

Figure 3PFE-2

3FE-3 Find the current I_x in the 4-ohm resistor in the circuit in Fig. 3PFE-3. **cs**

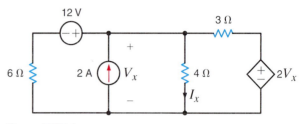

Figure 3PFE-3

3FE-4 Determine the voltage V_o in the circuit in Fig. 3PFE-4.

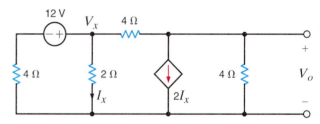

Figure 3PFE-4

Operational Amplifiers

In this chapter we discuss a very important commercially available circuit known as the operational amplifier, or op-amp. Op-amps are used in literally thousands of applications, including such things as compact disk (CD) players, random access memories (RAMs), analog-to-digital (A/D) and digital-to-analog (D/A) converters, headphone amplifiers, and electronic instrumentation of all types. Finally, we discuss the terminal characteristics of this circuit and demonstrate its use in practical applications as well as circuit design. ●

Access Problem-Solving Videos **PSV** *and Circuit Solutions* **CS** *at:*
http://www.justask4u.com/irwin
using the registration code on the inside cover and see a website with answers and more!

4.1   Introduction

It can be argued that the operational amplifier, or op-amp as it is commonly known, is the single most important integrated circuit for analog circuit design. It is a versatile interconnection of transistors and resistors that vastly expands our capabilities in circuit design, from engine control systems to cellular phones. Early op-amps were built of vacuum tubes, making them bulky and power hungry. The invention of the transistor at Bell Labs in 1947 allowed engineers to create op-amps that were much smaller and more efficient. Still, the op-amp itself consisted of individual transistors and resistors interconnected on a printed circuit board (PCB). When the manufacturing process for integrated circuits (ICs) was developed around 1970, engineers could finally put all of the op-amp transistors and resistors onto a single IC chip. Today, it is common to find as many as four high-quality op-amps on a single IC for as little as $0.40. A sample of commercial op-amps is shown in Fig. 4.1.

Figure 4.1

A selection of op-amps. On the left in (a) is a discrete op-amp assembled on a printed circuit board (PCB). On the right, top-down, a LM324 DIP, LMC6492 DIP, and MAX4240 in a SO-5 package (small outline/5 pins). The APEX PA03 with its lid removed (b) showing individual transistors and resistors.

(a)

(b)

Why are they called operational amplifiers? Originally, the op-amp was designed to perform mathematical operations such as addition, subtraction, differentiation, and integration. By adding simple networks to the op-amp, we can create these "building blocks" as well as voltage scaling, current-to-voltage conversion, and a myriad of more complex applications.

4.2   Op-Amp Models

How can we, understanding only sources and resistors, hope to comprehend the performance of the op-amp? The answer lies in modeling. When the bells and whistles are removed, an op-amp is just a really good voltage amplifier. In other words, the output voltage is a scaled replica of the input voltage. Modern op-amps are such good amplifiers that it is easy to create an accurate, first-order model. As mentioned earlier, the op-amp is very popular and is used extensively in circuit design at all levels. We should not be surprised to find that op-amps are available for every application—low voltage, high voltage, micro-power, high speed, high current, and so forth. Fortunately, the topology of our model is independent of these issues.

We start with the general-purpose LM324 quad (four in a pack) op-amp from National Semiconductor, shown in the upper right corner of Fig. 4.1a. The pinout for the LM324 is shown in Fig. 4.2 for a DIP (Dual Inline Pack) style package with dimensions in inches. Recognizing there are four identical op-amps in the package, we will focus on amplifier 1. Pins 3 and 2 are the input pins, *IN* 1+ and *IN* 1−, and are called the noninverting and inverting inputs, respectively.

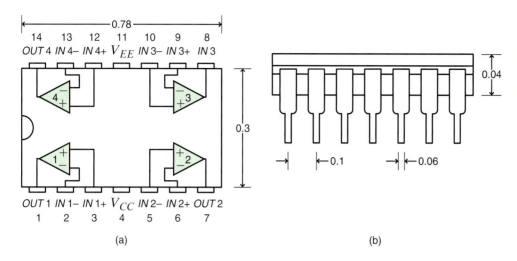

The output is at pin 1. A relationship exists between the output and input voltages,

$$V_o = A_o(IN_+ - IN_-) \qquad \textbf{4.1}$$

where all voltages are measured with respect to ground and A_o is the gain of the op-amp. (The location of the ground terminal will be discussed shortly.) From Eq. (4.1), we see that when IN_+ increases, so will V_o. However, if IN_- increases, then V_o will decrease—hence the names noninverting and inverting inputs. We mentioned earlier that op-amps are very good voltage amplifiers. How good? Typical values for A_o are between 10,000 and 1,000,000!

Amplification requires power that is provided by the dc voltage sources connected to pins 4 and 11, called V_{CC} and V_{EE}, respectively. Figure 4.3 shows how the power supplies, or rails, are connected for both dual- and single-supply applications and defines the ground node to which all input and output voltages are referenced. Traditionally, V_{CC} is a positive dc voltage with respect to ground, and V_{EE} is either a negative voltage or ground itself. Actual values for these power supplies can vary widely depending on the application, from as little as one volt up to several hundred.

How can we model the op-amp? A dependent voltage source can produce V_o! What about the currents into and out of the op-amp terminals (pins 3, 2, and 1)? Fortunately for us, the currents are fairly proportional to the pin voltages. That sounds like Ohm's law. So, we model the *I-V* performance with two resistors, one at the input terminals (R_i) and another at the output (R_o). The circuit in Fig. 4.4 brings everything together.

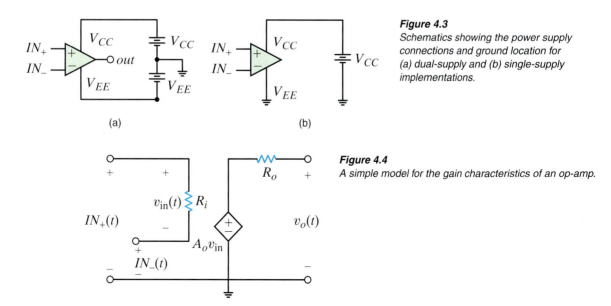

Figure 4.3
Schematics showing the power supply connections and ground location for (a) dual-supply and (b) single-supply implementations.

Figure 4.4
A simple model for the gain characteristics of an op-amp.

Figure 4.5

A network that depicts an op-amp circuit. V_S and R_{Th1} model the driving circuit, while the load is modeled by R_L. The circuit in Fig. 4.4 is the op-amp model.

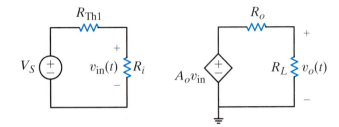

What values can we expect for A_o, R_i, and R_o? We can reason through this issue with the help of Fig. 4.5 where we have drawn an equivalent for the circuitry that drives the input nodes and we have modeled the circuitry connected to the output with a single resistor, R_L. Since the op-amp is supposed to be a great voltage amplifier, let's write an equation for the overall gain of the circuit V_o/V_S. Using voltage division at the input and again at the output, we quickly produce the expression

$$\frac{V_o}{V_S} = \left[\frac{R_i}{R_i + R_{Th1}}\right] A_o \left[\frac{R_L}{R_o + R_L}\right]$$

To maximize the gain regardless of the values of R_{Th1} and R_L, we make the voltage division ratios as close to unity as possible. The ideal scenario requires that A_o be infinity, R_i be infinity, and R_o be zero, yielding a large overall gain of A_o. Table 4.1 shows the actual values of A_o, R_i, and R_o for a sampling of commercial op-amps intended for very different applications. While A_o, R_i, and R_o are not ideal, they do have the correct tendencies.

The power supplies affect performance in two ways. First, each op-amp has minimum and maximum supply ranges over which the op-amp is guaranteed to function. Second, for proper operation, the input and output voltages are limited to no more than the supply voltages.* If the inputs/output can reach within a few dozen millivolts of the supplies, then the inputs/output are called rail-to-rail. Otherwise, the inputs/output voltage limits are more severe—usually a volt or so away from the supply values. Combining the model in Fig. 4.4, the values in Table 4.1, and these I/O limitations, we can produce the graph in Fig. 4.6 showing the output–input relation for each op-amp in Table 4.1. From the graph we see that LMC6492 and MAX4240 have rail-to-rail outputs while the LM324 and PA03 do not.

Table 4.1 A list of commercial op-amps and their model values

Manufacturer	Part No.	A_o (V/V)	R_i (MΩ)	R_o (Ω)	Comments
National	LM324	100,000	1.0	20	General purpose, up to ±16 V supplies, very inexpensive
National	LMC6492	50,000	10^7	150	Low voltage, rail-to-rail inputs and outputs*
Maxim	MAX4240	20,000	45	160	Micro-power (1.8 V supply @ 10 μA), rail-to-rail inputs and outputs
Apex	PA03	125,000	10^5	2	High-voltage, ±75 V, and high-output current capability, 30 A. That's 2 kW!

*Rail-to-rail is a trademark of Motorola Corporation. This feature is discussed further in the following paragraphs.

Even though the op-amp can function within the minimum and maximum supply voltages, because of the circuit configuration, an increase in the input voltage may not yield a corresponding increase in the output voltage. In this case, the op-amp is said to be in saturation. The following example addresses this issue.

---

*Op-amps are available that have input and/or output voltage ranges beyond the supply rails. However, these devices constitute a very small percentage of the op-amp market and will not be discussed here.

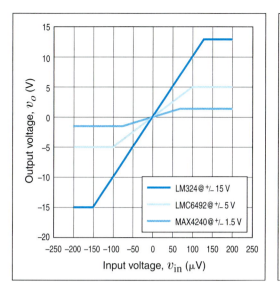
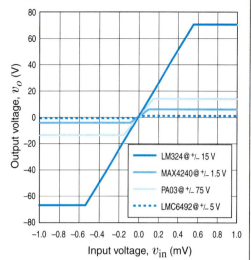

Figure 4.6
Transfer plots for the op-amps listed in Table 4.1. The supply voltages are listed in the plot legends. Note that the LMC6492 and MAX4240 have rail-to-rail output voltages (output voltage range extends to power supply values), while the LM324 and PA03 do not.

Example 4.1

The input and output signals for an op-amp circuit are shown in Fig. 4.7. We wish to determine (a) if the op-amp circuit is linear and (b) the circuit's gain.

SOLUTION

a. We know that if the circuit is linear, the output must be linearly related, that is, proportional, to the input. An examination of the input and output waveforms in Fig. 4.7 clearly indicates that in the region $t = 1.25$ to 2.5 and 4 to 6 ms the output is constant while the input is changing. In this case, the op-amp circuit is in saturation and therefore not linear.

b. In the region where the output is proportional to the input, that is, $t = 0$ to 1 ms, the input changes by 1 V and the output changes by 3.3 V. Therefore, the circuit's gain is 3.3.

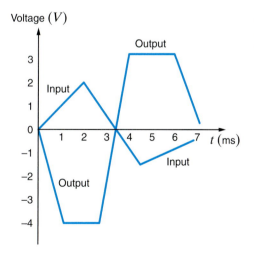

Figure 4.7
An op-amp input–output characteristic.

To introduce the performance of the op-amp in a practical circuit, consider the network in Fig. 4.8a called a unity gain buffer. Notice that the op-amp schematic symbol includes the power supplies. Substituting the model in Fig. 4.4 yields the circuit in Fig. 4.8b, containing just resistors and dependent sources, which we can easily analyze. Writing loop equations, we have

$$V_S = IR_i + IR_o + A_o V_{in}$$
$$V_{out} = IR_o + A_o V_{in}$$
$$V_{in} = IR_i$$

Figure 4.8
Circuit (a) and model (b) for the unity
gain buffer.

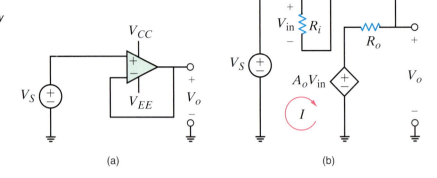

(a)　　　　　　　　　　(b)

Solving for the gain, V_o/V_S, we find

$$\frac{V_o}{V_S} = \frac{1}{1 + \dfrac{R_i}{R_o + A_o R_i}}$$

For $R_o \ll R_i$, we have

$$\frac{V_o}{V_S} \approx \frac{1}{1 + \dfrac{1}{A_o}}$$

And, if A_o is indeed $\gg 1$,

$$\frac{V_o}{V_S} \approx 1$$

The origin of the name *unity gain buffer* should be apparent. Table 4.2 shows the actual gain values for $V_S = 1$ V using the op-amps listed in Table 4.1. Notice how close the gain is to unity and how small the input voltage and current are. These results lead us to simplify the op-amp in Fig. 4.4 significantly. We introduce the *ideal op-amp model*, where A_o and R_i are infinite and R_o is zero. This produces two important results for analyzing op-amp circuitry, listed in Table 4.3.

Table 4.2 Unity gain buffer performance for the op-amps listed in Table 4.1

Op-Amp	Buffer Gain	$V_{in}(\mu V)$	$I(pA)$
LM324	0.999990	9.9999	9.9998
LMC6492	0.999980	19.999	1.9999×10^{-6}
MAX4240	0.999950	49.998	1.1111
PA05	0.999992	7.9999	7.9999×10^{-5}

Table 4.3 Consequences of the ideal op-amp model on input terminal I/V values

Model Assumption	Terminal Result
$A_o \to \infty$	input voltage $\to 0$ V
$R_i \to \infty$	input current $\to 0$ A

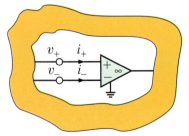

Figure 4.9
Ideal model for an operational
amplifier. Model parameters:
$i_+ = i_- = 0, v_+ = v_-.$

From Table 4.3 we find that the ideal model for the op-amp is reduced to that shown in Fig. 4.9. The important characteristics of the model are as follows: (1) Since R_i is extremely large, the input currents to the op-amp are approximately zero (i.e., $i_+ \approx i_- \approx 0$); and (2) if the output voltage is to remain bounded, then as the gain becomes very large and approaches infinity, the voltage across the input terminals must simultaneously become infinitesimally small so that as $A_o \to \infty$, $v_+ - v_- \to 0$ (i.e., $v_+ - v_- = 0$ or $v_+ = v_-$). The difference between these input voltages is often called the *error signal* for the op-amp (i.e., $v_+ - v_- = v_e$).

The ground terminal $\perp$ shown on the op-amp is necessary for signal current return, and it guarantees that Kirchhoff's current law is satisfied at both the op-amp and the ground node in the circuit.

In summary, then, our ideal model for the op-amp is simply stated by the following conditions:

$$i_+ = i_- = 0 \qquad \textbf{4.2}$$

$$v_+ = v_-$$

These simple conditions are extremely important because they form the basis of our analysis of op-amp circuits.

Let's use the ideal model to reexamine the unity gain buffer, drawn again in Fig. 4.10, where the input voltage and currents are shown as zero. Given V_{in} is zero, the voltage at both op-amp inputs is V_S. Since the inverting input is physically connected to the output, V_o is also V_S—unity gain!

Armed with the ideal op-amp model, let's change the circuit in Fig. 4.10 slightly as shown in Fig. 4.11 where V_S and R_S are an equivalent for the circuit driving the buffer and R_L models the circuitry connected to the output. There are three main points here. First, the gain is still unity. Second, the op-amp requires no current from the driving circuit. Third, the output current $(I_o = V_o/R_L)$ comes from the power supplies, through the op-amp and out of the output pin. In other words, the load current comes from the power supplies, which have plenty of current output capacity, rather than the driving circuit, which may have very little. This isolation of current is called buffering.

An obvious question at this point is this: If $V_o = V_S$, why not just connect V_S to V_o via two parallel connection wires; why do we need to place an op-amp between them? The answer to this question is fundamental and provides us with some insight that will aid us in circuit analysis and design.

Consider the circuit shown in Fig. 4.12a. In this case V_o is not equal to V_S because of the voltage drop across R_S:

$$V_o = V_S - IR_S$$

However, in Fig. 4.12b, the input current to the op-amp is zero and, therefore, V_S appears at the op-amp input. Since the gain of the op-amp configuration is 1, $V_o = V_S$. In Fig. 4.12a the resistive network's interaction with the source caused the voltage v_o to be less than V_S. In other words, the resistive network loads the source voltage. However, in Fig. 4.12b the op-amp isolates the source from the resistive network; therefore, the voltage follower is referred to as a *buffer amplifier* because it can be used to isolate one circuit from another. The energy supplied to the resistive network in the first case must come from the source V_S, whereas in the second case it comes from the power supplies that supply the amplifier, and little or no energy is drawn from V_S.

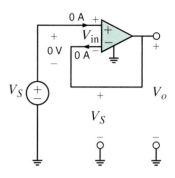

Figure 4.10
An ideal op-amp configured as a unity gain buffer.

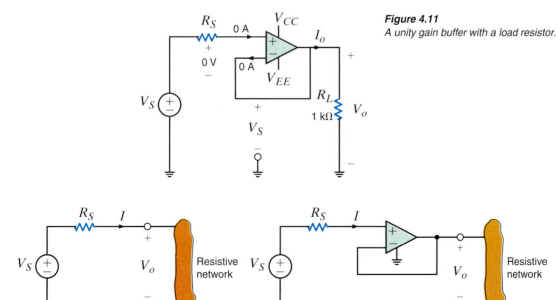

Figure 4.11
A unity gain buffer with a load resistor.

Figure 4.12
Illustration of the isolation capability of a voltage follower.

4.3 Fundamental Op-Amp Circuits

As a general rule, when analyzing op-amp circuits we write nodal equations at the op-amp input terminals, using the ideal op-amp model conditions. Thus, the technique is straight-forward and simple to implement.

Example 4.2

Let us determine the gain of the basic inverting op-amp configuration shown in Fig. 4.13a using both the nonideal and ideal op-amp models.

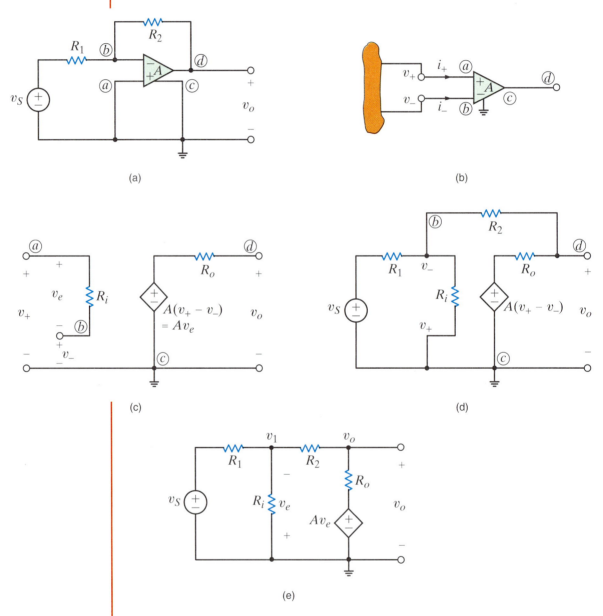

Figure 4.13 Op-amp circuit.

SOLUTION Our model for the op-amp is shown generically in Fig. 4.13b and specifically in terms of the parameters R_i, A, and R_o in Fig. 4.13c. If the model is inserted in the network in Fig. 4.13a, we obtain the circuit shown in Fig. 4.13d, which can be redrawn as shown in Fig. 4.13e.

The node equations for the network are

$$\frac{v_1 - v_S}{R_1} + \frac{v_1}{R_i} + \frac{v_1 - v_o}{R_2} = 0$$

$$\frac{v_o - v_1}{R_2} + \frac{v_o - Av_e}{R_o} = 0$$

where $v_e = -v_1$. The equations can be written in matrix form as

$$\begin{bmatrix} \dfrac{1}{R_1} + \dfrac{1}{R_i} + \dfrac{1}{R_2} & -\left(\dfrac{1}{R_2}\right) \\ -\left(\dfrac{1}{R_2} - \dfrac{A}{R_o}\right) & \dfrac{1}{R_2} + \dfrac{1}{R_o} \end{bmatrix} \begin{bmatrix} v_1 \\ v_o \end{bmatrix} = \begin{bmatrix} \dfrac{v_S}{R_1} \\ 0 \end{bmatrix}$$

Solving for the node voltages, we obtain

$$\begin{bmatrix} v_1 \\ v_o \end{bmatrix} = \frac{1}{\Delta} \begin{bmatrix} \dfrac{1}{R_2} + \dfrac{1}{R_o} & \dfrac{1}{R_2} \\ \dfrac{1}{R_2} - \dfrac{A}{R_o} & \dfrac{1}{R_1} + \dfrac{1}{R_i} + \dfrac{1}{R_o} \end{bmatrix} \begin{bmatrix} \dfrac{v_S}{R_1} \\ 0 \end{bmatrix}$$

where

$$\Delta = \left(\frac{1}{R_1} + \frac{1}{R_i} + \frac{1}{R_2}\right)\left(\frac{1}{R_2} + \frac{1}{R_o}\right) - \left(\frac{1}{R_2}\right)\left(\frac{1}{R_2} - \frac{A}{R_o}\right)$$

Hence,

$$v_o = \frac{\left(\dfrac{1}{R_2} - \dfrac{A}{R_o}\right)\left(\dfrac{v_S}{R_1}\right)}{\left(\dfrac{1}{R_1} + \dfrac{1}{R_i} + \dfrac{1}{R_2}\right)\left(\dfrac{1}{R_2} + \dfrac{1}{R_o}\right) - \left(\dfrac{1}{R_2}\right)\left(\dfrac{1}{R_2} - \dfrac{A}{R_o}\right)}$$

which can be written as

$$\frac{v_o}{v_S} = \frac{-(R_2/R_1)}{1 - \left[\left(\dfrac{1}{R_1} + \dfrac{1}{R_i} + \dfrac{1}{R_2}\right)\left(\dfrac{1}{R_2} + \dfrac{1}{R_o}\right) \bigg/ \left(\dfrac{1}{R_2}\right)\left(\dfrac{1}{R_2} - \dfrac{A}{R_o}\right)\right]}$$

If we now employ typical values for the circuit parameters (e.g., $A = 10^5$, $R_i = 10^8 \ \Omega$, $R_o = 10 \ \Omega$, $R_1 = 1 \ \text{k}\Omega$, and $R_2 = 5 \ \text{k}\Omega$), the voltage gain of the network is

$$\frac{v_o}{v_S} = -4.9996994 \approx -5.000$$

However, the ideal op-amp has infinite gain. Therefore, if we take the limit of the gain equation as $A \to \infty$, we obtain

$$\lim_{A \to \infty} \left(\frac{v_o}{v_S}\right) = -\frac{R_2}{R_1} = -5.000$$

Note that the ideal op-amp yielded a result accurate to within four significant digits of that obtained from an exact solution of a typical op-amp model. These results are easily repeated for the vast array of useful op-amp circuits.

We now analyze the network in Fig. 4.13a using the ideal op-amp model. In this model

$$i_+ = i_- = 0$$

$$v_+ = v_-$$

As shown in Fig. 4.13a, $v_+ = 0$ and, therefore, $v_- = 0$. If we now write a node equation at the negative terminal of the op-amp, we obtain

$$\frac{v_S - 0}{R_1} + \frac{v_o - 0}{R_2} = 0$$

or

$$\frac{v_o}{v_S} = -\frac{R_2}{R_1}$$

and we have immediately obtained the results derived previously.

Notice that the gain is a simple resistor ratio. This fact makes the amplifier very versatile in that we can control the gain accurately and alter its value by changing only one resistor. Also, the gain is essentially independent of op-amp parameters. Since the precise values of A_o, R_i, and R_o are sensitive to such factors as temperature, radiation, and age, their elimination results in a gain that is stable regardless of the immediate environment. Since it is much easier to employ the ideal op-amp model rather than the nonideal model, unless otherwise stated we will use the ideal op-amp assumptions to analyze circuits that contain operational amplifiers.

PROBLEM-SOLVING STRATEGY

Op-Amp Circuits

Step 1. Use the ideal op-amp model: $A_o = \infty$, $R_i = \infty$, $R_o = 0$.
- $i_+ = i_- = 0$
- $v_+ = v_-$

Step 2. Apply nodal analysis to the resulting circuit.

Step 3. Solve nodal equations to express the output voltage in terms of the op-amp input signals.

Example 4.3

Let us now determine the gain of the basic noninverting op-amp configuration shown in Fig. 4.14.

Figure 4.14
The noninverting op-amp configuration.

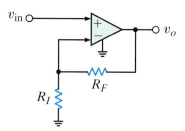

SOLUTION Once again we employ the ideal op-amp model conditions, that is, $v_- = v_+$ and $i_- = i_+$. Using the fact that $i_- = 0$ and $v_- = V_{in}$, the KCL equation at the negative terminal of the op-amp is

$$\frac{v_{in}}{R_I} = \frac{v_o - v_{in}}{R_F}$$

or

$$v_{in}\left(\frac{1}{R_I} + \frac{1}{R_F}\right) = \frac{v_o}{R_F}$$

Thus

$$\frac{v_o}{v_{in}} = 1 + \frac{R_F}{R_I}$$

Note the similarity of this case to the inverting op-amp configuration in the previous example. We find that the gain in this configuration is also controlled by a simple resistor ratio, but is not inverted; that is, the gain ratio is positive.

The remaining examples, though slightly more complicated, are analyzed in exactly the same manner as those outlined above.

Example 4.4

Gain error in an amplifier is defined as

$$GE = \left[\frac{\text{actual gain} - \text{ideal gain}}{\text{ideal gain}}\right] \times 100\%$$

We wish to show that for a standard noninverting configuration with finite gain A_o, the gain error is

$$GE = \frac{-100\%}{1 + A_o\beta}$$

where $\beta = R_1/(R_1 + R_2)$.

SOLUTION The standard noninverting configuration and its equivalent circuit are shown in Fig. 4.15a and b, respectively. The circuit equations for the network in Fig. 4.15b are

$$v_S = v_{\text{in}} + v_1, \qquad v_{\text{in}} = \frac{v_o}{A_o} \quad \text{and} \quad v_1 = \frac{R_1}{R_1 + R_2}v_o = \beta v_o$$

The expression that relates the input and output is

$$v_S = v_o\left[\frac{1}{A_o} + \beta\right] = v_o\left[\frac{1 + A_o\beta}{A_o}\right]$$

and thus the actual gain is

$$\frac{v_o}{v_S} = \frac{A_o}{1 + A_o\beta}$$

Recall that the ideal gain for this circuit is $(R_1 + R_2)/R_1 = 1/\beta$. Therefore, the gain error is

$$GE = \left[\frac{\dfrac{A_o}{1 + A_o\beta} - \dfrac{1}{\beta}}{1/\beta}\right]100\%$$

which when simplified yields

$$GE = \frac{-100\%}{1 + A_o\beta}$$

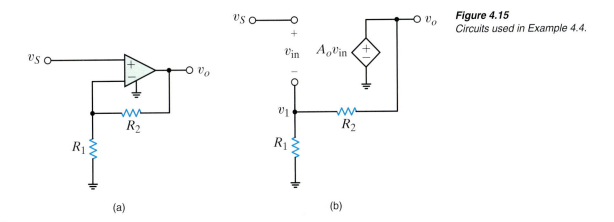

Figure 4.15
Circuits used in Example 4.4.

(a)

(b)

Example 4.5

Consider the op-amp circuit shown in Fig. 4.16. Let us determine an expression for the output voltage.

Figure 4.16
Differential amplifier operational amplifier circuit.

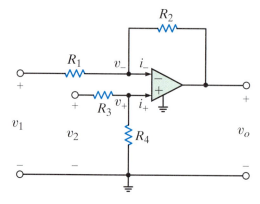

SOLUTION   The node equation at the inverting terminal is

$$\frac{v_1 - v_-}{R_1} + \frac{v_o - v_-}{R_2} = i_-$$

At the noninverting terminal KCL yields

$$\frac{v_2 - v_+}{R_3} = \frac{v_+}{R_4} + i_+$$

However, $i_+ = i_- = 0$ and $v_+ = v_-$. Substituting these values into the two preceding equations yields

$$\frac{v_1 - v_-}{R_1} + \frac{v_o - v_-}{R_2} = 0$$

and

$$\frac{v_2 - v_-}{R_3} = \frac{v_-}{R_4}$$

Solving these two equations for v_o results in the expression

$$v_o = \frac{R_2}{R_1}\left(1 + \frac{R_1}{R_2}\right)\frac{R_4}{R_3 + R_4}v_2 - \frac{R_2}{R_1}v_1$$

Note that if $R_4 = R_2$ and $R_3 = R_1$, the expression reduces to

$$v_o = \frac{R_2}{R_1}(v_2 - v_1)$$

Therefore, this op-amp can be employed to subtract two input voltages.

Example 4.6

The circuit shown in Fig. 4.17a is a precision differential voltage-gain device. It is used to provide a single-ended input for an analog-to-digital converter. We wish to derive an expression for the output of the circuit in terms of the two inputs.

SOLUTION   To accomplish this, we draw the equivalent circuit shown in Fig. 4.17b. Recall that the voltage across the input terminals of the op-amp is approximately zero and the currents into the op-amp input terminals are approximately zero. Note that we can write node equations for node voltages v_1 and v_2 in terms of v_o and v_a. Since we are interested in an expression for v_o in terms of the voltages v_1 and v_2, we simply eliminate the v_a terms from the two node equations. The node equations are

$$\frac{v_1 - v_o}{R_2} + \frac{v_1 - v_a}{R_1} + \frac{v_1 - v_2}{R_G} = 0$$

$$\frac{v_2 - v_a}{R_1} + \frac{v_2 - v_1}{R_G} + \frac{v_2}{R_2} = 0$$

Combining the two equations to eliminate v_a, and then writing v_o in terms of v_1 and v_2, yields

$$v_o = (v_1 - v_2)\left(1 + \frac{R_2}{R_1} + \frac{2R_2}{R_G}\right)$$

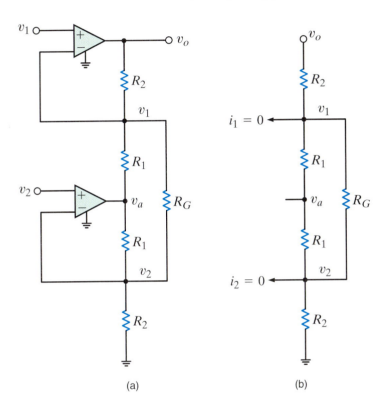

(a)

(b)

Figure 4.17

Instrumentation amplifier circuit.

E4.1 Find I_o in the network in Fig. E4.1.   **PSV**

ANSWER: $I_o = 8.4$ mA.

Figure E4.1

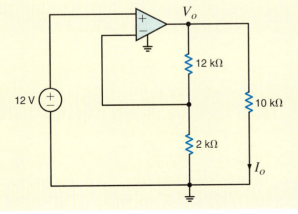

(continues on the next page)

E4.2 Determine the gain of the op-amp circuit in Fig. E4.2. **PSV**

ANSWER: $\dfrac{V_o}{V_S} = 1 + \dfrac{R_2}{R_1}$.

Figure E4.2

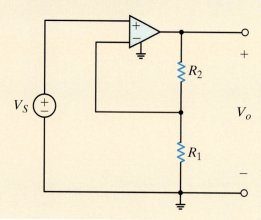

E4.3 Determine both the gain and the output voltage of the op-amp configuration shown in Fig. E4.3.

ANSWER: $V_o = 0.101$ V; gain = 101.

Figure E4.3

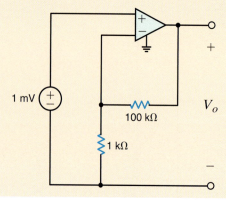

Example 4.7

The two op-amp circuits shown in Fig. 4.18 produce an output given by the equation

$$V_o = 8V_1 - 4V_2$$

where

$$1\text{ V} \le V_1 \le 2\text{ V} \quad \text{and} \quad 2\text{ V} \le V_2 \le 3\text{ V}$$

We wish to determine (a) the range of V_o and (b) if both of the circuits will produce the full range of V_o given that the dc supplies are ±10 V.

SOLUTION

a. Given that $V_o = 8V_1 - 4V_2$ and the range for both V_1 and V_2 as $1\text{ V} \le V_1 \le 2\text{ V}$ and $2\text{ V} \le V_2 \le 3\text{ V}$, we find that

$$V_{o\,\text{max}} = 8(2) - 4(2) = 8\text{ V} \quad \text{and} \quad V_{o\,\text{min}} = 8(1) - 4(3) = -4\text{ V}$$

and thus the range of V_o is -4 V to $+8$ V.

b. Consider first the network in Fig. 4.18a. The signal at V_x, which can be derived using the network in Example 4.5, is given by the equation $V_x = 2V_1 - V_2$. V_x is a maximum when $V_1 = 2$ V and $V_2 = 2$ V, that is, $V_{x\,\text{max}} = 2(2) - 2 = 2$ V. The minimum value for V_x occurs when $V_1 = 1$ V and $V_2 = 3$ V, that is $V_{x\,\text{min}} = 2(1) - 3 = -1$ V. Since both the max and min values are within the supply range of ±10 V, the first op-amp in Fig. 4.18a will not saturate. The output of the second op-amp in this circuit is given by the expression

$V_o = 4V_x$. Therefore, the range of V_o is $-4\ \text{V} \le V_o \le 8\ \text{V}$. Since this range is also within the power supply voltages, the second op-amp will not saturate, and this circuit will produce the full range of V_o.

Next, consider the network in Fig. 4.18b. The signal $V_y = -8V_1$ and so the range of V_y is $-16\ \text{V} \le V_y \le -8\ \text{V}$ and the range of V_y is outside the power supply limits. This circuit will saturate and fail to produce the full range of V_o.

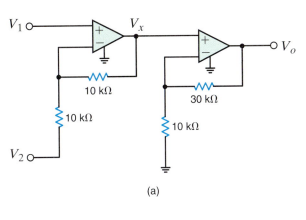

(a)

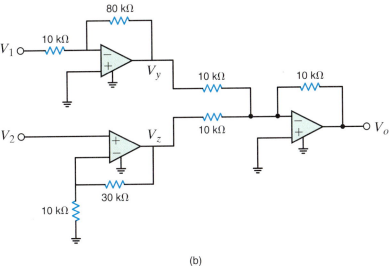

(b)

Figure 4.18
Circuits used in Example 4.7.

4.4   Comparators

A comparator, a variant of the op-amp, is designed to compare the noninverting and inverting input voltages. As shown in Fig. 4.19, when the noninverting input voltage is greater, the output goes as high as possible, at or near V_{CC}. On the other hand, if the inverting input voltage is greater, the output goes as low as possible, at or near V_{EE}. Of course, an ideal op-amp can do the same thing, that is, swing the output voltage as far as possible. However, op-amps are not designed to operate with the outputs saturated, whereas comparators are. As a result, comparators are faster and less expensive than op-amps.

We will present two very different quad comparators in this text, National Semiconductor's LM339 and Maxim's MAX917. Note that the LM339 requires a resistor, called a pull-up resistor, connected between the output pin and V_{CC}. The salient features of these products are listed in Table 4.4. From Table 4.4, it is easy to surmise that the LM339 is a general-purpose comparator, whereas the MAX917 is intended for low-power applications such as hand-held products.

Figure 4.19
(a) An ideal comparator and
(b) its transfer curve.

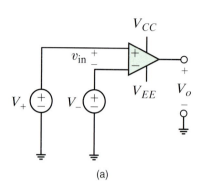

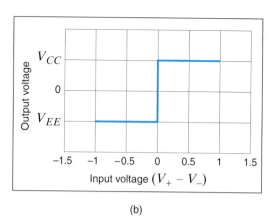

(a)

(b)

A common comparator application is the zero-crossing detector, shown in Fig. 4.20a using a LM339 with ±5 V supplies. As seen in Fig. 4.20b, when V_S is positive, V_o should be near +5 V and when V_S is negative, V_o should be near −5 V. The output changes value on every zero crossing!

Table 4.4 A listing of some of the features of the LM339 and MAX917 comparators

Product	Min. Supply	Max. Supply	Supply Current	Max. Output Current	Typical $R_{pull-up}$
LM339	2 V	36 V	3 mA	50 mA	3 kΩ
MAX919	1.8 V	5.5 V	0.8 μA	8 mA	NA

Figure 4.20
(a) A zero-crossing
detector and (b) the
corresponding
input/output waveforms.

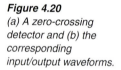

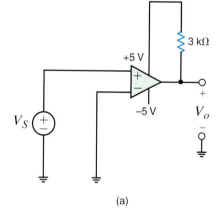

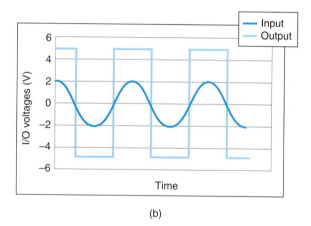

(a)

(b)

4.5 Application Examples

At this point, we have a new element, the op-amp, which we can effectively employ in both applications and circuit design. This device is an extremely useful element that vastly expands our capability in these areas. Because of its ubiquitous nature, the addition of the op-amp to our repertoire of circuit elements permits us to deal with a wide spectrum of practical circuits. Thus, we will employ it here, and also use it throughout this text.

Application Example 4.8

In a light meter, a sensor produces a current proportional to the intensity of the incident radiation. We wish to obtain a voltage proportional to the light's intensity using the circuit in Fig. 4.21. Thus, we select a value of R that will produce an output voltage of 1 V for each 10 μA of sensor current. Assume the sensor has zero resistance.

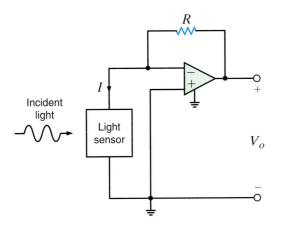

Figure 4.21
Light intensity to voltage converter.

SOLUTION   Applying KCL at the op-amp input,

$$I = V_o/R$$

Since V_o/I is 10^5,

$$R = 100 \text{ k}\Omega$$

Application Example 4.9

The circuit in Fig. 4.22 is an electronic ammeter. It operates as follows: The unknown current, I through R_I produces a voltage, V_I. V_I is amplified by the op-amp to produce a voltage, V_o, which is proportional to I. The output voltage is measured with a simple voltmeter. We want to find the value of R_2 such that 10 V appears at V_o for each milliamp of unknown current.

SOLUTION   Since the current into the op-amp + terminal is zero, the relationship between V_I and I is

$$V_I = IR_I$$

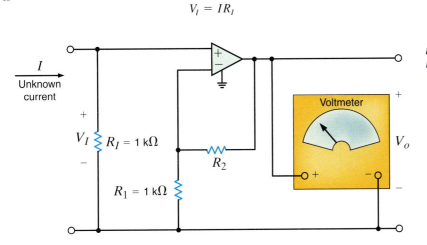

Figure 4.22
Electronic ammeter.

The relationship between the input and output voltages is

$$V_o = V_I\left(1 + \frac{R_2}{R_1}\right)$$

or, solving the equation for V_o/I, we obtain

$$\frac{V_o}{I} = R_I\left(1 + \frac{R_2}{R_1}\right)$$

Using the required ratio V_o/I of 10^4 and resistor values from Fig. 4.22, we can find that

$$R_2 = 9\,\text{k}\Omega$$

Application Example 4.10

Let us return to the dc motor control example in Chapter 3 (Example 3.22). We want to define the form of the power amplifier that reads the speed control signal, V_{speed} and outputs the dc motor voltage with sufficient current to drive the motor as shown in Fig. 4.23. Let us make our selection under the condition that the total power dissipation in the amplifier should not exceed 100 mW.

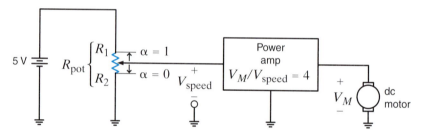

Figure 4.23 *The dc motor example from Chapter 3.*

SOLUTION From Table 4.1 we find that the only op-amp with sufficient output voltage—that is, a maximum output voltage of $(4)(5) = 20$ V—for this application is the PA03 from APEX. Since the required gain is +4, we can employ the standard noninverting amplifier configuration shown in Fig. 4.24. If the PA03 is assumed to be ideal, then

$$V_M = V_{speed}\left[1 + \frac{R_B}{R_A}\right] = 4V_{speed}$$

There are, of course, an infinite number of solutions that will satisfy this equation.

In order to select reasonable values, we should consider the possibility of high currents in R_A and R_B when V_M is at its peak value of 20 V. Assuming that R_{in} for the PA03 is much greater than R_A, the currents in R_B and R_A essentially determine the total power dissipated.

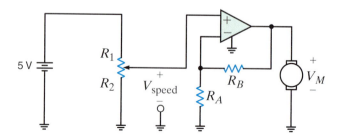

Figure 4.24 *The power amplifier configuration using the PA03 op-amp.*

The total power dissipated in R_A and R_B is

$$P_{total} = \frac{V_M^2}{R_A + R_B} \leq \frac{20^2}{R_A + R_B} = \frac{400}{R_A + R_B}$$

Since the total power should not exceed 100 mW, we can use 1/4 W resistors—an inexpensive industry standard—with room to spare. With this power specification, we find that

$$R_A + R_B = \frac{V_M^2}{P_{total}} = \frac{400}{0.1} = 4000$$

Also, since

$$1 + \frac{R_B}{R_A} = 4$$

then $R_B = 3\,R_A$. Combining this result with the power specification yields $R_A = 1\ \text{k}\Omega$ and $R_B = 3\ \text{k}\Omega$. Both are standard 5% tolerance values.

Application Example 4.11

An instrumentation amplifier of the form shown in Fig. 4.25 has been suggested. This amplifier should have high-input resistance, achieve a voltage gain $V_o/(V_1 - V_2)$ of 10, employ the MAX4240 op-amp listed in Table 4.1, and operate from two 1.5 V AA cell batteries in series. Let us analyze this circuit, select the resistor values, and explore the validity of this configuration.

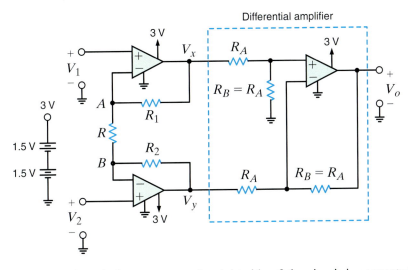

Figure 4.25
An instrumentation amplifier using the MAX4240 op-amp.

SOLUTION   As indicated, the op-amp on the right side of the circuit is connected in the traditional differential amplifier configuration. Example 4.5 indicates that the voltage gain for this portion of the network is

$$V_o = (V_x - V_y)\left[\frac{R_B}{R_A}\right]$$

And if $R_A = R_B$, the equation reduces to

$$V_o = V_x - V_y$$

If we can find a relationship between V_1, V_2, and V_x and V_y, then an expression for the overall voltage can be written. Applying KCL at node A yields

$$\frac{V_1 - V_2}{R} = \frac{V_x - V_1}{R_1}$$

or

$$V_x = V_1\left[1 + \frac{R_1}{R}\right] - V_2\left[\frac{R_1}{R}\right]$$

In a similar manner, at node B we obtain

$$\frac{V_1 - V_2}{R} = \frac{V_2 - V_y}{R_2}$$

or

$$V_y = -V_1 \left[\frac{R_2}{R} \right] + V_2 \left[1 + \frac{R_2}{R} \right]$$

By combining these equations, the output voltage can be expressed as

$$V_o = V_x - V_y = V_1 \left[1 + \frac{R_1}{R} \right] - V_2 \left[\frac{R_1}{R} \right] + V_1 \left[\frac{R_2}{R} \right] - V_2 \left[1 + \frac{R_2}{R} \right]$$

If the resistors are selected such that $R_1 = R_2$, then the voltage gain is

$$\frac{V_o}{V_1 - V_2} = 1 + \frac{2R_1}{R}$$

For a gain of +10, we set $R_1 = 4.5\,R$. To maintain low power, we will use fairly large values for these resistors. We somewhat arbitrarily choose $R = 100\ \text{k}\Omega$ and $R_1 = R_2 = 450\ \text{k}\Omega$. We can use $100\ \text{k}\Omega$ resistors in the differential amplifier stage as well.

Note that the voltage gain of the instrumentation amplifier is essentially the same as that of a generic differential amplifier. So why add the additional cost of two more op-amps? In this configuration the inputs V_1 and V_2 are directly connected to op-amp input terminals; therefore, the input resistance of the intrumentation amplifier is extremely large. From Table 4.1 we see that R_{in} for the MAX4240 is 45 MΩ. This is not the case in the traditional differential amplifier where the external resistor can significantly decrease the input resistance.

4.6   Design Examples

Design Example 4.12

We are asked to construct an amplifier that will reduce a very large input voltage (i.e., V_{in} ranges between ± 680 V) to a small output voltage in the range ∓ 5 V. Using only two resistors, we wish to design the best possible amplifier.

SOLUTION   Since we must reduce +680 V to −5 V, the use of an inverting amplifier seems to be appropriate. The input/output relationship for the circuit shown in Fig. 4.26 is

$$\frac{V_o}{V_{\text{in}}} = -\frac{R_2}{R_1}$$

Since the circuit must reduce the voltage, R_1 must be much larger than R_2. By trial and error, one excellent choice for the resistor pair, selected from the standard Table 2.1, is $R_1 = 27\ \text{k}\Omega$ and $R_2 = 200\ \Omega$. For a $V_{\text{in}} = 680$ V, the resulting output voltage is 5.037 V, resulting in a percent error of only 0.74%.

Figure 4.26
A standard inverting amplifier stage.

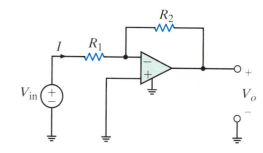

Design Example 4.13

There is a requirement to design a noninverting op-amp configuration with two resistors under the following conditions: the gain must be +10, the input range is ±2 V, and the total power consumed by the resistors must be less than 100 mW.

SOLUTION   For the standard noninverting configuration in Fig. 4.27a, the gain is

$$\frac{V_o}{V_{in}} = 1 + \frac{R_2}{R_1}$$

For a gain of 10, we find $R_2/R_1 = 9$. If $R_1 = 3\,k\Omega$ and $R_2 = 27\,k\Omega$, then the gain requirement is met exactly. Obviously, a number of other choices can be made, from the standard Table 2.1, with a 3/27 ratio. The power limitation can be formalized by referring to Fig. 4.27b where the maximum input voltage (2 V) is applied. The total power dissipated by the resistors is

$$P_R = \frac{2^2}{R_1} + \frac{(20-2)^2}{R_2} = \frac{4}{R_1} + \frac{324}{9R_1} < 0.1$$

The minimum value for R_1 is 400 Ω.

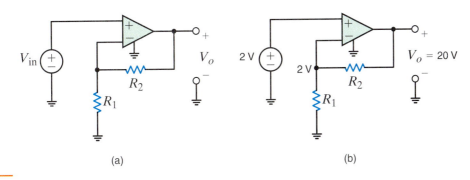

(a)   (b)

Figure 4.27
The noninverting op-amp configuration employed in Example 4.13.

Design Example 4.14

We wish to design a weighted-summer circuit that will produce the output

$$V_o = -0.9V_1 - 0.1V_2$$

The design specifications call for use of one op-amp and no more than three resistors. Furthermore, we wish to minimize power while using resistors no larger than 10 kΩ.

SOLUTION   A standard weighted-summer configuration is shown in Fig. 4.28. Our problem is reduced to finding values for the three resistors in the network. Using KCL, we can write

$$I_1 + I_2 = -\frac{V_o}{R}$$

where

$$I_1 = \frac{V_1}{R_1} \quad \text{and} \quad I_2 = \frac{V_2}{R_2}$$

Combining there relationships yields

$$V_o = -\left[\frac{R}{R_1}\right]V_1 - \left[\frac{R}{R_2}\right]V_2$$

Therefore, we require

$$\frac{R}{R_1} = 0.9 \quad \text{and} \quad \frac{R}{R_2} = 0.1$$

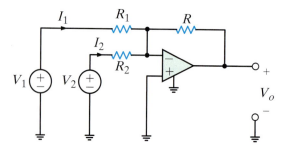

Figure 4.28 *A standard weighted-summer configuration.*

From these requirements, we see that the largest resistor is R_2 and that R is the smallest. Also, note that the R/R_1 ratio can be expressed as 27/30. Finally, to minimize power, we should use the largest possible resistor values. Based on this information, the best resistor values are $R = 270\ \Omega$, $R_1 = 300\ \Omega$, and $R_2 = 2.7\ \text{k}\Omega$, which yield the desired performance exactly.

Design Example 4.15

In Example 2.36, a 250-Ω resistor was used to convert a current in the 4- to 20-mA range to a voltage such that a 20-mA input produced a 5-V output. In this case, the minimum current (4 mA) produces a resistor voltage of 1 V. Unfortunately, many control systems operate on a 0- to 5-V range rather than a 1- to 5-V range. Let us design a new converter that will output 0 V at 4 mA and 5 V at 20 mA.

SOLUTION The simple resistor circuit we designed in Example 2.36 is a good start. However, the voltage span is only 4 V rather than the required 5 V, and the minimum value is not zero. These facts imply that a new resistor value is needed and the output voltage should be shifted down so that the minimum is zero. We begin by computing the necessary resistor value.

$$R = \frac{V_{\max} - V_{\min}}{I_{\max} - I_{\min}} = \frac{5 - 0}{0.02 - 0.004} = 312.5\ \Omega$$

The resistor voltage will now range from $(0.004)(312.5)$ to $(0.02)(312.5)$ or 1.25 to 6.25 V. We must now design a circuit that shifts these voltage levels so that the range is 0 to 5 V. One possible option for the level shifter circuit is the differential amplifier shown in Fig. 4.29. Recall that the output voltage of this device is

$$V_o = (V_I - V_{\text{shift}})\frac{R_2}{R_1}$$

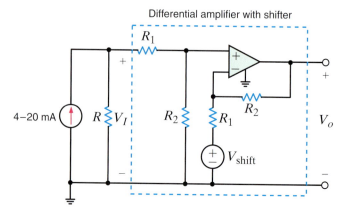

Figure 4.29 *A 4–20 mA to 0–5 V converter circuit.*

Since we have already chosen R for a voltage span of 5 V, the gain of the amplifier should be 1 (i.e., $R_1 = R_2$). Clearly, the value of the required shift voltage is 1.25 V. However, we can verify this value by inserting the minimum values into this last equation

$$0 = [(312.5)(0.004) - V_{\text{shift}}] \frac{R_2}{R_1}$$

and find

$$V_{\text{shift}} = (312.5)(0.004) = 1.25 \text{ V}$$

There is one caveat to this design. We don't want the converter resistor, R, to affect the differential amplifier, or vice versa. This means that the vast majority of the 4–20 mA current should flow entirely through R and not through the differential amplifier resistors. If we choose R_1 and $R_2 \gg R$, this requirement will be met. Therefore, we might select $R_1 = R_2 = 100 \text{ k}\Omega$ so that their resistance values are more than 300 times that of R.

SUMMARY

- Op-amps are characterized by

  High-input resistance

  Low-output resistance

  Very high gain

- The ideal op-amp is modeled using

$$i_+ = i_- = 0$$
$$v_+ = v_-$$

- Op-amp problems are typically analyzed by writing node equations at the op-amp input terminals

- The output of a comparator is dependent on the difference in voltage at the input terminals

PROBLEMS

PSV **CS** **both available on the web at: http://www.justask4u.com/irwin**

SECTION 4.2

4.1 An amplifier has a gain of 15 and the input waveform shown in Fig. P4.1. Draw the output waveform.

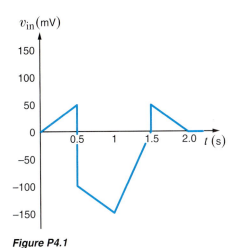

Figure P4.1

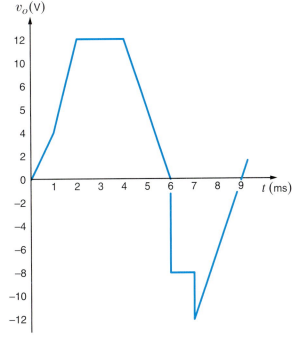

Figure P4.2

4.2 An amplifier has a gain of −5 and the output waveform shown in Fig. P4.2. Sketch the input waveform.

4.3 An op-amp base amplifier has supply voltages of ±5 V and a gain of 20.

   (a) Sketch the input waveform from the output waveform in Fig. P4.3.

   (b) Double the amplitude of your results in (a) and sketch the new output waveform.

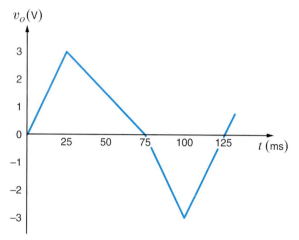

$v_o(V)$

Figure P4.3

4.4 For an ideal op-amp, the voltage gain and input resistance are infinite while the output resistance is zero. What are the consequences for

   (a) the op-amp's input voltage?

   (b) the op-amp's input currents?

   (c) the op-amp's output current?

4.5 Revisit your answers in Problem 4.4 under the following nonideal scenarios. **CS**

   (a) $R_{in} = \infty$, $R_{out} = 0$, $A_o \neq \infty$.

   (b) $R_{in} = \infty$, $R_{out} > 0$, $A_o = \infty$.

   (c) $R_{in} \neq \infty$, $R_{out} = 0$, $A_o = \infty$.

4.6 Revisit the exact analysis of the inverting configuration in Section 4.3.

   (a) Find an expression for the gain if $R_{in} = \infty$, $R_{out} = 0$, $A_o \neq \infty$.

   (b) Plot the ratio of the gain in (a) to the ideal gain versus A_o for $1 \leq A_o \leq 1000$ for an ideal gain of -10.

   (c) From your plot, does the actual gain approach the ideal value as A_o increases or decreases?

   (d) From your plot, what is the minimum value of A_o if the actual gain is within 5% of the ideal case?

4.7 Revisit the exact analysis of the inverting amplifier in Section 4.3.

   (a) Find an expression for the voltage gain if $R_{in} \neq \infty$, $R_{out} = 0$, $A_o \neq \infty$.

   (b) For $R_2 = 27$ kΩ and $R_1 = 3$ kΩ, plot the ratio of the actual gain to the ideal gain for $A_o = 1000$ and 1 k$\Omega \leq R_{in} \leq 100$ kΩ.

   (c) From your plot, does the ratio approach unity as R_{in} increases or decreases?

   (d) From your plot in (b), what is the minimum value of R_{in} if the gain ratio is to be at least 0.98?

4.8 An op-amp based amplifier has ±18 V supplies and a gain of -80. Over what input range is the amplifier linear?

SECTION 4.3

4.9 Determine the gain of the amplifier in Fig. P4.9. What is the value of I_o? **CS**

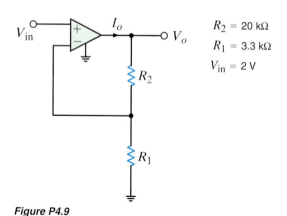

$R_2 = 20$ kΩ
$R_1 = 3.3$ kΩ
$V_{in} = 2$ V

Figure P4.9

4.10 For the amplifier in Fig. P4.10, find the gain and I_o.

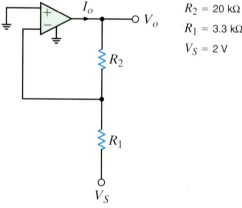

$R_2 = 20$ kΩ
$R_1 = 3.3$ kΩ
$V_S = 2$ V

Figure P4.10

4.11 Using the ideal op-amp assumptions, determine the values of V_o and I_1 in Fig. P4.11.

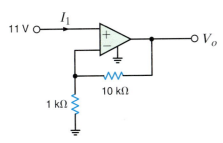

Figure P4.11

4.12 Using the ideal op-amp assumptions, determine I_1, I_2, and I_3 in Fig. P4.12. **PSV**

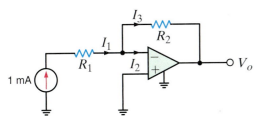

Figure P4.12

4.13 In a useful application, the amplifier drives a load. The circuit in Fig. P4.13 models this scenario. **CS**

(a) Sketch the gain V_o/V_S for $10\ \Omega \le R_L \le \infty$.

(b) Sketch I_o for $10\ \Omega \le R_L \le \infty$ if $V_S = 0.1$ V.

(c) Repeat (b) if $V_S = 1.0$ V.

(d) What is the minimum value of R_L if $|I_o|$ must be less than 100 mA for $|V_S| < 0.5$ V?

(e) What is the current I_S if R_L is 100 Ω? Repeat for $R_L = 10$ kΩ.

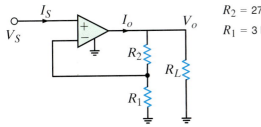

$R_2 = 27$ kΩ
$R_1 = 3$ kΩ

Figure P4.13

4.14 Repeat Problem 4.13 for the circuit in Fig. P4.14.

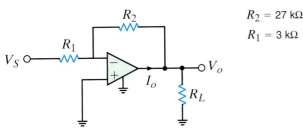

$R_2 = 27$ kΩ
$R_1 = 3$ kΩ

Figure P4.14

4.15 The op-amp in the amplifier in Fig. P4.15 operates with ± 15 V supplies and can output no more than 200 mA. What is the maximum gain allowable for the amplifier if the maximum value of V_S is 1 V?

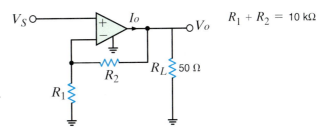

$R_1 + R_2 = 10$ kΩ

Figure P4.15

4.16 For the amplifier in Fig. P4.16, the maximum value of V_S is 2 V and the op-amp can deliver no more than 100 mA.

(a) If ± 10 V supplies are used, what is the maximum allowable value of R_2?

(b) Repeat for ± 3 V supplies.

(c) Discuss the impact of the supplies on the maximum allowable gain.

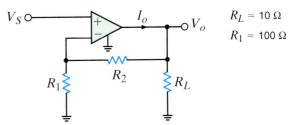

$R_L = 10\ \Omega$
$R_1 = 100\ \Omega$

Figure P4.16

4.17 For the circuit in Fig. P4.17,

(a) find V_o in terms of V_1 and V_2.

(b) If $V_1 = 2$ V and $V_2 = 6$ V, find V_o.

(c) If the op-amp supplies are ± 12 V, and $V_1 = 4$ V, what is the allowable range of V_2? **PSV**

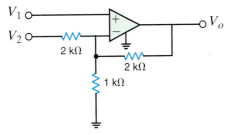

Figure P4.17

4.18 Find V_o in the circuit in Fig. P4.18 assuming the op-amp is ideal.

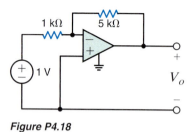

Figure P4.18

4.19 The network in Fig. P4.19 is a current-to-voltage converter or transconductance amplifier. Find v_o/i_S for this network.

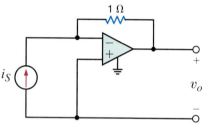

Figure P4.19

4.20 Calculate the transfer function i_o/v_1 for the network shown in Fig. P4.20.

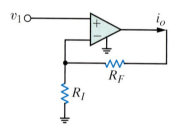

Figure P4.20

4.21 Determine the relationship between v_1 and i_o in the circuit shown in Fig. P4.21. **cs**

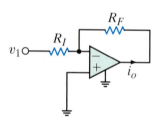

Figure P4.21

4.22 Find V_o in the network in Fig. P4.22 and explain what effect R_1 has on the output.

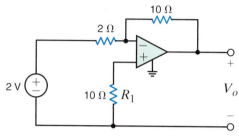

Figure P4.22

4.23 Determine the expression for v_o in the network in Fig. P4.23.

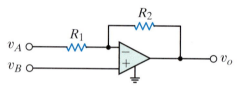

Figure P4.23

4.24 Show that the output of the circuit in Fig. P4.24 is

$$V_o = \left[1 + \frac{R_2}{R_1}\right]V_1 - \frac{R_2}{R_1}V_2$$

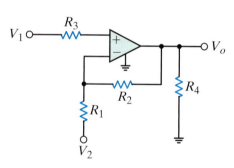

Figure P4.24

4.25 Find V_o in the network in Fig. P4.25. **cs**

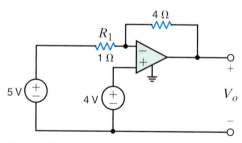

Figure P4.25

4.26 Find the voltage gain of the op-amp circuit shown in Fig. P4.26.

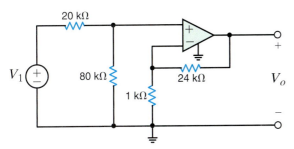

Figure P4.26

4.27 For the circuit in Fig. 4.27 find the value of R_1 that produces a voltage gain of 10.

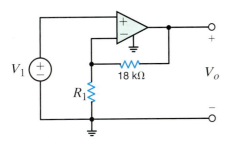

Figure P4.27

4.28 Determine the relationship between v_o and v_{in} in the circuit in Fig. P4.28.

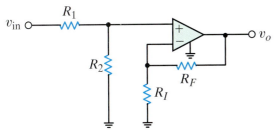

Figure P4.28

4.29 In the network in Fig. P4.29 derive the expression for v_o in terms of the inputs v_1 and v_2. **CS**

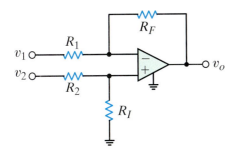

Figure P4.29

4.30 Find V_o in the circuit in Fig. P4.30.

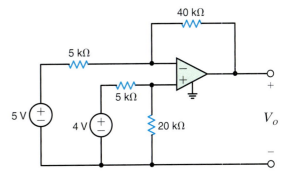

Figure P4.30

4.31 Find V_o in the circuit in Fig. P4.31.

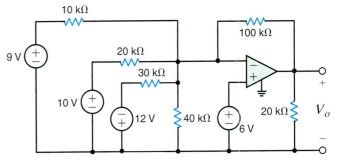

Figure P4.31

4.32 Determine the expression for the output voltage, v_o, of the inverting summer circuit shown in Fig. P4.32.

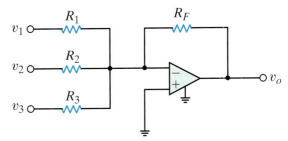

Figure P4.32

4.33 Determine the output voltage, v_o, of the noninverting averaging circuit shown in Fig. P4.33. **CS**

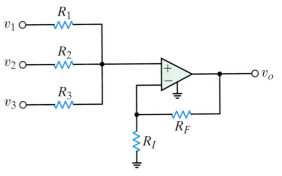

Figure P4.33

4.34 Find the input/output relationship for the current amplifier shown in Fig. P4.34.

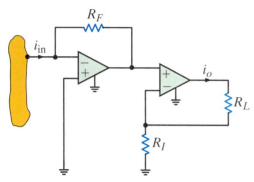

Figure P4.34

4.35 Find V_o in the circuit in Fig. P4.35. **PSV**

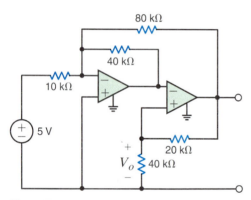

Figure P4.35

4.36 Find v_o in the circuit in Fig. P4.36.

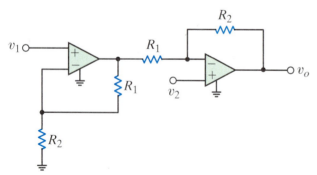

Figure P4.36

4.37 Find the expression for v_o in the differential amplifier circuit shown in Fig. P4.37. **CS**

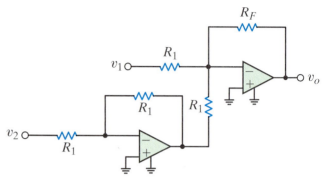

Figure P4.37

4.38 Find v_o in the circuit in Fig. P4.38. **PSV**

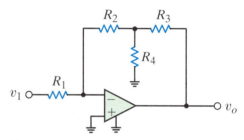

Figure P4.38

4.39 Find the output voltage, v_o, in the circuit in Fig. P4.39.

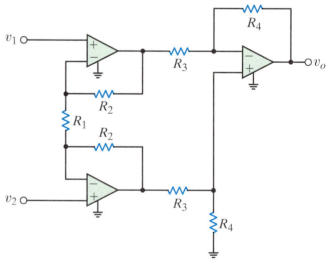

Figure P4.39

4.40 The electronic ammeter in Example 4.9 has been modified and is shown in Fig. P4.40. The selector switch allows the user to change the range of the meter. Using values for R_1 and R_2 from Example 4.9, find the values of R_A and R_B that will yield a 10-V output when the current being measured is 100 mA and 10 mA, respectively.

4.47 Design an op-amp-based circuit to produce the function

$$V_o = 5\,V_1 - 4\,V_2$$

4.48 Design an op-amp-based circuit to produce the function

$$V_o = 5\,V_1 - 7\,V_2$$

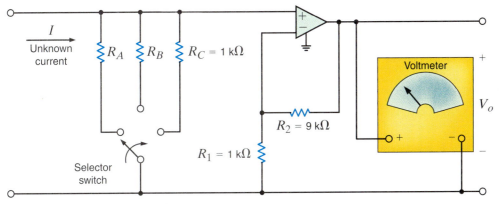

Figure P4.40

4.41 Given a box of 10-kΩ resistors and an op-amp, design a circuit that will have an output voltage of

$$V_o = -2V_1 - 4V_2 \quad \text{CS}$$

4.42 Design an op-amp circuit that has a gain of -50 using resistors no smaller than 1 kΩ.

4.43 Design a two-stage op-amp network that has a gain of $-50{,}000$ while drawing no current into its input terminal. Use no resistors smaller than 1 kΩ.

4.44 Design an op-amp circuit that has the following input/output relationship:

$$V_o = -5\,V_1 + 0.5\,V_2$$

4.45 A voltage waveform with a maximum value of 200 mV must be amplified to a maximum of 10 V and inverted. However, the circuit that produces the waveform can provide no more than 100 μA. Design the required amplifier. **CS**

4.46 An amplifier with a gain of $\pi \pm 1\%$ is needed. Using resistor values from Table 2.1, design the amplifier. Use as few resistors as possible.

4.49 Show that the circuit in Fig. P4.49 can produce the output

$$V_o = K_1 V_1 - K_2 V_2$$

only for $0 \leq K_1 \leq K_2 + 1$. **CS**

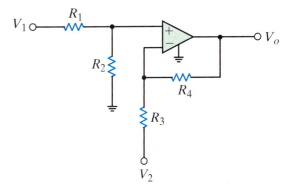

Figure P4.49

4.50 A 170°C maximum temperature digester is used in a paper mill to process wood chips that will eventually become paper. As shown in Fig. P4.50a, three electronic thermometers are placed along its length. Each thermometer outputs 0 V at 0°C, and the voltage changes 25 mV/°C. We will use the average of the three thermometer voltages to find an aggregate digester temperature. Furthermore, 1 volt should appear at V_o for every 10°C of average temperature. Design such an averaging circuit using the op-amp configuration shown in Fig. 4.50b if the final output voltage must be positive.

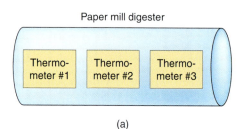

Paper mill digester

(a)

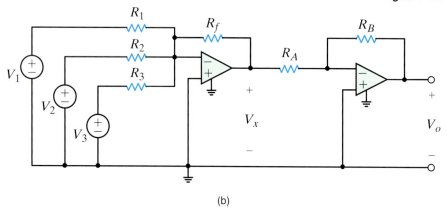

(b)

Figure P4.50

4.51 A 0.1-Ω shunt resistor is used to measure current in a fuel-cell circuit. The voltage drop across the shunt resistor is to be used to measure the current in the circuit. The maximum current is 20 A. Design the network shown in Fig. P4.51 so that a voltmeter attached to the output will read 0 volts when the current is 0 A and 20 V when the current is 20 A. Be careful not to load the shunt resistor, since loading will cause an inaccurate reading.

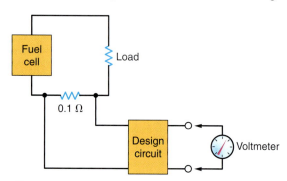

Figure P4.51

4.52 Wood pulp is used to make paper in a paper mill. The amount of lignin present in pulp is called the kappa number. A very sophisticated instrument is used to measure kappa, and the output of this instrument ranges from 1 to 5 volts, where 1 volt represents a kappa number of 12 and 5 volts represents a kappa number of 20. The pulp mill operator has asked to have a kappa meter installed on his console. Design a circuit that will employ as input the 1- to 5-volt signal and output the kappa number. An electronics engineer in the plant has suggested the circuit shown in Fig. P4.52.

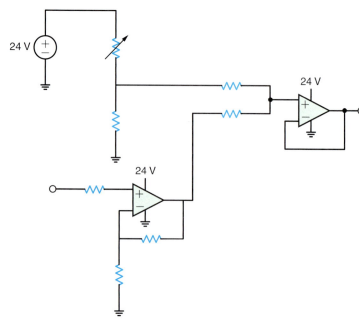

Figure P4.52

4.53 An operator in a chemical plant would like to have a set of indicator lights that indicate when a certain chemical flow is between certain specific values. The operator wants a RED light to indicate a flow of at least 10 GPM (gallons per minute), RED and YELLOW lights to indicate a flow of 60 GPM, and RED, YELLOW, and GREEN lights to indicate a flow rate of 80 GPM. The 4–20 mA flow meter instrument outputs 4 mA when the flow is zero and 20 mA when the flow rate is 100 GPM.

An experienced engineer has suggested the circuit shown in Fig. P4.53. The 4–20 mA flow meter and 250 Ω resistor provide a 1–5 V signal, which serves as one input for the three comparators. The light bulbs will turn on when the negative input to a comparator is higher than the positive input. Using this network, design a circuit that will satisfy the operator's requirements. **CS**

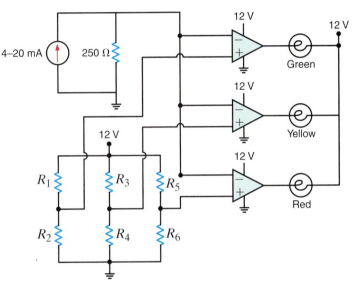

Figure P4.53

4.54 An industrial plant has a requirement for a circuit that uses as input the temperature of a vessel and outputs a voltage proportional to the vessel's temperature. The vessel's temperature ranges from 0°C to 500°C, and the corresponding output of the circuit should range from 0 to 12 V. A RTD (resistive thermal device), which is a linear device whose resistance changes with temperature

according to the plot in Fig. P4.54a, is available. The problem then is to use this RTD to design a circuit that employs this device as an input and produces a 0- to 12-V signal at the output, where 0 V corresponds to 0°C and 12 V corresponds to 500°C. An engineer familiar with this problem suggests the use of the circuit shown in Fig. P4.54b in which the RTD bridge circuit provides the input to a standard instrumentation amplifier. Determine the component values in this network needed to satisfy the design requirements.

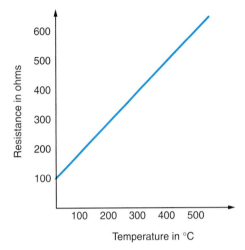

Figure P4.54a

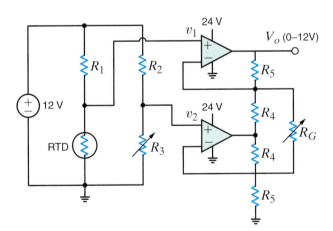

Figure P4.54b

4FE-1 Given the summing amplifier shown in Fig. 4PFE-1, select the values of R_2 that will produce an output voltage of -3 V. **CS**

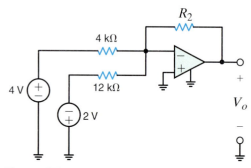

Figure 4PFE-1

4FE-2 Determine the output voltage V_o of the summing op-amp circuit shown in Fig. 4PFE-2.

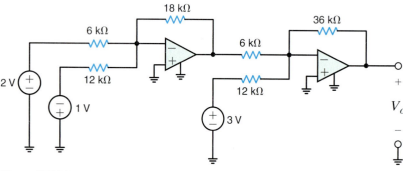

Figure 4PFE-2

Additional Analysis Techniques

At this point, we have mastered the ability to solve networks containing both independent and dependent sources using either nodal or loop analysis. In this chapter we introduce several new analysis techniques that bolster our arsenal of circuit analysis tools. We will find that in some situations these techniques lead to a quick solution and in other cases they do not. However, these new techniques in many cases do provide an insight into the circuit's operation that cannot be gained from a nodal or loop analysis.

In many practical situations we are interested in the analysis of some portion of a much larger network. If we can model the remainder of the network with a simple equivalent circuit, then our task will be much simpler. For example, consider the problem of analyzing some simple electronic device that is connected to the ac wall plug in our house. In this case the complete circuit includes not only the electronic device but the utility's power grid, which is connected to the device through the circuit breakers in the home. However, if we can accurately model everything outside the device with a simple equivalent circuit, then our analysis will be tractable. Two of the theorems that we present in this chapter will permit us to do just that. •

Access Problem-Solving Videos **PSV** *and Circuit Solutions* **CS** *at:*
http://www.justask4u.com/irwin
using the registration code on the inside cover and see a website with answers and more!

5.1   Introduction

Before introducing additional analysis techniques, let us review some of the topics we have used either explicitly or implicitly in our analyses thus far.

EQUIVALENCE   Table 5.1 is a short compendium of some of the equivalent circuits that have been employed in our analyses. This listing serves as a quick review as we begin to look at other techniques that can be used to find a specific voltage or current somewhere in a network and provide additional insight into the network's operation. In addition to the forms listed in the table, it is important to note that a series connection of current sources or a parallel connection of voltage sources is forbidden unless the sources are pointing in the same direction and have exactly the same values.

Table 5.1   Equivalent circuit forms

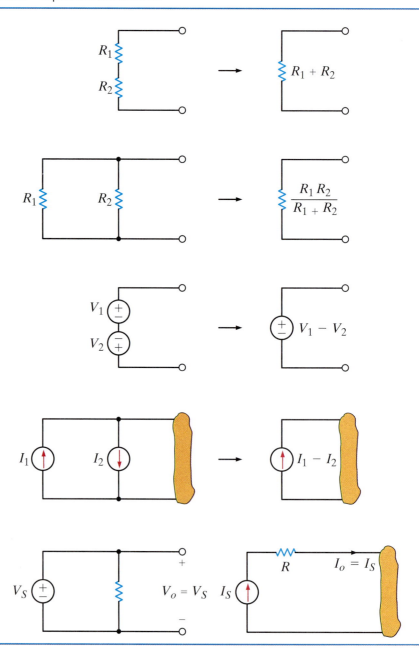

LINEARITY   All the circuits we have analyzed thus far have been linear circuits, which are described by a set of linear algebraic equations. Most of the circuits we will analyze in the remainder of the book will also be linear circuits, and any deviation from this type of network will be specifically identified as such.

Linearity requires both additivity and homogeneity (scaling). It can be shown that the circuits that we are examining satisfy this important property. The following example illustrates one way in which this property can be used.

Example 5.1

For the circuit shown in Fig. 5.1, we wish to determine the output voltage V_{out}. However, rather than approach the problem in a straightforward manner and calculate I_o, then I_1, then I_2, and so on, we will use linearity and simply assume that the output voltage is $V_{out} = 1$ V. This assumption will yield a value for the source voltage. We will then use the actual value of the source voltage and linearity to compute the actual value of V_{out}.

SOLUTION   If we assume that $V_{out} = V_2 = 1$ V , then

$$I_2 = \frac{V_2}{2k} = 0.5 \text{ mA}$$

V_1 can then be calculated as

$$V_1 = 4kI_2 + V_2$$
$$= 3 \text{ V}$$

Hence,

$$I_1 = \frac{V_1}{3k} = 1 \text{ mA}$$

Now, applying KCL,

$$I_o = I_1 + I_2 = 1.5 \text{ mA}$$

Then

$$V_o = 2kI_o + V_1$$
$$= 6 \text{ V}$$

Therefore, the assumption that $V_{out} = 1$ V produced a source voltage of 6 V. However, since the actual source voltage is 12 V, the actual output voltage is 1 V$(12/6) = 2$ V.

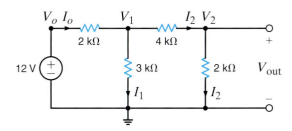

Figure 5.1
Circuit used in Example 5.1.

LEARNING EXTENSION

E5.1 Use linearity and the assumption that $I_o = 1$ mA to compute the correct current I_o in the circuit in Fig. E5.1 if $I = 6$ mA.   **ANSWER:** $I_o = 3$ mA.

Figure E5.1

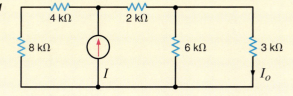

5.2 Superposition

Example 5.2

To provide motivation for this subject, let us examine the simple circuit of Fig. 5.2a in which two sources contribute to the current in the network. The actual values of the sources are left unspecified so that we can examine the concept of superposition.

SOLUTION   The mesh equations for this network are

$$6ki_1(t) - 3ki_2(t) = v_1(t)$$
$$-3ki_1(t) + 9ki_2(t) = -v_2(t)$$

Solving these equations for $i_1(t)$ yields

$$i_1(t) = \frac{v_1(t)}{5k} - \frac{v_2(t)}{15k}$$

In other words, the current $i_1(t)$ has a component due to $v_1(t)$ and a component due to $v_2(t)$. In view of the fact that $i_1(t)$ has two components, one due to each independent source, it would be interesting to examine what each source acting alone would contribute to $i_1(t)$. For $v_1(t)$ to act alone, $v_2(t)$ must be zero. As we pointed out in Chapter 2, $v_2(t) = 0$ means that the source $v_2(t)$ is replaced with a short circuit. Therefore, to determine the value of $i_1(t)$ due to $v_1(t)$ only, we employ the circuit in Fig. 5.2b and refer to this value of $i_1(t)$ as $i_1'(t)$.

$$i_1'(t) = \frac{v_1(t)}{3k + \dfrac{(3k)(6k)}{3k + 6k}} = \frac{v_1(t)}{5k}$$

Let us now determine the value of $i_1(t)$ due to $v_2(t)$ acting alone and refer to this value as $i_1''(t)$. Using the network in Fig. 5.2c,

$$i_2''(t) = -\frac{v_2(t)}{6k + \dfrac{(3k)(3k)}{3k + 3k}} = \frac{-2v_2(t)}{15k}$$

Then, using current division, we obtain

$$i_1''(t) = \frac{-2v_2(t)}{15k}\left(\frac{3k}{3k + 3k}\right) = \frac{-v_2(t)}{15k}$$

Now, if we add the values of $i_1'(t)$ and $i_1''(t)$, we obtain the value computed directly; that is,

$$i_1(t) = i_1'(t) + i_1''(t) = \frac{v_1(t)}{5k} - \frac{v_2(t)}{15k}$$

Note that we have *superposed* the value of $i_1'(t)$ on $i_1''(t)$, or vice versa, to determine the unknown current.

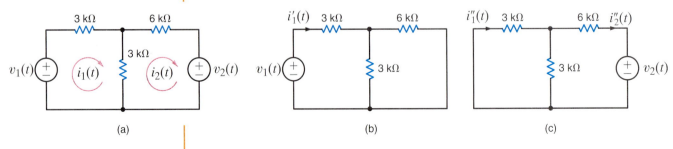

(a)   (b)   (c)

Figure 5.2 *Circuits used to illustrate superposition.*

What we have demonstrated in Example 5.2 is true in general for linear circuits and is a direct result of the property of linearity. *The principle of superposition*, which provides us with this ability to reduce a complicated problem to several easier problems—each containing only a single independent source—states that

In any linear circuit containing multiple independent sources, the current or voltage at any point in the network may be calculated as the algebraic sum of the individual contributions of each source acting alone.

When determining the contribution due to an independent source, any remaining voltage sources are made zero by replacing them with short circuits, and any remaining current sources are made zero by replacing them with open circuits.

Although superposition can be used in linear networks containing dependent sources, it is not useful in this case since the dependent source is never made zero.

As the previous example indicates, superposition provides some insight in determining the contribution of each source to the variable under investigation.

We will now demonstrate superposition with two examples and then provide a problem-solving strategy for the use of this technique. For purposes of comparison, we will also solve the networks using both node and loop analyses. Furthermore, we will employ these same networks when demonstrating subsequent techniques, if applicable.

Example 5.3

Let us use superposition to find V_o in the circuit in Fig. 5.3a.

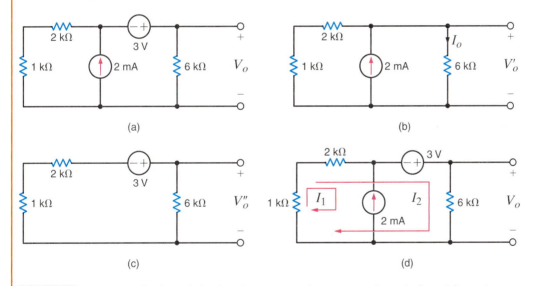

Figure 5.3
Circuits used in Example 5.3.

(a)   (b)   (c)   (d)

SOLUTION   The contribution of the 2-mA source to the output voltage is found from the network in Fig. 5.3b, using current division

$$I_o = (2 \times 10^{-3})\left(\frac{1k + 2k}{1k + 2k + 6k}\right) = \frac{2}{3} \text{ mA}$$

and

$$V'_o = I_o(6k) = 4 \text{ V}$$

The contribution of the 3-V source to the output voltage is found from the circuit in Fig. 5.3c. Using voltage division,

$$V''_o = 3\left(\frac{6k}{1k + 2k + 6k}\right)$$

$$= 2 \text{ V}$$

Therefore,

$$V_o = V'_o + V''_o = 6 \text{ V}$$

Although we used two separate circuits to solve the problem, both were very simple.

If we use nodal analysis and Fig. 5.3a to find V_o and recognize that the 3-V source and its connecting nodes form a supernode, V_o can be found from the node equation

$$\frac{V_o - 3}{1k + 2k} - 2 \times 10^{-3} + \frac{V_o}{6k} = 0$$

which yields $V_o = 6$ V. In addition, loop analysis applied as shown in Fig. 5.3d produces the equations

$$I_1 = -2 \times 10^{-3}$$

and

$$3k(I_1 + I_2) - 3 + 6kI_2 = 0$$

which yield $I_2 = 1$ mA and hence $V_o = 6$ V.

Example 5.4

Consider now the network in Fig. 5.4a. Let us use superposition to find V_o.

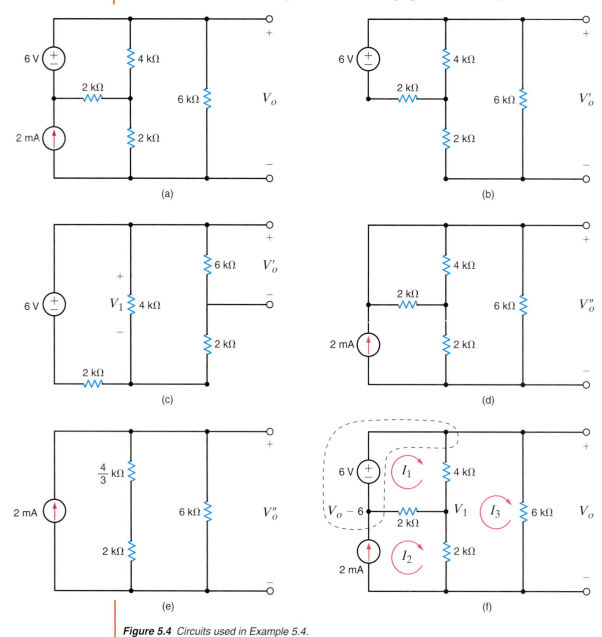

Figure 5.4 Circuits used in Example 5.4.

SOLUTION The contribution of the 6-V source to V_o is found from the network in Fig. 5.4b, which is redrawn in Fig. 5.4c. The 2 kΩ + 6 kΩ = 8-kΩ resistor and 4-kΩ resistor are in parallel, and their combination is an 8/3-kΩ resistor. Then, using voltage division,

$$V_1 = 6\left(\frac{\frac{8}{3}\text{k}}{\frac{8}{3}\text{k} + 2\text{k}}\right) = \frac{24}{7}\text{ V}$$

Applying voltage division again,

$$V'_o = V_1\left(\frac{6\text{k}}{6\text{k} + 2\text{k}}\right) = \frac{18}{7}\text{ V}$$

The contribution of the 2-mA source is found from Fig. 5.4d, which is redrawn in Fig. 5.4e. V''_o is simply equal to the product of the current source and the parallel combination of the resistors; that is,

$$V''_o = \left(2 \times 10^{-3}\right)\left(\frac{10}{3}\text{k}//6\text{k}\right) = \frac{30}{7}\text{ V}$$

Then

$$V_o = V'_o + V''_o = \frac{48}{7}\text{ V}$$

A nodal analysis of the network can be performed using Fig. 5.4f. The equation for the supernode is

$$-2 \times 10^{-3} + \frac{(V_o - 6) - V_1}{2\text{k}} + \frac{V_o - V_1}{4\text{k}} + \frac{V_o}{6\text{k}} = 0$$

The equation for the node labeled V_1 is

$$\frac{V_1 - V_o}{4\text{k}} + \frac{V_1 - (V_o - 6)}{2\text{k}} + \frac{V_1}{2\text{k}} = 0$$

Solving these two equations, which already contain the constraint equation for the supernode, yields $V_o = 48/7$ V.

Once again, referring to the network in Fig. 5.4f, the mesh equations for the network are

$$-6 + 4\text{k}(I_1 - I_3) + 2\text{k}(I_1 - I_2) = 0$$
$$I_2 = 2 \times 10^{-3}$$
$$2\text{k}(I_3 - I_2) + 4\text{k}(I_3 - I_1) + 6\text{k}I_3 = 0$$

Solving these equations, we obtain $I_3 = 8/7$ mA and, hence, $V_o = 48/7$ V.

Example 5.5

Let us demonstrate the power of superposition in the analysis of op-amp circuits by determining the input/output relationship for the op-amp configuration shown in Fig. 5.5a.

SOLUTION The contribution of V_1 to the output V_o is derived from the network in Fig. 5.5b where V_2 is set to zero. This circuit is the basic inverting gain configuration and

$$\frac{V_{o1}}{V_1} = -\frac{R_2}{R_1}$$

The contribution due to V_2 is shown in Fig. 5.5c where V_1 is set to zero. This circuit is the basic noninverting configuration and

$$\frac{V_{o2}}{V_2} = 1 + \frac{R_2}{R_1}.$$

Therefore, using superposition,

$$V_o = \left[1 + \frac{R_2}{R_1}\right]V_2 - \left[\frac{R_2}{R_1}\right]V_1$$

Thus, in this case, we have used what we learned in Chapter 4, via superposition, to immediately derive the input/output relationship for the network in Fig. 5.5a.

Figure 5.5
(a) a superposition example circuit; (b) the circuit with V_2 set to zero; (c) the circuit with V_1 set to zero.

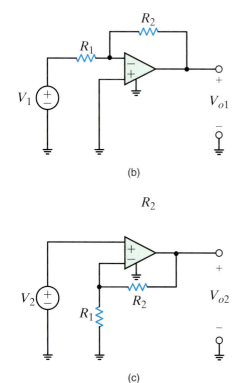

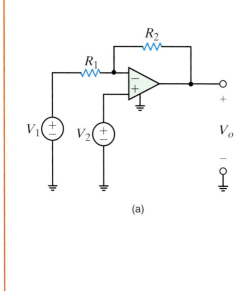

PROBLEM-SOLVING STRATEGY

Applying Superposition

Step 1. In a network containing multiple independent sources, each source can be applied independently with the remaining sources turned off.

Step 2. To turn off a voltage source, replace it with a short circuit, and to turn off a current source, replace it with an open circuit.

Step 3. When the individual sources are applied to the circuit, all the circuit laws and techniques we have learned, or will soon learn, can be applied to obtain a solution.

Step 4. The results obtained by applying each source independently are then added together algebraically to obtain a solution.

Superposition can be applied to a circuit with any number of dependent and independent sources. In fact, superposition can be applied to such a network in a variety of ways. For example, a circuit with three independent sources can be solved using each source acting alone, as we have just demonstrated, or we could use two at a time and sum the result with that obtained from the third acting alone. In addition, the independent sources do not have to assume their actual value or zero. However, it is mandatory that the sum of the different values chosen add to the total value of the source.

Superposition is a fundamental property of linear equations and, therefore, can be applied to any effect that is linearly related to its cause. In this regard it is important to point out that although superposition applies to the current and voltage in a linear circuit, it cannot be used to determine power because power is a nonlinear function.

LEARNING EXTENSION

E5.2 Compute V_o in the circuit in Fig. E5.2 using superposition. **PSV**

ANSWER: $V_o = \frac{4}{3}$ V.

Figure E5.2

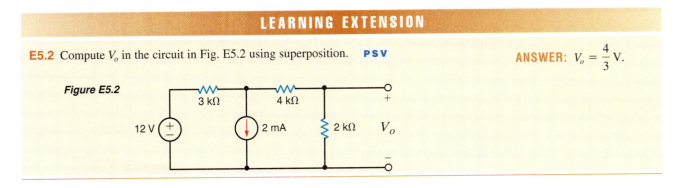

5.3   Thévenin's and Norton's Theorems

Thus far we have presented a number of techniques for circuit analysis. At this point we will add two theorems to our collection of tools that will prove to be extremely useful. The theorems are named after their authors, M. L. Thévenin, a French engineer, and E. L. Norton, a scientist formerly with Bell Telephone Laboratories.

Suppose that we are given a circuit and that we wish to find the current, voltage, or power that is delivered to some resistor of the network, which we will call the load. *Thévenin's theorem* tells us that we can replace the entire network, exclusive of the load, by an equivalent circuit that contains only an independent voltage source in series with a resistor in such a way that the current–voltage relationship at the load is unchanged. *Norton's theorem* is identical to the preceding statement except that the equivalent circuit is an independent current source in parallel with a resistor.

Note that this is a very important result. It tells us that if we examine any network from a pair of terminals, we know that with respect to those terminals, the entire network is equivalent to a simple circuit consisting of an independent voltage source in series with a resistor or an independent current source in parallel with a resistor.

In developing the theorems, we will assume that the circuit shown in Fig. 5.6a can be split into two parts, as shown in Fig. 5.6b. In general, circuit B is the load and may be linear or nonlinear. Circuit A is the balance of the original network exclusive of the load and must be linear. As such, circuit A may contain independent sources, dependent sources and resistors, or any other linear element. We require, however, that a dependent source and its control variable appear in the same circuit.

Circuit A delivers a current i to circuit B and produces a voltage v_o across the input terminals of circuit B. From the standpoint of the terminal relations of circuit A, we can replace circuit B by a voltage source of v_o volts (with the proper polarity), as shown in Fig. 5.6c. Since the terminal voltage is unchanged and circuit A is unchanged, the terminal current i is unchanged.

Figure 5.6
Concepts used to develop Thévenin's theorem.

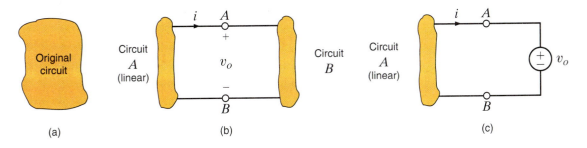

Now, applying the principle of superposition to the network shown in Fig. 5.6c, the total current i shown in the figure is the sum of the currents caused by all the sources in circuit A and the source v_o that we have just added. Therefore, via superposition the current i can be written

$$i = i_o + i_{sc} \qquad \textbf{5.1}$$

where i_o is the current due to v_o with all independent sources in circuit A made zero (i.e., voltage sources replaced by short circuits and current sources replaced by open circuits), and i_{sc} is the short-circuit current due to all sources in circuit A with v_o replaced by a short circuit.

The terms i_o and v_o are related by the equation

$$i_o = \frac{-v_o}{R_{Th}} \qquad \textbf{5.2}$$

where R_{Th} is the equivalent resistance looking back into circuit A from terminals A-B with all independent sources in circuit A made zero.

Substituting Eq. (5.2) into Eq. (5.1) yields

$$i = -\frac{v_o}{R_{Th}} + i_{sc} \qquad \textbf{5.3}$$

This is a general relationship and, therefore, must hold for any specific condition at terminals A-B. As a specific case, suppose that the terminals are open circuited. For this condition, $i = 0$ and v_o is equal to the open-circuit voltage v_{oc}. Thus, Eq. (5.3) becomes

$$i = 0 = \frac{-v_{oc}}{R_{Th}} + i_{sc} \qquad \textbf{5.4}$$

Hence,

$$v_{oc} = R_{Th} i_{sc} \qquad \textbf{5.5}$$

This equation states that the open-circuit voltage is equal to the short-circuit current times the equivalent resistance looking back into circuit A with all independent sources made zero. We refer to R_{Th} as the Thévenin equivalent resistance.

Substituting Eq. (5.5) into Eq. (5.3) yields

$$i = \frac{-v_o}{R_{Th}} + \frac{v_{oc}}{R_{Th}}$$

or

$$v_o = v_{oc} - R_{Th} i \qquad \textbf{5.6}$$

Let us now examine the circuits that are described by these equations. The circuit represented by Eq. (5.6) is shown in Fig. 5.7a. The fact that this circuit is equivalent at terminals A-B to circuit A in Fig. 5.6 is a statement of *Thévenin's theorem*. The circuit represented by Eq. (5.3) is shown in Fig. 5.7b. The fact that this circuit is equivalent at terminals A-B to circuit A in Fig. 5.6 is a statement of *Norton's theorem*.

Figure 5.7
Thévenin and Norton equivalent circuits.

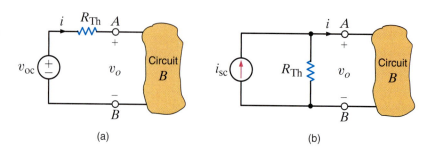

(a)   (b)

Having demonstrated that there is an inherent relationship between the Thévenin equivalent circuit and the Norton equivalent circuit, we now proceed to apply these two important and useful theorems. The manner in which these theorems are applied depends on the structure of the original network under investigation. For example, if only independent sources are present, we can calculate the open-circuit voltage or short-circuit current and the Thévenin equivalent resistance. However, if dependent sources are also present, the Thévenin equivalent will be determined by calculating v_{oc} and i_{sc}, since this is normally the best approach for determining R_{Th} in a network containing dependent sources. Finally, if circuit A contains no *independent* sources, then both v_{oc} and i_{sc} will necessarily be zero. (Why?) Thus, we cannot determine R_{Th} by v_{oc}/i_{sc}, since the ratio is indeterminate. We must look for another approach. Notice that if $v_{oc} = 0$, then the equivalent circuit is merely the unknown resistance R_{Th}. If we apply an external source to circuit A—a test source v_t—and determine the current, i_t, which flows into circuit A from v_t, then R_{Th} can be determined from $R_{Th} = v_t/i_t$. Although the numerical value of v_t need not be specified, we could let $v_t = 1$ V and then $R_{Th} = 1/i_t$. Alternatively, we could use a current source as a test source and let $i_t = 1$ A; then $v_t = (1)R_{Th}$.

Before we begin our analysis of several examples that will demonstrate the utility of these theorems, remember that these theorems, in addition to being another approach, often permit us to solve several small problems rather than one large one. They allow us to replace a network, no matter how large, *at a pair of terminals* with a Thévenin or Norton equivalent circuit. In fact, we could represent the entire U.S. power grid at a pair of terminals with one of the equivalent circuits. Once this is done we can quickly analyze the effect of different loads on a network. Thus, these theorems provide us with additional insight into the operation of a specific network.

CIRCUITS CONTAINING ONLY INDEPENDENT SOURCES

Example 5.6

Let us use Thévenin's and Norton's theorems to find V_o in the network in Example 5.3.

SOLUTION   The circuit is redrawn in Fig. 5.8a. To determine the Thévenin equivalent, we break the network at the 6-kΩ load as shown in Fig. 5.8b. KVL indicates that the open-circuit voltage, V_{oc}, is equal to 3 V plus the voltage V_1, which is the voltage across the current source. The 2 mA from the current source flows through the two resistors (where else could it possibly go!) and, therefore, $V_1 = (2 \times 10^{-3})(1k + 2k) = 6$ V. Therefore, $V_{oc} = 9$ V. By making both sources zero, we can find the Thévenin equivalent resistance, R_{Th}, using the circuit in Fig. 5.8c. Obviously, $R_{Th} = 3$ kΩ. Now our Thévenin equivalent circuit, consisting of V_{oc} and R_{Th}, is connected back to the original terminals of the load, as shown in Fig. 5.8d. Using a simple voltage divider, we find that $V_o = 6$ V.

Figure 5.8
Circuits used in Example 5.6.

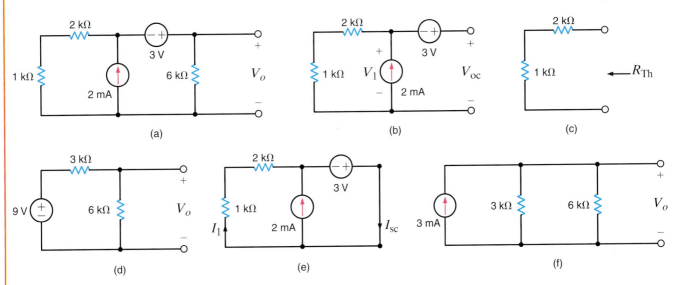

To determine the Norton equivalent circuit at the terminals of the load, we must find the short-circuit current as shown in Fig. 5.8e. Note that the short circuit causes the 3-V source to be directly across (i.e., in parallel with) the resistors and the current source. Therefore, $I_1 = 3/(1k + 2k) = 1$ mA. Then, using KCL, $I_{sc} = 3$ mA. We have already determined R_{Th}, and, therefore, connecting the Norton equivalent to the load results in the circuit in Fig. 5.8f. Hence, V_o is equal to the source current multiplied by the parallel resistor combination, which is 6 V.

Consider for a moment some salient features of this example. Note that in applying the theorems there is no point in breaking the network to the left of the 3-V source, since the resistors in parallel with the current source are already a Norton equivalent. Furthermore, once the network has been simplified using a Thévenin or Norton equivalent, we simply have a new network with which we can apply the theorems again. The following example illustrates this approach.

Example 5.7

Let us use Thévenin's theorem to find V_o in the network in Fig. 5.9a.

SOLUTION If we break the network to the left of the current source, the open-circuit voltage V_{oc_1} is as shown in Fig. 5.9b. Since there is no current in the 2-kΩ resistor and therefore no voltage across it, V_{oc_1} is equal to the voltage across the 6-kΩ resistor, which can be determined by voltage division as

$$V_{oc_1} = 12\left(\frac{6k}{6k + 3k}\right) = 8 \text{ V}$$

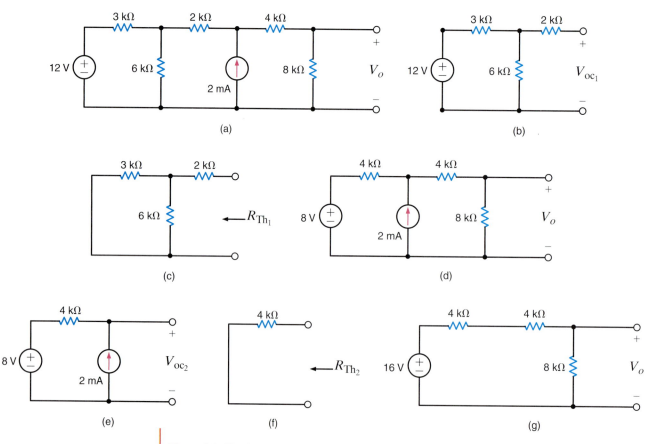

Figure 5.9 Circuits used in Example 5.7.

The Thévenin equivalent resistance, R_{Th_1}, is found from Fig. 5.9c as

$$R_{Th_1} = 2k + \frac{(3k)(6k)}{3k + 6k} = 4\ k\Omega$$

Connecting this Thévenin equivalent back to the original network produces the circuit shown in Fig. 5.9d. We can now apply Thévenin's theorem again, and this time we break the network to the right of the current source as shown in Fig. 5.9e. In this case V_{oc_2} is

$$V_{oc_2} = (2 \times 10^{-3})(4k) + 8 = 16\ V$$

and R_{Th_2} obtained from Fig. 5.9f is 4 kΩ. Connecting this Thévenin equivalent to the remainder of the network produces the circuit shown in Fig. 5.9g. Simple voltage division applied to this final network yields $V_o = 8$ V. Norton's theorem can be applied in a similar manner to solve this network; however, we save that solution as an exercise.

Example 5.8

It is instructive to examine the use of Thévenin's and Norton's theorems in the solution of the network in Fig. 5.4a, which is redrawn in Fig. 5.10a.

SOLUTION If we break the network at the 6-kΩ load, the open-circuit voltage is found from Fig. 5.10b. The equations for the mesh currents are

$$-6 + 4kI_1 + 2k(I_1 - I_2) = 0$$

and

$$I_2 = 2 \times 10^{-3}$$

from which we easily obtain $I_1 = 5/3$ mA. Then, using KVL, V_{oc} is

$$V_{oc} = 4kI_1 + 2kI_2$$

$$= 4k\left(\frac{5}{3} \times 10^{-3}\right) + 2k(2 \times 10^{-3})$$

$$= \frac{32}{3}\ V$$

R_{Th} is derived from Fig. 5.10c and is

$$R_{Th} = (2k//4k) + 2k = \frac{10}{3}\ k\Omega$$

Attaching the Thévenin equivalent to the load produces the network in Fig. 5.10d. Then using voltage division, we obtain

$$V_o = \frac{32}{3}\left(\frac{6k}{6k + \frac{10}{3}k}\right)$$

$$= \frac{48}{7}\ V$$

In applying Norton's theorem to this problem, we must find the short-circuit current shown in Fig. 5.10e. At this point the quick-thinking reader stops immediately! Three mesh equations applied to the original circuit will immediately lead to the solution, but the three mesh equations in the circuit in Fig. 5.10e will provide only part of the answer, specifically the short-circuit current. Sometimes the use of the theorems is more complicated than a straightforward attack using node or loop analysis. This would appear to be one of those situations.

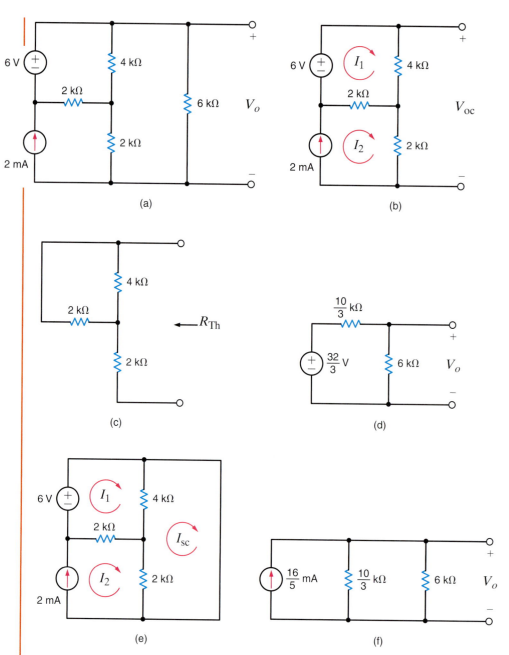

Figure 5.10 *Circuits used in Example 5.8.*

Interestingly, it is not. We can find I_{sc} from the network in Fig. 5.10e without using the mesh equations. The technique is simple, but a little tricky, and so we ignore it at this time. Having said all these things, let us now finish what we have started. The mesh equations for the network in Fig. 5.10e are

$$-6 + 4k(I_1 - I_{sc}) + 2k(I_1 - 2 \times 10^{-3}) = 0$$

$$2k(I_{sc} - 2 \times 10^{-3}) + 4k(I_{sc} - I_1) = 0$$

where we have incorporated the fact that $I_2 = 2 \times 10^{-3}$ A. Solving these equations yields $I_{sc} = 16/5$ mA. R_{Th} has already been determined in the Thévenin analysis. Connecting the Norton equivalent to the load results in the circuit in Fig. 5.10f. Applying Ohm's law to this circuit yields $V_o = 48/7$ V.

E5.3 Use Thévenin's theorem to find V_o in the network in Fig. E5.3.

ANSWER: $V_o = -3\text{V}$.

Figure E5.3

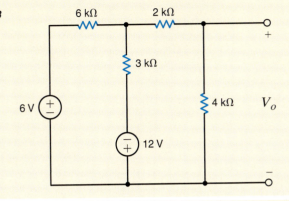

E5.4 Find V_o in the circuit in Fig. E5.2 using both Thévenin's and Norton's theorems. When deriving the Norton equivalent circuit, break the network to the left of the 2-kΩ resistor. Why?

ANSWER: $V_o = \dfrac{4}{3}\text{V}$.

CIRCUITS CONTAINING ONLY DEPENDENT SOURCES
As we have stated earlier, the Thévenin or Norton equivalent of a network containing only dependent sources is R_{Th}. The following examples will serve to illustrate how to determine this Thévenin equivalent resistance.

── **Example 5.9**

We wish to determine the Thévenin equivalent of the network in Fig. 5.11a at the terminals *A-B*.

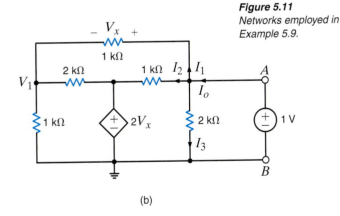

Figure 5.11
Networks employed in Example 5.9.

(a)   (b)

SOLUTION   Our approach to this problem will be to apply a 1-V source at the terminals as shown in Fig. 5.11b and then compute the current I_o and $R_{\text{Th}} = 1/I_o$.

The equations for the network in Fig. 5.11b are as follows. KVL around the outer loop specifies that

$$V_1 + V_x = 1$$

The KCL equation at the node labeled V_1 is

$$\frac{V_1}{1\text{k}} + \frac{V_1 - 2V_x}{2\text{k}} + \frac{V_1 - 1}{1\text{k}} = 0$$

Solving the equations for V_x yields $V_x = 3/7$ V. Knowing V_x, we can compute the currents I_1, I_2, and I_3. Their values are

$$I_1 = \frac{V_x}{1k} = \frac{3}{7} \text{ mA}$$

$$I_2 = \frac{1 - 2V_x}{1k} = \frac{1}{7} \text{ mA}$$

$$I_3 = \frac{1}{2k} = \frac{1}{2} \text{ mA}$$

Therefore,

$$I_o = I_1 + I_2 + I_3$$
$$= \frac{15}{14} \text{ mA}$$

and

$$R_{Th} = \frac{1}{I_o}$$
$$= \frac{14}{15} \text{ k}\Omega$$

Example 5.10

Let us determine R_{Th} at the terminals A-B for the network in Fig. 5.12a.

SOLUTION Our approach to this problem will be to apply a 1-mA current source at the terminals A-B and compute the terminal voltage V_2 as shown in Fig. 5.12b. Then $R_{Th} = V_2/0.001$.
 The node equations for the network are

$$\frac{V_1 - 2000I_x}{2k} + \frac{V_1}{1k} + \frac{V_1 - V_2}{3k} = 0$$

$$\frac{V_2 - V_1}{3k} + \frac{V_2}{2k} = 1 \times 10^{-3}$$

and

$$I_x = \frac{V_1}{1k}$$

Solving these equations yields

$$V_2 = \frac{10}{7} \text{ V}$$

and hence,

$$R_{Th} = \frac{V_2}{1 \times 10^{-3}}$$
$$= \frac{10}{7} \text{ k}\Omega$$

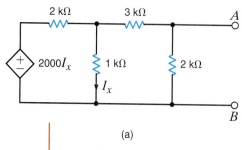

(a)

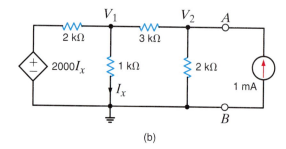

(b)

Figure 5.12 *Networks used in Example 5.10.*

CIRCUITS CONTAINING BOTH INDEPENDENT AND DEPENDENT SOURCES   In these types of circuits we must calculate both the open-circuit voltage and short-circuit current to calculate the Thévenin equivalent resistance. Furthermore, we must remember that we cannot split the dependent source and its controlling variable when we break the network to find the Thévenin or Norton equivalent.

We now illustrate this technique with a circuit containing a current-controlled voltage source.

Example 5.11

Let us use Thévenin's theorem to find V_o in the network in Fig. 5.13a.

SOLUTION   To begin, we break the network at points A-B. Could we break it just to the right of the 12-V source? No! Why? The open-circuit voltage is calculated from the network in Fig. 5.13b. Note that we now use the source $2000I'_x$ because this circuit is different from that in Fig. 5.13a. KCL for the supernode around the 12-V source is

$$\frac{(V_{oc} + 12) - (-2000I'_x)}{1k} + \frac{V_{oc} + 12}{2k} + \frac{V_{oc}}{2k} = 0$$

where

$$I'_x = \frac{V_{oc}}{2k}$$

yielding $V_{oc} = -6\,V$.

I_{sc} can be calculated from the circuit in Fig. 5.13c. Note that the presence of the short circuit forces I''_x to zero and, therefore, the network is reduced to that shown in Fig. 5.13d.

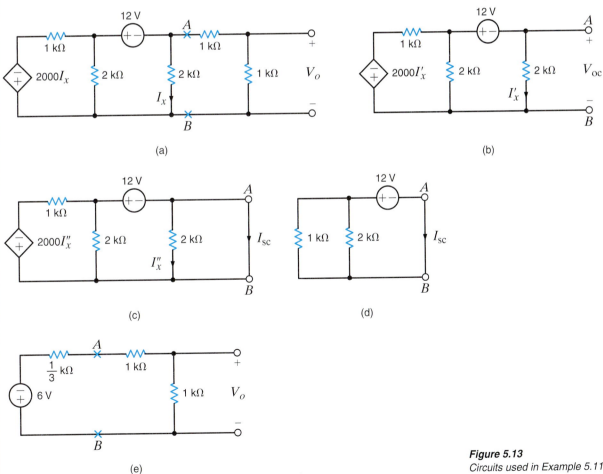

(a)

(b)

(c)

(d)

(e)

Figure 5.13

Circuits used in Example 5.11.

Therefore,

$$I_{sc} = \frac{-12}{\frac{2}{3}k} = -18 \text{ mA}$$

Then

$$R_{Th} = \frac{V_{oc}}{I_{sc}} = \frac{1}{3} k\Omega$$

Connecting the Thévenin equivalent circuit to the remainder of the network at terminals *A-B* produces the circuit in Fig. 5.13e. At this point, simple voltage division yields

$$V_o = (-6)\left(\frac{1k}{1k + 1k + \frac{1}{3}k}\right) = \frac{-18}{7} \text{ V}$$

Example 5.12

Let us find V_o in the network in Fig. 5.14a using Thévenin's theorem.

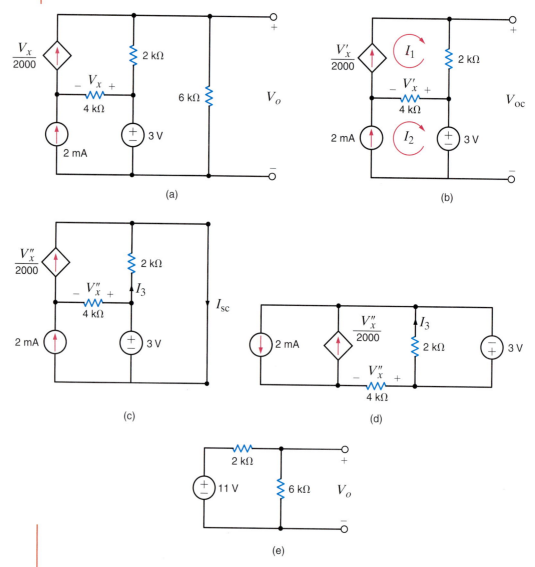

Figure 5.14 *Circuits used in Example 5.12.*

SOLUTION V_{oc} is determined from the network in Fig. 5.14b. Note that

$$I_1 = \frac{V'_x}{2k}$$

$$I_2 = 2 \text{ mA}$$

and

$$V'_x = 4k\left(\frac{V'_x}{2k} - 2 \times 10^{-3}\right)$$

Solving these equations yields $I_1 = 4$ mA and, hence,

$$V_{oc} = 2kI_1 + 3 = 11 \text{ V}$$

I_{sc} is derived from the circuit in Fig. 5.14c. Note that if we collapse the short circuit, the network is reduced to that in Fig. 5.14d. Although we have temporarily lost sight of I_{sc}, we can easily find the branch currents and they, in turn, will yield I_{sc}. KCL at the node at the bottom left of the network is

$$\frac{V''_x}{4k} = \frac{V''_x}{2000} - 2 \times 10^{-3}$$

or

$$V''_x = 8 \text{ V}$$

Then since

$$I_3 = \frac{3}{2k} = \frac{3}{2} \text{ mA}$$

as shown in Fig. 5.14c

$$I_{sc} = \frac{V''_x}{2000} + I_3$$

$$= \frac{11}{2} \text{ mA}$$

Then

$$R_{\text{Th}} = \frac{V_{oc}}{I_{sc}} = 2 \text{ k}\Omega$$

Connecting the Thévenin equivalent circuit to the remainder of the original network produces the circuit in Fig. 5.14e. Simple voltage division yields

$$V_o = 11\left(\frac{6k}{2k + 6k}\right)$$

$$= \frac{33}{4} \text{ V}$$

Example 5.13

We will now reexamine a problem that was solved earlier using both nodal and loop analysis. The circuit used in Examples 3.10 and 3.20 is redrawn in Fig. 5.15a. Since a dependent source is present, we will have to find the open-circuit voltage and the short-circuit current in order to employ Thévenin's theorem to determine the output voltage V_o.

SOLUTION   As we begin the analysis, we note that the circuit can be somewhat simplified by first forming a Thévenin equivalent for the leftmost and rightmost branches. Note that these two branches are in parallel and neither branch contains the control variable. Thus, we can simplify the network by reducing these two branches to one via a Thévenin equivalent. For the circuit shown in Fig. 5.15b, the open-circuit voltage is

$$V_{oc_1} = \frac{2}{k}(1k) + 4 = 6 \text{ V}$$

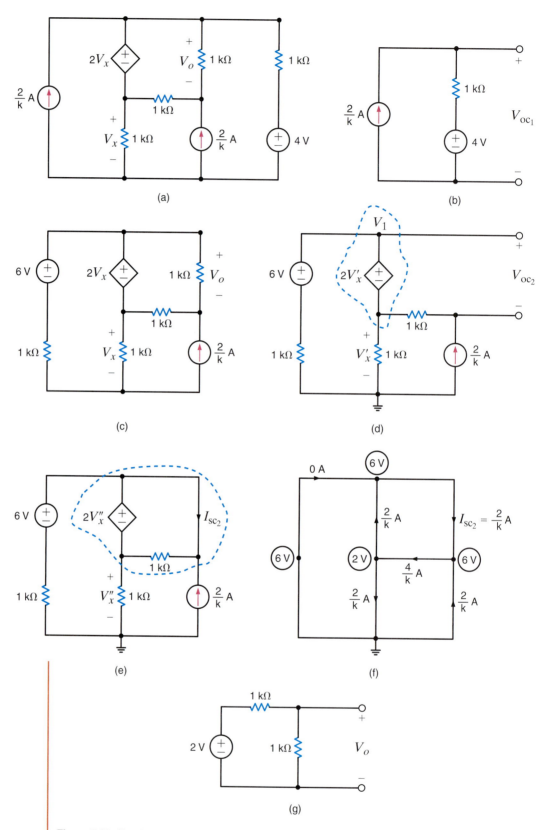

Figure 5.15 *Circuits used in Example 5.13.*

And the Thévenin equivalent resistance at the terminals, obtained by looking into the terminals with the sources made zero, is

$$R_{Th_1} = 1 \text{ k}\Omega$$

The resultant Thévenin equivalent circuit is now connected to the remaining portion of the circuit producing the network in Fig. 5.15c.

Now we break the network shown in Fig. 5.15c at the output terminals to determine the open-circuit voltage V_{oc_2} as shown in Fig. 5.15d. Because of the presence of the voltage sources, we will use a nodal analysis to find the open-circuit voltage with the help of a supernode. The node equations for this network are

$$V_1 = 3V'_x$$

$$\frac{V_1 - 6}{1k} + \frac{V_1 - 2V'_x}{1k} = \frac{2}{k}$$

and thus $V'_x = 2$ V and $V_1 = 6$ V. Then, the open-circuit voltage, obtained using the KVL equation

$$-2V'_x + V_{oc_2} + \frac{2}{k}(1k) = 0$$

is

$$V_{oc_2} = 2 \text{ V}$$

The short-circuit current is derived from the network shown in Fig. 5.15e. Once again we employ the supernode, and the network equations are

$$V_2 = 3V''_x$$

$$\frac{V_2 - 6}{1k} + \frac{V_2 - 2V''_x}{1k} = \frac{2}{k}$$

The node voltages obtained from these equations are $V''_x = 2$ V and $V_2 = 6$ V. The line diagram shown in Fig. 5.15f displays the node voltages and the resultant branch currents. (Node voltages are shown in the circles, and branch currents are identified with arrows.) The node voltages and resistors are used to compute the resistor currents, while the remaining currents are derived by KCL. As indicated, the short-circuit current is

$$I_{sc_2} = 2 \text{ mA}$$

Then, the Thévenin equivalent resistance is

$$R_{Th_2} = \frac{V_{oc_2}}{I_{sc_2}} = 1 \text{ k}\Omega$$

The Thévenin equivalent circuit now consists of a 2-V source in series with a 1-kΩ resistor. Connecting this Thévenin equivalent circuit to the load resistor yields the network shown in Fig. 5.15g. A simple voltage divider indicates that $V_o = 1$ V.

PROBLEM-SOLVING STRATEGY

Applying Thévenin's Theorem

Step 1. Remove the load and find the voltage across the open-circuit terminals, V_{oc}. All the circuit analysis techniques presented here can be used to compute this voltage.

Step 2. Determine the Thévenin equivalent resistance of the network at the open terminals with the load removed. Three different types of circuits may be encountered in determining the resistance, R_{Th}.

  (a) If the circuit contains only independent sources, they are made zero by replacing the voltage sources with short circuits and the current sources with open circuits. R_{Th} is then found by computing the resistance of the purely resistive network at the open terminals.

(continues on the next page)

(b) If the circuit contains only dependent sources, an independent voltage or current source is applied at the open terminals and the corresponding current or voltage at these terminals is measured. The voltage/current ratio at the terminals is the Thévenin equivalent resistance. Since there is no energy source, the open-circuit voltage is zero in this case.

(c) If the circuit contains both independent and dependent sources, the open-circuit terminals are shorted and the short-circuit current between these terminals is determined. The ratio of the open-circuit voltage to the short-circuit current is the resistance R_{Th}.

Step 3. If the load is now connected to the Thévenin equivalent circuit, consisting of V_{oc} in series with R_{Th}, the desired solution can be obtained.

The problem-solving strategy for Norton's theorem is essentially the same as that for Thévenin's theorem with the exception that we are dealing with the short-circuit current instead of the open-circuit voltage.

LEARNING EXTENSION

E5.5 Find V_o in the circuit in Fig. E5.5 using Thévenin's theorem. **PSV**

ANSWER: $V_o = \dfrac{36}{13}$ V.

Figure E5.5

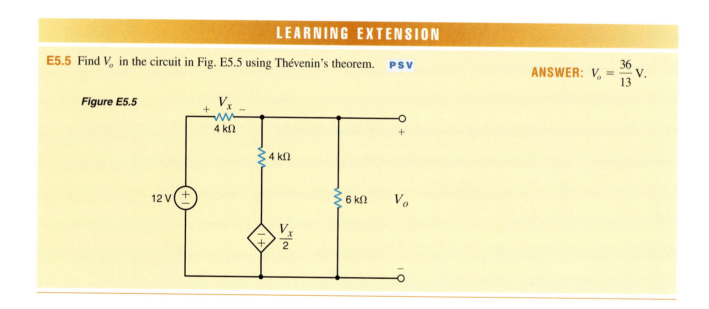

Having examined the use of Thévenin's and Norton's theorems in a variety of different types of circuits, it is instructive to look at yet one other aspect of these theorems that we find useful in circuit analysis and design. This additional aspect can be gleaned from the Thévenin equivalent and Norton equivalent circuits.

The relationships specified in Fig. 5.7 and Eq. (5.5) have special significance because they represent what is called a *source transformation* or *source exchange*. What these relationships tell us is that if we have embedded within a network a current source i in parallel with a resistor R, we can replace this combination with a voltage source of value $v = iR$ in series with the resistor R. The reverse is also true; that is, a voltage source v in series with a resistor R can be replaced with a current source of value $i = v/R$ in parallel with the resistor R. Parameters within the circuit (e.g., an output voltage) are unchanged under these transformations.

We must emphasize that the two equivalent circuits in Fig. 5.7 are *equivalent only at the two external nodes*. For example, if we disconnect circuit B from both networks in Fig. 5.7, the equivalent circuit in Fig. 5.7b dissipates power, but the one in Fig. 5.7a does not.

Example 5.14

We will now demonstrate how to find V_o in the circuit in Fig. 5.16a using the repeated application of source transformation.

SOLUTION   If we begin at the left end of the network in Fig. 5.16a, the series combination of the 12-V source and 3-kΩ resistor is converted to a 4-mA current source in parallel with the 3-kΩ resistor. If we combine this 3-kΩ resistor with the 6-kΩ resistor, we obtain the circuit in Fig. 5.16b. Note that at this point we have eliminated one circuit element. Continuing the reduction, we convert the 4-mA source and 2-kΩ resistor into an 8-V source in series with this same 2-kΩ resistor. The two 2-kΩ resistors that are in series are now combined to produce the network in Fig. 5.16c. If we now convert the combination of the 8-V source and 4-kΩ resistor into a 2-mA source in parallel with the 4-kΩ resistor and combine the resulting current source with the other 2-mA source, we arrive at the circuit shown in Fig. 5.16d. At this point, we can simply apply current division to the two parallel resistance paths and obtain

$$I_o = (4 \times 10^{-3})\left(\frac{4k}{4k + 4k + 8k}\right) = 1 \text{ mA}$$

and hence,

$$V_o = (1 \times 10^{-3})(8k) = 8 \text{ V}$$

The reader is encouraged to consider the ramifications of working this problem using any of the other techniques we have presented.

Figure 5.16
Circuits used in Example 5.14.

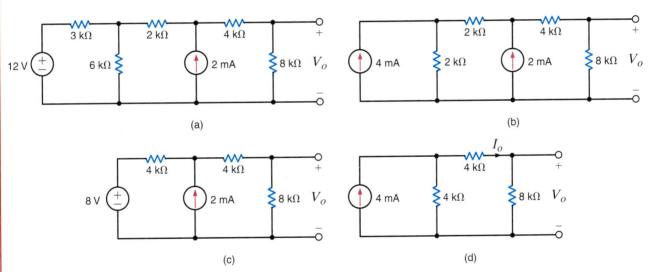

Note that this systematic, sometimes tedious, transformation allows us to reduce the network methodically to a simpler equivalent form with respect to some other circuit element. However, we should also realize that this technique is worthless for circuits of the form shown in Fig. 5.4. Furthermore, although applicable to networks containing dependent sources, it is not as useful as other techniques, and care must be taken not to transform the part of the circuit that contains the control variable.

LEARNING EXTENSION

E5.6 Find V_o in the circuit in Fig. E5.2 using source exchange.

ANSWER: $V_o = \dfrac{4}{3} \text{V}$.

At this point let us pause for a moment and reflect on what we have learned; that is, let us compare the use of node or loop analysis with that of the theorems discussed in this chapter. When we examine a network for analysis, one of the first things we should do is count the number of nodes and loops. Next we consider the number of sources. For example, are there a number of voltage sources or current sources present in the network? All these data, together with the information that we expect to glean from the network, give a basis for selecting the simplest approach. With the current level of computational power available to us, we can solve the node or loop equations that define the network in a flash.

With regard to the theorems, we have found that in some cases the theorems do not necessarily simplify the problem and a straightforward attack using node or loop analysis is as good an approach as any. This is a valid point provided that we are simply looking for some particular voltage or current. However, the real value of the theorems is the insight and understanding that they provide about the physical nature of the network. For example, superposition tells us what each source contributes to the quantity under investigation. However, a computer solution of the node or loop equations does not tell us the effect of changing certain parameter values in the circuit. It does not help us understand the concept of loading a network or the ramifications of interconnecting networks or the idea of matching a network for maximum power transfer. The theorems help us to understand the effect of using a transducer at the input of an amplifier with a given input resistance. They help us explain the effect of a load, such as a speaker, at the output of an amplifier. We derive none of this information from a node or loop analysis. In fact, as a simple example, suppose that a network at a specific pair of terminals has a Thévenin equivalent circuit consisting of a voltage source in series with a 2-kΩ resistor. If we connect a 2-Ω resistor to the network at these terminals, the voltage across the 2-Ω resistor will be essentially nothing. This result is fairly obvious using the Thévenin theorem approach; however, a node or loop analysis gives us no clue as to why we have obtained this result.

We have studied networks containing only dependent sources. This is a very important topic because all electronic devices, such as transistors, are modeled in this fashion. Motors in power systems are also modeled in this way. We use these amplification devices for many different purposes, such as speed control for automobiles.

In addition, it is interesting to note that when we employ source transformation as we did in Example 5.14, we are simply converting back and forth between a Thévenin equivalent circuit and a Norton equivalent circuit.

Finally, we have a powerful tool at our disposal that can be used to provide additional insight and understanding for both circuit analysis and design. That tool is Microsoft EXCEL, and it permits us to study the effects, on a network, of varying specific parameters. The following example will illustrate the simplicity of this approach.

Example 5.15

We wish to use Microsoft EXCEL to plot the Thévenin equivalent parameters V_{oc} and R_{Th} for the circuit in Fig. 5.17 over the R_x range 0 to 10 kΩ.

Figure 5.17
Circuit used in
Example 5.15.

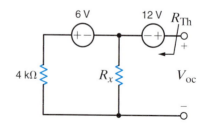

SOLUTION The Thévenin resistance is easily found by replacing the voltage sources with short circuits. The result is

$$R_{Th} = 4//R_x = \frac{4R_x}{4 + R_x}$$

5.7

where R_x and R_{Th} are in kΩ. Superposition can be used effectively to find V_{oc}. If the 12-V source is replaced by a short circuit

$$V_{oc_1} = -6\left[\frac{R_x}{R_x + 4}\right]$$

Applying this same procedure for the 6-V source yields

$$V_{oc_2} = 12$$

and the total open-circuit voltage is

$$V_{oc} = 12 - 6\left[\frac{R_x}{R_x + 4}\right]$$

5.8

In EXCEL we wish to (1) vary R_x between 0 and 10 kΩ, (2) calculate R_{Th} and V_{oc} at each R_x value, and (3) plot V_{oc} and R_{Th} versus R_x. We begin by opening EXCEL and entering column headings as shown in Fig. 5.18a. Next, we enter a zero in the first cell of the R_x column at column-row location A4. To automatically fill the column with values, go to the Edit menu and select Fill/Series to open the window shown in Fig. 5.18b, which has already been edited appropriately for 101 data points. The result is a series of R_x values from 0 to 10 kΩ in 100 Ω steps. To enter Eq. (5.8), go to location B4 (right under the V_{oc} heading). Enter the following text and do not forget the equal sign:

`=12-6*A4/(A4+4)`

This is Eq. (5.8) with R_x replaced by the first value for R_x, which is at column-row location A4. Similarly for R_{Th}, enter the following expression at C4.

`=4*A4/(A4+4)`

Figure 5.18
(a) The EXCEL spreadsheet for Example 5.15 showing the desired column headings. (b) The Fill/Series window edited for varying R_x and (c) the final plot of V_{oc} and R_{Th}.

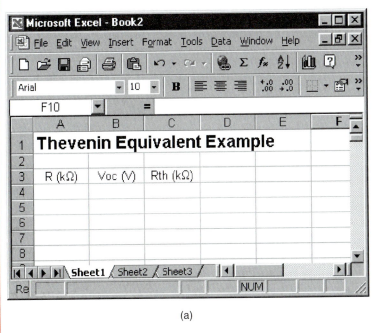

(a)

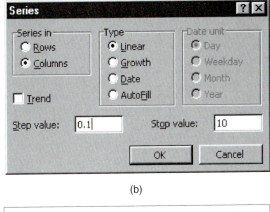

(b)

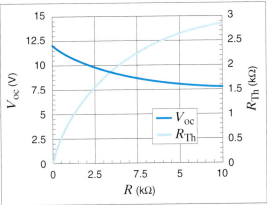

(c)

To replicate the expression in cell B4 for all R_x values, select cell B4, grab the lower right corner of the cell, hold and drag down to cell B104, and release. Repeat for R_{Th} by replicating cell C4.

To plot the data, first drag the cursor across all cells between A4 and C104. Next, from the Insert menu, select Chart. We recommend strongly that you choose the XY (Scatter) chart type. EXCEL will take you step by step through the basic formatting of your chart, which, after some manipulations, might look similar to the chart in Fig. 5.18c.

5.4 Maximum Power Transfer

In circuit analysis we are sometimes interested in determining the maximum power that can be delivered to a load. By employing Thévenin's theorem, we can determine the maximum power that a circuit can supply and the manner in which to adjust the load to effect maximum power transfer.

Suppose that we are given the circuit shown in Fig. 5.19. The power that is delivered to the load is given by the expression

$$P_{\text{load}} = i^2 R_L = \left(\frac{v}{R + R_L}\right)^2 R_L$$

We want to determine the value of R_L that maximizes this quantity. Hence, we differentiate this expression with respect to R_L and equate the derivative to zero.

$$\frac{dP_{\text{load}}}{dR_L} = \frac{(R + R_L)^2 v^2 - 2v^2 R_L (R + R_L)}{(R + R_L)^4} = 0$$

which yields

$$R_L = R$$

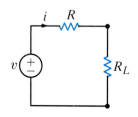

Figure 5.19

Equivalent circuit for examining maximum power transfer.

In other words, maximum power transfer takes place when the load resistance $R_L = R$. Although this is a very important result, we have derived it using the simple network in Fig. 5.19. However, we should recall that v and R in Fig. 5.19 could represent the Thévenin equivalent circuit for any linear network.

Example 5.16

Let us find the value of R_L for maximum power transfer in the network in Fig. 5.20a and the maximum power that can be transferred to this load.

SOLUTION To begin, we derive the Thévenin equivalent circuit for the network exclusive of the load. V_{oc} can be calculated from the circuit in Fig. 5.20b. The mesh equations for the network are

$$I_1 = 2 \times 10^{-3}$$

$$3k(I_2 - I_1) + 6kI_2 + 3 = 0$$

Solving these equations yields $I_2 = 1/3$ mA and, hence,

$$V_{oc} = 4kI_1 + 6kI_2$$

$$= 10 \text{ V}$$

R_{Th}, shown in Fig. 5.20c, is 6 kΩ; therefore, $R_L = R_{Th} = 6$ kΩ for maximum power transfer. The maximum power transferred to the load is

$$P_L = \left(\frac{10}{12k}\right)^2 (6k) = \frac{25}{6} \text{ mW}$$

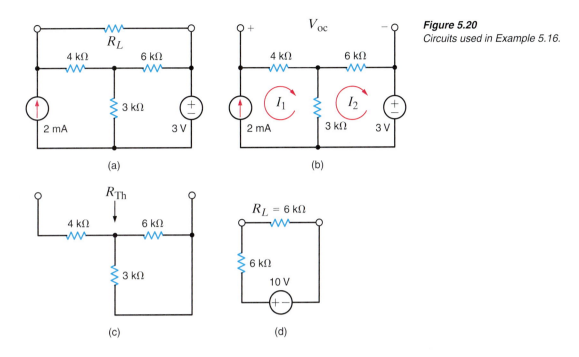

Figure 5.20
Circuits used in Example 5.16.

Example 5.17

Let us find R_L for maximum power transfer and the maximum power transferred to this load in the circuit in Fig. 5.21a.

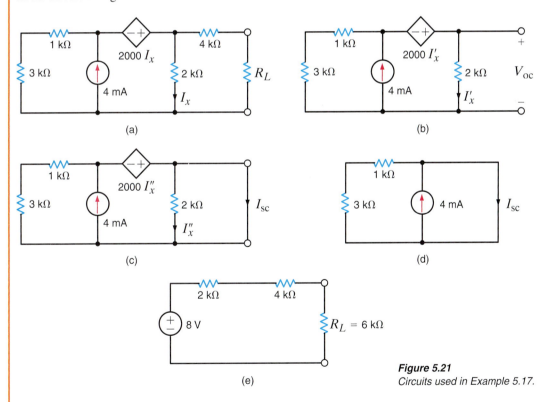

Figure 5.21
Circuits used in Example 5.17.

SOLUTION We wish to reduce the network to the form shown in Fig. 5.19. We could form the Thévenin equivalent circuit by breaking the network at the load. However, close examination of the network indicates that our analysis will be simpler if we break the network to the left of the 4-kΩ resistor. When we do this, however, we must realize that for maximum power transfer $R_L = R_{Th} + 4$ kΩ. V_{oc} can be calculated from the network in Fig. 5.21b.

Forming a supernode around the dependent source and its connecting nodes, the KCL equation for this supernode is

$$\frac{V_{oc} - 2000I'_x}{1k + 3k} + (-4 \times 10^{-3}) + \frac{V_{oc}}{2k} = 0$$

where

$$I'_x = \frac{V_{oc}}{2k}$$

These equations yield $V_{oc} = 8$ V. The short-circuit current can be found from the network in Fig. 5.21c. It is here that we find the advantage of breaking the network to the left of the 4-kΩ resistor. The short circuit shorts the 2-kΩ resistor and, therefore, $I''_x = 0$. Hence, the circuit is reduced to that in Fig. 5.21d, where clearly $I_{sc} = 4$ mA. Then

$$R_{Th} = \frac{V_{oc}}{I_{sc}} = 2 \text{ k}\Omega$$

Connecting the Thévenin equivalent to the remainder of the original circuit produces the network in Fig. 5.21e. For maximum power transfer $R_L = R_{Th} + 4 \text{ k}\Omega = 6 \text{ k}\Omega$, and the maximum power transferred is

$$P_L = \left(\frac{8}{12k}\right)^2 (6k) = \frac{8}{3} \text{ mW}$$

LEARNING EXTENSION

E5.7 Given the circuit in Fig. E5.7, find R_L for maximum power transfer and the maximum power transferred.

ANSWER: $R_L = 6 \text{ k}\Omega$;

$$P_L = \frac{2}{3} \text{ mW}.$$

Figure E5.7

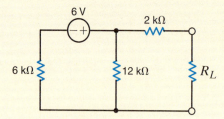

Example 5.18

Given the network in Fig. 5.22 with $V_{in} = 5$ V and $R_1 = 2 \ \Omega$, let us graphically examine a variety of aspects of maximum power transfer by plotting the parameters V_{out}, I, P_{out}, P_{in} and the efficiency $= P_{out}/P_{in}$ as a function of the resistor ratio R_2/R_1.

SOLUTION The parameters to be plotted can be determined by simple circuit analysis techniques. By voltage division

$$V_{out} = \left[\frac{R_2}{R_1 + R_2}\right] V_{in} = \left[\frac{R_2}{2 + R_2}\right](5)$$

From Ohm's law

$$I = \frac{V_{in}}{R_1 + R_2} = \frac{5}{2 + R_2}$$

Figure 5.22
Circuit used in maximum power transfer analysis.

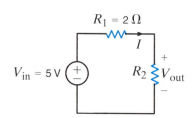

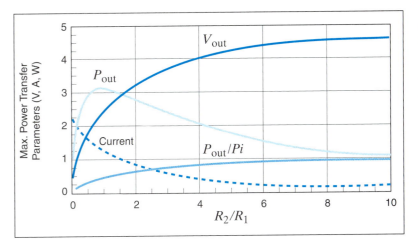

Figure 5.23
Maximum power transfer parameter plots for the network in Fig. 5.22. (The units for voltage, current, and power are volts, amperes, and watts, respectively.)

The input and output powers are

$$P_{in} = IV_{in} = \frac{V_{in}^2}{R_1 + R_2} = \frac{25}{2 + R_2} \qquad P_{out} = IV_{out} = R_2\left[\frac{V_{in}}{R_1 + R_2}\right]^2 = R_2\left[\frac{5}{2 + R_2}\right]^2$$

Finally, the efficiency is

$$\text{efficiency} = \frac{P_{out}}{P_{in}} = \frac{R_2}{R_1 + R_2} = \frac{R_2}{2 + R_2}$$

The resulting plots of the various parameters are shown in Fig. 5.23 for R_2 ranging from $0.1R_1$ to $10R_1$. Note that as R_2 increases, V_{out} increases toward V_{in} (5 V) as dictated by voltage division. Also, the current decreases in accordance with Ohm's law. Thus, for small values of R_2, V_{out} is small, and when R_2 is large, I is small. As a result, the output power (the product of these two parameters) has a maximum at $R_2/R_1 = 1$ as predicted by maximum power transfer theory.

Maximum power does not correspond to maximum output voltage, current, or efficiency. In fact, at maximum power transfer, the efficiency is always 0.5, or 50%. If you are an electric utility supplying energy to your customers, do you want to operate at maximum power transfer? The answer to this question is an obvious "No" because the efficiency is only 50%. The utility would only be able to charge its customers for one-half of the energy produced. It is not uncommon for a large electric utility to spend billions of dollars every year to produce electricity. The electric utility is more interested in operating at maximum efficiency.

5.5   *dc SPICE Analysis Using* Schematic *Capture*

INTRODUCTION   The original version of SPICE (<u>S</u>imulation <u>P</u>rogram with <u>I</u>ntegrated <u>C</u>ircuit <u>E</u>mphasis) was developed at the University of California at Berkeley. It quickly became an industry standard for simulating integrated circuits. With the advent and development of the PC industry, several companies began selling PC- and Macintosh-compatible versions of SPICE. One company, ORCAD Corporation, a division of Cadence Design Systems, Inc., produces a PC-compatible version called PSPICE, which we will discuss in some detail in this text.

In SPICE, circuit information such as the names and values of resistors and sources as well as how they are interconnected is input using data statements with a specific format. The particulars for every element must be typed in a precise order. This makes debugging difficult since you must know the proper format to recognize formatting errors.

PSPICE, as well as other SPICE-based simulators, now employs a feature known as schematic capture. Using an editor, we bypass the cryptic formatting and simply draw the circuit diagram, assigning element values via dialog boxes. The editor then converts the circuit diagram

Figure 5.24
In the student version of Release 9.1, you can choose to install either or both schematic capture editors.

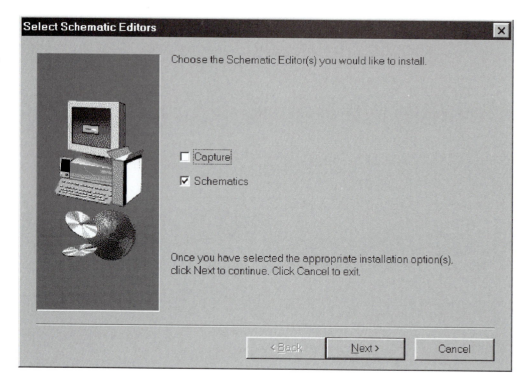

to an original SPICE format for actual simulation. In the student version of Release 9.1, the most current software available from ORCAD, there are two editors, CAPTURE and *Schematics*. CAPTURE is intended for printed circuit board (PCB) design, while *Schematics* is more generic. When installing PSPICE, the window in Fig. 5.24 will appear, giving you the option of choosing either or both editors. In this text, we will use *Schematics* exclusively.

When elements such as resistors and voltage sources are given values, it is convenient to use scale factors. PSPICE supports the following list.

T	tera	10^{12}	K	kilo	10^{3}	N	nano	10^{-9}
G	giga	10^{9}	M	milli	10^{-3}	P	pico	10^{-12}
MEG	mega	10^{6}	U	micro	10^{-6}	F	femto	10^{-15}

Component values must be immediately followed by a character—spaces are not allowed. Any text can follow the scale factor. Finally, PSPICE is case insensitive, so there is no difference between 1 Mohm and 1 mohm in PSPICE. Fig. 5.25 shows the five-step procedure for simulating circuits using *Schematics* and PSPICE. We will demonstrate this procedure in the following example.

A dc SIMULATION EXAMPLE   The following font conventions will be used throughout this tutorial. Uppercase text refers to PSPICE programs, menus, dialog boxes, and utilities. All boldface text denotes keyboard or mouse inputs. In each instance, the case in boldface text matches that in PSPICE. Let us simulate the circuit in Fig. 5.26 using PSPICE. Following the flowchart procedure in Fig. 5.25, the first step is to open *Schematics* using the **Start/Programs/Pspice Student/Schematics** sequence of pop-up menus. When *Schematics* opens, our screen will change to the *Schematics* editor window shown in Fig. 5.27.

Step 1: Drawing the Schematic

Next, we obtain and place the required parts: three resistors, two dc voltage sources, and a dc current source. To get the voltage sources, left-click on the **Draw** menu and select **Get New Part** as shown in Fig. 5.28. The **Part Browser** in Fig. 5.29 appears, listing all the parts available in PSPICE.

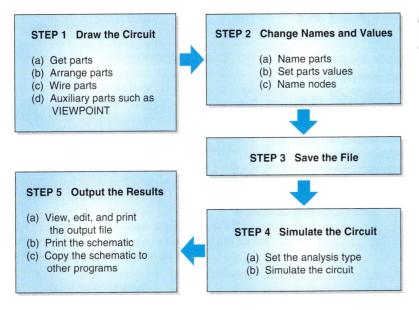

Figure 5.25
The five-step procedure for PSPICE simulations.

STEP 1   Draw the Circuit

(a) Get parts
(b) Arrange parts
(c) Wire parts
(d) Auxiliary parts such as VIEWPOINT

STEP 2   Change Names and Values

(a) Name parts
(b) Set parts values
(c) Name nodes

STEP 3   Save the File

STEP 5   Output the Results

(a) View, edit, and print the output file
(b) Print the schematic
(c) Copy the schematic to other programs

STEP 4   Simulate the Circuit

(a) Set the analysis type
(b) Simulate the circuit

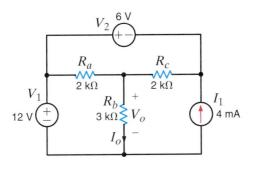

Figure 5.26
A dc circuit used for simulation.

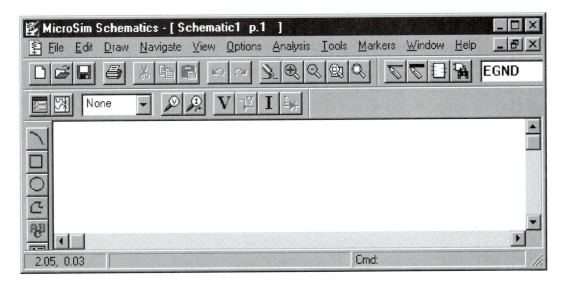

Figure 5.27
The Schematics *editor window.*

Since we do not know the *Schematics* name for a dc voltage source, select **Libraries**, and the **Library Browser** in Fig. 5.30 appears. This browser lists all the parts libraries available to us. The dc voltage source is called VDC and is located in the SOURCE.slb library, as seen in Fig. 5.30. Thus, we select the part VDC and click OK. The box in Fig. 5.29 reappears.

If we now select **Place & Close**, we revert to the *Schematics* editor in Fig. 5.27 with one difference: the mouse pointer has become a dc voltage source symbol. The source can be positioned within the drawing area by moving the mouse and then placed by left-clicking *once*.

Figure 5.29
The Part Browser window.

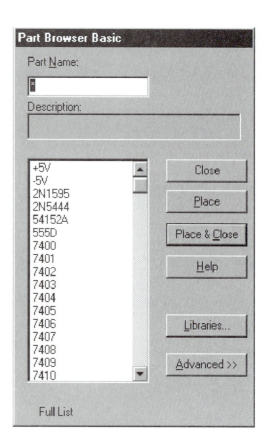

Figure 5.28
Getting a new part
in Schematics.

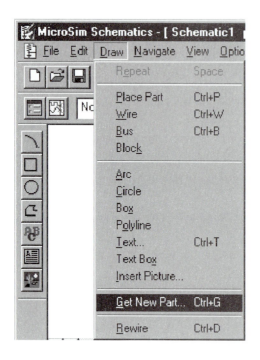

Figure 5.30
The Library Browser window.

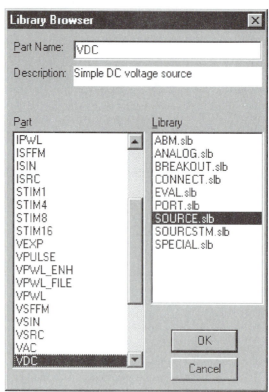

To place the second source, *move the mouse some distance*, and left-click again to place V2. To stop adding sources, we right-click once. Moving the mouse between part placements keeps the parts from stacking atop one another in the diagram. In Fig. 5.26, V2 is oriented horizontally.

To rotate V2 in your schematic, select it by clicking on it once. The part should turn red. In the **Edit** menu, choose **Rotate**. This causes V2 to spin 90° counterclockwise. Next, we click on V2 and drag it to the desired location. A diagram similar to that shown in Fig. 5.31 should result.

Next, we place the resistors. Repeat the process of **Draw/Get New Part**, except this time, when the **Part Browser Basic** dialog box appears, we type in **R** and select **Place & Close**. The mouse pointer then becomes a resistor. Place each resistor and right-click when done. Note that the resistors are automatically assigned default values of 1 kΩ. Current sources are in the SOURCE.slb library and are called IDC. Get one, place it, and rotate it twice. The resulting schematic is shown in Fig. 5.32.

The parts can now be interconnected. Go to the **Draw** menu and select **Wire**. The mouse pointer will turn into a symbolic pencil. To connect the top of V1 to R1, point the pencil at the end of the wire stub protruding from V1, click once and release. Next, we move the mouse up and over to the left end of R1. A line is drawn up and over at a 90° angle, appearing dashed as shown in Fig. 5.33. *Dashed lines are not yet wires!* We must left-click again to complete and "cut" the connection. The dashed lines become solid and the wiring connection is made.

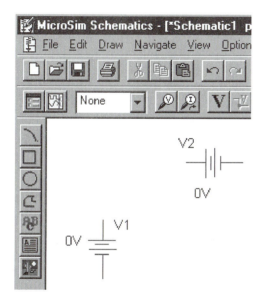

Figure 5.31
The Schematics *editor after placing the dc voltage sources.*

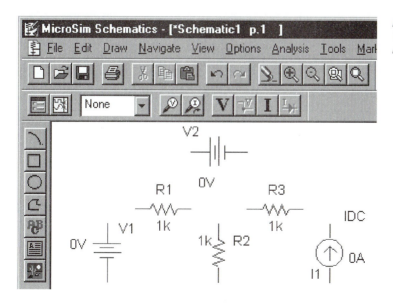

Figure 5.32
The schematic after part placement and ready for wiring.

Figure 5.33
Caught in the act of connecting V1 to R1.

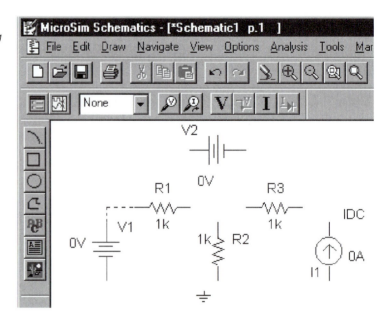

Excess wire fragments (extended dashed lines) can be removed by selecting **Re<u>d</u>raw** from the **View** menu. The remaining wires can be drawn using a *Schematics* shortcut. To reactivate the wiring pencil, double right-click. This shortcut reactivates the most recent mouse use. We simply repeat the steps listed above to complete the wiring.

PSPICE requires that all schematics have a ground or reference terminal. The ground node voltage will be zero, and all other node voltages are referenced to it. We can use either the analog ground (AGND) or the earth ground (EGND) part from the PORT.sIb library shown in Fig. 5.30. Get this part and place it at the bottom of the schematic as shown in Fig. 5.34. Make sure that the part touches the bottom wire in the diagram so that the node dot appears in your schematic. The wiring is now completed.

Step 2: Changing Component Names and Values

To change the name of R1 to Ra, double-click on the text "R1." The **Edit Reference Designator** dialog box appears as shown in Fig. 5.35. Simply type in the new name for the resistor, **Ra**, and select **OK**. Next we will change the resistor's value by double-clicking

Figure 5.34
Schematic with all parts and wiring completed.

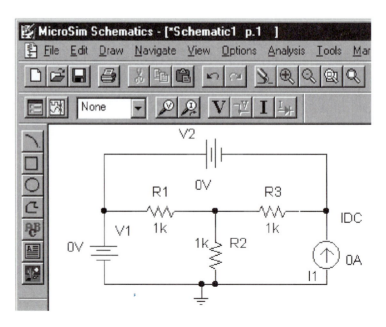

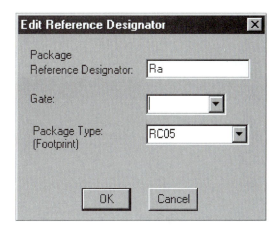

Figure 5.35
Changing the name of R1 to Ra.

on the value, "1k." Now the **Attributes** dialog box in Fig. 5.36 appears. Type in **2k** and select **OK**. In a similar manner, edit the names and values of the other parts. The circuit shown in Fig. 5.37 is now ready to be saved.

Step 3: Saving the Schematic

To save the schematic, simply go to the **File** menu and select **Save**. All *Schematics* files are automatically given the extension .sch.

   The Netlist   The netlist is the old-fashioned SPICE code listing for circuit diagrams drawn in *Schematics*. A netlist can be created directly or as part of the simulation process. To create the netlist directly, go to the **Analysis** menu in Fig. 5.27 and select **Create Netlist**. You should receive either the message *Netlist Created* or a dialog box informing you of netlist errors.

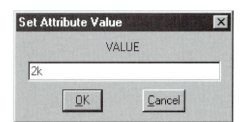

Figure 5.36
Changing the value of Ra from 1k to 2k.

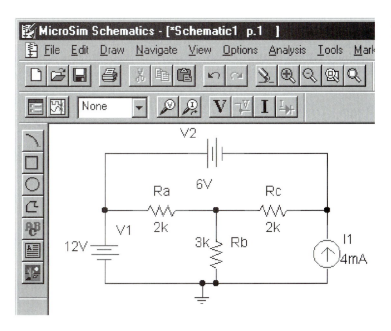

Figure 5.37
The finished schematic, ready for simulation.

Figure 5.38
The netlist for our circuit.

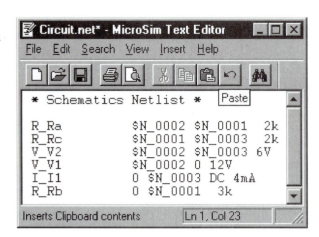

```
* Schematics  Netlist  *      Paste

R_Ra          $N_0002 $N_0001   2k
R_Rc          $N_0001 $N_0003   2k
V_V2          $N_0002 $N_0003  6V
V_V1          $N_0002 0 12V
I_I1          0 $N_0003 DC 4mA
R_Rb          0 $N_0001   3k
```

To view the netlist, return to the **Analysis** menu and select **Examine Netlist**, which opens the file shown in Fig. 5.38. All six elements appear along with their proper values. The text $N_0001 $N_0002 and so on are the node numbers that *Schematics* created when it converted the diagram to a netlist.

By tracing through the node numbers, we can be assured that our circuit is properly connected. The source V1 is 12 volts positive at node 2 with respect to node 0. EGND is always node number zero. Ra connects node 2 to node 1 and so forth.

How did PSPICE generate these node numbers in this particular order? In PSPICE, the terminals of a part symbol are called pins and are numbered. For example, the pin numbers for resistor, dc voltage, and current source symbols are shown in Fig. 5.39. None of the parts in Fig. 5.39 have been rotated. When the **Get New Part** sequence generates a part with a horizontal orientation, like the resistor, pin 1 is on the left. All vertically oriented parts (the sources) have pin 1 at the top. In the netlist, the order of the node numbers is always pin 1 then pin 2 for each component. Furthermore, when a part is rotated, the pin numbers also rotate. The most critical consequence of the pin numbers is current direction. PSPICE simulations always report the current flowing into pin 1 and out of pin 2.

To change the node number order of a part in the netlist, we must rotate that part 180°. Note that the order of node numbers for Rb in Fig. 5.38 is node 0 (EGND) then node 1. Accordingly, simulation results for the current in Rb will be the negative of Io as defined in Fig. 5.26. To force the netlist to agree with Fig. 5.26, we must rotate Rb twice. To do this, remove the wiring, click on Rb and select **Rotate** in the **Edit** menu twice, rewire and resave the file.

Step 4: Simulating the Circuit

The node voltages in the circuit are typically of interest to us, and *Schematics* permits us to identify each node with a unique name. For example, if we wish to call the node at Rb, Vo, we simply double-click on the wiring at the output node, and the dialog box in Fig. 5.40 will appear. Then type **Vo** in the space shown and select **OK**.

Simulation results for dc node voltages and branch currents can be displayed directly on the schematic. In the **Analysis/Display Results on Schematic** menu, choose **Enable Voltage Display** and **Enable Current Display**. This will display all voltages and currents. Unwanted voltage and current displays can be deleted by selecting the data and pressing the DELETE key.

Figure 5.39
A selection of parts showing the pin-numbering format used in Schematics.

pin 1 —\/\/\— pin 2
R1
1k

pin 1
+
V1
OV
-
pin 2

pin 1
I1
0A
pin 2

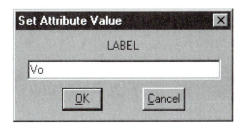

Figure 5.40
Creating a custom node name.

Individual node voltages can also be displayed using the VIEWPOINT (dc voltmeter) from the SPECIAL library. VIEWPOINT parts are placed at nodes and display dc node voltages with respect to ground. We will use a VIEWPOINT part to display Vo and **Analysis/Display/Enable Current Display** for the current, Io. The completed diagram will appear as shown in Fig. 5.41.

Simulation begins by choosing the type of analysis we wish to perform. This is done by selecting **Setup** from the **Analysis** menu. The SETUP dialog box is shown in Fig. 5.42. A *Schematics* dc analysis is requested by selecting **Bias Point Detail** then **Close**. The simulation results will include all node voltages, the currents through all voltage sources, and the total power dissipation. These data will be found in the output text file, accessible at **Analysis/Examine Output**. All VIEWPOINT voltages and currents will also be displayed on the schematic page.

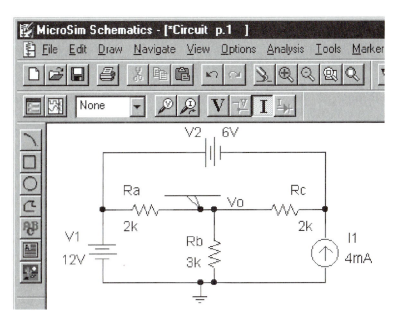

Figure 5.41
The finalized circuit, ready for simulation.

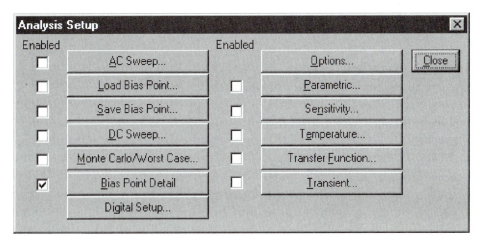

Figure 5.42
The ANALYSIS SETUP window showing the kinds of simulation we can request.

Figure 5.43
The simulation results: Vo = 6.75 V and Io = 2.25 mA.

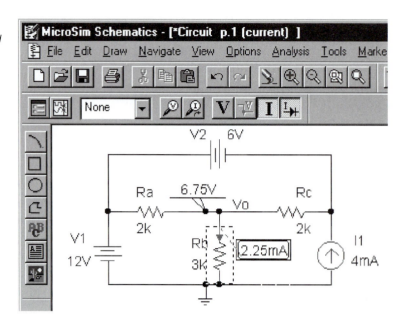

Clearly, a number of different analyses could be requested—for example, a **DC Sweep**. In this case, *Schematics* will ask for the dc source's value you wish to sweep, the start and stop values, and the increment. The simulation results will contain all dc node voltages and branch currents as a function of the varying source value. Two additional analyses, **Transient** and **AC Sweep**, will be discussed in subsequent chapters.

After exiting the SETUP dialog box, select **Simulate** from the **Analysis** menu. When the simulation is finished, your schematic should look much like that shown in Fig. 5.43, where Vo and Io are displayed automatically.

Step 5: Viewing and Printing the Results

All node voltages, voltage source currents, and total power dissipation are contained in the output file. To view these data, select **Examine Output** from the **Analysis** menu. Within the output file is a section containing the node voltages, voltage source currents, and total power dissipation as shown in Fig. 5.44. Indeed, Vo is 6.75 V and the total power dissipation is 5.25 mW. The current bears closer inspection. PSPICE says the currents through V1 and V2 are 1.75 mA and −4.375 mA, respectively. Recall from the netlist discussion that in PSPICE, current flows from a part's pin number 1 to pin 2. As seen in Fig. 5.39, pin 1 is at the positive end of the voltage source. PSPICE is telling us that the current flowing top-down through V1 in Fig. 5.26 is 1.75 mA, while 4.375 mA flows through V2 left-to-right. Based on the passive sign convention, V1 is consuming power while V2 generates power! Since the output file is simply a text editor, the file can be edited, saved, printed, or copied and pasted to other programs.

To print the circuit diagram, point the mouse pointer above and left of the upper left-hand corner of the diagram, click left, hold, and drag the mouse beyond the lower right edge of the drawing. A box will grow as you drag, eventually surrounding the circuit. Go to the **File** menu and select **Print**. The dialog box in Fig. 5.45 will appear. Select the options **Only Print Selected Area** and **User Definable Zoom Factor**. For most of the small schematics you will create, a scale of 125% to 200% will do fine. Other options are self-explanatory. Finally, select **OK** to print.

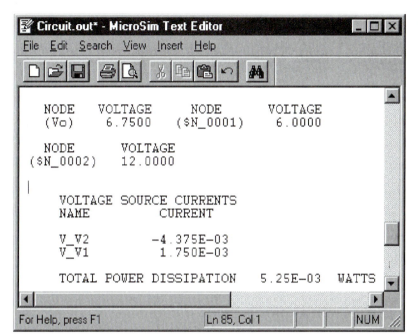

Figure 5.44
The output file simulation results.

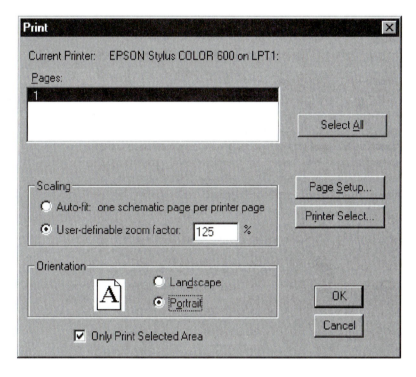

Figure 5.45
The Schematics *Print window.*

If you prefer that the grid dots not appear in the printout, go to the **Options/Display Options** menu and de-select the GRID ON option. To change the grid color, go to the **Options/Display Preferences** menu, select **Grid** and select a color from the drop-down edit box.

To incorporate your schematic into other applications such as text processors, draw a box around the diagram as described above, then under the **Edit** menu, select **Copy to Clipboard**. You can now paste the circuit into other programs.

Example 5.19

Let us use PSPICE to find the voltage V_o and the current I_x in the circuit in Fig. 5.46.

Figure 5.46
*Circuit used in
Example 5.19.*

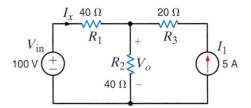

SOLUTION The PSPICE *Schematics* diagram is shown in Fig. 5.47. From the diagram, we find that $V_o = 150$ V and $I_x = -1.25$ A.

Figure 5.47
The Schematics
*diagram for the
network in Fig. 5.46.*

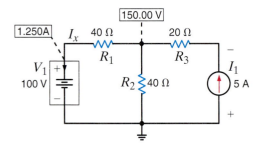

Example 5.20

Let us use both PSPICE and MATLAB to determine the voltage V_o in the network in Fig. 5.48.

Figure 5.48
*Circuit used in
Example 5.20.*

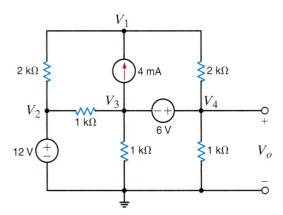

SOLUTION The PSPICE Schematics diagram is shown in Fig. 5.49. From this diagram we find that $V_o = 7.692$ V.

The MATLAB solution is obtained from the node equations of the network. The KCL equations at nonreference nodes V_1, V_2, and the supernode, including nodes V_3 and V_4, are

$$\frac{V_1 - V_2}{2k} + \frac{V_1 - V_4}{2k} = \frac{4}{k}$$

$$V_2 = 12$$

$$V_4 - V_3 = 6$$

$$\frac{4}{k} + \frac{V_3 - V_2}{1k} + \frac{V_3}{1k} + \frac{V_4 - V_1}{2k} + \frac{V_4}{1k} = 0$$

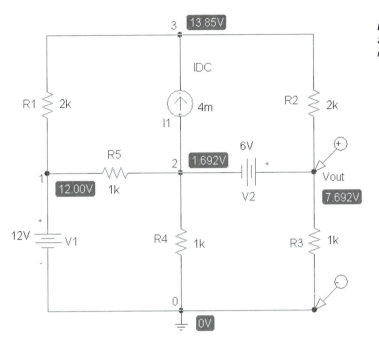

Figure 5.49
Schematics diagram for network in Fig. 5.48.

The simplified equations and MATLAB analysis are as follows:

```
   2v1 -  v2 + 0v3 -  v4 =  8
   0v1 +  v2 + 0v3 + 0v4 = 12
   0v1 + 0v2 -  v3 +  v4 =  6
  -v1  - 2v2 + 4v3 + 3v4 = -8
```

```
>> G = [2 -1 0 -1; 0 1 0 0; 0 0 -1 1; -1 -2 4 3]

G =

    2 -1  0 -1
    0  1  0  0
    0  0 -1  1
   -1 -2  4  3

>> i = [8; 12; 6; -8]

i =

    8
   12
    6
   -8

>> v=inv(G)*i

v =

   13.8462
   12.0000
    1.6923
    7.6923

>> vout=v(4)

vout =
    7.6923
```

Once again, we find that $V_4 = V_o = 7.692$ V.

Example 5.21

As a final example in this section on PSPICE, we will examine a circuit containing dependent sources that was analyzed in Chapter 3 using both node and loop equations. The network is shown in Fig. 5.50. We streamline this presentation by skipping the steps, already described in detail earlier, that are used to set up resistors and independent sources, and concentrate on the method employed to place dependent sources in the schematic.

Figure 5.50

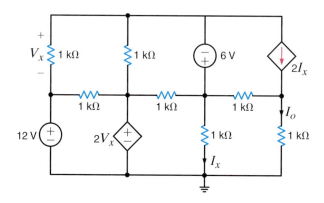

SOLUTION Using the techniques introduced earlier, the complete PSPICE schematic, shown in Fig. 5.51, can be created. Of particular interest are the dependent sources. The voltage-controlled voltage source part (VCVS) is called E in PSPICE. Conceptually, we consider the part to consist of a virtual voltmeter and a virtual voltage source. This "virtual meter" is connected across the dependent voltage, V_x, and the voltage across the "virtual source" is the output of the VCVS. To input the gain factor for the dependent source, simply double click on the E part to open the window shown in Fig. 5.52 and input the gain constant, 2 in this case.

Similarly, the current-controlled current source part (CCCS) is called F in PSPICE and should be viewed as a virtual ammeter-current source combination. Note how the dependent current, I_x, flows through the virtual meter. The CCCS gain is set just as it was in the VCVS case.

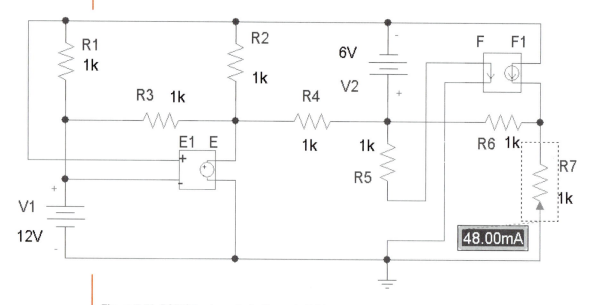

Figure 5.51 *PSPICE schematic for Example 5.21.*

Be aware that the direction of the metered current and output current of the dependent source do not necessarily have to match those in the original circuit since the sign of the gain can be used to compensate for any changes.

Finally, using the **Dipslay Results on Schematic** option, we find I_o is -48 mA, in agreement with our earlier analyses.

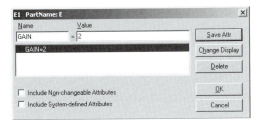

Figure 5.52
The Attribute window used to set the dependent source gain.

5.6   Application Example

Application Example 5.22

On Monday afternoon, Connie suddenly remembers that she has a term paper due Tuesday morning. When she sits at her computer to start typing, she discovers that the computer mouse doesn't work. After disassembly and some inspection, she finds that the mouse contains a printed circuit board that is powered by a 5-V supply contained inside the computer case. Furthermore, the board is found to contain several resistors, some op-amps, and one unidentifiable device, which is connected directly to the computer's 5-V supply as shown in Fig. 5.53a. Using a voltmeter to measure the node voltages, Connie confirms that all resistors and op-amps are functioning properly and the power supply voltage reaches the mouse board. However, without knowing the mystery device's function within the circuit, she cannot determine its condition. A phone call to the manufacturer reveals that the device is indeed linear but is also proprietary. With some persuasion, the manufacturer's representative agrees that if Connie can find the Thévenin equivalent circuit for the element at nodes A-B with the computer on, he will tell her if it is functioning properly. Armed with a single 1-kΩ resistor and a voltmeter, Connie attacks the problem.

SOLUTION   To find the Thévenin equivalent for the unknown device, together with the 5-V source, Connie first isolates nodes A and B from the rest of the devices on the board to measure the open-circuit voltage. The resulting voltmeter reading is $V_{AB} = 2.4$ V. Thus, the

Figure 5.53
Networks used in Example 5.22.

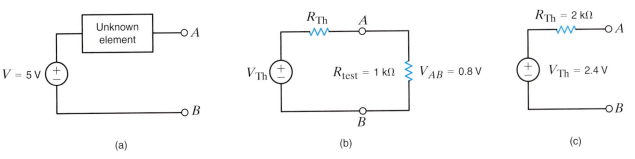

(a)   (b)   (c)

Thévenin equivalent voltage is 2.4 V. Then she connects the 1-kΩ resistor at nodes A-B as shown in Fig. 5.53b. The voltmeter reading is now $V_{AB} = 0.8$ V. Using voltage division to express V_{AB} in terms of V_{Th}, R_{Th}, and R_{test} in Fig. 5.43b yields the expression

$$0.8 = V_{Th}\left(\frac{1k}{1k + R_{Th}}\right)$$

Solving the equations for R_{Th}, we obtain

$$R_{Th} = 2.0 \text{ k}\Omega$$

Therefore, the unknown device and the 5-V source can be represented at the terminals A-B by the Thévenin equivalent circuit shown in Fig. 5.53c. When Connie phones the manufacturer with the data, the representative informs her that the device has indeed failed.

5.7 Design Examples

Design Example 5.23

We often find that in the use of electronic equipment, there is a need to adjust some quantity such as voltage, frequency, contrast, or the like. For very accurate adjustments, it is most convenient if coarse and fine-tuning can be separately adjusted. Therefore, let us design a circuit in which two inputs (i.e., coarse and fine voltages) are combined to produce a new voltage of the form

$$V_{tune} = \left[\frac{1}{2}\right]V_{coarse} + \left[\frac{1}{20}\right]V_{fine}$$

SOLUTION Because the equation to be realized is the sum of two terms, the solution appears to be an excellent application for superposition. Since the gain factors in the equation (i.e., 1/2 and 1/20) are both less than one, a voltage divider with two inputs would appear to be a logical choice. A typical circuit for this application is shown in Fig. 5.54a. The two superposition subcircuits are shown in Figs. 5.54b and c. Employing voltage division in the network in Fig. 5.54b yields

$$\frac{V_{tune\_C}}{V_{coarse}} = \left[\frac{R//R_2}{(R//R_2) + R_1}\right] = \frac{1}{2}$$

and therefore,

$$R//R_2 = R_1$$

In a similar manner, we find that

$$\frac{V_{tune\_F}}{V_{fine}} = \left[\frac{R//R_1}{(R//R_1) + R_2}\right] = \frac{1}{20}$$

which requires that

$$R_2 = 19(R//R_1)$$

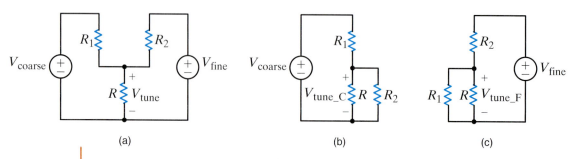

(a) (b) (c)

Figure 5.54 *(a) The coarse/fine adjustment circuit, (b) with V_{fine} set to zero and (c) with V_{coarse} set to zero.*

Note that the two constraint equations for the resistors have three unknowns—R, R_1, and R_2. Thus, we must choose one resistor value and then solve for the two remaining values. If we arbitrarily select $R = 1 \text{ k}\Omega$, then $R_1 = 900 \ \Omega$ and $R_2 = 9 \text{ k}\Omega$. This completes the design of the circuit. This example indicates that superposition is not only a useful analysis tool but provides insight into the design of new circuits.

Design Example 5.24

Coaxial cable is often used in very-high-frequency systems. For example, it is commonly used for signal transmission with cable television. In these systems resistance matching, the kind we use for maximum power transfer, is critical. In the laboratory, a common apparatus used in high-frequency research and development is the attenuator pad. The attenuator pad is basically a voltage divider, but the equivalent resistance at both its input ports is carefully designed for resistance matching. Given the network in Fig. 5.55 in which a source, modeled by V_S and R_S (50 Ω), drives an attenuator pad, which is connected to an equivalent load. Let us design the pad so that it has an equivalent resistance of 50 Ω and divides (i.e., attenuates) the input voltage by a factor of 10.

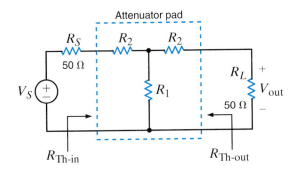

Figure 5.55
The model circuit for the attenuator pad design.

SOLUTION   Since the attenuator or "T-Network" must have an equivalent resistance of 50 Ω, we require that $R_{\text{Th-in}}$ and $R_{\text{Th-out}}$ be 50 Ω. Since these Thévenin resistance values are the same and the circuit is symmetric, we can use the label R_2 twice to indicate that those resistors will be the same value.

$$R_{\text{Th-in}} = R_2 + \left[R_1 // (R_2 + 50) \right] = 50$$

$$R_{\text{Th-out}} = R_2 + \left[R_1 // (R_2 + 50) \right] = 50$$

Since the equations are identical, we refer to both Thévenin equivalent resistance parameters simply as R_{Th}. The Thévenin equivalent voltage, V_{Th}, can be easily derived from the circuit in Fig. 5.56 a using voltage division.

$$V_{\text{Th}} = V_S \left[\frac{R_1}{R_1 + R_2 + 50} \right]$$

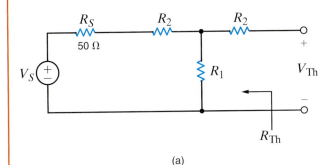

(a)

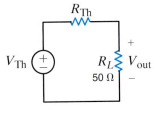

(b)

Figure 5.56
(a) The circuit used in finding V_{Th} and (b) the resulting model.

From the Thévenin equivalent circuit in Fig. 5.56b, we find

$$V_{\text{out}} = V_{\text{Th}}\left[\frac{50}{R_{\text{Th}} + 50}\right] = \frac{V_{\text{Th}}}{2}$$

Combining these equations yields the attenuation from V_S to V_{out}

$$\frac{V_{\text{out}}}{V_S} = \left[\frac{V_{\text{out}}}{V_{\text{Th}}}\right]\left[\frac{V_{\text{Th}}}{V_S}\right] = \frac{1}{2}\left[\frac{R_1}{R_1 + R_2 + 50}\right] = \frac{1}{10}$$

The Thévenin equivalent resistance equation and this attenuation equation provide us with two equations in the two unknowns R_1 and R_2. Solving these equations yields $R_1 = 20.83\ \Omega$ and $R_2 = 33.33\ \Omega$. For precise resistance matching, these resistors must be very accurate.

With such low resistor values, the power dissipation can become significant as V_S is increased. For example, if $V_S = 10$ V, $V_{\text{out}} = 1$ V and the power dissipated in the R_2 resistor connected to the input source is 333 mW. To keep the temperature of that resistor at reasonable levels, the power rating of that resistor should be at least 0.5 W.

Design Example 5.25

Let us design a circuit that will realize the following equation.

$$V_o = -3V_S - 2000I_S$$

SOLUTION   An examination of this equation indicates that we need to add two terms, one of which is from a voltage source and the other from a current source. Since the terms have negative signs, it would appear that the use of an inverting op-amp stage would be useful. Thus, one possible circuit for this application appears to be that shown in Fig. 5.57.

Figure 5.57
*Circuit used in
Example 5.25.*

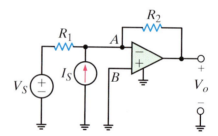

The Norton equivalent circuit at the terminals A-B will provide a composite view of the op-amp's input. Superposition can also be used in conjunction with the Norton equivalent to simplify the analysis. Using the network in Fig. 5.58a, we can determine the contribution of V_S to the short-circuit current, I_{sc}, which we call I_{sc_1}.

$$I_{\text{sc}_1} = \frac{V_S}{R_1}$$

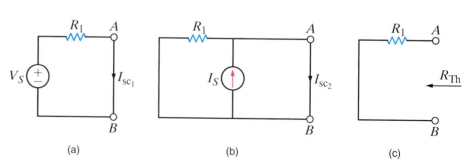

(a)   (b)   (c)

Figure 5.58 *Circuits used in deriving a Norton equivalent circuit.*

In a similar manner, using Fig. 5.58b, we find that the contribution of I_S to the short-circuit current is

$$I_{sc_2} = I_S$$

Employing superposition, the sum of these two currents yields the actual short-circuit current

$$I_{sc} = \frac{V_S}{R_1} + I_S$$

The Thévenin equivalent resistance at nodes *A-B* is obtained from the network in Fig. 5.58c as

$$R_{Th} = R_1$$

The equivalent circuit is now redrawn in Fig. 5.59 where we have employed the ideal op-amp conditions (i.e., $V_{in} = 0$), and the current into the op-amp terminals is zero. Since V_{in} is directly across R_{Th}, the current in this resistor is also zero. Hence, all the current I_{sc} will flow through R_2, producing the voltage

$$V_o = -R_2 \left[\frac{V_S}{R_1} + I_S \right] = -\frac{R_2}{R_1} V_S - I_S R_2$$

A comparison of this equation with the design requirement specifies that

$$\frac{R_2}{R_1} = 3 \quad \text{and} \quad R_2 = 2000 \ \Omega$$

which yields $R_1 = 667 \ \Omega$. Combining a 1-k Ω and a 2-kΩ resistor in parallel will yield the necessary 667 Ω exactly.

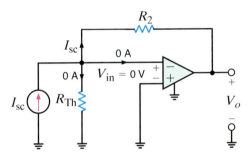

Figure 5.59
The required circuit containing the Norton equivalent.

Design Example 5.26

Fans are frequently needed to keep electronic circuits cool. They vary in size, power requirement, input voltage, and air-flow rate. In a particular application, three fans are connected in parallel to a 24-V source as shown in Fig. 5.60. A number of tests were run on this configuration, and it was found that the air flow, fan current, and input voltage are related by the following equations.

$$F_{CFM} = 200 I_F \qquad V_F = 100 I_F$$

where F_{CFM} represents the air-flow rate in cubic feet per minute, V_F is the fan voltage in volts, and I_F is the fan current in amperes. Note that fan current is related to fan speed, which in turn is related to air flow. A popular and inexpensive method for monitoring currents in applications where high accuracy is not critical involves placing a low-value sense resistor in series with the fan to "sense" the current by measuring the sense-resistor's voltage.

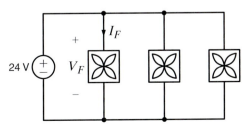

Figure 5.60
A trio of 24 V fans.

We wish to design a circuit that will measure the air flow in this three-fan system. Specifically, we want to

a. determine the value of the sense resistor, placed in series with the fan, such that its voltage is 2% of the nominal 24 V fan voltage, and specify a particular 1% component that can be obtained from the Digikey Corporation (Website: www.digikey.com).

b. design an op-amp circuit that will produce an output voltage proportional to total air flow, in which 1 V corresponds to 50 CFM.

Figure 5.61

The equivalent circuit for one fan and its sense resistor.

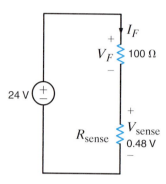

SOLUTION   The fan's voltage–current relationship specifies that the fan has a resistance of 100 Ω. Since the voltage across the sense resistor should be 2% of 24 V, or 0.48 V, the fan current, derived from the network in Fig. 5.61, is

$$I_F = \frac{24 - 0.48}{100} = 235.2 \text{ mA}$$

and the required value of the sense resistor is

$$R_{\text{sense}} = \frac{0.48}{0.2352} = 2.04 \ \Omega$$

The power dissipation in this component is only

$$P_{\text{sense}} = I_F^2 R_{\text{sense}} = 0.11 \text{ W}$$

And thus a standard 1/4 W 2-Ω resistor will satisfy the specifications.

The op-amp circuit must be capable of adding the air-flow contributions of all three fans and scaling the result such that 1 V corresponds to 50 CFM. A summing op-amp circuit would appear to be a logical choice in this situation, and thus we select the circuit shown in Fig. 5.62 where the second stage is simply an inverter that corrects for the negative sign resulting from the summer output. In order to determine the summer's gain, we calculate the volts/CFM at the sense resistors. For a single fan, the air flow is

$$F_{CFM} = 200I_F = 47.04 \text{ CFM}$$

And the volts per CFM at the input to the summer are

$$\frac{0.48 \text{ V}}{47.04 \text{ CFM}} = 0.0102 \text{ V/CFM}$$

Hence, the gain of the summer op-amp must be

$$\frac{V_o}{V_{\text{sense}}} = \frac{1 \text{ V/50 CFM}}{0.0102 \text{ V/CFM}} = 1.96 \text{ V/V}$$

This is a gain close to 2, and therefore we will use resistors that produce a 2:1 ratio, that is, very close to 1.96. At this point, one additional consideration must be addressed. Note that the resistors at the summer input are essentially connected in parallel with the sense resistors. To ensure that all the fan current flows in the sense resistors, we select very large values for the op-amp resistors. Let us choose $R_1 = R_2 = R_3 = 100$ kΩ and then $R_4 = 200$ kΩ.

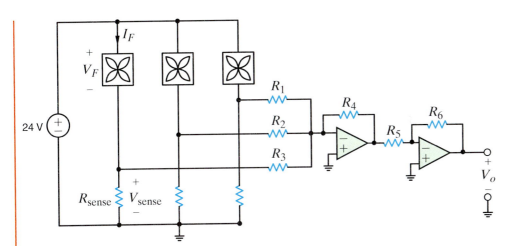

Figure 5.62
The complete air-flow measurement system.

Finally, the values for R_5 and R_6 can be somewhat arbitrary, as long as they are equal. If we select a value of 100 kΩ, then only two different resistor values are needed for the entire op-amp circuit.

SUMMARY

- Linearity: This property requires both additivity and homogeneity. Using this property, we can determine the voltage or current somewhere in a network by assuming a specific value for the variable and then determining what source value is required to produce it. The ratio of the specified source value to that computed from the assumed value of the variable, together with the assumed value of the variable, can be used to obtain a solution.

- In a linear network containing multiple independent sources, the principle of superposition allows us to compute any current or voltage in the network as the algebraic sum of the individual contributions of each source acting alone.

- Superposition is a linear property and does not apply to nonlinear functions such as power.

- Using Thévenin's theorem, we can replace some portion of a network at a pair of terminals with a voltage source V_{oc} in series with a resistor R_{Th}. V_{oc} is the open-circuit voltage at

the terminals, and R_{Th} is the Thévenin equivalent resistance obtained by looking into the terminals with all independent sources made zero.

- Using Norton's theorem, we can replace some portion of a network at a pair of terminals with a current source I_{sc} in parallel with a resistor R_{Th}. I_{sc} is the short-circuit current at the terminals and R_{Th} is the Thévenin equivalent resistance.

- Source transformation permits us to replace a voltage source V in series with a resistance R by a current source $I = V/R$ in parallel with the resistance R. The reverse is also true. This is an interchange relationship between Thévenin and Norton equivalent circuits.

- Maximum power transfer can be achieved by selecting the load R_L to be equal to R_{Th} found by looking into the network from the load terminals.

- dc PSPICE with Schematic Capture is an effective tool in analyzing dc circuits.

PROBLEMS

PSV **CS** both available on the web at: **http://www.justask4u.com/irwin**

SECTION 5.1

5.1 Find I_o in the circuit in Fig. P5.1 using linearity and the assumption that $I_o = 1$ mA. **CS**

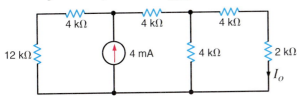

Figure P5.1

5.2 Find V_o in the network in Fig. P5.2 using linearity and the assumption that $V_o = 1$ V.

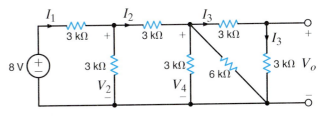

Figure P5.2

5.3 Find I_o in the network in Fig. P5.3 using linearity and the assumption that $I_o = 1$ mA. **PSV**

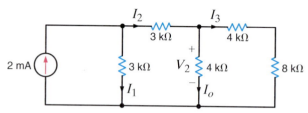

Figure P5.3

5.4 Find I_o in the network in Fig. P5.4 using linearity and the assumption that $I_o = 1$ mA.

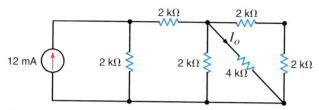

Figure P5.4

SECTION 5.2

5.5 In the network in Fig. P5.5, find I_o using superposition. **CS**

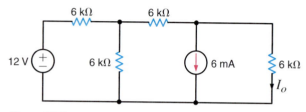

Figure P5.5

5.6 Find V_o in the network in Fig. P5.6 using superposition.

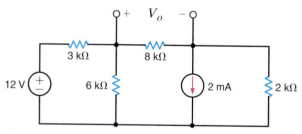

Figure P5.6

5.7 Find I_o in the network in Fig. P5.7 using superposition. **CS**

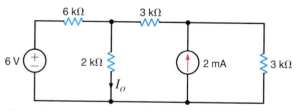

Figure P5.7

5.8 Use superposition to find V_o in the circuit in Fig. P5.8. **PSV**

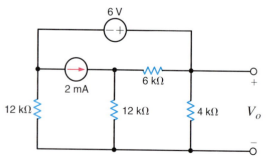

Figure P5.8

5.9 Find I_o in the circuit in Fig. P5.9 using superposition. **CS**

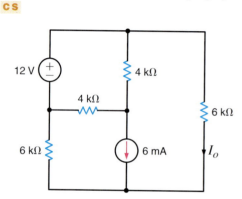

Figure P5.9

5.10 Use superposition to find I_o in the circuit in Fig. P5.10.

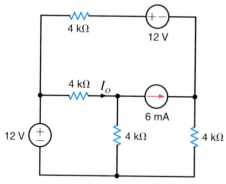

Figure P5.10

5.11 Find I_o in the network in Fig. P5.11 using superposition. **CS**

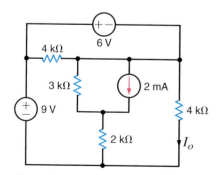

Figure P5.11

5.12 Find V_o in the circuit in Fig. P5.12 using superposition.

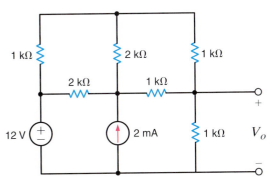

Figure P5.12

5.13 Given the network in Fig. P5.13, use superposition to find V_o.

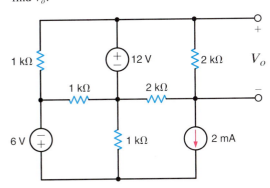

Figure P5.13

5.14 Use superposition to find V_o in the circuit in Fig. P5.14.

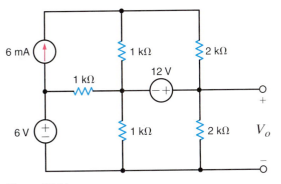

Figure P5.14

5.15 Find V_o in the circuit in Fig. P5.15 using superposition.

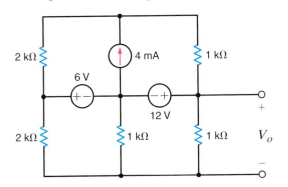

Figure P5.15

5.16 Find I_o in the circuit in Fig. P5.16 using superposition.

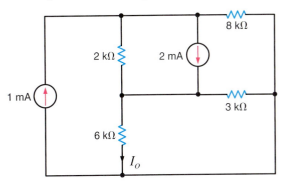

Figure P5.16

5.17 Use superposition to find I_o in the network in Fig. P5.17.

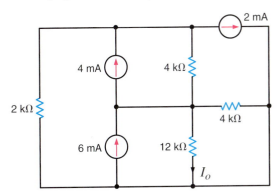

Figure P5.17

5.18 Use superposition to find I_o in the circuit in Fig. P5.18.

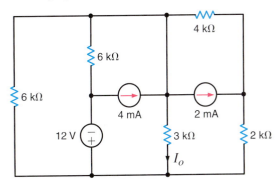

Figure P5.18

5.19 Find I_o in the circuit in Fig. P5.19 using superposition.

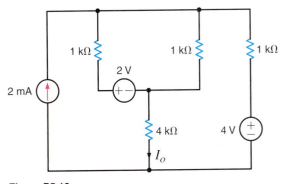

Figure P5.19

5.20 Use superposition to find I_o in the circuit in Fig. P5.20.

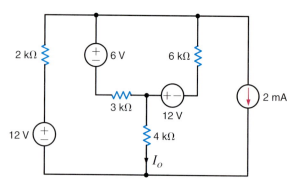

Figure P5.20

5.21 Find I_o in the circuit in Fig. P5.21 using superposition.

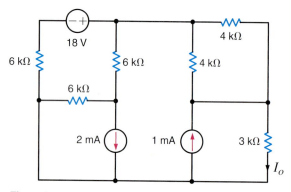

Figure P5.21

5.22 Use superposition to find I_o in the network in Fig. P5.22.
PSV

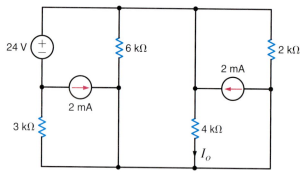

Figure P5.22

5.23 The loop equations for a two-loop network are

$$I_1 R_{11} + I_2 R_{12} = V_1$$
$$I_1 R_{21} + I_2 R_{22} = V_2$$

What is the relationship among V_1, V_2, and R_{ij} for $I_1 = 0$?

5.24 Use the results of Problem 5.23 to find the value of I_x that yields a $V_1 = 0$ V in the network in Fig. P5.24.

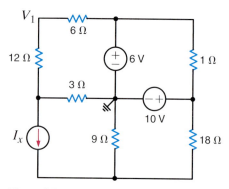

Figure P5.24

5.25 A three-node circuit is described by the following equations:

$$G_{11}V_1 + G_{12}V_2 + G_{13}V_3 = I_1$$
$$G_{21}V_1 + G_{22}V_2 + G_{23}V_3 = I_2$$
$$G_{31}V_1 + G_{32}V_2 + G_{33}V_3 = I_3$$

Show that for $V_2 = 0$,

$$I_1 \begin{vmatrix} G_{21} & G_{23} \\ G_{31} & G_{33} \end{vmatrix} - I_2 \begin{vmatrix} G_{11} & G_{13} \\ G_{31} & G_{33} \end{vmatrix} + I_3 \begin{vmatrix} G_{11} & G_{13} \\ G_{21} & G_{23} \end{vmatrix} = 0$$

5.26 Use the results of Problem 5.25 to determine the value of V_B such that V_o is zero in the network in Fig. P5.26.

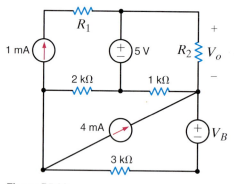

Figure P5.26

Explain why the values of R_1 and R_2 have no impact on your analysis.

SECTION 5.3

5.27 **(a)** Given the network in Fig. P5.27, find the value of R_2 such that $V_o = 0$ V. (b) Then find the Thévenin and Norton equivalent circuits at A-B as seen by R_3 using the results of (a).

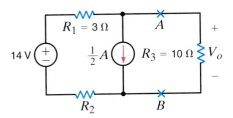

Figure P5.27

5.28 Use Thévenin's theorem to find V_o in the network in Fig. P5.28. **CS**

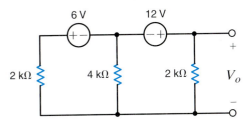

Figure P5.28

5.29 Find I_o in the network in Fig. P5.29 using Thévenin's theorem.

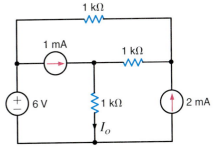

Figure P5.29

5.30 Find I_o in the circuit in Fig. P5.30 using Thévenin's theorem. **CS**

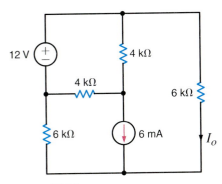

Figure P5.30

5.31 Find V_o in the network in Fig. P5.31 using Thévenin's theorem. **PSV**

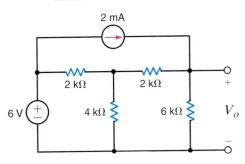

Figure P5.31

5.32 Find V_o in the circuit in Fig. P5.32 using Thévenin's theorem.

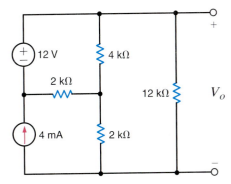

Figure P5.32

5.33 Find I_o in the network in Fig. P5.33 using Thévenin's theorem. **CS**

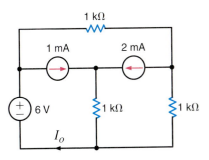

Figure P5.33

5.34 Find I_o in the network in Fig. P5.34 using Thévenin's theorem.

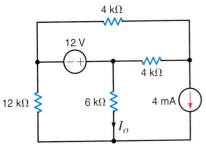

Figure P5.34

5.35 Find V_o in the circuit in Fig. P5.35 using Thévenin's theorem.

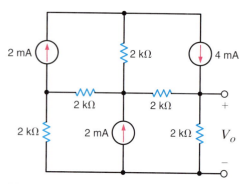

Figure P5.35

5.36 Find V_o in the network in Fig. P5.36 using Thévenin's theorem. **CS**

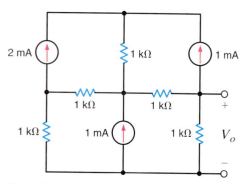

Figure P5.36

5.37 Find V_o in the network in Fig. P5.37 using Thévenin's theorem.

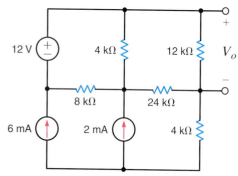

Figure P5.37

5.38 Use a combination of Thévenin's theorem and superposition to find V_o in the circuit in Fig. P5.38.

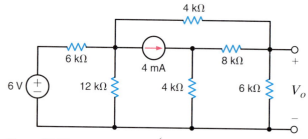

Figure P5.38

5.39 Find V_o in the network in Fig. P5.39 using Thévenin's theorem.

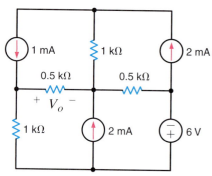

Figure P5.39

5.40 Find V_o in the network in Fig. P5.40 using Thévenin's theorem.

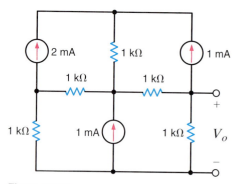

Figure P5.40

5.41 Solve Problem 5.12 using Thévenin's theorem.

5.42 Solve Problem 5.13 using Thévenin's theorem.

5.43 Use Thévenin's theorem to solve Problem 5.21.

5.44 Use Thévenin's theorem to solve Problem 5.22.

5.45 Given the linear circuit in Fig. P5.45, it is known that when a 2-kΩ load is connected to the terminals A-B, the load current is 10 mA. If a 10-kΩ load is connected to the terminals, the load current is 6 mA. Find the current in a 20-kΩ load.

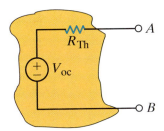

Figure P5.45

5.46 If an 8-kΩ load is connected to the terminals of the network in Fig. P5.46, $V_{AB} = 16$ V. If a 2-kΩ load is connected to the terminals, $V_{AB} = 8$ V. Find V_{AB} if a 20-kΩ load is connected to the terminals. **CS**

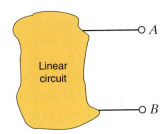

Figure P5.46

5.47 Find I_o in the network in Fig. P5.47 using Norton's theorem.

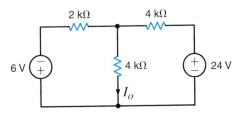

Figure P5.47

5.48 Find I_o in the network in Fig. P5.48 using Norton's theorem.

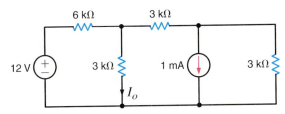

Figure P5.48

5.49 Use Norton's theorem to find I_o in the circuit in Fig. P5.49.

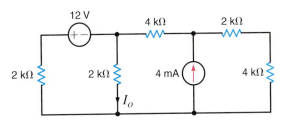

Figure P5.49

5.50 Find I_o in the network in Fig. P5.50 using Norton's theorem. **CS**

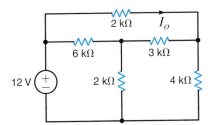

Figure P5.50

5.51 Use Norton's theorem to find V_o in the network in Fig. P5.51.

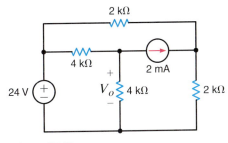

Figure P5.51

5.52 Find V_o in the network in Fig. P5.52 using Norton's theorem.

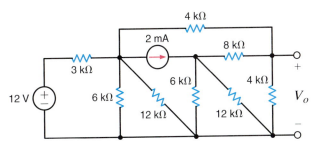

Figure P5.52

5.53 Solve Problem 5.14 using Norton's theorem.

5.54 Solve Problem 5.15 using Norton's theorem.

5.55 Use Norton's theorem to solve Problem 5.19.

5.56 Use Norton's theorem to solve Problem 5.21.

5.57 Use Norton's theorem to solve Problem 5.22.

5.58 Find I_o in the circuit in Fig. P5.58 using Norton's theorem. **PSV**

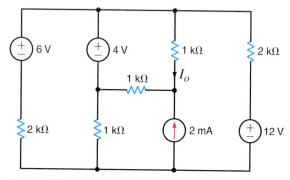

Figure P5.58

5.59 Find the Thévenin equivalent of the network in Fig. P5.59 at the terminals A-B. **CS**

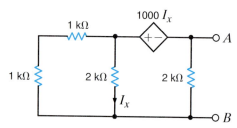

Figure P5.59

5.60 Find the Thévenin equivalent of the network in Fig. P5.60 at the terminals A-B. **PSV**

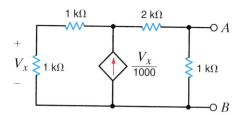

Figure P5.60

5.61 Find V_o in the network in Fig. P5.61 using Thévenin's theorem.

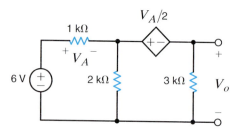

Figure P5.61

5.62 Use Thévenin's theorem to find V_o in the network in Fig. P5.62. **CS**

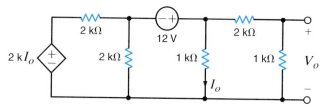

Figure P5.62

5.63 Use Thévenin's theorem to find I_o in the circuit in Fig. P5.63. **PSV**

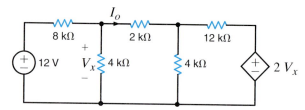

Figure P5.63

5.64 Find V_o in the network in Fig. P5.64 using Thévenin's theorem.

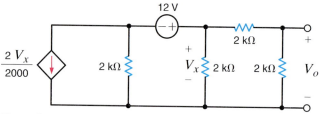

Figure P5.64

5.65 Find V_o in the circuit in Fig. P5.65 using Thévenin's theorem. **CS**

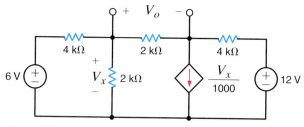

Figure P5.65

5.66 Find V_o in the network in Fig. P5.66 using Thévenin's theorem.

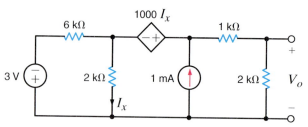

Figure P5.66

5.67 Use Thévenin's theorem to find V_o in the circuit in Fig. P5.67.

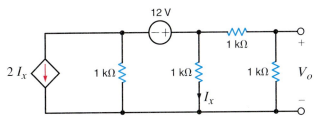

Figure P5.67

5.68 Use Norton's theorem to find V_o in the network in Fig. P5.68. **cs**

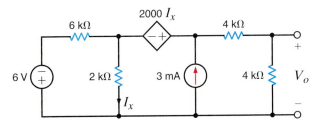

Figure P5.68

5.69 Find V_o in the network in Fig. P5.69 using Thévenin's theorem.

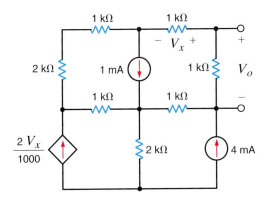

Figure P5.69

5.70 Find V_o in the circuit in Fig. P5.70 using Thévenin's theorem.

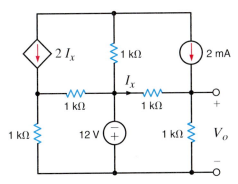

Figure P5.70

5.71 Find V_o in the network in Fig. P5.71 using Thévenin's theorem.

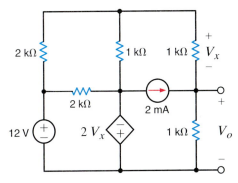

Figure P5.71

5.72 Use Thévenin's theorem to find V_o in the network in Fig. P5.72.

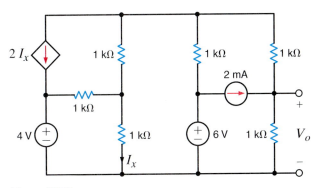

Figure P5.72

5.73 Use Thévenin's theorem to find I_o in the network in Fig. P5.73.

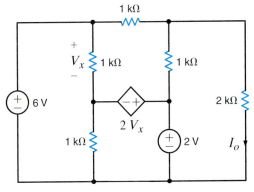

Figure P5.73

5.74 Using Thévenin's theorem find I_o in the circuit in Fig. P5.74.

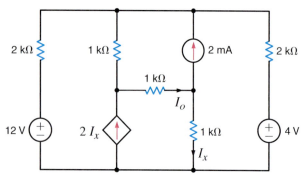

Figure P5.74

5.75 Find I_o in the network in Fig. P5.75 using Thévenin's theorem.

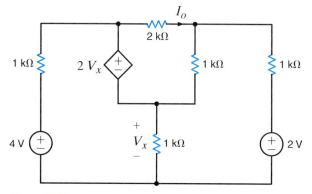

Figure P5.75

5.76 Use Thévenin's theorem to find the power supplied by the 2-mA source in the network in Fig. P5.74. **PSV**

5.77 Use Thévenin's theorem to find the power supplied by the 2-V source in the circuit in Fig. P5.75.

SECTION 5.4

5.78 Use source transformation to find V_o in the network in Fig. P5.78. **CS**

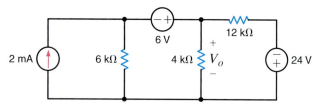

Figure P5.78

5.80 Use source transformation to find I_o in the network in Fig. P5.80. **PSV**

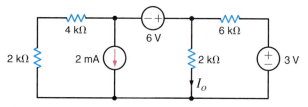

Figure P5.80

5.79 Find V_o in the network in Fig. P5.79 using source transformation.

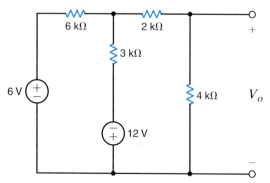

Figure P5.79

5.81 Use source transformation to find V_o in the network in Fig. P5.81.

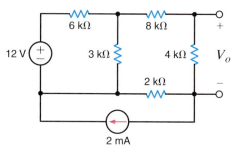

Figure P5.81

5.82 Find I_o in the network in Fig. P5.82 using source transformation.

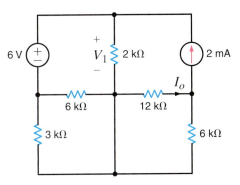

Figure P5.82

5.83 Find I_o in the network in Fig. P5.83.

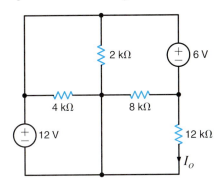

Figure P5.83

5.84 Find I_o in the network in Fig. P5.84 using source transformation.

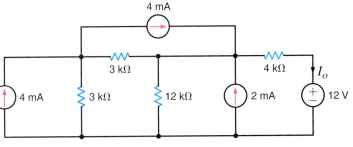

Figure P5.84

5.85 Use source transformation to find I_o in the circuit in Fig. P5.85. **CS**

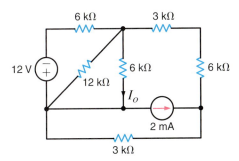

Figure P5.85

5.86 Find V_o in the network in Fig. P5.86 using source transformation.

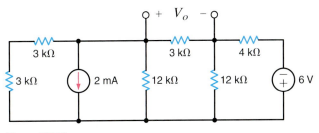

Figure P5.86

5.87 Find I_o in the circuit in Fig. P5.87 using source transformation.

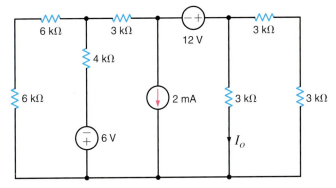

Figure P5.87

5.88 Find I_o in the network in Fig. P5.88 using source transformation. **CS**

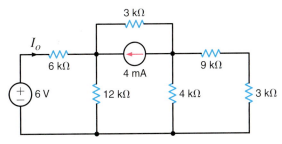

Figure P5.88

5.89 Solve Problem 5.16 using source transformation.

5.90 Solve Problem 5.17 using source transformation.

5.91 Use source transformation to solve Problem 5.18.

5.92 Use source transformation to solve Problem 5.20.

5.93 Use source transformation to solve Problem 5.21.

5.94 Solve Problem 5.22 using source transformation.

5.95 Find R_L in the network in Fig. P5.95 in order to achieve maximum power transfer.

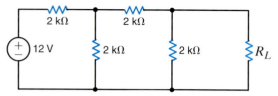

Figure P5.95

5.96 In the network in Fig. P5.96, find R_L, for maximum power transfer and the maximum power transferred to this load. **PSV**

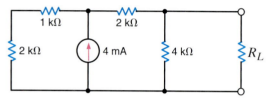

Figure P5.96

5.97 Find R_L for maximum power transfer and the maximum power that can be transferred to the load in Fig. P5.97.

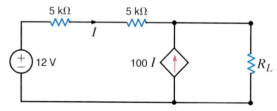

Figure P5.97

5.98 Choose R_L in Fig. P5.98 for maximum power transfer.

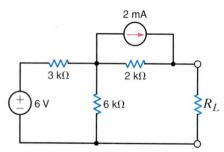

Figure P5.98

5.99 Find the value of R_L in the network in Fig. P5.99 for maximum power transfer. **PSV**

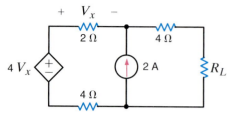

Figure P5.99

5.100 Calculate the maximum power that can be transferred to R_L in the circuit in Fig. P5.100.

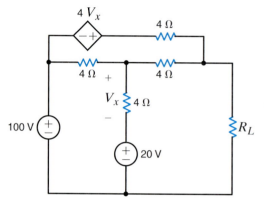

Figure P5.100

5.101 A cell phone antenna picks up a call. If the antenna and cell phone are modeled as shown in Fig. P5.101,

(a) Find R_{cell} for maximum output power.

(b) Determine the value of P_{out}.

(c) Determine the corresponding value of P_{ant}.

(d) Find v_o/v_{ant}.

(e) Determine the amount of power lost in R_{ant}.

(f) Calculate the efficiency $\eta = P_{out}/P_{ant}$.

(g) Determine the value of R_{cell} such that the efficiency is 90%.

(h) Given the change in (g), what is the new value of P_{ant}?

(i) Given the change in (g), what is the new value of P_{out}?

(j) Comment on the results obtained in (i) and (b).

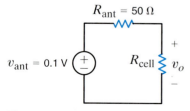

Figure P5.101

5.102 Some young engineers at the local electrical utility are debating ways to lower operating costs. They know that if they can reduce losses, they can lower operating costs. The question is whether they should design for maximum power transfer or maximum efficiency, where efficiency is defined as the ratio of customer power to power generated. Use the model in Fig. P5.102 to analyze this issue and justify your conclusions. Assume that both the generated voltage and the customer load are constant.

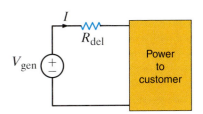

Figure P5.102

5.103 Find R_L for maximum power transfer and the maximum power that can be transferred in the network in Fig. P5.103.

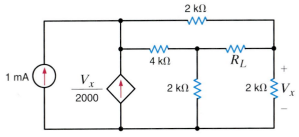

Figure P5.103

SECTION 5.5

5.104 Solve Problem 5.5 using PSPICE.

5.105 Solve Problem 5.78 using PSPICE.

5.106 Solve Problem 5.21 using PSPICE.

5.107 Solve Problem 5.71 using PSPICE.

TYPICAL PROBLEMS FOUND ON THE FE EXAM

5FE-1 Determine the maximum power that can be delivered to the load R_L in the network in Fig. 5PFE-1. **CS**

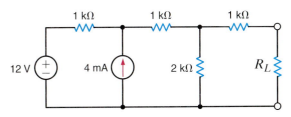

Fig. 5PFE-1

5FE-2 Find the value of the load R_L in the network in Fig. 5PFE-2 that will achieve maximum power transfer, and determine the value of the maximum power.

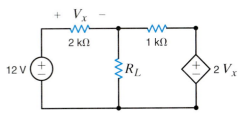

Fig. 5PFE-2

5FE-3 Find the value of R_L in the network in Fig. 5PFE-3 for maximum power transfer to this load.

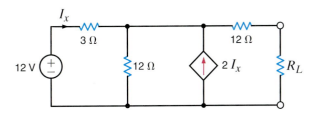

Fig. 5PFE-3

6

Capacitance and Inductance

Access Problem-Solving Videos PSV **and Circuit Solutions** CS **at:**
http://www.justask4u.com/irwin
using the registration code on the inside cover and see a website with answers and more!

Have you ever wondered how a tiny camera battery is able to produce a blinding flash or how a hand-held "stun gun" can deliver 50,000 V? The answer is energy storage, and in this chapter we introduce two elements that possess this property: the capacitor and the inductor. Both capacitors and inductors are linear elements; however, unlike the resistor, their terminal characteristics are described by linear differential equations. Another distinctive feature of these elements is their ability to absorb energy from the circuit, store it temporarily, and later return it. Elements that possess this energy storage capability are referred to simply as *storage elements*.

Capacitors are capable of storing energy when a voltage is present across the element. The energy is actually stored in an electric field not unlike that produced by sliding across a car seat on a dry winter day. Conversely, inductors are capable of storing energy when a current is passing through them, causing a magnetic field to form. This phenomenon can be demonstrated by placing a needle compass in the vicinity of a current. The current causes a magnetic field whose energy deflects the compass needle.

A very important circuit, which employs a capacitor in a vital role, is also introduced. This circuit, known as an op-amp integrator, produces an output voltage that is proportional to the integral of the input voltage. The significance of this circuit is that any system (for example, electrical, mechanical, hydraulic, biological, social, economic, and so on) that can be described by a set of linear differential equations with constant coefficients can be modeled by a network consisting of op-amp integrators. Thus, very complex and costly systems can be tested safely and inexpensively prior to construction and implementation.

Finally, we examine some practical circuits where capacitors and inductors are normally found or can be effectively used in circuit design. ●

6.1 Capacitors

A *capacitor* is a circuit element that consists of two conducting surfaces separated by a non-conducting, or *dielectric*, material. A simplified capacitor and its electrical symbol are shown in Fig. 6.1.

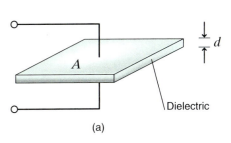

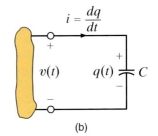

$$i = \frac{dq}{dt}$$

HINT
Note the use of the passive sign convention.

Figure 6.1
Capacitor and its electrical symbol.

(a)   (b)

There are many different kinds of capacitors, and they are categorized by the type of dielectric material used between the conducting plates. Although any good insulator can serve as a dielectric, each type has characteristics that make it more suitable for particular applications.

For general applications in electronic circuits (e.g., coupling between stages of amplification) the dielectric material may be paper impregnated with oil or wax, mylar, polystyrene, mica, glass, or ceramic.

Ceramic dielectric capacitors constructed of barium titanates have a large capacitance-to-volume ratio because of their high dielectric constant. Mica, glass, and ceramic dielectric capacitors will operate satisfactorily at high frequencies.

Aluminum electrolytic capacitors, which consist of a pair of aluminum plates separated by a moistened borax paste electrolyte, can provide high values of capacitance in small volumes. They are typically used for filtering, bypassing, and coupling, and in power supplies and motor-starting applications. Tantalum electrolytic capacitors have lower losses and more stable characteristics than those of aluminum electrolytic capacitors. Figure 6.2 shows a variety of typical discrete capacitors.

In addition to these capacitors, which we deliberately insert in a network for specific applications, stray capacitance is present any time there is a difference in potential between two conducting materials separated by a dielectric. Because this stray capacitance can cause unwanted coupling between circuits, extreme care must be exercised in the layout of electronic systems on printed circuit boards.

Figure 6.2
Some typical capacitors.

Figure 6.3
A 100-F double-layer capacitor and a 68,000-μF electrolytic capacitor.

100 F   68,000 μF

Capacitance is measured in coulombs per volt or farads. The unit *farad* (F) is named after Michael Faraday, a famous English physicist. Capacitors may be fixed or variable and typically range from thousands of microfarads (μF) to a few picofarads (pF).

Capacitor technology, initially driven by the modern interest in electric vehicles, is rapidly changing, however. For example, the capacitor on the left in the photograph in Fig. 6.3 is a double-layer capacitor, which is rated at 2.5 V and 100 F. An aluminum electrolytic capacitor, rated at 25 V and 68,000 μF, is shown on the right in this photograph. The electrolytic capacitor can store $0.5 * 6.8 \times 10^{-2} * 25^2 = 21.25$ joules (J). The double-layer capacitor can store $0.5 * 100 * 2.5^2 = 312.5$ J. Let's connect ten of the 100-F capacitors in series for an equivalent 25-V capacitor. The energy stored in this equivalent capacitor is 3125 J. We would need to connect 147 electrolytic capacitors in parallel to store that much energy.

It is interesting to calculate the dimensions of a simple equivalent capacitor consisting of two parallel plates each of area A, separated by a distance d as shown in Fig. 6.1. We learned in basic physics that the capacitance of two parallel plates of area A, separated by distance d, is

$$C = \frac{\varepsilon_o A}{d}$$

where ε_o, the permitivity of free space, is 8.85×10^{-12} F/m. If we assume the plates are separated by a distance in air of the thickness of one sheet of oil-impregnated paper, which is about 1.016×10^{-4} m, then

$$100 \text{ F} = \frac{(8.85 \times 10^{-12})A}{1.016 \times 10^{-4}}$$

$$A = 1.148 \times 10^9 \text{ m}^2$$

and since 1 square mile is equal to 2.59×10^6 square meters, the area is

$$A \approx 443 \text{ square miles}$$

which is the area of a medium-sized city! It would now seem that the double-layer capacitor in the photograph is much more impressive than it originally appeared. This capacitor is actually constructed using a high surface area material such as powdered carbon which is adhered to a metal foil. There are literally millions of pieces of carbon employed to obtain the required surface area.

Suppose now that a source is connected to the capacitor shown in Fig. 6.1; then positive charges will be transferred to one plate and negative charges to the other. The charge on the capacitor is proportional to the voltage across it such that

$$q = Cv \qquad \qquad \textbf{6.1}$$

where C is the proportionality factor known as the capacitance of the element in farads.

The charge differential between the plates creates an electric field that stores energy. Because of the presence of the dielectric, the conduction current that flows in the wires that connect the capacitor to the remainder of the circuit cannot flow internally between the plates. However, via electromagnetic field theory it can be shown that this conduction current is equal to the displacement current that flows between the plates of the capacitor and is present any time that an electric field or voltage varies with time.

Our primary interest is in the current–voltage terminal characteristics of the capacitor. Since the current is

$$i = \frac{dq}{dt}$$

then for a capacitor

$$i = \frac{d}{dt}(Cv)$$

which for constant capacitance is

$$i = C \frac{dv}{dt} \qquad \textbf{6.2}$$

Equation (6.2) can be rewritten as

$$dv = \frac{1}{C} i \, dt$$

Now integrating this expression from $t = -\infty$ to some time t and assuming $v(-\infty) = 0$ yields

$$v(t) = \frac{1}{C} \int_{-\infty}^{t} i(x) \, dx \qquad \textbf{6.3}$$

where $v(t)$ indicates the time dependence of the voltage. Equation (6.3) can be expressed as two integrals, so that

$$v(t) = \frac{1}{C} \int_{-\infty}^{t_0} i(x) \, dx + \frac{1}{C} \int_{t_0}^{t} i(x) \, dx$$
$$= v(t_0) + \frac{1}{C} \int_{t_0}^{t} i(x) \, dx \qquad \textbf{6.4}$$

where $v(t_0)$ is the voltage due to the charge that accumulates on the capacitor from time $t = -\infty$ to time $t = t_0$.

The energy stored in the capacitor can be derived from the power that is delivered to the element. This power is given by the expression

$$p(t) = v(t)i(t) = Cv(t) \frac{dv(t)}{dt} \qquad \textbf{6.5}$$

and hence the energy stored in the electric field is

$$w_C(t) = \int_{-\infty}^{t} Cv(x) \frac{dv(x)}{dx} \, dx = C \int_{-\infty}^{t} v(x) \frac{dv(x)}{dx} \, dx$$
$$= C \int_{v(-\infty)}^{v(t)} v(x) \, dv(x) = \frac{1}{2} Cv^2(x) \Big|_{v(-\infty)}^{v(t)}$$

$$= \frac{1}{2} Cv^2(t) \text{ J} \qquad \textbf{6.6}$$

since $v(t = -\infty) = 0$. The expression for the energy can also be written using Eq. (6.1) as

$$w_C(t) = \frac{1}{2} \frac{q^2(t)}{C} \qquad \textbf{6.7}$$

Equations (6.6) and (6.7) represent the energy stored by the capacitor, which, in turn, is equal to the work done by the source to charge the capacitor.

Now let's consider the case of a dc voltage applied across a capacitor. From Eq. (6.2), we see that the current flowing through the capacitor is directly proportional to the time rate of change of the voltage across the capacitor. A dc voltage does not vary with time, so the current flowing through the capacitor is zero. We can say that a capacitor "is an open circuit to dc" or "blocks dc." Capacitors are often utilized to remove or filter out an unwanted dc voltage. In analyzing a circuit containing dc voltage sources and capacitors, we can replace the capacitors with an open circuit and calculate voltages and currents in the circuit using our many analysis tools.

Note that the power absorbed by a capacitor, given by Eq. (6.5), is directly proportional to the time rate of change of the voltage across the capacitor. What if we had an instantaneous change in the capacitor voltage? This would correspond to $dv/dt = \infty$ and infinite power. Back in Chapter 1, we ruled out the possibility of any sources of infinite power. Since we only have finite power sources, the voltage across a capacitor cannot change instantaneously. This will be a particularly helpful idea in the next chapter when we encounter circuits containing switches. This idea of "continuity of voltage" for a capacitor tells us that the voltage across the capacitor just after a switch moves is the same as the voltage across the capacitor just before that switch moves.

The polarity of the voltage across a capacitor being charged is shown in Fig. 6.1b. In the ideal case, the capacitor will hold the charge for an indefinite period of time if the source is removed. If at some later time an energy-absorbing device (e.g., a flash bulb) is connected across the capacitor, a discharge current will flow from the capacitor and, therefore, the capacitor will supply its stored energy to the device.

Example 6.1

If the charge accumulated on two parallel conductors charged to 12 V is 600 pC, what is the capacitance of the parallel conductors?

SOLUTION Using Eq. (6.1), we find that

$$C = \frac{Q}{V} = \frac{(600)(10^{-12})}{12} = 50 \text{ pF}$$

Example 6.2

The voltage across a 5-μF capacitor has the waveform shown in Fig. 6.4a. Determine the current waveform.

SOLUTION Note that

$$v(t) = \frac{24}{6 \times 10^{-3}} t \qquad 0 \le t \le 6 \text{ ms}$$
$$= \frac{-24}{2 \times 10^{-3}} t + 96 \qquad 6 \le t < 8 \text{ ms}$$
$$= 0 \qquad 8 \text{ ms} \le t$$

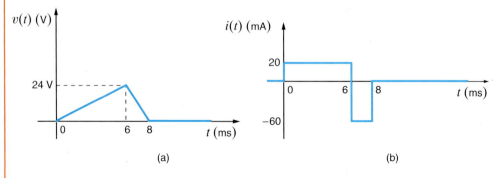

(a)   (b)

Figure 6.4 Voltage and current waveforms for a 5-μF capacitor.

Using Eq. (6.2), we find that

$$i(t) = C\frac{dv(t)}{dt}$$

$$= 5 \times 10^{-6}(4 \times 10^3) \qquad 0 \le t \le 6 \text{ ms}$$

$$= 20 \text{ mA} \qquad 0 \le t \le 6 \text{ ms}$$

$$i(t) = 5 \times 10^{-6}(-12 \times 10^3) \qquad 6 \le t \le 8 \text{ ms}$$

$$= -60 \text{ mA} \qquad 6 \le t < 8 \text{ ms}$$

and

$$i(t) = 0 \qquad 8 \text{ ms} \le t$$

Therefore, the current waveform is as shown in Fig. 6.4b and $i(t) = 0$ for $t > 8$ ms.

Example 6.3

Determine the energy stored in the electric field of the capacitor in Example 6.2 at $t = 6$ ms.

SOLUTION   Using Eq. (6.6), we have

$$w(t) = \frac{1}{2}Cv^2(t)$$

At $t = 6$ ms,

$$w(6 \text{ ms}) = \frac{1}{2}(5 \times 10^{-6})(24)^2$$

$$= 1440 \text{ μJ}$$

LEARNING EXTENSION

E6.1 A 10-μF capacitor has an accumulated charge of 500 nC. Determine the voltage across   **ANSWER:** 0.05 V.
the capacitor.

Example 6.4

The current in an initially uncharged 4-μF capacitor is shown in Fig. 6.5a. Let us derive the waveforms for the voltage, power, and energy and compute the energy stored in the electric field of the capacitor at $t = 2$ ms.

SOLUTION   The equations for the current waveform in the specific time intervals are

$$i(t) = \frac{16 \times 10^{-6}t}{2 \times 10^{-3}} \qquad 0 \le t \le 2 \text{ ms}$$

$$= -8 \times 10^{-6} \qquad 2 \text{ ms} \le t \le 4 \text{ ms}$$

$$= 0 \qquad 4 \text{ ms} < t$$

Since $v(0) = 0$, the equation for $v(t)$ in the time interval $0 \le t \le 2$ ms is

$$v(t) = \frac{1}{(4)(10^{-6})}\int_0^t 8(10^{-3})x \, dx = 10^3 t^2$$

and hence,

$$v(2 \text{ ms}) = 10^3(2 \times 10^{-3})^2 = 4 \text{ mV}$$

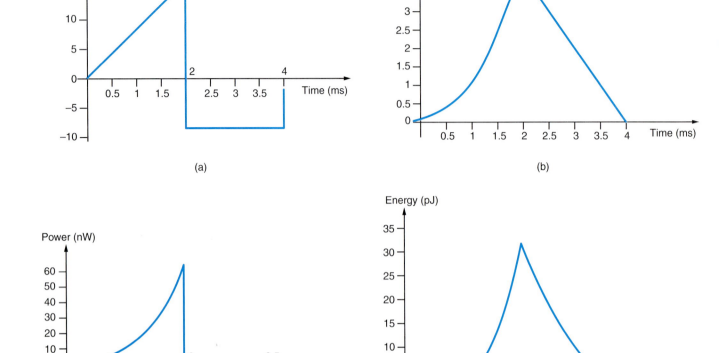

(a)

(b)

(c)

(d)

Figure 6.5 *Waveforms used in Example 6.4.*

In the time interval 2 ms $\leq t \leq$ 4 ms,

$$v(t) = \frac{1}{(4)(10^{-6})} \int_{2(10^{-3})}^{t} -(8)(10^{-6})dx + (4)(10^{-3})$$

$$= -2t + 8 \times 10^{-3}$$

The waveform for the voltage is shown in Fig. 6.5b.

Since the power is $p(t) = v(t)i(t)$, the expression for the power in the time interval $0 \leq t \leq 2$ ms is $p(t) = 8t^3$. In the time interval 2 ms $\leq t \leq 4$ ms, the equation for the power is

$$p(t) = -(8)(10^{-6})(-2t + 8 \times 10^{-3})$$

$$= 16(10^{-6})t - 64(10^{-9})$$

The power waveform is shown in Fig. 6.5c. Note that during the time interval $0 \leq t \leq 2$ ms, the capacitor is absorbing energy and during the interval 2 ms $\leq t \leq 4$ ms, it is delivering energy.

The energy is given by the expression

$$w(t) = \int_{t_0}^{t} p(x) \, dx + w(t_0)$$

In the time interval $0 \leq t \leq 2$ ms,

$$w(t) = \int_0^t 8x^3 \, dx = 2t^4$$

Hence,

$$w(2 \text{ ms}) = 32 \text{ pJ}$$

In the time interval $2 \leq t \leq 4$ ms,

$$w(t) = \int_{2\times10^{-3}}^t \left[(16 \times 10^{-6})x - (64 \times 10^{-9})\right] dx + 32 \times 10^{-12}$$

$$= \left[(8 \times 10^{-6})x^2 - (64 \times 10^{-9})x\right]_{2\times10^{-3}}^t + 32 \times 10^{-12}$$

$$= (8 \times 10^{-6})t^2 - (64 \times 10^{-9})t + 128 \times 10^{-12}$$

From this expression we find that $w(2 \text{ ms}) = 32$ pJ and $w(4 \text{ ms}) = 0$. The energy waveform is shown in Fig. 6.5d.

LEARNING EXTENSIONS

E6.2 The voltage across a 2-μF capacitor is shown in Fig. E6.2. Determine the waveform for the capacitor current. **PSV**

ANSWER:

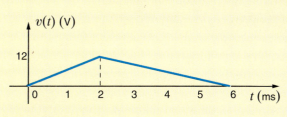

Figure E6.2

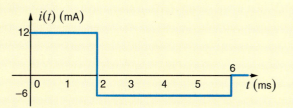

E6.3 Compute the energy stored in the electric field of the capacitor in Extension Exercise E6.2 at $t = 2$ ms.

ANSWER: $w = 144$ μJ.

6.2 Inductors

An *inductor* is a circuit element that consists of a conducting wire usually in the form of a coil. Two typical inductors and their electrical symbols are shown in Fig. 6.6. Inductors are typically categorized by the type of core on which they are wound. For example, the core material may be air or any nonmagnetic material, iron, or ferrite. Inductors made with air or nonmagnetic materials are widely used in radio, television, and filter circuits. Iron-core inductors are used in electrical power supplies and filters. Ferrite-core inductors are widely used in high-frequency applications. Note that in contrast to the magnetic core that confines the flux, as shown in Fig. 6.6b, the flux lines for nonmagnetic inductors extend beyond the inductor itself, as illustrated in Fig. 6.6a. Like stray capacitance, stray inductance can result from any element carrying current surrounded by flux linkages. Figure 6.7 shows a variety of typical inductors.

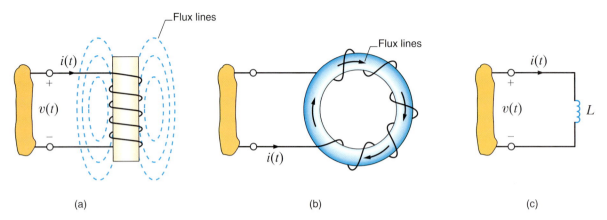

(a)   (b)   (c)

Figure 6.6
*Two inductors and their
electrical symbol.*

Figure 6.7
*Some typical
inductors.*

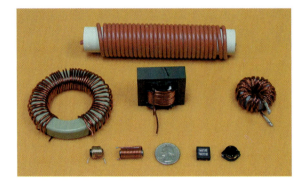

HINT

Note the use of the passive
sign convention, as
illustrated in Fig. 6.6.

From a historical standpoint, developments that led to the mathematical model we employ to represent the inductor are as follows. It was first shown that a current-carrying conductor would produce a magnetic field. It was later found that the magnetic field and the current that produced it were linearly related. Finally, it was shown that a changing magnetic field produced a voltage that was proportional to the time rate of change of the current that produced the magnetic field; that is,

$$v(t) = L \frac{di(t)}{dt} \qquad \textbf{6.8}$$

The constant of proportionality L is called the inductance and is measured in the unit *henry*, named after the American inventor Joseph Henry, who discovered the relationship. As seen in Eq. (6.8), 1 henry (H) is dimensionally equal to 1 volt-second per ampere.

Following the development of the mathematical equations for the capacitor, we find that the expression for the current in an inductor is

$$i(t) = \frac{1}{L} \int_{-\infty}^{t} v(x)\,dx \qquad \textbf{6.9}$$

which can also be written as

$$i(t) = i(t_0) + \frac{1}{L} \int_{t_0}^{t} v(x)\,dx \qquad \textbf{6.10}$$

The power delivered to the inductor can be used to derive the energy stored in the element.

This power is equal to

$$p(t) = v(t)i(t)$$

$$= \left[L \frac{di(t)}{dt} \right] i(t) \qquad \textbf{6.11}$$

Therefore, the energy stored in the magnetic field is

$$w_L(t) = \int_{-\infty}^{t} \left[L \frac{di(x)}{dx} \right] i(x)\, dx$$

Following the development of Eq. (6.6), we obtain

$$w_L(t) = \frac{1}{2} L i^2(t) \text{ J} \qquad \textbf{6.12}$$

Now let's consider the case of a dc current flowing through an inductor. From Eq. (6.8), we see that the voltage across the inductor is directly proportional to the time rate of change of the current flowing through the inductor. A dc current does not vary with time, so the voltage across the inductor is zero. We can say that an inductor "is a short circuit to dc." In analyzing a circuit containing dc sources and inductors, we can replace any inductors with short circuits and calculate voltages and currents in the circuit using our many analysis tools.

Note from Eq. (6.11) that an instantaneous change in inductor current would require infinite power. Since we don't have any infinite power sources, the current flowing through an inductor cannot change instantaneously. This will be a particularly helpful idea in the next chapter when we encounter circuits containing switches. This idea of "continuity of current" for an inductor tells us that the current flowing through an inductor just after a switch moves is the same as the current flowing through an inductor just before that switch moves.

Example 6.5

Find the total energy stored in the circuit of Fig. 6.8a.

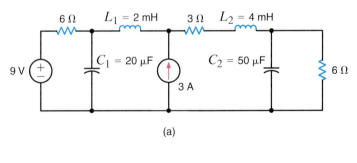

(a)

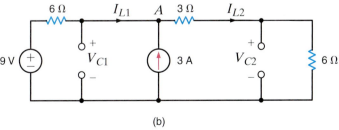

(b)

Figure 6.8
Circuits used in Example 6.5.

SOLUTION This circuit has only dc sources. Based on our earlier discussions about capacitors and inductors and constant sources, we can replace the capacitors with open circuits and the inductors with short circuits. The resulting circuit is shown in Fig. 6.8b.

This resistive circuit can now be solved using any of the techniques we have learned in earlier chapters. If we apply KCL at node A, we get

$$I_{L2} = I_{L1} + 3$$

Applying KVL around the outside of the circuit yields

$$4I_{L1} + 3I_{L2} + 6I_{L2} = 9$$

Solving these equations yields $I_{L1} = -1.2$ A and $I_{L2} = 1.8$ A. The voltages V_{C1} and V_{C2} can be calculated from the currents.

$$V_{C1} = -6I_{L1} + 9 = 16.2 \text{ V}$$
$$V_{C2} = 6I_{L2} = 6(1.8) = 10.8 \text{ V}$$

The total energy stored in the circuit is the sum of the energy stored in the two inductors and two capacitors.

$$w_{L1} = \frac{1}{2}(2 \times 10^{-3})(-1.2)^2 = 1.44 \text{ mJ}$$

$$w_{L2} = \frac{1}{2}(4 \times 10^{-3})(1.8)^2 = 6.48 \text{ mJ}$$

$$w_{C1} = \frac{1}{2}(20 \times 10^{-6})(16.2)^2 = 2.62 \text{ mJ}$$

$$w_{C2} = \frac{1}{2}(50 \times 10^{-6})(10.8)^2 = 2.92 \text{ mJ}$$

The total stored energy is 13.46 mJ.

The inductor, like the resistor and capacitor, is a passive element. The polarity of the voltage across the inductor is shown in Fig. 6.6.

Practical inductors typically range from a few microhenrys to tens of henrys. From a circuit design standpoint it is important to note that inductors cannot be easily fabricated on an integrated circuit chip, and therefore chip designs typically employ only active electronic devices, resistors, and capacitors that can be easily fabricated in microcircuit form.

Example 6.6

The current in a 10-mH inductor has the waveform shown in Fig. 6.9a. Determine the voltage waveform.

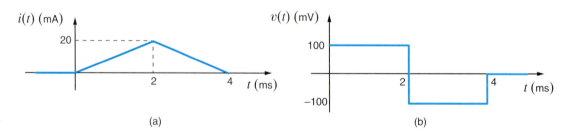

(a)   (b)

Figure 6.9 Current and voltage waveforms for a 10-mH inductor.

SOLUTION Using Eq. (6.8) and noting that

$$i(t) = \frac{20 \times 10^{-3}t}{2 \times 10^{-3}} \qquad 0 \le t \le 2 \text{ ms}$$

$$i(t) = \frac{-20 \times 10^{-3}t}{2 \times 10^{-3}} + 40 \times 10^{-3} \qquad 2 \le t \le 4 \text{ ms}$$

and

$$i(t) = 0 \qquad 4 \text{ ms} < t$$

we find that

$$v(t) = \left(10 \times 10^{-3}\right) \frac{20 \times 10^{-3}}{2 \times 10^{-3}} \qquad 0 \le t \le 2 \text{ ms}$$

$$= 100 \text{ mV}$$

and

$$v(t) = \left(10 \times 10^{-3}\right) \frac{-20 \times 10^{-3}}{2 \times 10^{-3}} \qquad 2 \le t \le 4 \text{ ms}$$

$$= -100 \text{ mV}$$

and $v(t) = 0$ for $t > 4$ ms. Therefore, the voltage waveform is shown in Fig. 6.9b.

Example 6.7

The current in a 2-mH inductor is

$$i(t) = 2 \sin 377t \text{ A}$$

Determine the voltage across the inductor and the energy stored in the inductor.

SOLUTION   From Eq. (6.8), we have

$$v(t) = L \frac{di(t)}{dt}$$

$$= \left(2 \times 10^{-3}\right) \frac{d}{dt} \left(2 \sin 377t\right)$$

$$= 1.508 \cos 377t \text{ V}$$

and from Eq. (6.12),

$$w_L(t) = \frac{1}{2} L i^2(t)$$

$$= \frac{1}{2} \left(2 \times 10^{-3}\right) \left(2 \sin 377t\right)^2$$

$$= 0.004 \sin^2 377t \text{ J}$$

Example 6.8

The voltage across a 200-mH inductor is given by the expression

$$v(t) = (1 - 3t)e^{-3t} \text{mV} \quad t \ge 0$$

$$= 0 \qquad\qquad t < 0$$

Let us derive the waveforms for the current, energy, and power.

SOLUTION   The waveform for the voltage is shown in Fig. 6.10a. The current is derived from Eq. (6.10) as

$$i(t) = \frac{10^3}{200} \int_0^t (1 - 3x)e^{-3x} \, dx$$

$$= 5 \left\{ \int_0^t e^{-3x} \, dx - 3 \int_0^t x e^{-3x} \, dx \right\}$$

$$= 5 \left\{ \frac{e^{-3x}}{-3} \Big|_0^t - 3 \left[-\frac{e^{-3x}}{9} (3x + 1) \right]_0^t \right\}$$

$$= 5t e^{-3t} \text{ mA} \quad t \ge 0$$

$$= 0 \qquad\qquad t < 0$$

A plot of the current waveform is shown in Fig. 6.10b.
The power is given by the expression

$$p(t) = v(t)i(t)$$
$$= 5t(1 - 3t)e^{-6t}\ \mu W \qquad t \geq 0$$
$$= 0 \qquad\qquad t < 0$$

The equation for the power is plotted in Fig. 6.10c.
The expression for the energy is

$$w(t) = \frac{1}{2}Li^2(t)$$
$$= 2.5t^2e^{-6t}\ \mu J \qquad t \geq 0$$
$$= 0 \qquad\qquad t < 0$$

This equation is plotted in Fig. 6.10d.

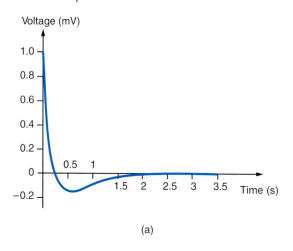

(a)

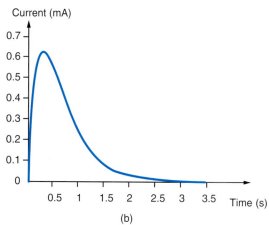

(b)

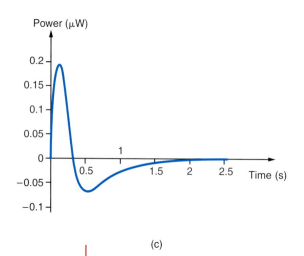

(c)

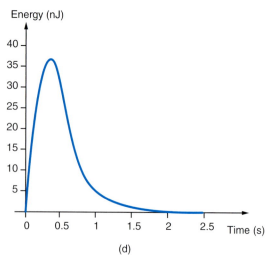

(d)

Figure 6.10 *Waveforms used in Example 6.8.*

LEARNING EXTENSIONS

E6.4 The current in a 5-mH inductor has the waveform shown in Fig. E6.4. Compute the waveform for the inductor voltage. **PSV**

ANSWER:

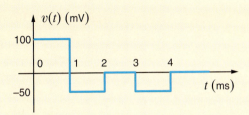

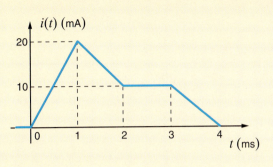

Figure E6.4

E6.5 Compute the energy stored in the magnetic field of the inductor in Extension Exercise E6.4 at $t = 1.5$ ms.

ANSWER: $W = 562.5$ nJ.

CAPACITOR AND INDUCTOR SPECIFICATIONS

There are a couple of important parameters that are used to specify capacitors and inductors. In the case of capacitors, the capacitance value, working voltage, and tolerance are issues that must be considered in their application. Standard capacitor values range from a few pF to about 50 mF. Capacitors larger than 1 F are available but will not be discussed here. Table 6.1 is a list of standard capacitor values, which are typically given in picofarads or microfarads. Although both smaller and larger ratings are available, the standard working voltage, or dc voltage rating, is typically between 6.3 V and 500 V. Manufacturers specify this working voltage since it is critical to keep the applied voltage below the breakdown point of the dielectric. Tolerance is an adjunct to the capacitance value and is usually listed as a percentage of the nominal value. Standard tolerance values are $\pm 5\%$, $\pm 10\%$, and $\pm 20\%$. Occasionally, tolerances for single-digit pF capacitors are listed in pF. For example, 5 pF $\pm$ 0.25 pF.

Table 6.1 Standard capacitor values

pF	pF	pF	pF	μF	μF	μF	μF	μF	μF	μF
1	10	100	1000	0.010	0.10	1.0	10	100	1000	10,000
	12	120	1200	0.012	0.12	1.2	12	120	1200	12,000
1.5	15	150	1500	0.015	0.15	1.5	15	150	1500	15,000
	18	180	1800	0.018	0.18	1.8	18	180	1800	18,000
2	20	200	2000	0.020	0.20	2.0	20	200	2000	20,000
	22	220	2200	0.022	0.22	2.2	22	220	2200	22,000
	27	270	2700	0.027	0.27	2.7	27	270	2700	27,000
3	33	330	3300	0.033	0.33	3.3	33	330	3300	33,000
4	39	390	3900	0.039	0.39	3.9	39	390	3900	39,000
5	47	470	4700	0.047	0.47	4.7	47	470	4700	47,000
6	51	510	5100	0.051	0.51	5.1	51	510	5100	51,000
7	56	560	5600	0.056	0.56	5.6	56	560	5600	56,000
8	68	680	6800	0.068	0.68	6.8	68	680	6800	68,000
9	82	820	8200	0.082	0.82	8.2	82	820	8200	82,000

There are two principal inductor specifications, inductance and resistance. Standard commercial inductances range from about 1 nH to around 100 mH. Larger inductances can, of course, be custom built for a price. Table 6.2 lists the standard inductor values. The current rating for inductors typically extends from a few dozen mA's to about 1 A. Tolerances are typically 5% or 10% of the specified value.

Table 6.2 Standard inductor values

nH	nH	nH	μH	μH	μH	mH	mH	mH
1	10	100	1.0	10	100	1.0	10	100
1.2	12	120	1.2	12	120	1.2	12	
1.5	15	150	1.5	15	150	1.5	15	
1.8	18	180	1.8	18	180	1.8	18	
2	20	200	2.0	20	200	2.0	20	
2.2	22	220	2.2	22	220	2.2	22	
2.7	27	270	2.7	27	270	2.7	27	
3	33	330	3.3	33	330	3.3	33	
4	39	390	3.9	39	390	3.9	39	
5	47	470	4.7	47	470	4.7	47	
6	51	510	5.1	51	510	5.1	51	
7	56	560	5.6	56	560	5.6	56	
8	68	680	6.8	68	680	6.8	68	
9	82	820	8.2	82	820	8.2	82	

As indicated in Chapter 2, wire-wound resistors are simply coils of wire, and therefore it is only logical that inductors will have some resistance. The major difference between wire-wound resistors and inductors is the wire material. High-resistance materials such as Nichrome are used in resistors, and low-resistance copper is used in inductors. The resistance of the copper wire is dependent on the length and diameter of the wire. Table 6.3 lists the American Wire Gauge (AWG) standard wire diameters and the resulting resistance per foot for copper wire.

Table 6.3 Resistance per foot of solid copper wire

AWG No.	Diameter (in.)	$m\Omega/ft$
12	0.0808	1.59
14	0.0641	2.54
16	0.0508	4.06
18	0.0400	6.50
20	0.0320	10.4
22	0.0253	16.5
24	0.0201	26.2
26	0.0159	41.6
28	0.0126	66.2
30	0.0100	105
32	0.0080	167
34	0.0063	267
36	0.0049	428
38	0.0039	684
40	0.0031	1094

Example 6.9

We wish to find the possible range of capacitance values for a 51-mF capacitor that has a tolerance of 20%.

SOLUTION The minimum capacitor value is $0.8C = 40.8$ mF, and the maximum capacitor value is $1.2C = 61.2$ mF.

Example 6.10

The capacitor in Fig. 6.11a is a 100-nF capacitor with a tolerance of 20%. If the voltage waveform is as shown in Fig. 6.11b, let us graph the current waveform for the minimum and maximum capacitor values.

SOLUTION The maximum capacitor value is $1.2C = 120$ nF, and the minimum capacitor value is $0.8C = 80$ nF. The maximum and minimum capacitor currents, obtained from the equation

$$i(t) = C\frac{dv(t)}{dt}$$

are shown in Fig. 6.11c.

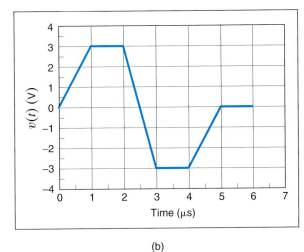

Figure 6.11
Circuit and graphs used in Example 6.10.

(a)

(b)

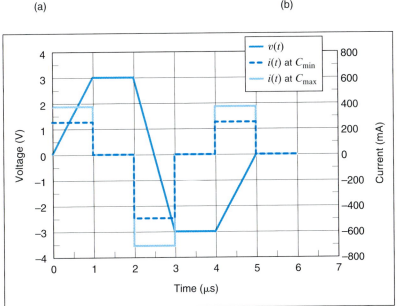

(c)

Example 6.11

The inductor in Fig. 6.12a is a 100-µH inductor with a tolerance of 10%. If the current waveform is as shown in Fig. 6.12b, let us graph the voltage waveform for the minimum and maximum inductor values.

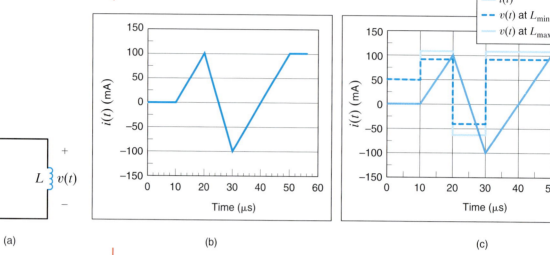

(a) (b) (c)

Figure 6.12 *Circuit and graphs used in Example 6.11.*

SOLUTION The maximum inductor value is $1.1L = 110$ µH, and the minimum inductor value is $0.9L = 90$ µH. The maximum and minimum inductor voltages, obtained from the equation

$$v(t) = L \frac{di(t)}{dt}$$

are shown in Fig. 6.12c.

6.3 Capacitor and Inductor Combinations

SERIES CAPACITORS If a number of capacitors are connected in series, their equivalent capacitance can be calculated using KVL. Consider the circuit shown in Fig. 6.13a. For this circuit

$$v(t) = v_1(t) + v_2(t) + v_3(t) + \cdots + v_N(t) \qquad \textbf{6.13}$$

but

$$v_i(t) = \frac{1}{C_i} \int_{t_0}^{t} i(t)\, dt + v_i(t_0) \qquad \textbf{6.14}$$

Figure 6.13
Equivalent circuit for N series-connected capacitors.

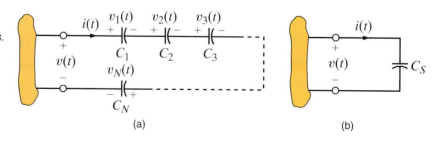

(a) (b)

Therefore, Eq. (6.13) can be written as follows using Eq. (6.14):

$$v(t) = \left(\sum_{i=1}^{N} \frac{1}{C_i} \right) \int_{t_0}^{t} i(t)\, dt + \sum_{i=1}^{N} v_i(t_0) \qquad \textbf{6.15}$$

$$= \frac{1}{C_S} \int_{t_0}^{t} i(t)\, dt + v(t_0) \qquad \textbf{6.16}$$

where

$$v(t_0) = \sum_{i=1}^{N} v_i(t_0)$$

and

$$\frac{1}{C_S} = \sum_{i=1}^{N} \frac{1}{C_i} = \frac{1}{C_1} + \frac{1}{C_2} + \cdots + \frac{1}{C_N} \qquad \textbf{6.17}$$

HINT

Capacitors in series combine like resistors in parallel.

Thus, the circuit in Fig. 6.13b is equivalent to that in Fig. 6.13a under the conditions stated previously.

It is also important to note that since the same current flows in each of the series capacitors, each capacitor gains the same charge in the same time period. The voltage across each capacitor will depend on this charge and the capacitance of the element.

Example 6.12

Determine the equivalent capacitance and the initial voltage for the circuit shown in Fig. 6.14.

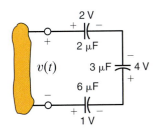

Figure 6.14
Circuit containing multiple capacitors with initial voltages.

SOLUTION   Note that these capacitors must have been charged before they were connected in series or else the charge of each would be equal and the voltages would be in the same direction.

The equivalent capacitance is

$$\frac{1}{C_S} = \frac{1}{2} + \frac{1}{3} + \frac{1}{6}$$

where all capacitance values are in microfarads.

Therefore, $C_S = 1\ \mu\text{F}$ and, as seen from the figure, $v(t_0) = -3$ V. Note that the total energy stored in the circuit is

$$w(t_0) = \frac{1}{2}\left[2 \times 10^{-6}(2)^2 + 3 \times 10^{-6}(-4)^2 + 6 \times 10^{-6}(-1)^2\right]$$

$$= 31\ \mu\text{J}$$

However, the energy recoverable at the terminals is

$$w_C(t_0) = \frac{1}{2} C_S v^2(t)$$

$$= \frac{1}{2}\left[1 \times 10^{-6}(-3)^2\right]$$

$$= 4.5\ \mu\text{J}$$

Example 6.13

Two previously uncharged capacitors are connected in series and then charged with a 12-V source. One capacitor is 30 μF and the other is unknown. If the voltage across the 30-μF capacitor is 8 V, find the capacitance of the unknown capacitor.

SOLUTION   The charge on the 30-μF capacitor is

$$Q = CV = (30 \text{ μF})(8 \text{ V}) = 240 \text{ μC}$$

Since the same current flows in each of the series capacitors, each capacitor gains the same charge in the same time period.

$$C = \frac{Q}{V} = \frac{240 \text{ μC}}{4 \text{V}} = 60 \text{ μF}$$

PARALLEL CAPACITORS   To determine the equivalent capacitance of N capacitors connected in parallel, we employ KCL. As can be seen from Fig. 6.15a,

$$i(t) = i_1(t) + i_2(t) + i_3(t) + \cdots + i_N(t) \qquad \textbf{6.18}$$

$$= C_1 \frac{dv(t)}{dt} + C_2 \frac{dv(t)}{dt} + C_3 \frac{dv(t)}{dt} + \cdots + C_N \frac{dv(t)}{dt}$$

$$= \left(\sum_{i=1}^{N} C_i \right) \frac{dv(t)}{dt}$$

$$= C_p \frac{dv(t)}{dt} \qquad \textbf{6.19}$$

where

$$C_p = C_1 + C_2 + C_3 + \cdots + C_N \qquad \textbf{6.20}$$

HINT

Capacitors in parallel combine like resistors in series.

Figure 6.15
Equivalent circuit for N capacitors connected in parallel.

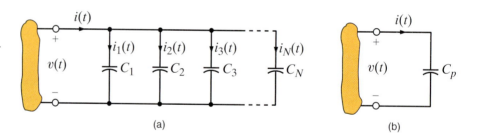

(a)   (b)

Example 6.14

Determine the equivalent capacitance at terminals A-B of the circuit shown in Fig. 6.16.

SOLUTION $C_p = 15 \text{ μF}$

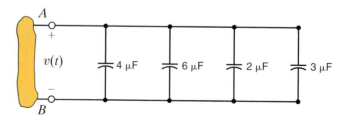

Figure 6.16 *Circuit containing multiple capacitors in parallel.*

E6.6 Two initially uncharged capacitors are connected as shown in Fig. E6.6. After a period of time, the voltage reaches the value shown. Determine the value of C_1.

ANSWER: $C_1 = 4\ \mu\text{F}$.

Figure E6.6

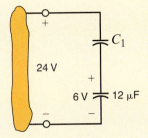

E6.7 Compute the equivalent capacitance of the network in Fig. E6.7.

ANSWER: $C_{eq} = 1.5\ \mu\text{F}$.

Figure E6.7

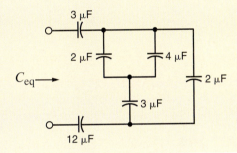

SERIES INDUCTORS   If N inductors are connected in series, the equivalent inductance of the combination can be determined as follows. Referring to Fig. 6.17a and using KVL, we see that

$$v(t) = v_1(t) + v_2(t) + v_3(t) + \cdots + v_N(t) \qquad \textbf{6.21}$$

and therefore,

$$v(t) = L_1 \frac{di(t)}{dt} + L_2 \frac{di(t)}{dt} + L_3 \frac{di(t)}{dt} + \cdots + L_N \frac{di(t)}{dt} \qquad \textbf{6.22}$$

$$= \left(\sum_{i=1}^{N} L_i \right) \frac{di(t)}{dt}$$

$$= L_S \frac{di(t)}{dt} \qquad \textbf{6.23}$$

where

$$L_S = \sum_{i=1}^{N} L_i = L_1 + L_2 + \cdots + L_N \qquad \textbf{6.24}$$

> **HINT**
>
> Inductors in series combine like resistors in series.

Therefore, under this condition the network in Fig. 6.17b is equivalent to that in Fig. 6.17a.

Figure 6.17
Equivalent circuit for N series-connected inductors.

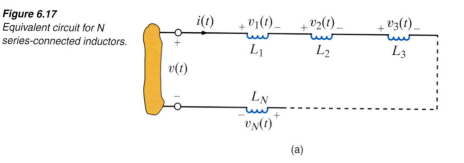

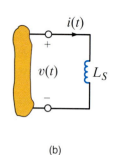

(a)　　　　　　　　　　　　　　　　　(b)

Example 6.15

Find the equivalent inductance of the circuit shown in Fig. 6.18.

Figure 6.18
Circuit containing multiple inductors.

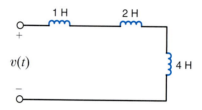

SOLUTION　　The equivalent inductance of the circuit shown in Fig. 6.18 is

$$L_S = 1H + 2H + 4H$$
$$= 7H$$

PARALLEL INDUCTORS　　Consider the circuit shown in Fig. 6.19a, which contains N parallel inductors. Using KCL, we can write

$$i(t) = i_1(t) + i_2(t) + i_3(t) + \cdots + i_N(t) \qquad \textbf{6.25}$$

However,

$$i_j(t) = \frac{1}{L_j} \int_{t_0}^{t} v(x)\,dx + i_j(t_0) \qquad \textbf{6.26}$$

Substituting this expression into Eq. (6.25) yields

$$i(t) = \left(\sum_{j=1}^{N} \frac{1}{L_j} \right) \int_{t_0}^{t} v(x)\,dx + \sum_{j=1}^{N} i_j(t_0) \qquad \textbf{6.27}$$

$$= \frac{1}{L_p} \int_{t_0}^{t} v(x)\,dx + i(t_0) \qquad \textbf{6.28}$$

where

$$\frac{1}{L_p} = \frac{1}{L_1} + \frac{1}{L_2} + \frac{1}{L_3} + \cdots + \frac{1}{L_N} \qquad \textbf{6.29}$$

and $i(t_0)$ is equal to the current in L_p at $t = t_0$. Thus, the circuit in Fig. 6.19b is equivalent to that in Fig. 6.19a under the conditions stated previously.

Figure 6.19
Equivalent circuits for N inductors connected in parallel.

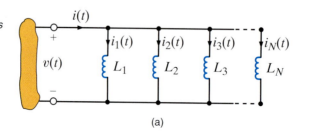

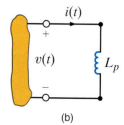

(a)　　　　　　　　　　　　　　　　　(b)

Example 6.16

Determine the equivalent inductance and the initial current for the circuit shown in Fig. 6.20.

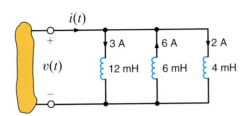

Figure 6.20
Circuit containing multiple inductors with initial currents.

SOLUTION   The equivalent inductance is

$$\frac{1}{L_p} = \frac{1}{12} + \frac{1}{6} + \frac{1}{4}$$

where all inductance values are in millihenrys

$$L_p = 2 \text{ mH}$$

and the initial current is $i(t_0) = -1$ A.

The previous material indicates that capacitors combine like conductances, whereas inductances combine like resistances.

LEARNING EXTENSION

E6.8 Determine the equivalent inductance of the network in Fig. E6.8 if all inductors are 6 mH.   **ANSWER:** 9.429 mH.

Figure E6.8

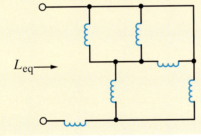

$L_{eq} \longrightarrow$

6.4   *RC* Operational Amplifier Circuits

Two very important *RC* op-amp circuits are the differentiator and the integrator. These circuits are derived from the circuit for an inverting op-amp by replacing the resistors R_1 and R_2, respectively, by a capacitor. Consider, for example, the circuit shown in Fig. 6.21a. The circuit equations are

$$C_1 \frac{d}{dt}(v_1 - v_-) + \frac{v_o - v_-}{R_2} = i_-$$

HINT

The properties of the ideal op-amp are $v_+ = v_-$ and $i_+ = i_- = 0$.

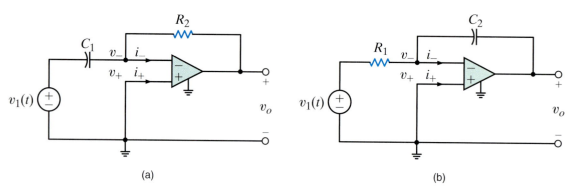

Figure 6.21
Differentiator and integrator operational amplifier circuits.

However, $v_- = 0$ and $i_- = 0$. Therefore,

$$v_o(t) = -R_2 C_1 \frac{dv_1(t)}{dt} \qquad \textbf{6.30}$$

Thus, the output of the op-amp circuit is proportional to the derivative of the input.
The circuit equations for the op-amp configuration in Fig. 6.21b are

$$\frac{v_1 - v_-}{R_1} + C_2 \frac{d}{dt}(v_o - v_-) = i_-$$

but since $v_- = 0$ and $i_- = 0$, the equation reduces to

$$\frac{v_1}{R_1} = -C_2 \frac{dv_o}{dt}$$

or

$$v_o(t) = \frac{-1}{R_1 C_2} \int_{-\infty}^{t} v_1(x)\, dx$$

$$= \frac{-1}{R_1 C_2} \int_{0}^{t} v_1(x)\, dx + v_o(0) \qquad \textbf{6.31}$$

If the capacitor is initially discharged, then $v_o(0) = 0$; hence,

$$v_o(t) = \frac{-1}{R_1 C_2} \int_{0}^{t} v_1(x)\, dx \qquad \textbf{6.32}$$

Thus, the output voltage of the op-amp circuit is proportional to the integral of the input voltage.

Example 6.17

The waveform in Fig. 6.22a is applied at the input of the differentiator circuit shown in Fig. 6.21a. If $R_2 = 1$ kΩ and $C_1 = 2$ μF, determine the waveform at the output of the op-amp.

Figure 6.22
Input and output waveforms for a differentiator circuit.

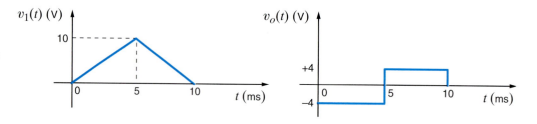

SOLUTION   Using Eq. (6.30), we find that the op-amp output is

$$v_o(t) = -R_2 C_1 \frac{dv_1(t)}{dt}$$

$$= -(2)10^{-3} \frac{dv_1(t)}{dt}$$

$dv_1(t)/dt = (2)10^3$ for $0 \le t < 5$ ms, and therefore,

$$v_o(t) = -4 \text{ V} \qquad 0 \le t < 5 \text{ ms}$$

$dv_1(t)/dt = -(2)10^3$ for $5 \le t < 10$ ms, and therefore,

$$v_o(t) = 4 \text{ V} \qquad 5 \le t < 10 \text{ ms}$$

Hence, the output waveform of the differentiator is shown in Fig. 6.22b.

Example 6.18

If the integrator shown in Fig. 6.21b has the parameters $R_1 = 5$ kΩ and $C_2 = 0.2$ µF, determine the waveform at the op-amp output if the input waveform is given as in Fig. 6.23a and the capacitor is initially discharged.

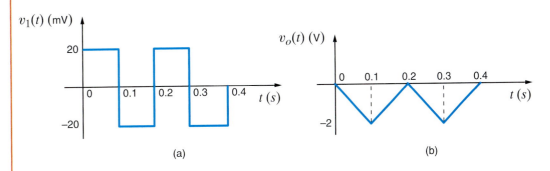

Figure 6.23
Input and output
waveforms for an
integrator circuit.

(a)   (b)

SOLUTION   The integrator output is given by the expression

$$v_o(t) = \frac{-1}{R_1 C_2} \int_0^t v_1(x)\, dx$$

which with the given circuit parameters is

$$v_o(t) = -10^3 \int_0^t v_1(x)\, dx$$

In the interval $0 \le t < 0.1$ s, $v_1(t) = 20$ mV. Hence,

$$v_o(t) = -10^3 (20) 10^{-3} t \qquad 0 \le t < 0.1 \text{ s}$$

$$= -20t$$

At $t = 0.1$ s, $v_o(t) = -2$ V. In the interval from 0.1 to 0.2 s, the integrator produces a positive slope output of $20t$ from $v_o(0.1) = -2$ V to $v_o(0.2) = 0$ V. This waveform from $t = 0$ to $t = 0.2$ s is repeated in the interval $t = 0.2$ to $t = 0.4$ s, and therefore, the output waveform is shown in Fig. 6.23b.

E6.9 The waveform in Fig. E6.9 is applied to the input terminals of the op-amp differentiator circuit. Determine the differentiator output waveform if the op-amp circuit parameters are $C_1 = 2$ F and $R_2 = 2 \Omega$.

ANSWER:

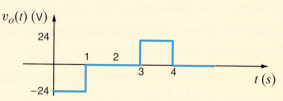

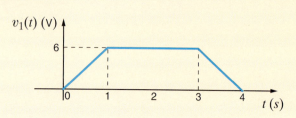

Figure E6.9

6.5 Application Examples

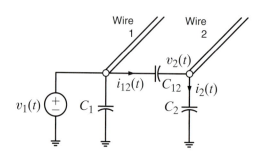

Figure 6.24
SEM Image (Tom Way/Ginger Conly. Courtesy of International Business Machines Corporation. Unauthorized use not permitted.)

Application Example 6.19

In integrated circuits, wires carrying high-speed signals are closely spaced as shown by the micrograph in Fig. 6.24. As a result, a signal on one conductor can "mysteriously" appear on a different conductor. This phenomenon is called crosstalk. Let us examine this condition and propose some methods for reducing it.

SOLUTION   The origin of crosstalk is capacitance. In particular, it is undesired capacitance, often called *parasitic capacitance*, that exists between wires that are closely spaced. The simple model in Fig. 6.25 can be used to investigate crosstalk between two long parallel wires. A signal is applied to wire 1. Capacitances C_1 and C_2 are the parasitic capacitances of the conductors with respect to ground, while C_{12} is the capacitance between the conductors. Recall that we introduced the capacitor as two closely spaced conducting plates. If we stretch those plates into thin wires, certainly the geometry of the conductors would change and thus the amount of capacitance. However, we should still expect some capacitance between the wires.

In order to quantify the level of crosstalk, we want to know how much of the voltage on wire 1 appears on wire 2. A nodal analysis at wire 2 yields

$$i_{12}(t) = C_{12}\left[\frac{dv_1(t)}{dt} - \frac{dv_2(t)}{dt}\right] = i_2(t) = C_2\left[\frac{dv_2(t)}{dt}\right]$$

Solving for $dv_2(t)/dt$, we find that

$$\frac{dv_2(t)}{dt} = \left[\frac{C_{12}}{C_{12} + C_2}\right]\frac{dv_1(t)}{dt}$$

Integrating both sides of this equation yields

$$v_2(t) = \left[\frac{C_{12}}{C_{12} + C_2}\right]v_1(t)$$

Figure 6.25
A simple model for investigating crosstalk.

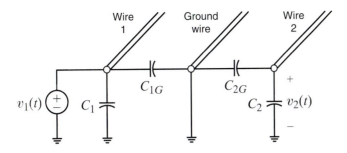

Figure 6.26

Use of a ground wire in the crosstalk model.

Note that it is a simple capacitance ratio that determines how effectively $v_1(t)$ is "coupled" into wire 2. Clearly, ensuring that C_{12} is much less than C_2 is the key to controlling crosstalk. How is this done? First, we can make C_{12} as small as possible by increasing the spacing between wires. Second, we can increase C_2 by putting it closer to the ground wiring. Unfortunately, the first option takes up more real estate, and the second one slows down the voltage signals in wire 1. At this point, we seem to have a typical engineering tradeoff: to improve one criterion, that is, decreased crosstalk, we must sacrifice another, space or speed. One way to address the space issue would be to insert a ground connection between the signal-carrying wires as shown in Fig. 6.26. However, any advantage achieved with grounded wires must be traded off against the increase in space, since inserting grounded wires between adjacent conductors would nearly double the width consumed without them.

Redrawing the circuit in Fig. 6.27 immediately indicates that wires 1 and 2 are now electrically isolated and there should be no crosstalk whatsoever—a situation that is highly unlikely. Thus, we are prompted to ask the question, "Is our model accurate enough to model crosstalk?" A more accurate model for the crosstalk reduction scheme is shown in Fig. 6.28 where the capacitance between signal wires 1 and 2 is no longer ignored. Once again, we will determine the amount of crosstalk by examining the ratio $v_2(t)/v_1(t)$. Employing nodal analysis at wire 2 in the circuit in Fig. 6.29 yields

$$i_{12}(t) = C_{12}\left[\frac{dv_1(t)}{dt} - \frac{dv_2(t)}{dt}\right] = i_2(t) = (C_2 + C_{2G})\left[\frac{dv_2(t)}{dt}\right]$$

Solving for $dv_2(t)/dt$, we obtain

$$\frac{dv_2(t)}{dt} = \left[\frac{C_{12}}{C_{12} + C_2 + C_{2G}}\right]\frac{dv_1(t)}{dt}$$

Integrating both sides of this equation yields

$$v_2(t) = \left[\frac{C_{12}}{C_{12} + C_2 + C_{2G}}\right]v_1(t)$$

Note that this result is very similar to our earlier result with the addition of the C_{2G} term. Two benefits in this situation reduce crosstalk. First, C_{12} is smaller because adding the ground wire moves wires 1 and 2 farther apart. Second, C_{2G} makes the denominator of the crosstalk equation bigger. If we assume that $C_{2G} = C_2$ and that C_{12} has been halved by the extra spacing, we can expect the crosstalk to be reduced by a factor of roughly 4.

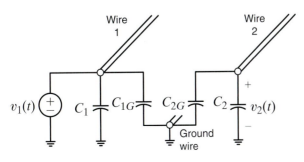

Figure 6.27

Electrical isolation using a ground wire in crosstalk model.

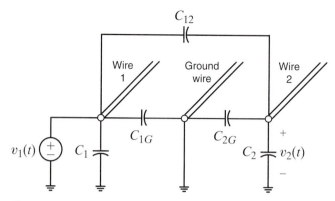

Figure 6.28 A more accurate crosstalk model.

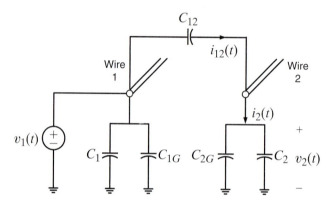

Figure 6.29 A redrawn version of the more accurate crosstalk model.

Application Example 6.20

An excellent example of capacitor operation is the memory inside a personal computer. This memory, called dynamic random access memory (DRAM), contains as many as two billion data storage sites called cells (circa 2005). Expect this number to roughly double every two years for the next decade or two. Let us examine in some detail the operation of a single DRAM cell.

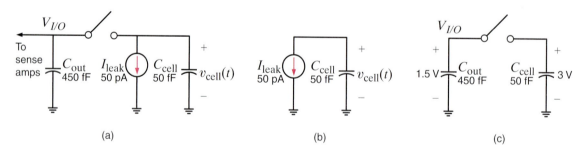

Figure 6.30 A simple circuit model showing (a) the DRAM memory cell, (b) the effect of charge leakage from the cell capacitor, and (c) cell conditions at the beginning of a read operation.

SOLUTION Figure 6.30a shows a simple model for a DRAM cell. Data are stored on the cell capacitor in true/false (or 1/0) format, where a large-capacitor voltage represents a true condition and a low voltage represents a false condition. The switch closes to allow access from the processor to the DRAM cell. Current source I_{leak} is an unintentional, or parasitic, current that models charge leakage from the capacitor. Another parasitic model element is the capacitance, C_{out}, the capacitance of the wiring connected to the output side of the cell. Both I_{leak} and C_{out} have enormous impacts on DRAM performance and design.

Consider storing a true condition in the cell. A high voltage of 3.0 V is applied at node I/O and the switch is closed, causing the voltage on C_{cell} to quickly rise to 3.0 V. We open the switch and the data are stored. During the store operation the charge, energy and number of electrons, n, used are

$$Q = CV = (50 \times 10^{-15})(3) = 150 \text{ fC}$$

$$W = \frac{1}{2}CV^2 = (0.5)(50 \times 10^{-15})(3^2) = 225 \text{ fJ}$$

$$n = Q/q = 150 \times 10^{-15}/(1.6 \times 10^{-19}) = 937,500 \text{ electrons}$$

Once data are written, the switch opens and the capacitor begins to discharge through I_{leak}. A measure of DRAM quality is the time required for the data voltage to drop by half, from 3.0 V to 1.5 V. Let us call this time t_H. For the capacitor, we know

$$v_{cell}(t) = \frac{1}{C_{cell}} \int i_{cell} \, dt \text{ V}$$

where, from Fig. 6.30b, $i_{cell}(t) = -I_{leak}$. Performing the integral yields

$$v_{cell}(t) = \frac{1}{C_{cell}} \int (-I_{leak}) \, dt = -\frac{I_{leak}}{C_{cell}} t + K$$

We know that at $t = 0$, $v_{cell} = 3$ V. Thus, $K = 3$ and the cell voltage is

$$v_{cell}(t) = 3 - \frac{I_{leak}}{C_{cell}} t \text{ V} \qquad\qquad \textbf{6.33}$$

Substituting $t = t_H$ and $v_{cell}(t_H) = 1.5$ V into Eq. (6.33) and solving for t_H yields $t_H = 15$ ms. Thus, the cell data are gone in only a few milliseconds! The solution is rewriting the data before it can disappear. This technique, called *refresh*, is a must for all DRAM using this one-transistor cell.

To see the effect of C_{out}, consider reading a fully charged $(v_{cell} = 3.0 \text{ V})$ true condition. The I/O line is usually precharged to half the data voltage. In this example, that would be 1.5 V as seen in Fig. 6.30c. (To isolate the effect of C_{out}, we have removed I_{leak}.) Next, the switch is closed. What happens next is best viewed as a conservation of charge. Just before the switch closes, the total stored charge in the circuit is

$$Q_T = Q_{out} + Q_{cell} = V_{I/O}C_{out} + V_{cell}C_{cell} = (1.5)(450 \times 10^{-15}) + (3)(50 \times 10^{-15}) = 825 \text{ fC}$$

When the switch closes, the capacitor voltages are the same (let us call it V_o) and the total charge is unchanged.

$$Q_T = 825 \text{ fC} = V_o C_{out} + V_o C_{cell} = V_o(450 \times 10^{-15} + 50 \times 10^{-15})$$

and

$$V_o = 1.65 \text{ V}$$

Thus, the change in voltage at $V_{I/O}$ during the read operation is only 0.15 V. A very sensitive amplifier is required to quickly detect such a small change. In DRAMs, these amplifiers are called *sense amps*. How can v_{cell} change instantaneously when the switch closes? It cannot. In an actual DRAM cell, a transistor, which has a small equivalent resistance, acts as the switch. The resulting RC time constant is very small, indicating a very fast circuit. Recall that we are not analyzing the cell's speed—only the final voltage value, V_o. As long as the power lost in the switch is small compared to the capacitor energy, we can be comfortable in neglecting the switch resistance. By the way, if a false condition (zero volts) were read from the cell, then V_o would drop from its precharged value of 1.5 V to 1.35 V—a negative change of 0.15 V. This symmetric voltage change is the reason for precharging the I/O node to half the data voltage. Review the effects of I_{leak} and C_{out}. You will find that eliminating them would greatly simplify the refresh requirement and improve the voltage swing at node I/O when reading data. DRAM designers earn a very good living trying to do just that.

6.6 Design Examples

— **Design Example 6.21**

We have all undoubtedly experienced a loss of electrical power in our office or our home. When this happens, even for a second, we typically find that we have to reset all of our digital alarm clocks. Let's assume that such a clock's internal digital hardware requires a current of 1 mA at a typical voltage level of 3.0 V, but the hardware will function properly down to 2.4 V. Under these assumptions, we wish to design a circuit that will "hold" the voltage level for a short duration, for example, 1 second.

SOLUTION   We know that the voltage across a capacitor cannot change instantaneously, and hence its use appears to be viable in this situation. Thus, we model this problem using the circuit in Fig. 6.31 where the capacitor is employed to hold the voltage and the 1-mA source represents the 1-mA load.

As the circuit indicates, when the power fails, the capacitor must provide all the power for the digital hardware. The load, represented by the current source, will discharge the capacitor linearly in accordance with the expression

$$v(t) = 3.0 - \frac{1}{C} \int i(t)\, dt$$

After 1 second, $v(t)$ should be at least 2.4 V, that is, the minimum functioning voltage, and hence

$$2.4 = 3.0 - \frac{1}{C} \int_o^1 (0.001)\, dt$$

Solving this equation for C yields

$$C = 1670\ \mu F$$

Thus, from the standard capacitor values in Table 6.1, connecting three 560-μF capacitors in parallel produces 1680 μF. Although three 560-μF capacitors in parallel will satisfy the design requirements, this solution may require more space than is available. An alternate solution involves the use of "double-layer capacitors" or what are known as Supercaps. A web search of this topic will indicate that a company by the name of Elna America, Inc. is a major supplier of double-layer capacitors. An investigation of their product listing indicates that their DCK series of small coin-shaped supercaps is a possible alternative in this situation. In particular, the DCK3R3224 supercap is a 220-mF capacitor rated at 3.3 V with a diameter of 7 mm, or about 1/4 inch, and a thickness of 2.1 mm. Since only one of these items is required, this is a very compact solution from a space standpoint. However, there is yet another factor of importance and that is cost. To minimize this latter item, we may need to look for yet another alternate solution.

Figure 6.31
A simple model for a power outage ride-through circuit.

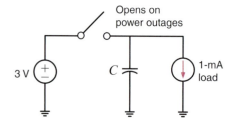

— **Design Example 6.22**

Let us design an op-amp circuit in which the relationship between the output voltage and two inputs is

$$v_o(t) = 5 \int v_1(t)\, dt - 2v_2(t)$$

SOLUTION In order to satisfy the output voltage equation, we must add two inputs, one of which must be integrated. Thus, the design equation calls for an integrator and a summer as shown in Fig. 6.32.

Using the known equations for both the integrator and summer, we can express the output voltage as

$$v_o(t) = -v_2(t)\left[\frac{R_4}{R_3}\right] - \left[\frac{R_4}{R_2}\right]\left\{-\frac{1}{R_1C}\int v_1(t)\,dt\right\} = \frac{R_4}{R_1R_2C}\int v_1(t)\,dt - \left[\frac{R_4}{R_3}\right]v_2(t)$$

If we now compare this equation to our design requirement, we find that the following equalities must hold.

$$\frac{R_4}{R_1R_2C} = 5 \qquad \frac{R_4}{R_3} = 2$$

Note that we have five variables and two constraint equations. Thus, we have some flexibility in our choice of components. First, we select $C = 2\,\mu F$, a value that is neither large nor small. If we arbitrarily select $R_4 = 20\,k\Omega$, then R_3 must be $10\,k\Omega$ and furthermore

$$R_1R_2 = 2 \times 10^9$$

If our third choice is $R_1 = 100\,k\Omega$, then $R_2 = 20\,k\Omega$. If we employ standard op-amps with supply voltages of approximately ± 10 V, then all currents will be less than 1 mA, which are, reasonable values.

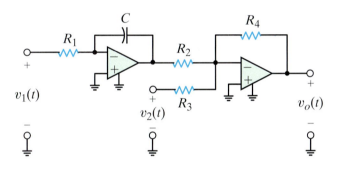

Figure 6.32
Op-amp circuit with integrator and summer.

SUMMARY

- The important (dual) relationships for capacitors and inductors are as follows:

$$q = Cv$$

$$i(t) = C\frac{dv(t)}{dt} \qquad v(t) = L\frac{di(t)}{dt}$$

$$v(t) = \frac{1}{C}\int_{-\infty}^{t} i(x)\,dx \qquad i(t) = \frac{1}{L}\int_{-\infty}^{t} v(x)\,dx$$

$$p(t) = Cv(t)\frac{dv(t)}{dt} \qquad p(t) = Li(t)\frac{di(t)}{dt}$$

$$w_C(t) = \frac{1}{2}Cv^2(t) \qquad w_L(t) = \frac{1}{2}Li^2(t)$$

- The passive sign convention is used with capacitors and inductors.

- In dc steady state, a capacitor looks like an open circuit and an inductor looks like a short circuit.

- The voltage across a capacitor and the current flowing through an inductor cannot change instantaneously.

- Leakage resistance is present in practical capacitors.

- When capacitors are interconnected, their equivalent capacitance is determined as follows: Capacitors in series combine like resistors in parallel, and capacitors in parallel combine like resistors in series.

- When inductors are interconnected, their equivalent inductance is determined as follows: Inductors in series combine like resistors in series, and inductors in parallel combine like resistors in parallel.

- RC operational amplifier circuits can be used to differentiate or integrate an electrical signal.

PROBLEMS

PSV **CS** both available on the web at: http://www.justask4u.com/irwin

SECTION 6.1

6.1 A 6-μF capacitor was charged to 12 V. Find the charge accumulated in the capacitor.

6.2 A capacitor has an accumulated charge of 600 μC with 5 V across it. What is the value of capacitance?

6.3 An uncharged 100-μF capacitor is charged by a constant current of 1 mA. Find the voltage across the capacitor after 4 s. **CS**

6.4 A 10-μF capacitor is charged by a constant current source, and its voltage is increased to 2 V in 5 s. Find the value of the constant current source.

6.5 A 50-μF capacitor initially charged to −12 V is charged by a constant current of 2.5 μA. Find the voltage across the capacitor after 3 min.

6.6 The energy that is stored in a 25-μF capacitor is $w(t) = 12 \sin^2 377t$ J. Find the current in the capacitor. **CS**

6.7 The voltage across a 150-μF capacitor is given by the expression $v(t) = 60 \sin 377t$ V. Find (a) the current in the capacitor and (b) the expression for the energy stored in the element.

6.8 The voltage across a 12-μF capacitor is shown in Fig. P6.8. Compute the waveform for the current in the capacitor.

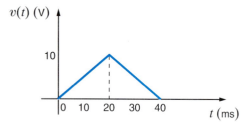

Figure P6.8

6.9 The current in a 100-μF capacitor is shown in Fig. P6.9. Determine the waveform for the voltage across the capacitor if it is initially uncharged. **CS**

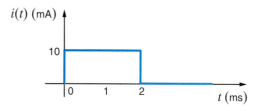

Figure P6.9

6.10 The voltage across a 50-μF capacitor is shown in Fig. P6.10. Compute the waveform for the current in the capacitor.

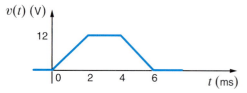

Figure P6.10

6.11 The voltage across a 25-μF capacitor is shown in Fig. P6.11. Determine the current waveform. **PSV**

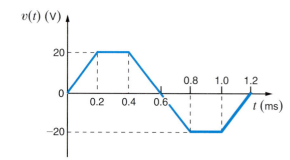

Figure P6.11

6.12 The voltage across a 2-F capacitor is given by the waveform in Fig. P6.12. Find the waveform for the current in the capacitor.

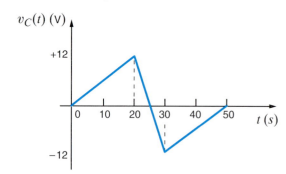

Figure P6.12

6.13 The voltage across a 2-μF capacitor is given by the waveform in Fig. P6.13. Compute the current waveform. **CS**

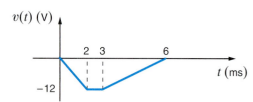

Figure P6.13

6.14 Draw the waveform for the current in a 24-μF capacitor when the capacitor voltage is as described in Fig. P6.14.

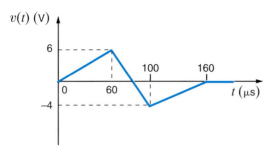

Figure P6.14

6.15 Draw the waveform for the current in a 3-μF capacitor when the voltage across the capacitor is given in Fig. P6.15. **CS**

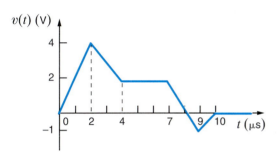

Figure P6.15

6.16 The voltage across a 10-μF capacitor is given by the waveform in Fig. P6.16. Plot the waveform for the capacitor current.

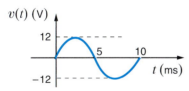

Figure P6.16

6.17 The waveform for the current in a 50-μF capacitor is shown in Fig. P6.17. Determine the waveform for the capacitor voltage. **PSV**

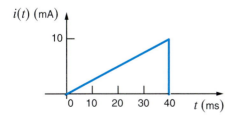

Figure P6.17

6.18 The waveform for the current in a 100-μF initially uncharged capacitor is shown in Fig. P6.18. Determine the waveform for the capacitor's voltage.

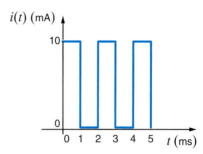

Figure P6.18

6.19 The waveform for the current in a 50-μF initially uncharged capacitor is shown in Fig. P6.19. Determine the waveform for the capacitor's voltage.

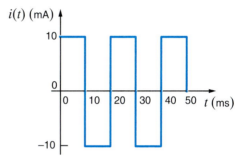

Figure P6.19

6.20 If $v_C(t = 2\text{ s}) = 10$ V in the circuit in Fig. P6.20, find the energy stored in the capacitor and the power supplied by the source at $t = 6$ s.

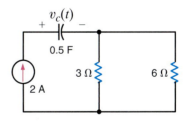

Figure P6.20

SECTION 6.2

6.21 The current in an inductor changes from 0 to 200 mA in 4 ms and induces a voltage of 100 mV. What is the value of the inductor?

6.22 The current in a 100-mH inductor is $i(t) = 2 \sin 377t$ A. Find (a) the voltage across the inductor and (b) the expression for the energy stored in the element. **CS**

6.23 If the current $i(t) = 1.5t$ A flows through a 2-H inductor, find the energy stored at $t = 2s$.

6.24 The current in a 25-mH inductor is given by the expressions

$$i(t) = 0 \qquad\qquad t < 0$$

$$i(t) = 10(1 - e^{-t})\text{mA} \quad t > 0$$

Find (a) the voltage across the inductor and (b) the expression for the energy stored in it.

6.25 Given the data in the previous problem, find the voltage across the inductor and the energy stored in it after 1 s. **CS**

6.26 The current in a 10-mH inductor is shown in Fig. P6.26. Determine the waveform for the voltage across the inductor.

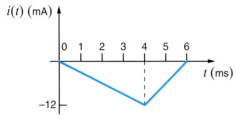

Figure P6.26

6.27 The current in a 50-mH inductor is given in Fig. P6.27. Sketch the inductor voltage. **CS**

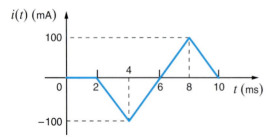

Figure P6.27

6.28 The current in a 50-mH inductor is shown in Fig. P6.28. Find the voltage across the inductor. **PSV**

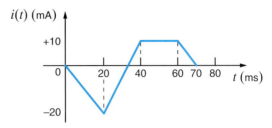

Figure P6.28

6.29 Draw the waveform for the voltage across a 24-mH inductor when the inductor current is given by the waveform shown in Fig. P6.29.

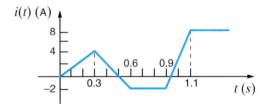

Figure P6.29

6.30 The current in a 24-mH inductor is given by the waveform in Fig. P6.30. Find the waveform for the voltage across the inductor.

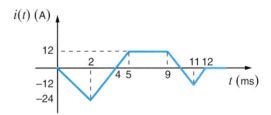

Figure P6.30

6.31 The current in a 4-mH inductor is given by the waveform in Fig. P6.31. Plot the voltage across the inductor.

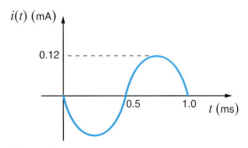

Figure P6.31

6.32 The voltage across a 2-H inductor is given by the waveform shown in Fig. P6.32. Find the waveform for the current in the inductor.

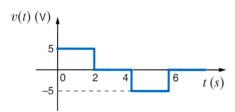

Figure P6.32

6.33 The waveform for the voltage across a 20-mH inductor is shown in Fig. P6.33. Compute the waveform for the inductor current. $v(t) = 0, t < 0$. **CS**

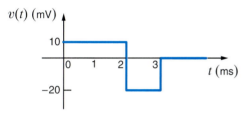

Figure P6.33

6.34 The voltage across a 4-H inductor is given by the waveform shown in Fig. P6.34. Find the waveform for the current in the inductor. $v(t) = 0, t < 0$. **PSV**

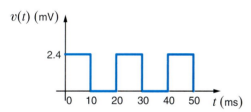

Figure P6.34

6.35 The voltage across a 24-mH inductor is shown in Fig. P6.35. Determine the waveform for the inductor current. $v(t) = 0, t < 0$.

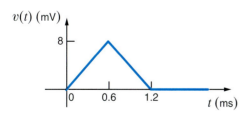

Figure P6.35

6.36 Find the possible capacitance range of the following capacitors.

(a) 0.068 μF with a tolerance of 10%.

(b) 120 pF with a tolerance of 20%.

(c) 39 μF with a tolerance of 20%.

6.37 The capacitor in Fig. P6.37a is 51 nF with a tolerance of 10%. Given the voltage waveform in Fig. P6.37b, graph the current $i(t)$ for the minimum and maximum capacitor values.

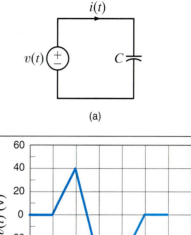

(a)

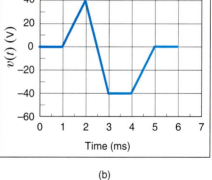

(b)

Figure P6.37

6.38 Find the possible inductance range of the following inductors. **CS**

(a) 10 mH with a tolerance of 10%.

(b) 2.0 nH with a tolerance of 5%.

(c) 68 μH with a tolerance of 10%.

6.39 The inductor in Fig. P6.39a is 330 µH with a tolerance of 5%. Given the current waveform in Fig. P6.39b, graph the voltage $v(t)$ for the minimum and maximum inductor values.

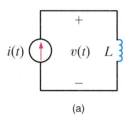

(a)

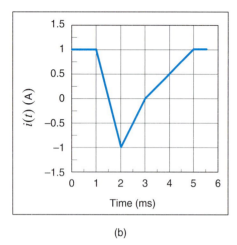

(b)

Figure P6.39

6.40 The inductor in Fig. P6.40a is 4.7 µH with a tolerance of 20%. Given the current waveform in Fig. 6.40b, graph the voltage $v(t)$ for the minimum and maximum inductor values.

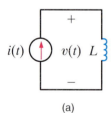

(a)

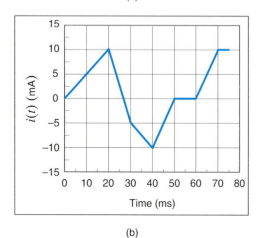

(b)

Figure P6.40

6.41 If the total energy stored in the circuit in Fig. P6.41 is 80 mJ, what is the value of L?

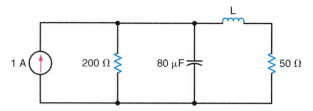

Figure P6.41

6.42 Find the value of C if the energy stored in the capacitor in Fig. P6.42 equals the energy stored in the inductor.

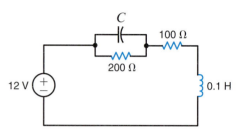

Figure P6.42

6.43 Given the network in Fig. P6.43, find the power dissipated in the 3-Ω resistor and the energy stored in the capacitor. **PSV**

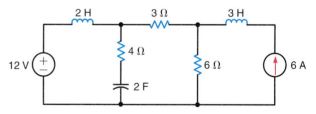

Figure P6.43

SECTION 6.3

6.44 What values of capacitance can be obtained by interconnecting a 2-μF capacitor, a 4-μF capacitor, and an 8-μF capacitor?

6.45 Given a 1-, 3-, and 4-μF capacitor, can they be interconnected to obtain an equivalent 2-μF capacitor? **cs**

6.46 Given four 6-μF capacitors, find the maximum value and minimum value that can be obtained by interconnecting the capacitors in series/parallel combinations.

6.47 The two capacitors in Fig. P6.47 were charged and then connected as shown. Determine the equivalent capacitance, the initial voltage at the terminals, and the total energy stored in the network.

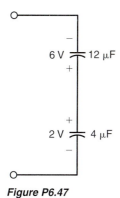

Figure P6.47

6.48 The two capacitors shown in Fig. P6.48 have been connected for some time and have reached their present values. Find V_o.

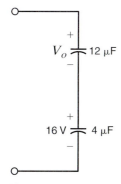

Figure P6.48

6.49 The three capacitors shown in Fig. P6.49 have been connected for some time and have reached their present values. Find V_1 and V_2. **cs**

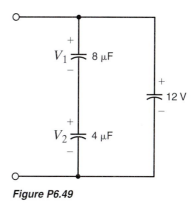

Figure P6.49

6.50 Select the value of C to produce the desired total capacitance of $C_T = 10$ μF in the circuit in Fig. P6.50.

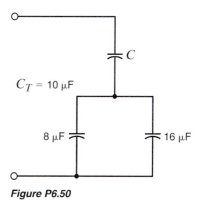

Figure P6.50

6.51 Select the value of C to produce the desired total capacitance of $C_T = 1$ μF in the circuit in Fig. P6.51.

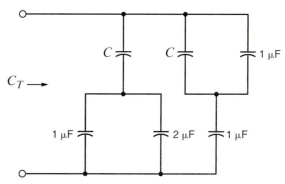

Figure P6.51

6.52 Find C_T in the network in Fig. P6.52 if (a) the switch is open and (b) the switch is closed.

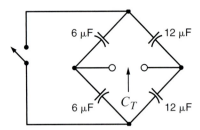

Figure P6.52

6.53 Find the equivalent capacitance at terminals *A-B* in Fig. P6.53. **PSV**

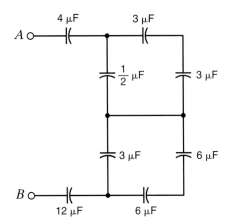

Figure P6.53

6.54 Determine the total capacitance of the network in Fig. P6.54. **CS**

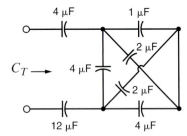

Figure P6.54

6.55 Find the total capacitance C_T shown in the network in Fig. P6.55.

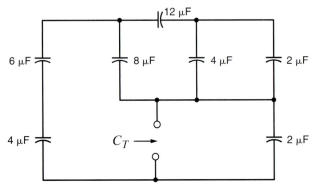

Figure P6.55

6.56 Find the total capacitance C_T of the network in Fig. P6.56.

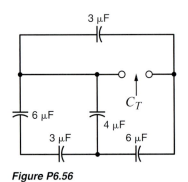

Figure P6.56

6.57 Find the total capacitance C_T of the network in Fig. P6.57.

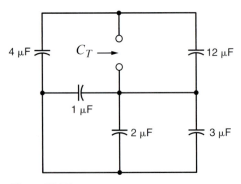

Figure P6.57

6.58 In the network in Fig. P6.58, find the capacitance C_T if (a) the switch is open and (b) the switch is closed.

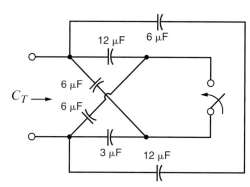

Figure P6.58

6.59 Compute the equivalent capacitance of the network in Fig. P6.59 if all the capacitors are 6 μF.

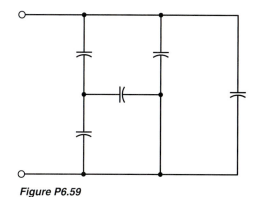

Figure P6.59

6.60 If all the capacitors in Fig. P6.60 are 6 μF, find C_{eq}. **c s**

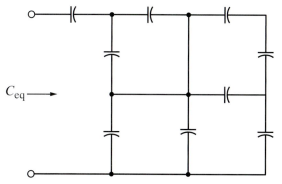

Figure P6.60

6.61 Given the capacitors in Fig. P6.61 are $C_1 = 2.0$ μF with a tolerance of 2% and $C_2 = 2.0$ μF with a tolerance of 20%, find the following.

(a) The nominal value of C_{eq}.

(b) The minimum and maximum possible values of C_{eq}.

(c) The percent errors of the minimum and maximum values.

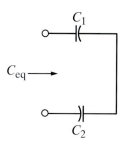

Figure P6.61

6.62 The capacitor values for the network in Fig. P6.62 are $C_1 = 0.1$ μF with a tolerance of 10%, $C_2 = 0.33$ μF with a tolerance of 20%, and $C_3 = 1$ μF with a tolerance of 10%. Find the following.

(a) The nominal value of C_{eq}.

(b) The minimum and maximum possible values of C_{eq}.

(c) The percent errors of the minimum and maximum values.

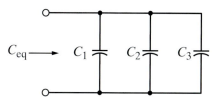

Figure P6.62

6.63 A 20-mH inductor and a 12-mH inductor are connected in series with a 1-A current source. Find (a) the equivalent inductance and (b) the total energy stored.

6.64 Two inductors are connected in parallel, as shown in Fig. P6.64. Find i.

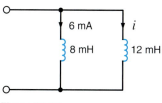

Figure P6.64

6.65 Find the value of L in the network in Fig. P6.65 so that the total inductance L_T will be 2 mH. **CS**

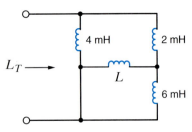

Figure P6.65

6.66 Determine the inductance at terminals A-B in the network in Fig. P6.66. **PSV**

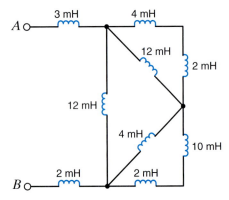

Figure P6.66

6.67 Determine the inductance at terminals A-B in the network in Fig P6.67. **CS**

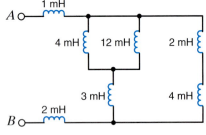

Figure P6.67

6.68 Given the network shown in Fig. P6.68, find (a) the equivalent inductance at terminals A-B with terminals C-D short circuited, and (b) the equivalent inductance at terminals C-D with terminals A-B open circuited.

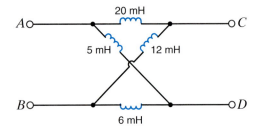

Figure P6.68

6.69 Find the total inductance at the terminals of the network in Fig. P6.69.

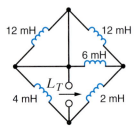

Figure P6.69

6.70 Compute the equivalent inductance of the network in Fig. P6.70 if all inductors are 12 mH.

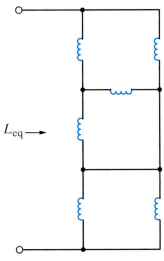

Figure P6.70

6.71 Find L_T in the network in Fig. P6.71 (a) with the switch open and (b) with the switch closed. All inductors are 12 mH.

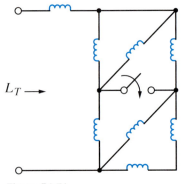

Figure P6.71

SECTION 6.4

6.72 For the network in Fig. P6.72, $v_S(t) = 120 \cos 377t$ V. Find $v_o(t)$.

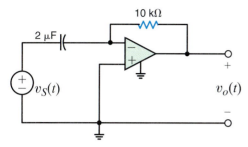

Figure P6.72

6.73 For the network in Fig. P6.73, $v_S(t) = 115 \sin 377t$ V. Find $v_o(t)$.

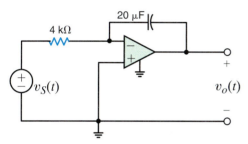

Figure P6.73

6.74 For the network in Fig. P6.74, $v_{S_1}(t) = 80 \cos 377t$ V and $v_{S_2}(t) = 40 \cos 377t$ V. Find $v_o(t)$.

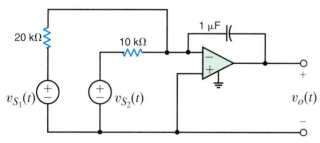

Figure P6.74

6.75 For the network in Fig. P6.75, choose C such that

$$v_o = -10 \int v_S \, dt$$

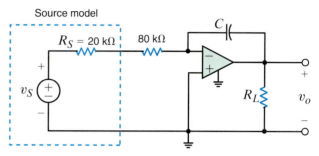

Figure P6.75

6.76 An integrator is required that has the following performance

$$v_o(t) = 10^6 \int v_s \, dt$$

where the capacitor values must be greater than 10 nF and the resistor values must be greater than 10 kΩ.

(a) Design the integrator.

(b) If ±10-V supplies are used, what are the maximum and minimum values of v_o?

(c) Suppose $v_S = 1$ V. What is the rate of change of v_o?

6.77 The circuit shown in Fig. P6.77 is known as a "Deboo" integrator.

(a) Express the output voltage in terms of the input voltage and circuit parameters.

(b) How is the Deboo integrator's performance different from that of a standard integrator?

(c) What kind of application would justify the use of this device?

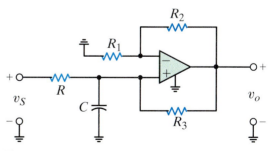

Figure P6.77

6.78 A driverless automobile is under development. One critical issue is braking, particularly at red lights. It is decided that the braking effort should depend on distance to the light (if you're close, you better stop now) and speed (if you're going fast, you'll need more brakes). The resulting design equation is

$$\text{braking effort} = K_1\left[\frac{dx(t)}{dt}\right] + K_2 x(t)$$

where x, the distance from the vehicle to the intersection, is measured by a sensor whose output is proportional to x, $v_{\text{sense}} = \alpha x$. Use superposition to show that the circuit in Fig. P6.78 can produce the braking effort signal.

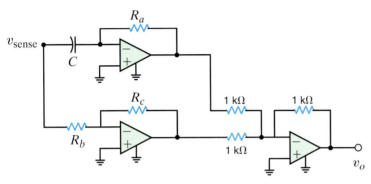

Figure P6.78

TYPICAL PROBLEMS FOUND ON THE FE EXAM

6FE-1 Given three capacitors with values 2 μF, 4 μF, and 6 μF, can the capacitors be interconnected so that the combination is an equivalent 3 μF? **CS**

6FE-2 The current pulse shown in Fig. 6PFE-2 is applied to a 1-μF capacitor. Determine the charge on the capacitor and the energy stored.

6FE-3 The two capacitors shown in Fig. 6PFE-3 have been connected for some time and have reached their present values. Determine the energy stored in the unknown capacitor C_x. **CS**

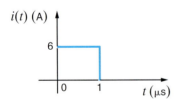

Figure 6PFE-2

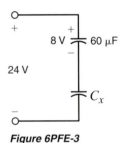

Figure 6PFE-3

First- and Second-Order Transient Circuits

We now consider circuits that are in transition from one state to another; that is, a circuit may have no forcing function and we suddenly apply one, or we instantly remove from a circuit the source of energy. We call the study of the circuit behavior in this transition phase a transient analysis. This transition is affected by the presence of a capacitor or an inductor, or both, since these two elements are capable of storing energy and releasing it over some interval of time. Our analysis includes both first-order circuits, those containing a single capacitor or inductor, and second-order circuits, those in which both a capacitor and inductor are present. ●

LEARNING GOALS

Access Problem-Solving Videos **PSV** *and Circuit Solutions* **CS** *at:* **http://www.justask4u.com/irwin** using the registration code on the inside cover and see a website with answers and more!

7.1 Introduction

In this chapter we perform what is normally referred to as a transient analysis. We begin our analysis with first-order circuits—that is, those that contain only a single storage element. When only a single storage element is present in the network, the network can be described by a first-order differential equation.

Our analysis involves an examination and description of the behavior of a circuit as a function of time after a sudden change in the network occurs due to switches opening or closing. Because of the presence of one or more storage elements, the circuit response to a sudden change will go through a transition period prior to settling down to a steady-state value. It is this transition period that we will examine carefully in our transient analysis.

One of the important parameters that we will examine in our transient analysis is the circuit's time constant. This is a very important network parameter because it tells us how fast the circuit will respond to changes. We can contrast two very different systems to obtain a feel for the parameter. For example, consider the model for a room air-conditioning system and the model for a single-transistor stage of amplification in a computer chip. If we change the setting for the air conditioner from 70 degrees to 60 degrees, the unit will come on and the room will begin to cool. However, the temperature measured by a thermometer in the room will fall very slowly and, thus, the time required to reach the desired temperature is long. However, if we send a trigger signal to a transistor to change state, the action may take only a few nanoseconds. These two systems will have vastly different time constants.

Our analysis of first-order circuits begins with the presentation of two techniques for performing a transient analysis: the differential equation approach, in which a differential equation is written and solved for each network, and a step-by-step approach, which takes advantage of the known form of the solution in every case. In the second-order case, both an inductor and a capacitor are present simultaneously, and the network is described by a second-order differential equation. Although the *RLC* circuits are more complicated than the first-order single storage circuits, we will follow a development similar to that used in the first-order case.

Our presentation will deal only with very simple circuits, since the analysis can quickly become complicated for networks that contain more than one loop or one nonreference node. Furthermore, we will demonstrate a much simpler method for handling these circuits when we cover the Laplace transform later in this book. We will analyze several networks in which the parameters have been chosen to illustrate the different types of circuit response. In addition, we will extend our PSPICE analysis techniques to the analysis of transient circuits. Finally, a number of application-oriented examples are presented and discussed.

We begin our discussion by recalling that in Chapter 6 we found that capacitors and inductors were capable of storing electric energy. In the case of a charged capacitor, the energy is stored in the electric field that exists between the positively and negatively charged plates. This stored energy can be released if a circuit is somehow connected across the capacitor that provides a path through which the negative charges move to the positive charges. As we know, this movement of charge constitutes a current. The rate at which the energy is discharged is a direct function of the parameters in the circuit that is connected across the capacitor's plates.

As an example, consider the flash circuit in a camera. Recall that the operation of the flash circuit, from a user standpoint, involves depressing the push button on the camera that triggers both the shutter and the flash and then waiting a few seconds before repeating the process to take the next picture. This operation can be modeled using the circuit in Fig. 7.1a. The voltage source and resistor R_S model the batteries that power the camera and flash. The capacitor models the energy storage, the switch models the push button, and finally the resistor R models the xenon flash lamp. Thus, if the capacitor is charged, when the switch is closed, the capacitor voltage drops and energy is released through the xenon lamp, producing the flash. In practice this energy release takes about a millisecond, and the discharge time is a function of the elements in the circuit. When the push button is released and the switch is then opened, the battery begins to recharge the capacitor. Once again, the time required to charge the capacitor is a function of the circuit elements. The discharge and charge cycles are graphically illustrated in Fig. 7.1b. Although the discharge time is very fast, it is not instantaneous.

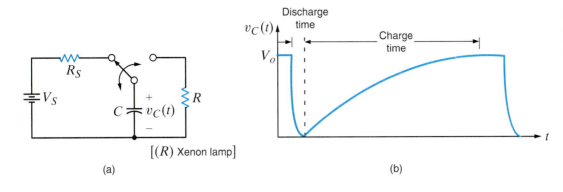

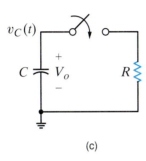

(c)

Figure 7.1
*Diagrams used to
describe a camera's
flash circuit.*

To provide further insight into this phenomenon, consider what we might call a *free-body diagram* of the right half of the network in Fig. 7.1a as shown in Fig. 7.1c (that is, a charged capacitor that is discharged through a resistor). When the switch is closed, KCL for the circuit is

$$C \frac{dv_C(t)}{dt} + \frac{v_C(t)}{R} = 0$$

or

$$\frac{dv_C(t)}{dt} + \frac{1}{RC} v_C(t) = 0$$

We will demonstrate in the next section that the solution of this equation is

$$v_C(t) = V_o e^{-t/RC}$$

Note that this function is a decaying exponential and the rate at which it decays is a function of the values of R and C. The product RC is a very important parameter, and we will give it a special name in the following discussions.

7.2   First-Order Circuits

GENERAL FORM OF THE RESPONSE EQUATIONS
In our study of first-order transient circuits we will show that the solution of these circuits (i.e., finding a voltage or current) requires us to solve a first-order differential equation of the form

$$\frac{dx(t)}{dt} + ax(t) = f(t) \qquad \textbf{7.1}$$

Although a number of techniques may be used for solving an equation of this type, we will obtain a general solution that we will then employ in two different approaches to transient analysis.

A fundamental theorem of differential equations states that if $x(t) = x_p(t)$ is any solution to Eq. (7.1), and $x(t) = x_c(t)$ is any solution to the homogeneous equation

$$\frac{dx(t)}{dt} + ax(t) = 0 \qquad \textbf{7.2}$$

then

$$x(t) = x_p(t) + x_c(t) \qquad \textbf{7.3}$$

is a solution to the original Eq. (7.1). The term $x_p(t)$ is called *the particular integral solution*, or forced response, and $x_c(t)$ is called the *complementary solution*, or natural response.

At the present time we confine ourselves to the situation in which $f(t) = A$ (i.e., some constant). The general solution of the differential equation then consists of two parts that are obtained by solving the two equations

$$\frac{dx_p(t)}{dt} + ax_p(t) = A \qquad \textbf{7.4}$$

$$\frac{dx_c(t)}{dt} + ax_c(t) = 0 \qquad \textbf{7.5}$$

Since the right-hand side of Eq. (7.4) is a constant, it is reasonable to assume that the solution $x_p(t)$ must also be a constant. Therefore, we assume that

$$x_p(t) = K_1 \qquad \textbf{7.6}$$

Substituting this constant into Eq. (7.4) yields

$$K_1 = \frac{A}{a} \qquad \textbf{7.7}$$

Examining Eq. (7.5), we note that

$$\frac{dx_c(t)/dt}{x_c(t)} = -a \qquad \textbf{7.8}$$

This equation is equivalent to

$$\frac{d}{dt}\left[\ln x_c(t)\right] = -a$$

Hence,

$$\ln x_c(t) = -at + c$$

and therefore,

$$x_c(t) = K_2 e^{-at} \qquad \textbf{7.9}$$

Thus, a solution of Eq. (7.1) is

$$x(t) = x_p(t) + x_c(t)$$

$$= \frac{A}{a} + K_2 e^{-at} \qquad \textbf{7.10}$$

The constant K_2 can be found if the value of the independent variable $x(t)$ is known at one instant of time.

Equation (7.10) can be expressed in general in the form

$$x(t) = K_1 + K_2 e^{-t/\tau} \qquad \textbf{7.11}$$

Once the solution in Eq. (7.11) is obtained, certain elements of the equation are given names that are commonly employed in electrical engineering. For example, the term K_1 is referred to as the *steady-state solution*: the value of the variable $x(t)$ as $t \rightarrow \infty$ when the second term becomes negligible. The constant τ is called the *time constant* of the circuit.

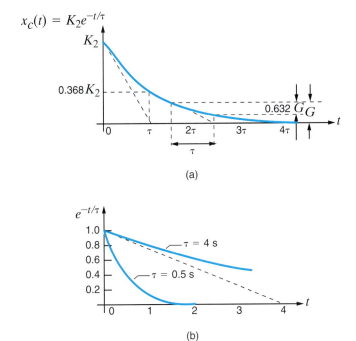

Figure 7.2
Time-constant illustrations.

Note that the second term in Eq. (7.11) is a decaying exponential that has a value, if $\tau > 0$, of K_2 for $t = 0$ and a value of 0 for $t = \infty$. The rate at which this exponential decays is determined by the time constant τ. A graphical picture of this effect is shown in Fig. 7.2a. As can be seen from the figure, the value of $x_c(t)$ has fallen from K_2 to a value of $0.368K_2$ in one time constant, a drop of 63.2%. In two time constants the value of $x_c(t)$ has fallen to $0.135K_2$, a drop of 63.2% from the value at time $t = \tau$. This means that the gap between a point on the curve and the final value of the curve is closed by 63.2% each time constant. Finally, after five time constants, $x_c(t) = 0.0067K_2$, which is less than 1%.

An interesting property of the exponential function shown in Fig. 7.2a is that the initial slope of the curve intersects the time axis at a value of $t = \tau$. In fact, we can take any point on the curve, not just the initial value, and find the time constant by finding the time required to close the gap by 63.2%. Finally, the difference between a small time constant (i.e., fast response) and a large time constant (i.e., slow response) is shown in Fig. 7.2b. These curves indicate that if the circuit has a small time constant, it settles down quickly to a steady-state value. Conversely, if the time constant is large, more time is required for the circuit to settle down or reach steady state. In any case, note that the circuit response essentially reaches steady state within five time constants (i.e., 5τ).

Note that the previous discussion has been very general in that no particular form of the circuit has been assumed—except that it results in a first-order differential equation.

ANALYSIS TECHNIQUES

The Differential Equation Approach   Equation (7.11) defines the general form of the solution of first-order transient circuits; that is, it represents the solution of the differential equation that describes an unknown current or voltage *anywhere in the network*. One of the ways that we can arrive at this solution is to solve the equations that describe the network behavior using what is often called the *state-variable approach*. In this technique we write the equation for the voltage across the capacitor and/or the equation for the current through the inductor. Recall from Chapter 6 that these quantities cannot change instantaneously. Let us first illustrate this technique in the general sense and then examine two specific examples.

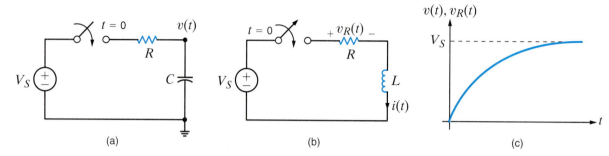

Figure 7.3
(a) RC *circuit*, (b) RL *circuit*,
(c) *plot of the capacitor voltage
in (a) and resistor voltage in (b).*

Consider the circuit shown in Fig. 7.3a. At time $t = 0$, the switch closes. The KCL equation that describes the capacitor voltage for time $t > 0$ is

$$C\frac{dv(t)}{dt} + \frac{v(t) - V_S}{R} = 0$$

or

$$\frac{dv(t)}{dt} + \frac{v(t)}{RC} = \frac{V_S}{RC}$$

From our previous development, we assume that the solution of this first-order differential equation is of the form

$$v(t) = K_1 + K_2 e^{-t/\tau}$$

Substituting this solution into the differential equation yields

$$-\frac{K_2}{\tau}e^{-t/\tau} + \frac{K_1}{RC} + \frac{K_2}{RC}e^{-t/\tau} = \frac{V_S}{RC}$$

Equating the constant and exponential terms, we obtain

$$K_1 = V_S$$
$$\tau = RC$$

Therefore,

$$v(t) = V_S + K_2 e^{-t/RC}$$

where V_S is the steady-state value and RC is the network's time constant. K_2 is determined by the initial condition of the capacitor. For example, if the capacitor is initially uncharged (that is, the voltage across the capacitor is zero at $t = 0$), then

$$0 = V_S + K_2$$

or

$$K_2 = -V_S$$

Hence, the complete solution for the voltage $v(t)$ is

$$v(t) = V_S - V_S e^{-t/RC}$$

The circuit in Fig. 7.3b can be examined in a similar manner. The KVL equation that describes the inductor current for $t > 0$ is

$$L\frac{di(t)}{dt} + Ri(t) = V_S$$

A development identical to that just used yields

$$i(t) = \frac{V_S}{R} + K_2 e^{-\left(\frac{R}{L}\right)t}$$

where V_S/R is the steady-state value and L/R is the circuit's time constant. If there is no initial current in the inductor, then at $t = 0$

$$0 = \frac{V_S}{R} + K_2$$

and

$$K_2 = \frac{-V_S}{R}$$

Hence,

$$i(t) = \frac{V_S}{R} - \frac{V_S}{R} e^{-\frac{R}{L}t}$$

is the complete solution. Note that if we wish to calculate the voltage across the resistor, then

$$v_R(t) = Ri(t)$$

$$= V_S\left(1 - e^{-\frac{R}{L}t}\right)$$

Therefore, we find that the voltage across the capacitor in the *RC* circuit and the voltage across the resistor in the *RL* circuit have the same general form. A plot of these functions is shown in Fig. 7.3c.

Example 7.1

Consider the circuit shown in Fig. 7.4a. Assuming that the switch has been in position 1 for a long time, at time $t = 0$ the switch is moved to position 2. We wish to calculate the current $i(t)$ for $t > 0$.

Figure 7.4
Analysis of RC circuits.

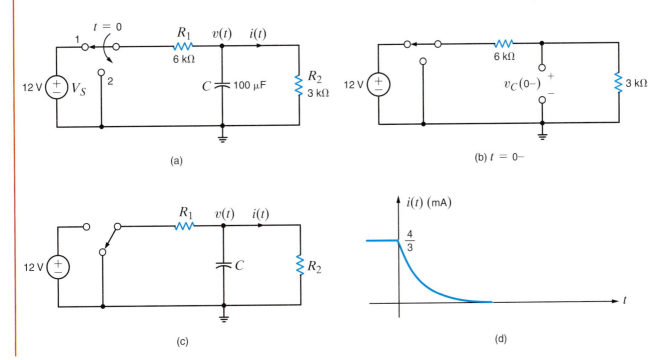

(a)

(b) $t = 0-$

(c)

(d)

SOLUTION At $t = 0-$ the capacitor is fully charged and conducts no current since the capacitor acts like an open circuit to dc. The initial voltage across the capacitor can be found using voltage division. As shown in Fig. 7.4b,

$$v_C(0-) = 12\left(\frac{3k}{6k + 3k}\right) = 4 \text{ V}$$

The network for $t > 0$ is shown in Fig. 7.4c. The KCL equation for the voltage across the capacitor is

$$\frac{v(t)}{R_1} + C\frac{dv(t)}{dt} + \frac{v(t)}{R_2} = 0$$

Using the component values, the equation becomes

$$\frac{dv(t)}{dt} + 5v(t) = 0$$

The form of the solution to this homogeneous equation is

$$v(t) = K_2 e^{-t/\tau}$$

If we substitute this solution into the differential equation, we find that $\tau = 0.2$ s. Thus,

$$v(t) = K_2 e^{-t/0.2} \text{ V}$$

Using the initial condition $v_C(0-) = v_C(0+) = 4$ V, we find that the complete solution is

$$v(t) = 4e^{-t/0.2} \text{ V}$$

Then $i(t)$ is simply

$$i(t) = \frac{v(t)}{R_2}$$

or

$$i(t) = \frac{4}{3}e^{-t/0.2} \text{ mA}$$

An ordinary differential equation (ODE) can also be solved using MATLAB symbolic operations. For example, given an ordinary differential equation of the form

$$\frac{dx^2}{dt^2} + a\frac{dx}{dt} + bx = f, \ x(0) = A, \ \frac{dx}{dt}(0) = B$$

The MATLAB solution is generated from the equation

$$x = d\text{solve } ('D2x + a*Dx + b*x = f', \ 'x(0) = A', \ 'Dx(0) = B')$$

Note that the variable x cannot be i, since i is used to represent the square root of -1. In Example 7.1 the ODE is

$$\frac{dv}{dt} + 5v = 0, \quad v(0) = 4$$

The equation, in symbolic notation, and its solution are as follows.

```
dsolve('Dv + 5*v = 0','v(0) = 4')
ans =
4*exp(-5*t)
```

Example 7.2

The switch in the network in Fig. 7.5a opens at $t = 0$. Let us find the output voltage $v_o(t)$ for $t > 0$.

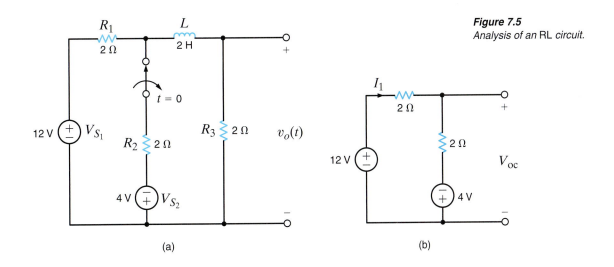

Figure 7.5
Analysis of an RL circuit.

(a)   (b)

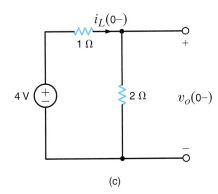

(c)

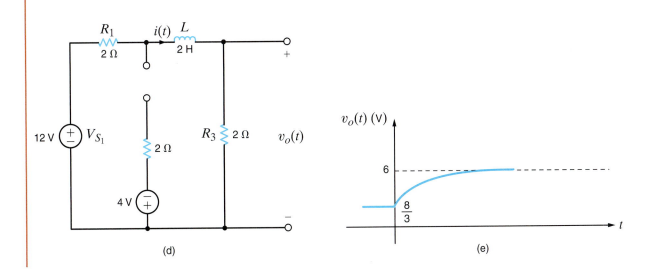

(d)   (e)

SOLUTION At $t = 0-$ the circuit is in steady state and the inductor acts like a short circuit. The initial current through the inductor can be found in many ways; however, we will form a Thévenin equivalent for the part of the network to the left of the inductor, as shown in Fig. 7.5b. From this network we find that $I_1 = 4$ A and $V_{oc} = 4$ V. In addition, $R_{Th} = 1\ \Omega$. Hence, $i_L(0-)$ obtained from Fig. 7.5c is $i_L(0-) = 4/3$ A.

The network for $t > 0$ is shown in Fig. 7.5d. Note that the 4-V independent source and the 2-ohm resistor in series with it no longer have any impact on the resulting circuit. The KVL equation for the circuit is

$$-V_{S_1} + R_1 i(t) + L\frac{di(t)}{dt} + R_3 i(t) = 0$$

which with the component values reduces to

$$\frac{di(t)}{dt} + 2i(t) = 6$$

The solution to this equation is of the form

$$i(t) = K_1 + K_2 e^{-t/\tau}$$

which when substituted into the differential equation yields

$$K_1 = 3$$

$$\tau = 1/2$$

Therefore,

$$i(t) = \left(3 + K_2 e^{-2t}\right) \text{A}$$

Evaluating this function at the initial condition, which is

$$i_L(0-) = i_L(0+) = i(0) = 4/3 \text{ A}$$

we find that

$$K_2 = \frac{-5}{3}$$

Hence,

$$i(t) = \left(3 - \frac{5}{3} e^{-2t}\right) \text{A}$$

and then

$$v_o(t) = 6 - \frac{10}{3} e^{-2t} \text{ V}$$

A plot of the voltage $v_o(t)$ is shown in Fig. 7.5e.

LEARNING EXTENSIONS

E7.1 Find $v_C(t)$ for $t > 0$ in the circuit shown in Fig. E7.1.

ANSWER:
$v_C(t) = 8e^{-t/0.6}$ V.

Figure E7.1

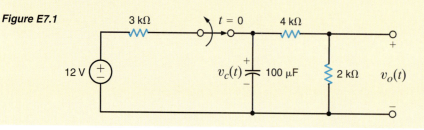

E7.2 In the circuit shown in Fig. E7.2, the switch opens at $t = 0$. Find $i_1(t)$ for $t > 0$.   **ANSWER:** $i_1(t) = 1e^{-9t}$ A.

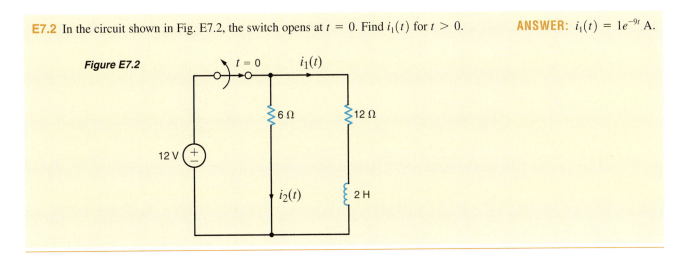

Figure E7.2

The Step-by-Step Approach   In the previous analysis technique, we derived the differential equation for the capacitor voltage or inductor current, solved the differential equation, and used the solution to find the unknown variable in the network. In the very methodical technique that we will now describe, we will use the fact that Eq. (7.11) is the form of the solution and we will employ circuit analysis to determine the constants K_1, K_2, and τ.

From Eq. (7.11) we note that as $t \to \infty$, $e^{-at} \to 0$ and $x(t) = K_1$. Therefore, if the circuit is solved for the variable $x(t)$ in steady state (i.e., $t \to \infty$) with the capacitor replaced by an open circuit [v is constant and therefore $i = C(dv/dt) = 0$] or the inductor replaced by a short circuit [i is constant and therefore $v = L(di/dt) = 0$], then the variable $x(t) = K_1$. Note that since the capacitor or inductor has been removed, the circuit is a dc circuit with constant sources and resistors, and therefore only dc analysis is required in the steady-state solution.

The constant K_2 in Eq. (7.11) can also be obtained via the solution of a dc circuit in which a capacitor is replaced by a voltage source or an inductor is replaced by a current source. The value of the voltage source for the capacitor or the current source for the inductor is a known value at one instant of time. In general, we will use the initial condition value since it is generally the one known, but the value at any instant could be used. This value can be obtained in numerous ways and is often specified as input data in a statement of the problem. However, a more likely situation is one in which a switch is thrown in the circuit and the initial value of the capacitor voltage or inductor current is determined from the previous circuit (i.e., the circuit before the switch is thrown). It is normally assumed that the previous circuit has reached steady state, and therefore the voltage across the capacitor or the current through the inductor can be found in exactly the same manner as was used to find K_1.

Finally, the value of the time constant can be found by determining the Thévenin equivalent resistance at the terminals of the storage element. Then $\tau = R_{Th}C$ for an RC circuit, and $\tau = L/R_{Th}$ for an RL circuit.

Let us now reiterate this procedure in a step-by-step fashion.

PROBLEM-SOLVING STRATEGY

Using the Step-by-Step Approach

Step 1. We assume a solution for the variable $x(t)$ of the form $x(t) = K_1 + K_2 e^{-t/\tau}$.

Step 2. Assuming that the original circuit has reached steady state before a switch was thrown (thereby producing a new circuit), draw this previous circuit with the capacitor replaced by an open circuit or the inductor replaced by a short circuit. Solve for the voltage across the capacitor, $v_C(0-)$, or the current through the inductor, $i_L(0-)$, prior to switch action.

(continues on the next page)

Step 3. Recall from Chapter 6 that voltage across a capacitor and the current flowing through an inductor cannot change in zero time. Draw the circuit valid for $t = 0+$ with the switches in their new positions. Replace a capacitor with a voltage source $v_C(0+) = v_C(0-)$ or an inductor with a current source of value $i_L(0+) = i_L(0-)$. Solve for the initial value of the variable $x(0+)$.

Step 4. Assuming that steady state has been reached after the switches are thrown, draw the equivalent circuit, valid for $t > 5\tau$, by replacing the capacitor by an open circuit or the inductor by a short circuit. Solve for the steady-state value of the variable

$$x(t)|_{t>5\tau} \doteq x(\infty)$$

Step 5. Since the time constant for all voltages and currents in the circuit will be the same, it can be obtained by reducing the entire circuit to a simple series circuit containing a voltage source, resistor, and a storage element (i.e., capacitor or inductor) by forming a simple Thévenin equivalent circuit at the terminals of the storage element. This Thévenin equivalent circuit is obtained by looking into the circuit from the terminals of the storage element. The time constant for a circuit containing a capacitor is $\tau = R_{Th}\,C$, and for a circuit containing an inductor it is $\tau = L/R_{Th}$.

Step 6. Using the results of steps 3, 4, and 5, we can evaluate the constants in step 1 as

$$x(0+) = K_1 + K_2$$
$$x(\infty) = K_1$$

Therefore, $K_1 = x(\infty)$, $K_2 = x(0+) - x(\infty)$, and hence the solution is

$$x(t) = x(\infty) + [x(0+) - x(\infty)]e^{-t/\tau}$$

Keep in mind that this solution form applies only to a first-order circuit having constant, dc sources. If the sources are not dc, the forced response will be different. Generally, the forced response is of the same form as the forcing functions (sources) and their derivatives.

Example 7.3

Consider the circuit shown in Fig. 7.6a. The circuit is in steady state prior to time $t = 0$, when the switch is closed. Let us calculate the current $i(t)$ for $t > 0$.

SOLUTION

Step 1. $i(t)$ is of the form $K_1 + K_2 e^{-t/\tau}$.

Step 2. The initial voltage across the capacitor is calculated from Fig. 7.6b as

$$v_C(0-) = 36 - (2)(2)$$
$$= 32 \text{ V}$$

Step 3. The new circuit, valid only for $t = 0+$, is shown in Fig. 7.6c. The value of the voltage source that replaces the capacitor is $v_C(0-) = v_C(0+) = 32$ V. Hence,

$$i(0+) = \frac{32}{6k}$$
$$= \frac{16}{3} \text{ mA}$$

Step 4. The equivalent circuit, valid for $t > 5\tau$, is shown in Fig. 7.6d. The current $i(\infty)$ caused by the 36-V source is

$$i(\infty) = \frac{36}{2k + 6k}$$
$$= \frac{9}{2} \text{ mA}$$

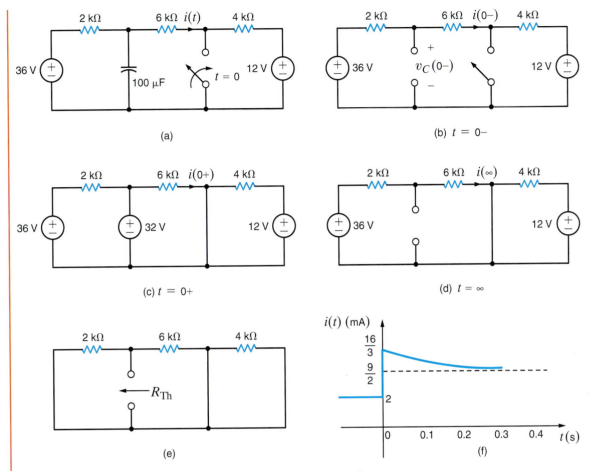

Figure 7.6 *Analysis of an* RC *transient circuit with a constant forcing function.*

Step 5. The Thévenin equivalent resistance, obtained by looking into the open-circuit terminals of the capacitor in Fig. 7.6e, is

$$R_{Th} = \frac{(2k)(6k)}{2k + 6k} = \frac{3}{2} k\Omega$$

Therefore, the circuit time constant is

$$\tau = R_{Th}C$$

$$= \left(\frac{3}{2}\right)(10^3)(100)(10^{-6})$$

$$= 0.15 \text{ s}$$

Step 6.

$$K_1 = i(\infty) = \frac{9}{2} \text{ mA}$$

$$K_2 = i(0+) - i(\infty) = i(0+) - K_1$$

$$= \frac{16}{3} - \frac{9}{2}$$

$$= \frac{5}{6} \text{ mA}$$

Therefore,

$$i(t) = \frac{36}{8} + \frac{5}{6} e^{-t/0.15} \text{ mA}$$

Let us now employ MATLAB to plot the function. First, an interval for the variable t must be specified. The beginning of the interval will be chosen to be $t = 0$. The end of the interval will be chosen to be 10 times the time constant. This is realized in MATLAB as follows:

```
>>tau = 0.15
>>tend = 10*tau
```

Once the time interval has been specified, we can use MATLAB's linspace function to generate an array of evenly spaced points in the interval. The linspace command has the following syntax: linspace(x1,x2,N) where x1 and x2 denote the beginning and ending points in the interval and N represents the number of points. Thus, to generate an array containing 150 points in the interval [0, tend], we execute the following command:

```
>>t = linspace(0, tend, 150)
```

The MATLAB program for generating a plot of the function is

```
>>tau = 0.15;
>>tend = 10*tau;
>>t = linspace(0, tend, 150);
>>i = 9/2 + (5/6)*exp(-t/tau);
>>plot(t,i)
>>xlabel('Time (s)')
>>ylabel('Current (mA)')
```

The MATLAB plot is shown in Fig. 7.7 and can be compared to the sketch in Fig. 7.6(f). Examination of Fig. 7.6(f) indicates once again that although the voltage across the capacitor is continuous at $t = 0$, the current $i(t)$ in the 6-kΩ resistor jumps at $t = 0$ from 2 mA to 5 1/3 mA, and finally decays to 4 1/2 mA.

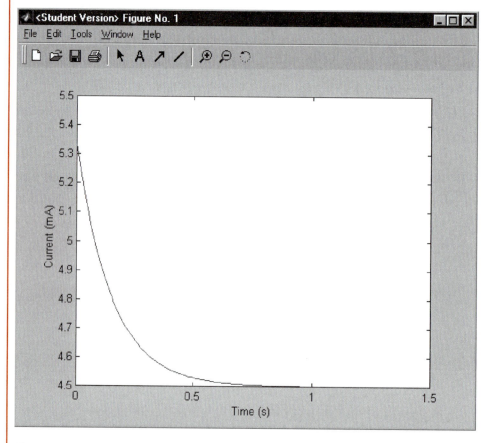

Figure 7.7 *MATLAB plot for Example 7.3.*

Example 7.4

The circuit shown in Fig. 7.8a is assumed to have been in a steady-state condition prior to switch closure at $t = 0$. We wish to calculate the voltage $v(t)$ for $t > 0$.

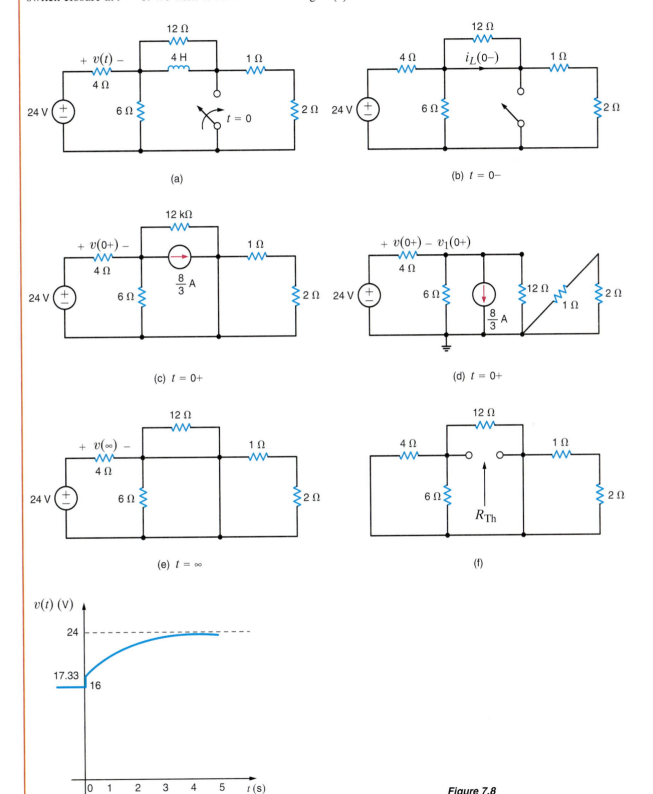

(a)

(b) $t = 0-$

(c) $t = 0+$

(d) $t = 0+$

(e) $t = \infty$

(f)

(g)

Figure 7.8
Analysis of an RL transient circuit with a constant forcing function.

SOLUTION

Step 1. $v(t)$ is of the form $K_1 + K_2 e^{-t/\tau}$.

Step 2. In Fig. 7.8b we see that

$$i_L(0-) = \frac{24}{4 + \dfrac{(6)(3)}{6+3}}\left(\frac{6}{6+3}\right)$$

$$= \frac{8}{3}\text{ A}$$

Step 3. The new circuit, valid only for $t = 0+$, is shown in Fig. 7.8c, which is equivalent to the circuit shown in Fig. 7.8d. The value of the current source that replaces the inductor is $i_L(0-) = i_L(0+) = 8/3$ A. The node voltage $v_1(0+)$ can be determined from the circuit in Fig. 7.8d using a single-node equation, and $v(0+)$ is equal to the difference between the source voltage and $v_1(0+)$. The equation for $v_1(0+)$ is

$$\frac{v_1(0+) - 24}{4} + \frac{v_1(0+)}{6} + \frac{8}{3} + \frac{v_1(0+)}{12} = 0$$

or

$$v_1(0+) = \frac{20}{3}\text{ V}$$

Then

$$v(0+) = 24 - v_1(0+)$$

$$= \frac{52}{3}\text{ V}$$

Step 4. The equivalent circuit for the steady-state condition after switch closure is given in Fig. 7.8e. Note that the 6-, 12-, 1-, and 2-Ω resistors are shorted, and therefore $v(\infty) = 24$ V.

Step 5. The Thévenin equivalent resistance is found by looking into the circuit from the inductor terminals. This circuit is shown in Fig. 7.8f. Note carefully that R_{Th} is equal to the 4-, 6-, and 12-Ω resistors in parallel. Therefore, $R_{\text{Th}} = 2\ \Omega$, and the circuit time constant is

$$\tau = \frac{L}{R_{\text{Th}}} = \frac{4}{2} = 2\text{ s}$$

Step 6. From the previous analysis we find that

$$K_1 = v(\infty) = 24$$

$$K_2 = v(0+) - v(\infty) = -\frac{20}{3}$$

and hence that

$$v(t) = 24 - \frac{20}{3}e^{-t/2}\text{ V}$$

From Fig. 7.8b we see that the value of $v(t)$ before switch closure is 16 V. This value jumps to 17.33 V at $t = 0$. The MATLAB program for generating the plot (shown in Fig. 7.9) of this function for $t > 0$ is listed next.

```
>>tau = 2;
>>tend = 10*tau;
>>t = linspace(0, tend, 150);
```

```
>>v = 24 - (20/3)*exp(-t/tau);
>>plot (t,v)
>>xlabel('Time (s)')
>>ylabel('Voltage (V)')
```

This plot can be compared to the sketch shown in Fig. 7.8g.

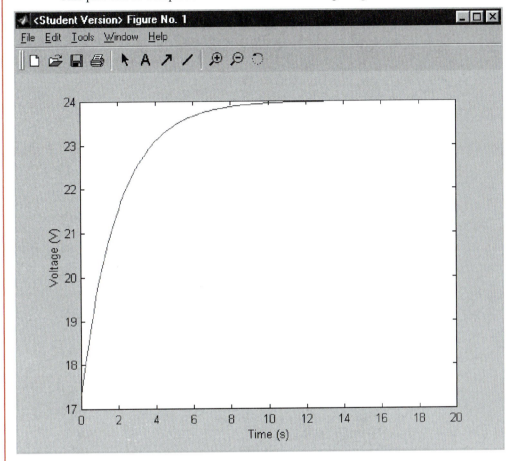

Figure 7.9
*MATLAB plot for
Example 7.4.*

E7.3 Consider the network in Fig. E7.3. The switch opens at $t = 0$. Find $v_o(t)$ for $t > 0$. **PSV**   **ANSWER:**
$$v_o(t) = \frac{24}{5} + \frac{1}{5}e^{-(5/8)t} \text{ V}.$$

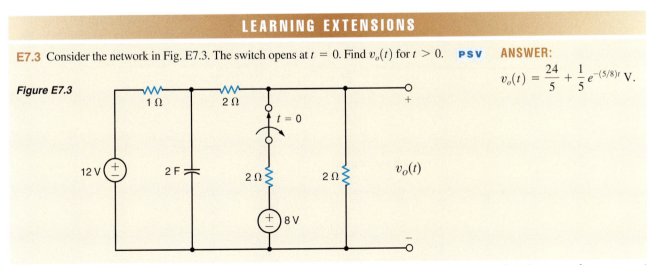

Figure E7.3

(continues on the next page)

E7.4 Consider the network in Fig. E7.4. If the switch opens at $t = 0$, find the output voltage $v_o(t)$ for $t > 0$.

ANSWER:

$$v_o(t) = 6 - \frac{10}{3} e^{-2t} \text{ V}.$$

Figure E7.4

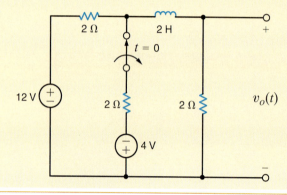

Example 7.5

The circuit shown in Fig. 7.10a has reached steady state with the switch in position 1. At time $t = 0$ the switch moves from position 1 to position 2. We want to calculate $v_o(t)$ for $t > 0$.

SOLUTION

Step 1. $v_o(t)$ is of the form $K_1 + K_2 e^{-t/\tau}$.

Step 2. Using the circuit in Fig. 7.10b, we can calculate $i_L(0-)$

$$i_A = \frac{12}{4} = 3 \text{ A}$$

Then

$$i_L(0-) = \frac{12 + 2i_A}{6} = \frac{18}{6} = 3 \text{ A}$$

Step 3. The new circuit, valid only for $t = 0+$, is shown in Fig. 7.10c. The value of the current source that replaces the inductor is $i_L(0-) = i_L(0+) = 3$ A. Because of the current source

$$v_o(0+) = (3)(6) = 18 \text{ V}$$

Step 4. The equivalent circuit, for the steady-state condition after switch closure, is given in Fig. 7.10d. Using the voltages and currents defined in the figure, we can compute $v_o(\infty)$ in a variety of ways. For example, using node equations we can find $v_o(\infty)$ from

$$\frac{v_B - 36}{2} + \frac{v_B}{4} + \frac{v_B + 2i'_A}{6} = 0$$

$$i'_A = \frac{v_B}{4}$$

$$v_o(\infty) = v_B + 2i'_A$$

or, using loop equations,

$$36 = 2(i_1 + i_2) + 4i_1$$

$$36 = 2(i_1 + i_2) + 6i_2 - 2i_1$$

$$v_o(\infty) = 6i_2$$

Using either approach, we find that $v_o(\infty) = 27$ V.

Step 5. The Thévenin equivalent resistance can be obtained via v_{oc} and i_{sc} because of the presence of the dependent source. From Fig. 7.10e we note that

$$i_A'' = \frac{36}{2+4} = 6 \text{ A}$$

Therefore,

$$v_{oc} = (4)(6) + 2(6)$$
$$= 36 \text{ V}$$

Figure 7.10
Analysis of an RL transient circuit containing a dependent source.

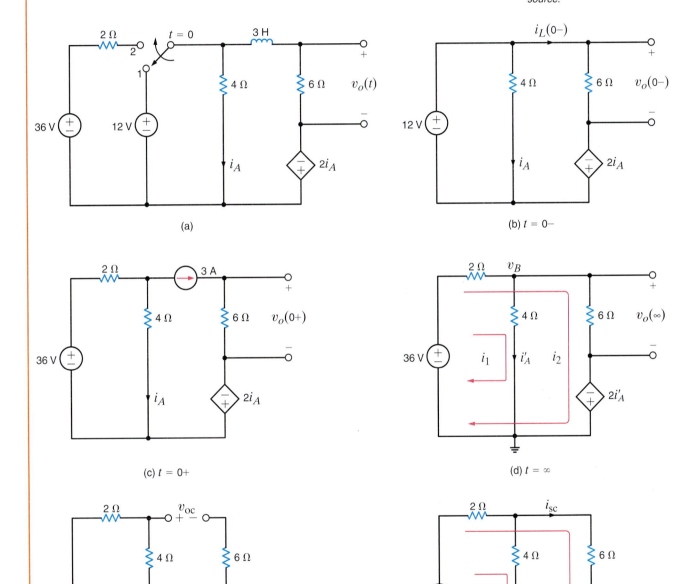

(a)

(b) $t = 0-$

(c) $t = 0+$

(d) $t = \infty$

(e)

(f)

From Fig. 7.10f we can write the following loop equations:

$$36 = 2(i_A''' + i_{sc}) + 4i_A'''$$

$$36 = 2(i_A''' + i_{sc}) + 6i_{sc} - 2i_A'''$$

Solving these equations for i_{sc} yields

$$i_{sc} = \frac{9}{2} \text{ A}$$

Therefore,

$$R_{Th} = \frac{v_{oc}}{i_{sc}} = \frac{36}{9/2} = 8 \ \Omega$$

Hence, the circuit time constant is

$$\tau = \frac{L}{R_{Th}} = \frac{3}{8} \text{ s}$$

Step 6. Using the information just computed, we can derive the final equation for $v_o(t)$:

$$K_1 = v_o(\infty) = 27$$

$$K_2 = v_o(0+) - v_o(\infty) = 18 - 27 = -9$$

Therefore,

$$v_o(t) = 27 - 9e^{-t/(3/8)} \text{ V}$$

LEARNING EXTENSION

E7.5 If the switch in the network in Fig. E7.5 closes at $t = 0$, find $v_o(t)$ for $t > 0$. **PSV**

ANSWER:
$$v_o(t) = 24 + 36e^{-(t/12)} \text{ V}.$$

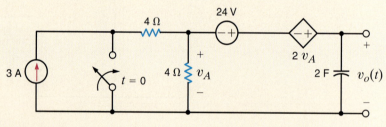

Figure E7.5

At this point, it is appropriate to state that not all switch action will always occur at time $t = 0$. It may occur at any time t_0. In this case the results of the step-by-step analysis yield the following equations:

$$x(t_0) = K_1 + K_2$$

$$x(\infty) = K_1$$

and

$$x(t) = x(\infty) + [x(t_0) - x(\infty)]e^{-(t-t_0)/\tau} \qquad t > t_0$$

The function is essentially time-shifted by t_0 seconds.

Finally, note that if more than one independent source is present in the network, we can simply employ superposition to obtain the total response.

PULSE RESPONSE   Thus far we have examined networks in which a voltage or current source is suddenly applied. As a result of this sudden application of a source, voltages or currents in the circuit are forced to change abruptly. A forcing function whose value changes in a discontinuous manner or has a discontinuous derivative is called a *singular function*. Two such singular functions that are very important in circuit analysis are the unit impulse function and the unit step function. We will defer a discussion of the unit impulse function until a later chapter and concentrate on the unit step function.

The *unit step function* is defined by the following mathematical relationship:

$$u(t) = \begin{cases} 0 & t < 0 \\ 1 & t > 0 \end{cases}$$

In other words, this function, which is dimensionless, is equal to zero for negative values of the argument and equal to 1 for positive values of the argument. It is undefined for a zero argument where the function is discontinuous. A graph of the unit step is shown in Fig. 7.11a. The unit step is dimensionless, and therefore a voltage step of V_o volts or a current step of I_o amperes is written as $V_o u(t)$ and $I_o u(t)$, respectively. Equivalent circuits for a voltage step are shown in Figs. 7.11b and c. Equivalent circuits for a current step are shown in Figs. 7.11d and e. If we use the definition of the unit step, it is easy to generalize this function by replacing the argument t by $t - t_0$. In this case

$$u(t - t_0) = \begin{cases} 0 & t < t_0 \\ 1 & t > t_0 \end{cases}$$

A graph of this function is shown in Fig. 7.11f. Note that $u(t - t_0)$ is equivalent to delaying $u(t)$ by t_0 seconds, so that the abrupt change occurs at time $t = t_0$.

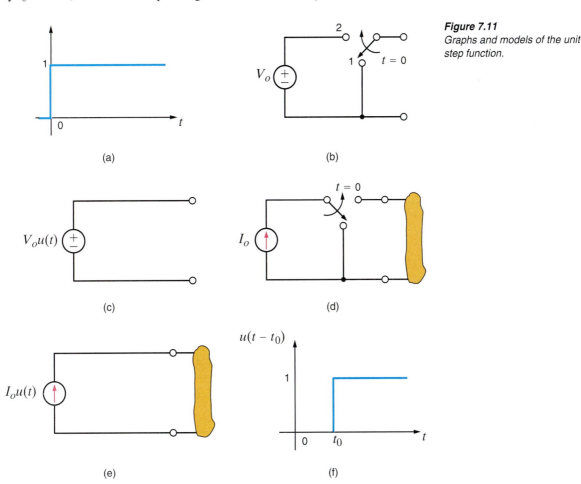

Figure 7.11
Graphs and models of the unit step function.

(a)

(b)

(c)

(d)

(e)

(f)

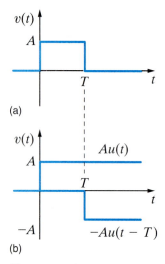

Figure 7.12 *Construction of a pulse via two step functions.*

Step functions can be used to construct one or more pulses. For example, the voltage pulse shown in Fig. 7.12a can be formulated by initiating a unit step at $t = 0$ and subtracting one that starts at $t = T$, as shown in Fig. 7.12b. The equation for the pulse is

$$v(t) = A\big[u(t) - u(t - T)\big]$$

If the pulse is to start at $t = t_0$ and have width T, the equation would be

$$v(t) = A\{u(t - t_0) - u[t - (t_0 + T)]\}$$

Using this approach, we can write the equation for a pulse starting at any time and ending at any time. Similarly, using this approach, we could write the equation for a series of pulses, called a *pulse train*, by simply forming a summation of pulses constructed in the manner illustrated previously.

The following example will serve to illustrate many of the concepts we have just presented.

Example 7.6

Consider the circuit shown in Fig. 7.13a. The input function is the voltage pulse shown in Fig. 7.13b. Since the source is zero for all negative time, the initial conditions for the network are zero $\big[$i.e., $v_C(0-) = 0\big]$. The response $v_o(t)$ for $0 < t < 0.3$ s is due to the application of the constant source at $t = 0$ and is not influenced by any source changes that will occur later. At $t = 0.3$ s the forcing function becomes zero, and therefore $v_o(t)$ for $t > 0.3$ s is the source-free or natural response of the network.

Let us determine the expression for the voltage $v_o(t)$.

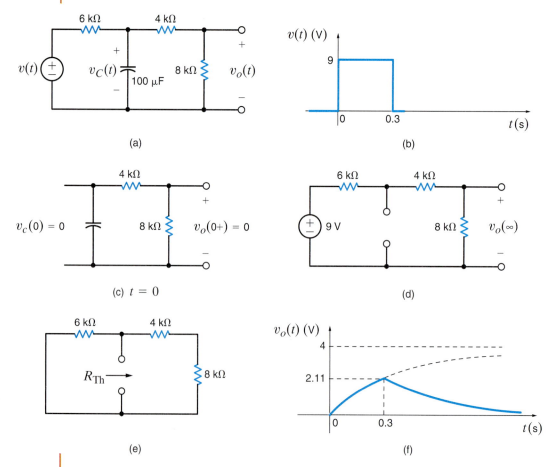

Figure 7.13 *Pulse response of a network.*

SOLUTION Since the output voltage $v_o(t)$ is a voltage division of the capacitor voltage, and the initial voltage across the capacitor is zero, we know that $v_o(0+) = 0$, as shown in Fig. 7.13c.

If no changes were made in the source after $t = 0$, the steady-state value of $v_o(t)$ [i.e., $v_o(\infty)$] due to the application of the unit step at $t = 0$ would be

$$v_o(\infty) = \frac{9}{6k + 4k + 8k}(8k)$$
$$= 4 \text{ V}$$

as shown in Fig. 7.13d.

The Thévenin equivalent resistance is

$$R_{\text{Th}} = \frac{(6k)(12k)}{6k + 12k}$$
$$= 4 \text{ k}\Omega$$

as illustrated in Fig. 7.13e.

Therefore, the circuit time constant τ is

$$\tau = R_{\text{Th}}C$$
$$= (4)(10^3)(100)(10^{-6})$$
$$= 0.4 \text{ s}$$

Therefore, the response $v_o(t)$ for the period $0 < t < 0.3$ s is

$$v_o(t) = 4 - 4e^{-t/0.4} \text{ V} \quad 0 < t < 0.3 \text{ s}$$

The capacitor voltage can be calculated by realizing that using voltage division, $v_o(t) = 2/3\, v_C(t)$. Therefore,

$$v_C(t) = \frac{3}{2}\left(4 - 4e^{-t/0.4}\right) \text{ V}$$

Since the capacitor voltage is continuous,

$$v_C(0.3-) = v_C(0.3+)$$

and therefore,

$$v_o(0.3+) = \frac{2}{3}v_C(0.3-)$$
$$= 4\left(1 - e^{-0.3/0.4}\right)$$
$$= 2.11 \text{ V}$$

Since the source is zero for $t > 0.3$ s, the final value for $v_o(t)$ as $t \to \infty$ is zero. Therefore, the expression for $v_o(t)$ for $t > 0.3$ s is

$$v_o(t) = 2.11e^{-(t-0.3)/0.4} \text{ V} \quad t > 0.3 \text{ s}$$

The term $e^{-(t-0.3)/0.4}$ indicates that the exponential decay starts at $t = 0.3$ s. The complete solution can be written by means of superposition as

$$v_o(t) = 4\left(1 - e^{-t/0.4}\right)u(t) - 4\left(1 - e^{-(t-0.3)/0.4}\right)u(t - 0.3) \text{ V}$$

or, equivalently, the complete solution is

$$v_o(t) = \begin{cases} 0 & t < 0 \\ 4\left(1 - e^{-t/0.4}\right) \text{ V} & 0 < t < 0.3 \text{ s} \\ 2.11e^{-(t-0.3)/0.4} \text{ V} & 0.3 \text{ s} < t \end{cases}$$

which in mathematical form is

$$v_o(t) = 4\left(1 - e^{-t/0.4}\right)\left[u(t) - u(t - 0.3)\right] + 2.11e^{-(t-0.3)/0.4}u(t - 0.3) \text{ V}$$

Note that the term $\left[u(t) - u(t - 0.3)\right]$ acts like a gating function that captures only the part of the step response that exists in the time interval $0 < t < 0.3$ s. The output as a function of time is shown in Fig. 7.13f.

LEARNING EXTENSION

E7.6 The voltage source in the network in Fig. E7.6a is shown in Fig. E7.6b. The initial current in the inductor must be zero. (Why?) Determine the output voltage $v_o(t)$ for $t > 0$.

ANSWER: $v_o(t) = 0$ for $t < 0$, $4(1 - e^{-(3/2)t})$ V for $0 \leq t \leq 1$ s, and $3.11e^{-(3/2)(t-1)}$ V for 1 s $< t$.

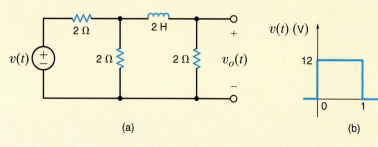

(a)   (b)

Figure E7.6

7.3 Second-Order Circuits

THE BASIC CIRCUIT EQUATION   To begin our development, let us consider the two basic *RLC* circuits shown in Fig. 7.14. We assume that energy may be initially stored in both the inductor and capacitor. The node equation for the parallel *RLC* circuit is

$$\frac{v}{R} + \frac{1}{L}\int_{t_0}^{t} v(x)\,dx + i_L(t_0) + C\frac{dv}{dt} = i_S(t)$$

Similarly, the loop equation for the series *RLC* circuit is

$$Ri + \frac{1}{C}\int_{t_0}^{t} i(x)\,dx + v_C(t_0) + L\frac{di}{dt} = v_S(t)$$

Note that the equation for the node voltage in the parallel circuit is of the same form as that for the loop current in the series circuit. Therefore, the solution of these two circuits is dependent on solving one equation. If the two preceding equations are differentiated with respect to time, we obtain

$$C\frac{d^2v}{dt^2} + \frac{1}{R}\frac{dv}{dt} + \frac{v}{L} = \frac{di_S}{dt}$$

and

$$L\frac{d^2i}{dt^2} + R\frac{di}{dt} + \frac{i}{C} = \frac{dv_S}{dt}$$

Since both circuits lead to a second-order differential equation with constant coefficients, we will concentrate our analysis on this type of equation.

Figure 7.14
Parallel and series RLC circuits.

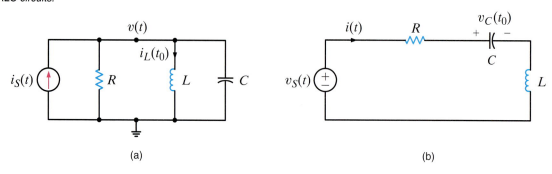

(a)   (b)

THE RESPONSE EQUATIONS In concert with our development of the solution of a first-order differential equation that results from the analysis of either an *RL* or an *RC* circuit as outlined earlier, we will now employ the same approach here to obtain the solution of a second-order differential equation that results from the analysis of *RLC* circuits. As a general rule, for this case we are confronted with an equation of the form

$$\frac{d^2x(t)}{dt^2} + a_1\frac{dx(t)}{dt} + a_2x(t) = f(t) \qquad \textbf{7.12}$$

Once again we use the fact that if $x(t) = x_p(t)$ is a solution to Eq. (7.12), and if $x(t) = x_c(t)$ is a solution to the homogeneous equation

$$\frac{d^2x(t)}{dt^2} + a_1\frac{dx(t)}{dt} + a_2x(t) = 0$$

then

$$x(t) = x_p(t) + x_c(t)$$

is a solution to the original Eq. (7.12). If we again confine ourselves to a constant forcing function $\big[$i.e., $f(t) = A\big]$, the development at the beginning of this chapter shows that the solution of Eq. (7.12) will be of the form

$$x(t) = \frac{A}{a_2} + x_c(t) \qquad \textbf{7.13}$$

Let us now turn our attention to the solution of the homogeneous equation

$$\frac{d^2x(t)}{dt^2} + a_1\frac{dx(t)}{dt} + a_2x(t) = 0$$

where a_1 and a_2 are constants. For simplicity we will rewrite the equation in the form

$$\frac{d^2x(t)}{dt^2} + 2\zeta\omega_0\frac{dx(t)}{dt} + \omega_0^2 x(t) = 0 \qquad \textbf{7.14}$$

where we have made the following simple substitutions for the constants $a_1 = 2\zeta\omega_0$ and $a_2 = \omega_0^2$.

Following the development of a solution for the first-order homogeneous differential equation earlier in this chapter, the solution of Eq. (7.14) must be a function whose first- and second-order derivatives have the same form, so that the left-hand side of Eq. (7.14) will become identically zero for all t. Again we assume that

$$x(t) = Ke^{st}$$

Substituting this expression into Eq. (7.14) yields

$$s^2 Ke^{st} + 2\zeta\omega_0 s Ke^{st} + \omega_0^2 Ke^{st} = 0$$

Dividing both sides of the equation by Ke^{st} yields

$$s^2 + 2\zeta\omega_0 s + \omega_0^2 = 0 \qquad \textbf{7.15}$$

This equation is commonly called the *characteristic equation*; ζ is called the *exponential damping ratio*, and ω_0 is referred to as the *undamped natural frequency*. The importance of this terminology will become clear as we proceed with the development. If this equation is satisfied, our assumed solution $x(t) = Ke^{st}$ is correct. Employing the quadratic formula, we find that Eq. (7.15) is satisfied if

$$s = \frac{-2\zeta\omega_0 \pm \sqrt{4\zeta^2\omega_0^2 - 4\omega_0^2}}{2}$$

$$= -\zeta\omega_0 \pm \omega_0\sqrt{\zeta^2 - 1} \qquad \textbf{7.16}$$

Therefore, two values of s, s_1 and s_2, satisfy Eq. (7.15):

$$s_1 = -\zeta\omega_0 + \omega_0\sqrt{\zeta^2 - 1}$$
$$s_2 = -\zeta\omega_0 - \omega_0\sqrt{\zeta^2 - 1}$$

7.17

In general, then, the complementary solution of Eq. (7.14) is of the form

$$x_c(t) = K_1 e^{s_1 t} + K_2 e^{s_2 t}$$

7.18

K_1 and K_2 are constants that can be evaluated via the initial conditions $x(0)$ and $dx(0)/dt$. For example, since

$$x(t) = K_1 e^{s_1 t} + K_2 e^{s_2 t}$$

then

$$x(0) = K_1 + K_2$$

and

$$\left.\frac{dx(t)}{dt}\right|_{t=0} = \frac{dx(0)}{dt} = s_1 K_1 + s_2 K_2$$

Hence, $x(0)$ and $dx(0)/dt$ produce two simultaneous equations, which when solved yield the constants K_1 and K_2.

Close examination of Eqs. (7.17) and (7.18) indicates that the form of the solution of the homogeneous equation is dependent on the value ζ. For example, if $\zeta > 1$, the roots of the characteristic equation, s_1 and s_2, also called the *natural frequencies* because they determine the natural (unforced) response of the network, are real and unequal; if $\zeta < 1$, the roots are complex numbers; and finally, if $\zeta = 1$, the roots are real and equal.

Let us now consider the three distinct forms of the unforced response—that is, the response due to an initial capacitor voltage or initial inductor current.

Case 1, $\zeta > 1$ This case is commonly called *overdamped*. The natural frequencies s_1 and s_2 are real and unequal; therefore, the natural response of the network described by the second-order differential equation is of the form

$$x_c(t) = K_1 e^{-(\zeta\omega_0 - \omega_0\sqrt{\zeta^2-1})t} + K_2 e^{-(\zeta\omega_0 + \omega_0\sqrt{\zeta^2-1})t}$$

7.19

where K_1 and K_2 are found from the initial conditions. This indicates that the natural response is the sum of two decaying exponentials.

Case 2, $\zeta < 1$ This case is called *underdamped*. Since $\zeta < 1$, the roots of the characteristic equation given in Eq. (7.17) can be written as

$$s_1 = -\zeta\omega_0 + j\omega_0\sqrt{1 - \zeta^2} = -\sigma + j\omega_d$$
$$s_2 = -\zeta\omega_0 - j\omega_0\sqrt{1 - \zeta^2} = -\sigma - j\omega_d$$

where $j = \sqrt{-1}$, $\sigma = \zeta\omega_0$, and $\omega_d = \omega_0\sqrt{1 - \zeta^2}$. Thus, the natural frequencies are complex numbers (briefly discussed in the Appendix). The natural response is then of the form

$$x_c(t) = e^{-\zeta\omega_0 t}\left(A_1 \cos\omega_0\sqrt{1 - \zeta^2}\,t + A_2 \sin\omega_0\sqrt{1 - \zeta^2}\,t\right)$$

7.20

where A_1 and A_2, like K_1 and K_2, are constants, which are evaluated using the initial conditions $x(0)$ and $dx(0)/dt$. This illustrates that the natural response is an exponentially damped oscillatory response.

Case 3, $\zeta = 1$ This case, called *critically damped*, results in

$$s_1 = s_2 = -\zeta\omega_0$$

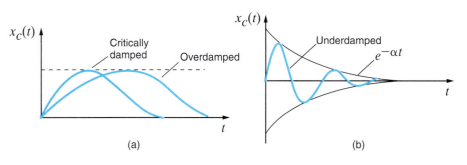

Figure 7.15
Comparison of overdamped, critically damped, and underdamped responses.

In the case where the characteristic equation has repeated roots, the general solution is of the form

$$x_c(t) = B_1 e^{-\zeta\omega_0 t} + B_2 t e^{-\zeta\omega_0 t} \qquad \textbf{7.21}$$

where B_1 and B_2 are constants derived from the initial conditions.

It is informative to sketch the natural response for the three cases we have discussed: over-damped, Eq. (7.19); underdamped, Eq. (7.20); and critically damped, Eq. (7.21). Figure 7.15 graphically illustrates the three cases for the situations in which $x_c(0) = 0$. Note that the critically damped response peaks and decays faster than the overdamped response. The underdamped response is an exponentially damped sinusoid whose rate of decay is dependent on the factor ζ. Actually, the terms $\pm e^{-\zeta\omega_0 t}$ define what is called the *envelope* of the response, and the damped oscillations (i.e., the oscillations of decreasing amplitude) exhibited by the waveform in Fig. 7.15b are called *ringing*.

LEARNING EXTENSIONS

E7.7 A parallel *RLC* circuit has the following circuit parameters: $R = 1\ \Omega$, $L = 2$ H, and $C = 2$ F. Compute the damping ratio and the undamped natural frequency of this network.

ANSWER: $\zeta = 0.5$; $\omega_0 = 0.5$ rad/s.

E7.8 A series *RLC* circuit consists of $R = 2\ \Omega$, $L = 1$ H, and a capacitor. Determine the type of response exhibited by the network if (a) $C = 1/2$ F, (b) $C = 1$ F, and (c) $C = 2$ F.

ANSWER: (a) underdamped; (b) critically damped; (c) overdamped.

THE NETWORK RESPONSE We will now analyze a number of simple *RLC* networks that contain both nonzero initial conditions and constant forcing functions. Circuits that exhibit overdamped, underdamped, and critically damped responses will be considered.

PROBLEM-SOLVING STRATEGY

Second-Order Transient Circuits

Step 1. Write the differential equation that describes the circuit.

Step 2. Derive the characteristic equation, which can be written in the form $s^2 + 2\zeta\omega_0 s + \omega_0^2 = 0$, where ζ is the damping ratio and ω_0 is the undamped natural frequency.

Step 3. The two roots of the characteristic equation will determine the type of response. If the roots are real and unequal (i.e., $\zeta > 1$), the network response is overdamped. If the roots are real and equal (i.e., $\zeta = 1$), the network response is critically damped. If the roots are complex (i.e., $\zeta < 1$), the network response is underdamped.

(continues on the next page)

Step 4. The damping condition and corresponding response for the aforementioned three cases outlined are as follows:

Overdamped: $x(t) = K_1 e^{-(\zeta\omega_0 - \omega_0\sqrt{\zeta^2-1})t} + K_2 e^{-(\zeta\omega_0 + \omega_0\sqrt{\zeta^2-1})t}$

Critically damped: $x(t) = B_1 e^{-\zeta\omega_0 t} + B_2 t e^{-\zeta\omega_0 t}$

Underdamped: $x(t) = e^{-\sigma t}(A_1 \cos\omega_d t + A_2 \sin\omega_d t)$, where $\sigma = \zeta\omega_0$, and $\omega_d = \omega_0\sqrt{1-\zeta^2}$

Step 5. Two initial conditions, either given or derived, are required to obtain the two unknown coefficients in the response equation.

The following examples will demonstrate the analysis techniques.

Example 7.7

Consider the parallel *RLC* circuit shown in Fig. 7.16. The second-order differential equation that describes the voltage $v(t)$ is

$$\frac{d^2v}{dt^2} + \frac{1}{RC}\frac{dv}{dt} + \frac{v}{LC} = 0$$

Figure 7.16
Parallel RLC *circuit.*

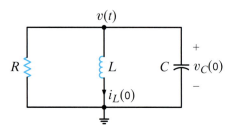

A comparison of this equation with Eqs. (7.14) and (7.15) indicates that for the parallel *RLC* circuit the damping term is $1/2\,RC$ and the undamped natural frequency is $1/\sqrt{LC}$. If the circuit parameters are $R = 2\,\Omega$, $C = 1/5$ F, and $L = 5$ H, the equation becomes

$$\frac{d^2v}{dt^2} + 2.5\frac{dv}{dt} + v = 0$$

Let us assume that the initial conditions on the storage elements are $i_L(0) = -1$ A and $v_C(0) = 4$ V. Let us find the node voltage $v(t)$ and the inductor current.

SOLUTION The characteristic equation for the network is

$$s^2 + 2.5s + 1 = 0$$

and the roots are

$$s_1 = -2$$
$$s_2 = -0.5$$

Since the roots are real and unequal, the circuit is overdamped, and $v(t)$ is of the form

$$v(t) = K_1 e^{-2t} + K_2 e^{-0.5t}$$

The initial conditions are now employed to determine the constants K_1 and K_2. Since $v(t) = v_C(t)$,

$$v_C(0) = v(0) = 4 = K_1 + K_2$$

The second equation needed to determine K_1 and K_2 is normally obtained from the expression

$$\frac{dv(t)}{dt} = -2K_1 e^{-2t} - 0.5K_2 e^{-0.5t}$$

However, the second initial condition is not $dv(0)/dt$. If this were the case, we would simply evaluate the equation at $t = 0$. This would produce a second equation in the unknowns K_1 and K_2.

We can, however, circumvent this problem by noting that the node equation for the circuit can be written as

$$C \frac{dv(t)}{dt} + \frac{v(t)}{R} + i_L(t) = 0$$

or

$$\frac{dv(t)}{dt} = \frac{-1}{RC} v(t) - \frac{i_L(t)}{C}$$

At $t = 0$,

$$\frac{dv(0)}{dt} = \frac{-1}{RC} v(0) - \frac{1}{C} i_L(0)$$

$$= -2.5(4) - 5(-1)$$

$$= -5$$

However, since

$$\frac{dv(t)}{dt} = -2K_1 e^{-2t} - 0.5K_2 e^{-0.5t}$$

then when $t = 0$

$$-5 = -2K_1 - 0.5K_2$$

This equation, together with the equation

$$4 = K_1 + K_2$$

produces the constants $K_1 = 2$ and $K_2 = 2$. Therefore, the final equation for the voltage is

$$v(t) = 2e^{-2t} + 2e^{-0.5t} \text{ V}$$

Note that the voltage equation satisfies the initial condition $v(0) = 4$ V. The response curve for this voltage $v(t)$ is shown in Fig. 7.17.

The inductor current is related to $v(t)$ by the equation

$$i_L(t) = \frac{1}{L} \int v(t)\, dt$$

Substituting our expression for $v(t)$ yields

$$i_L(t) = \frac{1}{5} \int \left[2e^{-2t} + 2e^{-0.5t} \right] dt$$

or

$$i_L(t) = -\frac{1}{5} e^{-2t} - \frac{4}{5} e^{-0.5t} \text{ A}$$

Note that in comparison with the *RL* and *RC* circuits, the response of this *RLC* circuit is controlled by two time constants. The first term has a time constant of 1/2 s, and the second term has a time constant of 2 s.

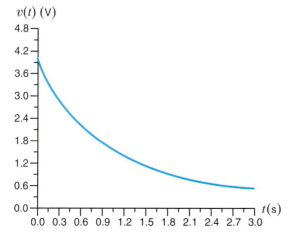

Figure 7.17
Overdamped response.

Once again, the MATLAB symbolic operations can be employed to solve the ordinary differential equation (ODE) that result from the analysis of these second-order circuits.

In Example 7.7, the ODE and its initial conditions are

$$\frac{d^2v}{dt^2} + 2.5\frac{dv}{dt} + v = 0, \quad v(0) = 4, \quad i(0) = -1$$

However,

$$C\frac{dv(0)}{dt} + \frac{v(0)}{R} + i(0) = 0$$

Therefore,

$$\frac{dv(0)}{dt} = 5\left(1 - \frac{4}{2}\right) = -5$$

Following the MATLAB symbolic operation development used in conjunction with Example 7.1, we can list the symbolic notation and solution for this equation as follows.

```
dsolve('D2v + 2.5*Dv + v = 0', 'v(0) = 4', 'Dv(0) = -5')
ans =
2*exp(-2*t)+2*exp(-1/2*t)
```

Example 7.8

The series *RLC* circuit shown in Fig. 7.18 has the following parameters: $C = 0.04$ F, $L = 1$ H, $R = 6\ \Omega$, $i_L(0) = 4$ A, and $v_C(0) = -4$ V. The equation for the current in the circuit is given by the expression

$$\frac{d^2i}{dt^2} + \frac{R}{L}\frac{di}{dt} + \frac{i}{LC} = 0$$

A comparison of this equation with Eqs. (7.14) and (7.15) illustrates that for a series *RLC* circuit the damping term is $R/2L$ and the undamped natural frequency is $1/\sqrt{LC}$. Substituting the circuit element values into the preceding equation yields

$$\frac{d^2i}{dt^2} + 6\frac{di}{dt} + 25i = 0$$

Let us determine the expression for both the current and the capacitor voltage.

Figure 7.18
Series RLC circuit.

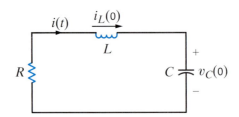

SOLUTION The characteristic equation is then

$$s_2 + 6s + 25 = 0$$

and the roots are

$$s_1 = -3 + j4$$
$$s_2 = -3 - j4$$

Since the roots are complex, the circuit is underdamped, and the expression for $i(t)$ is

$$i(t) = K_1 e^{-3t}\cos 4t + K_2 e^{-3t}\sin 4t$$

Using the initial conditions, we find that

$$i(0) = 4 = K_1$$

and

$$\frac{di}{dt} = -4K_1 e^{-3t} \sin 4t - 3K_1 e^{-3t} \cos 4t + 4K_2 e^{-3t} \cos 4t - 3K_2 e^{-3t} \sin 4t$$

and thus

$$\frac{di(0)}{dt} = -3K_1 + 4K_2$$

Although we do not know $di(0)/dt$, we can find it via KVL. From the circuit we note that

$$Ri(0) + L\frac{di(0)}{dt} + v_C(0) = 0$$

or

$$\frac{di(0)}{dt} = -\frac{R}{L}i(0) - \frac{v_C(0)}{L}$$

$$= -\frac{6}{1}(4) + \frac{4}{1}$$

$$= -20$$

Therefore,

$$-3K_1 + 4K_2 = -20$$

and since $K_1 = 4$, $K_2 = -2$, the expression then for $i(t)$ is

$$i(t) = 4e^{-3t} \cos 4t - 2e^{-3t} \sin 4t \text{ A}$$

Note that this expression satisfies the initial condition $i(0) = 4$. The voltage across the capacitor could be determined via KVL using this current:

$$Ri(t) + L\frac{di(t)}{dt} + v_C(t) = 0$$

or

$$v_C(t) = -Ri(t) - L\frac{di(t)}{dt}$$

Substituting the preceding expression for $i(t)$ into this equation yields

$$v_C(t) = -4e^{-3t} \cos 4t + 22e^{-3t} \sin 4t \text{ V}$$

Note that this expression satisfies the initial condition $v_C(0) = -4$ V.

The MATLAB program for plotting this function in the time interval $t > 0$ is as follows:

```
>>tau = 1/3;
>>tend = 10*tau;
>>t = linspace(0, tend, 150);
>>v1 = -4*exp(-3*t).*cos(4*t);
>>v2 = 22*exp(-3*t).*sin(4*t);
>>v = v1+v2;
>>plot (t,v)
>>xlabel('Time (s)')
>>ylabel('Voltage (V)')
```

Note that the functions `exp(-3*t)` and `cos(4*t)` produce arrays the same size as `t`. Hence, when we multiply these functions together, we need to specify that we want the arrays to be multiplied element by element. Element-by-element multiplication of an array is denoted in MATLAB by the `.*` notation. Multiplying an array by a scalar `(-4*exp(-3*t)`, for example, can be performed by using the `*` by itself.

The plot generated by this program is shown in Fig. 7.19.

Figure 7.19
Underdamped response.

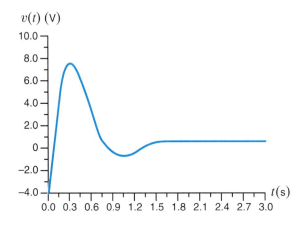

$v(t)$ (V)

Example 7.9

Let us examine the circuit in Fig. 7.20, which is slightly more complicated than the two we have considered earlier. The two equations that describe the network are

$$L\frac{di(t)}{dt} + R_1 i(t) + v(t) = 0$$

$$i(t) = C\frac{dv(t)}{dt} + \frac{v(t)}{R_2}$$

Figure 7.20
Series-parallel RLC circuit.

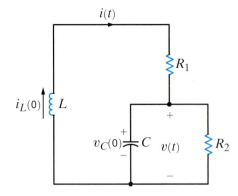

Substituting the second equation into the first yields

$$\frac{d^2v}{dt^2} + \left(\frac{1}{R_2 C} + \frac{R_1}{L}\right)\frac{dv}{dt} + \frac{R_1 + R_2}{R_2 LC}v = 0$$

If the circuit parameters and initial conditions are

$$R_1 = 10\ \Omega \qquad C = \frac{1}{8}\text{F} \qquad v_C(0) = 1\ \text{V}$$

$$R_2 = 8\ \Omega \qquad L = 2\ \text{H} \qquad i_L(0) = \frac{1}{2}\ \text{A}$$

the differential equation becomes

$$\frac{d^2v}{dt^2} + 6\frac{dv}{dt} + 9v = 0$$

We wish to find expressions for the current $i(t)$ and the voltage $v(t)$.

SOLUTION   The characteristic equation is then

$$s^2 + 6s + 9 = 0$$

and hence the roots are

$$s_1 = -3$$
$$s_2 = -3$$

Since the roots are real and equal, the circuit is critically damped. The term $v(t)$ is then given by the expression

$$v(t) = K_1 e^{-3t} + K_2 t e^{-3t}$$

Since $v(t) = v_C(t)$,

$$v(0) = v_C(0) = 1 = K_1$$

In addition,

$$\frac{dv(t)}{dt} = -3K_1 e^{-3t} + K_2 e^{-3t} - 3K_2 t e^{-3t}$$

However,

$$\frac{dv(t)}{dt} = \frac{i(t)}{C} - \frac{v(t)}{R_2 C}$$

Setting these two expressions equal to one another and evaluating the resultant equation at $t = 0$ yields

$$\frac{1/2}{1/8} - \frac{1}{1} = -3K_1 + K_2$$

$$3 = -3K_1 + K_2$$

Since $K_1 = 1$, $K_2 = 6$ and the expression for $v(t)$ is

$$v(t) = e^{-3t} + 6t e^{-3t} \text{ V}$$

Note that the expression satisfies the initial condition $v(0) = 1$.

The current $i(t)$ can be determined from the nodal analysis equation at $v(t)$.

$$i(t) = C\frac{dv(t)}{dt} + \frac{v(t)}{R_2}$$

Substituting $v(t)$ from the preceding equation, we find

$$i(t) = \frac{1}{8}\left[-3e^{-3t} + 6e^{-3t} - 18t e^{-3t}\right] + \frac{1}{8}\left[e^{-3t} + 6t e^{-3t}\right]$$

or

$$i(t) = \frac{1}{2}e^{-3t} - \frac{3}{2}t e^{-3t} \text{ A}$$

If this expression for the current is employed in the circuit equation,

$$v(t) = -L\frac{di(t)}{dt} - R_1 i(t)$$

we obtain

$$v(t) = e^{-3t} + 6t e^{-3t} \text{ V}$$

which is identical to the expression derived earlier.

Figure 7.21
Critically damped response.

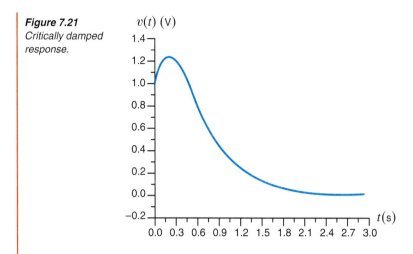

The MATLAB program for generating a plot of this function, shown in Fig. 7.21, is listed here.

```
>>tau = 1/3;
>>tend = 10*tau;
>>t = linspace(0, tend, 150);
>>v = exp(-3*t) + 6*t.*exp(-3*t);
>>plot(t,v)
>>xlabel('Time (s)')
>>ylabel('Voltage (V)')
```

LEARNING EXTENSIONS

E7.9 The switch in the network in Fig. E7.9 opens at $t = 0$. Find $i(t)$ for $t > 0$. **PSV**

ANSWER:
$i(t) = -2e^{-t/2} + 4e^{-t}$ A.

Figure E7.9

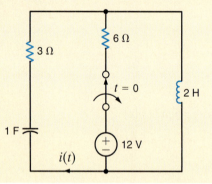

E7.10 The switch in the network in Fig. E7.10 moves from position 1 to position 2 at $t = 0$. Find $v_o(t)$ for $t > 0$.

ANSWER:
$v_o(t) = 2(e^{-t} - 3e^{-3t})$ V.

Figure E7.10

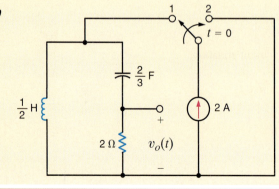

Example 7.10

Consider the circuit shown in Fig. 7.22. This circuit is the same as that analyzed in Example 7.8, except that a constant forcing function is present. The circuit parameters are the same as those used in Example 7.8:

$$C = 0.04 \text{ F} \qquad i_L(0) = 4 \text{ A}$$
$$L = 1 \text{ H} \qquad v_C(0) = -4 \text{ V}$$
$$R = 6 \ \Omega$$

We want to find an expression for $v_C(t)$ for $t > 0$.

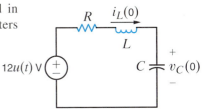

Figure 7.22
Series RLC circuit with a step function input.

SOLUTION   From our earlier mathematical development we know that the general solution of this problem will consist of a particular solution plus a complementary solution. From Example 7.8 we know that the complementary solution is of the form $K_3 e^{-3t} \cos 4t + K_4 e^{-3t} \sin 4t$. The particular solution is a constant, since the input is a constant and therefore the general solution is

$$v_C(t) = K_3 e^{-3t} \cos 4t + K_4 e^{-3t} \sin 4t + K_5$$

An examination of the circuit shows that in the steady state the final value of $v_C(t)$ is 12 V, since in the steady-state condition, the inductor is a short circuit and the capacitor is an open circuit. Thus, $K_5 = 12$. The steady-state value could also be immediately calculated from the differential equation. The form of the general solution is then

$$v_C(t) = K_3 e^{-3t} \cos 4t + K_4 e^{-3t} \sin 4t + 12$$

The initial conditions can now be used to evaluate the constants K_3 and K_4.

$$v_C(0) = -4 = K_3 + 12$$
$$-16 = K_3$$

Since the derivative of a constant is zero, the results of Example 7.8 show that

$$\frac{dv_C(0)}{dt} = \frac{i(0)}{C} = 100 = -3K_3 + 4K_4$$

and since $K_3 = -16$, $K_4 = 13$. Therefore, the general solution for $v_C(t)$ is

$$v_C(t) = 12 - 16e^{-3t} \cos 4t + 13e^{-3t} \sin 4t \text{ V}$$

Note that this equation satisfies the initial condition $v_C(0) = -4$ and the final condition $v_C(\infty) = 12$ V.

Example 7.11

Let us examine the circuit shown in Fig. 7.23. A close examination of this circuit will indicate that it is identical to that shown in Example 7.9 except that a constant forcing function is present. We assume the circuit is in steady state at $t = 0-$. The equations that describe the circuit for $t > 0$ are

$$L \frac{di(t)}{dt} + R_1 i(t) + v(t) = 24$$

$$i(t) = C \frac{dv(t)}{dt} + \frac{v(t)}{R_2}$$

Figure 7.23
Series-parallel RLC circuit with a constant forcing function.

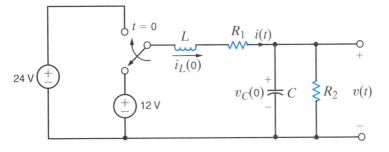

Combining these equations, we obtain

$$\frac{d^2v(t)}{dt^2} + \left(\frac{1}{R_2C} + \frac{R_1}{L}\right)\frac{dv(t)}{dt} + \frac{R_1 + R_2}{R_2LC}v(t) = \frac{24}{LC}$$

If the circuit parameters are $R_1 = 10\ \Omega$, $R_2 = 2\ \Omega$, $L = 2$ H, and $C = 1/4$ F, the differential equation for the output voltage reduces to

$$\frac{d^2v(t)}{dt^2} + 7\frac{dv(t)}{dt} + 12v(t) = 48$$

Let us determine the output voltage $v(t)$.

SOLUTION The characteristic equation is

$$s^2 + 7s + 12 = 0$$

and hence the roots are

$$s_1 = -3$$
$$s_2 = -4$$

The circuit response is overdamped, and therefore the general solution is of the form

$$v(t) = K_1e^{-3t} + K_2e^{-4t} + K_3$$

The steady-state value of the voltage, K_3, can be computed from Fig. 7.24a. Note that

$$v(\infty) = 4\ \text{V} = K_3$$

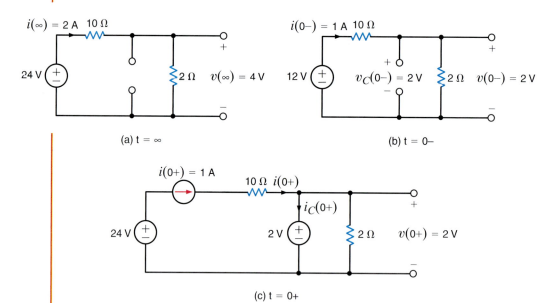

(a) t = ∞

(b) t = 0−

(c) t = 0+

Figure 7.24 Equivalent circuits at $t = \infty$, $t = 0-$, and $t = 0+$ for the circuit in Fig. 7.23.

The initial conditions can be calculated from Figs. 7.24b and c, which are valid at $t = 0-$ and $t = 0+$, respectively. Note that $v(0+) = 2$ V and, hence, from the response equation

$$v(0+) = 2\ \text{V} = K_1 + K_2 + 4$$
$$-2 = K_1 + K_2$$

Figure 7.24c illustrates that $i(0+) = 1$. From the response equation we see that

$$\frac{dv(0)}{dt} = -3K_1 - 4K_2$$

and since

$$\frac{dv(0)}{dt} = \frac{i(0)}{C} - \frac{v(0)}{R_2 C}$$

$$= 4 - 4$$

$$= 0$$

then

$$0 = -3K_1 - 4K_2$$

Solving the two equations for K_1 and K_2 yields $K_1 = -8$ and $K_2 = 6$. Therefore, the general solution for the voltage response is

$$v(t) = 4 - 8e^{-3t} + 6e^{-4t} \text{ V}$$

Note that this equation satisfies both the initial and final values of $v(t)$.

LEARNING EXTENSION

E7.11 The switch in the network in Fig. E7.11 moves from position 1 to position 2 at $t = 0$. Compute $i_o(t)$ for $t > 0$ and use this current to determine $v_o(t)$ for $t > 0$.

ANSWER:

$$i_o(t) = -\frac{11}{6}e^{-3t} + \frac{14}{6}e^{-6t} \text{ A};$$

$$v_o(t) = 12 + 18i_o(t) \text{ V}.$$

Figure E7.11

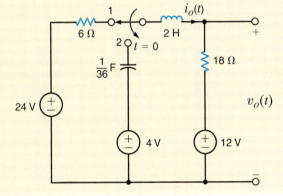

7.4   *Transient PSPICE Analysis Using* Schematic *Capture*

INTRODUCTION   In transient analyses, we determine voltages and currents as functions of time. Typically, the time dependence is demonstrated by plotting the waveforms using time as the independent variable. PSPICE can perform this kind of analysis, called a *Transient* simulation, in which all voltages and currents are determined over a specified time duration. To facilitate plotting, PSPICE uses what is known as the PROBE utility, which will be described later. As an introduction to transient analysis, let us simulate the circuit in Fig. 7.25, plot the voltage $v_C(t)$ and the current $i(t)$, and extract the time constant.

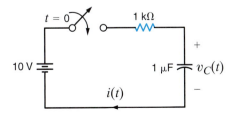

Figure 7.25
A circuit used for Transient simulation.

Figure 7.26
The Schematics *circuit.*

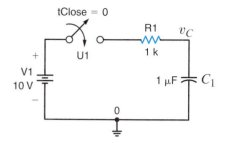

Although we will introduce some new PSPICE topics in this section, *Schematics* fundamentals such as getting parts, wiring, and editing part names and values have already been covered in Chapter 5. Also, uppercase text refers to PSPICE utilities and dialog boxes, whereas boldface denotes keyboard or mouse inputs.

THE SWITCH PARTS The inductor and capacitor parts are called L and C, respectively, and are in the ANALOG library. The switch, called SW_TCLOSE, is in the EVAL library. There is also a SW_TOPEN part that models an opening switch. After placing and wiring the switch along with the other parts, the *Schematics* circuit appears as that shown in Fig. 7.26.

To edit the switch's attributes, double-click on the switch symbol and the ATTRIBUTES box in Fig. 7.27 will appear. Deselecting the **Include Non-changeable Attributes** and **Include System-defined Attributes** fields limits the attribute list to those we can edit and is highly recommended.

The attribute **tClose** is the time at which the switch begins to close, and **ttran** is the time required to complete the closure. Switch attributes **Rclosed** and **Ropen** are the switch's resistance in the closed and open positions, respectively. During simulations, the resistance of the switch changes linearly from **Ropen** at $t = $ **tClose** to **Rclosed** at $t = $ **tClose + ttran.**

When using the SW_TCLOSE and SW_TOPEN parts to simulate ideal switches, care should be taken to ensure that the values for **ttran**, **Rclosed**, and **Ropen** are appropriate for valid simulation results. In our present example, we see that the switch and R_1 are in series; thus, their resistances add. Using the default values listed in Fig. 7.27, we find that when the switch is closed, the switch resistance, **Rclosed**, is 0.01 Ω, 100,000 times smaller than that of the resistor. The resulting series-equivalent resistance is essentially that of the resistor. Alternatively, when the switch is open, the switch resistance is 1 MΩ, 1,000 times larger than that of the resistor. Now, the equivalent resistance is much larger than that of the resistor. Both are desirable scenarios.

Figure 7.27
The switch's ATTRIBUTES box.

U1 PartName: Sw_tClose		✕
Name	Value	
tClose	= 0	Save Attr
tClose=0		Change Display
ttran=1u		
Rclosed=0.01		Delete
Ropen=1Meg		
☐ Include Non-changeable Attributes		OK
☐ Include System-defined Attributes		Cancel

To determine a reasonable value for **ttran**, we first estimate the duration of the transient response. The component values yield a time constant of 1 ms, and thus all voltages and currents will reach steady state in about 5 ms. For accurate simulations, **ttran** should be much less than 5 ms. Therefore, the default value of 1 µs is viable in this case.

THE IMPORTANCE OF PIN NUMBERS

As mentioned in Chapter 5, each component within the various Parts libraries has two or more terminals. Within PSPICE, these terminals are called pins and are numbered sequentially starting with pin 1, as shown in Fig. 7.28 for several two-terminal parts. The significance of the pin numbers is their effect on currents plotted using the PROBE utility. PROBE always plots the current entering pin 1 and exiting pin 2. Thus, if the current through an element is to be plotted, the part should be oriented in the *Schematics* diagram such that the defined current direction enters the part at pin 1. This can be done by using the ROTATE command in the EDIT menu. ROTATE causes the part to spin 90° counterclockwise. In our example, we will plot the current $i(t)$ by plotting the current through the capacitor, I(C1). Therefore, when the *Schematics* circuit in Fig. 7.26 was created, the capacitor was rotated 270°. As a result, pin 1 is at the top of the diagram and the assigned current direction in Fig. 7.25 matches the direction presumed by PROBE. If a component's current direction in PROBE is opposite the desired direction, simply go to the *Schematics* circuit, rotate the part in question 180°, and re-simulate.

SETTING INITIAL CONDITIONS

To set the initial condition of the capacitor voltage, double-click on the capacitor symbol in Fig. 7.26 to open its ATTRIBUTE box, as shown in Fig. 7.29. Click on the **IC** field and set the value to the desired voltage, 0 V in this example. Setting the initial condition on an inductor current is done in a similar fashion. Be forewarned that the initial condition for a capacitor voltage is positive at pin 1 versus pin 2. Similarly, the initial condition for an inductor's current will flow into pin 1 and out of pin 2.

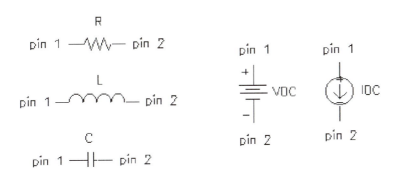

Figure 7.28
Pin numbers for common PSPICE parts.

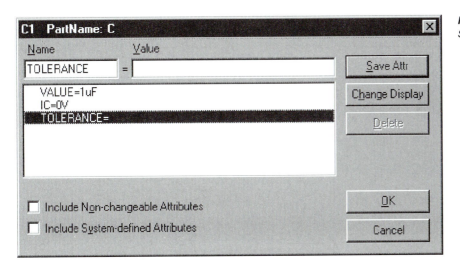

Figure 7.29
Setting the capacitor initial condition.

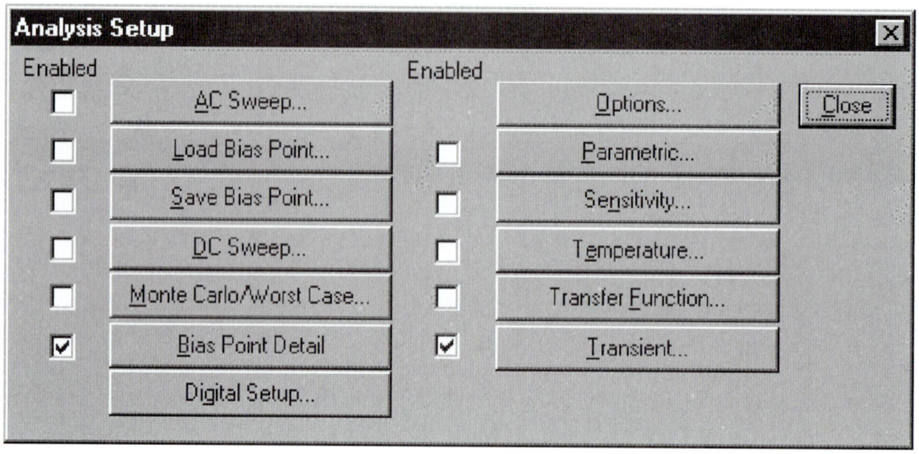

Figure 7.30
The ANALYSIS SETUP window.

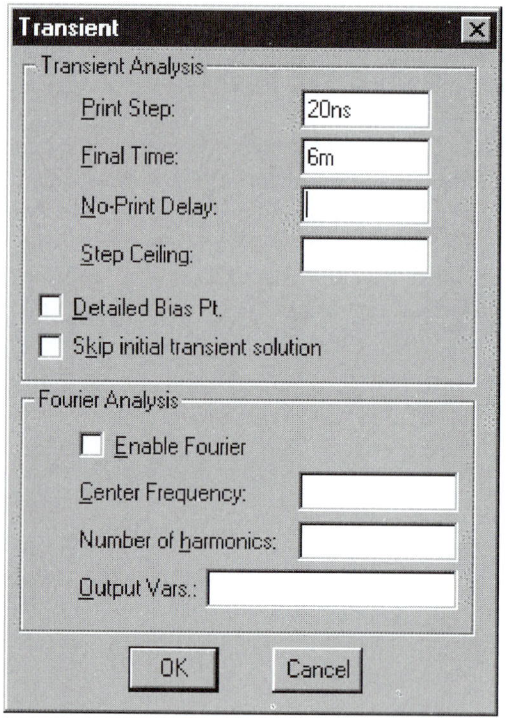

Figure 7.31
The TRANSIENT window.

SETTING UP A TRANSIENT ANALYSIS   The simulation duration is selected using **SETUP** from the **ANALYSIS** menu. When the SETUP window shown in Fig. 7.30 appears, double-click on the text **TRANSIENT** and the TRANSIENT window in Fig. 7.31 will appear. The simulation period described by **Final time** is selected as 6 milliseconds. All simulations start at $t = 0$. The **No-Print Delay** field sets the time the simulation runs before data collection begins. **Print Step** is the interval used for printing data to the output file. **Print Step** has no effect on the data used to create PROBE plots. The **Detailed Bias Pt.** option is useful when simulating circuits containing transistors and diodes, and thus will not be used here. When **Skip initial transient solution** is enabled, all capacitors and inductors that do not have specific initial condition values in their ATTRIBUTES boxes will use zero initial conditions.

   Sometimes, plots created in PROBE are not smooth. This is caused by an insufficient number of data points. More data points can be requested by inputting a **Step Ceiling** value. A reasonable first guess would be a hundredth of the **Final Time**. If the resulting PROBE plots are still unsatisfactory, reduce the **Step Ceiling** further. As soon as the TRANSIENT window is complete, simulate the circuit by selecting **Simulate** from the **Analysis** menu.

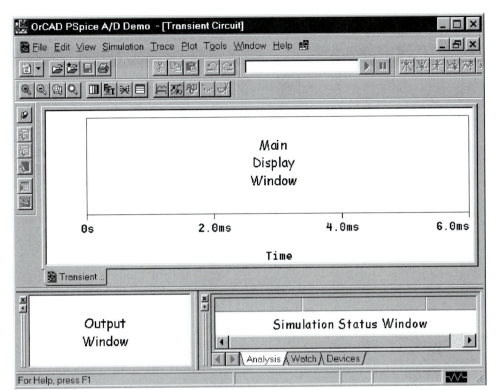

Figure 7.32
The PROBE window.

PLOTTING IN PROBE   When the PSPICE simulation is finished, the PROBE window shown in Fig. 7.32 will open. If not, select **Run Probe** from the **Analysis** menu. In Fig. 7.32, we see three subwindows: the main display window, the output window, and the simulation status window. The waveforms we choose to plot appear in the main display window. The output window shows messages from PSPICE about the success or failure of the simulation. Run-time information about the simulation appears in the simulation status window. Here we will focus on the main display window.

To plot the voltage, $v_C(t)$, select **Add Trace** from the **Trace** menu. The ADD TRACES window is shown in Fig. 7.33. Note that the options **Alias Names** and **Subcircuit Nodes** have been deselected, which greatly simplifies the ADD TRACES window. The capacitor voltage is obtained by clicking on **V(Vc)** in the left column. The PROBE window should look like that shown in Fig. 7.34.

Before adding the current $i(t)$ to the plot, we note that the dc source is 10 V and the resistance is 1 kΩ, which results in a loop current of a few milliamps. Since the capacitor voltage span is much greater, we will plot the current on a second y axis. From the **Plot** menu, select **Add Y Axis**. To add the current to the plot, select **Add Trace** from the **Trace** menu, then select **I(C1)**. Figure 7.35 shows the PROBE plot for $v_C(t)$ and $i(t)$.

FINDING THE TIME CONSTANT   Given that the final value of $v_C(t)$ is 10 V, we can write

$$v_C(t) = 10\left[1 - e^{-t/\tau}\right] \text{ V}$$

When $t = \tau$,

$$v_C(\tau) = 10\left(1 - e^{-1}\right) \text{ V} = 6.32 \text{ V}$$

To determine the time at which the capacitor voltage is 6.32 V, we activate the cursors by selecting **Cursor/Display** in the **Trace** menu. Two cursors can be used to extract x-y data from the plots. Use the $\leftarrow$ and $\rightarrow$ arrow keys to move the first cursor. By holding the SHIFT key down, the arrow keys move the second cursor. Moving the first cursor along the voltage plot, as shown in Fig. 7.36, we find that 6.32 V occurs at a time of 1 ms. Therefore, the time constant is 1 ms—exactly the *RC* product.

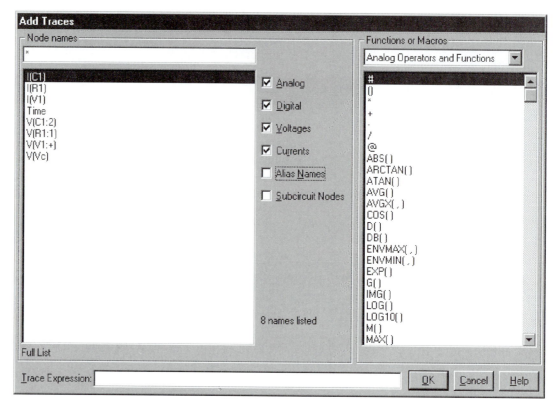

Figure 7.33
The Add Traces window.

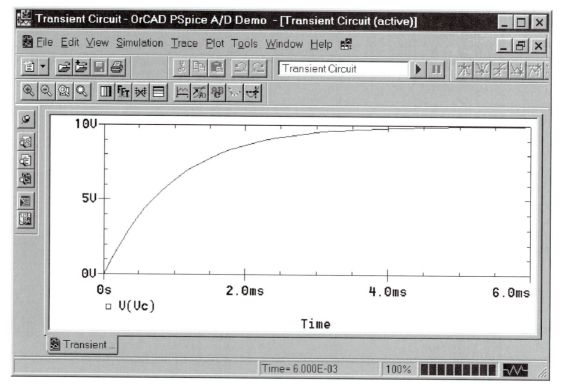

Figure 7.34
The capacitor voltage.

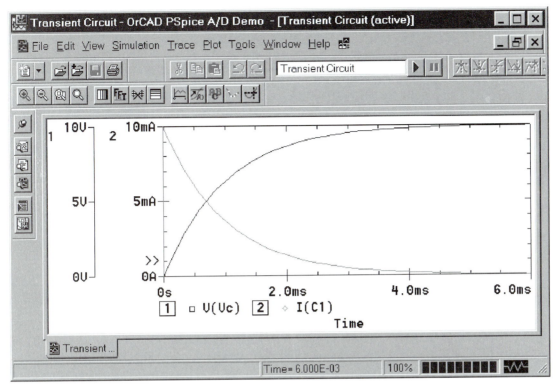

Figure 7.35
The capacitor voltage and the clockwise loop current.

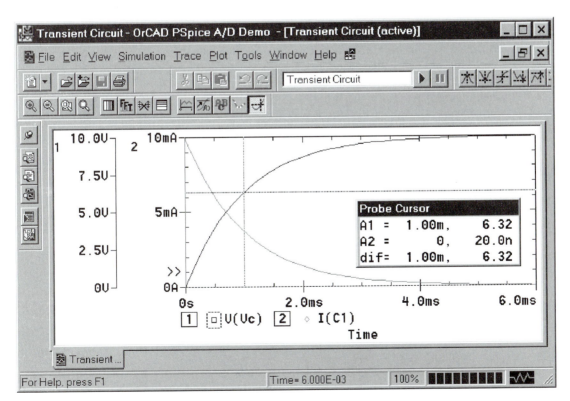

Figure 7.36
PROBE plot for $v_C(t)$ and time-constant extraction.

Figure 7.37
The SAVE/RESTORE DISPLAY window used in PROBE to save plots.

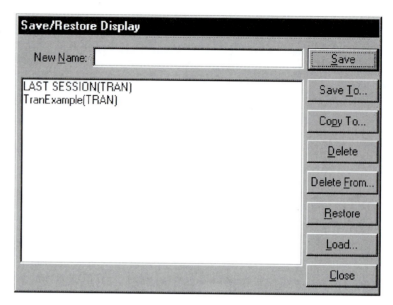

SAVING AND PRINTING PROBE PLOTS Saving plots within PROBE requires the use of the **Display Control** command in the **Windows** menu. Using the SAVE/RESTORE DISPLAY window shown in Fig. 7.37, we simply name the plot and click on **Save**. Figure 7.37 shows that one plot has already been saved, TranExample. This procedure saves the plot attributes such as axes settings, additional text, and cursor settings in a file with a .prb extension. In addition, the .prb file contains a reference to the appropriate data file, which has a .dat extension and contains the actual simulation results. Therefore, in using **Display Control** to save a plot, we do not save the plot itself, only the plot settings and the .dat file's name. To access an old PROBE plot, enter PROBE, and from the **File** menu, open the appropriate .dat file. Next, access the DISPLAY CONTROL window, select the file of interest, and click on **RESTORE**. Use the **Save As** option in Fig. 7.37 to save the .prb file to any directory on any disk, hard or floppy.

To copy the PROBE plot to other documents such as word processors, select the **Copy to Clipboard** command in the **Window** menu. The window in Fig. 7.38 will appear showing several options. If your PROBE display screen background is black, it is recommended that

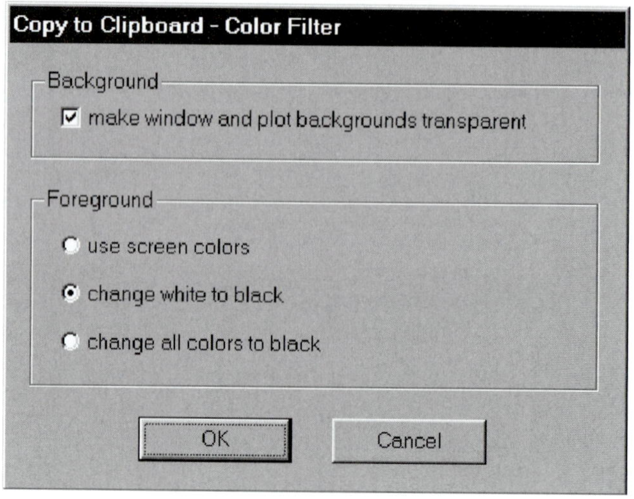

Figure 7.38
The COPY TO CLIPBOARD window used in PROBE to copy plots to other documents.

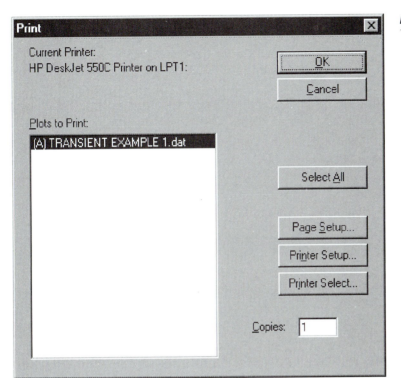

Figure 7.39
The PRINT window used in PROBE.

you choose the **make window and plot backgrounds transparent** and **change white to black** options. When the plot is pasted, it will have a white background with black text—a better scenario for printing. To print a PROBE plot, select **Print** from the **File** menu and the PRINT window in Fig. 7.39 will open. The options in this window are self-explanatory. Note that printed plots have white backgrounds with black text.

OTHER PROBE FEATURES   There are several features within PROBE for plot manipulation and data extraction. Within the **Plot** menu resides commands for editing the plot itself. These include altering the axes, adding axes, and adding more plots to the page. Also in the **Tools** menu is the **Label** command, which allows one to add marks (data point values), explanatory text, lines, and shapes to the plot.

Example 7.12

Using the PSPICE *Schematics* editor, draw the circuit in Fig. 7.40 and use the PROBE utility to find the time at which the capacitor and inductor current are equal.

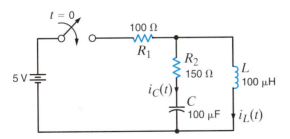

Figure 7.40
Circuit used in Example 7.12.

SOLUTION   Figure 7.41a shows the *Schematics* circuit, and the simulation results are shown in Fig. 7.41b. Based on the PROBE plot, the currents are equal at 561.8 ns.

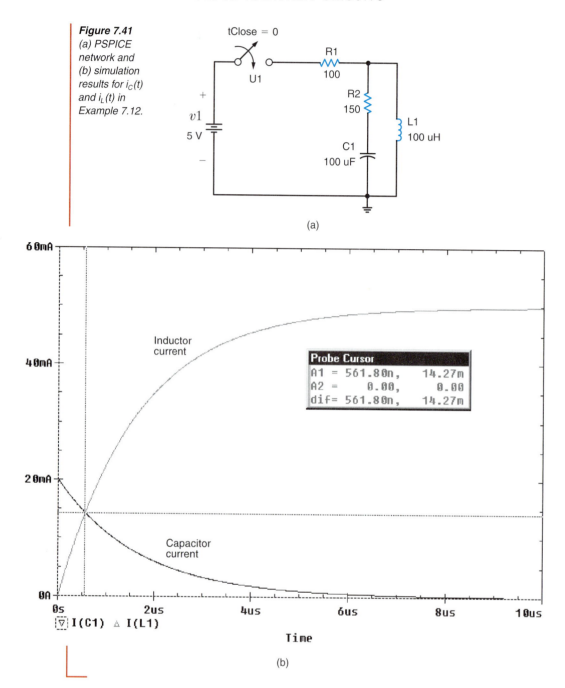

Figure 7.41
(a) PSPICE network and (b) simulation results for $i_C(t)$ and $i_L(t)$ in Example 7.12.

(a)

(b)

7.5 Application Examples

There are a wide variety of applications for transient circuits. The following examples will demonstrate some of them.

Application Example 7.13

Let us return to the camera flash circuit, redrawn in Fig. 7.42, which was discussed in the introduction of this chapter. The Xenon flash has the following specifications:

Voltage required for successful flash: $\begin{cases} \text{minimum } 50 \text{ V} \\ \text{maximum } 70 \text{ V} \end{cases}$

Equivalent resistance: 80 Ω

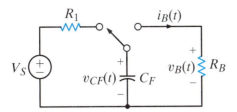

Figure 7.42
A model for a camera flash charging circuit.

For this particular application, a time constant of 1 ms is required during flash time. In addition, to minimize the physical size of the circuit, the resistor R_1 must dissipate no more than 100-mW peak power. We want to determine values for V_S, C_F, and R_1. Furthermore, we wish to determine the recharge time, the flash bulb's voltage, current, power, and total energy dissipated during the flash.

SOLUTION   We begin by selecting the source voltage, V_S. Since the capacitor is applied directly to the Xenon bulb during flash, and since at least 50 V is required to flash, we should set V_S higher than 50 V. We will somewhat arbitrarily split the difference in the bulb's required voltage range and select 60 V for V_S.

Now we consider the time constant during the flash time. From Fig. 7.42, during flash, the time constant is simply

$$\tau_F = R_B C_F \qquad\qquad \textbf{7.22}$$

Given $t_F = 1$ ms and $R_B = 80\ \Omega$, we find $C_F = 12.5\ \mu\text{F}$.

Next, we turn to the value of R_1. At the beginning of the charge time, the capacitor voltage is zero and both the current and power in R_1 are at their maximum values. Setting the power to the maximum allowed value of 100 mW, we can write

$$P_{R\text{max}} = \frac{V_S^2}{R_1} = \frac{3600}{R_1} = 0.1 \qquad\qquad \textbf{7.23}$$

and find that $R_1 = 36$ kΩ. The recharge time is the time required for the capacitor to charge from zero up to at least 50 V. At that point the flash can be successfully discharged. We will define $t = 0$ as the point at which the switch moves from the bulb back to R_1. At $t = 0$, the capacitor voltage is zero, and at $t = \infty$, the capacitor voltage is 60 V; the time constant is simply $R_1 C_F$. The resulting equation for the capacitor voltage during recharge is

$$v_{CF}(t) = K_1 + K_2 e^{-t/\tau} = 60 - 60 e^{-t/R_1 C_F}\ \text{V} \qquad\qquad \textbf{7.24}$$

At $t = t_{\text{charge}}$, $v_{CF}(t) = 50$ V. Substituting this and the values of R_1 and C_F into Eq. (7.24) yields a charge time of $t_{\text{charge}} = 806$ ms—just less than a second. As a point of interest, let us reconsider our choice for V_S. What happens if V_S is decreased to only 51 V? First, from Eq. (7.23), R_1 changes to 26.01 kΩ. Second, from Eq. (7.24), the charge time increases only slightly to 1.28 s. Therefore, it appears that selection of V_S will not have much effect on the flash unit's performance, and thus there exists some flexibility in the design.

Finally, we consider the waveforms for the flash bulb itself. The bulb and capacitor voltage are the same during flash and are given by the decaying exponential function

$$v_B(t) = 60 e^{-1000t}\ \text{V} \qquad\qquad \textbf{7.25}$$

where the time constant is defined in Eq. (7.22), and we have assumed that the capacitor is allowed to charge fully to V_S (i.e., 60 V). Since the bulb's equivalent resistance is 80 Ω, the bulb current must be

$$i_B(t) = \frac{60 e^{-1000t}}{80} = 750 e^{-1000t}\ \text{mA} \qquad\qquad \textbf{7.26}$$

As always, the power is the *v-i* product.

$$p_B(t) = v_B(t)i_B(t) = 45e^{-2000t} \text{ W} \qquad \textbf{7.27}$$

Finally, the total energy consumed by the bulb during flash is only

$$w_B(t) = \int_0^\infty p_B(t)\, dt = \int_0^\infty 45e^{-2000t}\, dt = \frac{45}{2000}\, e^{-2000t}\Big|_\infty^0 = \frac{45}{2000} = 22.5 \text{ mJ} \quad \textbf{7.28}$$

Application Example 7.14

One very popular application for inductors is storing energy in the present for release in the future. This energy is in the form of a magnetic field, and current is required to maintain the field. In an analogous situation, the capacitor stores energy in an electric field, and a voltage across the capacitor is required to maintain it. As an application of the inductor's energy storage capability, let us consider the high-voltage pulse generator circuit shown in Fig. 7.43. This circuit is capable of producing high-voltage pulses from a small dc voltage. Let's see if this circuit can produce an output voltage peak of 500 V every 2 ms, that is, 500 times per second.

Figure 7.43
A simple high-voltage pulse generator.

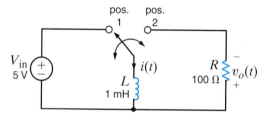

SOLUTION At the heart of this circuit is a single-pole, double-throw switch, that is, a single switch (single-pole) with two electrically connected positions (double-throw) 1 and 2. As shown in Fig. 7.44a, when in position 1, the inductor current grows linearly in accordance with the equation

$$i(t) = \frac{1}{L}\int_0^{T_1} V_{in}\, dt$$

Then the switch moves from position 1 to position 2 at time T_1. The peak inductor current is

$$i_p(t) = \frac{V_{in}T_1}{L}$$

While at position 1, the resistor is isolated electrically, and therefore its voltage is zero.

At time $t > T_1$, when the switch is in position 2 as shown in Fig. 7.44b, the inductor current flows into the resistor producing the voltage

$$v_o(t - T_1) = i(t - T_1)R \qquad t > T_1$$

At this point, we know that the form of the voltage $v_o(t)$, in the time interval $t > T_1$, is

$$v_o(t) = Ke^{-(t-T_1)/\tau} \qquad t > T_1$$

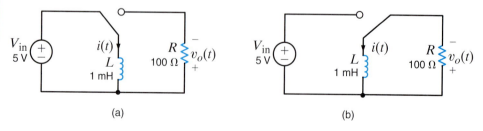

(a) (b)

Figure 7.44 *(a) Pulse generator with switch in position 1. Inductor is energized. (b) Switch in position 2. As energy is drained from the inductor, the voltage and current decay toward zero.*

And $\tau = L/R$. The initial value of $v_o(t)$ in the time interval $t > T_1$ is K since at time T_1 the exponential term is 1. According to the design specifications, this initial value is 500 and therefore $K = 500$.

Since this voltage is created by the peak inductor current I_P flowing in R,

$$K = 500 = (V_{\text{in}} T_1 R)/L = 5T_1(100)/10^{-3}$$

and thus T_1 is 1 ms and I_P is 5 A.

The equation for the voltage in the time interval $t > T_1$, or $t > 1$ ms, is

$$v_o(t - 1 \text{ ms}) = 500e^{-100,000(t-1\text{ ms})} \text{ V}$$

At the end of the 2-ms period, that is, at $t = 2$ ms, the voltage is $500e^{-100}$ or essentially zero. The complete waveform for the voltage is shown in Fig. 7.45.

It is instructive at this point to consider the ratings of the various components used in this pulse generator circuit. First, 500 V is a rather high voltage, and thus each component's voltage rating should be at least 600 V in order to provide some safety margin. Second, the inductor's peak current rating should be at least 6 A. Finally, at peak current, the power losses in the resistor are 2500 W! This resistor will have to be physically large to handle this power load without getting too hot. Fortunately, the resistor power is pulsed rather than continuous; thus, a lower power rated resistor will work fine, perhaps 500 W. In later chapters we will address the issue of power in much more detail.

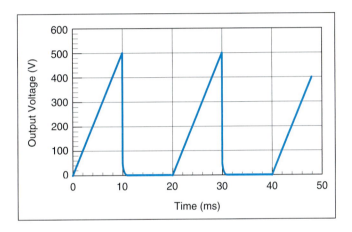

Figure 7.45
The output voltage of the pulse generator.

Application Example 7.15

A heart pacemaker circuit is shown in Fig. 7.46. The SCR (silicon-controlled rectifier) is a solid-state device that has two distinct modes of operation. When the voltage across the SCR is increasing but less than 5 V, the SCR behaves like an open circuit, as shown in Fig. 7.47a. Once the voltage across the SCR reaches 5 V, the device functions like a current source, as shown in Fig. 7.47b. This behavior will continue as long as the SCR voltage remains above 0.2 V. At this voltage, the SCR shuts off and again becomes an open circuit.

Assume that at $t = 0$, $v_C(t)$ is 0 V and the 1-μF capacitor begins to charge toward the 6-V source voltage. Find the resistor value such that $v_C(t)$ will equal 5 V (the SCR firing voltage) at 1 s. At $t = 1$ s, the SCR fires and begins discharging the capacitor. Find the time required for $v_C(t)$ to drop from 5 V to 0.2 V. Finally, plot $v_C(t)$ for the three cycles.

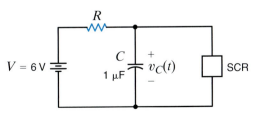

Figure 7.46
Heart pacemaker equivalent circuit.

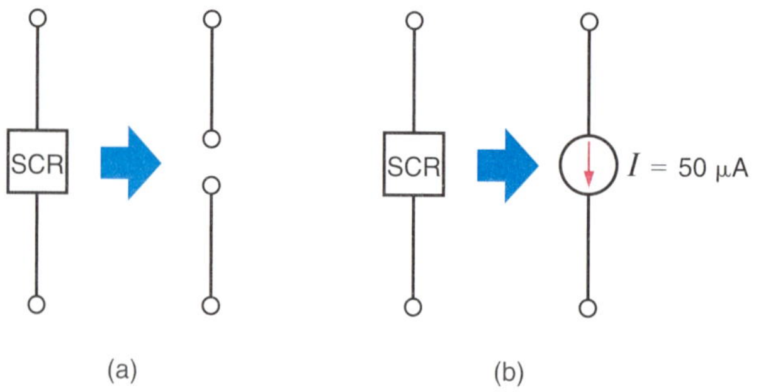

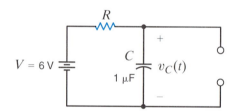

Figure 7.47 *Equivalent circuits for silicon-controlled rectifer.*

SOLUTION For $t < 1$ s, the equivalent circuit for the pacemaker is shown in Fig. 7.48. As indicated earlier, the capacitor voltage has the form

$$v_C(t) = 6 - 6e^{-t/RC} \text{ V}$$

A voltage of 0.2 V occurs at

$$t_1 = 0.034RC$$

whereas a voltage of 5 V occurs at

$$t_2 = 1.792RC$$

Figure 7.48
*Pacemaker
equivalent network
during capacitor
charge cycle.*

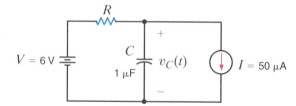

We desire that $t_2 - t_1 = 1$ s. Therefore,

$$t_2 - t_1 = 1.758RC = 1 \text{ s}$$

and

$$RC = 0.569 \text{ s} \quad \text{and} \quad R = 569 \text{ k}\Omega$$

At $t = 1$ s the SCR fires and the pacemaker is modeled by the circuit in Fig. 7.49. The form of the discharge waveform is

$$v(t) = K_1 + K_2 e^{-(t-1)/RC}$$

Figure 7.49
*Pacemaker
equivalent network
during capacitor
discharge cycle.*

The term $(t - 1)$ appears in the exponential to shift the function 1 s, since during that time the capacitor was charging. Just after the SCR fires at $t = 1^+$ s, $v_C(t)$ is still 5 V, whereas at $t = \infty$, $v_C(t) = 6 - IR$. Therefore,

$$K_1 + K_2 = 5 \quad \text{and} \quad K_1 = 6 - IR$$

Our solution, then, is of the form

$$v_C(t) = 6 - IR + (IR - 1)e^{-(t-1)/RC}$$

Let T be the time beyond 1 s necessary for $v(t)$ to drop to 0.2 V. We write

$$v_C(T + 1) = 6 - IR + (IR - 1)e^{-T/RC} = 0.2$$

Substituting for I, R, and C, we find

$$T = 0.11 \text{ s}$$

The output waveform is shown in Fig. 7.50.

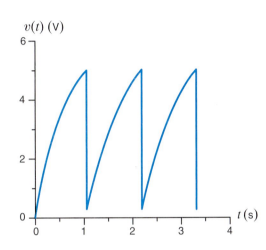

Figure 7.50
Heart pacemaker output voltage waveform.

Application Example 7.16

Consider the simple *R-L* circuit, shown in Fig. 7.51, which forms the basis for essentially every dc power supply in the world. The switch opens at $t = 0$. V_S and R have been chosen simply to create a 1-A current in the inductor prior to switching. Let us find the peak voltage across the inductor and across the switch.

SOLUTION We begin the analysis with an expression for the inductor current. At $t = 0$, the inductor current is 1 A. At $t = \infty$, the current is 0. The time constant is simply L/R, but when the switch is open, R is infinite and the time constant is zero! As a result, the inductor current is

$$i_L(t) = 1e^{-\alpha t} \text{ A} \qquad \textbf{7.29}$$

where α is infinite. The resulting inductor voltage is

$$v_L(t) = L\frac{di_L(t)}{dt} = -\alpha e^{-\alpha t} \qquad \textbf{7.30}$$

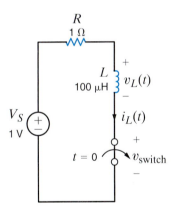

Figure 7.51
The switched inductor network at the heart of modern power supplies.

At $t = 0$, the peak inductor voltage is negative infinity! This voltage level is caused by the attempt to disrupt the inductor current instantaneously, driving di/dt through the roof. Employing KVL, the peak switch voltage must be positive infinity (give or take the supply voltage). This phenomenon is called *inductive kick*, and it is the nemesis of power supply designers.

Given this situation, we naturally look for a way to reduce this excessive voltage and, more importantly, predict and control it. Let's look at what we have and what we know. We have a transient voltage that grows very quickly without bound. We also have an initial current in the inductor that must go somewhere. We know that capacitor voltages cannot change quickly and resistors consume energy. Therefore, let's put an *RC* network around the switch, as shown in Fig. 7.52, and examine the performance that results from this change.

Figure 7.52
Conversion of a switched inductor circuit to an RLC network in an attempt to control inductive kick.

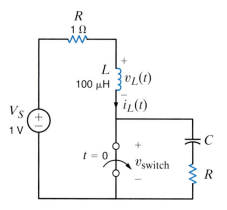

The addition of the *RC* network yields a series circuit. We need the characteristic equation of this series *RLC* network when the switch is open. From Eq. (7.15), we know that the characteristic equation for the series *RLC* circuit is

$$s^2 + 2\zeta\omega_0 s + \omega_0^2 = s^2 + \left[\frac{R+1}{L}\right]s + \frac{1}{LC} = 0 \qquad \textbf{7.31}$$

To retain some switching speed, we will somewhat arbitrarily choose a critically damped system where $\zeta = 1$ and $\omega_0 = 10^6$ rad/s. This choice for ω_0 should allow the system to stabilize in a few microseconds. From Eq. (7.31) we can now write expressions for *C* and *R*.

$$\omega_0^2 = 10^{12} = \frac{1}{LC} = \frac{1}{10^{-4}C} \qquad 2\zeta\omega_0 = 2 \times 10^6 = \frac{R+1}{L} = \frac{R+1}{10^{-4}} \qquad \textbf{7.32}$$

Solving these equations yields the parameter values $C = 10$ nF and $R = 199\ \Omega$. Now we can focus on the peak switch voltage. When the switch opens, the inductor current, set at 1 A by the dc source and the 1-Ω resistor, flows through the *RC* circuit. Since the capacitor was previously discharged by the closed switch, its voltage cannot change immediately and its voltage remains zero for an instant. The resistor voltage is simply $I_L R$ where I_L is the initial inductor current. Given our I_L and R values, the resistor voltage just after opening the switch is 199 V. The switch voltage is then just the sum of the capacitor and resistor voltages (i.e., 199 V). This is a tremendous improvement over the first scenario!

A plot of the switch voltage, shown in Fig. 7.53, clearly agrees with our analysis. This plot illustrates the effectiveness of the *RC* network in reducing the inductive kick generated by opening the switch. Note that the switch voltage is controlled at a 199-V peak value and the system is critically damped; that is, there is little or no overshoot, having stabilized in less than 5 μs. Because of its importance, this *R-C* network is called a *snubber* and is the engineer's solution of choice for controlling inductive kick.

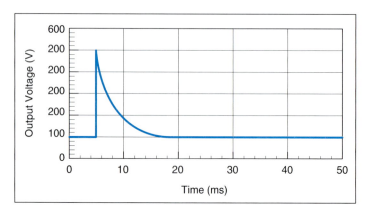

Figure 7.53 Plot of the switch voltage when the snubber circuit is employed to reduce inductive kick.

Application Example 7.17

One of the most common and necessary subcircuits that appears in a wide variety of electronic systems—for example, stereos, TVs, radios, and computers—is a quality dc voltage source or power supply. The standard wall socket supplies an alternating current (ac) voltage waveform shown in Fig. 7.54a, and the conversion of this voltage to a desired dc level is done as illustrated in Fig. 7.54b. The ac waveform is converted to a quasi-dc voltage by an inexpensive ac–dc converter whose output contains remnants of the ac input and is unregulated. A higher quality dc output is created by a switching dc–dc converter. Of the several versions of dc–dc converters, we will focus on a topology called the boost converter, shown in Fig. 7.55. Let us develop an equation relating the output voltage to the switching characteristics.

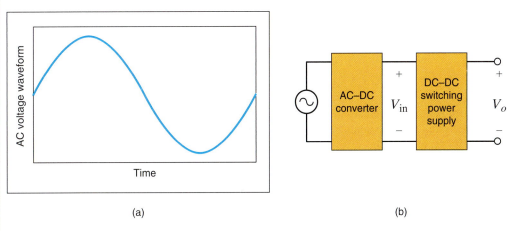

(a)

(b)

Figure 7.54
(a) The ac voltage waveform at a standard wall outlet and (b) a block diagram of a modern dc power supply.

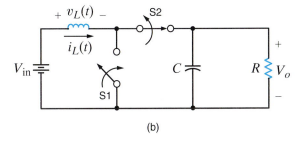

(a)

Figure 7.55
The boost converter with switch settings for time intervals (a) t_{on} and (b) t_{off}.

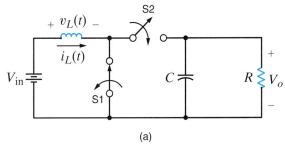

(b)

SOLUTION   Consider the boost converter in Fig. 7.55a, where switch 1 (S1) is closed and S2 is open for a time interval t_{on}. This isolates the inductor from the capacitor, creating two subcircuits that can be analyzed independently. Note that during t_{on} the inductor current and stored energy are increasing while at the output node, the capacitor voltage discharges exponentially into the load. If the capacitor's time constant ($\tau = RC$) is large, then the output voltage will decrease slowly. Thus, during t_{on} energy is stored in the inductor and the capacitor provides energy to the load.

Next, we change both switch positions so that S1 is open and S2 is closed for a time interval t_{off}, as seen in Fig. 7.55b. Since the inductor current cannot change instantaneously, current flows into the capacitor and the load, recharging the capacitor. During t_{off} the energy that was added to the inductor during t_{on} is used to recharge the capacitor and drive the load. When t_{off} has elapsed, the cycle is repeated.

Note that the energy added to the inductor during t_{on} must go to the capacitor and load during t_{off}; otherwise, the inductor energy would increase to the point that the inductor would fail. This requires that the energy stored in the inductor must be the same at the end of each switching cycle. Recalling that the inductor energy is related to the current by

$$w(t) = \frac{1}{2} L i^2(t)$$

we can state that the inductor current must also be the same at the end of each switching cycle, as shown in Fig. 7.56. The inductor current during t_{on} and t_{off} can be written as

$$i_L(t) = \frac{1}{L} \int_0^{t_{on}} v_L(t)\, dt = \frac{1}{L} \int_0^{t_{on}} V_{in}\, dt = \left[\frac{V_{in}}{L}\right] t_{on} + I_0 \qquad 0 < t < t_{on}$$

$$i_L(t) = \frac{1}{L} \int_{t_{on}}^{t_{on}+t_{off}} v_L(t)\, dt = \frac{1}{L} \int_{t_{on}}^{t_{on}+t_{off}} (V_{in} - V_o)\, dt =$$

$$\left[\frac{V_{in} - V_o}{L}\right] t_{off} + I_0 \qquad t_{on} < t < t_{off} \qquad\qquad \textbf{7.33}$$

where I_0 is the initial current at the beginning of each switching cycle. If the inductor current is the same at the beginning and end of each switching cycle, then the integrals in Eq. (7.33) must sum to zero. Or,

$$V_{in} t_{on} = (V_o - V_{in}) t_{off} = (V_o - V_{in})(T - t_{on})$$

where T is the period $(T = t_{on} + t_{off})$. Solving for V_o yields

$$V_o = V_{in}\left[\frac{T}{T - t_{on}}\right] = V_{in}\left[\frac{1}{(T - t_{on})/T}\right] = V_{in}\left[\frac{1}{(1 - t_{on}/T)}\right] = V_{in}\left[\frac{1}{1 - D}\right]$$

where D is the duty cycle $(D = t_{on}/T)$. Thus, by controlling the duty cycle, we control the output voltage. Since D is always a positive fraction, V_o is always bigger than V_{in}—thus the name, boost converter. A plot of V_o/V_{in} versus duty cycle is shown in Fig. 7.57.

Figure 7.56
Waveform sketches for the inductor voltage and current.

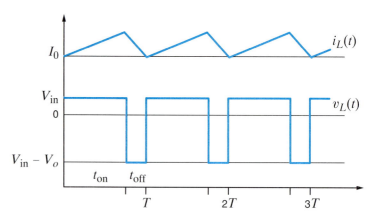

Figure 7.57
Effect of duty cycle on boost converter gain.

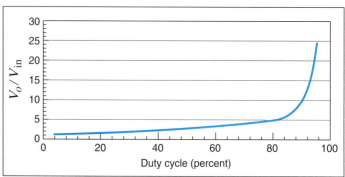

Application Example 7.18

An experimental schematic for a railgun is shown in Fig. 7.58. With switch Sw-2 open, switch Sw-1 is closed and the power supply charges the capacitor bank to 10 kV. Then switch Sw-1 is opened. The railgun is fired by closing switch Sw-2. When the capacitor discharges, the current causes the foil at the end of the gun to explode, creating a hot plasma that accelerates down the tube. The voltage drop in vaporizing the foil is negligible, and, therefore, more than 95% of the energy remains available for accelerating the plasma. The current flow establishes a magnetic field, and the force on the plasma caused by the magnetic field, which is proportional to the square of the current at any instant of time, accelerates the plasma. A higher initial voltage will result in more acceleration.

The circuit diagram for the discharge circuit is shown in Fig. 7.59. The resistance of the bus (a heavy conductor) includes the resistance of the switch. The resistance of the foil and resultant plasma is negligible; therefore, the current flowing between the upper and lower conductors is dependent on the remaining circuit components in the closed path, as specified in Fig. 7.58.

The differential equation for the natural response of the current is

$$\frac{d^2 i(t)}{dt^2} + \frac{R_{\text{bus}}}{L_{\text{bus}}}\frac{di(t)}{dt} + \frac{i(t)}{L_{\text{bus}}C} = 0$$

Let us use the characteristic equation to describe the current waveform.

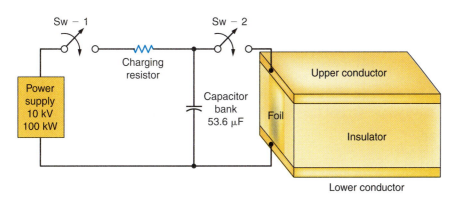

Figure 7.58
Experimental schematic for a railgun.

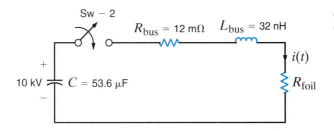

Figure 7.59
Railgun discharge circuit.

SOLUTION   Using the circuit values, the characteristic equation is

$$s^2 + 37.5 \times 10^4 s + 58.3 \times 10^{10} = 0$$

and the roots of the equation are

$$s_1, s_2 = (-18.75 \pm j74) \times 10^4$$

and hence the network is underdamped.

The roots of the characteristic equation illustrate that the damped resonant frequency is

$$\omega_d = 740 \text{ krad/s}$$

Therefore,

$$f_d = 118 \text{ kHz}$$

and the period of the waveform is

$$T = \frac{1}{f_d} = 8.5 \text{ μs}$$

An actual plot of the current is shown in Fig. 7.60 and this plot verifies that the period of the damped response is indeed 8.5 μs.

Figure 7.60
Load current with a
capacitor bank
charged to 10 kV.

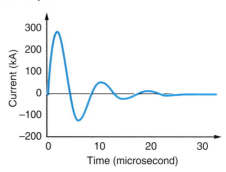

7.6 Design Examples

Design Example 7.19

We wish to design an efficient electric heater that operates from a 24-V dc source and creates heat by driving a 1-Ω resistive heating element. For the temperature range of interest, the power absorbed by the heating element should be between 100 and 400 W. An experienced engineer has suggested that two quite different techniques be examined as possible solutions: a simple voltage divider and a switched inductor circuit.

Figure 7.61
A simple circuit for
varying the temperature
of a heating element.

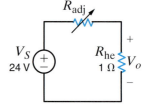

SOLUTION In the first case, the required network is shown in Fig. 7.61. The variable resistance element is called a rheostat. Potentiometers are variable resistors that are intended for low power (i.e., less than 1 W) operation. Rheostats, on the other hand, are devices used at much higher power levels.

We know from previous work that the heating element voltage is

$$V_o = V_S \frac{R_{\text{he}}}{R_{\text{he}} + R_{\text{adj}}} \qquad \textbf{7.34}$$

Changing R_{adj} will change the voltage across, and the power dissipated by, the heating element. The power can be expressed as

$$P_o = \frac{V_o^2}{R_{\text{he}}} = V_S^2 \frac{R_{\text{he}}}{\left(R_{\text{he}} + R_{\text{adj}}\right)^2} \qquad \textbf{7.35}$$

By substituting the maximum and minimum values of the output power into Eq. (7.35) we can determine the range of resistance required for the rheostat.

$$R_{\text{adj,min}} = \sqrt{\frac{V_S^2 R_{\text{he}}}{P_{o,\text{max}}}} - R_{\text{he}} = \sqrt{\frac{(24^2)(1)}{400}} - 1 = 0.2 \text{ Ω}$$

$$R_{\text{adj,max}} = \sqrt{\frac{V_S^2 R_{\text{he}}}{P_{o,\text{min}}}} - R_{\text{he}} = \sqrt{\frac{(24^2)(1)}{100}} - 1 = 1.4 \text{ Ω} \qquad \textbf{7.36}$$

So, a 2-Ω rheostat should work just fine. But what about the efficiency of our design? How much power is lost in the rheostat? The rheostat power can be expressed as

$$P_{adj} = \frac{(V_S - V_o)^2}{R_{adj}} = V_S^2 \frac{R_{adj}}{(R_{he} + R_{adj})^2} \qquad 7.37$$

We know from our studies of maximum power transfer that the value of R_{adj} that causes maximum power loss, and thus the worst-case efficiency for the circuit, occurs when $R_{adj} = R_{he} = 1\ \Omega$. Obviously, the resistances consume the same power, and the efficiency is only 50%.

Now that we understand the capability of this voltage-divider technique, let's explore the alternative solution. At this point, it would at least appear that the use of a switched inductor is a viable alternative since this element consumes no power. So, if we could set up a current in an inductor, switch it into the heating element, and repeat this operation fast enough, the heating element would respond to the average power delivered to it and maintain a constant temperature.

Consider the circuit in Fig. 7.62 where the switch moves back and forth, energizing the inductor with current, then directing that current to the heating element. Let's examine this concept to determine its effectiveness. We begin by assuming the inductor current is zero and the switch has just moved to position 1. The inductor current will begin to grow *linearly* in accordance with the fundamental equation

$$i_L(t) = \frac{1}{L}\int v_L(t)\,dt = \frac{1}{L}\int V_S\,dt = \frac{V_S}{L}t \qquad 7.38$$

Note that V_S/L is the slope of the linear growth. Since V_S is set at 24 V, we can control the slope with our selection for L. The inductor current increases until the switch moves at time $t = t_1$ at which point the peak current is

$$I_{peak} = \frac{V_S}{L}t_1 \qquad 7.39$$

This inductor current will discharge exponentially through the heating element according to the equation

$$i_L(t') = I_{peak}e^{-t'/\tau} \qquad 7.40$$

where t' is zero when the switch moves to position 2 and $\tau = L/R_{he}$. If the switch is maintained in position 2 for about 5 time constants, the inductor current will essentially reach zero and the switch can return to position 1 under the initial conditions—zero inductor current. A sketch of the inductor current over a single switching cycle is shown in Fig. 7.63. Repeated switching cycles will transfer power to the heating element. If the switching period is much shorter than the element's thermal time constant—a measure of how quickly the element heats up—then the element's temperature will be determined by the average power. This is a concept we don't understand at this point. However, we will present Average Power in Chapter 9 and this example provides at least some motivation for its examination.

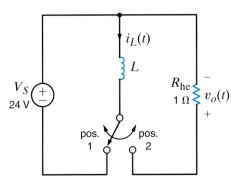

Figure 7.62
A switched inductor solution to varying the heating element's temperature.

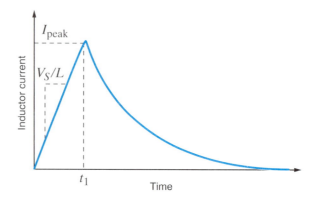

Figure 7.63 *A single switching cycle for the inductor-based solution. The value of I_{peak} is directly proportional to t_1.*

Nevertheless, we should recognize two things: the load current is just the exponential decaying portion of the inductor current, and the initial value of that exponential is I_{peak} as defined in Eq. (7.39). Increasing I_{peak} will increase the element's power and temperature. As Eq. (7.39) indicates, this is easily done by controlling t_1! It is impossible to proceed with the design until we can accurately predict the average power at the load.

Returning to our original concern, have we improved the efficiency at all? Note that there are no power-consuming components in our new circuit other than the heating element itself. Therefore, ignoring resistance in the inductor and the switch, we find that our solution is 100% efficient! In actuality, efficiencies approaching 95% are attainable. This is a drastic improvement over the other alternative, which employs a rheostat.

Design Example 7.20

Consider the circuit in Fig. 7.64a where a dc power supply, typically fed from a wall outlet, is modeled as a dc voltage source in series with a resistor R_S. The load draws a constant current and is modeled as a current source. We wish to design the simplest possible circuit that will isolate the load device from disturbances in the power supply voltage. In effect, our task is to improve the performance of the power supply at very little additional cost.

A standard solution to this problem involves the use of a capacitor C_D as shown in Fig. 7.64b. The two voltage sources and the single-pole double-throw switch model the input disturbance sketched in Fig. 7.64c. Engineers call C_D a *decoupling capacitor* since it

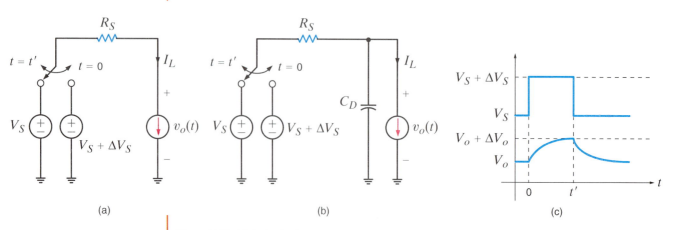

Figure 7.64 *(a) A simple dc circuit that models disturbances in the source voltage. (b) The use of a decoupling capacitor to reduce disturbances in the load voltage. (c) Definitions of the input and output voltage disturbances.*

decouples disturbances in the input voltage from the output voltage. In typical electronic circuits, we find liberal use of these decoupling capacitors. Thus, our task is to develop a design equation for C_D in terms of R_S, V_S, V_o, ΔV_S, ΔV_o, and t'. Our result will be applicable to any scenario that can be modeled by the circuit in Fig. 7.64b.

SOLUTION   The voltage across C_D can be expressed in the standard form as

$$v_o(t) = K_1 + K_2 e^{-t/\tau} \qquad \textbf{7.41}$$

Equivalent circuits for $t = 0$ and $t = \infty$ are shown in Fig. 7.65a and b, respectively. At these two time extremes we find

$$v_o(0) = K_1 + K_2 = V_S - I_L R_S$$

$$v_o(\infty) = K_1 = V_S + \Delta V_S - I_L R_S \qquad \textbf{7.42}$$

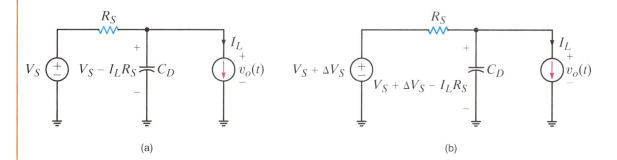

(a)   (b)

Figure 7.65
The circuit in Fig. 7.64b at t = 0 just after the switch has moved. (b) The same circuit at t = ∞.

To determine the time constant's equivalent resistance, we return to the circuit in Fig. 7.65b, reduce all independent sources to zero, and view the resulting circuit from the capacitor's terminals. It is easy to see that the time constant is simply $R_S C_D$. Thus,

$$v_o(t) = V_S + \Delta V_S - I_L R_S - \Delta V_S e^{-t/R_S C_D} \qquad \textbf{7.43}$$

At exactly $t = t'$, the output voltage is its original value, $V_o(0)$ plus ΔV_o. Substituting this condition into Eq. (7.43) yields

$$v_o(t') = V_S - I_L R_S + \Delta V_o = V_S + \Delta V_S - I_L R_S - \Delta V_S e^{-t'/R_S C_D}$$

which can be reduced to the expression

$$\Delta V_S - \Delta V_o = \Delta V_S e^{-t'/R_S C_D} \qquad \textbf{7.44}$$

Notice that the value of C_D depends not on the input and output voltages, but rather on the *changes* in those voltages! This makes Eq. (7.44) very versatile indeed. A simple algebraic manipulation of Eq. (7.44) yields the design equation for C_D.

$$C_D = \frac{t'}{R_S \ln\left[\dfrac{\Delta V_S}{\Delta V_S - \Delta V_o}\right]} \qquad \textbf{7.45}$$

Examining this expression, we see that C_D is directly related to t' and inversely related to R_S. If t' doubles or if R_S is halved, then C_D will double as well. This result is not very surprising. The dependence on the voltage changes is more complex. Let us isolate this term and express it as

$$f = \frac{1}{\ln\left[\dfrac{\Delta V_S}{\Delta V_S - \Delta V_o}\right]} = \frac{1}{\ln\left[\dfrac{1}{1 - \Delta V_o/\Delta V_S}\right]} = \frac{-1}{\ln\left[1 - \dfrac{\Delta V_o}{\Delta V_S}\right]} \qquad \textbf{7.46}$$

Figure 7.66 shows a plot of this term versus the ratio $\Delta V_o / \Delta V_S$. Note that for very small ΔV_o (i.e., a large degree of decoupling) this term is very large. Since this term is multiplied with t' in Eq. (7.45), we find that the price for excellent decoupling is a very large capacitance.

Finally, as an example, consider the scenario in which V_S is 5 V, R_S is 20 Ω, and the input disturbance is characterized by $\Delta V_S = 1$ V and $t' = 0.5$ ms. If the output changes are to be limited to only 0.2 V, the required capacitance would be $C_D = 112.0$ μF. Such a capacitor rated for operation at up to 16 V costs less than $0.20 and should be slightly smaller than a peanut M&M.

This very simple, but very important, application demonstrates how an engineer can apply his or her basic circuit analysis skills to attack and describe a practical application in such a way that the result is broadly applicable. Remember, the key to this entire exercise was the creation of a circuit model for the decoupling scenario.

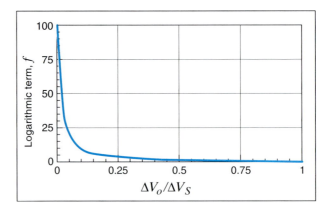

Figure 7.66 *A plot of the function, f, versus $\Delta V_o / \Delta V_S$.*

Design Example 7.21

The network in Fig. 7.67 models an automobile ignition system. The voltage source represents the standard 12-V battery. The inductor is the ignition coil, which is magnetically coupled to the starter (not shown). The inductor's internal resistance is modeled by the resistor, and the switch is the keyed ignition switch. Initially, the switch connects the ignition circuitry to the battery, and thus the capacitor is charged to 12 V. To start the motor, we close the switch, thereby discharging the capacitor through the inductor. Assuming that optimum starter operation requires an overdamped response for $i_L(t)$ that reaches at least 1 A within 100 ms after switching and remains above 1 A for between 1 and 1.5 s, let us find a value for the capacitor that will produce such a current waveform. In addition, let us plot the response, including the time interval just prior to moving the switch, and verify our design.

Figure 7.67
Circuit model for ignition system.

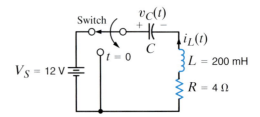

SOLUTION Before the switch is moved at $t = 0$, the capacitor looks like an open circuit, and the inductor acts like a short circuit. Thus,

$$i_L(0^-) = i_L(0^+) = 0 \text{ A} \quad \text{and} \quad v_C(0^-) = v_C(0^+) = 12 \text{ V}$$

After switching, the circuit is a series *RLC* unforced network described by the characteristic equation

$$s^2 + \frac{R}{L}s + \frac{1}{LC} = 0$$

with roots at $s = -s_1$ and $-s_2$. The characteristic equation is of the form

$$(s + s_1)(s + s_2) = s^2 + (s_1 + s_2)s + s_1 s_2 = 0$$

Comparing the two expressions, we see that

$$\frac{R}{L} = s_1 + s_2 = 20$$

and

$$\frac{1}{LC} = s_1 s_2$$

Since the network must be overdamped, the inductor current is of the form

$$i_L(t) = K_1 e^{-s_1 t} + K_2 e^{-s_2 t}$$

Just after switching,

$$i_L(0^+) = K_1 + K_2 = 0$$

or

$$K_2 = -K_1$$

Also, at $t = 0^+$, the inductor voltage equals the capacitor voltage because $i_L = 0$ and therefore $i_L R = 0$. Thus, we can write

$$v_L(0^+) = L\frac{di_L(0^+)}{dt} \Rightarrow -s_1 K_1 + s_2 K_1 = \frac{12}{L}$$

or

$$K_1 = \frac{60}{s_2 - s_1}$$

At this point, let us arbitrarily choose $s_1 = 3$ and $s_2 = 17$, which satisfies the condition $s_1 + s_2 = 20$, and furthermore,

$$K_1 = \frac{60}{s_2 - s_1} = \frac{60}{14} = 4.29$$

$$C = \frac{1}{L s_1 s_2} = \frac{1}{(0.2)(3)(17)} = 98 \text{ mF}$$

Hence, $i_L(t)$ is

$$i_L(t) = 4.29\left[e^{-3t} - e^{-17t}\right] \text{A}$$

Figure 7.68a shows a plot of $i_L(t)$. At 100 ms the current has increased to 2.39 A, which meets the initial magnitude specifications. However, one second later at $t = 1.1$ s, $i_L(t)$ has fallen

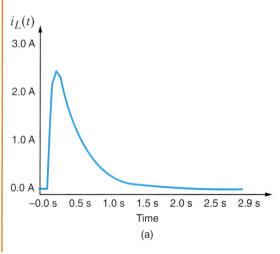

(a)

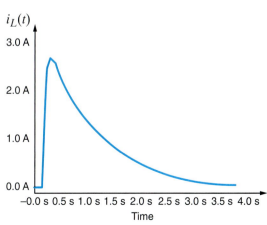

(b)

Figure 7.68

Ignition current as a function of time.

to only 0.16 A—well below the magnitude-over-time requirement. Simply put, the current falls too quickly. To make an informed estimate for s_1 and s_2, let us investigate the effect the roots exhibit on the current waveform when $s_2 > s_1$.

Since $s_2 > s_1$, the exponential associated with s_2 will decay to zero faster than that associated with s_1. This causes $i_L(t)$ to rise—the larger the value of s_2, the faster the rise. After $5(1/s_2)$ seconds have elapsed, the exponential associated with s_2 is approximately zero and $i_L(t)$ decreases exponentially with a time constant of $\tau = 1/s_1$. Thus, to slow the fall of $i_L(t)$ we should reduce s_1. Hence, let us choose $s_1 = 1$. Since $s_1 + s_2$ must equal 20, $s_2 = 19$. Under these conditions

$$C = \frac{1}{Ls_1s_2} = \frac{1}{(0.2)(1)(19)} = 263 \text{ mF}$$

and

$$K_1 = \frac{60}{s_2 - s_1} = \frac{60}{18} = 3.33$$

Thus, the current is

$$i_L(t) = 3.33[e^{-t} - e^{-19t}] \text{ A}$$

which is shown in Fig. 7.68b. At 100 ms the current is 2.52 A. Also, at $t = 1.1$ s, the current is 1.11 A—above the 1-A requirement. Therefore, the choice of $C = 263$ mF meets all starter specifications.

Design Example 7.22

A defibrillator is a device that is used to stop heart fibrillations—erratic uncoordinated quivering of the heart muscle fibers—by delivering an electric shock to the heart. The Lown defibrillator was developed by Dr. Bernard Lown in 1962. Its key feature, shown in Fig. 7.69a, is its voltage waveform. A simplified circuit diagram that is capable of producing the Lown waveform is shown in Fig. 7.69b. Let us find the necessary values for the inductor and capacitor.

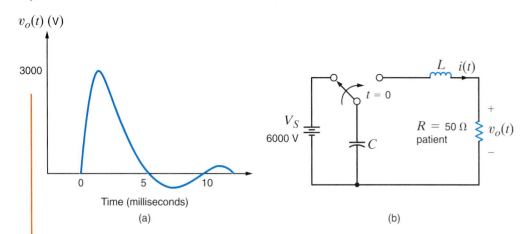

Figure 7.69 *Lown defibrillator waveform and simplified circuit. Reprinted with permission from John Wiley & Sons, Inc.,* Introduction to Biomedical Equipment Technology.

SOLUTION Since the Lown waveform is oscillatory in nature, we know that the circuit is underdamped ($\zeta < 1$) and the voltage applied to the patient is of the form

$$v_o(t) = K_1 e^{-\zeta\omega_o t} \sin[\omega t]$$

where

$$\zeta\omega_o = \frac{R}{2L}$$

$$\omega = \omega_o \sqrt{1 - \zeta^2}$$

and

$$\omega_o = \frac{1}{\sqrt{LC}}$$

for the series *RLC* circuit. From Fig. 7.69a, we see that the period of the sine function is

$$T = 10 \text{ ms}$$

Thus, we have one expression involving ω_o and ζ

$$\omega = \frac{2\pi}{T} = \omega_o \sqrt{1 - \zeta^2} = 200\pi \text{ rad/s}$$

A second expression can be obtained by solving for the ratio of $v_o(t)$ at $t = T/4$ to that at $t = 3T/4$. At these two instants of time the sine function is equal to $+1$ and -1, respectively. Using values from Fig. 7.69a, we can write

$$\frac{v_o(t/4)}{-v_o(3T/4)} = \frac{K_1 e^{-\zeta\omega_o(T/4)}}{K_1 e^{-\zeta\omega_o(3T/4)}} = e^{\zeta\omega_o(T/2)} \approx \frac{3000}{250} = 12$$

or

$$\zeta\omega_o = 497.0$$

Given $R = 50 \ \Omega$, the necessary inductor value is

$$L = 50.3 \text{ mH}$$

Using our expression for ω,

$$\omega^2 = (200\pi)^2 = \omega_o^2 - (\zeta\omega_o)^2$$

or

$$= (200\pi)^2 = \frac{1}{LC} - (497.0)^2$$

Solving for the capacitor value, we find

$$C = 31.0 \ \mu\text{F}$$

Let us verify our design using PSPICE. The PSPICE circuit is shown in Fig. 7.70a. The output voltage plot shown in Fig. 7.70b matches the Lown waveform in Fig. 7.69a; thus, we can consider the design to be a success.

It is important to note that while this solution is a viable one, it is not the only one. Like many design problems, there are often various ways to satisfy the design specifications.

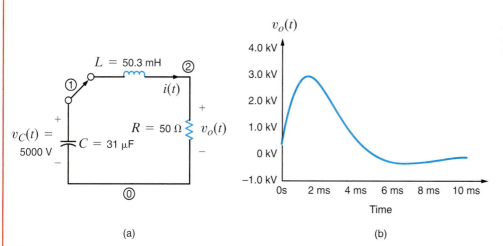

Figure 7.70
PSPICE circuit and output plot for the Lown defibrillator.

(a)   (b)

SUMMARY

First-Order Circuits

- An RC or RL transient circuit is said to be first order if it contains only a single capacitor or single inductor. The voltage or current anywhere in the network can be obtained by solving a first-order differential equation.

- The form of a first-order differential equation with a constant forcing function is

$$\frac{dx(t)}{dt} + \frac{x(t)}{\tau} = A$$

and the solution is

$$x(t) = A\tau + K_2 e^{-t/\tau}$$

where $A\tau$ is referred to as the steady-state solution and τ is called the time constant.

- The function $e^{-t/\tau}$ decays to a value that is less than 1% of its initial value after a period of 5τ. Therefore, the time constant, τ, determines the time required for the circuit to reach steady state.

- The time constant for an RC circuit is $R_{Th}C$ and for an RL circuit is L/R_{Th}, where R_{Th} is the Thévenin equivalent resistance looking into the circuit at the terminals of the storage element (i.e., capacitor or inductor).

- The two approaches proposed for solving first-order transient circuits are the differential equation approach and the step-by-step method. In the former case, the differential equation that describes the dynamic behavior of the circuit is solved to determine the desired solution. In the latter case, the initial conditions and the steady-state value of the voltage across the capacitor or current in the inductor are used in conjunction with the circuit's time constant and the known form of the desired variable to obtain a solution.

- The response of a first-order transient circuit to an input pulse can be obtained by treating the pulse as a combination of two step-function inputs.

Second-Order Circuits

- The voltage or current in an RLC transient circuit can be described by a constant coefficient differential equation of the form

$$\frac{d^2 x(t)}{dt^2} + 2\zeta\omega_0 \frac{dx(t)}{dt} + \omega_0^2 x(t) = f(t)$$

where $f(t)$ is the network forcing function.

- The characteristic equation for a second-order circuit is $s^2 + 2\zeta\omega_0 s + \omega_0^2 = 0$, where ζ is the damping ratio and ω_0 is the undamped natural frequency.

- If the two roots of the characteristic equation are
  - real and unequal, then $\zeta > 1$ and the network response is overdamped
  - real and equal, then $\zeta = 1$ and the network response is critically damped
  - complex conjugates, then $\zeta < 1$ and the network response is underdamped

- The three types of damping together with the corresponding network response are as follows:

  1. Overdamped:
     $$x(t) = K_1 e^{-(\zeta\omega_0 - \omega_0\sqrt{\zeta^2-1})t} + K_2 e^{-(\zeta\omega_0 + \omega_0\sqrt{\zeta^2-1})t}$$
  2. Critically damped: $x(t) = B_1 e^{-\zeta\omega_0 t} + B_2 t e^{-\zeta\omega_0 t}$
  3. Underdamped: $x(t) = e^{-\sigma t}(A_1 \cos \omega_d t + A_2 \sin \omega_d t)$, where $\sigma = \zeta\omega_0$ and $\omega_d = \omega_0 \sqrt{1 - \zeta^2}$

- Two initial conditions are required to derive the two unknown coefficients in the network response equations.

PROBLEMS

PSV **CS** both available on the web at: http://www.justask4u.com/irwin

SECTION 7.2

7.1 Use the differential equation approach to find $v_o(t)$ for $t > 0$ in the circuit in Fig. P7.1 and plot the response including the time interval just prior to switch action.

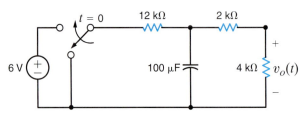

Figure P7.1

7.2 Use the differential equation approach to find $v_C(t)$ for $t > 0$ in the circuit in Fig. P7.2 and plot the response including the time interval just prior to closing the switch.

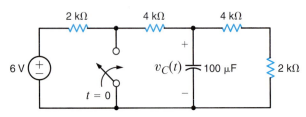

Figure P7.2

7.3 Use the differential equation approach to find $v_C(t)$ for $t > 0$ in the circuit in Fig. P7.3. **CS**

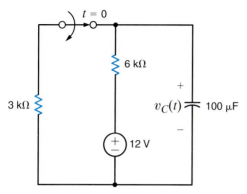

Figure P7.3

7.4 Use the differential equation approach to find $v_C(t)$ for $t > 0$ in the circuit in Fig. P7.4.

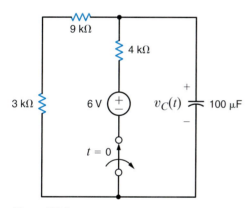

Figure P7.4

7.5 Use the differential equation approach to find $v_C(t)$ for $t > 0$ in the circuit in Fig. P7.5 and plot the response including the time interval just prior to opening the switch. **CS**

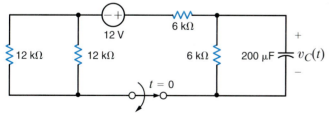

Figure P7.5

7.6 Use the differential equation approach to find $v_o(t)$ for $t > 0$ in the circuit in Fig. P7.6 and plot the response including the time interval just prior to opening the switch. **CS**

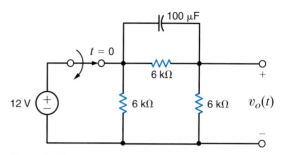

Figure P7.6

7.7 Use the differential equation approach to find $i_o(t)$ for $t > 0$ in the circuit in Fig. P7.7 and plot the response including the time interval just prior to closing the switch.

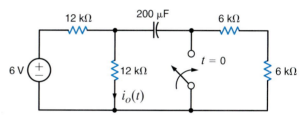

Figure P7.7

7.8 Use the differential equation approach to find $v_o(t)$ for $t > 0$ in the circuit in Fig. P7.8 and plot the response including the time interval just prior to closing the switch. **CS**

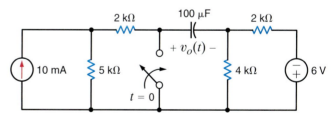

Figure P7.8

7.9 Use the differential equation approach to find $v_C(t)$ for $t > 0$ in the circuit in Fig. P7.9 and plot the response including the time interval just prior to opening the switch.

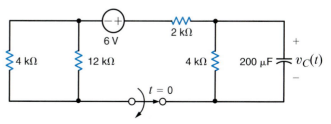

Figure P7.9

7.10 Use the differential equation approach to find $i_o(t)$ for $t > 0$ in the circuit in Fig. P7.10 and plot the response including the time interval just prior to opening the switch. **CS**

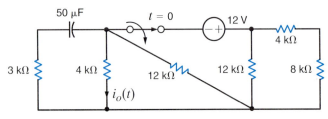

Figure P7.10

7.11 Use the differential equation approach to find $i_o(t)$ for $t > 0$ in the circuit in Fig. P7.11 and plot the response including the time interval just prior to opening the switch.

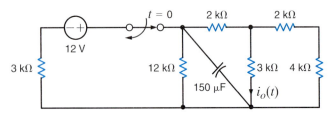

Figure P7.11

7.12 Use the differential equation approach to find $v_o(t)$ for $t > 0$ in the circuit in Fig. P7.12 and plot the response including the time interval just prior to opening the switch.

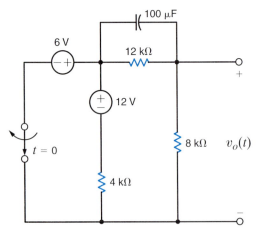

Figure P7.12

7.13 Use the differential equation approach to find $v_o(t)$ for $t > 0$ in the circuit in Fig. P7.13 and plot the response including the time interval just prior to opening the switch.

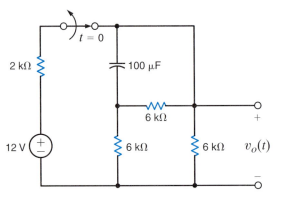

Figure P7.13

7.14 Use the differential equation approach to find $v_o(t)$ for $t > 0$ in the circuit in Fig. P7.14 and plot the response including the time interval just prior to closing the switch. **PSV**

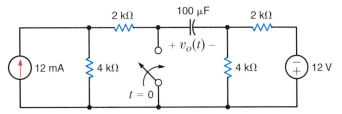

Figure P7.14

7.15 Use the differential equation approach to find $i(t)$ for $t > 0$ in the network in Fig. P7.15.

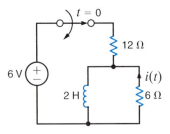

Figure P7.15

7.16 Use the differential equation approach to find $i(t)$ for $t > 0$ in the circuit in Fig. P7.16 and plot the response including the time interval just prior to switch movement.

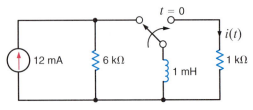

Figure P7.16

7.17 In the circuit in Fig. 7.17, find $i_o(t)$ for $t > 0$ using the differential equation approach.

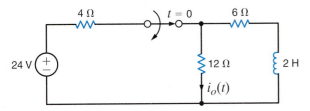

Figure P7.17

7.18 Use the differential equation approach to find $i(t)$ for $t > 0$ in the circuit in Fig. P7.18 and plot the response including the time interval just prior to opening the switch.

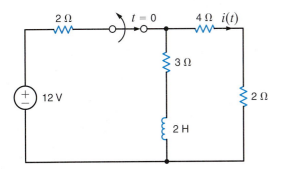

Figure P7.18

7.19 In the network in Fig. 7.19, find $i_o(t)$ for $t > 0$ using the differential equation approach. **CS**

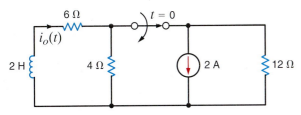

Figure P7.19

7.20 Use the differential equation approach to find $i(t)$ for $t > 0$ in the circuit in Fig. P7.20 and plot the response including the time interval just prior to switch movement. **PSV**

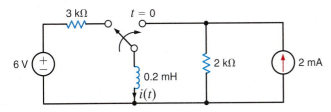

Figure P7.20

7.21 Use the differential equation approach to find $i_L(t)$ for $t > 0$ in the circuit in Fig. P7.21 and plot the response including the time interval just prior to opening the switch.

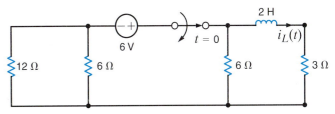

Figure P7.21

7.22 Use the differential equation approach to find $i_o(t)$ for $t > 0$ in the circuit in Fig. P7.22 and plot the response including the time interval just prior to opening the switch.

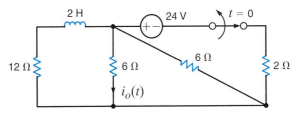

Figure P7.22

7.23 Using the differential equation approach, find $i_o(t)$ for $t > 0$ in the circuit in Fig. P7.23 and plot the response including the time interval just prior to opening the switch.

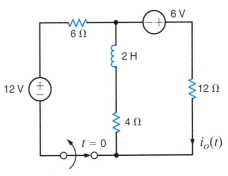

Figure P7.23

7.24 Use the differential equation approach to find $v_o(t)$ for $t > 0$ in the circuit in Fig. P7.24 and plot the response including the time interval just prior to opening the switch.

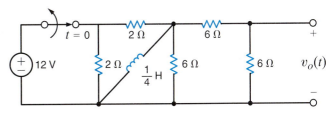

Figure P7.24

7.25 Use the differential equation approach to find $i(t)$ for $t > 0$ in the circuit in Fig. P7.25 and plot the response including the time interval just prior to opening the switch.

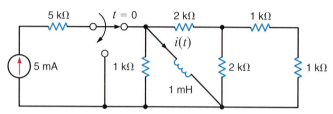

Figure P7.25

7.26 Find $v_C(t)$ for $t > 0$ in the network in Fig. P7.26 using the step-by-step method. **CS**

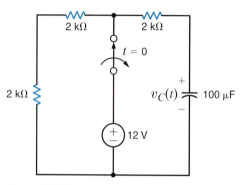

Figure P7.26

7.27 Use the step-by-step method to find $i_o(t)$ for $t > 0$ in the circuit in Fig. P7.27.

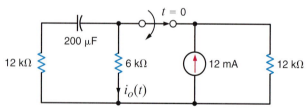

Figure P7.27

7.28 Use the step-by-step technique to find $i_o(t)$ for $t > 0$ in the network in Fig. P7.28. **CS**

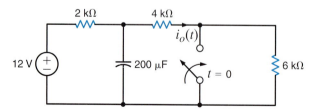

Figure P7.28

7.29 Use the step-by-step method to find $v_o(t)$ for $t > 0$ in the network in Fig. P7.29.

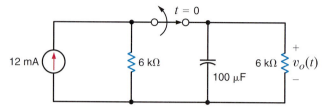

Figure P7.29

7.30 Use the step-by-step method to find $i_o(t)$ for $t > 0$ in the circuit in Fig. P7.30.

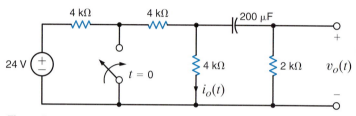

Figure P7.30

7.31 Find $v_o(t)$ for $t > 0$ in the network in Fig. P7.30 using the step-by-step technique.

7.32 Use the step-by-step technique to find $i_o(t)$ for $t > 0$ in the network in Fig. P7.32.

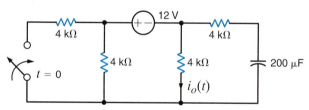

Figure P7.32

7.33 Find $v_o(t)$ for $t > 0$ in the network in Fig. P7.33 using the step-by-step method. **CS**

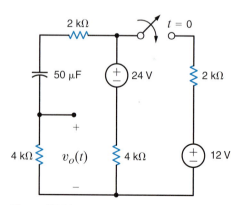

Figure P7.33

7.34 Find $v_o(t)$ for $t > 0$ in the circuit in Fig. P7.34 using the step-by-step method. **PSV**

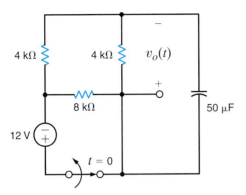

Figure P7.34

7.35 Use the step-by-step method to find $i_o(t)$ for $t > 0$ in the network in Fig. P7.35. **CS**

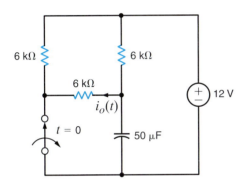

Figure P7.35

7.36 Use the step-by-step technique to find $v_o(t)$ for $t > 0$ in the circuit in Fig. P7.36.

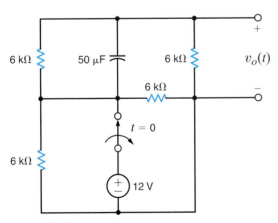

Figure P7.36

7.37 Find $i_o(t)$ for $t > 0$ in the network in Fig. P7.37 using the step-by-step method. **CS**

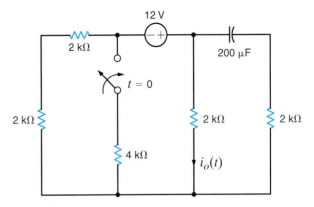

Figure P7.37

7.38 Use the step-by-step technique to find $i_o(t)$ for $t > 0$ in the network in Fig. P7.38.

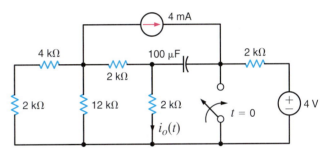

Figure P7.38

7.39 Use the step-by-step method to find $i_o(t)$ for $t > 0$ in the circuit in Fig. P7.39.

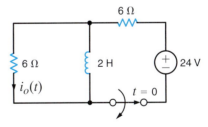

Figure P7.39

7.40 Find $i_o(t)$ for $t > 0$ in the network in Fig. P7.40 using the step-by-step method.

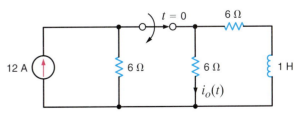

Figure P7.40

7.41 Find $i_o(t)$ for $t > 0$ in the network in Fig. P7.41 using the step-by-step method. **CS**

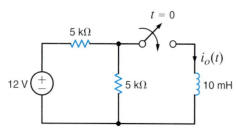

Figure P7.41

7.42 Find $v_o(t)$ for $t > 0$ in the network in Fig. P7.42 using the step-by-step method.

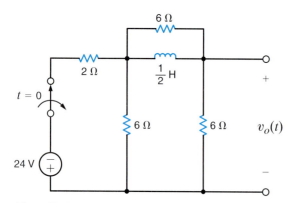

Figure P7.42

7.43 Use the step-by-step method to find $v_o(t)$ for $t > 0$ in the network in Fig. P7.43. **PSV**

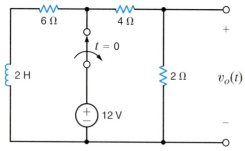

Figure P7.43

7.44 Find $i_o(t)$ for $t > 0$ in the network in Fig. P7.44 using the step-by-step method.

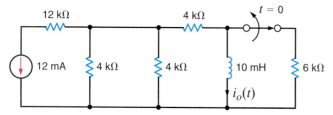

Figure P7.44

7.45 Use the step-by-step technique to find $v_o(t)$ for $t > 0$ in the circuit in Fig. P7.45.

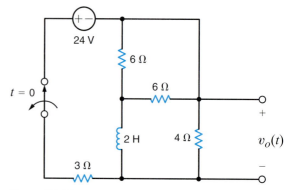

Figure P7.45

7.46 Use the step-by-step method to find $v_o(t)$ for $t > 0$ in the network in Fig. P7.46.

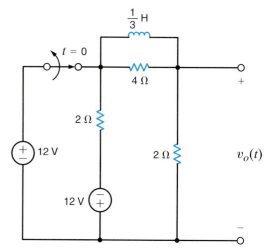

Figure P7.46

7.47 Find $v_o(t)$ for $t > 0$ in the circuit in Fig. P7.47 using the step-by-step method.

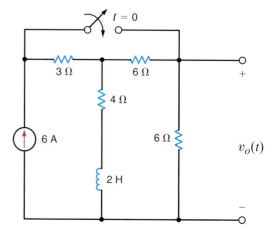

Figure P7.47

7.48 Find $v_o(t)$ for $t > 0$ in the network in Fig. P7.48 using the step-by-step technique.

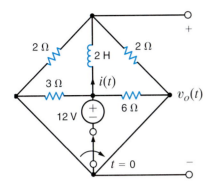

Figure P7.48

7.49 Use the step-by-step method to find $v_o(t)$ for $t > 0$ in the circuit in Fig. P7.49.

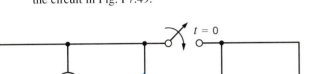
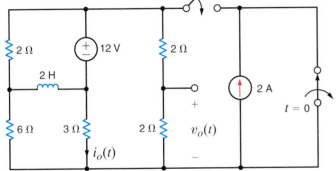

Figure P7.49

7.50 Find $i(t)$ for $t > 0$ in the circuit of Fig. P7.50 using the step-by-step method.

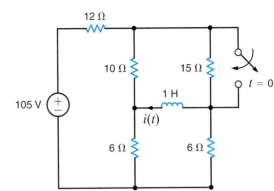

Figure P7.50

7.51 Find $v_C(t)$ for $t > 0$ in the circuit of Fig. P7.51 using the step-by-step method.

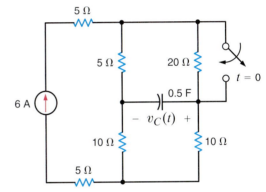

Figure P7.51

7.52 Find $i(t)$ for $t > 0$ in the circuit of Fig. P7.52 using the step-by-step method.

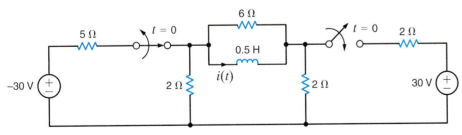

Figure P7.52

7.53 Find $i_o(t)$ for $t > 0$ in the circuit in Fig. P7.53 using the step-by-step method.

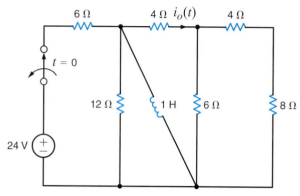

Figure P7.53

7.54 Find $v_o(t)$ for $t > 0$ in the network in Fig. P7.54 using the step-by-step method.

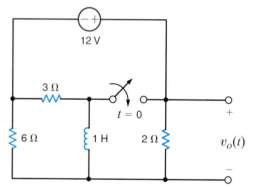

Figure P7.54

7.55 Find $i_o(t)$ for $t > 0$ in the network in Fig. P7.55 using the step-by-step method. **PSV**

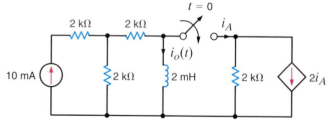

Figure P7.55

7.56 Find $i_L(t)$ for $t > 0$ in the circuit of Fig. P7.56 using the step-by-step method.

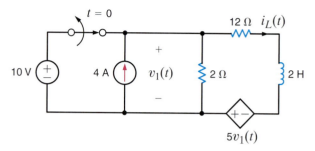

Figure P7.56

7.57 Use the step-by-step technique to find $v_o(t)$ for $t > 0$ in the network in Fig. P7.57.

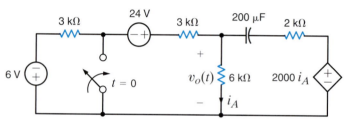

Figure P7.57

7.58 Use the step-by-step method to find $v_o(t)$ for $t > 0$ in the network in Fig. P7.58. **CS**

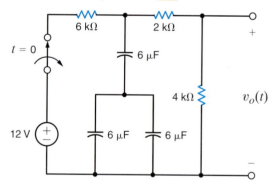

Figure P7.58

7.59 Find $i_o(t)$ for $t > 0$ in the circuit in Fig. P7.59 using the step-by-step method. **CS**

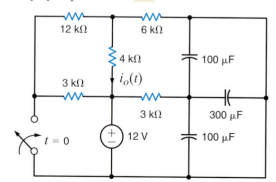

Figure P7.59

7.60 Find $v_o(t)$ for $t > 0$ in the network in Fig. P7.60 using the step-by-step method.

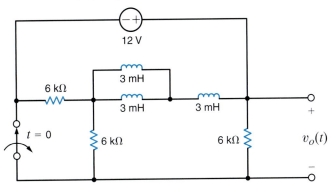

Figure P7.60

7.61 Use the step-by-step method to find $i_o(t)$ for $t > 0$ in the network in Fig. P7.61.

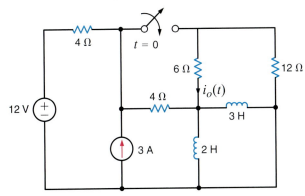

Figure P7.61

7.62 The current source in the network in Fig. P7.62a is defined in Fig. P7.62b. The initial voltage across the capacitor must be zero. (Why?) Determine the current $i_o(t)$ for $t > 0$.

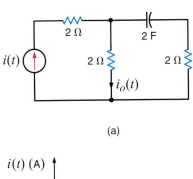

(a)

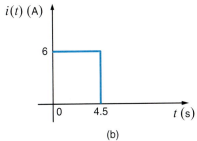

(b)

Figure P7.62

7.63 Determine the equation for the voltage $v_o(t)$ for $t > 0$, in Fig. P7.63a when subjected to the input pulse shown in Fig. P7.63b.

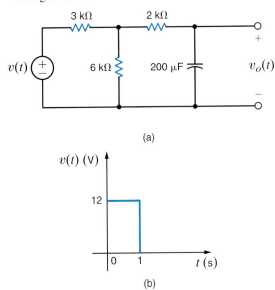

(a)

(b)

Figure P7.63

7.64 Find the output voltage $v_o(t)$ in the network in Fig. P7.64 if the input voltage is
$$v_i(t) = 5(u(t) - u(t - 0.05)) \text{ V}.$$

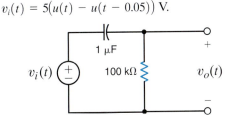

Figure P7.64

7.65 The voltage $v(t)$ shown in Fig. P7.65a is given by the graph shown in Fig. P7.65b. If $i_L(0) = 0$, answer the following questions: (a) how much energy is stored in the inductor at $t = 3$ s?, (b) how much power is supplied by the source at $t = 4$ s?, (c) what is $i(t = 6 \text{ s})$?, and (d) how much power is absorbed by the inductor at $t = 3$ s?

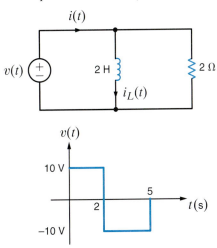

Figure P7.65

7.66 In the circuit in Fig. P7.66, $v_R(t) = 100e^{-400t}$ V for $t < 0$. Find $v_R(t)$ for $t > 0$.

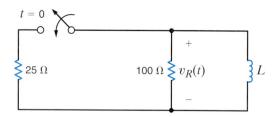

Figure P7.66

7.67 Given that $v_{C1}(0-) = -10$ V and $v_{C2}(0-) = 20$ V in the circuit in Fig. P7.67, find $i(0+)$.

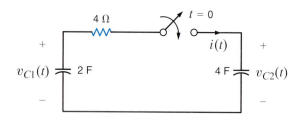

Figure P7.67

7.68 The switch in the circuit in Fig. P7.68 is closed at $t = 0$. If $i_1(0-) = 2$ A, determine $i_2(0+)$, $v_R(0+)$, and $i_1(t = \infty)$.

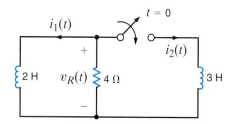

Figure P7.68

7.69 In the network in Fig. P7.69 find $i(t)$ for $t > 0$. If $v_{C1}(0-) = -10$ V, calculate $v_{C2}(0-)$.

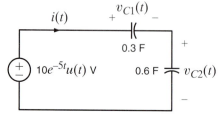

Figure P7.69

7.70 The switch in the circuit in Fig. P7.70 has been closed for a long time and is opened at $t = 0$. If $v_C(t) = 20 - 8e^{-0.05t}$ V, find R_1, R_2, and C.

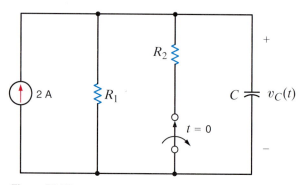

Figure P7.70

7.71 Given that $i(t) = 13.33e^{-t} - 8.33e^{-0.5t}$ A for $t > 0$ in the network in Fig. P7.71, find the following: (a) $v_C(0)$, (b) $v_C(t = 1 \text{ s})$, and (c) the capacitance C.

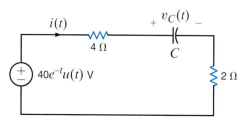

Figure P7.71

7.72 Given that $i(t) = 2.5 + 1.5e^{-4t}$ A for $t > 0$ in the circuit in Fig. P7.72, find R_1, R_2, and L.

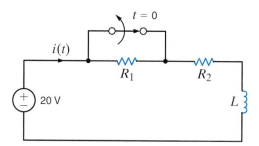

Figure P7.72

SECTION 7.3

7.73 The differential equation that describes the current $i_o(t)$ in a network is

$$\frac{d^2i_o(t)}{dt^2} + 6\left[\frac{di_o(t)}{dt}\right] + 4i_o(t) = 0$$

Find **(a)** the characteristic equation of the network, **(b)** the network's natural frequencies, and **(c)** the expression for $i_o(t)$.

7.74 The terminal current in a network is described by the equation

$$\frac{d^2i_o(t)}{dt^2} + 8\left[\frac{di_o(t)}{dt}\right] + 16i_o(t) = 0$$

Find **(a)** the characteristic equation of the network, **(b)** the network's natural frequencies, and **(c)** the equation for $i_o(t)$.

7.75 The voltage $v_1(t)$ in a network is defined by the equation

$$\frac{d^2v_1(t)}{dt^2} + 2\left[\frac{dv_1(t)}{dt}\right] + 5v_1(t) = 0$$

Find

(a) the characteristic equation of the network.

(b) the circuit's natural frequencies.

(c) the expression for $v_1(t)$. **CS**

7.76 The output voltage of a circuit is described by the differential equation

$$\frac{d^2v_o(t)}{dt^2} + 8\left[\frac{dv_o(t)}{dt}\right] + 10v_o(t) = 0$$

Find **(a)** the characteristic equation of the circuit, **(b)** the network's natural frequencies, and **(c)** the equation for $v_o(t)$.

7.77 The parameters for a parallel RLC circuit are $R = 1\ \Omega$, $L = 1/2$ H, and $C = 1/2$ F. Determine the type of damping exhibited by the circuit.

7.78 A series RLC circuit contains a resistor $R = 2\ \Omega$ and a capacitor $C = 1/2$ F. Select the value of the inductor so that the circuit is critically damped.

7.79 A parallel RLC circuit contains a resistor $R = 1\ \Omega$ and an inductor $L = 2$ H. Select the value of the capacitor so that the circuit is critically damped.

7.80 For the underdamped circuit shown in Fig. P7.80, determine the voltage $v(t)$ if the initial conditions on the storage elements are $i_L(0) = 1$ A and $v_C(0) = 10$ V. **CS**

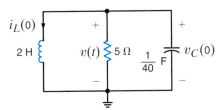

Figure P7.80

7.81 In the critically damped circuit shown in Fig. P7.81, the initial conditions on the storage elements are $i_L(0) = 2$ A and $v_C(0) = 5$ V. Determine the voltage $v(t)$.

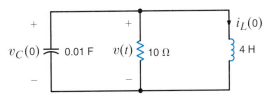

Figure P7.81

7.82 Given the circuit and the initial conditions from Problem 7.81, determine the current $i_L(t)$ that is flowing through the inductor.

7.83 Find $v_C(t)$ for $t > 0$ in the circuit in Fig. P7.83. **CS**

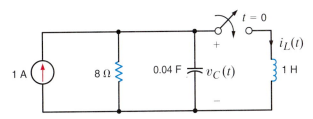

Figure P7.83

7.84 Find $v_C(t)$ for $t > 0$ in the circuit in Fig. P7.84 if $v_C(0) = 0$.

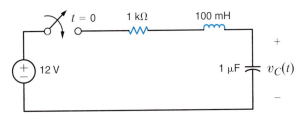

Figure P7.84

7.85 Find $v_o(t)$ for $t > 0$ in the circuit in Fig. P7.85 and plot the response including the time interval just prior to closing the switch. **PSV**

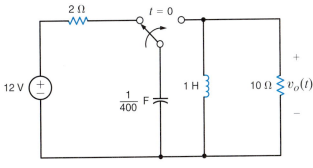

Figure P7.85

7.86 Find $v_o(t)$ for $t > 0$ in the circuit in Fig. P7.86 and plot the response including the time interval just prior to closing the switch.

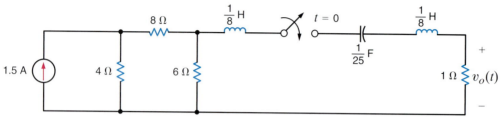

Figure P7.86

7.87 In the circuit shown in Fig. P7.87, switch action occurs at $t = 0$. Determine the voltage $v_o(t)$, $t > 0$.

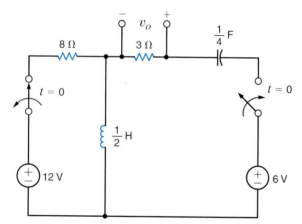

Figure P7.87

7.89 The switch in the circuit in Fig. P7.89 has been closed for a long time and is opened at $t = 0$. Find $i(t)$ for $t > 0$. **PSV**

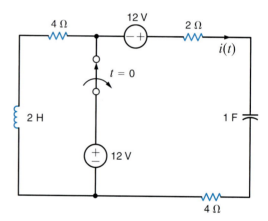

Figure P7.89

7.88 Find $v_o(t)$ for $t > 0$ in the circuit in Fig. P7.88 and plot the response including the time interval just prior to moving the switch.

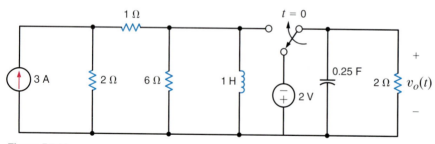

Figure P7.88

PROBLEMS **343**

7.90 The switch in the circuit in Fig. P7.90 has been closed for a long time and is opened at $t = 0$. Solve for $i(t)$ for $t > 0$.

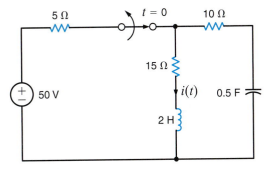

Figure P7.90

7.91 The switch in the circuit in Fig. P7.91 has been closed for a long time and is opened at $t = 0$. Solve for $i(t)$ for $t > 0$.

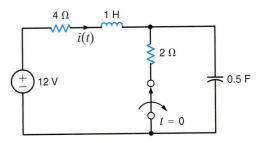

Figure P7.91

7.92 The switch in the circuit in Fig. P7.92 has been closed for a long time and is opened at $t = 0$. Find $i(t)$ for $t > 0$.

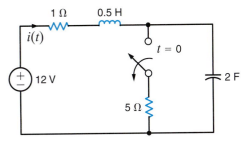

Figure P7.92

7.93 The switch in the circuit in Fig. P7.93 has been closed for a long time and is opened at $t = 0$. Find $i(t)$ for $t > 0$.

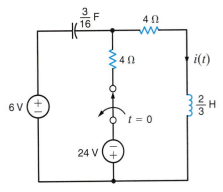

Figure P7.93

7.94 The switch in the circuit in Fig. P7.94 has been closed for a long time and is opened at $t = 0$. Find $i(t)$ for $t > 0$.

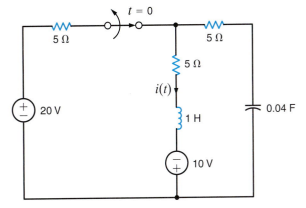

Figure P7.94

7.95 The switch in the circuit in Fig. P7.95 has been closed for a long time and is opened at $t = 0$. Find $i(t)$ for $t > 0$.

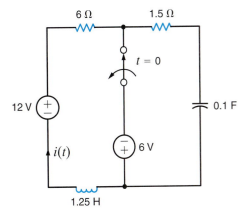

Figure P7.95

SECTION 7.4

7.96 Using the PSPICE *Schematics* editor, draw the circuit in Fig. P7.96, and use the PROBE utility to plot $v_C(t)$ and determine the time constants for $0 < t < 1$ ms and 1 ms $< t < \infty$. Also, find the maximum voltage on the capacitor.

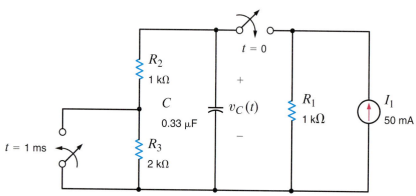

Figure P7.96

7.97 Using the PSPICE *Schematics* editor, draw the circuit in Fig. P7.97, and use the PROBE utility to find the maximum values of $v_L(t)$, $i_C(t)$, and $i(t)$.

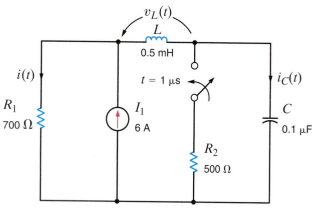

Figure P7.97

7.98 Design a parallel *RLC* circuit with $R \geq 1$ kΩ that has the characteristic equation

$$s^2 + 4 \times 10^7 s + 4 \times 10^{14} = 0$$

7.99 Design a parallel *RLC* circuit with $R \geq 1$ kΩ that has the characteristic equation

$$s^2 + 4 \times 10^7 s + 3 \times 10^{14} = 0$$

7.100 The curve shown in Fig. P7.100 is used to model the pressure in a vessel located in a chemical plant. We wish to design a circuit to realize this function so that we can study various parameters in the vessel, such as volume.

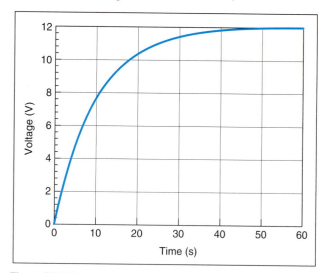

Figure P7.100

7.101 Let us redesign the pulse generator in Example 7.14 such that a voltage with the following characteristics is created across a 10-kΩ resistor: a peak value of 250 V, a cycle time of 10,000 pulses/second, and a T_1 value of one-half the cycle time.

7FE-1 In the circuit in Fig. 7PFE-1, the switch, which has been closed for a long time, opens at $t = 0$. Find the value of the capacitor voltage $v_C(t)$ at $t = 2$ s. **CS**

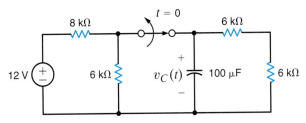

Figure 7PFE-1

7FE-2 In the network in Fig. 7PFE-2, the switch closes at $t = 0$. Find $v_o(t)$ at $t = 1$ s.

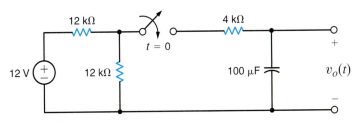

Figure 7PFE-2

7FE-3 Assume that the switch in the network in Fig. 7PFE-3 has been closed for some time. At $t = 0$ the switch opens. Determine the time required for the capacitor voltage to decay to one-half of its initially charged value. **CS**

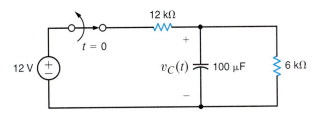

Figure 7PFE-3

AC Steady-State Analysis

LEARNING GOALS

Access Problem-Solving Videos **PSV** *and Circuit Solutions* **CS** *at:* **http://www.justask4u.com/irwin** using the registration code on the inside cover and see a website with answers and more!

In the preceding chapters we have considered in some detail both the natural and forced response of a network. We found that the natural response was a characteristic of the network and was independent of the forcing function. The forced response, however, depends directly on the type of forcing function, which until now has generally been a constant. At this point we will diverge from this tack to consider an extremely important excitation: the *sinusoidal forcing function*. Nature is replete with examples of sinusoidal phenomena, and although this is important to us as we examine many different types of physical systems, one reason that we can appreciate at this point for studying this forcing function is that it is the dominant waveform in the electric power industry. The signal present at the ac outlets in our home, office, laboratory, and so on is sinusoidal. In addition, we will demonstrate in Chapter 15 that via Fourier analysis we can represent any periodic electrical signal by a sum of sinusoids.

In this chapter we concentrate on the steady-state forced response of networks with sinusoidal driving functions. We will ignore the initial conditions and the transient or natural response, which will eventually vanish for the type of circuits with which we will be dealing. We refer to this as an *ac steady-state analysis*.

Our approach will be to begin by first studying the characteristics of a sinusoidal function as a prelude to using it as a forcing function for a circuit. We will mathematically relate this sinusoidal forcing function to a complex forcing function, which will lead us to define a phasor. By employing phasors, we effectively transform a set of differential equations with sinusoidal forcing functions in the time domain into a set of algebraic equations containing complex numbers in the frequency domain. We will show that in this frequency domain Kirchhoff's laws are valid, and, thus, all the analysis techniques that we have learned for dc analysis are applicable in ac steady-state analysis. Finally, we demonstrate the power of MATLAB and PSPICE in the solution of ac steady-state circuits. ●

8.1   Sinusoids

Let us begin our discussion of sinusoidal functions by considering the sine wave

$$x(t) = X_M \sin \omega t \qquad \textbf{8.1}$$

where $x(t)$ could represent either $v(t)$ or $i(t)$. X_M is the *amplitude, maximum value* or peak value, ω is the *radian* or *angular frequency*, and ωt is the *argument* of the sine function. A plot of the function in Eq. (8.1) as a function of its argument is shown in Fig. 8.1a. Obviously, the function repeats itself every 2π radians. This condition is described mathematically as $x(\omega t + 2\pi) = x(\omega t)$ or in general for period T as

$$x\big[\omega(t + T)\big] = x(\omega t) \qquad \textbf{8.2}$$

meaning that the function has the same value at time $t + T$ as it does at time t.

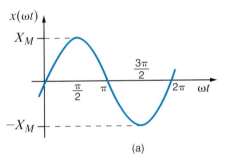

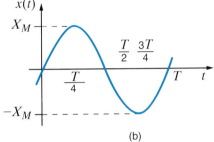

(a)   (b)

Figure 8.1
Plots of a sine wave as a function of both ωt and t.

The waveform can also be plotted as a function of time, as shown in Fig. 8.1b. Note that this function goes through one period every T seconds, or in other words, in 1 second it goes through $1/T$ periods or cycles. The number of cycles per second, called Hertz, is the frequency f, where

$$f = \frac{1}{T} \qquad \textbf{8.3}$$

HINT
The relationship between frequency and period

Now since $\omega T = 2\pi$, as shown in Fig. 8.1a, we find that

$$\omega = \frac{2\pi}{T} = 2\pi f \qquad \textbf{8.4}$$

HINT
The relationship between frequency, period, and radian frequency

which is, of course, the general relationship among period in seconds, frequency in Hertz, and radian frequency.

Now that we have discussed some of the basic properties of a sine wave, let us consider the following general expression for a sinusoidal function:

$$x(t) = X_M \sin(\omega t + \theta) \qquad \textbf{8.5}$$

In this case $(\omega t + \theta)$ is the argument of the sine function, and θ is called the *phase angle*. A plot of this function is shown in Fig. 8.2, together with the original function in Eq. (8.1) for comparison. Because of the presence of the phase angle, any point on the waveform $X_M \sin(\omega t + \theta)$ occurs θ radians earlier in time than the corresponding point on the waveform $X_M \sin \omega t$. Therefore, we say that $X_M \sin \omega t$ *lags* $X_M \sin(\omega t + \theta)$ by θ radians. In the more general situation, if

$$x_1(t) = X_{M_1} \sin(\omega t + \theta)$$

HINT
Phase lag defined

and

$$x_2(t) = X_{M_2} \sin(\omega t + \phi)$$

HINT
In phase and out of phase defined

HINT

Phase lead graphically
illustrated

Figure 8.2
Graphical illustration of
$X_M \sin(\omega t + \theta)$ *leading*
$X_M \sin \omega t$ *by θ radians.*

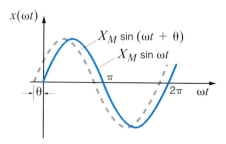

then $x_1(t)$ *leads* $x_2(t)$ by $\theta - \phi$ radians and $x_2(t)$ lags $x_1(t)$ by $\theta - \phi$ radians. If $\theta = \phi$, the waveforms are identical and the functions are said to be *in phase*. If $\theta \neq \phi$, the functions are *out of phase*.

The phase angle is normally expressed in degrees rather than radians. Therefore, we will simply state at this point that we will use the two forms interchangeably; that is,

$$x(t) = X_M \sin\left(\omega t + \frac{\pi}{2}\right) = X_M \sin(\omega t + 90°) \qquad \textbf{8.6}$$

HINT

A very important point

Rigorously speaking, since ωt is in radians, the phase angle should be as well. However, it is common practice and convenient to use degrees for phase; therefore, that will be our practice in this text.

In addition, it should be noted that adding to the argument integer multiples of either 2π radians or 360° does not change the original function. This can easily be shown mathematically but is visibly evident when examining the waveform, as shown in Fig. 8.2.

Although our discussion has centered on the sine function, we could just as easily have used the cosine function, since the two waveforms differ only by a phase angle; that is,

$$\cos \omega t = \sin\left(\omega t + \frac{\pi}{2}\right) \qquad \textbf{8.7}$$

HINT

Some trigonometric
identities that are useful in
phase angle calculations

$$\sin \omega t = \cos\left(\omega t - \frac{\pi}{2}\right) \qquad \textbf{8.8}$$

We are often interested in the phase difference between two sinusoidal functions. Three conditions must be satisfied before we can determine the phase difference: (1) the frequency of both sinusoids must be the same, (2) the amplitude of both sinusoids must be positive, and (3) both sinusoids must be written as sine waves or cosine waves. Once in this format, the phase angle between the functions can be computed as outlined previously. Two other trigonometric identities that normally prove useful in phase angle determination are:

$$-\cos(\omega t) = \cos(\omega t \pm 180°) \qquad \textbf{8.9}$$

$$-\sin(\omega t) = \sin(\omega t \pm 180°) \qquad \textbf{8.10}$$

Finally, the angle-sum and angle-difference relationships for sines and cosines may be useful in the manipulation of sinusoidal functions. These relations are

$$\sin(\alpha + \beta) = \sin\alpha \cos\beta + \cos\alpha \sin\beta$$

$$\cos(\alpha + \beta) = \cos\alpha \cos\beta - \sin\alpha \sin\beta$$

$$\sin(\alpha - \beta) = \sin\alpha \cos\beta - \cos\alpha \sin\beta \qquad \textbf{8.11}$$

$$\cos(\alpha - \beta) = \cos\alpha \cos\beta + \sin\alpha \sin\beta$$

Example 8.1

We wish to plot the waveforms for the following functions:

a. $v(t) = 1 \cos(\omega t + 45°)$,

b. $v(t) = 1 \cos(\omega t + 225°)$, and

c. $v(t) = 1 \cos(\omega t - 315°)$.

SOLUTION   Figure 8.3a shows a plot of the function $v(t) = 1 \cos \omega t$. Figure 8.3b is a plot of the function $v(t) = 1 \cos(\omega t + 45°)$. Figure 8.3c is a plot of the function $v(t) = 1 \cos(\omega t + 225°)$. Note that since

$$v(t) = 1 \cos(\omega t + 225°) = 1 \cos(\omega t + 45° + 180°)$$

this waveform is 180° out of phase with the waveform in Fig. 8.3b; that is, $\cos(\omega t + 225°) = -\cos(\omega t + 45°)$, and Fig. 8.3c is the negative of Fig. 8.3b. Finally, since the function

$$v(t) = 1 \cos(\omega t - 315°) = 1 \cos(\omega t - 315° + 360°) = 1 \cos(\omega t + 45°)$$

this function is identical to that shown in Fig. 8.3b.

Figure 8.3
Cosine waveforms with various phase angles.

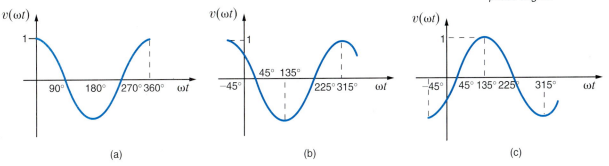

(a)   (b)   (c)

Example 8.2

Determine the frequency and the phase angle between the two voltages $v_1(t) = 12 \sin(1000t + 60°)$ V and $v_2(t) = -6 \cos(1000t + 30°)$ V.

SOLUTION   The frequency in Hertz (Hz) is given by the expression

$$f = \frac{\omega}{2\pi} = \frac{1000}{2\pi} = 159.2 \text{ Hz}$$

Using Eq. (8.9), $v_2(t)$ can be written as

$$v_2(t) = -6 \cos(\omega t + 30°) = 6 \cos(\omega t + 210°) \text{ V}$$

Then employing Eq. (8.7), we obtain

$$6 \sin(\omega t + 300°) \text{ V} = 6 \sin(\omega t - 60°) \text{ V}$$

Now that both voltages of the same frequency are expressed as sine waves with positive amplitudes, the phase angle between $v_1(t)$ and $v_2(t)$ is $60° - (-60°) = 120°$; that is, $v_1(t)$ leads $v_2(t)$ by 120° or $v_2(t)$ lags $v_1(t)$ by 120°.

E8.1 Given the voltage $v(t) = 120 \cos(314t + \pi/4)$ V, determine the frequency of the voltage in Hertz and the phase angle in degrees.

ANSWER: $f = 50$ Hz; $\theta = 45°$.

E8.2 Three branch currents in a network are known to be

$$i_1(t) = 2 \sin(377t + 45°) \text{ A}$$

$$i_2(t) = 0.5 \cos(377t + 10°) \text{ A}$$

$$i_3(t) = -0.25 \sin(377t + 60°) \text{ A}$$

Determine the phase angles by which $i_1(t)$ leads $i_2(t)$ and $i_1(t)$ leads $i_3(t)$.

ANSWER: i_1 leads i_2 by $-55°$; i_1 leads i_3 by $165°$.

8.2 Sinusoidal and Complex Forcing Functions

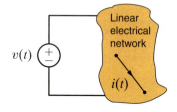

$v(t)$

Figure 8.4
Current response to an applied voltage in an electrical network.

In the preceding chapters we applied a constant forcing function to a network and found that the steady-state response was also constant.

In a similar manner, if we apply a sinusoidal forcing function to a linear network, the steady-state voltages and currents in the network will also be sinusoidal. This should also be clear from the KVL and KCL equations. For example, if one branch voltage is a sinusoid of some frequency, the other branch voltages must be sinusoids of the same frequency if KVL is to apply around any closed path. This means, of course, that the forced solutions of the differential equations that describe a network with a sinusoidal forcing function are sinusoidal functions of time. For example, if we assume that our input function is a voltage $v(t)$ and our output response is a current $i(t)$ as shown in Fig. 8.4, then if $v(t) = A \sin(\omega t + \theta)$, $i(t)$ will be of the form $i(t) = B \sin(\omega t + \phi)$. The critical point here is that we know the form of the output response, and therefore the solution involves simply determining the values of the two parameters B and ϕ.

Example 8.3

Consider the circuit in Fig. 8.5. Let us derive the expression for the current.

Figure 8.5
A simple RL circuit.

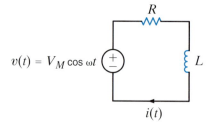

$v(t) = V_M \cos \omega t$

$i(t)$

SOLUTION The KVL equation for this circuit is

$$L \frac{di(t)}{dt} + Ri(t) = V_M \cos \omega t$$

Since the input forcing function is $V_M \cos \omega t$, we assume that the forced response component of the current $i(t)$ is of the form

$$i(t) = A \cos(\omega t + \phi)$$

which can be written using Eq. (8.11) as

$$i(t) = A \cos \phi \cos \omega t - A \sin \phi \sin \omega t$$
$$= A_1 \cos \omega t + A_2 \sin \omega t$$

Note that this is, as we observed in Chapter 7, of the form of the forcing function $\cos \omega t$ and its derivative $\sin \omega t$. Substituting this form for $i(t)$ into the preceding differential equation yields

$$L \frac{d}{dt} \left(A_1 \cos \omega t + A_2 \sin \omega t \right) + R \left(A_1 \cos \omega t + A_2 \sin \omega t \right) = V_M \cos \omega t$$

Evaluating the indicated derivative produces

$$-A_1 \omega L \sin \omega t + A_2 \omega L \cos \omega t + RA_1 \cos \omega t + RA_2 \sin \omega t = V_M \cos \omega t$$

By equating coefficients of the sine and cosine functions, we obtain

$$-A_1 \omega L + A_2 R = 0$$
$$A_1 R + A_2 \omega L = V_M$$

that is, two simultaneous equations in the unknowns A_1 and A_2. Solving these two equations for A_1 and A_2 yields

$$A_1 = \frac{R V_M}{R^2 + \omega^2 L^2}$$

$$A_2 = \frac{\omega L V_M}{R^2 + \omega^2 L^2}$$

Therefore,

$$i(t) = \frac{R V_M}{R^2 + \omega^2 L^2} \cos \omega t + \frac{\omega L V_M}{R^2 + \omega^2 L^2} \sin \omega t$$

which, using the last identity in Eq. (8.11), can be written as

$$i(t) = A \cos(\omega t + \phi)$$

where A and ϕ are determined as follows:

$$A \cos \phi = \frac{R V_M}{R^2 + \omega^2 L^2}$$

$$A \sin \phi = \frac{-\omega L V_M}{R^2 + \omega^2 L^2}$$

Hence,

$$\tan \phi = \frac{A \sin \phi}{A \cos \phi} = -\frac{\omega L}{R}$$

and therefore,

$$\phi = -\tan^{-1} \frac{\omega L}{R}$$

and since

$$(A \cos \phi)^2 + (A \sin \phi)^2 = A^2(\cos^2 \phi + \sin^2 \phi) = A^2$$

$$A^2 = \frac{R^2 V_M^2}{\left(R^2 + \omega^2 L^2 \right)^2} + \frac{(\omega L)^2 V_M^2}{\left(R^2 + \omega^2 L^2 \right)^2}$$

$$= \frac{V_M^2}{R^2 + \omega^2 L^2}$$

$$A = \frac{V_M}{\sqrt{R^2 + \omega^2 L^2}}$$

Hence, the final expression for $i(t)$ is

$$i(t) = \frac{V_M}{\sqrt{R^2 + \omega^2 L^2}} \cos\left(\omega t - \tan^{-1}\frac{\omega L}{R}\right)$$

The preceding analysis indicates that ϕ is zero if $L = 0$ and hence $i(t)$ is in phase with $v(t)$. If $R = 0$, $\phi = -90°$, and the current lags the voltage by $90°$. If L and R are both present, the current lags the voltage by some angle between $0°$ and $90°$.

This example illustrates an important point—solving even a simple one-loop circuit containing one resistor and one inductor is very complicated when compared to the solution of a single-loop circuit containing only two resistors. Imagine for a moment how laborious it would be to solve a more complicated circuit using the procedure we have employed in Example 8.3. To circumvent this approach, we will establish a correspondence between sinusoidal time functions and complex numbers. We will then show that this relationship leads to a set of algebraic equations for currents and voltages in a network (e.g., loop currents or node voltages) in which the coefficients of the variables are complex numbers. Hence, once again we will find that determining the currents or voltages in a circuit can be accomplished by solving a set of algebraic equations; however, in this case, their solution is complicated by the fact that variables in the equations have complex, rather than real, coefficients.

The vehicle we will employ to establish a relationship between time-varying sinusoidal functions and complex numbers is Euler's equation, which for our purposes is written as

$$e^{j\omega t} = \cos \omega t + j \sin \omega t \qquad \textbf{8.12}$$

This complex function has a real part and an imaginary part:

$$\text{Re}(e^{j\omega t}) = \cos \omega t$$
$$\text{Im}(e^{j\omega t}) = \sin \omega t \qquad \textbf{8.13}$$

where $\text{Re}(\cdot)$ and $\text{Im}(\cdot)$ represent the real part and the imaginary part, respectively, of the function in the parentheses. Recall that $j = \sqrt{-1}$.

Now suppose that we select as our forcing function in Fig. 8.4 the nonrealizable voltage

$$v(t) = V_M e^{j\omega t} \qquad \textbf{8.14}$$

which because of Euler's identity can be written as

$$v(t) = V_M \cos \omega t + j V_M \sin \omega t \qquad \textbf{8.15}$$

The real and imaginary parts of this function are each realizable. We think of this complex forcing function as two forcing functions, a real one and an imaginary one, and as a consequence of linearity, the principle of superposition applies and thus the current response can be written as

$$i(t) = I_M \cos(\omega t + \phi) + j I_M \sin(\omega t + \phi) \qquad \textbf{8.16}$$

where $I_M \cos(\omega t + \phi)$ is the response due to $V_M \cos \omega t$ and $j I_M \sin(\omega t + \phi)$ is the response due to $j V_M \sin \omega t$. This expression for the current containing both a real and an imaginary term can be written via Euler's equation as

$$i(t) = I_M e^{j(\omega t + \phi)} \qquad \textbf{8.17}$$

Because of the preceding relationships, we find that rather than applying the forcing function $V_M \cos \omega t$ and calculating the response $I_M \cos(\omega t + \phi)$, we can apply the complex forcing function $V_M e^{j\omega t}$ and calculate the response $I_M e^{j(\omega t + \phi)}$, the real part of which is the desired response $I_M \cos(\omega t + \phi)$. Although this procedure may initially appear to be more complicated, it is not. It is through this technique that we will convert the differential equation to an algebraic equation that is much easier to solve.

Example 8.4

Once again, let us determine the current in the *RL* circuit examined in Example 8.3. However, rather than applying $V_M \cos \omega t$ we will apply $V_M e^{j\omega t}$.

SOLUTION The forced response will be of the form

$$i(t) = I_M e^{j(\omega t + \phi)}$$

where only I_M and ϕ are unknown. Substituting $v(t)$ and $i(t)$ into the differential equation for the circuit, we obtain

$$RI_M e^{j(\omega t + \phi)} + L \frac{d}{dt} \left(I_M e^{j(\omega t + \phi)} \right) = V_M e^{j\omega t}$$

Taking the indicated derivative, we obtain

$$RI_M e^{j(\omega t + \phi)} + j\omega L I_M e^{j(\omega t + \phi)} = V_M e^{j\omega t}$$

Dividing each term of the equation by the common factor $e^{j\omega t}$ yields

$$RI_M e^{j\phi} + j\omega L I_M e^{j\phi} = V_M$$

which is an algebraic equation with complex coefficients. This equation can be written as

$$I_M e^{j\phi} = \frac{V_M}{R + j\omega L}$$

Converting the right-hand side of the equation to exponential or polar form produces the equation

$$I_M e^{j\phi} = \frac{V_M}{\sqrt{R^2 + \omega^2 L^2}} e^{j[-\tan^{-1}(\omega L/R)]}$$

(A quick refresher on complex numbers is given in the Appendix for readers who need to sharpen their skills in this area.) The preceding form clearly indicates that the magnitude and phase of the resulting current are

$$I_M = \frac{V_M}{\sqrt{R^2 + \omega^2 L^2}}$$

and

$$\phi = -\tan^{-1} \frac{\omega L}{R}$$

However, since our actual forcing function was $V_M \cos \omega t$ rather than $V_M e^{j\omega t}$, our actual response is the real part of the complex response:

$$i(t) = I_M \cos(\omega t + \phi)$$

$$= \frac{V_M}{\sqrt{R^2 + \omega^2 L^2}} \cos\left(\omega t - \tan^{-1} \frac{\omega L}{R} \right)$$

Note that this is identical to the response obtained in the previous example by solving the differential equation for the current $i(t)$.

HINT

Summary of complex number relationships:

$$x + jy = r e^{j\theta}$$
$$r = \sqrt{x^2 + y^2}$$
$$\theta = \tan^{-1} \frac{y}{x}$$
$$x = r \cos\theta$$
$$y = r \sin\theta$$
$$\frac{1}{e^{j\theta}} = e^{-j\theta}$$

8.3 Phasors

Once again let us assume that the forcing function for a linear network is of the form

$$v(t) = V_M e^{j\omega t}$$

Then every steady-state voltage or current in the network will have the same form and the same frequency ω; for example, a current $i(t)$ will be of the form $i(t) = I_M e^{j(\omega t + \phi)}$.

As we proceed in our subsequent circuit analyses, we will simply note the frequency and then drop the factor $e^{j\omega t}$ since it is common to every term in the describing equations. Dropping the term $e^{j\omega t}$ indicates that every voltage or current can be fully described by a magnitude and phase. For example, a voltage $v(t)$ can be written in exponential form as

$$v(t) = V_M \cos(\omega t + \theta) = \text{Re}\left[V_M e^{j(\omega t + \theta)}\right] \qquad \textbf{8.18}$$

or as a complex number

$$v(t) = \text{Re}\left(V_M \underline{/\theta}\; e^{j\omega t}\right) \qquad \textbf{8.19}$$

HINT

If $v(t) = V_M \cos(\omega t + \theta)$ and $i(t) = I_M \cos(\omega t + \phi)$, then in phasor notation

$$\mathbf{V} = V_M \underline{/\theta}$$

and

$$\mathbf{I} = I_M \underline{/\phi}$$

Since we are working with a complex forcing function, the real part of which is the desired answer, and each term in the equation will contain $e^{j\omega t}$, we can drop $\text{Re}(\cdot)$ and $e^{j\omega t}$ and work only with the complex number $V_M \underline{/\theta}$. This complex representation is commonly called a *phasor*. As a distinguishing feature, phasors will be written in boldface type. In a completely identical manner a voltage $v(t) = V_M \cos(\omega t + \theta) = \text{Re}\left[V_M e^{j(\omega t + \theta)}\right]$ and a current $i(t) = I_M \cos(\omega t + \phi) = \text{Re}\left[I_M e^{j(\omega t + \phi)}\right]$ are written in phasor notation as $\mathbf{V} = V_M \underline{/\theta}$ and $\mathbf{I} = I_M \underline{/\phi}$, respectively. Note that it is common practice to express phasors with positive magnitudes.

Example 8.5

Again, we consider the RL circuit in Example 8.3. Let us use phasors to determine the expression for the current.

SOLUTION   The differential equation is

$$L \frac{di(t)}{dt} + Ri(t) = V_M \cos \omega t$$

The forcing function can be replaced by a complex forcing function that is written as $\mathbf{V}e^{j\omega t}$ with phasor $\mathbf{V} = V_M \underline{/0°}$. Similarly, the forced response component of the current $i(t)$ can be replaced by a complex function that is written as $\mathbf{I}e^{j\omega t}$ with phasor $\mathbf{I} = I_M \underline{/\phi}$. From our previous discussions we recall that the solution of the differential equation is the real part of this current.

Using the complex forcing function, we find that the differential equation becomes

$$L \frac{d}{dt}\left(\mathbf{I}e^{j\omega t}\right) + R\mathbf{I}e^{j\omega t} = \mathbf{V}e^{j\omega t}$$

$$j\omega L \mathbf{I}e^{j\omega t} + R\mathbf{I}e^{j\omega t} = \mathbf{V}e^{j\omega t}$$

Note that $e^{j\omega t}$ is a common factor and, as we have already indicated, can be eliminated, leaving the phasors; that is,

$$j\omega L \mathbf{I} + R\mathbf{I} = \mathbf{V}$$

Therefore,

$$\mathbf{I} = \frac{\mathbf{V}}{R + j\omega L} = I_M \underline{/\phi} = \frac{V_M}{\sqrt{R^2 + \omega^2 L^2}} \underline{\bigg/ -\tan^{-1}\frac{\omega L}{R}}$$

Thus,

HINT

The differential equation is reduced to a phasor equation.

$$i(t) = \frac{V_M}{\sqrt{R^2 + \omega^2 L^2}} \cos\left(\omega t - \tan^{-1}\frac{\omega L}{R}\right)$$

which once again is the function we obtained earlier.

We define relations between phasors after the $e^{j\omega t}$ term has been eliminated as "phasor, or frequency domain, analysis." Thus, we have transformed a set of differential equations with sinusoidal forcing functions in the time domain into a set of algebraic equations containing complex numbers in the frequency domain. In effect, we are now faced with solving a set of algebraic equations for the unknown phasors. The phasors are then simply transformed back to the time domain to yield the solution of the original set of differential equations. In addition, we note that the solution of sinusoidal steady-state circuits would be relatively simple if we could write the phasor equation directly from the circuit description. In Section 8.4 we will lay the groundwork for doing just that.

Note that in our discussions we have tacitly assumed that sinusoidal functions would be represented as phasors with a phase angle based on a cosine function. Therefore, if sine functions are used, we will simply employ the relationship in Eq. (8.7) to obtain the proper phase angle.

In summary, while $v(t)$ represents a voltage in the time domain, the phasor **V** represents the voltage in the frequency domain. The phasor contains only magnitude and phase information, and the frequency is implicit in this representation. The transformation from the time domain to the frequency domain, as well as the reverse transformation, is shown in Table 8.1.

Table 8.1 Phasor representation

Time Domain	Frequency Domain
$A\cos(\omega t \pm \theta)$	$A\,\underline{/\pm\theta}$
$A\sin(\omega t \pm \theta)$	$A\,\underline{/\pm\theta - 90°}$

PROBLEM-SOLVING STRATEGY

Phasor analysis

Step 1. Using phasors, transform a set of differential equations in the time domain into a set of algebraic equations in the frequency domain.

Step 2. Solve the algebraic equations for the unknown phasors.

Step 3. Transform the now-known phasors back to the time domain.

Recall that the phase angle is based on a cosine function and, therefore, if a sine function is involved, a 90° shift factor must be employed, as shown in the table.

It is important to note, however, that if a network contains only sine sources, there is no need to perform the 90° shift. We simply perform the normal phasor analysis, and then the *imaginary* part of the time-varying complex solution is the desired response. Simply put, cosine sources generate a cosine response, and sine sources generate a sine response.

LEARNING EXTENSIONS

E8.3 Convert the following voltage functions to phasors.

$$v_1(t) = 12\cos(377t - 425°)\ \text{V}$$
$$v_2(t) = 18\sin(2513t + 4.2°)\ \text{V}$$

ANSWER: $\mathbf{V_1} = 12\,\underline{/-425°}$ V; $\mathbf{V_2} = 18\,\underline{/-85.8°}$ V.

E8.4 Convert the following phasors to the time domain if the frequency is 400 Hz.

$$\mathbf{V_1} = 10\,\underline{/20°}\ \text{V}$$
$$\mathbf{V_2} = 12\,\underline{/-60°}\ \text{V}$$

ANSWER:
$$v_1(t) = 10\cos(800\pi t + 20°)\ \text{V};$$
$$v_2(t) = 12\cos(800\pi t - 60°)\ \text{V}.$$

8.4 Phasor Relationships for Circuit Elements

As we proceed in our development of the techniques required to analyze circuits in the sinusoidal steady-state, we are now in a position to establish the phasor relationships between voltage and current for the three passive elements R, L, and C.

In the case of a resistor as shown in Fig. 8.6a, the voltage–current relationship is known to be

HINT

Current and voltage are in phase.

$$v(t) = Ri(t) \qquad \textbf{8.20}$$

Applying the complex voltage $V_M e^{j(\omega t + \theta_v)}$ results in the complex current $I_M e^{j(\omega t + \theta_i)}$, and therefore Eq. (8.20) becomes

$$V_M e^{j(\omega t + \theta_v)} = R I_M e^{j(\omega t + \theta_i)}$$

which reduces to

$$V_M e^{j\theta_v} = R I_M e^{j\theta_i} \qquad \textbf{8.21}$$

Equation (8.21) can be written in phasor form as

$$\mathbf{V} = R\mathbf{I} \qquad \textbf{8.22}$$

where

$$\mathbf{V} = V_M e^{j\theta_v} = V_M \underline{/\theta_v} \quad \text{and} \quad \mathbf{I} = I_M e^{j\theta_i} = I_M \underline{/\theta_i}$$

From Eq. (8.21) we see that $\theta_v = \theta_i$ and thus the current and voltage for this circuit are *in phase*.

Historically, complex numbers have been represented as points on a graph in which the x-axis represents the real axis and the y-axis the imaginary axis. The line segment connecting the origin with the point provides a convenient representation of the magnitude and angle when the complex number is written in a polar form. A review of the Appendix will indicate how these complex numbers or line segments can be added, subtracted, and so on. Since phasors are complex numbers, it is convenient to represent the phasor voltage and current graphically as line segments. A plot of the line segments representing the phasors is called a *phasor diagram*. This pictorial representation of phasors provides immediate information on the relative magnitude of one phasor with another, the angle between two phasors, and the relative position of one phasor with respect to another (i.e., leading or lagging). A phasor diagram and the sinusoidal waveforms for the resistor are shown in Figs. 8.6c and d, respectively. A phasor diagram will be drawn for each of the other circuit elements in the remainder of this section.

Figure 8.6
Voltage–current relationships for a resistor.

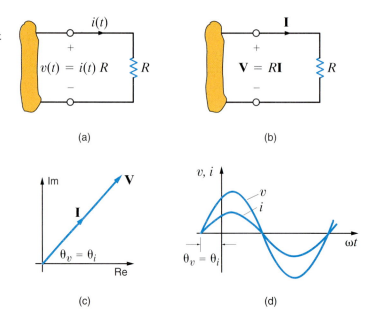

(a)

(b)

(c)

(d)

Example 8.6

If the voltage $v(t) = 24\cos(377t + 75°)$ V is applied to a 6-Ω resistor as shown in Fig. 8.6a, we wish to determine the resultant current.

SOLUTION   Since the phasor voltage is

$$\mathbf{V} = 24 \,\underline{/75°}\ \text{V}$$

the phasor current from Eq. (8.22) is

$$\mathbf{I} = \frac{24 \,\underline{/75°}}{6} = 4 \,\underline{/75°}\ \text{A}$$

which in the time domain is

$$i(t) = 4\cos(377t + 75°)\ \text{A}$$

LEARNING EXTENSION

E8.5 The current in a 4-Ω resistor is known to be $\mathbf{I} = 12 \,\underline{/60°}$ A. Express the voltage across the resistor as a time function if the frequency of the current is 4 kHz.

ANSWER:
$v(t) = 48\cos(8000\pi t + 60°)$ V.

The voltage–current relationship for an inductor, as shown in Fig. 8.7a, is

$$v(t) = L\frac{di(t)}{dt} \qquad\qquad \textbf{8.23}$$

Substituting the complex voltage and current into this equation yields

$$V_M e^{j(\omega t + \theta_v)} = L\frac{d}{dt} I_M e^{j(\omega t + \theta_i)}$$

which reduces to

$$V_M e^{j\theta_v} = j\omega L I_M e^{j\theta_i} \qquad\qquad \textbf{8.24}$$

Equation (8.24) in phasor notation is

$$\mathbf{V} = j\omega L\mathbf{I} \qquad\qquad \textbf{8.25}$$

HINT

The derivative process yields a frequency-dependent function.

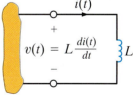
(a)

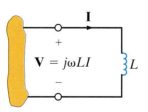
(b)

Figure 8.7
Voltage–current relationships for an inductor.

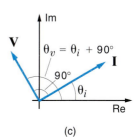
(c)

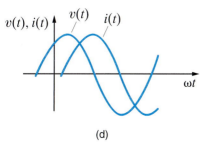
(d)

Note that the differential equation in the time domain (8.23) has been converted to an algebraic equation with complex coefficients in the frequency domain. This relationship is shown in Fig. 8.7b. Since the imaginary operator $j = 1e^{j90°} = 1\underline{/90°} = \sqrt{-1}$, Eq. (8.24) can be written as

$$V_M e^{j\theta_v} = \omega L I_M e^{j(\theta_i + 90°)} \qquad \textbf{8.26}$$

HINT

The voltage leads the current or the current lags the voltage.

Therefore, the voltage and current are *90° out of phase*, and in particular the voltage leads the current by 90° or the current lags the voltage by 90°. The phasor diagram and the sinusoidal waveforms for the inductor circuit are shown in Figs. 8.7c and d, respectively.

Example 8.7

The voltage $v(t) = 12\cos(377t + 20°)$ V is applied to a 20-mH inductor as shown in Fig. 8.7a. Find the resultant current.

SOLUTION The phasor current is

$$I = \frac{V}{j\omega L} = \frac{12\underline{/20°}}{\omega L \underline{/90°}}$$

HINT

Applying $V = j\omega LI$

$$\frac{x_1 \underline{/\theta_1}}{x_2 \underline{/\theta_2}} = \frac{x_1}{x_2}\underline{/\theta_1 - \theta_2}$$

$$= \frac{12\underline{/20°}}{(377)(20 \times 10^{-3})\underline{/90°}}$$

$$= 1.59\underline{/-70°} \text{ A}$$

or

$$i(t) = 1.59\cos(377t - 70°) \text{ A}$$

LEARNING EXTENSION

E8.6 The current in a 0.05-H inductor is $I = 4\underline{/-30°}$ A. If the frequency of the current is 60 Hz, determine the voltage across the inductor.

ANSWER:
$v_L(t) = 75.4\cos(377t + 60°)$ V.

HINT

The current leads the voltage or the voltage lags the current.

The voltage–current relationship for our last passive element, the capacitor, as shown in Fig. 8.8a, is

$$i(t) = C\frac{dv(t)}{dt} \qquad \textbf{8.27}$$

Once again employing the complex voltage and current, we obtain

$$I_M e^{j(\omega t + \theta_i)} = C\frac{d}{dt}V_M e^{j(\omega t + \theta_v)}$$

which reduces to

$$I_M e^{j\theta_i} = j\omega C V_M e^{j\theta_v} \qquad \textbf{8.28}$$

In phasor notation this equation becomes

$$I = j\omega CV \qquad \textbf{8.29}$$

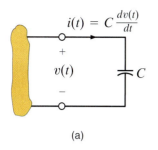 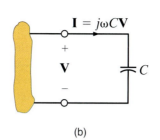

(a)   (b)

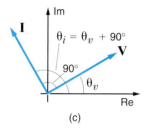 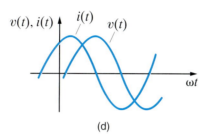

(c)   (d)

Figure 8.8
Voltage–current relationships for a capacitor.

Equation (8.27), a differential equation in the time domain, has been transformed into Eq. (8.29), an algebraic equation with complex coefficients in the frequency domain. The phasor relationship is shown in Fig. 8.8b. Substituting $j = 1e^{j90°}$ into Eq. (8.28) yields

$$I_M e^{j\theta_i} = \omega C V_M e^{j(\theta_v + 90°)} \qquad \textbf{8.30}$$

Note that the voltage and current are *90° out of phase*. Equation (8.30) states that the current leads the voltage by 90° or the voltage lags the current by 90°. The phasor diagram and the sinusoidal waveforms for the capacitor circuit are shown in Figs. 8.8c and d, respectively.

Example 8.8

The voltage $v(t) = 100 \cos(314t + 15°)$ V is applied to a 100-μF capacitor as shown in Fig. 8.8a. Find the current.

SOLUTION   The resultant phasor current is

$$\mathbf{I} = j\omega C \left(100 \underline{/15°}\right)$$

$$= (314)\left(100 \times 10^{-6} \underline{/90°}\right)\left(100 \underline{/15°}\right)$$

$$= 3.14 \underline{/105°} \text{ A}$$

Therefore, the current written as a time function is

$$i(t) = 3.14 \cos(314t + 105°) \text{ A}$$

HINT

Applying $\mathbf{I} = j\omega C \mathbf{V}$

LEARNING EXTENSION

E8.7 The current in a 150-μF capacitor is $\mathbf{I} = 3.6 \underline{/-145°}$ A. If the frequency of the current is 60 Hz, determine the voltage across the capacitor.

ANSWER:
$v_C(t) = 63.66 \cos(377t - 235°)$ V.

8.5 Impedance and Admittance

We have examined each of the circuit elements in the frequency domain on an individual basis. We now wish to treat these passive circuit elements in a more general fashion. We define the two-terminal input *impedance* **Z**, also referred to as the driving point impedance, in exactly the same manner in which we defined resistance earlier. Later we will examine another type of impedance, called transfer impedance.

Impedance is defined as the ratio of the phasor voltage **V** to the phasor current **I**:

$$\mathbf{Z} = \frac{\mathbf{V}}{\mathbf{I}} \qquad \qquad \textbf{8.31}$$

at the two terminals of the element related to one another by the passive sign convention, as illustrated in Fig. 8.9. Since **V** and **I** are complex, the impedance **Z** is complex and

$$\mathbf{Z} = \frac{V_M \underline{/\theta_v}}{I_M \underline{/\theta_i}} = \frac{V_M}{I_M} \underline{/\theta_v - \theta_i} = Z\underline{/\theta_z} \qquad \qquad \textbf{8.32}$$

Since **Z** is the ratio of **V** to **I**, the units of **Z** are ohms. Thus, impedance in an ac circuit is analogous to resistance in a dc circuit. In rectangular form, impedance is expressed as

$$\mathbf{Z}(\omega) = R(\omega) + jX(\omega) \qquad \qquad \textbf{8.33}$$

where $R(\omega)$ is the real, or resistive, component and $X(\omega)$ is the imaginary, or reactive, component. In general, we simply refer to R as the resistance and X as the reactance. It is important to note that R and X are real functions of ω and therefore $\mathbf{Z}(\omega)$ is frequency dependent. Equation (8.33) clearly indicates that **Z** is a complex number; however, it is not a phasor, since phasors denote sinusoidal functions.

Equations (8.32) and (8.33) indicate that

$$Z\underline{/\theta_z} = R + jX \qquad \qquad \textbf{8.34}$$

Therefore,

$$Z = \sqrt{R^2 + X^2}$$

$$\theta_z = \tan^{-1}\frac{X}{R} \qquad \qquad \textbf{8.35}$$

where

$$R = Z\cos\theta_z$$

$$X = Z\sin\theta_z$$

Figure 8.9
General impedance relationship.

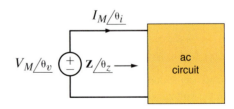

For the individual passive elements the impedance is as shown in Table 8.2. However, just as it was advantageous to know how to determine the equivalent resistance in dc circuits, we want to learn how to determine the equivalent impedance in ac circuits.

Table 8.2 Passive element impedance

Passive element	Impedance
R	$\mathbf{Z} = R$
L	$\mathbf{Z} = j\omega L = jX_L = \omega L \underline{/90°}, X_L = \omega L$
C	$\mathbf{Z} = \dfrac{1}{j\omega C} = jX_C = -\dfrac{1}{\omega C} \underline{/90°}, X_C = -\dfrac{1}{\omega C}$

KCL and KVL are both valid in the frequency domain. We can use this fact, as was done in Chapter 2 for resistors, to show that impedances can be combined using the same rules that we established for resistor combinations. That is, if $\mathbf{Z}_1, \mathbf{Z}_2, \mathbf{Z}_3, \dots, \mathbf{Z}_n$ are connected in series, the equivalent impedance $\mathbf{Z}_s$ is

$$\mathbf{Z}_s = \mathbf{Z}_1 + \mathbf{Z}_2 + \mathbf{Z}_3 + \cdots + \mathbf{Z}_n \qquad \text{8.36}$$

and if $\mathbf{Z}_1, \mathbf{Z}_2, \mathbf{Z}_3, \dots, \mathbf{Z}_n$ are connected in parallel, the equivalent impedance is given by

$$\frac{1}{\mathbf{Z}_p} = \frac{1}{\mathbf{Z}_1} + \frac{1}{\mathbf{Z}_2} + \frac{1}{\mathbf{Z}_3} + \cdots + \frac{1}{\mathbf{Z}_n} \qquad \text{8.37}$$

Example 8.9

Determine the equivalent impedance of the network shown in Fig. 8.10 if the frequency is $f = 60$ Hz. Then compute the current $i(t)$ if the voltage source is $v(t) = 50 \cos(\omega t + 30°)$ V. Finally, calculate the equivalent impedance if the frequency is $f = 400$ Hz.

SOLUTION   The impedances of the individual elements at 60 Hz are

$$\mathbf{Z}_R = 25 \ \Omega$$

$$\mathbf{Z}_L = j\omega L = j(2\pi \times 60)(20 \times 10^{-3}) = j7.54 \ \Omega$$

$$\mathbf{Z}_C = \frac{-j}{\omega C} = \frac{-j}{(2\pi \times 60)(50 \times 10^{-6})} = -j53.05 \ \Omega$$

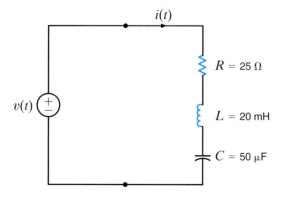

Figure 8.10
Series ac circuit.

Since the elements are in series,

$$\mathbf{Z} = \mathbf{Z}_R + \mathbf{Z}_L + \mathbf{Z}_C$$
$$= 25 - j45.51 \ \Omega$$

The current in the circuit is given by

$$\mathbf{I} = \frac{\mathbf{V}}{\mathbf{Z}} = \frac{50 \ \underline{/30°}}{25 - j45.51} = \frac{50 \ \underline{/30°}}{51.93 \ \underline{/-61.22°}} = 0.96 \ \underline{/91.22°} \ \text{A}$$

or in the time domain, $i(t) = 0.96 \cos(377t + 91.22°)$ A.

If the frequency is 400 Hz, the impedance of each element is

$$\mathbf{Z}_R = 25 \ \Omega$$
$$\mathbf{Z}_L = j\omega L = j50.27 \ \Omega$$
$$\mathbf{Z}_C = \frac{-j}{\omega C} = -j7.96 \ \Omega$$

The total impedance is then

$$\mathbf{Z} = 25 + j42.31 = 49.14 \ \underline{/59.42°} \ \Omega$$

It is important to note that at the frequency $f = 60$ Hz, the reactance of the circuit is capacitive; that is, if the impedance is written as $R + jX$, $X < 0$; however, at $f = 400$ Hz the reactance is inductive since $X > 0$.

PROBLEM-SOLVING STRATEGY

Basic AC Analysis

Step 1. Express $v(t)$ as a phasor and determine the impedance of each passive element.

Step 2. Combine impedances and solve for the phasor $\mathbf{I}$.

Step 3. Convert the phasor $\mathbf{I}$ to $i(t)$.

LEARNING EXTENSION

E8.8 Find the current $i(t)$ in the network in Fig. E8.8. **PSV**

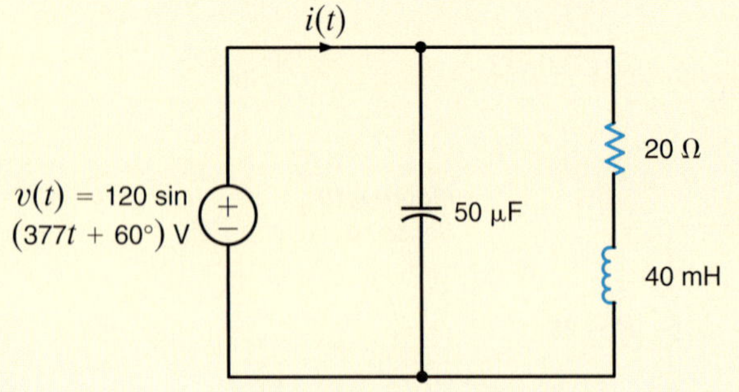

ANSWER:
$i(t) = 3.88 \cos(377t - 39.2°)$ A.

Figure E8.8

Another quantity that is very useful in the analysis of ac circuits is the two-terminal input *admittance*, which is the reciprocal of impedance; that is,

$$\mathbf{Y} = \frac{1}{\mathbf{Z}} = \frac{\mathbf{I}}{\mathbf{V}} \qquad \qquad 8.38$$

The units of $\mathbf{Y}$ are siemens, and this quantity is analogous to conductance in resistive dc circuits. Since $\mathbf{Z}$ is a complex number, $\mathbf{Y}$ is also a complex number.

$$\mathbf{Y} = Y_M \ \underline{/\theta_y} \qquad \qquad 8.39$$

which is written in rectangular form as

$$\mathbf{Y} = G + jB \qquad \text{8.40}$$

where G and B are called *conductance* and *susceptance*, respectively. Because of the relationship between $\mathbf{Y}$ and $\mathbf{Z}$, we can express the components of one quantity as a function of the components of the other. From the expression

$$G + jB = \frac{1}{R + jX} \qquad \text{8.41}$$

we can show that

$$G = \frac{R}{R^2 + X^2}, \qquad B = \frac{-X}{R^2 + X^2} \qquad \text{8.42}$$

and in a similar manner, we can show that

$$R = \frac{G}{G^2 + B^2}, \qquad X = \frac{-B}{G^2 + B^2} \qquad \text{8.43}$$

It is very important to note that in general R and G are *not* reciprocals of one another. The same is true for X and B. The purely resistive case is an exception. In the purely reactive case, the quantities are negative reciprocals of one another.

The admittance of the individual passive elements are

$$\mathbf{Y}_R = \frac{1}{R} = G$$

$$\mathbf{Y}_L = \frac{1}{j\omega L} = -\frac{1}{\omega L}\underline{/90^\circ} \qquad \text{8.44}$$

$$\mathbf{Y}_C = j\omega C = \omega C\underline{/90^\circ}$$

> **HINT**
>
> Technique for taking the reciprocal:
>
> $$\frac{1}{R + jX} = \frac{R - jX}{(R + jX)(R - jX)}$$
> $$= \frac{R - jX}{R^2 + X^2}$$

Once again, since KCL and KVL are valid in the frequency domain, we can show, using the same approach outlined in Chapter 2 for conductance in resistive circuits, that the rules for combining admittances are the same as those for combining conductances; that is, if $\mathbf{Y}_1, \mathbf{Y}_2, \mathbf{Y}_3, \ldots, \mathbf{Y}_n$ are connected in parallel, the equivalent admittance is

$$\mathbf{Y}_p = \mathbf{Y}_1 + \mathbf{Y}_2 + \cdots + \mathbf{Y}_n \qquad \text{8.45}$$

and if $\mathbf{Y}_1, \mathbf{Y}_2, \ldots, \mathbf{Y}_n$ are connected in series, the equivalent admittance is

$$\frac{1}{\mathbf{Y}_S} = \frac{1}{\mathbf{Y}_1} + \frac{1}{\mathbf{Y}_2} + \cdots + \frac{1}{\mathbf{Y}_n} \qquad \text{8.46}$$

Example 8.10

Calculate the equivalent admittance $\mathbf{Y}_p$ for the network in Fig. 8.11 and use it to determine the current $\mathbf{I}$ if $\mathbf{V}_S = 60\underline{/45^\circ}$ V.

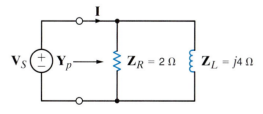

Figure 8.11
An example parallel circuit.

SOLUTION   From Fig. 8.11 we note that

$$\mathbf{Y}_R = \frac{1}{\mathbf{Z}_R} = \frac{1}{2}\,\text{S}$$

$$\mathbf{Y}_L = \frac{1}{\mathbf{Z}_L} = \frac{-j}{4}\,\text{S}$$

HINT

Admittances add in parallel.

Therefore,

$$\mathbf{Y}_p = \frac{1}{2} - j\frac{1}{4}\,\mathrm{S}$$

and hence,

$$\mathbf{I} = \mathbf{Y}_p\mathbf{V}_S$$

$$= \left(\frac{1}{2} - j\frac{1}{4}\right)(60\,\underline{/45°})$$

$$= 33.5\,\underline{/18.43°}\,\mathrm{A}$$

LEARNING EXTENSION

E8.9 Find the current **I** in the network in Fig. E8.9.

ANSWER: $\mathbf{I} = 9.01\,\underline{/53.7°}\,\mathrm{A}$.

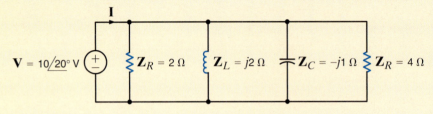

Figure E8.9

As a prelude to our analysis of more general ac circuits, let us examine the techniques for computing the impedance or admittance of circuits in which numerous passive elements are interconnected. The following example illustrates that our technique is analogous to our earlier computations of equivalent resistance.

Example 8.11

Consider the network shown in Fig. 8.12a. The impedance of each element is given in the figure. We wish to calculate the equivalent impedance of the network $\mathbf{Z}_{\mathrm{eq}}$ at terminals A-B.

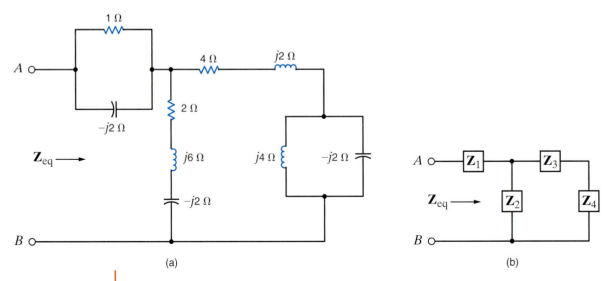

(a)   (b)

Figure 8.12 Example circuit for determining equivalent impedance in two steps.

SOLUTION   The equivalent impedance $\mathbf{Z}_{eq}$ could be calculated in a variety of ways; we could use only impedances, or only admittances, or a combination of the two. We will use the latter. We begin by noting that the circuit in Fig. 8.12a can be represented by the circuit in Fig. 8.12b.

Note that

$$\mathbf{Y}_4 = \mathbf{Y}_L + \mathbf{Y}_C$$

$$= \frac{1}{j4} + \frac{1}{-j2}$$

$$= j\frac{1}{4}\,\text{S}$$

Therefore,

$$\mathbf{Z}_4 = -j4\ \Omega$$

Now

$$\mathbf{Z}_{34} = \mathbf{Z}_3 + \mathbf{Z}_4$$

$$= (4 + j2) + (-j4)$$

$$= 4 - j2\ \Omega$$

and hence,

$$\mathbf{Y}_{34} = \frac{1}{\mathbf{Z}_{34}}$$

$$= \frac{1}{4 - j2}$$

$$= 0.20 + j0.10\ \text{S}$$

Since

$$\mathbf{Z}_2 = 2 + j6 - j2$$

$$= 2 + j4\ \Omega$$

then

$$\mathbf{Y}_2 = \frac{1}{2 + j4}$$

$$= 0.10 - j0.20\ \text{S}$$

$$\mathbf{Y}_{234} = \mathbf{Y}_2 + \mathbf{Y}_{34}$$

$$= 0.30 - j0.10\ \text{S}$$

The reader should carefully note our approach—we are adding impedances in series and adding admittances in parallel.

From $\mathbf{Y}_{234}$ we can compute $\mathbf{Z}_{234}$ as

$$\mathbf{Z}_{234} = \frac{1}{\mathbf{Y}_{234}}$$

$$= \frac{1}{0.30 - j0.10}$$

$$= 3 + j1\ \Omega$$

Now

$$\mathbf{Y}_1 = \mathbf{Y}_R + \mathbf{Y}_C$$

$$= \frac{1}{1} + \frac{1}{-j2}$$

$$= 1 + j\frac{1}{2}\,\text{S}$$

and then

$$\mathbf{Z}_1 = \frac{1}{1 + j\dfrac{1}{2}}$$

$$= 0.8 - j0.4\ \Omega$$

Therefore,

$$\mathbf{Z}_{eq} = \mathbf{Z}_1 + \mathbf{Z}_{234}$$
$$= 0.8 - j0.4 + 3 + j1$$
$$= 3.8 + j0.6 \ \Omega$$

PROBLEM-SOLVING STRATEGY

Combining Impedances and Admittances

Step 1. Add the admittances of elements in parallel.

Step 2. Add the impedances of elements in series.

Step 3. Convert back and forth between admittance and impedance in order to combine neighboring elements.

LEARNING EXTENSION

E8.10 Compute the impedance $\mathbf{Z}_T$ in the network in Fig. E8.10.   **PSV**

ANSWER:
$\mathbf{Z}_T = 3.38 + j1.08 \ \Omega$.

Figure E8.10

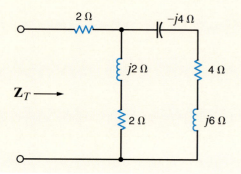

8.6 Phasor Diagrams

Impedance and admittance are functions of frequency, and therefore their values change as the frequency changes. These changes in $\mathbf{Z}$ and $\mathbf{Y}$ have a resultant effect on the current–voltage relationships in a network. This impact of changes in frequency on circuit parameters can be easily seen via a phasor diagram. The following examples will serve to illustrate these points.

Example 8.12

Let us sketch the phasor diagram for the network shown in Fig. 8.13.

Figure 8.13
Example parallel circuit.

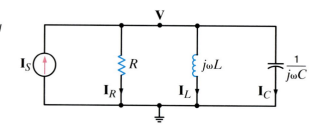

SOLUTION The pertinent variables are labeled on the figure. For convenience in forming a phasor diagram, we select **V** as a reference phasor and arbitrarily assign it a 0° phase angle. We will, therefore, measure all currents with respect to this phasor. We suffer no loss of generality by assigning **V** a 0° phase angle, since if it is actually 30°, for example, we will simply rotate the entire phasor diagram by 30° because all the currents are measured with respect to this phasor.

At the upper node in the circuit KCL is

$$\mathbf{I}_S = \mathbf{I}_R + \mathbf{I}_L + \mathbf{I}_C = \frac{\mathbf{V}}{R} + \frac{\mathbf{V}}{j\omega L} + \frac{\mathbf{V}}{1/j\omega C}$$

Since $\mathbf{V} = V_M \underline{/0°}$, then

$$\mathbf{I}_S = \frac{V_M \underline{/0°}}{R} + \frac{V_M \underline{/-90°}}{\omega L} + V_M \omega C \underline{/90°}$$

The phasor diagram that illustrates the phase relationship between **V**, $\mathbf{I}_R$, $\mathbf{I}_L$, and $\mathbf{I}_C$ is shown in Fig. 8.14a. For small values of ω such that the magnitude of $\mathbf{I}_L$ is greater than that of $\mathbf{I}_C$, the phasor diagram for the currents is shown in Fig. 8.14b. In the case of large values of ω— that is, those for which $\mathbf{I}_C$ is greater than $\mathbf{I}_L$—the phasor diagram for the currents is shown in Fig. 8.14c. Note that as ω increases, the phasor $\mathbf{I}_S$ moves from $\mathbf{I}_{S_1}$ to $\mathbf{I}_{S_n}$ along a locus of points specified by the dashed line shown in Fig. 8.14d.

Note that $\mathbf{I}_S$ is in phase with **V** when $\mathbf{I}_C = \mathbf{I}_L$ or, in other words, when $\omega L = 1/\omega C$. Hence, the node voltage **V** is in phase with the current source $\mathbf{I}_S$ when

$$\omega = \frac{1}{\sqrt{LC}}$$

This can also be seen from the KCL equation

$$\mathbf{I} = \left[\frac{1}{R} + j\left(\omega C - \frac{1}{\omega L} \right) \right] \mathbf{V}$$

HINT

From a graphical standpoint, phasors can be manipulated like vectors.

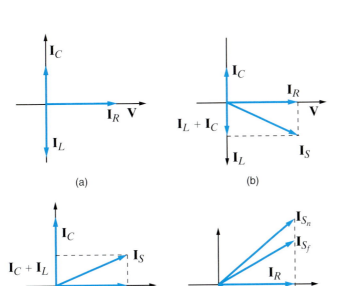

Figure 8.14
Phasor diagrams for the circuit in Fig. 8.13.

(a)

(b)

(c)

(d)

Example 8.13

Let us determine the phasor diagram for the series circuit shown in Fig. 8.15a.

SOLUTION KVL for this circuit is of the form

$$\mathbf{V}_S = \mathbf{V}_R + \mathbf{V}_L + \mathbf{V}_C$$

$$= IR + \omega L\mathbf{I}\underline{/90°} + \frac{\mathbf{I}}{\omega C}\underline{/-90°}$$

If we select $\mathbf{I}$ as a reference phasor so that $\mathbf{I} = I_M\underline{/0°}$, then if $\omega LI_M > I_M/\omega C$, the phasor diagram will be of the form shown in Fig. 8.15b. Specifically, if $\omega = 377$ rad/s (i.e., $f = 60$ Hz), then $\omega L = 6$ and $1/\omega C = 2$. Under these conditions, the phasor diagram is as shown in Fig. 8.15c. If, however, we select $\mathbf{V}_S$ as reference with, for example,

$$v_S(t) = 12\sqrt{2}\cos(377t + 90°)\text{ V}$$

then

$$\mathbf{I} = \frac{\mathbf{V}}{\mathbf{Z}} = \frac{12\sqrt{2}\underline{/90°}}{4 + j6 - j2}$$

$$= \frac{12\sqrt{2}\underline{/90°}}{4\sqrt{2}\underline{/45°}}$$

$$= 3\underline{/45°}\text{ A}$$

and the entire phasor diagram, as shown in Figs. 8.15b and c, is rotated 45°, as shown in Fig. 8.15d.

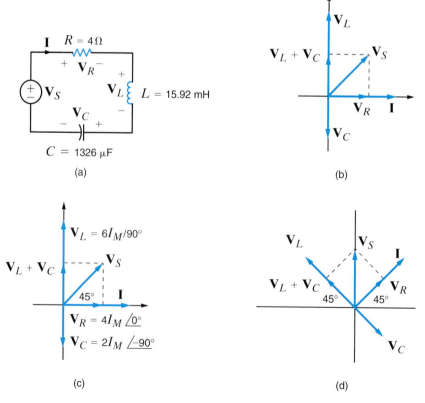

Figure 8.15 *Series circuit and certain specific phasor diagrams (plots are not drawn to scale).*

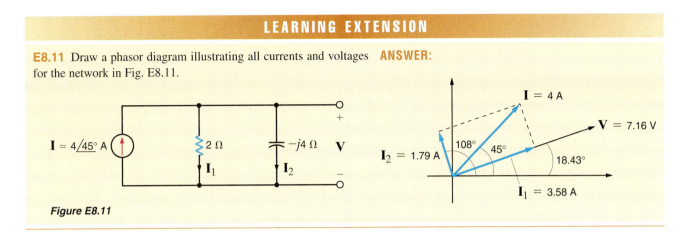

E8.11 Draw a phasor diagram illustrating all currents and voltages for the network in Fig. E8.11.   **ANSWER:**

Figure E8.11

8.7   Basic Analysis Using Kirchhoff's Laws

We have shown that Kirchhoff's laws apply in the frequency domain, and therefore they can be used to compute steady-state voltages and currents in ac circuits. This approach involves expressing these voltages and currents as phasors, and once this is done, the ac steady-state analysis employing phasor equations is performed in an identical fashion to that used in the dc analysis of resistive circuits. Complex number algebra is the tool that is used for the mathematical manipulation of the phasor equations, which, of course, have complex coefficients. We will begin by illustrating that the techniques we have applied in the solution of dc resistive circuits are valid in ac circuit analysis also—the only difference being that in steady-state ac circuit analysis the algebraic phasor equations have complex coefficients.

PROBLEM-SOLVING STRATEGY

AC Steady-State Analysis

- For relatively simple circuits (e.g., those with a single source), use
  - Ohm's law for ac analysis, i.e., $\mathbf{V} = \mathbf{IZ}$
  - The rules for combining $\mathbf{Z}_s$ and $\mathbf{Y}_p$
  - KCL and KVL
  - Current and voltage division
- For more complicated circuits with multiple sources, use
  - Nodal analysis
  - Loop or mesh analysis
  - Superposition
  - Source exchange
  - Thévenin's and Norton's theorems
  - MATLAB
  - PSPICE

   At this point, it is important for the reader to understand that in our manipulation of algebraic phasor equations with complex coefficients we will, for the sake of simplicity, normally carry only two digits to the right of the decimal point. In doing so, we will introduce round-off errors in our calculations. Nowhere are these errors more evident than when two or more approaches are used to solve the same problem, as is done in the following example.

HINT

Technique

1. Compute $\mathbf{I}_1$.

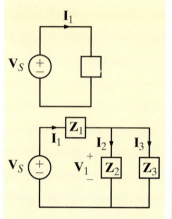

2. Determine $\mathbf{V}_1 = \mathbf{V}_S - \mathbf{I}_1\mathbf{Z}_1$.

Then $\mathbf{I}_2 = \dfrac{\mathbf{V}_1}{\mathbf{Z}_2}$ and $\mathbf{I}_3 = \dfrac{\mathbf{V}_1}{\mathbf{Z}_3}$

Current and voltage division are also applicable.

Example 8.14

We wish to calculate all the voltages and currents in the circuit shown in Fig. 8.16a.

SOLUTION Our approach will be as follows. We will calculate the total impedance seen by the source $\mathbf{V}_S$. Then we will use this to determine $\mathbf{I}_1$. Knowing $\mathbf{I}_1$, we can compute $\mathbf{V}_1$ using KVL. Knowing $\mathbf{V}_1$, we can compute $\mathbf{I}_2$ and $\mathbf{I}_3$, and so on.

The total impedance seen by the source $\mathbf{V}_S$ is

$$
\begin{aligned}
\mathbf{Z}_{eq} &= 4 + \frac{(j6)(8 - j4)}{j6 + 8 - j4} \\
&= 4 + \frac{24 + j48}{8 + j2} \\
&= 4 + 4.24 + j4.94 \\
&= 9.61\,\underline{/30.94°}\ \Omega
\end{aligned}
$$

Then

$$
\begin{aligned}
\mathbf{I}_1 &= \frac{\mathbf{V}_S}{\mathbf{Z}_{eq}} = \frac{24\,\underline{/60°}}{9.61\,\underline{/30.94°}} \\
&= 2.5\,\underline{/29.06°}\ \text{A}
\end{aligned}
$$

$\mathbf{V}_1$ can be determined using KVL:

$$
\begin{aligned}
\mathbf{V}_1 &= \mathbf{V}_S - 4\mathbf{I}_1 \\
&= 24\,\underline{/60°} - 10\,\underline{/29.06°} \\
&= 3.26 + j15.93 \\
&= 16.26\,\underline{/78.43°}\ \text{V}
\end{aligned}
$$

Note that $\mathbf{V}_1$ could also be computed via voltage division:

$$
\mathbf{V}_1 = \frac{\mathbf{V}_S\,\dfrac{(j6)(8 - j4)}{j6 + 8 - j4}}{4 + \dfrac{(j6)(8 - j4)}{j6 + 8 - 4}}\ \text{V}
$$

which from our previous calculation is

$$
\begin{aligned}
\mathbf{V}_1 &= \frac{(24\,\underline{/60°})(6.51\,\underline{/49.36°})}{9.61\,\underline{/30.94°}} \\
&= 16.26\,\underline{/78.42°}\ \text{V}
\end{aligned}
$$

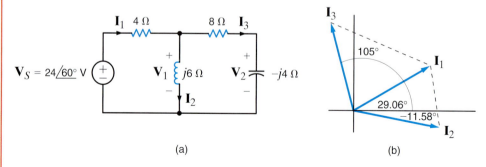

(a)   (b)

Figure 8.16 (a) Example ac circuit, (b) phasor diagram for the currents (plots are not drawn to scale).

Knowing $\mathbf{V}_1$, we can calculate both $\mathbf{I}_2$ and $\mathbf{I}_3$:

$$\mathbf{I}_2 = \frac{\mathbf{V}_1}{j6} = \frac{16.26 \,\underline{/78.43^\circ}}{6 \,\underline{/90^\circ}}$$

$$= 2.71 \,\underline{/-11.58^\circ} \text{ A}$$

and

$$\mathbf{I}_3 = \frac{\mathbf{V}_1}{8 - j4}$$

$$= 1.82 \,\underline{/105^\circ} \text{ A}$$

Note that $\mathbf{I}_2$ and $\mathbf{I}_3$ could have been calculated by current division. For example, $\mathbf{I}_2$ could be determined by

$$\mathbf{I}_2 = \frac{\mathbf{I}_1(8 - j4)}{8 - j4 + j6}$$

$$= \frac{(2.5 \,\underline{/29.06^\circ})(8.94 \,\underline{/-26.57^\circ})}{8 + j2}$$

$$= 2.71 \,\underline{/-11.55^\circ} \text{ A}$$

Finally, $\mathbf{V}_2$ can be computed as

$$\mathbf{V}_2 = \mathbf{I}_3(-j4)$$

$$= 7.28 \,\underline{/15^\circ} \text{ V}$$

This value could also have been computed by voltage division. The phasor diagram for the currents $\mathbf{I}_1$, $\mathbf{I}_2$, and $\mathbf{I}_3$ is shown in Fig. 8.16b and is an illustration of KCL.

Finally, the reader is encouraged to work the problem in reverse; that is, given $\mathbf{V}_2$, find $\mathbf{V}_S$. Note that if $\mathbf{V}_2$ is known, $\mathbf{I}_3$ can be computed immediately using the capacitor impedance. Then $\mathbf{V}_2 + \mathbf{I}_3(8)$ yields $\mathbf{V}_1$. Knowing $\mathbf{V}_1$ we can find $\mathbf{I}_2$. Then $\mathbf{I}_2 + \mathbf{I}_3 = \mathbf{I}_1$, and so on. Note that this analysis, which is the subject of Learning Extensions Exercise E8.12, involves simply a repeated application of Ohm's law, KCL, and KVL.

LEARNING EXTENSION

E8.12 In the network in Fig. E8.12, $\mathbf{V}_o$ is known to be $8 \,\underline{/45^\circ}$ V. Compute $\mathbf{V}_S$.   **PSV**

ANSWER:
$\mathbf{V}_S = 17.89 \,\underline{/-18.43^\circ}$ V.

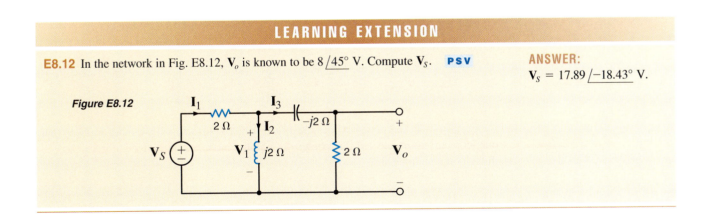

Figure E8.12

8.8   Analysis Techniques

In this section we revisit the circuit analysis methods that were successfully applied earlier to dc circuits and illustrate their applicability to ac steady-state analysis. The vehicle we employ to present these techniques is examples in which all the theorems, together with nodal analysis and loop analysis, are used to obtain a solution.

Example 8.15

Let us determine the current $\mathbf{I}_o$ in the network in Fig. 8.17a using nodal analysis, loop analysis, superposition, source exchange, Thévenin's theorem, and Norton's theorem.

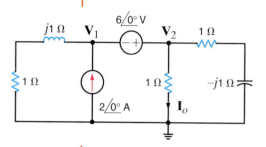 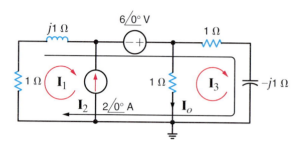

Figure 8.17 *Circuits used in Example 8.15 for node and loop analysis.*

SOLUTION

HINT

Summing the current, leaving the supernode. Outbound currents have a positive sign.

1. *Nodal Analysis* We begin with a nodal analysis of the network. The KCL equation for the supernode that includes the voltage source is

$$\frac{\mathbf{V}_1}{1+j} - 2\underline{/0^\circ} + \frac{\mathbf{V}_2}{1} + \frac{\mathbf{V}_2}{1-j} = 0$$

and the associated KVL constraint equation is

$$\mathbf{V}_1 + 6\underline{/0^\circ} = \mathbf{V}_2$$

Solving for $\mathbf{V}_1$ in the second equation and using this value in the first equation yields

$$\frac{\mathbf{V}_2 - 6\underline{/0^\circ}}{1+j} - 2\underline{/0^\circ} + \mathbf{V}_2 + \frac{\mathbf{V}_2}{1-j} = 0$$

or

$$\mathbf{V}_2\left[\frac{1}{1+j} + 1 + \frac{1}{1-j}\right] = \frac{6+2+2j}{1+j}$$

Solving for $\mathbf{V}_2$, we obtain

$$\mathbf{V}_2 = \left(\frac{4+j}{1+j}\right)\text{V}$$

Therefore,

$$\mathbf{I}_o = \frac{4+j}{1+j} = \left(\frac{5}{2} - \frac{3}{2}j\right)\text{A}$$

HINT

Just as in a dc analysis, the loop equations assume that a decrease in potential level is + and an increase is −.

2. *Loop Analysis* The network in Fig. 8.17b is used to perform a loop analysis. Note that one loop current is selected that passes through the independent current source. The three loop equations are

$$\mathbf{I}_1 = -2\underline{/0^\circ}$$

$$1(\mathbf{I}_1 + \mathbf{I}_2) + j1(\mathbf{I}_1 + \mathbf{I}_2) - 6\underline{/0^\circ} + 1(\mathbf{I}_2 + \mathbf{I}_3) - j1(\mathbf{I}_2 + \mathbf{I}_3) = 0$$

$$1\mathbf{I}_3 + 1(\mathbf{I}_2 + \mathbf{I}_3) - j1(\mathbf{I}_2 + \mathbf{I}_3) = 0$$

Combining the first two equations yields

$$\mathbf{I}_2(2) + \mathbf{I}_3(1-j) = 8 + 2j$$

The third loop equation can be simplified to the form

$$\mathbf{I}_2(1-j) + \mathbf{I}_3(2-j) = 0$$

Solving this last equation for $\mathbf{I}_2$ and substituting the value into the previous equation yields

$$\mathbf{I}_3\left[\frac{-4 + 2j}{1 - j} + 1 - j\right] = 8 + 2j$$

or

$$\mathbf{I}_3 = \frac{-10 + 6j}{4}$$

and finally

$$\mathbf{I}_o = -\mathbf{I}_3 = \left(\frac{5}{2} - \frac{3}{2}j\right)\text{ A}$$

3. *Superposition*   In using superposition, we apply one independent source at a time. The network in which the current source acts alone is shown in Fig. 8.18a. By combining the two parallel impedances on each end of the network, we obtain the circuit in Fig. 8.18b, where

$$\mathbf{Z}' = \frac{(1 + j)(1 - j)}{(1 + j) + (1 - j)} = 1\ \Omega$$

HINT

In applying superposition in this case, each source is applied independently and the results are added to obtain the solution.

Therefore, using current division

$$\mathbf{I}'_o = 1\ \underline{/0°}\text{ A}$$

The circuit in which the voltage source acts alone is shown in Fig. 8.18c. The voltage $\mathbf{V}''_1$ obtained using voltage division is

$$\mathbf{V}''_1 = \frac{(6\ \underline{/0°})\left[\dfrac{1(1 - j)}{1 + 1 - j}\right]}{1 + j + \left[\dfrac{1(1 - j)}{1 + 1 - j}\right]}$$

$$= \frac{6(1 - j)}{4}\text{ V}$$

and hence,

$$\mathbf{I}''_0 = \frac{6}{4}(1 - j)\text{ A}$$

Then

$$\mathbf{I}_o = \mathbf{I}'_o + \mathbf{I}''_o = 1 + \frac{6}{4}(1 - j) = \left(\frac{5}{2} - \frac{3}{2}j\right)\text{ A}$$

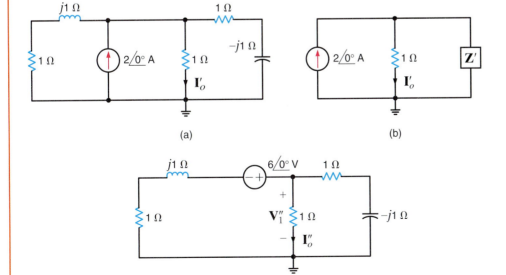

Figure 8.18
Circuits used in Example 8.15 for a superposition analysis.

(a)

(b)

(c)

Figure 8.19
Circuits used in Example 8.15 for a source exchange analysis.

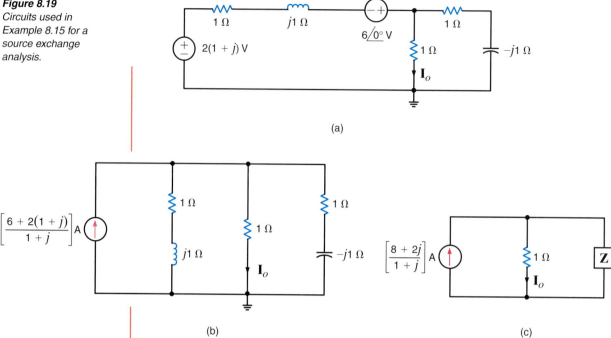

(a)

(b)

(c)

HINT

In source exchange, a voltage source in series with an impedance can be exchanged for a current source in parallel with the impedance, and vice versa. Repeated application systematically reduces the number of circuit elements.

4. *Source Exchange* As a first step in the source exchange approach, we exchange the current source and parallel impedance for a voltage source in series with the impedance, as shown in Fig. 8.19a.

Adding the two voltage sources and transforming them and the series impedance into a current source in parallel with that impedance are shown in Fig. 8.19b. Combining the two impedances that are in parallel with the 1-Ω resistor produces the network in Fig. 8.19c, where

$$\mathbf{Z} = \frac{(1 + j)(1 - j)}{1 + j + 1 - j} = 1\ \Omega$$

Therefore, using current division,

$$\mathbf{I}_o = \left(\frac{8 + 2j}{1 + j}\right)\left(\frac{1}{2}\right) = \frac{4 + j}{1 + j}$$

$$= \left(\frac{5}{2} - \frac{3}{2}j\right)\ \text{A}$$

HINT

In this Thévenin analysis,

1. Remove the 1-Ω load and find the voltage across the open terminals, $\mathbf{V}_{oc}$.

2. Determine the impedance $\mathbf{Z}_{Th}$ at the open terminals with all sources made zero.

3. Construct the following circuit and determine $\mathbf{I}_o$.

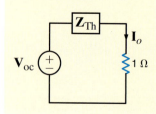

5. *Thévenin Analysis* In applying Thévenin's theorem to the circuit in Fig. 8.17a, we first find the open-circuit voltage, $\mathbf{V}_{oc}$, as shown in Fig. 8.20a. To simplify the analysis, we perform a source exchange on the left end of the network, which results in the circuit in Fig. 8.20b. Now using voltage division,

$$\mathbf{V}_{oc} = [6 + 2(1 + j)]\left[\frac{1 - j}{1 - j + 1 + j}\right]$$

or

$$\mathbf{V}_{oc} = (5 - 3j)\ \text{V}$$

The Thévenin equivalent impedance, $\mathbf{Z}_{Th}$, obtained at the open-circuit terminals when the current source is replaced with an open circuit and the voltage source is replaced with a short circuit is shown in Fig. 8.20c and calculated to be

$$\mathbf{Z}_{Th} = \frac{(1 + j)(1 - j)}{1 + j + 1 - j} = 1\ \Omega$$

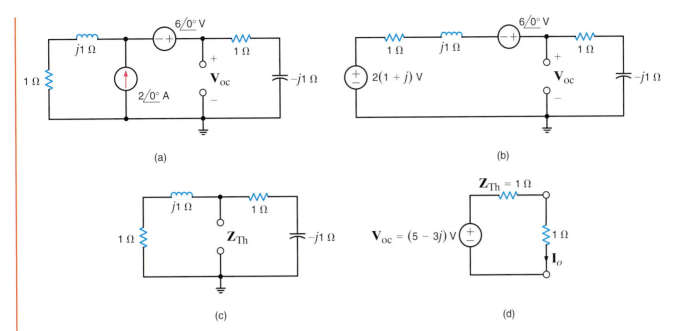

(a)   (b)

(c)   (d)

Figure 8.20
*Circuits used in Example 8.15
for a Thévenin analysis.*

Connecting the Thévenin equivalent circuit to the 1-Ω resistor containing $\mathbf{I}_o$ in the original network yields the circuit in Fig. 8.20d. The current $\mathbf{I}_o$ is then

$$\mathbf{I}_o = \left(\frac{5}{2} - \frac{3}{2}j\right) \text{A}$$

6. Norton Analysis   Finally, in applying Norton's theorem to the circuit in Fig. 8.17a, we calculate the short-circuit current, $\mathbf{I}_{sc}$, using the network in Fig. 8.21a. Note that because of the short circuit, the voltage source is directly across the impedance in the left-most branch. Therefore,

$$\mathbf{I}_1 = \frac{6\,\underline{/0^\circ}}{1+j}$$

Then using KCL,

$$\mathbf{I}_{sc} = \mathbf{I}_1 + 2\,\underline{/0^\circ} = 2 + \frac{6}{1+j}$$

$$= \left(\frac{8+2j}{1+j}\right) \text{A}$$

The Thévenin equivalent impedance, $\mathbf{Z}_{Th}$, is known to be 1 Ω and, therefore, connecting the Norton equivalent to the 1-Ω resistor containing $\mathbf{I}_o$ yields the network in Fig. 8.21b. Using current division, we find that

$$\mathbf{I}_o = \frac{1}{2}\left(\frac{8+2j}{1+j}\right)$$

$$= \left(\frac{5}{2} - \frac{3}{2}j\right) \text{A}$$

HINT

In this Norton analysis,

1. Remove the 1-Ω load and find the current $\mathbf{I}_{sc}$ through the short-circuited terminals.

2. Determine the impedance $\mathbf{Z}_{Th}$ at the open load terminals with all sources made zero.

3. Construct the following circuit and determine $\mathbf{I}_o$.

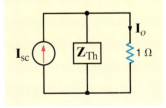

Figure 8.21
*Circuits used in Example 8.15
for a Norton analysis.*

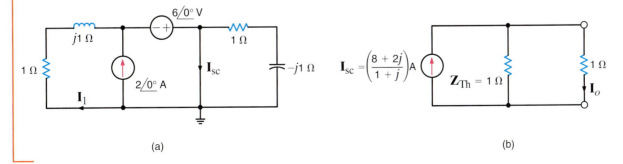

(a)   (b)

Let us now consider an example containing a dependent source.

Example 8.16

Let us determine the voltage $\mathbf{V}_o$ in the circuit in Fig. 8.22a. In this example we will use node equations, loop equations, Thévenin's theorem, and Norton's theorem. We will omit the techniques of superposition and source transformation. Why?

SOLUTION

1. *Nodal Analysis*   To perform a nodal analysis, we label the node voltages and identify the supernode as shown in Fig. 8.22b. The constraint equation for the supernode is

$$\mathbf{V}_3 + 12\,\underline{/0^\circ} = \mathbf{V}_1$$

and the KCL equations for the nodes of the network are

$$\frac{\mathbf{V}_1 - \mathbf{V}_2}{-j1} + \frac{\mathbf{V}_3 - \mathbf{V}_2}{1} - 4\,\underline{/0^\circ} + \frac{\mathbf{V}_3 - \mathbf{V}_o}{1} + \frac{\mathbf{V}_3}{j1} = 0$$

$$\frac{\mathbf{V}_2 - \mathbf{V}_1}{-j1} + \frac{\mathbf{V}_2 - \mathbf{V}_3}{1} - 2\left(\frac{\mathbf{V}_3 - \mathbf{V}_o}{1}\right) = 0$$

$$4\,\underline{/0^\circ} + \frac{\mathbf{V}_o - \mathbf{V}_3}{1} + \frac{\mathbf{V}_o}{1} = 0$$

At this point we can solve the foregoing equations using a matrix analysis or, for example, substitute the first and last equations into the remaining two equations, which yields

$$3\mathbf{V}_o - (1 + j)\mathbf{V}_2 = -(4 + j12)$$

$$-(4 + j2)\mathbf{V}_o + (1 + j)\mathbf{V}_2 = 12 + j16$$

HINT

How does the presence of a dependent source affect superposition and source exchange?

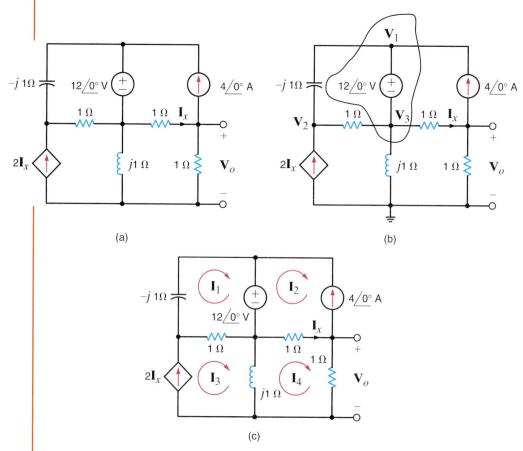

(a)

(b)

(c)

Figure 8.22 Circuits used in Example 8.16 for nodal and loop analysis.

Solving these equations for $\mathbf{V}_o$ yields

$$\mathbf{V}_o = \frac{-(8 + j4)}{1 + j2}$$

$$= +4\,\underline{/143.13°}\ \text{V}$$

2. **Loop Analysis** The mesh currents for the network are defined in Fig. 8.22c. The constraint equations for the circuit are

$$\mathbf{I}_2 = -4\,\underline{/0°}$$

$$\mathbf{I}_x = \mathbf{I}_4 - \mathbf{I}_2 = \mathbf{I}_4 + 4\,\underline{/0°}$$

$$\mathbf{I}_3 = 2\mathbf{I}_x = 2\mathbf{I}_4 + 8\,\underline{/0°}$$

The KVL equations for mesh 1 and mesh 4 are

$$-j1\mathbf{I}_1 + 1(\mathbf{I}_1 - \mathbf{I}_3) = -12\,\underline{/0°}$$

$$j1(\mathbf{I}_4 - \mathbf{I}_3) + 1(\mathbf{I}_4 - \mathbf{I}_2) + 1\mathbf{I}_4 = 0$$

Note that if the constraint equations are substituted into the second KVL equation, the only unknown in the equation is $\mathbf{I}_4$. This substitution yields

$$\mathbf{I}_4 = +4\,\underline{/143.13°}\ \text{A}$$

and hence,

$$\mathbf{V}_o = +4\,\underline{/143.13°}\ \text{V}$$

3. **Thévenin's Theorem** In applying Thévenin's theorem, we will find the open-circuit voltage and then determine the Thévenin equivalent impedance using a test source at the open-circuit terminals. We could determine the Thévenin equivalent impedance by calculating the short-circuit current; however, we will determine this current when we apply Norton's theorem.

The open-circuit voltage is determined from the network in Fig. 8.23a. Note that $\mathbf{I}'_x = 4\,\underline{/0°}$ A and since $2\mathbf{I}'_x$ flows through the inductor, the open-circuit voltage $\mathbf{V}_{\text{oc}}$ is

$$\mathbf{V}_{\text{oc}} = -1(4\,\underline{/0°}) + j1(2\mathbf{I}'_x)$$

$$= -4 + j8\ \text{V}$$

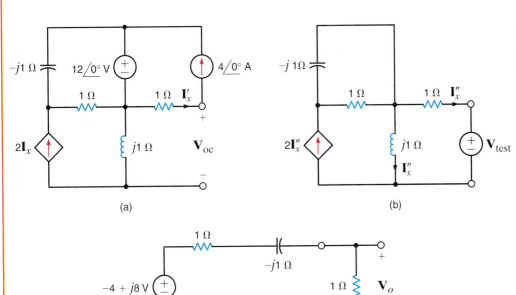

Figure 8.23
Circuits used in Example 8.16 when applying Thévenin's theorem.

To determine the Thévenin equivalent impedance, we turn off the independent sources, apply a test voltage source to the output terminals, and compute the current leaving the test source. As shown in Fig. 8.23b, since $\mathbf{I}''_x$ flows in the test source, KCL requires that the current in the inductor be $\mathbf{I}''_x$ also. KVL around the mesh containing the test source indicates that

$$j1\mathbf{I}''_x - 1\mathbf{I}''_x - \mathbf{V}_{\text{test}} = 0$$

Therefore,

$$\mathbf{I}''_x = \frac{-\mathbf{V}_{\text{test}}}{1 - j}$$

Then

$$\mathbf{Z}_{\text{Th}} = \frac{\mathbf{V}_{\text{test}}}{-\mathbf{I}''_x}$$

$$= 1 - j \; \Omega$$

If the Thévenin equivalent network is now connected to the load, as shown in Fig. 8.23c, the output voltage $\mathbf{V}_o$ is found to be

$$\mathbf{V}_o = \frac{-4 + 8j}{2 - j1} \quad (1)$$

$$= +4 \underline{/143.13°} \; \text{V}$$

4. Norton's Theorem   In using Norton's theorem, we will find the short-circuit current from the network in Fig. 8.24a. Once again, using the supernode, the constraint and KCL equations are

$$\mathbf{V}_3 + 12 \underline{/0°} = \mathbf{V}_1$$

$$\frac{\mathbf{V}_2 - \mathbf{V}_1}{-j1} + \frac{\mathbf{V}_2 - \mathbf{V}_3}{1} - 2\mathbf{I}'''_x = 0$$

$$\frac{\mathbf{V}_1 - \mathbf{V}_2}{-j1} + \frac{\mathbf{V}_3 - \mathbf{V}_2}{1} - 4\underline{/0°} + \frac{\mathbf{V}_3}{j1} + \mathbf{I}'''_x = 0$$

$$\mathbf{I}'''_x = \frac{\mathbf{V}_3}{1}$$

Substituting the first and last equations into the remaining equations yields

$$(1 + j)\mathbf{V}_2 - (3 + j)\mathbf{I}'''_x = j12$$

$$-(1 + j)\mathbf{V}_2 + (2)\mathbf{I}'''_x = 4 - j12$$

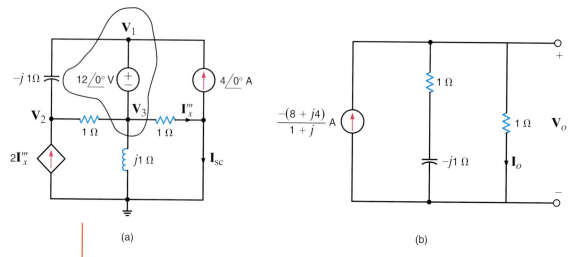

(a)   (b)

Figure 8.24 Circuits used in Example 8.16 when applying Norton's theorem.

Solving these equations for $\mathbf{I}_x'''$ yields

$$\mathbf{I}_x''' = \frac{-4}{1 + j}\,\text{A}$$

The KCL equation at the right-most node in the network in Fig. 8.24a is

$$\mathbf{I}_x''' = 4\,\underline{/0°} + \mathbf{I}_{sc}$$

Solving for $\mathbf{I}_{sc}$, we obtain

$$\mathbf{I}_{sc} = \frac{-(8 + j4)}{1 + j}\,\text{A}$$

The Thévenin equivalent impedance was found earlier to be

$$\mathbf{Z}_{Th} = 1 - j\,\Omega$$

Using the Norton equivalent network, the original network is reduced to that shown in Fig. 8.24b. The voltage $\mathbf{V}_o$ is then

$$\mathbf{V}_o = \frac{-(8 + j4)}{1 + j}\left[\frac{(1)(1 - j)}{1 + 1 - j}\right]$$

$$= -4\left[\frac{3 - j}{3 + j}\right]$$

$$= +4\,\underline{/+143.13°}\,\text{V}$$

E8.13 Use nodal analysis to find $\mathbf{V}_o$ in the network in Fig. E8.13.

ANSWER:
$\mathbf{V}_o = 2.12\,\underline{/75°}\,\text{V}$.

Figure E8.13

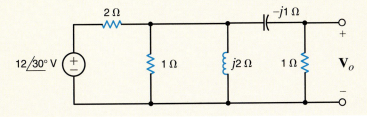

E8.14 Use (a) mesh equations and (b) Thévenin's theorem to find $\mathbf{V}_o$ in the network in Fig. E8.14. **PSV**

ANSWER:
$\mathbf{V}_o = 10.88\,\underline{/36°}\,\text{V}$.

Figure E8.14

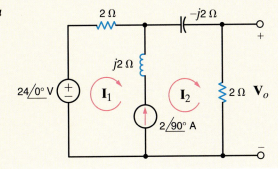

(continues on the next page)

E8.15 Use (a) superposition, (b) source transformation, and (c) Norton's theorem to find V_o in the network in Fig. E8.15.   **ANSWER:** $V_o = 12 \underline{/90°}$ V.

Figure E8.15

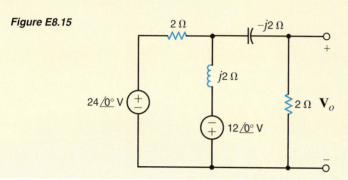

Example 8.17

Let's solve for the current $i(t)$ in the circuit in Fig. 8.25. At first glance, this appears to be a simple single-loop circuit. A more detailed observation reveals that the two sources operate at different frequencies. The radian frequency for the source on the left is 10 rad/s, while the source on the right operates at a radian frequency of 20 rad/s. If we draw a frequency-domain circuit, which frequency do we use? How can we solve this problem?

Figure 8.25
Circuit used in Example 8.17.

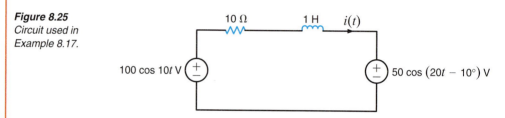

SOLUTION   Recall that the principle of superposition tells us that we can analyze the circuit with each source operating alone. The circuit responses to each source acting alone are then added together to give us the response with both sources active. Let's use the principle of superposition to solve this problem. First, calculate the response $i'(t)$ from the source on the left using the circuit shown in Fig. 8.26a. Now we can draw a frequency-domain circuit for $\omega = 10$ rad/s.

Figure 8.26
Circuits used to illustrate superposition.

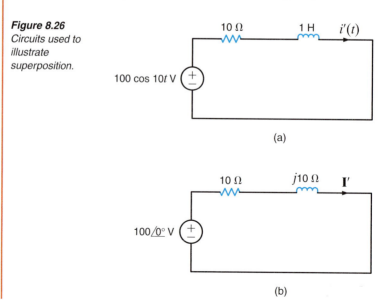

Then $\mathbf{I}' = \dfrac{100\,\underline{/0^\circ}}{10 + j10} = 7.07\,\underline{/-45^\circ}$ A. Therefore, $i'(t) = 7.07\cos(10t - 45^\circ)$ A.

The response due to the source on the right can be determined using the circuit in Fig. 8.27. Note that $i''(t)$ is defined in the opposite direction to $i(t)$ in the original circuit. The frequency-domain circuit for $\omega = 20$ rad/s is also shown in Fig. 8.27b.

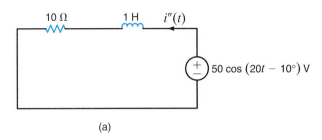

(a)

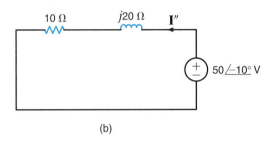

(b)

Figure 8.27
Circuits used to illustrate superposition.

The current $\mathbf{I}'' = \dfrac{50\,\underline{/-10^\circ}}{10 + j20} = 2.24\,\underline{/-73.43^\circ}$ A. Therefore, $i''(t) = 2.24\cos(20t - 73.43^\circ)$ A.

The current $i(t)$ can now be calculated as $i'(t) - i''(t) = 7.07\cos(10t - 45^\circ) - 2.24\cos(20t - 73.43^\circ)$ A.

MATLAB Analysis

When dealing with large networks it is impractical to determine the currents and voltages within the network without the help of a mathematical software or CAD package. Thus, we will once again demonstrate how to bring this computing power to bear when solving more complicated ac networks.

Earlier we used MATLAB to solve a set of simultaneous equations, which yielded the node voltages or loop currents in dc circuits. We now apply this technique to ac circuits. In the ac case where the number-crunching involves complex numbers, we use j to represent the imaginary part of a complex number (unless it has been previously defined as something else), and the complex number $x + jy$ is expressed in MATLAB as x + j*y. Although we use j in defining a complex number, MATLAB will list the complex number using i.

Complex sources are expressed in rectangular form, and we use the fact that 360° equals 2 pi radians. For example, a source V = $10\,\underline{/45^\circ}$ will be entered into MATLAB data as

        V = $10\,\underline{/45^\circ}$ = x+j*y
where the real and imaginary components are

        x = 10*cos(45*pi/180)
          = 7.07

and

        y = 10*sin(45*pi/180)
          = 7.07

When using MATLAB to determine the node voltages in ac circuits, we enter the Y matrix, the I vector, and then the solution equation

```
V = inv(Y)*I
```

as was done in the dc case. The following example will serve to illustrate the use of MATLAB in the solution of ac circuits.

Example 8.18

Consider the network in Fig. 8.28. We wish to find all the node voltages in this network. The five simultaneous equations describing the node voltages are

$$\mathbf{V}_1 = 12\,\underline{/30°}$$

$$\frac{\mathbf{V}_2 - \mathbf{V}_1}{1} + \frac{\mathbf{V}_2 - \mathbf{V}_5}{j2} + \frac{\mathbf{V}_2 - \mathbf{V}_3}{-j1} = 0$$

$$\frac{\mathbf{V}_3 - \mathbf{V}_2}{-j1} + \frac{\mathbf{V}_3 - \mathbf{V}_5}{2} + \frac{\mathbf{V}_3}{1} = 0$$

$$\frac{\mathbf{V}_4 - \mathbf{V}_1}{2} + \frac{\mathbf{V}_4 - \mathbf{V}_5}{1} + \frac{\mathbf{V}_4}{-j1} = 0$$

$$\frac{\mathbf{V}_5 - \mathbf{V}_2}{j2} + \frac{\mathbf{V}_5 - \mathbf{V}_3}{2} + \frac{\mathbf{V}_5}{-j1} + \frac{\mathbf{V}_5 - \mathbf{V}_4}{1} = 2\,\underline{/45°}$$

Figure 8.28
Circuit used in Example 8.18.

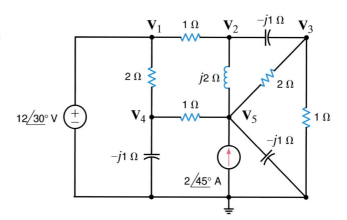

Expressing the equations in matrix form, we obtain

$$
\begin{Bmatrix}
1 & 0 & 0 & 0 & 0 \\
-1 & 1+j0.5 & -j1 & 0 & j0.5 \\
0 & -j1 & 1.5+j1 & 0 & -0.5 \\
-0.5 & 0 & 0 & 1.5+j1 & -1 \\
0 & j0.5 & -0.5 & -1 & 1.5+j0.5
\end{Bmatrix}
\begin{Bmatrix}
\mathbf{V}_1 \\ \mathbf{V}_2 \\ \mathbf{V}_3 \\ \mathbf{V}_4 \\ \mathbf{V}_5
\end{Bmatrix}
=
\begin{Bmatrix}
12\,\underline{/30°} \\ 0 \\ 0 \\ 0 \\ 2\,\underline{/45°}
\end{Bmatrix}
$$

The following MATLAB data consist of the conversion of the sources to rectangular form, the coefficient matrix Y, the vector *I*, the solution equation V = inv(Y)*I, and the solution vector **V**.

```
>> x = 12*cos(30*pi/180)

x =
  10.3923

>> y = 12*sin(30*pi/180)
```

```
y =
 6.0000

>> v1 = x+j*y

v1 =
 10.3923 + 6.0000i

>> I2 = 2*cos(45*pi/180) + j*2*sin(45*pi/180)

I2 =
 1.4142 + 1.4142i

>> Y = [1 0 0 0 0; -1 1+j*0.5 -j*1 0 j*0.5; 0 -j*1 1.5+j*1 0
      -0.5; -0.5 0 0 1.5+j*1 -1; 0 j*0.5 -0.5 -1 1.5+j*0.5]

Y =

Columns 1 through 4

  1.0000        0               0                  0
 -1.0000 1.0000 +0.5000i        0 -1.0000i         0
      0        0 -1.0000i  1.5000 +1.0000i          0
 -0.5000        0               0             1.5000 +1.0000i
      0        0 +0.5000i -0.5000             -1.0000

Column 5

      0
      0  +0.5000i
 -0.5000
 -1.0000
  1.5000  +0.5000i

>> I = [v1; 0; 0; 0; I2]

I =
 10.3923  +6.0000i
      0
      0
      0
  1.4142  +1.4142i

>> V = inv(Y) *I

V =
 10.3923  +6.0000i
  7.0766  +2.1580i
  1.4038  +2.5561i
  3.7661  -2.9621i
  3.4151  -3.6771i
```

Example 8.19

Consider the circuit shown in Fig. 8.29a. We wish to determine the current I_o. In contrast to the previous example, this network contains two dependent sources, one of which is a current source and the other a voltage source. We will work this problem using MATLAB here and then address it again later using PSPICE.

SOLUTION Our method of attack will be a nodal analysis. There are six nodes, and therefore we will need five linearly independent simultaneous equations to determine the nonreference node voltages. Actually, seven equations are needed, but two of them simply define the dependent variables. The network is redrawn in Fig. 8.29b where the node voltages are labeled. The five required equations, together with the two constraint equations, needed to determine the node voltages, and thus the current $\mathbf{I}_o$, are listed as follows:

$$\frac{\mathbf{V}_1 - \mathbf{V}_2}{1} - 2 + \frac{\mathbf{V}_1 - \mathbf{V}_4}{-j} + 2\mathbf{I}_x = 0$$

$$\mathbf{V}_2 = 2\mathbf{V}_x$$

$$\mathbf{V}_4 - \mathbf{V}_3 = 12$$

$$2 + \frac{\mathbf{V}_3 - \mathbf{V}_2}{1} + \frac{\mathbf{V}_3}{1} + \frac{\mathbf{V}_4}{1} + \frac{\mathbf{V}_4 - \mathbf{V}_5}{1} + \frac{\mathbf{V}_4 - \mathbf{V}_1}{-j} = 0$$

$$\frac{\mathbf{V}_5 - \mathbf{V}_4}{1} + \frac{\mathbf{V}_5}{j} = 2\mathbf{I}_x$$

$$\mathbf{V}_x = \mathbf{V}_3$$

$$\mathbf{I}_x = \frac{\mathbf{V}_4}{1}$$

Figure 8.29
Circuit used in
Example 8.19.

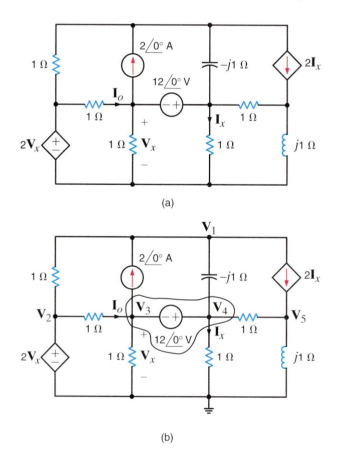

(a)

(b)

Combining these equations and expressing the results in matrix form yields

$$\begin{bmatrix} 1+j & -1 & 0 & 2-j & 0 \\ 0 & 1 & -2 & 0 & 0 \\ 0 & 0 & -1 & 1 & 0 \\ -j & -1 & 2 & 2+j & -1 \\ 0 & 0 & 0 & -3 & 1-j \end{bmatrix} \begin{bmatrix} \mathbf{V}_1 \\ \mathbf{V}_2 \\ \mathbf{V}_3 \\ \mathbf{V}_4 \\ \mathbf{V}_5 \end{bmatrix} = \begin{bmatrix} 2 \\ 0 \\ 12 \\ -2 \\ 0 \end{bmatrix}$$

The MATLAB solution is then

```
>> Y = [1+j*1 -1 0 2-j*10; 0 1 -2 0 0; 0 0 -1 1 0;
        -j*1 -1 2 2+j*1 -1; 0 0 0 -3 1-j*1]

Y =

Columns 1 through 4

  1.0000 +1.0000i    -1.0000          0      2.0000 -1.0000i
       0              1.0000    -2.0000           0
       0                   0    -1.0000      1.0000
       0 -1.0000i      -1.0000     2.0000      2.0000 +1.0000i
       0                   0          0     -3.0000

Column 5

       0
       0
       0
 -1.0000
  1.0000  -1.0000i

>> I = [2; 0; 12; -2; 0]

I =

     2
     0
    12
    -2
     0

>> V = inv(Y) *I

V =

    -5.5000    +4.5000i
   -26.0000   -24.0000i
   -13.0000   -12.0000i
    -1.0000   -12.0000i
    16.5000   -19.5000i

>>
```

As indicated in Fig. 8.29b, $\mathbf{I}_o$ is

$$\mathbf{I}_o = (\mathbf{V}_2 - \mathbf{V}_3)/1$$
$$= -13 - j12 \text{ A}$$

8.9 *AC PSPICE Analysis Using Schematic Capture*

INTRODUCTION In this chapter we found that an ac steady-state analysis is facilitated by the use of phasors. PSPICE can perform ac steady-state simulations, outputting magnitude and phase data for any voltage or current phasors of interest. In addition, PSPICE can perform an AC SWEEP in which the frequency of the sinusoidal sources is varied over a user-defined range. In this case, the simulation results are the magnitude and phase of every node voltage and branch current as a function of frequency.

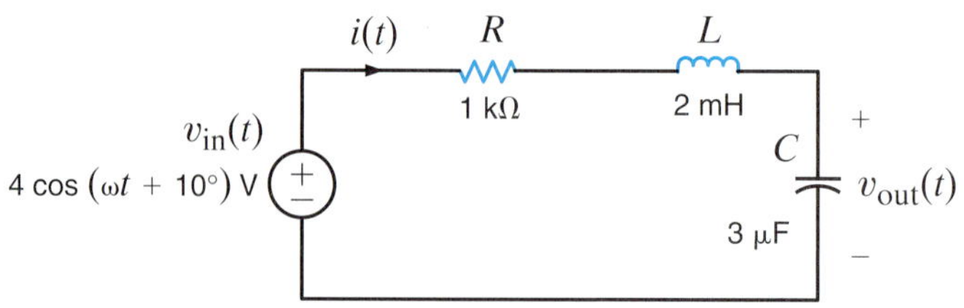

Figure 8.30
Circuit for ac simulation.

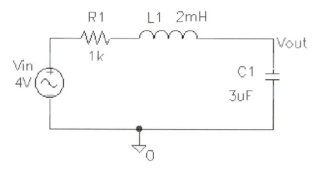

Figure 8.31
The Schematics diagram for the circuit in Fig. 8.30.

We will introduce five new *Schematics*/PSPICE topics in this section: defining AC sources, simulating at a single frequency, simulating over a frequency range, using the PROBE feature to create plots and, finally, saving and printing these plots. *Schematics* fundamentals such as getting parts, wiring, and changing part names and values were already discussed in Chapter 5. As in Chapter 5, we will use the following font conventions. Uppercase text refers to programs and utilities within PSPICE such as the AC SWEEP feature and PROBE graphing utility. All boldface text, whether upper or lowercase, denotes keyboard or mouse entries. For example, when placing a resistor into a circuit schematic, one must specify the resistor **VALUE** using the keyboard. The case of the boldface text matches that used in PSPICE.

DEFINING AC SOURCES   Figure 8.30 shows the circuit we will simulate at a frequency of 60 Hz. We will continue to follow the flowchart shown in Fig. 5.24 in performing this simulation. Inductor and capacitor parts are in the ANALOG library and are called L and C, respectively. The AC source, VAC, is in the SOURCE library. Figure 8.31 shows the resulting *Schematics* diagram after wiring and editing the part's names and values.

   To set up the AC source for simulation, double-click on the source symbol to open its ATTRIBUTES box, which is shown, after editing, in Fig. 8.32. As discussed in Chapter 7, we deselect the fields **Include Non-changeable Attributes** and **Include System-defined Attributes**. Each line in the ATTRIBUTES box is called an attribute of the ac source.

Figure 8.32
Setting the ac source phase angle.

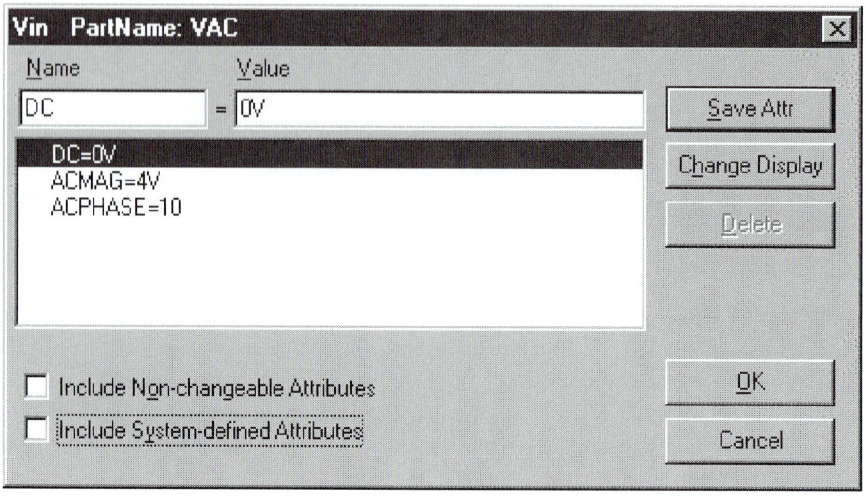

Vin PartName: VAC		✕
<u>N</u>ame	<u>V</u>alue	
DC	= 0V	<u>S</u>ave Attr
DC=0V ACMAG=4V ACPHASE=10		Change Display
		Delete
☐ Include No<u>n</u>-changeable Attributes		<u>O</u>K
☐ Include System-defined Attributes		Cancel

Analysis Setup

Enabled		Enabled		
☑	AC Sweep...		Options...	Close
☐	Load Bias Point...	☐	Parametric...	
☐	Save Bias Point...	☐	Sensitivity...	
☐	DC Sweep...	☐	Temperature...	
☐	Monte Carlo/Worst Case...	☐	Transfer Function...	
☑	Bias Point Detail	☐	Transient...	
	Digital Setup...			

Figure 8.33
The ANALYSIS SETUP window.

Each attribute has a name and a value. The **DC** attribute is the dc value of the source for dc analyses. The **ACMAG** and **ACPHASE** attributes set the magnitude and phase of the phasor representing Vin for ac analyses. Each of these attributes defaults to zero. The value of the **ACMAG** attribute was set to 4 V when we created the schematic in Fig. 8.31. To set the **ACPHASE** attribute to 10°, click on the **ACPHASE** attribute line, enter 10 in the **Value** field, press **Save Attr** and **OK**. When the ATTRIBUTE box looks like that shown in Fig. 8.32, the source is ready for simulation.

SINGLE-FREQUENCY AC SIMULATIONS   Next, we must specify the frequency for simulation. This is done by selecting **Setup** from the **Analysis** menu. The SETUP box in Fig. 8.33 should appear. If we double-click on the text **AC Sweep**, the AC SWEEP AND NOISE ANALYSIS window in Fig. 8.34 will open. All of the fields in Fig. 8.34 have been set for our 60-Hz simulation.

Since the simulation will be performed at only one frequency, 60 Hz, graphing the simulation results is not an attractive option. Instead, we will write the magnitude and phase of the phasors **Vout** and **I** to the output file using the VPRINT1 and IPRINT parts, from the SPECIAL library, which have been added to the circuit diagram as shown in Fig. 8.35. The VPRINT1 part acts as a voltmeter, measuring the voltage at any single node with respect to the ground node. There is also a VPRINT2 part, which measures the voltage between any two non-reference nodes. Similarly, the IPRINT part acts as an ammeter and must be placed in series with the branch current of interest. By convention, current in the IPRINT part is assumed to exit from its negatively marked terminal. To find the clockwise loop current, as defined in Fig. 8.30, the IPRINT part has been flipped. The FLIP command is in the EDIT menu.

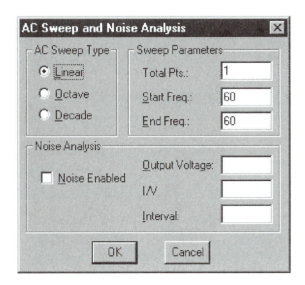

Figure 8.34
Setting the frequency range for a single frequency simulation.

Figure 8.35

A Schematics diagram ready for single-frequency ac simulation.

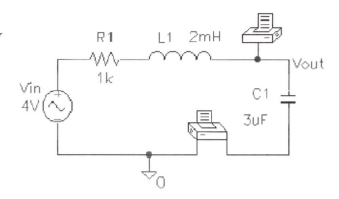

After placing the VPRINT1 part, double-click on it to open its ATTRIBUTES box, shown in Fig. 8.36. The VPRINT1 part can be configured to meter the node voltage in any kind of simulation: dc, ac, or transient. Since an ac analysis was specified in the SETUP window in Fig. 8.33, the values of the **AC**, **MAG**, and **PHASE** attributes are set to **Y**, where Y stands for YES. This process is repeated for the IPRINT part. When we return to *Schematics*, the simulation is ready to run.

When an AC SWEEP is performed, PSPICE, unless instructed otherwise, will attempt to plot the results using the PROBE plotting program. To turn off this feature, select **Probe Setup** in the **Analysis** menu. When the PROBE SETUP window shown in Fig. 8.37 appears, select **Do Not Auto-Run Probe** and **OK**.

The circuit is simulated by selecting **Simulate** from the **Analysis** menu. Since the results are in the output file, select **Examine Output** from the **Analysis** menu to view the data. At the bottom of the file, we find the results as seen in Fig. 8.38: **Vout** = 2.651 $\underline{/-38.54°}$ V and **I** = 2.998 $\underline{/51.46°}$ mA.

VARIABLE FREQUENCY AC SIMULATIONS To sweep the frequency over a range, 1 Hz to 10 MHz, for example, return to the AC SWEEP AND NOISE ANALYSIS box shown in Fig. 8.34. Change the fields to those shown in Fig. 8.39. Since the frequency range is so large, we have chosen a log axis for frequency with 50 data points in each decade. We can now plot the data using the PROBE utility. This procedure requires two steps. First, we remove the VPRINT1 and IPRINT parts in Fig. 8.35. Second, we return to the PROBE SETUP window shown in Fig. 8.37, and select **Automatically Run Probe After Simulation** and **OK**.

Figure 8.36

Setting the VPRINT1 measurements for ac magnitude and phase.

PRINT8   PartName: VPRINT1

Name	Value
DC	=

DC=
AC=y
TRAN=
MAG=y
PHASE=y
REAL=
IMAG=

Save Attr
Change Display
Delete

☐ Include Non-changeable Attributes
☐ Include System-defined Attributes

OK
Cancel

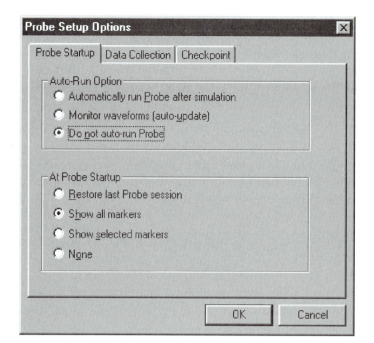

Figure 8.37
The PROBE SETUP window.

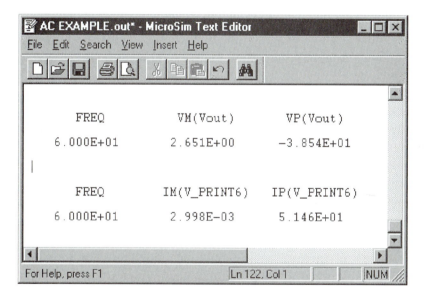

Figure 8.38
Magnitude and phase data for **Vout** *and* **I** *are at the bottom of the output file.*

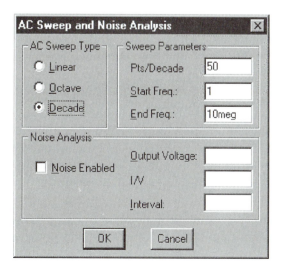

Figure 8.39
Setting the frequency range for a swept frequency simulation.

CREATING PLOTS IN PROBE When the PSPICE simulation is finished, the PROBE window shown in Fig. 8.40 will open. Actually, there are three windows here: the main display window, the output window, and the simulation status window. The latter two can be toggled off and on in the **View** menu. We will focus on the main display window, where the frequency is displayed on a log axis, as requested. To plot the magnitude and phase of **Vout**, select **Add** from the **Trace** menu. The ADD TRACES window shown in Fig. 8.41 will appear. To display the magnitude of **Vout**, we select **V(Vout)** from the left column. When a voltage or current is selected, the magnitude of the phasor will be plotted. Now the PROBE window should look like that shown in Fig. 8.42.

Before adding the phase to the plot, we note that **Vout** spans a small range—that is, 0 to 4 V. Since the phase change could span a much greater range, we will plot the phase on a second *y*-axis. From the **Plot** menu, select **Add Y Axis**. To add the phase to the plot, select **Add** from the **Trace** menu. On the right side of the ADD TRACES window in Fig. 8.41, scroll down to the entry, **P()**. Click on that, and then click on **V(Vout)** in the left column. The TRACE EXPRESSION line at the bottom of the window will contain the expression **P(V(Vout))**—the phase of **Vout**. Figure 8.43 shows the PROBE plot for both magnitude and phase of **Vout**.

To plot the current, **I**, on a new plot, we select **New** from the **Window** menu. Then, we add the traces for the magnitude and phase of the current through R1 (PSPICE calls it **I(R1)**) using the process described above for plotting the magnitude and phase of **Vout**. The results are shown in Fig. 8.44.

The procedures for saving and printing PROBE plots, as well as the techniques for plot manipulation and data extraction, are described in Chapter 7.

Figure 8.40
The PROBE window.

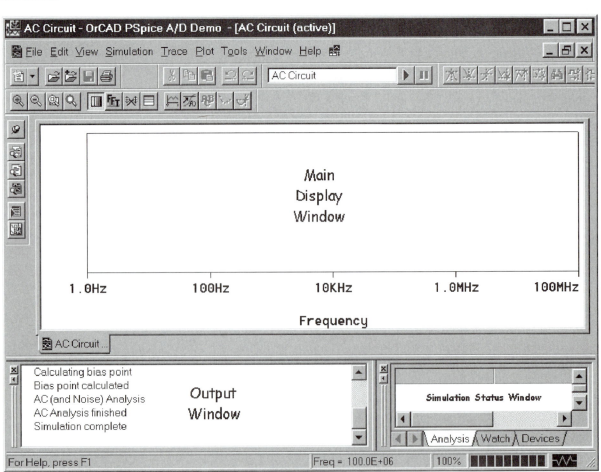

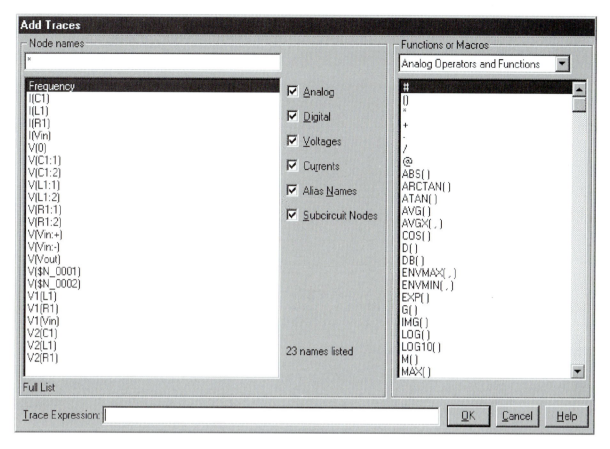

Figure 8.41
The Add Traces window.

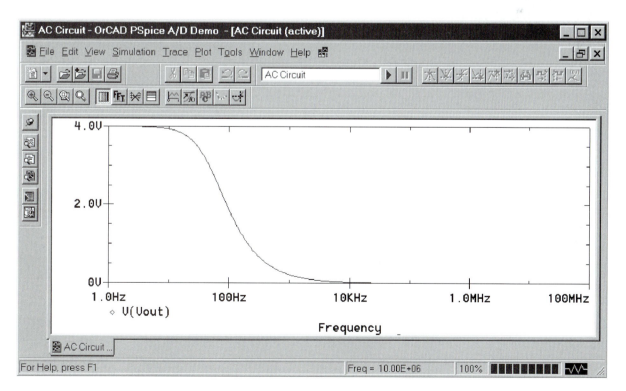

Figure 8.42
The magnitude of **Vout**.

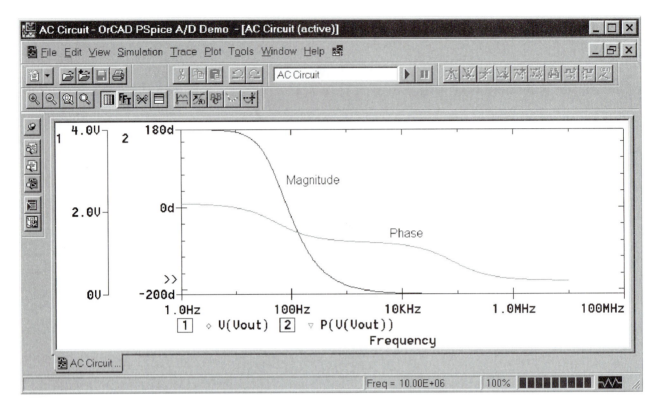

Figure 8.43
The magnitude and phase of **Vout**.

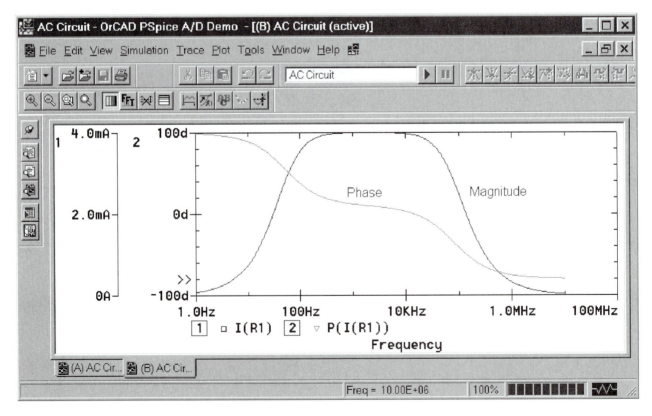

Figure 8.44
*The magnitude and phase of
the current,* **I**.

Example 8.20

Using the PSPICE *Schematics* editor, draw the circuit in Fig. 8.45, and use the PROBE utility to create plots for the magnitude and phase of $\mathbf{V}_{out}$ and $\mathbf{I}$. At what frequency does maximum $|\mathbf{I}|$ occur? What are the phasors $\mathbf{V}_{out}$ and $\mathbf{I}$ at that frequency?

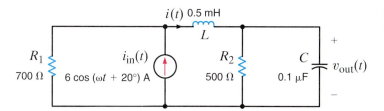

Figure 8.45
Circuit for Example 8.20.

SOLUTION   The *Schematics* diagram for the simulation is shown in Fig. 8.46, where an AC SWEEP has been set up for the frequency range 10 Hz to 10 MHz at 100 data points per decade. Plots for $\mathbf{V}_{out}$ and $\mathbf{I}$ magnitudes and phases are given in Figs. 8.47a and b, respectively. From Fig. 8.47b we see that the maximum inductor current magnitude occurs at 31.42 kHz. At that frequency, the phasors of interest are $\mathbf{I} = 5.94\,\underline{/16.06°}$ A and $\mathbf{V}_{out} = 299.5\,\underline{/-68.15°}$ V.

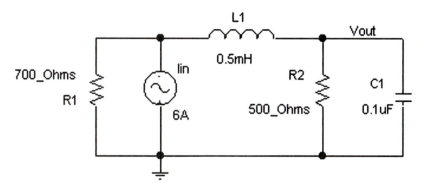

Figure 8.46
PSPICE Schematics diagram for Example 8.20.

Figure 8.47
Simulation results for Example 8.20, (a) **Vout** and (b) **I**.

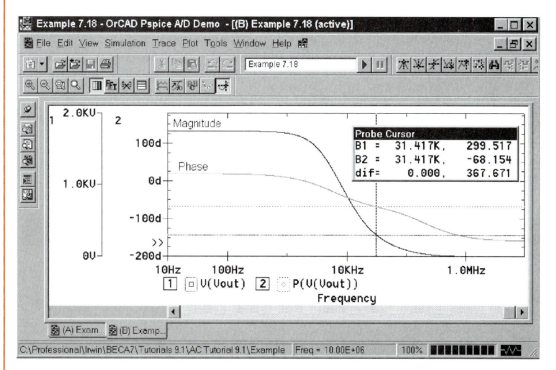

(a)

Figure 8.47
(continued)

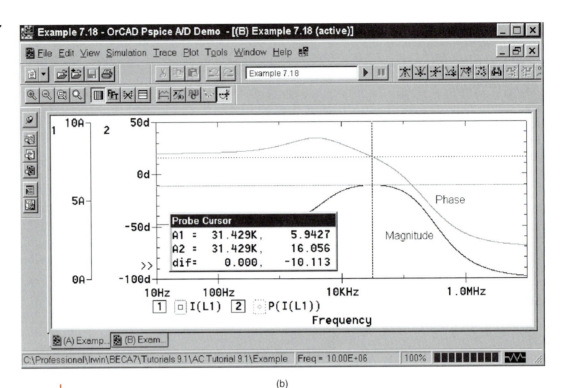

(b)

Example 8.21

Given the network in Fig. 8.48, let us determine the output voltage $\mathbf{V}_o$ using both PSPICE and MATLAB.

SOLUTION Since the component values are given in the frequency domain, in order to calculate the output voltage using PSPICE, we will simply assume that the frequency is $\omega = 1$ rad/s or $f = 0.159$ Hz. Then the capacitor value can be listed as $C = 0.5$ F and the inductor value as $L = 1$ H. The *Schematics* diagram for this circuit is shown in Fig. 8.49, and the output obtained with a frequency sweep in which the start and stop frequencies are both $f = 0.159$ is listed as

```
    FREQ           VM(Vout)       VP(Vout)

    1.590E-01      8.762E+00      -3.228E+01
```

Figure 8.48
Circuit used in
Example 8.21.

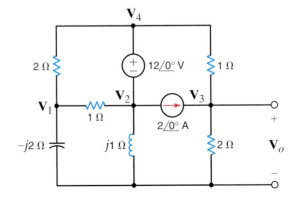

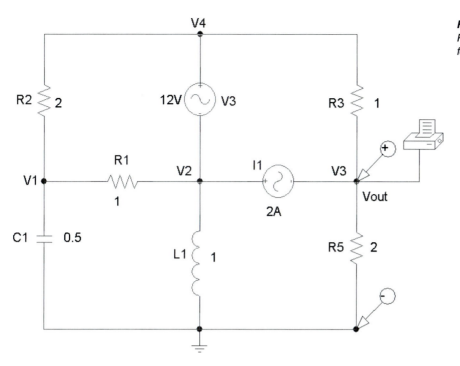

Figure 8.49
PSPICE Schematics *diagram*
for the network in Fig. 8.48.

The nodal equations for the network are

$$\mathbf{V}_4 - \mathbf{V}_2 = 12$$

$$\frac{\mathbf{V}_4 - \mathbf{V}_1}{2} + \frac{\mathbf{V}_4 - \mathbf{V}_3}{1} + \frac{\mathbf{V}_2 - \mathbf{V}_1}{1} + 2 + \frac{\mathbf{V}_2}{j1} = 0$$

$$\frac{\mathbf{V}_1 - \mathbf{V}_4}{2} + \frac{\mathbf{V}_1}{-j2} + \frac{\mathbf{V}_1 - \mathbf{V}_2}{1} = 0$$

$$\frac{\mathbf{V}_3 - \mathbf{V}_4}{1} + \frac{\mathbf{V}_3}{2} = 2$$

These equations can be placed in the following format, and the MATLAB solution is then listed as follows:

$$0v1 - v2 + 0v3 + v4 = 12$$

$$(-1 + j3)v1 - j2v2 + 0v3 - jv4 = 0$$

$$-j3v1 + (2 + j2)v2 - j2v3 + j3v4 = -j4$$

$$0v1 + 0v2 + 3v3 - 2v4 = 4$$

```
>> y = [0 -1 0 1; -1+(3*j) -2*j 0 -1*j;
      -3*j 2+2*j -2*j 3*j; 0 0 3 -2]
y =

      0              -1          0        1
   -1 + 3i         0 - 2i        0      0 - 1i
    0 - 3i         2 + 2i      0 - 2i   0 + 3i
      0              0           3       -2
>> i = [12; 0; -4*j; 4]

i =

      12
       0
     0 - 4i
       4
>> v = inv(y) *I
```

```
v =
    -1.1192 - 6.6528i
    -2.9016 - 7.0259i
     7.3990 - 4.6839i
     9.0984 - 7.0259i
>> vout=v(3)
vout =
     7.3990 - 4.6839i
```

which is identical to the PSPICE solution.

Example 8.22

As indicated earlier, we will now consider once again the network in Fig. 8.29, which has been redrawn in Fig. 8.50a, and determine the current I_o. In this final PSPICE example, which contains two dependent sources—one current source and one voltage source—we will, once again, outline in some detail the various issues that must be addressed in the solution.

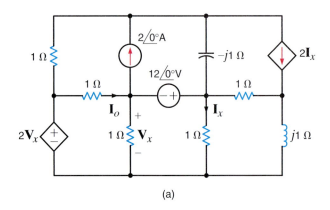

(a)

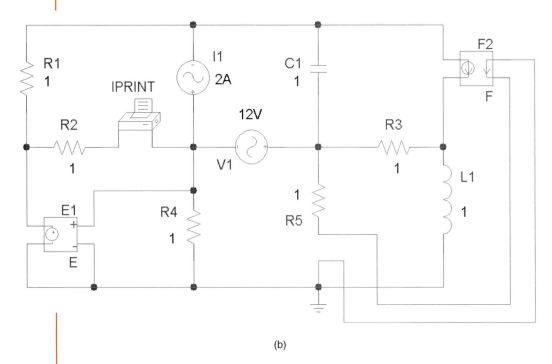

(b)

Figure 8.50 (a) The circuit diagram and (b) the PSPICE Schematics diagram for our introductory example.

Choosing the Frequency.   In PSPICE we must specify inductors in henries and capacitors in farads even though reactive impedances are given in ohms in the frequency-domain schematic in Fig. 8.50a. When the value of f or ω is not specified, we may choose any frequency we like, and a very smart choice is $\omega = 1$ rad/s ($f = 1/2\pi = 0.1592$ Hz). As a result, reactive impedances are related to the inductor and capacitor values by simple expressions.

$$Z_L = \omega L = L \qquad Z_C = 1/\omega C = 1/C$$

To set the frequency, select SETUP from the ANALYSIS menu and choose AC SWEEP. The window in Fig. 8.51, already edited for this circuit, should open. By limiting the frequency range to just one data point, making a plot of $\mathbf{I_o}$ versus frequency is silly. Instead, we will obtain a printout for $\mathbf{I_o}$.

Printing Currents and Voltages to the Output File.   The IPRINT part, utilized here, saves current values in the output file. The small negative sign marks the terminal where current exits the part. To set up the IPRINT part, double-click on it to open the window in Fig. 8.52. We must indicate what kind of data we want to acquire. In this case, we have used Y (short for "yes") to select the REAL and IMAGINARY components of the current that result from an AC simulation. When the simulation is complete, $\mathbf{I_o}$ will be at the bottom of the output file. To log voltages to the output file, use the VPRINT1 and VPRINT2 parts. VPRINT1 measures node voltage with respect to the ground node, while VPRINT2 measures voltages between two nonreference nodes.

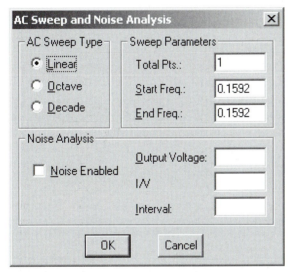

Figure 8.51
Setting the frequency sweep to a single frequency (0.1592 Hz) such that $\omega L = L$.

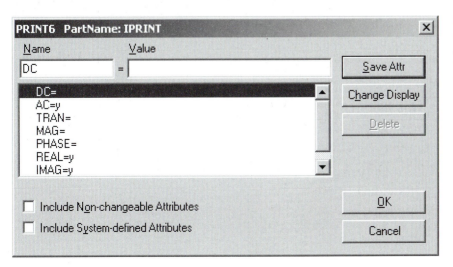

Figure 8.52
Initializing the IPRINT part to print real and imaginary current components from an AC simulation.

Current Direction in AC Current Sources. Dating back to the earliest forms of SPICE, the current through current sources was defined as flowing into the terminal marked with a + sign and out of the terminal marked with the − sign. Although this gives the appearance that the source is consuming power, note that +/− markings are just that and the voltage across the source will depend on the entire circuit. Look carefully at I1 and I2 in Fig. 8.50b. They are connected exactly as shown in Fig. 8.50a.

Setting Dependent Source Gains. We will use the gain of the CCCS as an example. Comparing Figs. 8.50a and b, it appears that the direction of I_x is reversed. This was done to simplify the wiring between the CCCS and the branch where I_x is defined. By setting the gain of the CCCS to *negative* 2, the circuits become identical. Now consider the VCVS, which is wired in agreement with Fig. 8.50a. Here the gain is entered as *positive* 2. To set the gain of a dependent source, simply double-click on it to open the attribute box shown in Fig. 8.53 and enter the gain.

Accessing the OUTPUT FILE. When you simulate the circuit, the simulation engine in PSPICE will open and the window in Fig. 8.54 will appear. When the simulation is complete, close this window to return to *Schematics* where you can access the output file through Examine Output File option in the ANALYSIS menu. When the OUTPUT FILE opens, scroll to the bottom to find $\mathbf{I_o}$. You might notice that the output file window is actually the NOTEPAD text editor common to Windows-based PCs. So, you can copy and paste the simulation results directly into other software.

When this circuit is simulated, the results in the output file are

```
FREQ          IR(V_PRINT1)     II(V_PRINT1)

1.592E-01     -1.300E+01       -1.200E+01
```

which correspond to $\mathbf{I_o}$ is $-13 - j12$ A, exactly the value calculated in Example 8.19. Note that the current is listed as flowing through a part called V_PRINT1. Again, dating back to its origins, SPICE has always produced the node voltages and the currents through all voltage sources. So, an ammeter was easily "constructed" by inserting a voltage source set to 0V. That is exactly what the IPRINT part is, a voltage source called V_PRINT set to 0V.

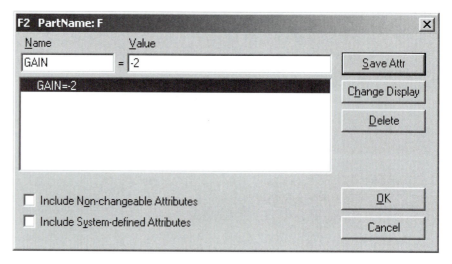

Figure 8.53 *Setting the gain of the CCCS to negative 2 offsets the wiring differences between Figs. 8.50a and b.*

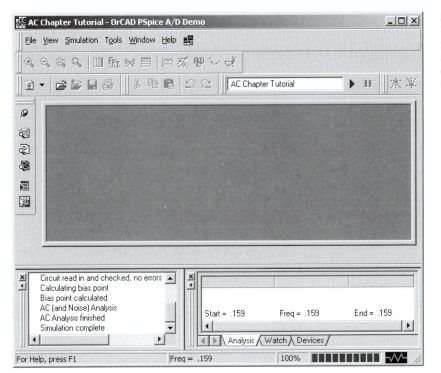

Figure 8.54
*The true PSPICE window
where simulation status is
reported in the lower window
cells and plots are displayed in
the main cell.*

8.10   Application Examples

Application Example 8.23

The network in Fig. 8.55 models an unfortunate situation that is all too common. Node A, which is the voltage $v_{in}(t)$ at the output of a temperature sensor, has "picked-up" a high-frequency voltage, $v_{noise}(t)$, caused by a nearby AM radio station. The noise frequency is 700 kHz. In this particular scenario, the sensor voltage, like temperature, tends to vary slowly. Our task then is to modify the circuit to reduce the noise at the output without disturbing the desired signal, $v_{in}(t)$.

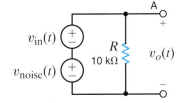

Figure 8.55
*Modeling radio frequency
noise pickup.*

SOLUTION   Consider the network in Fig. 8.56a. If component X has a high impedance, that is, much greater than R at 700 kHz, and an impedance of zero at dc, we should be able to alleviate the problem. Using voltage division to obtain $\mathbf{V}_o$ in Fig. 8.56b, we find

$$\mathbf{V}_o = \left[\frac{R}{R + j\omega L} \right] \mathbf{V}_l$$

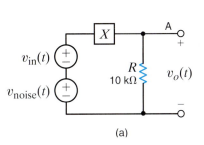

(a)

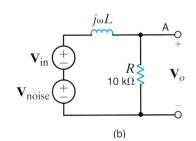

(b)

Figure 8.56
*(a) Model used to
reject $\mathbf{V}_{noise}$ and
(b) the required
component.*

where V_1 and ω are either $\mathbf{V}_{in}$ and 0, or $\mathbf{V}_{noise}$ and $2\pi(700 \times 10^3)$. At dc, $\omega = 0$, the inductor's impedance is zero, the voltage division ratio is unity, V_1 is $\mathbf{V}_{in}$ and $\mathbf{V}_o$ equals $\mathbf{V}_{in}$. But at 700 kHz, V_1 is $\mathbf{V}_{noise}$ and the desired voltage division ratio should be very small; that is, the inductor impedance must be much greater than R, so that $\mathbf{V}_o$ becomes nearly zero. If we choose to reduce the noise at the output by 90%, we find

$$\left| \frac{R}{R + j\omega L} \right| = \frac{1}{10} \quad \text{at } f = 700 \text{ kHz}$$

Solving this equation yields $L = 22.6$ mH, which is close to a standard inductor value.

Application Example 8.24

The circuit in Fig. 8.57 is called a General Impedance Converter or GIC. We wish to develop an expression for the impedance $\mathbf{Z}_{eq}$ in terms of $\mathbf{Z}_1, \mathbf{Z}_2, \mathbf{Z}_3, \mathbf{Z}_4$, and $\mathbf{Z}_5$, and then using resistors of equal value and a 1-μF capacitor, create a 1-H equivalent inductance.

Figure 8.57
The general impedance converter.

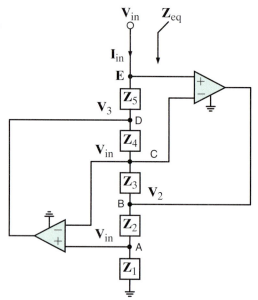

SOLUTION We simply employ the ideal op-amp assumptions; that is, there is no current entering any op-amp input, and the voltage across any op-amp's input terminals is zero. As a result, the voltage at both nodes A and C are $\mathbf{V}_{in}$. Next, we apply KCL at each op-amp input. At node A

$$\frac{\mathbf{V}_2 - \mathbf{V}_{in}}{\mathbf{Z}_2} = \frac{\mathbf{V}_{in}}{\mathbf{Z}_1}$$

which yields

$$\mathbf{V}_2 = \mathbf{V}_{in}\left[1 + \frac{\mathbf{Z}_2}{\mathbf{Z}_1} \right] \qquad\qquad \textbf{8.47}$$

At node C, we find

$$\frac{\mathbf{V}_3 - \mathbf{V}_{in}}{\mathbf{Z}_4} = \frac{\mathbf{V}_{in} - \mathbf{V}_2}{\mathbf{Z}_3}$$

Solving for $\mathbf{V}_3$ yields

$$\mathbf{V}_3 = \mathbf{V}_{in}\left[1 + \frac{\mathbf{Z}_4}{\mathbf{Z}_3} \right] - \mathbf{V}_2\left[\frac{\mathbf{Z}_4}{\mathbf{Z}_3} \right]$$

Substituting our results from Eq. 8.47, we can express $\mathbf{V}_3$ as

$$\mathbf{V}_3 = \mathbf{V}_{\text{in}}\left[1 - \frac{\mathbf{Z}_2\mathbf{Z}_4}{\mathbf{Z}_1\mathbf{Z}_3}\right]$$ **8.48**

At node E, we write

$$\mathbf{I}_{\text{in}} = \frac{\mathbf{V}_{\text{in}} - \mathbf{V}_3}{\mathbf{Z}_5}$$

Then using our expression for $\mathbf{V}_3$ in Eq. (8.48), we find that

$$\mathbf{I}_{\text{in}} = \mathbf{V}_{\text{in}}\left[\frac{\mathbf{Z}_2\mathbf{Z}_4}{\mathbf{Z}_1\mathbf{Z}_3\mathbf{Z}_5}\right]$$

Finally, the impedance of interest is

$$\mathbf{Z}_{\text{eq}} = \frac{\mathbf{V}_{\text{in}}}{\mathbf{I}_{\text{in}}} = \left[\frac{\mathbf{Z}_1\mathbf{Z}_3\mathbf{Z}_5}{\mathbf{Z}_2\mathbf{Z}_4}\right]$$ **8.49**

Now, if $\mathbf{Z}_1 = \mathbf{Z}_3 = \mathbf{Z}_5 = \mathbf{Z}_2 = R$ and $\mathbf{Z}_4 = 1/j\omega C$, then $\mathbf{Z}_{\text{eq}}$ becomes

$$\mathbf{Z}_{\text{eq}} = j\omega CR^2 = j\omega L_{\text{eq}}$$

Hence, the value of R necessary to yield a 1-H inductance is 1000 Ω. At this point, we must address the question: why go to all this trouble just to make inductance? The answer is size and weight. A 1-H inductor would be very large and heavy. The GIC is easy to construct with integrated circuit components, requires very little space, and weighs only a few grams!

8.11   Design Examples

Design Example 8.25

In Chapter 4 we found that the op-amp provided us with an easy and effective method of producing controllable voltage gain. From these earlier studies, we have come to expect gain from these "active" devices in a configuration like that shown in Fig. 8.58a. However, an experienced engineer has suggested that we could achieve some gain from the proper configuration of "passive" elements as illustrated in Fig. 8.58b, and has proposed the circuit in Fig. 8.58c. Therefore, let us use this suggested configuration in an attempt to design for a gain of 10 at 1 kHz if the load is 100 Ω.

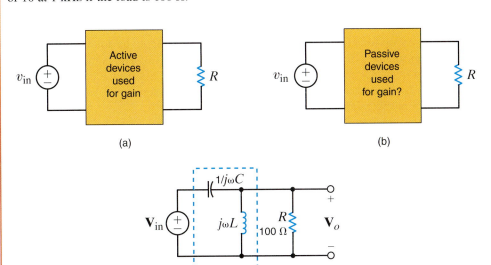

Figure 8.58
Circuit configurations used in Example 8.25.

SOLUTION   The voltage gain of the network in Fig. 8.58c can be expressed as

$$\frac{\mathbf{V}_o}{\mathbf{V}_{\text{in}}} = \left[\frac{\mathbf{Z}}{\mathbf{Z} + \dfrac{1}{j\omega C}} \right]$$

where

$$\mathbf{Z} = \frac{(j\omega L)R}{j\omega L + R}$$

Combining these two equations and rearranging the terms yields the expression

$$\frac{\mathbf{V}_o}{\mathbf{V}_{\text{in}}} = \left[\frac{j\omega L}{j\left[\omega L - \dfrac{1}{\omega C} \right] + \dfrac{L}{CR}} \right]$$

We know that in order to achieve amplification, the denominator must be less than the numerator. In addition, the denominator will be reduced if the reactances of the inductor and capacitor are made equal in magnitude, since they are opposite in sign. Thus, by selecting the parameters such that $\omega^2 LC = 1$, the reactance of the inductor will cancel that of the capacitor. Under this condition, the gain is reduced to

$$\frac{\mathbf{V}_o}{\mathbf{V}_{\text{in}}} = j\omega RC$$

For the given load and frequency values, a capacitor value of 15.9 μF will provide the required gain. The inductor value can then be obtained from the constraint

$$\omega^2 LC = 1$$

which yields $L = 1.59$ mH. It should be noted that if the frequency changes, the impedance of both the inductor and capacitor will also change, thus altering the gain. Finally, we will find, in a later chapter, that the equation $\omega^2 LC = 1$ is an extremely important expression and one that can have a dramatic effect on circuits.

Design Example 8.26

A sinusoidal signal, $v_1(t) = 2.5\cos(\omega t)$ when added to a dc level of $V_2 = 2.5$ V, provides a 0- to 5-V clock signal used to control a microprocessor. If the oscillation frequency of the signal is to be 1 GHz, let us design the appropriate circuit.

SOLUTION   As we found in Chapter 4, this application appears to be a natural application for an op-amp summer. However, the frequency of oscillation (1 GHz) is much higher than the maximum frequency most op-amps can handle—typically less than 200 MHz. Since no amplification is required in this case, we should be able to design an op-amp-less summer that, while not precise, should get the job done.

Consider the circuit in Fig. 8.59a where inputs $v_1(t)$ and V_2 are connected to yield the output $v_o(t)$. For this application, component A should block any dc component in $v_1(t)$ from reaching the output but permit the 1-GHz signal to pass right through. Similarly, component B should pass V_2 while blocking any high-frequency signal. Thus, the impedance of component A should be infinite at dc but very low at 1 GHz. And the impedance of component B should be zero at dc but very high at high frequency. Our earlier studies indicate that component A must be a capacitor and component B an inductor. The resulting circuit, called a *bias T*, is shown in Fig. 8.59b.

The values for C and L are dependent on both the signal frequency and the precision required in the summing operation, and can be easily seen by using superposition to investigate the contribution of each input to $v_o(t)$. In Fig. 8.60a, the dc voltage V_2 has been reduced

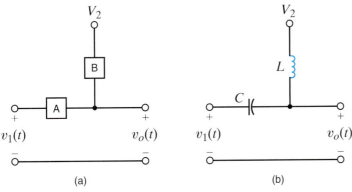

Figure 8.59
(a) A simple passive summer circuit; (b) a solution—the bias T.

to zero and an ac circuit has been drawn at a frequency of 1 GHz, that is, the frequency of $v_1(t)$. Using voltage division, we can express the output voltage as

$$\mathbf{V}_{o1} = \left[\frac{j\omega L}{j\omega L - \dfrac{j}{\omega C}}\right]\mathbf{V}_1 = \left[\frac{\omega^2 LC}{\omega^2 LC - 1}\right]\mathbf{V}_1 \qquad \textbf{8.50}$$

Note that in order to achieve a perfect summer, the voltage division ratio must be unity. However, such a voltage division ratio requires the impractical condition $\omega^2 LC$ equal infinity.

Instead, we will approach the problem by choosing values for the inductive and capacitive reactances. As stated earlier, the capacitive reactance should be small; we choose 1 Ω. And the inductive reactance should be large; let's say 10 kΩ. The resulting L and C values are

$$C = \frac{1}{\omega X_C} = \frac{1}{2\pi \times 10^9} = 159 \text{ pF}$$

and

$$L = \frac{X_L}{\omega} = 1.59 \text{ μH}$$

Now we consider V_2. In Fig. 8.60b, $v_1(t)$ has been reduced to zero and an ac circuit has been drawn at dc—the frequency of V_2. Again, voltage division could be used to express the output voltage. However, we see that the impedances of the capacitor and inductor are infinity and unity, respectfully. As a result, the output is EXACTLY equal to V_2 regardless of the values of C and L!

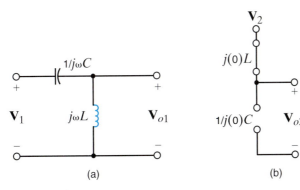

Figure 8.60
Exploring the bias-T by steady-state ac superposition with (a) $V_2 = 0$ and (b) $v_1(t) = 0$

Thus, the output voltage consists of two voltages, one at dc and the other at 1 GHz. The dc component is just $V_2 = 2.5$ V. From Eq. 8.50, the ac component at 1 GHz is

$$\mathbf{V}_{o1} = \left[\frac{j10{,}000}{j10{,}000 - j1}\right]2.5\underline{/0°} = 2.50025\underline{/0°} \text{ V}$$

Back in the time domain, the output voltage is, to three significant digits,

$$v_o(t) = 2.5 + 2.5\cos\left[2\pi(10^9)t\right] \text{ V}$$

SUMMARY

- **The sinusoidal function definition**
  The sinusoidal function $x(t) = X_M \sin(\omega t + \theta)$ has an amplitude of X_M, a radian frequency of ω, a period of $2\pi/\omega$, and a phase angle of θ.

- **The phase lead and phase lag definitions**
  If $x_1(t) = X_{M_1} \sin(\omega t + \theta)$ and $x_2(t) = X_{M_2} \sin(\omega t + \phi)$, $x_1(t)$ leads $x_2(t)$ by $\theta - \phi$ radians and $x_2(t)$ lags $x_1(t)$ by $\theta - \phi$ radians.

- **The phasor definition** The sinusoidal voltage $v(t) = V_M \cos(\omega t + \theta)$ can be written in exponential form as $v(t) = \mathrm{Re}\left[V_M e^{j(\omega t + \theta)}\right]$ and in phasor form as $\mathbf{V} = V_M \underline{/\theta}$.

- **The phase relationship in θ_v and θ_i for elements R, L, and C** If θ_v and θ_i represent the phase angles of the voltage across and the current through a circuit element, then $\theta_i = \theta_v$ if the element is a resistor, θ_i lags θ_v by 90° if the element is an inductor, θ_i leads θ_v by 90° if the element is a capacitor.

- **The impedances of R, L, and C** Impedance, $\mathbf{Z}$, is defined as the ratio of the phasor voltage, $\mathbf{V}$, to the phasor current, $\mathbf{I}$, where $\mathbf{Z} = R$ for a resistor, $\mathbf{Z} = j\omega L$ for an inductor, and $\mathbf{Z} = 1/j\omega C$ for a capacitor.

- **The phasor diagrams** Phasor diagrams can be used to display the magnitude and phase relationships of various voltages and currents in a network.

- **Frequency-domain analysis**
  1. Represent all voltages, $v_i(t)$, and all currents, $i_j(t)$, as phasors and represent all passive elements by their impedance or admittance.
  2. Solve for the unknown phasors in the frequency (ω) domain.
  3. Transform the now-known phasors back to the time domain.

- **Solution techniques for ac steady-state problems**
  Ohm's law
  KCL and KVL
  PSPICE
  MATLAB
  Nodal and loop analysis
  Superposition and source exchange
  Thévenin's theorem
  Norton's theorem

PROBLEMS

PSV **CS** both available on the web at: http://www.justask4u.com/irwin

SECTION 8.1

8.1 Given $i(t) = 5\cos(400t - 120°)$ A, determine the period of the current and the frequency in Hertz. **CS**

8.2 Determine the relative phase relationship of the two waves.

$$v_1(t) = 10\cos(377t - 30°) \text{ V}$$
$$v_2(t) = 10\cos(377t + 90°) \text{ V}$$

8.3 Given the following voltage and current

$$i(t) = 5\sin(377t - 20°) \text{ V}$$
$$v(t) = 10\cos(377t + 30°) \text{ V}$$

determine the phase relationship between $i(t)$ and $v(t)$.

8.4 Determine the phase angles by which $v_1(t)$ leads $i_1(t)$ and $v_1(t)$ leads $i_2(t)$, where

$$v_1(t) = 4\sin(377t + 25°) \text{ V}$$
$$i_1(t) = 0.05\cos(377t - 20°) \text{ A}$$
$$i_2(t) = -0.1\sin(377t + 45°) \text{ A}$$

8.5 Calculate the current in the resistor in Fig. P8.5 if the voltage input is

(a) $v_1(t) = 10\cos(377t + 180°)$ V.

(b) $v_2(t) = 12\sin(377t + 45°)$ V.

Give the answers in both the time and frequency domains. **CS**

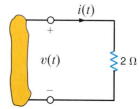

Figure P8.5

8.6 Calculate the current in the inductor shown in Fig. P8.6 if the voltage input is

(a) $v_1(t) = 10\cos(377t + 45°)$ V

(b) $v_2(t) = 5\sin(377t - 90°)$ V

Give the answers in both the time and frequency domains.

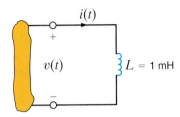

Figure P8.6

8.7 Calculate the current in the capacitor shown in Fig. P8.7 if the voltage input is

(a) $v_1(t) = 10\cos(377t - 30°)$ V

(b) $v_2(t) = 5\sin(377t + 60°)$ V

Give the answers in both the time and frequency domains.

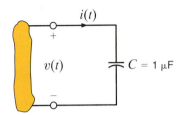

Figure P8.7

SECTION 8.5

8.8 Find the frequency-domain impedance, **Z**, as shown in Fig. P8.8. **CS**

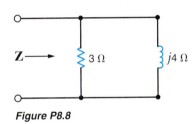

Figure P8.8

8.9 Find the impedance, **Z**, shown in Fig. P8.9 at a frequency of 60 Hz.

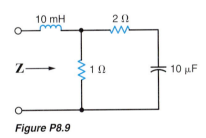

Figure P8.9

8.10 Find the impedance, **Z**, shown in Fig. P8.9 at a frequency of 400 Hz.

8.11 In the network in Fig. P8.11, find $\mathbf{Z}(j\omega)$ at a frequency of 60 Hz. **CS**

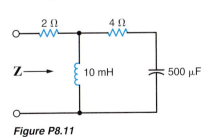

Figure P8.11

8.12 Find the frequency-domain impedance, **Z**, shown in Fig. P8.12.

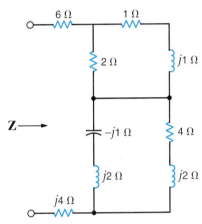

Figure P8.12

8.13 Find the frequency-domain impedance, **Z**, shown in Fig. P8.13. **PSV**

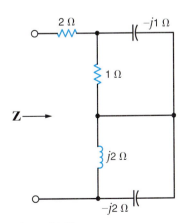

Figure P8.13

8.14 Find **Z** in the network in Fig. P8.14. **CS**

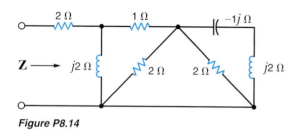

Figure P8.14

8.15 Find **Z** in the network in Fig. P8.15.

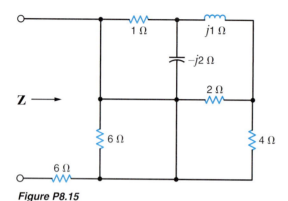

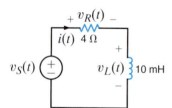

Figure P8.15

8.16 Draw the frequency-domain circuit and calculate $i(t)$ for the circuit shown in Fig. P8.16 if $v_S(t) = 2\cos(377t)$ V.

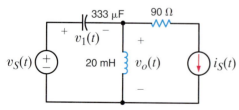

Figure P8.16

8.17 Draw the frequency-domain circuit and calculate $v(t)$ for the circuit shown in Fig. P8.17 if $i_S(t) = 10\cos(377t + 30°)$ A.

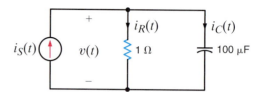

Figure P8.17

8.18 Draw the frequency-domain circuit and calculate $v(t)$ for the circuit shown in Fig. P8.18 if $i_S(t) = 20\cos(377t + 120°)$ A. **CS**

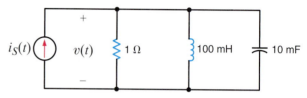

Figure P8.18

8.19 Draw the frequency-domain circuit and calculate $v(t)$ for the circuit shown in Fig. P8.19 if $i_S(t) = 2\cos(1000t + 120°)$ A.

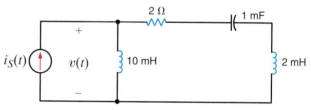

Figure P8.19

SECTION 8.6

8.20 Draw the frequency-domain network and calculate $v_o(t)$ in the circuit shown in Fig. P8.20 if $v_S(t)$ is $4\sin(500t + 45°)$ V and $i_S(t)$ is $1\cos(500t + 45°)$ A. Also, use a phasor diagram to determine $v_1(t)$.

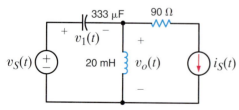

Figure P8.20

8.21 Draw the frequency-domain network and calculate $v_o(t)$ in the circuit shown in Fig. P8.21 if $i_1(t)$ is $200\cos(10^5t + 60°)$ mA, $i_2(t)$ is $100\sin(10^5t + 90°)$ mA, and $v_S(t) = 10\sin(10^5t)$ V. Also, use a phasor diagram to determine $v_C(t)$.

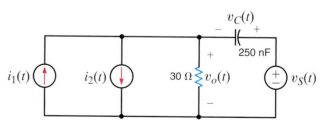

Figure P8.21

8.22 The impedance of the network in Fig. P8.22 is found to be purely real at $f = 400$ Hz. What is the value of C?

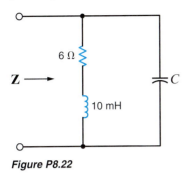

Figure P8.22

8.23 In the circuit shown in Fig. P8.23, determine the value of the inductance such that the current is in phase with the source voltage. **PSV**

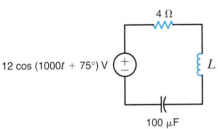

Figure P8.23

SECTION 8.7

8.26 The voltages $v_R(t)$ and $v_L(t)$ in the circuit shown in Fig. P8.16 can be drawn as phasors in a phasor diagram. Use a phasor diagram to show that $v_R(t) + v_L(t) = v_S(t)$.

8.27 The currents $i_R(t)$ and $i_C(t)$ in the circuit shown in Fig. P8.17 can be drawn as phasors in a phasor diagram. Use the diagram to show that $i_R(t) + i_C(t) = i_S(t)$.

8.28 The currents $i_R(t)$ and $i_C(t)$ in the circuit shown in Fig. P8.28 can be drawn as phasors in a phasor diagram. Use the diagram to show that $i_R(t) + i_C(t) = i_S(t)$. **CS**

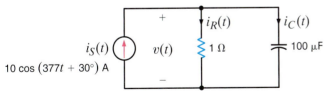

Figure P8.28

8.24 The impedance of the box in Fig. P8.24 is $5 + j4\ \Omega$ at 1000 rad/s. What is the impedance at 1300 rad/s?

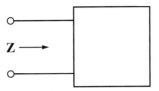

Figure P8.24

8.25 The admittance of the box in Fig. P8.25 is $0.1 + j0.2$ S at 500 rad/s. What is the impedance at 300 rad/s?

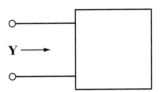

Figure P8.25

8.29 Draw the frequency-domain network and calculate $v_o(t)$ in the circuit shown in Fig. P8.29 if $i_S(t)$ is $300 \sin(10^4 t - 45°)$ mA. Also, using a phasor diagram, show that $i_1(t) + i_2(t) = i_S(t)$. **CS**

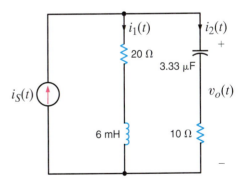

Figure P8.29

8.30 Find the value of C in the circuit shown in Fig. P8.30 so that $\mathbf{Z}$ is purely resistive at the frequency of 60 Hz.

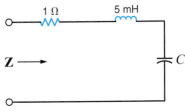

Figure P8.30

8.31 Find the frequency at which the circuit shown in Fig. P8.31 is purely resistive.

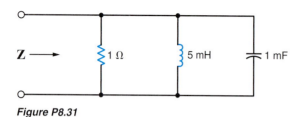

Figure P8.31

8.32 In the circuit shown in Fig. P8.32, determine the frequency at which $i(t)$ is in phase with $v_S(t)$.

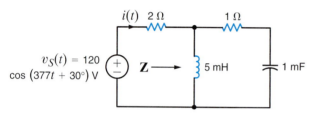

Figure P8.32

8.33 In the circuit shown in Fig. P8.33, determine the value of the inductance such that $v(t)$ is in phase with $i_S(t)$.

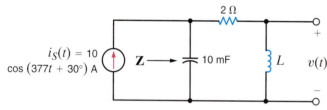

Figure P8.33

8.34 Find the current **I** shown in Fig. P8.34.

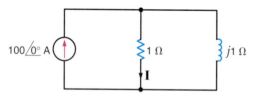

Figure P8.34

8.35 Find the voltage **V** shown in Fig. P8.35.

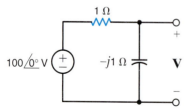

Figure P8.35

8.36 Find the frequency-domain voltage V_o, as shown in Fig. P8.36. **CS**

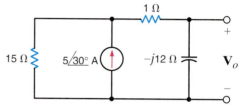

Figure P8.36

8.37 Find the voltage, V_o, shown in Fig. P8.37.

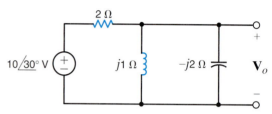

Figure P8.37

8.38 Given the network in Fig. P8.38, determine the value of V_o if $V_S = 24\,\underline{/0°}$ V.

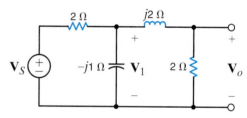

Figure P8.38

8.39 Find V_S in the network in Fig. P8.39, if $V_1 = 4\,\underline{/0°}$ V. **CS**

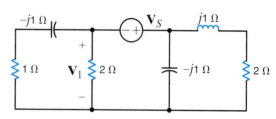

Figure P8.39

8.40 Find V_o in the network in Fig. P8.40. **PSV**

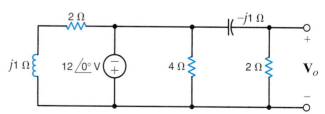

Figure P8.40

8.41 If $\mathbf{V}_1 = 4\underline{/0°}$ V, find $\mathbf{I}_o$ in Fig. P8.41. **CS**

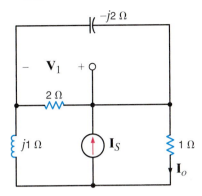

Figure P8.41

8.42 In the network in Fig. P8.42 $\mathbf{I}_o = 4\underline{/0°}$ A, find $\mathbf{I}_x$.

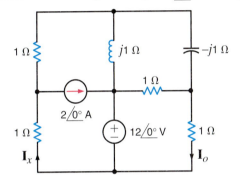

Figure P8.42

8.43 If $\mathbf{I}_o = 4\underline{/0°}$ A in the circuit in Fig. P8.43, find $\mathbf{I}_x$.

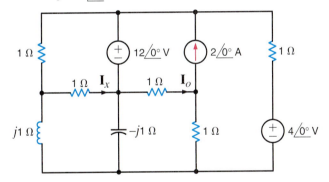

Figure P8.43

8.44 If $\mathbf{I}_o = 4\underline{/0°}$ A in the network in Fig. P8.44, find $\mathbf{I}_x$.

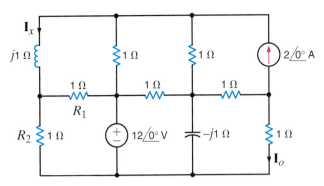

Figure P8.44

8.45 In the network in Fig. P8.45, $\mathbf{V}_o$ is known to be $4\underline{/45°}$ V. Find $\mathbf{Z}$.

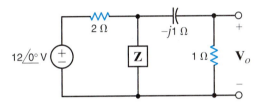

Figure P8.45

8.46 In the network in Fig. P8.46, $\mathbf{V}_1 = 2\underline{/45°}$ V. Find $\mathbf{Z}$.

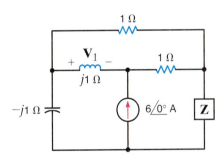

Figure P8.46

SECTION 8.8

8.47 Find $\mathbf{V}_o$ in the circuit in Fig. P8.47.

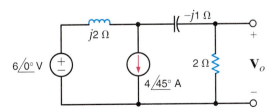

Figure P8.47

8.48 Using nodal analysis, find $\mathbf{I}_o$ in the circuit in Fig. P8.48. **CS**

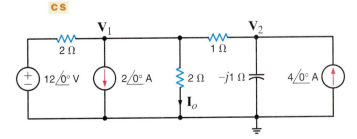

Figure P8.48

8.49 Determine $\mathbf{V}_o$ in the circuit in Fig. P8.49.

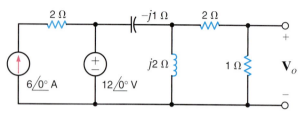

Figure P8.49

8.50 Using nodal analysis, find $\mathbf{I}_o$ in the circuit in Fig. P8.50.

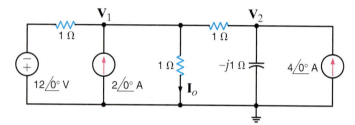

Figure P8.50

8.51 Use nodal analysis to find $\mathbf{I}_o$ in the circuit in Fig. P8.51.

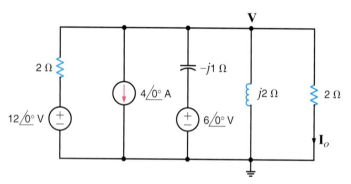

Figure P8.51

8.52 Find $\mathbf{V}_o$ in the network in Fig. P8.52. **CS**

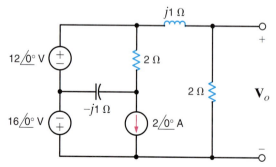

Figure P8.52

8.53 Find $\mathbf{V}_o$ in the network in Fig. P8.53 using nodal analysis. **PSV**

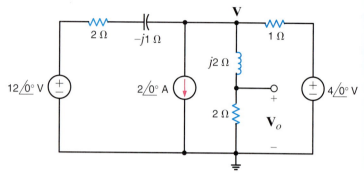

Figure P8.53

8.54 Find $\mathbf{I}_o$ in the circuit in Fig. P8.54 using nodal analysis.

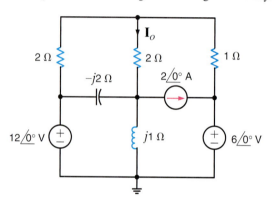

Figure P8.54

8.55 Use the supernode technique to find $\mathbf{I}_o$ in the circuit in Fig. P8.55.

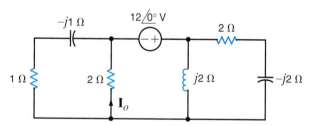

Figure P8.55

8.56 Find $\mathbf{I}_o$ in the network in Fig. P8.56.

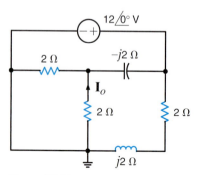

Figure P8.56

8.57 Find $\mathbf{V}_o$ in the network in Fig. P8.57.

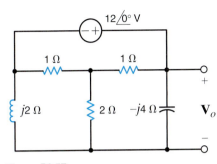

Figure P8.57

8.58 Use nodal analysis to find $\mathbf{V}_o$ in the circuit in Fig. P8.58.

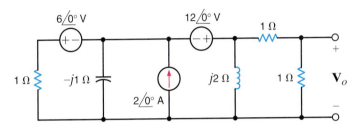

Figure P8.58

8.59 The low-frequency equivalent circuit for a common-emitter transistor amplifier is shown in Fig. P8.59. Compute the voltage gain $\mathbf{V}_o/\mathbf{V}_S$.

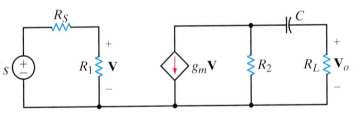

Figure P8.59

8.60 Use nodal analysis to find $\mathbf{V}_o$ in the circuit in Fig. P8.60.
PSV

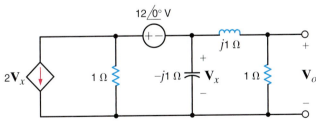

Figure P8.60

8.61 Find the voltage across the inductor in the circuit shown in Fig. P8.61 using nodal analysis.

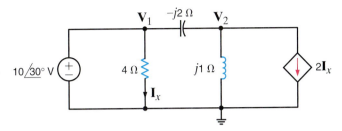

Figure P8.61

8.62 Use nodal analysis to find $\mathbf{I}_o$ in the circuit in Fig. P8.62.

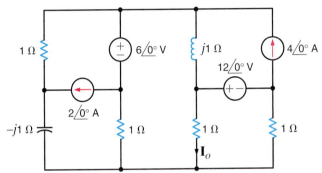

Figure P8.62

8.63 Use mesh analysis to find $\mathbf{V}_o$ in the circuit shown in Fig. P8.63.

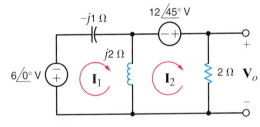

Figure P8.63

8.64 Use mesh analysis to find $\mathbf{V}_o$ in the circuit shown in Fig. P8.64.

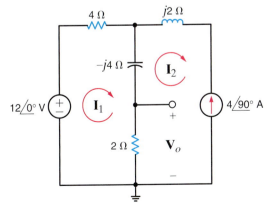

Figure P8.64

8.65 Find $\mathbf{V}_o$ in the circuit in Fig. P8.65 using mesh analysis.

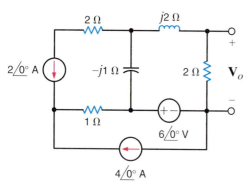

Figure P8.65

8.66 Use mesh analysis to find $\mathbf{V}_o$ in the circuit in Fig. P8.66.

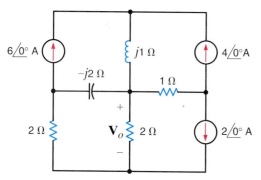

Figure P8.66

8.67 Using loop analysis and MATLAB, find $\mathbf{I}_o$ in the network in Fig. P8.67. **CS**

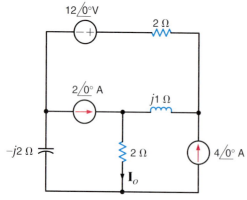

Figure P8.67

8.68 Find $\mathbf{V}_o$ in the network in Fig. P8.68.

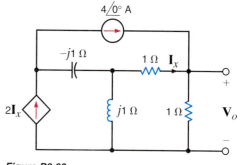

Figure P8.68

8.69 Find $\mathbf{V}_o$ in the network in Fig. P8.69. **PSV**

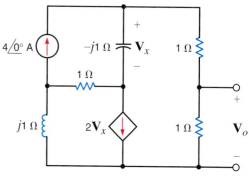

Figure P8.69

8.70 Use loop analysis to find $\mathbf{I}_o$ in the network in Fig. P8.70.

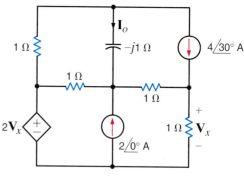

Figure P8.70

8.71 Use superposition to find $\mathbf{V}_o$ in the network in Fig. P8.71.

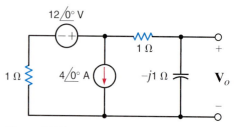

Figure P8.71

8.72 Using superposition, find $\mathbf{V}_o$ in the circuit in Fig. P8.72.

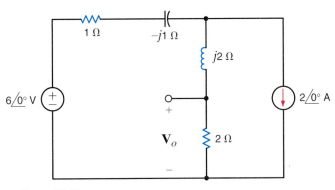

Figure P8.72

8.73 Find $\mathbf{V}_o$ in the network in Fig. P8.73 using superposition.
CS

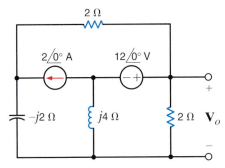

Figure P8.73

8.74 Use both superposition and MATLAB to determine $\mathbf{V}_o$ in the circuit in Fig. P8.74.

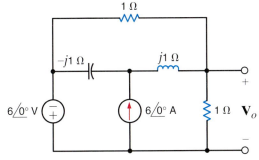

Figure P8.74

8.75 Find $\mathbf{V}_o$ in the network in Fig. P8.75 using superposition.
PSV

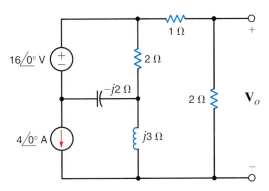

Figure P8.75

8.76 Use source exchange to determine $\mathbf{V}_o$ in the network in Fig. P8.76.

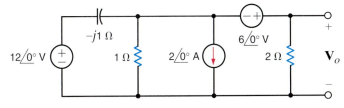

Figure P8.76

8.77 Use source exchange to find the current $\mathbf{I}_o$ in the network in Fig. P8.77. **CS**

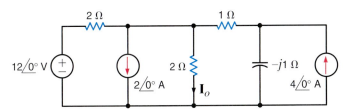

Figure P8.77

8.78 Use source transformation to determine $\mathbf{I}_o$ in the network in Fig. P8.50.

8.79 Use source transformation to determine $\mathbf{I}_o$ in the network in Fig. P8.51.

8.80 Use source transformation to determine $\mathbf{V}_o$ in the network in Fig. P8.76.

8.81 Using Thévenin's theorem, find $\mathbf{V}_o$ in the network in Fig. P8.63.

8.82 Use Thévenin's theorem to find $\mathbf{V}_o$ in the circuit in Fig. P8.64. **CS**

8.83 Solve Problem 8.52 using Thévenin's theorem.

8.84 Apply Thévenin's theorem twice to find $\mathbf{V}_o$ in the circuit in Fig. P8.84.

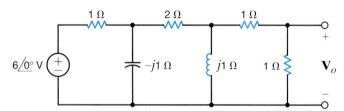

Figure P8.84

8.85 Solve Problem 8.49 using Thévenin's theorem.

8.86 Use Thévenin's theorem to find $\mathbf{V}_o$ in the network in Fig. P8.86.

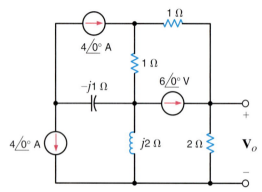

Figure P8.86

8.87 Find $\mathbf{V}_o$ in the network in Fig. P8.87 using Thévenin's theorem.

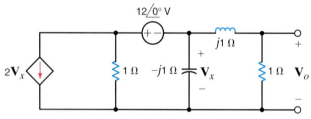

Figure P8.87

8.88 Find the Thévenin's equivalent for the network in Fig. P8.88 at the terminals *A-B*. **CS**

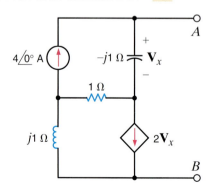

Figure P8.88

8.89 Given the network in Fig. P8.89, find the Thévenin's equivalent of the network at the terminals *A-B*.

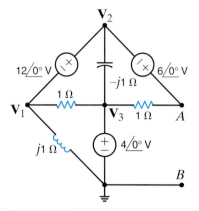

Figure P8.89

8.90 Find $\mathbf{V}_x$ in the circuit in Fig. P8.90 using Norton's theorem.

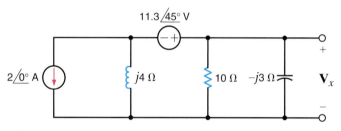

Figure P8.90

8.91 Find $\mathbf{I}_o$ in the network in Fig. P8.91 using Norton's theorem.

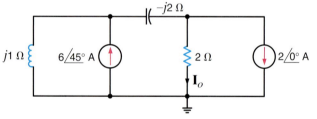

Figure P8.91

8.92 Apply both Norton's theorem and MATLAB to find $\mathbf{V}_o$ in the network in Fig. P8.92.

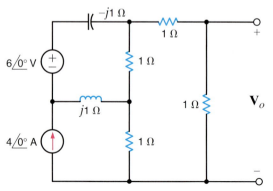

Figure P8.92

8.93 Find $\mathbf{V}_o$ using Norton's theorem for the circuit in Fig. P8.93. **CS**

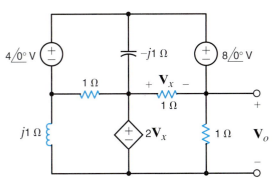

Figure P8.93

8.94 Use Norton's theorem to find $\mathbf{V}_o$ in the network in Fig. P8.94. **PSV**

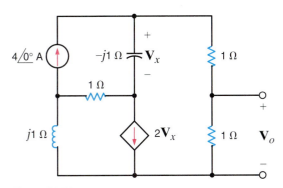

Figure P8.94

8.95 Use MATLAB to find the node voltages in the network in Fig. P8.95.

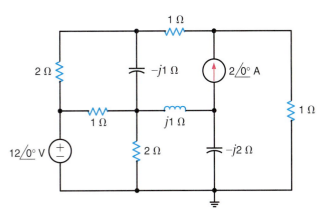

Figure P8.95

8.96 Use MATLAB to find $\mathbf{I}_o$ in the network in Fig. P8.96.

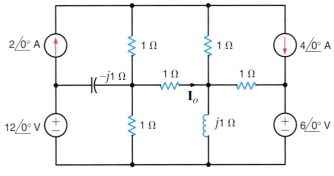

Figure P8.96

8.97 Find $\mathbf{V}_o$ in the circuit in Fig. P8.97 using MATLAB.

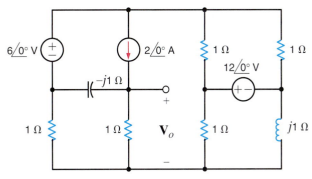

Figure P8.97

8.98 Determine $\mathbf{V}_o$ in the network in Fig. P8.98 using MATLAB.

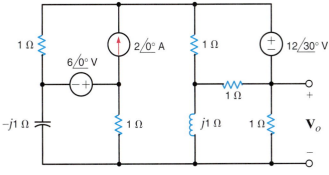

Figure P8.98

8.99 Find $\mathbf{I}_o$ in the network in Fig. P8.99 using MATLAB.

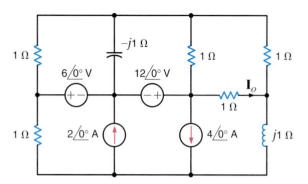

Figure P8.99

8.100 Use MATLAB to determine $\mathbf{I}_o$ in the network in Fig. P8.100.

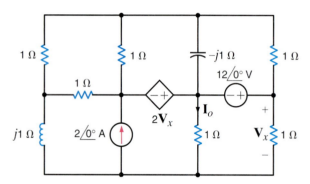

Figure P8.100

8.101 Find $\mathbf{I}_o$ in the circuit in Fig. P8.101 using MATLAB.

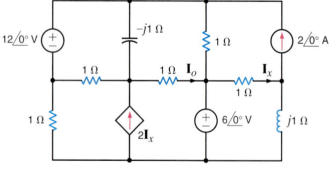

Figure P8.101

8.102 Use MATLAB to find $\mathbf{I}_o$ in the network in Fig. P8.102.

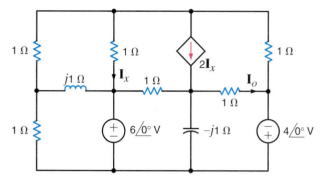

Figure P8.102

8.103 Use both a nodal analysis and a loop analysis, each in conjunction with MATLAB, to find $\mathbf{I}_o$ in the network in Fig. P8.103.

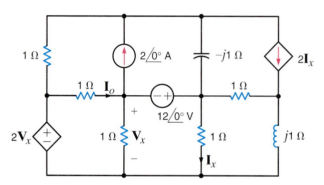

Figure P8.103

8.104 Use Thévenin's theorem, in conjunction with MATLAB, to determine $\mathbf{I}_o$ in the network in Fig. P8.104.

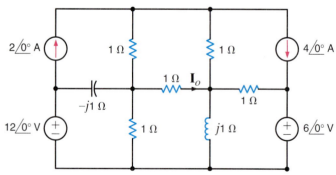

Figure P8.104

SECTION 8.9

8.105 Using the PSPICE *Schematics* editor, draw the circuit in Fig. P8.105. At what frequency are the magnitudes of $i_C(t)$ and $i_L(t)$ equal?

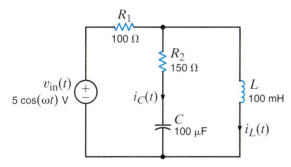

Figure P8.105

8.106 Using the PSPICE *Schematics* editor, draw the circuit in Fig. P8.106. At what frequency are the phases of $i_1(t)$ and $v_x(t)$ equal?

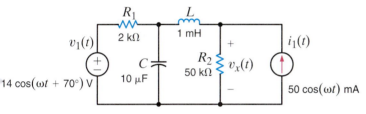

Figure P8.106

8.107 Solve Problem 8.48 using PSPICE.

8.108 Solve Problem 8.54 using PSPICE.

8.109 Solve Problem 8.70 using PSPICE.

8.110 Physical inductors are essentially coils of "long" pieces of wire, usually copper with a very thin enamel coating for insulation. Since copper has some resistivity, inductors have some resistance. In most inductors, the coils touch each other. As we learned earlier, conductors in close proximity have capacitance between them. Since the enamel insulation is so thin, the capacitance in an inductor is larger than you would expect for coils of plastic coated wire. Thus, one practical electrical model of a specific inductor is shown in Fig. 8.110. Develop an equation for the inductor's impedance and determine the frequency at which the impedance is real.

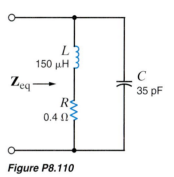

Figure P8.110

8.111 We have available a sinusoidal voltage, $v_1(t) = 10 \cos\left[2\pi(10^3 t)\right]$ V. We are asked to design a circuit that will produce a second voltage that has the same magnitude as $v_1(t)$ but leads it by 60°. A good starting point for the design is an *RC* voltage divider.

TYPICAL PROBLEMS FOUND ON THE FE EXAM

8FE-1 Find $\mathbf{V}_o$ in the network in Fig. 8PFE-1. **cs**

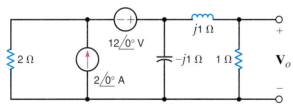

Figure 8PFE-1

8FE-2 Find $\mathbf{V}_o$ in the circuit in Fig. 8PFE-2.

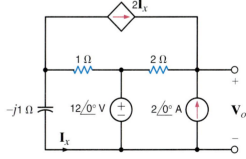

Figure 8PFE-2

8FE-3 Find $\mathbf{V}_o$ in the network in Fig. 8PFE-3. **CS**

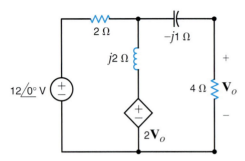

Figure 8PFE-3

8FE-4 Determine the midband (where the coupling capacitors can be ignored) gain of the single-stage transistor amplifier shown in Fig. 8PFE-4.

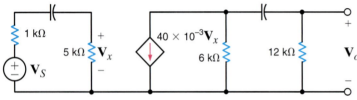

Figure 8PFE-4

Steady-State Power Analysis

9

D**o you understand the electrical wiring circuit in your** home that is used to power all the appliances and lights? Do you know the hazards associated with working with electrical power circuits? These are just two of the topics that will be examined in this chapter.

In the preceding chapters we have been concerned primarily with determining the voltage or current at some point in a network. Of equal importance to us in many situations is the power that is supplied or absorbed by some element. Typically, electrical and electronic devices have peak power or maximum instantaneous power ratings that cannot be exceeded without damaging the devices.

In electrical and electronic systems, power comes in all sizes. The power absorbed by some device on an integrated circuit chip may be in picowatts, whereas the power supplied by a large generating station may be in gigawatts. Note that the range between these two examples is phenomenally large (10^{21}).

In our previous work we defined instantaneous power to be the product of voltage and current. Average power, obtained by averaging the instantaneous power, is the average rate at which energy is absorbed or supplied. In the dc case, where both current and voltage are constant, the instantaneous power is equal to the average power. However, as we will demonstrate, this is not the case when the currents and voltages are sinusoidal functions of time.

In this chapter we explore the many ramifications of power in ac circuits. We examine instantaneous power, average power, maximum power transfer, average power for periodic nonsinusoidal waveforms, the power factor, and complex power.

Some very practical safety considerations will be introduced and discussed through a number of examples.

Finally, a wide variety of application-oriented examples is presented. The origin of many of these examples is the common household wiring circuit. ●

LEARNING GOALS

9.1 *Instantaneous Power*...*Page 420*

9.2 *Average Power* Average power is derived by averaging the instantaneous power over a period. The average power is defined as one-half the product of the maximum values of the voltage and current waveforms multiplied by cos θ, where θ is the phase angle between the voltage and current waveforms...*Page 421*

9.3 *Maximum Average Power Transfer* Maximum average power transfer to a load is achieved when the loading impedance is the complex conjugate of the source impedance...*Page 426*

9.4 *Effective or rms Values* The rms value of a periodic current is the dc value that as a current would deliver the same average power to a load resistor...*Page 430*

9.5 *The Power Factor* The power factor angle is the phase angle of the load impedance. Since the power factor is the cosine of this angle, the power factor is 1 for purely resistive loads and 0 for purely reactive loads...*Page 433*

9.6 *Complex Power* Complex power is defined as the product of the phasor voltage and the complex conjugate of the phasor current. Complex power **S** can be expressed as **S** = P + jQ, where P is the average power, Q is the reactive power, and the phase angle is the power factor angle...*Page 435*

9.7 *Power Factor Correction* The power factor of a load can be adjusted by connecting capacitors in parallel with the load...*Page 440*

9.8 *Single-Phase Three-Wire Circuits* The ac household circuit is typically a single-phase three-wire system that supplies 120 V rms to lights and small appliances and 240 V rms to large appliances...*Page 443*

9.9 *Safety Considerations* Safety is of the utmost importance in electric circuits, and readers are continuously cautioned to avoid any contact with them unless it is under the direction of a licensed professional...*Page 446*

Access Problem-Solving Videos **PSV** *and Circuit Solutions* **CS** *at:* **http://www.justask4u.com/irwin** using the registration code on the inside cover and see a website with answers and more!

9.1 Instantaneous Power

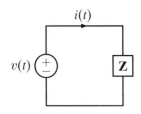

Figure 9.1
Simple ac network.

By employing the sign convention adopted in the earlier chapters, we can compute the instantaneous power supplied or absorbed by any device as the product of the instantaneous voltage across the device and the instantaneous current through it.

Consider the circuit shown in Fig. 9.1. In general, the steady-state voltage and current for the network can be written as

$$v(t) = V_M \cos(\omega t + \theta_v) \qquad \text{9.1}$$

$$i(t) = I_M \cos(\omega t + \theta_i) \qquad \text{9.2}$$

The instantaneous power is then

$$
\begin{aligned}
p(t) &= v(t)i(t) \\
&= V_M I_M \cos(\omega t + \theta_v) \cos(\omega t + \theta_i) \qquad \text{9.3}
\end{aligned}
$$

Employing the following trigonometric identity,

$$\cos\phi_1 \cos\phi_2 = \frac{1}{2}\left[\cos(\phi_1 - \phi_2) + \cos(\phi_1 + \phi_2)\right] \qquad \text{9.4}$$

we find that the instantaneous power can be written as

$$p(t) = \frac{V_M I_M}{2}\left[\cos(\theta_v - \theta_i) + \cos(2\omega t + \theta_v + \theta_i)\right] \qquad \text{9.5}$$

Note that the instantaneous power consists of two terms. The first term is a constant (i.e., it is time independent), and the second term is a cosine wave of twice the excitation frequency. We will examine this equation in more detail in Section 9.2.

HINT

Note that $p(t)$ contains a dc term and a cosine wave with twice the frequency of $v(t)$ and $i(t)$.

Example 9.1

The circuit in Fig. 9.1 has the following parameters: $v(t) = 4\cos(\omega t + 60°)$ V and $\mathbf{Z} = 2\,\underline{/30°}\ \Omega$. We wish to determine equations for the current and the instantaneous power as a function of time and plot these functions with the voltage on a single graph for comparison.

SOLUTION Since

$$\mathbf{I} = \frac{4\,\underline{/60°}}{2\,\underline{/30°}}$$

$$= 2\,\underline{/30°}\ \text{A}$$

then

$$i(t) = 2\cos(\omega t + 30°)\ \text{A}$$

From Eq. (9.5),

$$p(t) = 4[\cos(30°) + \cos(2\omega t + 90°)]$$

$$= 3.46 + 4\cos(2\omega t + 90°)\ \text{W}$$

A plot of this function, together with plots of the voltage and current, is shown in Fig. 9.2. As can be seen in this figure, the instantaneous power has a dc or constant term and a second term whose frequency is twice that of the voltage or current.

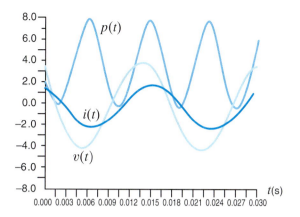

Figure 9.2

Plots of v(t), i(t), and p(t) for the circuit in Example 9.1 using f = 60 Hz.

9.2   Average Power

The average value of any periodic waveform (e.g., a sinusoidal function) can be computed by integrating the function over a complete period and dividing this result by the period. Therefore, if the voltage and current are given by Eqs. (9.1) and (9.2), respectively, the average power is

$$P = \frac{1}{T} \int_{t_0}^{t_0+T} p(t)\, dt$$

$$= \frac{1}{T} \int_{t_0}^{t_0+T} V_M I_M \cos(\omega t + \theta_v) \cos(\omega t + \theta_i)\, dt \qquad \textbf{9.6}$$

where t_0 is arbitrary, $T = 2\pi/\omega$ is the period of the voltage or current, and P is measured in watts. Actually, we may average the waveform over any integral number of periods so that Eq. (9.6) can also be written as

$$P = \frac{1}{nT} \int_{t_0}^{t_0+nT} V_M I_M \cos(\omega t + \theta_v) \cos(\omega t + \theta_i)\, dt \qquad \textbf{9.7}$$

where n is a positive integer.

Employing Eq. (9.5) for the expression in (9.6), we obtain

$$P = \frac{1}{T} \int_{t_0}^{t_0+T} \frac{V_M I_M}{2} \left[\cos(\theta_v - \theta_i) + \cos(2\omega t + \theta_v + \theta_i) \right] dt \qquad \textbf{9.8}$$

We could, of course, plod through the indicated integration; however, with a little forethought we can determine the result by inspection. The first term is independent of t, and therefore a constant in the integration. Integrating the constant over the period and dividing by the period simply results in the original constant. The second term is a cosine wave. It is well known that the average value of a cosine wave over one complete period or an integral number of periods is zero, and therefore the second term in Eq. (9.8) vanishes. In view of this discussion, Eq. (9.8) reduces to

$$P = \frac{1}{2} V_M I_M \cos(\theta_v - \theta_i) \qquad \textbf{9.9}$$

HINT

A frequently used equation for calculating the average power.

Note that since $\cos(-\theta) = \cos(\theta)$, the argument for the cosine function can be either $\theta_v - \theta_i$ or $\theta_i - \theta_v$. In addition, note that $\theta_v - \theta_i$ is the angle of the circuit impedance as shown in Fig. 9.1. Therefore, *for a purely resistive circuit,*

$$P = \frac{1}{2} V_M I_M \qquad \textbf{9.10}$$

and *for a purely reactive circuit,*

$$P = \frac{1}{2} V_M I_M \cos(90°)$$
$$= 0$$

Because purely reactive impedances absorb no average power, they are often called *lossless elements*. The purely reactive network operates in a mode in which it stores energy over one part of the period and releases it over another.

Example 9.2

We wish to determine the average power absorbed by the impedance shown in Fig. 9.3.

Figure 9.3
Example RL circuit.

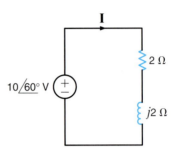

SOLUTION   From the figure we note that

$$\mathbf{I} = \frac{\mathbf{V}}{\mathbf{Z}} = \frac{V_M \,\underline{/\theta_v}}{2 + j2} = \frac{10\,\underline{/60°}}{2.83\,\underline{/45°}} = 3.53\,\underline{/15°}\ \text{A}$$

Therefore,

$$I_M = 3.53\ \text{A} \quad \text{and} \quad \theta_i = 15°$$

Hence,

$$P = \frac{1}{2} V_M I_M \cos(\theta_v - \theta_i)$$

$$= \frac{1}{2}(10)(3.53)\cos(60° - 15°)$$

$$= 12.5\ \text{W}$$

Since the inductor absorbs no power, we can employ Eq. (9.10) provided that V_M in that equation is the voltage across the resistor. Using voltage division, we obtain

$$\mathbf{V}_R = \frac{(10\,\underline{/60°})(2)}{2 + j2} = 7.07\,\underline{/15°}\ \text{V}$$

and therefore,

$$P = \frac{1}{2}(7.07)(3.53)$$

$$= 12.5\ \text{W}$$

In addition, using Ohm's law, we could also employ the expressions

$$P = \frac{1}{2}\frac{V_M^2}{R}$$

or

$$P = \frac{1}{2} I_M^2 R$$

where once again we must be careful that the V_M and I_M in these equations refer to the voltage across the resistor and the current through it, respectively.

Example 9.3

For the circuit shown in Fig. 9.4, we wish to determine both the total average power absorbed and the total average power supplied.

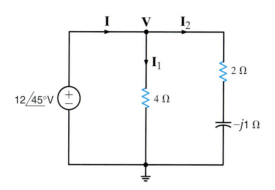

Figure 9.4
Example circuit for illustrating a power balance.

SOLUTION   From the figure we note that

$$\mathbf{I}_1 = \frac{12\,\underline{/45^\circ}}{4} = 3\,\underline{/45^\circ}\ \text{A}$$

$$\mathbf{I}_2 = \frac{12\,\underline{/45^\circ}}{2 - j1} = \frac{12\,\underline{/45^\circ}}{2.24\,\underline{/-26.57^\circ}} = 5.36\,\underline{/71.57^\circ}\ \text{A}$$

and therefore,

$$\mathbf{I} = \mathbf{I}_1 + \mathbf{I}_2$$
$$= 3\,\underline{/45^\circ} + 5.36\,\underline{/71.57^\circ}$$
$$= 8.15\,\underline{/62.10^\circ}\ \text{A}$$

The average power absorbed in the 4-Ω resistor is

$$P_4 = \frac{1}{2}V_M I_M = \frac{1}{2}(12)(3) = 18\ \text{W}$$

The average power absorbed in the 2-Ω resistor is

$$P_2 = \frac{1}{2}I_M^2 R = \frac{1}{2}(5.34)^2(2) = 28.7\ \text{W}$$

Therefore, the total average power absorbed is

$$P_A = 18 + 28.7 = 46.7\ \text{W}$$

Note that we could have calculated the power absorbed in the 2-Ω resistor using $1/2\,V_M^2/R$ if we had first calculated the voltage across the 2-Ω resistor.

The total average power supplied by the source is

$$P_S = \frac{1}{2}V_M I_M \cos(\theta_v - \theta_i)$$

$$= \frac{1}{2}(12)(8.15)\cos(45^\circ - 62.10^\circ)$$

$$= 46.7\ \text{W}$$

Thus, the total average power supplied is, of course, equal to the total average power absorbed.

E9.1 Find the average power absorbed by each resistor in the network in Fig. E9.1.

ANSWER: $P_{2\Omega} = 7.20$ W; $P_{4\Omega} = 7.20$ W.

Figure E9.1

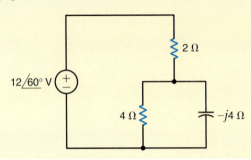

E9.2 Given the network in Fig. E9.2, find the average power absorbed by each passive circuit element and the total average power supplied by the current source.

ANSWER: $P_{3\Omega} = 56.60$ W; $P_{4\Omega} = 33.96$ W; $P_L = 0$; $P_{CS} = 90.50$ W.

Figure E9.2

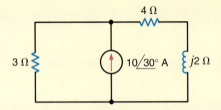

HINT

Superposition is not applicable to power. Why?

When determining average power, if more than one source is present in a network, we can use any of our network analysis techniques to find the necessary voltage and/or current to compute the power. However, we must remember that in general we cannot apply superposition to power.

Example 9.4

HINT

Under the following condition

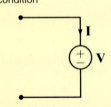

if $P = \mathbf{IV}$ is positive, power is being absorbed.
If $P = \mathbf{IV}$ is negative, power is being generated.

Consider the network shown in Fig. 9.5. We wish to determine the total average power absorbed and supplied by each element.

Figure 9.5
Example RL circuit with two sources.

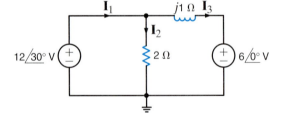

SOLUTION From the figure we note that

$$\mathbf{I}_2 = \frac{12\,\underline{/30^\circ}}{2} = 6\,\underline{/30^\circ} \text{ A}$$

and

$$\mathbf{I}_3 = \frac{12\,\underline{/30^\circ} - 6\,\underline{/0^\circ}}{j1} = \frac{4.39 + j6}{j1} = 7.44\,\underline{/-36.21^\circ} \text{ A}$$

The power absorbed by the 2-Ω resistor is

$$P_2 = \frac{1}{2}V_M I_M = \frac{1}{2}(12)(6) = 36 \text{ W}$$

According to the direction of $\mathbf{I}_3$, the $6\underline{/0°}$-V source is absorbing power. The power it absorbs is given by

$$P_{6\underline{/0°}} = \frac{1}{2} V_M I_M \cos(\theta_v - \theta_i)$$

$$= \frac{1}{2}(6)(7.44)\cos[0° - (-36.21°)]$$

$$= 18 \text{ W}$$

At this point an obvious question arises: How do we know whether the $6\underline{/0°}$-V source is supplying power to the remainder of the network or absorbing it? The answer to this question is actually straightforward. If we employ our passive sign convention that was adopted in the earlier chapters—that is, if the current reference direction enters the positive terminal of the source and the answer is positive—the source is absorbing power. If the answer is negative, the source is supplying power to the remainder of the circuit. A generator sign convention could have been used, and under this condition the interpretation of the sign of the answer would be reversed. Note that once the sign convention is adopted and used, the sign for average power will be negative only if the angle difference is greater than 90° (i.e., $|\theta_v - \theta_i| > 90°$).

To obtain the power supplied to the network, we compute $\mathbf{I}_1$ as

$$\mathbf{I}_1 = \mathbf{I}_2 + \mathbf{I}_3$$

$$= 6\underline{/30°} + 7.44\underline{/-36.21°}$$

$$= 11.29\underline{/-7.10°} \text{ A}$$

Therefore, the power supplied by the $12\underline{/30°}$-V source using the generator sign convention is

$$P_S = \frac{1}{2}(12)(11.29)\cos(30° + 7.10°)$$

$$= 54 \text{ W}$$

and hence the power absorbed is equal to the power supplied.

LEARNING EXTENSIONS

E9.3 Determine the total average power absorbed and supplied by each element in the network in Fig. E9.3.

Figure E9.3

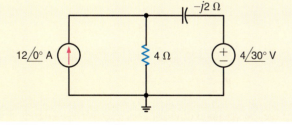

ANSWER:
$P_{CS} = -69.4$ W;
$P_{VS} = 19.8$ W;
$P_{4\Omega} = 49.6$ W; $P_C = 0$.

E9.4 Given the network in Fig. E9.4, determine the total average power absorbed or supplied by each element.

Figure E9.4

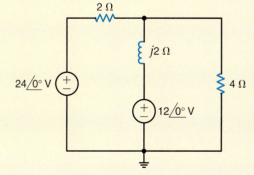

ANSWER:
$P_{24\underline{/0°}} = -55.4$ W;
$P_{12\underline{/0°}} = 5.5$ W;
$P_{2\Omega} = 22.2$ W;
$P_{4\Omega} = 27.7$ W; $P_L = 0$.

9.3 Maximum Average Power Transfer

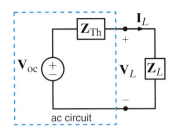

Figure 9.6
Circuit used to examine maximum average power transfer.

HINT

This impedance-matching concept is an important issue in the design of high-speed computer chips and motherboards. For today's high-speed chips with internal clocks running at about 1 GHz and mother-boards with a bus speed above 100 MHz, impedance matching is necessary in order to obtain the required speed for signal propagation. Although this high-speed transmission line is based on a distributed circuit (discussed later in electrical engineering courses), the impedance-matching technique for the transmission line is the same as that of the lumped parameter circuit for maximum average power transfer.

In our study of resistive networks, we addressed the problem of maximum power transfer to a resistive load. We showed that if the network excluding the load was represented by a Thévenin equivalent circuit, maximum power transfer would result if the value of the load resistor was equal to the Thévenin equivalent resistance $\left(\text{i.e., } R_L = R_{\text{Th}}\right)$. We will now reexamine this issue within the present context to determine the load impedance for the network shown in Fig. 9.6 that will result in maximum average power being absorbed by the load impedance $\mathbf{Z}_L$.

The equation for average power at the load is

$$P_L = \frac{1}{2} V_L I_L \cos\left(\theta_{v_L} - \theta_{i_L}\right) \qquad \textbf{9.11}$$

The phasor current and voltage at the load are given by the expressions

$$\mathbf{I}_L = \frac{\mathbf{V}_{oc}}{\mathbf{Z}_{\text{Th}} + \mathbf{Z}_L} \qquad \textbf{9.12}$$

$$\mathbf{V}_L = \frac{\mathbf{V}_{oc}\mathbf{Z}_L}{\mathbf{Z}_{\text{Th}} + \mathbf{Z}_L} \qquad \textbf{9.13}$$

where

$$\mathbf{Z}_{\text{Th}} = R_{\text{Th}} + jX_{\text{Th}} \qquad \textbf{9.14}$$

and

$$\mathbf{Z}_L = R_L + jX_L \qquad \textbf{9.15}$$

The magnitude of the phasor current and voltage are given by the expressions

$$I_L = \frac{V_{oc}}{\left[\left(R_{\text{Th}} + R_L\right)^2 + \left(X_{\text{Th}} + X_L\right)^2\right]^{1/2}} \qquad \textbf{9.16}$$

$$V_L = \frac{V_{oc}\left(R_L^2 + X_L^2\right)^{1/2}}{\left[\left(R_{\text{Th}} + R_L\right)^2 + \left(X_{\text{Th}} + X_L\right)^2\right]^{1/2}} \qquad \textbf{9.17}$$

The phase angles for the phasor current and voltage are contained in the quantity $\left(\theta_{v_L} - \theta_{i_L}\right)$. Note also that $\theta_{v_L} - \theta_{i_L} = \theta_{\mathbf{Z}_L}$ and, in addition,

$$\cos\theta_{\mathbf{Z}_L} = \frac{R_L}{\left(R_L^2 + X_L^2\right)^{1/2}} \qquad \textbf{9.18}$$

Substituting Eqs. (9.16) to (9.18) into Eq. (9.11) yields

$$P_L = \frac{1}{2}\frac{V_{oc}^2 R_L}{\left(R_{\text{Th}} + R_L\right)^2 + \left(X_{\text{Th}} + X_L\right)^2} \qquad \textbf{9.19}$$

which could, of course, be obtained directly from Eq. (9.16) using $P_L = \frac{1}{2}I_L^2 R_L$. Once again, a little forethought will save us some work. From the standpoint of maximizing P_L, V_{oc} is a constant. The quantity $\left(X_{\text{Th}} + X_L\right)$ absorbs no power, and therefore any nonzero value of this quantity only serves to reduce P_L. Hence, we can eliminate this term by selecting $X_L = -X_{\text{Th}}$. Our problem then reduces to maximizing

$$P_L = \frac{1}{2}\frac{V_{oc}^2 R_L}{\left(R_L + R_{\text{Th}}\right)^2} \qquad \textbf{9.20}$$

However, this is the same quantity we maximized in the purely resistive case by selecting $R_L = R_{\text{Th}}$. Therefore, for maximum average power transfer to the load shown in Fig. 9.6, $\mathbf{Z}_L$ should be chosen so that

$$\mathbf{Z}_L = R_L + jX_L = R_{\text{Th}} - jX_{\text{Th}} = \mathbf{Z}_{\text{Th}}^* \qquad \textbf{9.21}$$

Finally, if the load impedance is purely resistive (i.e., $X_L = 0$), the condition for maximum average power transfer can be derived via the expression

$$\frac{dP_L}{dR_L} = 0$$

where P_L is the expression in Eq. (9.19) with $X_L = 0$. *The value of R_L that maximizes P_L under the condition $X_L = 0$ is*

$$R_L = \sqrt{R_{\text{Th}}^2 + X_{\text{Th}}^2}$$ **9.22**

PROBLEM-SOLVING STRATEGY

Maximum Average Power Transfer

Step 1. Remove the load $\mathbf{Z}_L$ and find the Thévenin equivalent for the remainder of the circuit.

Step 2. Construct the circuit shown in Fig. 9.6.

Step 3. Select $\mathbf{Z}_L = \mathbf{Z}_{\text{Th}}^* = R_{\text{Th}} - jX_{\text{Th}}$, and then $\mathbf{I}_L = \mathbf{V}_{\text{oc}}/2\,R_{\text{Th}}$ and the maximum

average power transfer $= \dfrac{1}{2}\mathbf{I}_L^2 R_{\text{Th}} = \mathbf{V}_{\text{oc}}^2/8\,R_{\text{Th}}$.

Example 9.5

Given the circuit in Fig. 9.7a, we wish to find the value of $\mathbf{Z}_L$ for maximum average power transfer. In addition, we wish to find the value of the maximum average power delivered to the load.

SOLUTION   To solve the problem, we form a Thévenin equivalent at the load. The circuit in Fig. 9.7b is used to compute the open-circuit voltage

$$\mathbf{V}_{\text{oc}} = \frac{4\,\underline{/0^\circ}\,(2)}{6 + j1}\,(4) = 5.26\,\underline{/-9.46^\circ}\ \text{V}$$

The Thévenin equivalent impedance can be derived from the circuit in Fig. 9.7c. As shown in the figure,

$$\mathbf{Z}_{\text{Th}} = \frac{4(2 + j1)}{6 + j1} = 1.41 + j0.43\ \Omega$$

Therefore, $\mathbf{Z}_L$ for maximum average power transfer is

$$\mathbf{Z}_L = 1.41 - j0.43\ \Omega$$

With $\mathbf{Z}_L$ as given previously, the current in the load is

$$\mathbf{I} = \frac{5.26\,\underline{/-9.46^\circ}}{2.82} = 1.87\,\underline{/-9.46^\circ}\ \text{A}$$

Therefore, the maximum average power transferred to the load is

$$P_L = \frac{1}{2}I_M^2 R_L = \frac{1}{2}(1.87)^2(1.41) = 2.47\ \text{W}$$

HINT

In this Thévenin analysis,

1. Remove $\mathbf{Z}_L$ and find the voltage across the open terminals, $\mathbf{V}_{\text{oc}}$.

2. Determine the impedance $\mathbf{Z}_{\text{Th}}$ at the open terminals with all independent sources made zero.

3. Construct the following circuit and determine I and P_L.

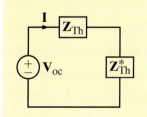

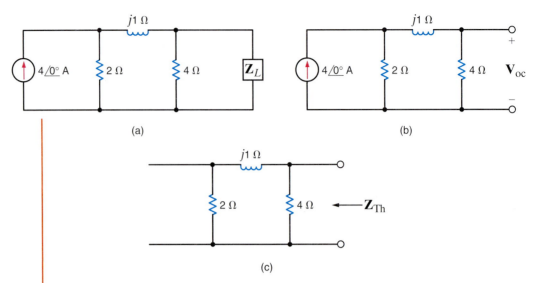

Figure 9.7 *Circuits for illustrating maximum average power transfer.*

Example 9.6

For the circuit shown in Fig. 9.8a, we wish to find the value of $\mathbf{Z}_L$ for maximum average power transfer. In addition, let us determine the value of the maximum average power delivered to the load.

HINT

When there is a dependent source, both $\mathbf{V}_{oc}$ and $\mathbf{I}_{sc}$ must be found and $\mathbf{Z}_{Th}$ computed from the equation

$$\mathbf{Z}_{Th} = \frac{\mathbf{V}_{oc}}{\mathbf{I}_{sc}}.$$

SOLUTION   We will first reduce the circuit, with the exception of the load, to a Thévenin equivalent circuit. The open-circuit voltage can be computed from Fig. 9.8b. The equations for the circuit are

$$\mathbf{V}'_x + 4 = (2 + j4)\mathbf{I}_1$$
$$\mathbf{V}'_x = -2\mathbf{I}_1$$

Solving for $\mathbf{I}_1$, we obtain

$$\mathbf{I}_1 = \frac{1\underline{/-45°}}{\sqrt{2}}$$

The open-circuit voltage is then

$$\mathbf{V}_{oc} = 2\mathbf{I}_1 - 4\underline{/0°}$$
$$= \sqrt{2}\underline{/-45°} - 4\underline{/0°}$$
$$= -3 - j1$$
$$= +3.16\underline{/-161.57°} \text{ V}$$

The short-circuit current can be derived from Fig. 9.8c. The equations for this circuit are

$$\mathbf{V}''_x + 4 = (2 + j4)\mathbf{I} - 2\mathbf{I}_{sc}$$
$$-4 = -2\mathbf{I} + (2 - j2)\mathbf{I}_{sc}$$
$$\mathbf{V}''_x = -2(\mathbf{I} - \mathbf{I}_{sc})$$

Solving these equations for $\mathbf{I}_{sc}$ yields

$$\mathbf{I}_{sc} = -(1 + j2) \text{ A}$$

The Thévenin equivalent impedance is then

$$\mathbf{Z}_{Th} = \frac{\mathbf{V}_{oc}}{\mathbf{I}_{sc}} = \frac{3 + j1}{1 + j2} = 1 - j1 \text{ } \Omega$$

Therefore, for maximum average power transfer the load impedance should be

$$\mathbf{Z}_L = 1 + j1 \; \Omega$$

The current in this load $\mathbf{Z}_L$ is then

$$\mathbf{I}_L = \frac{\mathbf{V}_{oc}}{\mathbf{Z}_{Th} + \mathbf{Z}_L} = \frac{-3 - j1}{2} = 1.58 \underline{/-161.57^\circ} \; \text{A}$$

Hence, the maximum average power transferred to the load is

$$P_L = \frac{1}{2}(1.58)^2(1)$$

$$= 1.25 \; \text{W}$$

Figure 9.8
Circuits for illustrating maximum average power transfer.

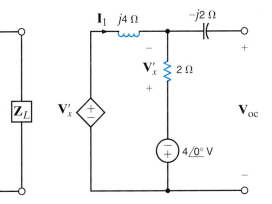

(a)

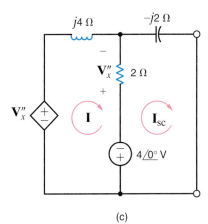

(b)

(c)

LEARNING EXTENSIONS

E9.5 Given the network in Fig. E9.5, find $\mathbf{Z}_L$ for maximum average power transfer and the maximum average power transferred to the load. **PSV**

ANSWER: $\mathbf{Z}_L = 1 + j1 \; \Omega$; $P_L = 45 \; \text{W}.$

Figure E9.5

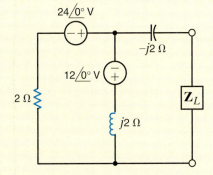

E9.6 Find $\mathbf{Z}_L$ for maximum average power transfer and the maximum average power transferred to the load in the network in Fig. E9.6.

ANSWER: $\mathbf{Z}_L = 2 - j2 \; \Omega$; $P_L = 45 \; \text{W}.$

Figure E9.6

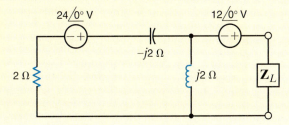

9.4 Effective or rms Values

In the preceding sections of this chapter, we have shown that the average power absorbed by a resistive load is directly dependent on the type, or types, of sources that are delivering power to the load. For example, if the source was dc, the average power absorbed was I^2R, and if the source was sinusoidal, the average power was $1/2\, I_M^2 R$. Although these two types of waveforms are extremely important, they are by no means the only waveforms we will encounter in circuit analysis. Therefore, a technique by which we can compare the *effectiveness* of different sources in delivering power to a resistive load would be quite useful.

To accomplish this comparison, we define what is called the *effective value of a periodic waveform*, representing either voltage or current. Although either quantity could be used, we will employ current in the definition. Hence, we define the effective value of a periodic current as a constant or dc value, which as current would deliver the same average power to a resistor R. Let us call the constant current I_{eff}. Then the average power delivered to a resistor as a result of this current is

$$P = I_{\text{eff}}^2 R$$

Similarly, the average power delivered to a resistor by a periodic current $i(t)$ is

$$P = \frac{1}{T} \int_{t_0}^{t_0+T} i^2(t)R\, dt$$

Equating these two expressions, we find that

$$I_{\text{eff}} = \sqrt{\frac{1}{T} \int_{t_0}^{t_0+T} i^2(t)\, dt} \qquad \textbf{9.23}$$

Note that this effective value is found by first determining the *square* of the current, then computing the average or *mean* value, and finally taking the square *root*. Thus, in "reading" the mathematical Eq. (9.23), we are determining the root mean square, which we abbreviate as rms, and therefore I_{eff} is called I_{rms}.

Since dc is a constant, the rms value of dc is simply the constant value. Let us now determine the rms value of other waveforms. The most important waveform is the sinusoid, and therefore, we address this particular one in the following example.

Example 9.7

We wish to compute the rms value of the waveform $i(t) = I_M \cos(\omega t - \theta)$, which has a period of $T = 2\pi/\omega$.

SOLUTION Substituting these expressions into Eq. (9.23) yields

$$I_{\text{rms}} = \left[\frac{1}{T} \int_0^T I_M^2 \cos^2(\omega t - \theta)\, dt \right]^{1/2}$$

Using the trigonometric identity

$$\cos^2\phi = \frac{1}{2} + \frac{1}{2}\cos 2\phi$$

we find that the preceding equation can be expressed as

$$I_{\text{rms}} = I_M \left\{ \frac{\omega}{2\pi} \int_0^{2\pi/\omega} \left[\frac{1}{2} + \frac{1}{2}\cos(2\omega t - 2\theta) \right] dt \right\}^{1/2}$$

Since we know that the average or mean value of a cosine wave is zero,

$$I_{\text{rms}} = I_M \left(\frac{\omega}{2\pi} \int_0^{2\pi/\omega} \frac{1}{2}\, dt \right)^{1/2}$$

$$= I_M \left[\frac{\omega}{2\pi} \left(\frac{t}{2} \right) \Big|_0^{2\pi/\omega} \right]^{1/2} = \frac{I_M}{\sqrt{2}} \qquad \textbf{9.24}$$

Therefore, the rms value of a sinusoid is equal to the maximum value divided by the $\sqrt{2}$. Hence, a sinusoidal current with a maximum value of I_M delivers the same average power to a resistor R as a dc current with a value of $I_M/\sqrt{2}$. Recall that earlier a phasor $\mathbf{X}$ was defined as $X_M\underline{/\theta}$ for a sinusoidal wave of the form $X_M\cos(\omega t + \theta)$. This phasor can also be represented as $X_M/\sqrt{2}\underline{/\theta}$ if the units are given in rms. For example, $120\underline{/30°}$ V rms is equivalent to $170\underline{/30°}$ V.

On using the rms values for voltage and current, the average power can be written, in general, as

$$P = V_{\text{rms}}I_{\text{rms}}\cos(\theta_v - \theta_i)$$ **9.25**

The power absorbed by a resistor R is

$$P = I_{\text{rms}}^2 R = \frac{V_{\text{rms}}^2}{R}$$ **9.26**

In dealing with voltages and currents in numerous electrical applications, it is important to know whether the values quoted are maximum, average, rms, or what. We are familiar with the 120-V ac electrical outlets in our home. In this case, the 120 V is the rms value of the voltage in our home. The maximum or peak value of this voltage is $120\sqrt{2} = 170$ V. The voltage at our electrical outlets could be written as $170\cos 377t$ V. The maximum or peak value must be given if we write the voltage in this form. There should be no question in our minds that this is the peak value. It is common practice to specify the voltage rating of ac electrical devices in terms of the rms voltage. For example, if you examine an incandescent light bulb, you will see a voltage rating of 120 V, which is the rms value. For now we will add an rms to our voltages and currents to indicate that we are using rms values in our calculations.

Example 9.8

We wish to compute the rms value of the voltage waveform shown in Fig. 9.9.

SOLUTION   The waveform is periodic with period $T = 3$ s. The equation for the voltage in the time frame $0 \le t \le 3$ s is

$$v(t) = \begin{bmatrix} 4t\text{ V} & 0 < t \le 1\text{ s} \\ 0\text{ V} & 1 < t \le 2\text{ s} \\ -4t + 8\text{ V} & 2 < t \le 3\text{ s} \end{bmatrix}$$

The rms value is

$$V_{\text{rms}} = \left\{ \frac{1}{3}\left[\int_0^1 (4t)^2\,dt + \int_1^2 (0)^2\,dt + \int_2^3 (8 - 4t)^2\,dt \right] \right\}^{1/2}$$

$$= \left[\frac{1}{3}\left(\frac{16t^3}{3}\Big|_0^1 + \left(64t - \frac{64t^2}{2} + \frac{16t^3}{3}\right)\Big|_2^3 \right) \right]^{1/2}$$

$$= 1.89\text{ V}$$

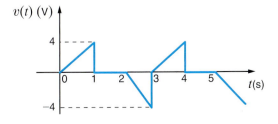

Figure 9.9
Waveform used to illustrate rms values.

Example 9.9

Determine the rms value of the current waveform in Fig. 9.10 and use this value to compute the average power delivered to a 2-Ω resistor through which this current is flowing.

SOLUTION The current waveform is periodic with a period of $T = 4$ s. The rms value is

$$I_{\text{rms}} = \left\{ \frac{1}{4} \left[\int_0^2 (4)^2 \, dt + \int_2^4 (-4)^2 \, dt \right] \right\}^{1/2}$$

$$= \left[\frac{1}{4} \left(16t \Big|_0^2 + 16t \Big|_2^4 \right) \right]^{1/2}$$

$$= 4 \text{ A}$$

The average power delivered to a 2-Ω resistor with this current is

$$P = I_{\text{rms}}^2 R = (4)^2 (2) = 32 \text{ W}$$

Figure 9.10
Waveform used to illustrate rms values.

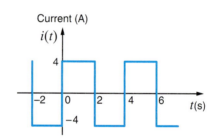

LEARNING EXTENSIONS

E9.7 Compute the rms value of the voltage waveform shown in Fig. E9.7.

ANSWER: $V_{\text{rms}} = 1.633$ V.

Figure E9.7

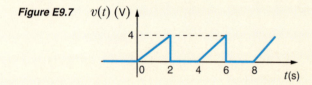

E9.8 The current waveform in Fig. E9.8 is flowing through a 4-Ω resistor. Compute the average power delivered to the resistor.

ANSWER: $P = 32$ W.

Figure E9.8

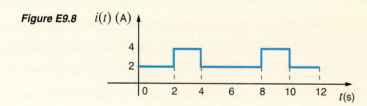

E9.9 The current waveform in Fig. E9.9 is flowing through a 10-Ω resistor. Determine the average power delivered to the resistor.

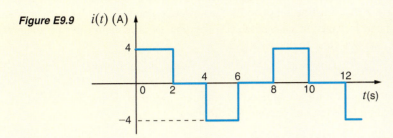

Figure E9.9

9.5   The Power Factor

The power factor is a very important quantity. Its importance stems in part from the economic impact it has on industrial users of large amounts of power. In this section we carefully define this term and then illustrate its significance via some practical examples.

In Section 9.4 we showed that a load operating in the ac steady state is delivered an average power of

$$P = V_{rms}I_{rms}\cos(\theta_v - \theta_i)$$

We will now further define the terms in this important equation. The product $V_{rms}I_{rms}$ is referred to as the *apparent power*. Although the term $\cos(\theta_v - \theta_i)$ is a dimensionless quantity, and the units of P are watts, apparent power is normally stated in volt-amperes (VA) or kilovolt-amperes (kVA) to distinguish it from average power.

We now define the *power factor* (pf) as the ratio of the average power to the apparent power; that is,

$$\text{pf} = \frac{P}{V_{rms}I_{rms}} = \cos(\theta_v - \theta_i) \qquad \textbf{9.27}$$

where

$$\cos(\theta_v - \theta_i) = \cos\theta_{\mathbf{Z}_L} \qquad \textbf{9.28}$$

The angle $\theta_v - \theta_i = \theta_{\mathbf{Z}_L}$ is the phase angle of the load impedance and is often referred to as the *power factor angle*. The two extreme positions for this angle correspond to a purely resistive load where $\theta_{\mathbf{Z}_L} = 0$ and the pf is 1, and the purely reactive load where $\theta_{\mathbf{Z}_L} = \pm 90°$ and the pf is 0. It is, of course, possible to have a unity pf for a load containing R, L, and C elements if the values of the circuit elements are such that a zero phase angle is obtained at the particular operating frequency.

There is, of course, a whole range of power factor angles between $\pm 90°$ and $0°$. If the load is an equivalent RC combination, then the pf angle lies between the limits $-90° < \theta_{\mathbf{Z}_L} < 0°$. On the other hand, if the load is an equivalent RL combination, then the pf angle lies between the limits $0 < \theta_{\mathbf{Z}_L} < 90°$. Obviously, confusion in identifying the type of load could result, due to the fact that $\cos\theta_{\mathbf{Z}_L} = \cos(-\theta_{\mathbf{Z}_L})$. To circumvent this problem, the pf is said to be either *leading* or *lagging*, where these two terms *refer to the phase of the current with respect to the voltage*. Since the current leads the voltage in an RC load, the load has a leading pf. In a similar manner, an RL load has a lagging pf; therefore, load impedances of $\mathbf{Z}_L = 1 - j1\ \Omega$ and $\mathbf{Z}_L = 2 + j1\ \Omega$ have power factors of $\cos(-45°) = 0.707$ leading and $\cos(26.57°) = 0.894$ lagging, respectively.

HINT

Technique

1. Given P_L, pf, and V_{rms}, determine I_{rms}.
2. Then $P_S = P_L + I^2_{rms} R_{line}$, where R_{line} is the line resistance.

Example 9.10

An industrial load consumes 88 kW at a pf of 0.707 lagging from a 480-V rms line. The transmission line resistance from the power company's transformer to the plant is 0.08 Ω. Let us determine the power that must be supplied by the power company (a) under present conditions and (b) if the pf is somehow changed to 0.90 lagging. (It is economically advantageous to have a power factor as close to one as possible.)

SOLUTION

a. The equivalent circuit for these conditions is shown in Fig. 9.11. Using Eq. (9.27), we obtain the magnitude of the rms current into the plant:

$$I_{rms} = \frac{P_L}{(\text{pf})(V_{rms})}$$

$$= \frac{(88)(10^3)}{(0.707)(480)}$$

$$= 259.3 \text{ A rms}$$

The power that must be supplied by the power company is

$$P_S = P_L + (0.08)I^2_{rms}$$

$$= 88,000 + (0.08)(259.3)^2$$

$$= 93.38 \text{ kW}$$

b. Suppose now that the pf is somehow changed to 0.90 lagging but the voltage remains constant at 480 V. The rms load current for this condition is

$$I_{rms} = \frac{P_L}{(\text{pf})(V_{rms})}$$

$$= \frac{(88)(10^3)}{(0.90)(480)}$$

$$= 203.7 \text{ A rms}$$

Under these conditions the power company must generate

$$P_S = P_L + (0.08)I^2_{rms}$$

$$= 88,000 + (0.08)(203.7)^2$$

$$= 91.32 \text{ kW}$$

Note carefully the difference between the two cases. A simple change in the pf of the load from 0.707 lagging to 0.90 lagging has had an interesting effect. Note that in the first case the power company must generate 93.38 kW in order to supply the plant with 88 kW of power because the low power factor means that the line losses will be high—5.38 kW. However, in the second case the power company need only generate 91.32 kW in order to supply the plant with its required power, and the corresponding line losses are only 3.32 kW.

Figure 9.11
Example circuit for examining changes in power factor.

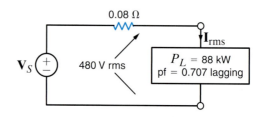

This example clearly indicates the economic impact of the load's power factor. The cost of producing electricity for a large electric utility can easily be in the billions of dollars. A low power factor at the load means that the utility generators must be capable of carrying more current at constant voltage, and they must also supply power for higher $I_{rms}^2 R_{line}$ losses than would be required if the load's power factor were high. Since line losses represent energy expended in heat and benefit no one, the utility will insist that a plant maintain a high pf, typically 0.9 lagging, and adjust the rate it charges a customer that does not conform to this requirement. We will demonstrate a simple and economical technique for achieving this power factor correction in a future section.

LEARNING EXTENSION

E9.10 An industrial load consumes 100 kW at 0.707 pf lagging. The 60-Hz line voltage at the load is $480\,\underline{/0°}$ V rms. The transmission-line resistance between the power company transformer and the load is 0.1 Ω. Determine the power savings that could be obtained if the pf is changed to 0.94 lagging. **PSV**

ANSWER: Power saved is 3.771 kW.

9.6 Complex Power

In our study of ac steady-state power, it is convenient to introduce another quantity, which is commonly called *complex power*. To develop the relationship between this quantity and others we have presented in the preceding sections, consider the circuit shown in Fig. 9.12.

The complex power is defined to be

$$\mathbf{S} = \mathbf{V}_{rms}\mathbf{I}_{rms}^* \qquad \textbf{9.29}$$

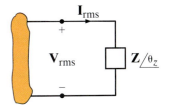

Figure 9.12
Circuit used to explain power relationships.

where $\mathbf{I}_{rms}^*$ refers to the complex conjugate of $\mathbf{I}_{rms}$; that is, if $\mathbf{I}_{rms} = I_{rms}\underline{/\theta_i} = I_R + jI_I$, then $\mathbf{I}_{rms}^* = I_{rms}\underline{/-\theta_i} = I_R - jI_I$. Complex power is then

$$\mathbf{S} = V_{rms}\underline{/\theta_v}\, I_{rms}\underline{/-\theta_i} = V_{rms}I_{rms}\underline{/\theta_v - \theta_i} \qquad \textbf{9.30}$$

or

$$\mathbf{S} = V_{rms}I_{rms}\cos(\theta_v - \theta_i) + jV_{rms}I_{rms}\sin(\theta_v - \theta_i) \qquad \textbf{9.31}$$

where, of course, $\theta_v - \theta_i = \theta_\mathbf{Z}$. We note from Eq. (9.31) that the real part of the complex power is simply the *real* or *average* power. The imaginary part of $\mathbf{S}$ we call the *reactive* or *quadrature power*. Therefore, complex power can be expressed in the form

$$\mathbf{S} = P + jQ \qquad \textbf{9.32}$$

where

$$P = \mathrm{Re}(\mathbf{S}) = V_{rms}I_{rms}\cos(\theta_v - \theta_i) \qquad \textbf{9.33}$$

$$Q = \mathrm{Im}(\mathbf{S}) = V_{rms}I_{rms}\sin(\theta_v - \theta_i) \qquad \textbf{9.34}$$

As shown in Eq. (9.31), the magnitude of the complex power is what we have called the *apparent power*, and the phase angle for complex power is simply the power factor angle. Complex power, like apparent power, is measured in volt-amperes, real power is measured in watts, and to distinguish Q from the other quantities, which in fact have the same dimensions, it is measured in volt-amperes reactive, or var.

Now let's examine the expressions in Eqs. (9.33) and (9.34) in more detail for our three basic circuit elements: R, L, C. For a resistor, $\theta_v - \theta_i = 0°$, $\cos(\theta_v - \theta_i) = 1$, and $\sin(\theta_v - \theta_i) = 0$. As a result, a resistor absorbs real power ($P > 0$) but does not absorb any reactive power ($Q = 0$). For an inductor, $\theta_v - \theta_i = 90°$ and

$$P = V_{\text{rms}} I_{\text{rms}} \cos(90°) = 0$$

$$Q = V_{\text{rms}} I_{\text{rms}} \sin(90°) > 0$$

An inductor absorbs reactive power but does not absorb real power. Repeating for a capacitor, we get $\theta_v - \theta_i = -90°$ and

$$P = V_{\text{rms}} I_{\text{rms}} \cos(-90°) = 0$$

$$Q = V_{\text{rms}} I_{\text{rms}} \sin(-90°) < 0$$

A capacitor does not absorb any real power; however, the reactive power is now negative. How do we interpret the negative reactive power? Referring back to Fig. 9.12, note that the voltage and current are specified such that they satisfy the passive sign convention. In this case, the product of the voltage and current gives us the power absorbed by the impedance in that figure. If the reactive power absorbed by the capacitor is negative, then the capacitor must be supplying reactive power. The fact that capacitors are a source of reactive power will be utilized in the next section on power factor correction.

We see that resistors absorb only real power, while inductors and capacitors absorb only reactive power. What is a fundamental difference between these elements? Resistors only absorb energy. On the other hand, capacitors and inductors store energy and then release it back to the circuit. Since inductors and capacitors absorb only reactive power and not real power, we can conclude that reactive power is related to energy storage in these elements.

Now let's substitute $\mathbf{V}_{\text{rms}} = \mathbf{I}_{\text{rms}}{}^*\mathbf{Z}$ into Eq. (9.29). Multiplying $\mathbf{I}_{\text{rms}} * \mathbf{I}_{\text{rms}}^* = I_{\text{rms}} \underline{/\theta_i} * I_{\text{rms}} \underline{/-\theta_i}$ yields I_{rms}^2. The complex power absorbed by an impedance can be obtained by multiplying the square of the rms magnitude of the current flowing through that impedance by the impedance.

$$\mathbf{S} = \mathbf{V}_{\text{rms}}\mathbf{I}_{\text{rms}}^* = (\mathbf{I}_{\text{rms}}\mathbf{Z})\mathbf{I}_{\text{rms}}^* = \mathbf{I}_{\text{rms}}\mathbf{I}_{\text{rms}}^*\mathbf{Z} = I_{\text{rms}}^2\mathbf{Z} = I_{\text{rms}}^2(R + jX) = P + jQ \quad \textbf{9.35}$$

Instead of substituting for $\mathbf{V}_{\text{rms}}$ in Eq. (9.29), let's substitute for $\mathbf{I}_{\text{rms}}$.

$$\mathbf{S} = \mathbf{V}_{\text{rms}}\mathbf{I}_{\text{rms}}^* = \mathbf{V}_{\text{rms}}\left(\frac{\mathbf{V}_{\text{rms}}}{\mathbf{Z}}\right)^* = \frac{V_{\text{rms}}^2}{\mathbf{Z}^*} = V_{\text{rms}}^2\mathbf{Y}^* = V_{\text{rms}}^2(G + jB)^* = P + jQ \quad \textbf{9.36}$$

This expression tells us that we can calculate the complex power absorbed by an admittance by multiplying the square of the rms magnitude of the voltage across the admittance by the conjugate of the admittance. Suppose the box in Fig. 9.12 contains a capacitor. The admittance for a capacitor is $j\omega C$. Plugging into the equation above yields

$$\mathbf{S} = V_{\text{rms}}^2(j\omega C)^* = -j\omega C V_{\text{rms}}^2 \quad \textbf{9.37}$$

Note the negative sign on the complex power. This agrees with our previous statement that a capacitor does not absorb real power but is a source of reactive power.

The diagrams in Fig. 9.13 further explain the relationships among the various quantities of power. As shown in Fig. 9.13a, the phasor current can be split into two components: one that is in phase with $\mathbf{V}_{\text{rms}}$ and one that is 90° out of phase with $\mathbf{V}_{\text{rms}}$. Equations (9.33) and (9.34)

Figure 9.13

Diagram for illustrating power relationships.

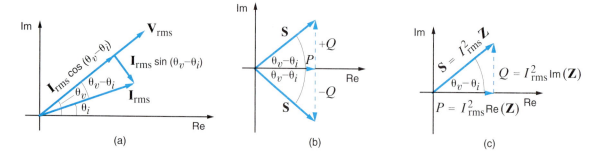

(a)   (b)   (c)

illustrate that the in-phase component produces the real power, and the 90° component, called the *quadrature component*, produces the reactive or quadrature power. In addition, Eqs. (9.33) and (9.34) indicate that

$$\tan(\theta_v - \theta_i) = \frac{Q}{P} \qquad \textbf{9.38}$$

which relates the pf angle to P and Q in what is called the *power triangle*.

The relationships among **S**, P, and Q can be expressed via the diagrams shown in Figs. 9.13b and c. In Fig. 9.13b we note the following conditions. If Q is positive, the load is inductive, the power factor is lagging, and the complex number **S** lies in the first quadrant. If Q is negative, the load is capacitive, the power factor is leading, and the complex number **S** lies in the fourth quadrant. If Q is zero, the load is resistive, the power factor is unity, and the complex number **S** lies along the positive real axis. Figure 9.13c illustrates the relationships expressed by Eqs. (9.35) to (9.37) for an inductive load.

Finally, it is important to state that complex power, like energy, is conserved; that is, the total complex power delivered to any number of individual loads is equal to the sum of the complex powers delivered to each individual load, regardless of how the loads are interconnected.

PROBLEM-SOLVING STRATEGY

Determining P or S

If $v(t)$ and $i(t)$ are known and we wish to find P given an impedance $\mathbf{Z}\underline{/\theta} = R + jX$, two viable approaches are as follows:

Step 1. Determine **V** and **I** and then calculate

$$P = V_{\text{rms}} I_{\text{rms}} \cos\theta \quad \text{or} \quad P = V_{\text{rms}} I_{\text{rms}} \cos(\theta_v - \theta_i)$$

Step 2. Use **I** to calculate the real part of **S**; that is,

$$P = R_e(\mathbf{S}) = I^2 R$$

The latter method may be easier to calculate than the former. However, if the imaginary part of the impedance, X, is not zero, then

$$P \neq \frac{V^2}{R}$$

which is a common mistake. Furthermore, the P and Q portions of **S** are directly related to $Z\underline{/\theta}$ and provide a convenient way in which to relate power, current, and impedance. That is,

$$\tan\theta = \frac{Q}{P}$$

$$\mathbf{S} = I^2\mathbf{Z}$$

The following example illustrates the usefulness of **S**.

Example 9.11

A load operates at 20 kW, 0.8 pf lagging. The load voltage is $220\underline{/0°}$ V rms at 60 Hz. The impedance of the line is $0.09 + j0.3\ \Omega$. We wish to determine the voltage and power factor at the input to the line.

SOLUTION The circuit diagram for this problem is shown in Fig. 9.14. As illustrated in Fig. 9.13,

$$S = \frac{P}{\cos\theta} = \frac{P}{\text{pf}} = \frac{20{,}000}{0.8} = 25{,}000 \text{ VA}$$

Therefore, at the load

$$\mathbf{S}_L = 25{,}000 \underline{/\theta} = 25{,}000 \underline{/36.87°} = 20{,}000 + j15{,}000 \text{ VA}$$

Since $\mathbf{S}_L = \mathbf{V}_L \mathbf{I}_L^*$

$$\mathbf{I}_L = \left[\frac{25{,}000 \underline{/36.87°}}{220 \underline{/0°}} \right]^*$$

$$= 113.64 \underline{/-36.87°} \text{ A rms}$$

The complex power losses in the line are

$$\mathbf{S}_{\text{line}} = I_L^2 \mathbf{Z}_{\text{line}}$$

$$= (113.64)^2 (0.09 + j0.3)$$

$$= 1162.26 + j3874.21 \text{ VA}$$

As stated earlier, complex power is conserved, and, therefore, the complex power at the generator is

$$\mathbf{S}_S = \mathbf{S}_L + \mathbf{S}_{\text{line}}$$

$$= 21{,}162.26 + j18{,}874.21$$

$$= 28{,}356.25 \underline{/41.73°} \text{ VA}$$

Hence, the generator voltage is

$$V_S = \frac{|\mathbf{S}_S|}{I_L} = \frac{28{,}356.25}{113.64}$$

$$= 249.53 \text{ V rms}$$

and the generator power factor is

$$\cos(41.73°) = 0.75 \text{ lagging}$$

We could have solved this problem using KVL. For example, we calculated the load current as

$$\mathbf{I}_L = 113.64 \underline{/-36.87°} \text{ A rms}$$

Hence, the voltage drop in the transmission line is

$$\mathbf{V}_{\text{line}} = (113.64 \underline{/-36.87°})(0.09 + j0.3)$$

$$= 35.59 \underline{/36.43°} \text{ V rms}$$

Therefore, the generator voltage is

$$\mathbf{V}_S = 220 \underline{/0°} + 35.59 \underline{/36.43°}$$

$$= 249.53 \underline{/4.86°} \text{ V rms}$$

Hence, the generator voltage is 249.53 V rms. In addition,

$$\theta_v - \theta_i = 4.86° - (-36.87°) = 41.73°$$

and therefore,

$$\text{pf} = \cos(41.73°) = 0.75 \text{ lagging}$$

HINT

1. Use the given P_L, $\cos\theta$, and V_L rms to obtain $\mathbf{S}_L$ and $\mathbf{I}_L$ based on Eqs. (9.33) and (9.29), respectively.

2. Use $\mathbf{I}_L$ and $\mathbf{Z}_{\text{line}}$ to obtain $\mathbf{S}_{\text{line}}$ using Eq. (9.35).

3. Use $\mathbf{S}_S = \mathbf{S}_{\text{line}} + \mathbf{S}_L$.

4. $\mathbf{V}_S = \mathbf{S}_S / \mathbf{I}_L^*$ yields V_S and θ_v. Since $\mathbf{V}_S = V_S \underline{/\theta_v}$ and θ_i is the phase of $\mathbf{I}_L$, $\text{pf} = \cos(\theta_v - \theta_i)$.

Figure 9.14
Example circuit for power analysis.

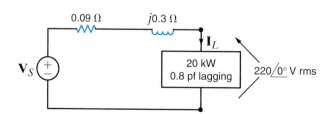

Example 9.12

Two networks A and B are connected by two conductors having a net impedance of $\mathbf{Z} = 0 + j1\ \Omega$, as shown in Fig. 9.15. The voltages at the terminals of the networks are $\mathbf{V}_A = 120\ \underline{/30°}$ V rms and $\mathbf{V}_B = 120\ \underline{/0°}$ V rms. We wish to determine the average power flow between the networks and identify which is the source and which is the load.

SOLUTION   As shown in Fig. 9.15,

$$\mathbf{I} = \frac{\mathbf{V}_A - \mathbf{V}_B}{\mathbf{Z}}$$

$$= \frac{120\ \underline{/30°} - 120\ \underline{/0°}}{j1}$$

$$= 62.12\ \underline{/15°}\ \text{A rms}$$

The power delivered by network A is

$$P_A = |\mathbf{V}_A|\,|\mathbf{I}|\cos\left(\theta_{\mathbf{V}_A} - \theta_{\mathbf{I}}\right)$$

$$= (120)(62.12)\cos(30° - 15°)$$

$$= 7200.4\ \text{W}$$

The power absorbed by network B is

$$P_B = |\mathbf{V}_B|\,|\mathbf{I}|\cos\left(\theta_{\mathbf{V}_B} - \theta_{\mathbf{I}}\right)$$

$$= (120)(62.12)\cos(0° - 15°)$$

$$= 7200.4\ \text{W}$$

If the power flow had actually been from network B to network A, the resultant signs on P_A and P_B would have been negative.

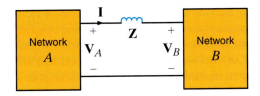

Figure 9.15
Network used in Example 9.12.

LEARNING EXTENSIONS

E9.11 An industrial load requires 40 kW at 0.84 pf lagging. The load voltage is $220\ \underline{/0°}$ V rms at 60 Hz. The transmission-line impedance is $0.1 + j0.25\ \Omega$. Determine the real and reactive power losses in the line and the real and reactive power required at the input to the transmission line.

ANSWER:
$P_{\text{line}} = 4.685\ \text{kW}$;
$Q_{\text{line}} = 11.713\ \text{kvar}$;
$P_S = 44.685\ \text{kW}$;
$Q_S = 37.55\ \text{kvar}$.

E9.12 A load requires 60 kW at 0.85 pf lagging. The 60-Hz line voltage at the load is $220\ \underline{/0°}$ V rms. If the transmission-line impedance is $0.12 + j0.18\ \Omega$, determine the line voltage and power factor at the input.   **PSV**

ANSWER:
$\mathbf{V}_{\text{in}} = 284.6\ \underline{/5.8°}$ V rms;
$\text{pf}_{\text{in}} = 0.792$ lagging.

9.7 Power Factor Correction

Industrial plants that require large amounts of power have a wide variety of loads. However, by nature the loads normally have a lagging power factor. In view of the results obtained in Example 9.10, we are naturally led to ask whether there is any convenient technique for raising the power factor of a load. Since a typical load may be a bank of induction motors or other expensive machinery, the technique for raising the pf should be an economical one to be feasible.

To answer the question we pose, consider the diagram in Fig. 9.16. A typical industrial load with a lagging pf is supplied by an electrical source. Also shown is the power triangle for the load. The load pf is $\cos(\theta_{old})$. If we want to improve the power factor, we need to reduce the angle shown on the power triangle in Fig. 9.16. From Eq. (9.38) we know that the tangent of this angle is equal to the ratio of Q to P. We could decrease the angle by increasing P. This is not an economically attractive solution because our increased power consumption would increase the monthly bill from the electric utility.

Figure 9.16
Diagram for power factor correction.

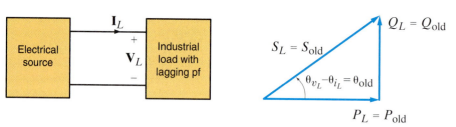

The other option we have to reduce this angle is to decrease Q. How can we decrease Q? Recall from a previous section that a capacitor is a source of reactive power and does not absorb real power. Suppose we connect a capacitor in parallel with our industrial load as shown in Fig. 9.17. The corresponding power triangles for this diagram are also shown in Fig. 9.17. Let's define

$$\mathbf{S}_{old} = P_{old} + jQ_{old} = |\mathbf{S}_{old}| \underline{/\theta_{old}} \quad \text{and} \quad \mathbf{S}_{new} = P_{old} + jQ_{new} = |\mathbf{S}_{new}| \underline{/\theta_{new}}$$

Then

$$\mathbf{S}_{new} - \mathbf{S}_{old} = \mathbf{S}_{cap} = (P_{old} + jQ_{new}) - (P_{old} + jQ_{old}) = j(Q_{new} - Q_{old}) = jQ_{cap}.$$

Equation (9.37) can be used to find the required value of C in order to achieve the new specified power factor, as illustrated in Fig. 9.17. Hence, we can obtain a particular power factor for the total load (industrial load and capacitor) simply by judiciously selecting a capacitor and placing it in parallel with the original load. In general, we want the power factor to be large, and therefore the power factor angle must be small (i.e., the larger the desired power factor, the smaller the angle $(\theta_{v_L} - \theta_{i_T})$).

Figure 9.17
Power factor correction diagram including capacitor.

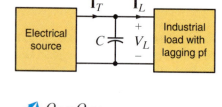

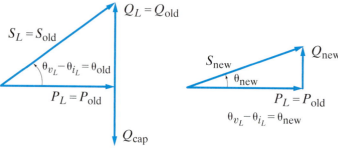

Example 9.13

Every month our electrical energy provider sends us a bill for the amount of electrical energy that we have consumed. The rate is often expressed in ¢ per kWh and consists of at least two components: (1) the demand charge, which covers the cost of lines, poles, transformers, and so on, and (2) the energy charge, which covers the cost to produce electric energy at power plants. The energy charge is the subject of the deregulation of the electric utility industry where you, as a customer, choose your energy provider.

It is common for an industrial facility operating at a poor power factor to be charged more by the electric utility providing electrical service. Let's suppose that our industrial facility operates at 277 V rms and requires 500 kW at a power factor of 0.75 lagging. Assume an energy charge of 2¢ per kWh and a demand charge of $3.50 per kW per month if the power factor is between 0.9 lagging and unity and $5 per kVA per month if the power factor is less than 0.9 lagging.

The monthly energy charge is $500 \times 24 \times 30 \times \$0.02 = \$7,200$. Let's calculate the monthly demand charge with the 0.75 lagging power factor. The complex power absorbed by the industrial facility is

$$\mathbf{S}_{old} = \frac{500}{0.75} \underline{/\cos^{-1}(0.75)} = 666.67 \underline{/41.4°} = 500 + j441 \text{ kVA}$$

The monthly demand charge is $666.67 \times \$5 = \$3,333.35$. The total bill from the energy provider is $\$7,200 + \$3,333.35 = \$10,533.35$ per month.

Let's consider installing a capacitor bank to correct the power factor to reduce our demand charge. The demand charge is such that we only need to correct the power factor to 0.9 lagging. The monthly demand charge will be the same whether the power factor is corrected to 0.9 or unity. The complex power absorbed by the industrial facility and capacitor bank will be

$$\mathbf{S}_{new} = \frac{500}{0.9} \underline{/\cos^{-1}(0.9)} = 555.6 \underline{/25.84°} = 500 + j242.2 \text{ kVA}$$

The monthly demand charge for our industrial facility with the capacitor bank is $500 \times \$3.50 = \$1,750$ per month. The average power absorbed by our capacitor bank is negligible compared to the average power absorbed by the industrial facility, so our monthly energy charge remains $7,200 per month. With the capacitor bank installed, the total bill from the energy provider is $\$7,200 + \$1,750 = \$8,950$ per month.

How many kvars of capacitance do we need to correct the power factor to 0.9 lagging?

$$\mathbf{S}_{new} - \mathbf{S}_{old} = \mathbf{S}_{cap} = (500 + j242.2) - (500 + j441) = -j198.8 \text{ kvar}$$

Let's assume that it costs $100 per kvar to install the capacitor bank at the industrial facility for an installation cost of $19,880. How long will it take to recover the cost of installing the capacitor bank? The difference in the monthly demand charge without the bank and with the bank is $\$3,333.35 - \$1,750 = \$1,583.35$. Dividing this value into the cost of installing the bank yields $\$19,880/\$1,583.35 = 12.56$ months.

Example 9.14

Plastic kayaks are manufactured using a process called rotomolding, which is diagrammed in Fig. 9.18. Molten plastic is injected into a mold, which is then spun on the long axis of the kayak until the plastic cools, resulting in a hollow one-piece craft. Suppose that the induction motors used to spin the molds consume 50 kW at a pf of 0.8 lagging from a $220\underline{/0°}$-V rms, 60-Hz line. We wish to raise the pf to 0.95 lagging by placing a bank of capacitors in parallel with the load.

Figure 9.18 *Rotomolding manufacturing process.*

SOLUTION The circuit diagram for this problem is shown in Fig. 9.19. $P_L = 50$ kW and since $\cos^{-1} 0.8 = 36.87°$, $\theta_{\text{old}} = 36.87°$. Therefore,

$$Q_{\text{old}} = P_{\text{old}} \tan \theta_{\text{old}} = (50)(10^3)(0.75) = 37.5 \text{ kvar}$$

Hence,

$$\mathbf{S}_{\text{old}} = P_{\text{old}} + jQ_{\text{old}} = 50,000 + j37,500$$

and

$$\mathbf{S}_{\text{cap}} = 0 + jQ_{\text{cap}}$$

Figure 9.19
*Example circuit
for power factor
correction.*

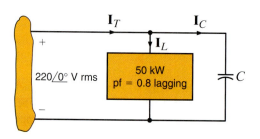

Since the required power factor is 0.95,

$$\theta_{\text{new}} = \cos^{-1}(\text{pf}_{\text{new}}) = \cos^{-1}(0.95)$$
$$= 18.19°$$

Then

$$Q_{\text{new}} = P_{\text{old}} \tan \theta_{\text{new}}$$
$$= 50,000 \tan(18.19°)$$
$$= 16,430 \text{ var}$$

Hence

$$Q_{\text{new}} - Q_{\text{old}} = Q_{\text{cap}} = -\omega CV^2 \text{ rms}$$

$$16,430 - 37,500 = -\omega CV^2 \text{ rms}$$

Solving the equation for C yields

$$C = \frac{21,070}{(377)(220)^2}$$
$$= 1155 \text{ } \mu\text{F}$$

By using a capacitor of this magnitude in parallel with the industrial load, we create, from the utility's perspective, a load pf of 0.95 lagging. However, the parameters of the actual load remain unchanged. Under these conditions, the current supplied by the utility to the kayak manufacturer is less and therefore they can use smaller conductors for the same amount of power. Or, if the conductor size is fixed, the line losses will be less since these losses are a function of the square of the current.

> ### PROBLEM-SOLVING STRATEGY
>
> ## Power Factor Correction
>
> **Step 1.** Find Q_{old} from P_L and θ_{old}, or the equivalent pf_{old}.
>
> **Step 2.** Find θ_{new} from the desired pf_{new}.
>
> **Step 3.** Determine $Q_{new} = P_{old} \tan \theta_{new}$.
>
> **Step 4.** $Q_{new} - Q_{old} = Q_{cap} = -\omega C V^2$ rms.

LEARNING EXTENSION

E9.13 Compute the value of the capacitor necessary to change the power factor in Extension Exercise E9.10 to 0.95 lagging. **PSV**

ANSWER: $C = 773 \,\mu\text{F}$.

9.8   Single-Phase Three-Wire Circuits

The single-phase three-wire ac circuit shown in Figure 9.20 is an important topic because it is the typical ac power network found in households. Note that the voltage sources are equal; that is, $\mathbf{V}_{an} = \mathbf{V}_{nb} = \mathbf{V}$. Thus, the magnitudes are equal and the phases are equal (single phase). The line-to-line voltage $\mathbf{V}_{ab} = 2\mathbf{V}_{an} = 2\mathbf{V}_{nb} = 2\mathbf{V}$. Within a household, lights and small appliances are connected from one line to *neutral n*, and large appliances such as hot water heaters and air conditioners are connected line to line. Lights operate at about 120 V rms and large appliances operate at approximately 240 V rms.

Let us now attach two identical loads to the single-phase three-wire voltage system using perfect conductors as shown in Fig. 9.20b. From the figure we note that

$$\mathbf{I}_{aA} = \frac{\mathbf{V}}{\mathbf{Z}_L}$$

and

$$\mathbf{I}_{bB} = -\frac{\mathbf{V}}{\mathbf{Z}_L}$$

KCL at point N is

$$\mathbf{I}_{nN} = -(\mathbf{I}_{aA} + \mathbf{I}_{bB})$$

$$= -\left(\frac{\mathbf{V}}{\mathbf{Z}_L} - \frac{\mathbf{V}}{\mathbf{Z}_L}\right)$$

$$= 0$$

Note that there is no current in the neutral wire, and therefore it could be removed without affecting the remainder of the system; that is, all the voltages and currents would be unchanged. One is naturally led to wonder just how far the simplicity exhibited by this system will extend. For example, what would happen if each line had a line impedance, if the neutral conductor had an impedance associated with it, and if there were a load tied from line to line? To explore these questions, consider the circuit in Fig. 9.20c. Although we could examine this circuit using many of the techniques we have employed in previous chapters, the symmetry of the network suggests that perhaps superposition may lead us to some conclusions without having to resort to a brute-force assault. Employing superposition, we consider the two circuits in Figs. 9.20d and e. The currents in Fig. 9.20d are labeled arbitrarily.

Figure 9.20

Single-phase three-wire system.

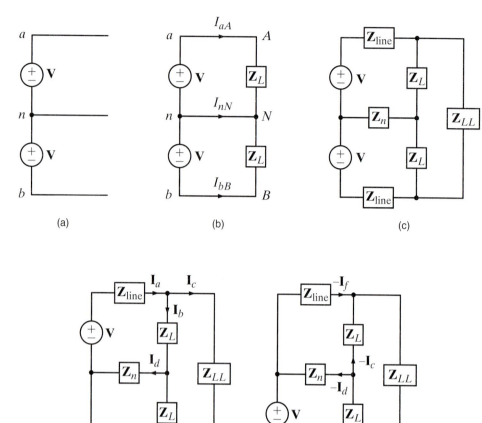

Because of the symmetrical relationship between Figs. 9.20d and e, the currents in Fig. 9.20e correspond directly to those in Fig. 9.20d. If we add the two *phasor* currents in each branch, we find that the neutral current is again zero. A neutral current of zero is a direct result of the symmetrical nature of the network. If either the line impedances $\mathbf{Z}_{\text{line}}$ or the load impedances $\mathbf{Z}_L$ are unequal, the neutral current will be nonzero. We will make direct use of these concepts when we study three-phase networks in Chapter 10.

Example 9.15

A three-wire single-phase household circuit is shown in Fig. 9.21a. The use of the lights, stereo, and range for a 24-hour period is demonstrated in Fig. 9.21b. Let us calculate the energy use over the 24 hours in kilowatt-hours. Assuming that this represents a typical day and that our utility rate is \$0.08/kWh, let us also estimate the power bill for a 30-day month.

SOLUTION Applying nodal analysis to Fig. 9.21a yields

$$\mathbf{I}_{aA} = \mathbf{I}_L + \mathbf{I}_R$$

$$\mathbf{I}_{bB} = -\mathbf{I}_S - \mathbf{I}_R$$

$$\mathbf{I}_{nN} = \mathbf{I}_S - \mathbf{I}_L$$

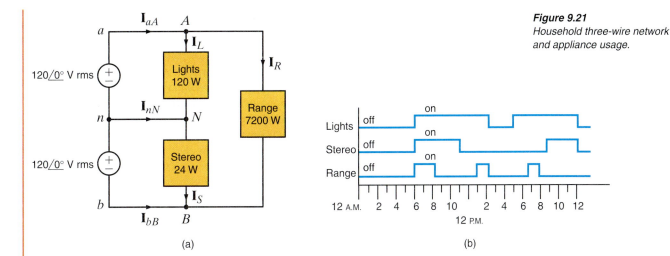

Figure 9.21
*Household three-wire network
and appliance usage.*

The current magnitudes for each load can be found from the corresponding power levels as follows.

$$I_L = \frac{P_L}{V_{an}} = \frac{120}{120} = 1 \text{ A rms}$$

$$I_S = \frac{P_S}{V_{nb}} = \frac{24}{120} = 0.2 \text{ A rms}$$

$$I_R = \frac{P_R}{V_{ab}} = \frac{7200}{240} = 30 \text{ A rms}$$

The energy used is simply the integral of the power delivered by the two sources over the 24-hour period. Since the voltage magnitudes are constants, we can express the energy delivered by the sources as

$$E_{an} = V_{an} \int I_{aA} \, dt$$

$$E_{nb} = V_{nb} \int -I_{bB} \, dt$$

The integrals of I_{aA} and I_{bB} can be determined graphically from Fig. 9.21b.

$$\int_{12\text{A.M.}}^{12\text{A.M.}} I_{aA} \, dt = 4I_R + 15I_L = 135$$

$$\int_{12\text{A.M.}}^{12\text{A.M.}} -I_{bB} \, dt = 8I_S + 4I_R = 121.6$$

Therefore, the daily energy for each source and the total energy is

$$E_{an} = 16.2 \text{ kWh}$$

$$E_{nb} = 14.6 \text{ kWh}$$

$$E_{\text{total}} = 30.8 \text{ kWh}$$

Over a 30-day month, a \$0.08/kWh utility rate results in a power bill of

$$\text{Cost} = (30.8)(30)(0.08) = \$73.92$$

9.9 Safety Considerations

Although this book is concerned primarily with the theory of circuit analysis, we recognize that most students, by this point in their study, will have begun to relate the theory to the electrical devices and systems that they encounter in the world around them. Thus, it seems advisable to depart briefly from the theoretical and spend some time discussing the very practical and important subject of safety. Electrical safety is a very broad and diverse topic that would require several volumes for a comprehensive treatment. Instead, we will limit our discussion to a few introductory concepts and illustrate them with examples.

It would be difficult to imagine that anyone in our society could have reached adolescence without having experienced some form of electrical shock. Whether that shock was from a harmless electrostatic discharge or from accidental contact with an energized electrical circuit, the response was probably the same—an immediate and involuntary muscular reaction. In either case, the cause of the reaction is current flowing through the body. The severity of the shock depends on several factors, the most important of which are the magnitude, the duration, and the pathway of the current through the body.

The effect of electrical shock varies widely from person to person. Figure 9.22 shows the general reactions that occur as a result of 60-Hz ac current flow through the body from hand to hand, with the heart in the conduction pathway. Observe that there is an intermediate range of current, from about 0.1 to 0.2 A, which is most likely to be fatal. Current levels in this range are apt to produce ventricular fibrillation, a disruption of the orderly contractions of the heart muscle. Recovery of the heartbeat generally does not occur without immediate medical intervention. Current levels above that fatal range tend to cause the heart muscle to contract severely, and if the shock is removed soon enough, the heart may resume beating on its own.

Figure 9.22

Effects of electrical shock. (From C. F. Dalziel and W. R. Lee, "Lethal Electric Currents," IEEE Spectrum, February 1969, pp. 44–50, and C. F. Dalziel, "Electric Shock Hazard," IEEE Spectrum, February 1972, pp. 41–50.)

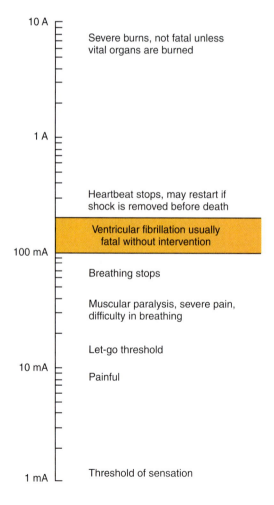

The voltage required to produce a given current depends on the quality of the contact to the body and the impedance of the body between the points of contact. The electrostatic voltage such as might be produced by sliding across a car seat on a dry winter day may be on the order of 20,000 to 40,000 V, and the current surge on touching the door handle, on the order of 40 A. However, the pathway for the current flow is mainly over the body surface, and its duration is for only a few microseconds. Although that shock could be disastrous for some electronic components, it causes nothing more than mild discomfort and aggravation to a human being.

Electrical appliances found about the home typically require 120 or 240 V rms for operation. Although the voltage level is small compared with that of the electrostatic shock, the potential for harm to the individual and to property is much greater. Accidental contact is more apt to result in current flow either from hand to hand or from hand to foot—either of which will subject the heart to shock. Moreover, the relatively slowly changing (low frequency) 60-Hz current tends to penetrate more deeply into the body as opposed to remaining on the surface as a rapidly changing (high frequency) current would tend to do. In addition, the energy source has the capability of sustaining a current flow without depletion. Thus, subsequent discussion will concentrate primarily on hazards associated with the 60-Hz ac power system.

The single-phase three-wire system introduced earlier is commonly, though not exclusively, used for electrical power distribution in residences. Two important aspects of this or any system that relate to safety were not mentioned earlier: circuit fusing and grounding.

Each branch circuit, regardless of the type of load it serves, is protected from excessive current flow by circuit breakers or fuses. Receptacle circuits are generally limited to 20 amps and lighting circuits to 15 amps. Clearly, these cannot protect persons from lethal shock. The primary purpose of these current-limiting devices is to protect equipment.

The neutral conductor of the power system is connected to ground (earth) at a multitude of points throughout the system and, in particular, at the service entrance to the residence. The connection to earth may be by way of a driven ground rod or by contact to a cold water pipe of a buried metallic water system. The 120-V branch circuits radiating from the distribution panel (fuse box) generally consist of three conductors rather than only two, as was shown in Fig. 9.20. The third conductor is the ground wire, as shown in Fig. 9.23.

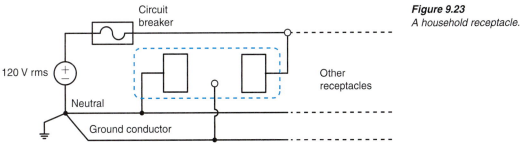

Figure 9.23
A household receptacle.

The ground conductor may appear to be redundant, since it plays no role in the normal operation of a load that might be connected to the receptacle. Its role is illustrated by the following example.

Example 9.16

Joe has a workshop in his basement where he uses a variety of power tools such as drills, saws, and sanders. The basement floor is concrete, and being below ground level, it is usually damp. Damp concrete is a relatively good conductor. Unknown to Joe, the insulation on a wire in his electric drill has been nicked and the wire is in contact with (or shorted to) the metal case of the drill, as shown in Fig. 9.24. Is Joe in any danger when using the drill?

SOLUTION Without the ground conductor connected to the metal case of the tool, Joe would receive a severe, perhaps fatal, shock when he attempts to use the drill. The voltage between his hand and his feet would be 120 V, and the current through his body would be limited by the resistance of his body and of the concrete floor. Typically, the circuit breakers would not operate. However, if the ground conductor is present and properly connected to the drill case, the case remains at ground potential, the 120-V source becomes shorted to ground, the circuit breaker operates, and Joe lives to drill another hole.

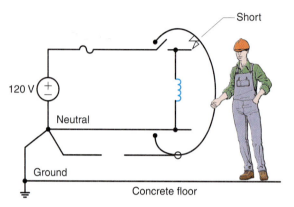

Figure 9.24 *Faulty circuit, when the case of the tool is not grounded through the power cord.*

It was mentioned earlier that the circuit breaker or fuse cannot provide effective protection against shock. There is, however, a special type of device called a ground-fault interrupter (GFI) that can provide protection for personnel. This device detects current flow outside the normal circuit. Consider the circuit of Fig. 9.24. In the normal safe operating condition, the current in the neutral conductor must be the same as that in the line conductor. If at any time the current in the line does not equal the current in the neutral, then a secondary path has somehow been established, creating an unsafe condition. This secondary path is called a fault. For example, the fault path in Fig. 9.24 is through Joe and the concrete floor. The GFI detects this fault and opens the circuit in response. Its principle of operation is illustrated by the following example.

Example 9.17

Let us describe the operation of a GFI.

SOLUTION Consider the action of the magnetic circuit in Fig. 9.25. Under normal operating conditions, i_1 and i_2 are equal, and if the coils in the neutral and line conductors are identical, as we learned in basic physics, the magnetic flux in the core will be zero. Consequently, no voltage will be induced in the sensing coil.

If a fault should occur at the load, current will flow in the ground conductor and perhaps in the earth; thus, i_1 and i_2 will no longer be equal, the magnetic flux will not be zero, and a voltage will be induced in the sensing coil. That voltage can be used to activate a circuit breaker. This is the essence of the GFI device.

Figure 9.25
Ground-fault interrupter circuit.

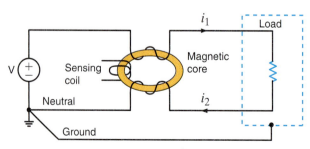

Ground-fault interrupters are available in the form of circuit breakers and also as receptacles. They are now required in branch circuits that serve outlets in areas such as bathrooms, basements, garages, and outdoor sites. The devices will operate at ground-fault currents on the order of a few milliamperes. Unfortunately, the GFI is a relatively new device and electrical code requirements are generally not retroactive. Thus few older residences have them.

Requirements for the installation and maintenance of electrical systems are meticulously defined by various codes that have been established to provide protection of personnel and property. Installation, alteration, or repair of electrical devices and systems should be undertaken only by qualified persons. The subject matter that we study in circuit analysis does not provide that qualification.

The following examples illustrate the potential hazards that can be encountered in a variety of everyday situations. We begin by revisiting a situation described in a previous example.

Example 9.18

Suppose that a man is working on the roof of a mobile home with a hand drill. It is early in the day, the man is barefoot, and dew covers the mobile home. The ground prong on the electrical plug of the drill has been removed. Will the man be shocked if the "hot" electrical line shorts to the case of the drill?

SOLUTION   To analyze this problem, we must construct a model that adequately represents the situation described. In his book *Medical Instrumentation* (Boston: Houghton Mifflin, 1978), John G. Webster suggests the following values for resistance of the human body: $R_{\text{skin}}(\text{dry}) = 15\ \text{k}\Omega$, $R_{\text{skin}}(\text{wet}) = 150\ \Omega$, $R_{\text{limb}}(\text{arm or leg}) = 100\ \Omega$, and $R_{\text{trunk}} = 200\ \Omega$.

The network model is shown in Fig. 9.26. Note that since the ground line is open-circuited, a closed path exists from the hot wire through the short, the human body, the mobile home, and the ground. For the conditions stated previously, we assume that the surface contact resistances R_{sc_1} and R_{sc_2} are 150 Ω each. The body resistance, R_{body}, consisting of arm, trunk, and leg, is 400 Ω. The mobile home resistance is assumed to be zero, and the ground resistance, R_{gnd}, from the mobile home ground to the actual source ground is assumed to be 1 Ω. Therefore, the magnitude of the current through the body from hand to foot would be

$$
\begin{aligned}
\mathbf{I}_{\text{body}} &= \frac{120}{R_{\text{sc}_1} + R_{\text{body}} + R_{\text{sc}_2} + R_{\text{gnd}}} \\
&= \frac{120}{701} \\
&= 171\ \text{mA}
\end{aligned}
$$

A current of this magnitude can easily cause heart failure.

It is important to note that additional protection would be provided if the circuit breaker were a ground-fault interrupter.

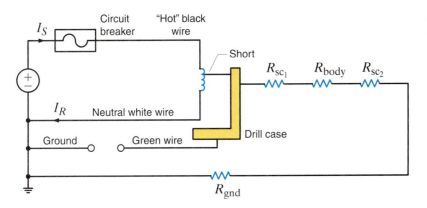

Figure 9.26
Model for Example 9.18.

Example 9.19

Two boys are playing basketball in their backyard. To cool off, they decide to jump into their pool. The pool has a vinyl lining, so the water is electrically insulated from the earth. Unknown to the boys, there is a ground fault in one of the pool lights. One boy jumps in and while standing in the pool with water up to his chest, reaches up to pull in the other boy, who is holding onto a grounded hand rail, as shown in Fig. 9.27a. What is the impact of this action?

SOLUTION The action in Fig. 9.27a is modeled as shown in Fig. 9.27b. Note that since a ground fault has occurred, there exists a current path through the two boys. Assuming that the fault, pool, and railing resistances are approximately zero, the magnitude of the current through the two boys would be

$$I = \frac{120}{\left(3R_{arm}\right) + 3\left(3R_{wet\ contact}\right) + R_{trunk}}$$

$$= \frac{120}{950}$$

$$= 126\ mA$$

This current level would cause severe shock in both boys. The boy outside the pool could experience heart failure.

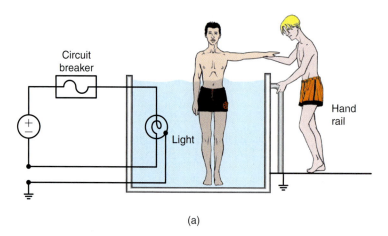

(a)

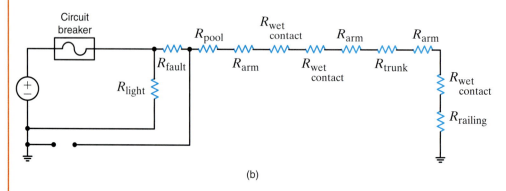

(b)

Figure 9.27 *Diagrams used in Example 9.19.*

Example 9.20

A patient in a medical laboratory has a muscle stimulator attached to her left forearm. Her heart rate is being monitored by an EKG machine with two differential electrodes over the heart and the ground electrode attached to her right ankle. This activity is illustrated in Fig. 9.28a. The stimulator acts as a current source that drives 150 mA through the muscle from the active electrode to the passive electrode. If the laboratory technician mistakenly decides to connect the passive electrode of the stimulator to the ground electrode of the EKG system to achieve a common ground, is there any risk?

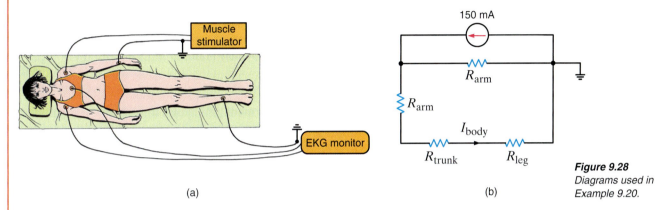

(a)   (b)

Figure 9.28
Diagrams used in
Example 9.20.

SOLUTION When the passive electrode of the stimulator is connected to the ground electrode of the EKG system, the equivalent network in Fig. 9.28b illustrates the two paths for the stimulator current: one through half an arm and the other through half an arm and the body. Using current division, the body current is

$$\mathbf{I}_{body} = \frac{(150)(10^{-3})(50)}{50 + 50 + 200 + 100}$$

$$= 19 \text{ mA}$$

Therefore, a dangerously high level of current will flow from the stimulator through the body to the EKG ground.

Example 9.21

A cardiac care patient with a pacing electrode has ignored the hospital rules and is listening to a cheap stereo. The stereo has an amplified 60-Hz hum that is very annoying. The patient decides to dismantle the stereo partially in an attempt to eliminate the hum. In the process, while he is holding one of the speaker wires, the other touches the pacing electrode. What are the risks in this situation?

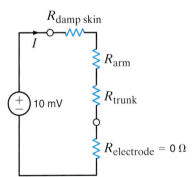

SOLUTION Let us suppose that the patient's skin is damp and that the 60-Hz voltage across the speaker wires is only 10 mV. Then the circuit model in this case would be shown in Fig. 9.29. The current through the heart would be

$$I = \frac{(10)(10^{-3})}{150 + 100 + 200}$$

$$= 22.2 \text{ μA}$$

It is known that 10 μA delivered directly to the heart is potentially lethal.

Figure 9.29
Circuit model for Example 9.21.

Example 9.22

While maneuvering in a muddy area, a crane operator accidentally touched a high-voltage line with the boom of the crane, as illustrated in Fig. 9.30a. The line potential was 7200 V. The neutral conductor was grounded at the pole. When the crane operator realized what had happened, he jumped from the crane and ran in the direction of the pole, which was approximately 10 m away. He was electrocuted as he ran. Can we explain this very tragic accident?

SOLUTION The conditions depicted in Fig. 9.30a can be modeled as shown in Fig. 9.30b. The crane was at 7200 V with respect to earth. Therefore, a gradient of 720 V/m existed along the earth between the crane and the power pole. This earth between the crane and the pole is modeled as a resistance. If the man's stride was about 1 m, the difference in potential between his feet was approximately 720 V. A man standing in the same area with his feet together was unharmed.

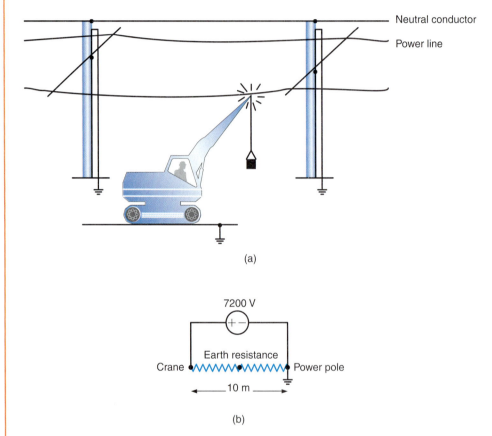

(a)

(b)

Figure 9.30 Illustrations used in Example 9.22.

The examples of this section have been provided in an attempt to illustrate some of the potential dangers that exist when working or playing around electric power. In the worst case, failure to prevent an electrical accident can result in death. However, even nonlethal electrical contacts can cause such things as burns or falls. Therefore, we must always be alert to ensure not only our own safety, but also that of others who work and play with us.

The following guidelines will help us minimize the chances of injury.

HINT
Safety guidelines

1. Avoid working on energized electrical systems.

2. Always assume that an electrical system is energized unless you can absolutely verify that it is not.

3. Never make repairs or alterations that are not in compliance with the provisions of the prevailing code.

4. Do not work on potentially hazardous electrical systems alone.

5. If another person is "frozen" to an energized electrical circuit, deenergize the circuit, if possible. If that cannot be done, use nonconductive material such as dry wooden boards, sticks, belts, and articles of clothing to separate the body from the contact. Act quickly but take care to protect yourself.

6. When handling long metallic equipment, such as ladders, antennas, and so on, outdoors, be continuously aware of overhead power lines and avoid any possibility of contact with them.

LEARNING EXTENSION

E9.14 A woman is driving her car in a violent rainstorm. While she is waiting at an intersection, a power line falls on her car and makes contact. The power line voltage is 7200 V.

(a) Assuming that the resistance of the car is negligible, what is the potential current through her body if, while holding the door handle with a dry hand, she steps out onto the wet ground?

(b) If she remained in the car, what would happen?

ANSWER: (a) $I = 463$ mA, extremely dangerous; (b) she should be safe.

Safety when working with electric power must always be a primary consideration. Regardless of how efficient or expedient an electrical network is for a particular application, it is worthless if it is also hazardous to human life.

The safety device shown in Fig. 9.31, which is also used for troubleshooting, is a proximity-type sensor that will indicate whether a circuit is energized by simply touching the conductor on the outside of the insulation. This device is typically carried by all electricians and is helpful when working on electric circuits.

Figure 9.31
A modern safety or troubleshooting device (courtesy of Fluke Corporation).

In addition to the numerous deaths that occur each year due to electrical accidents, fire damage that results from improper use of electrical wiring and distribution equipment amounts to millions of dollars per year.

To prevent the loss of life and damage to property, very detailed procedures and specifications have been established for the construction and operation of electrical systems to ensure their safe operation. The *National Electrical Code* ANSI C1 (ANSI—American National Standards Institute) is the primary guide. There are other codes, however: for example, the National Electric Safety Code, ANSI C2, which deals with safety requirements for public utilities. The Underwriters' Laboratory (UL) tests all types of devices and systems to ensure that they are safe for use by the general public. We find the UL label on all types of electrical equipment that is used in the home, such as appliances and extension cords.

Electric energy plays a central role in our lives. It is extremely important to our general health and well-being. However, if not properly used, it can be lethal.

9.10 Application Examples

The following application-oriented examples illustrate a practical use of the material studied in this chapter.

Application Example 9.23

For safety reasons the National Electrical Code restricts the circuit-breaker rating in a 120-V household lighting branch circuit to no more than 20 A. Furthermore, the code also requires a 25% safety margin for continuous-lighting loads. Under these conditions let us determine the number of 100-W lighting fixtures that can be placed in one branch circuit.

SOLUTION The model for the branch circuit is shown in Fig. 9.32. The current drawn by each 100-W bulb is

$$I_{bulb} = 100/120 = 0.833 \text{ A rms}$$

Using the safety margin recommendation, the estimated current drawn by each bulb is 25% higher or

$$I_{bulb} = (1.25)(0.83) = 1.04 \text{ A rms}$$

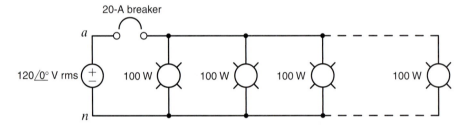

Figure 9.32 *20-A branch circuit for householding lighting.*

Therefore, the maximum number of fixtures on one circuit breaker is

$$n = 20/1.04 = 19 \text{ fixtures}$$

Application Example 9.24

An electric lawn mower requires 12 A rms at 120 V rms but will operate down to 110 V rms without damage. At 110 V rms, the current draw is 13.1 A rms, as shown in Fig. 9.33. Give the maximum length extension cord that can be used with a 120-V rms power source if the extension cord is made from

1. 16-gauge wire (4 mΩ/ft)

2. 14-gauge wire (2.5 mΩ/ft)

Figure 9.33
Circuit model in Example 9.24.

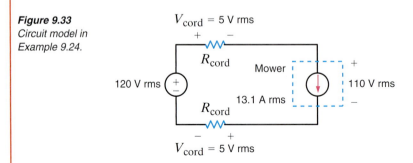

SOLUTION   The voltage drop across the extension cord is

$$V_{cord} = (2)(13.1)R_{cord} = 10 \text{ V rms}$$

or

$$R_{cord} = 0.382 \ \Omega$$

If ℓ_{cord} is the length of the extension cord, then for 16-gauge wire we find

$$\ell_{cord} = \frac{R_{cord}}{0.004} = 95.5 \text{ feet}$$

and for 14-gauge wire

$$\ell_{cord} = \frac{R_{cord}}{0.0025} = 152.8 \text{ feet}$$

Application Example 9.25

While sitting in a house reading a book, we notice that every time the air conditioner comes on, the lights momentarily dim. Let us investigate this phenomenon using the single-phase three-wire circuit shown in Fig. 9.34a and some typical current requirements for a 10,000-Btu/h air conditioner, assuming a line resistance of 0.5 Ω.

SOLUTION   The 60-W light bulb can be roughly modeled by its equivalent resistance:

$$P_{bulb} = \frac{V_{an}^2}{R_{bulb}}$$

or

$$R_{bulb} = 240 \ \Omega$$

When the air conditioner unit first turns on, the current requirement is 40 A, as shown in Fig. 9.34b. As the compressor motor comes up to speed, the current requirement drops quickly to a steady-state value of 10 A, as shown in Fig. 9.34c. We will compare the voltage across the light fixture, V_{AN}, both at turn-on and in steady state.

Using superposition, let us first find that portion of V_{AN} caused by the voltage sources. The appropriate circuit is shown in Fig. 9.34d. Using voltage division, we find that

$$V_{AN1} = V_{AN}\left(\frac{R_{bulb}}{R_{bulb} + 2R_L}\right)$$

or

$$V_{AN1} = 119.50 \text{ V rms}$$

Figure 9.34e will yield the contribution to V_{AN} caused by the 10-A steady-state current. Using current division to calculate the current through the light bulb, we find that

$$V_{AN2} = -\left\{I_{AB}\left(\frac{R_L}{R_{bulb} + 2R_L}\right)\right\}R_{bulb}$$

or

$$V_{AN2} = -4.98 \text{ V rms}$$

Therefore, the steady-state value of V_{AN} is

$$V_{AN} = V_{AN1} + V_{AN2} = 114.52 \text{ V rms}$$

At start-up, our expression for V_{AN2} can be used with $I_{AB} = 40$ A, which yields $V_{AN2} = -19.92$ V rms. The resulting value for V_{AN} is

$$V_{AN} = V_{AN1} + V_{AN2} = 119.50 - 19.92 = 99.58 \text{ V rms}$$

The voltage delivered to the light fixture at start-up is 13% lower than the steady-state value, resulting in a momentary dimming of the lights.

HINT

Technique

1. Find the resistance of the light bulb.

2. Use a large current source to represent the transient current of the air conditioner and a small current source to represent the steady-state current.

3. Find the voltage drop across the light bulb during both the transient and steady-state operations.

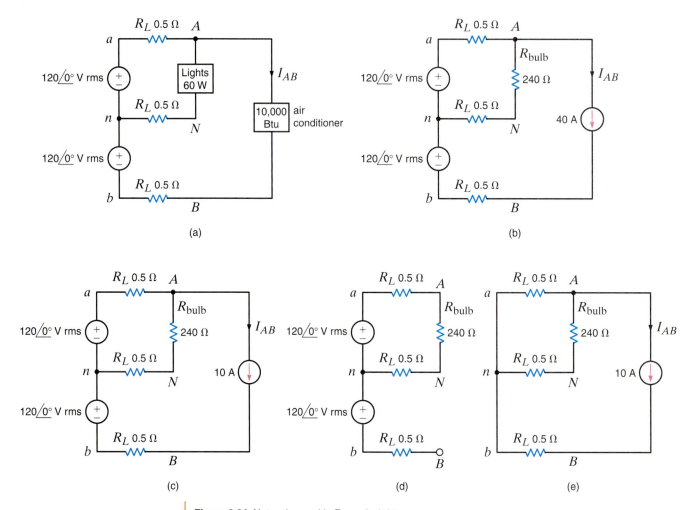

Figure 9.34 *Networks used in Example 9.25.*

Application Example 9.26

Most clothes dryers operate from a 240 V rms outlet and have several temperature settings such as low, medium, and high. Let's examine the manner in which the dryer creates heat for drying and the way in which it regulates the temperature.

SOLUTION   First we will consider the heat itself. A simple model for this situation is shown in Fig. 9.35a, where a resistive heating element is connected across the 240-V rms supply. For a particular dryer, the resistance of the element is roughly 11 Ω. The current through this element is

$$I_{he} = \frac{240}{11} = 21.81 \text{ A rms} \qquad \textbf{9.39}$$

Since the element is resistive, the voltage and current are in phase. The power dissipation is

$$P_{he} = I_{he}^2 R_{he} = (21.81)^2(11) = 5236 \text{ W}$$

This value, more than 5 kW, is a lot of power! Note, however, that this is the power dissipated under the assumption that the element is connected to the 240-V rms supply 100% of the time. If we manipulate the percentage of the time the element is energized, we can control the average power and thus the average temperature.

A fairly standard method for temperature control is shown schematically in Fig. 9.35b, where a standard residential single-phase, three-wire service powers the heating element at 240 V rms and the control circuit at 120 V rms. The temperature switch is connected to three resistors. Each temperature setting produces a different current through the thermostat heater, which is just another resistor. Each switch setting will alter the temperature of the thermostat heater that controls the temperature set point. We can calculate the power dissipated in the thermostat heater for each temperature setting. If we let R_S be the resistance that corresponds to the switch setting, then

$$P_{Th} = I_{Th}^2 R_{Th} = \left(\frac{120}{R_{Th} + R_S} \right)^2 R_{Th} \quad \begin{cases} = 1.3 \text{ W} & \text{for the low setting} \\ = 1.79 \text{ W} & \text{for the medium setting} \\ = 4.1 \text{ W} & \text{for the high setting} \end{cases} \quad \textbf{9.41}$$

We see that the thermostat heater power dissipation is lowest at the low setting and increases as the switch is moved to the high setting.

Now let us consider the critical issue in the temperature control system. The thermostat heater is located physically adjacent to the control thermostat (very similar to the ones used to control heat and cooling in your homes where you set the temperature manually). In the dryer, the thermostat heater acts as the desired setting. If the temperature at the thermostat exceeds the setting, then the thermostat switch opens, deenergizing the heating element and allowing it to cool off. When the thermostat determines the temperature is too low, the thermostat switch will close, energizing the element and increasing the temperature. In this way, the thermostat switch opens and closes throughout the drying cycle, maintaining the correct temperature as selected by the temperature switch.

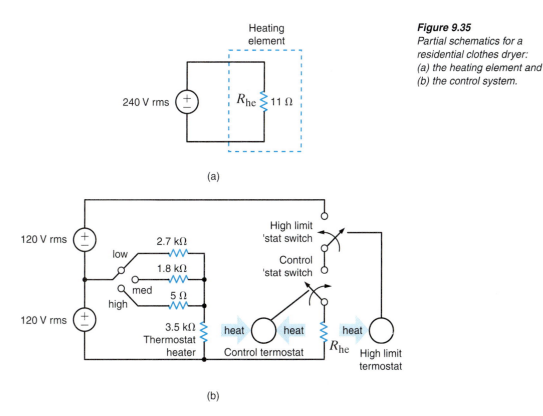

Figure 9.35
Partial schematics for a residential clothes dryer: (a) the heating element and (b) the control system.

Note that all of our calculations have been made for power levels, not temperatures. The exact temperatures of the heating element, the thermostat heater, and the thermostat itself depend on how heat moves about within the dryer—a thermodynamics issue that cannot be addressed with a simple circuit diagram.

Finally, note the high limit thermostat and its associated switch. This thermostat is mounted very close to the heating element. If the control thermostat fails, there is no temperature control and we can expect trouble. The high limit thermostat will detect these excessive temperatures and deenergize the heating element. Once the temperature drops, normal operation can resume. Thus, the high limit thermostat is used to protect the dryer and by extension your home.

9.11 Design Example

The following design example is presented in the context of a single-phase three-wire household circuit and serves as a good introduction to the material in Chapter 11.

Design Example 9.27

A light-duty commercial single-phase three-wire 60-Hz circuit serves lighting, heating, and motor loads, as shown in Fig. 9.36a. Lighting and heating loads are essentially pure resistance and, hence, unity power factor (pf), whereas motor loads have lagging pf.

We wish to design a balanced configuration for the network and determine its economic viability using the following procedure.

a. Compute the phase and neutral currents, and the complex power and pf for each source.

b. Now move the heating load (panel H) to phase b, as shown in Fig. 9.36b. This is called "balancing" the load. Repeat the analysis of (a).

c. Assume that the phase and neutral conductor resistances are each 0.05 Ω and have negligible effect on the results of (a). Evaluate the system line losses for (a) and (b). If the loads in question operate 24 hours per day and 365 days per year, at \$0.08/kWh, how much energy (and money) is saved by operating in the balanced mode?

SOLUTION (a) The magnitudes of the rms currents are

$$I_L = I_H = \frac{P}{V} = \frac{5000}{120} = 41.67 \text{ A rms}$$

and

$$I_m = \frac{10{,}000}{240} = 41.67 \text{ A rms}$$

In addition,

$$\theta_m = \cos^{-1}(0.8) = -36.9°$$

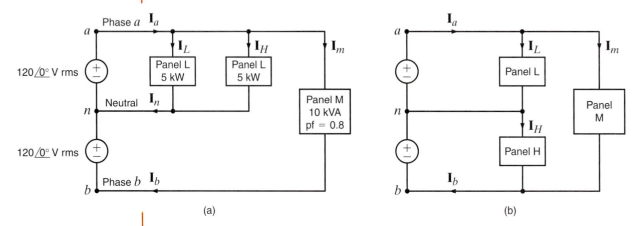

Figure 9.36 Single-phase three-wire power distribution system.

Therefore,

$$\mathbf{I}_a = \mathbf{I}_L + \mathbf{I}_H + \mathbf{I}_m$$
$$= 41.67\,\underline{/0°} + 41.67\,\underline{/0°} + 41.67\,\underline{/-36.9°}$$
$$= 119.4\,\underline{/-12.1°} \text{ A rms}$$

The currents in the neutral and phase b lines are

$$\mathbf{I}_n = \mathbf{I}_L + \mathbf{I}_H = 83.34\,\underline{/0°} \text{ A}$$
$$\mathbf{I}_b = 41.67\,\underline{/-36.9°} \text{ A rms}$$

The complex power and power factor for each source are

$$\mathbf{S}_a = \mathbf{V}_{an}\mathbf{I}_a^* = (120\,\underline{/0°})(119.4\,\underline{/+12.1°}) = 14 + j3 \text{ kVA}$$
$$\text{pf}_a = \cos(12.1°) = 0.9778 \text{ lagging}$$

and in a similar manner

$$\mathbf{S}_b = \mathbf{V}_{bn}\mathbf{I}_b^* = 4 + j3 \text{ kVA}$$
$$\text{pf}_b = 0.8 \text{ lagging}$$

(b) Under the balanced condition

$$\mathbf{I}_a = \mathbf{I}_L + \mathbf{I}_m = 41.67\,\underline{/0°} + 41.67\,\underline{/-36.9°}$$
$$= 79.06\,\underline{/-18.4°} \text{ A rms}$$

and

$$\mathbf{I}_b = 79.06\,\underline{/-18.4°} \text{ A rms}$$
$$\mathbf{I}_n = 0$$

Therefore,

$$\mathbf{S}_a = \mathbf{V}_{na}\mathbf{I}_a^* = 9 + j3 \text{ kVA}$$
$$\text{pf}_a = 0.9487 \text{ lagging}$$

and

$$\mathbf{S}_b = \mathbf{V}_{bn}\mathbf{I}_b^* = 9 + j3 \text{ kVA}$$
$$\text{pf}_b = 0.9487 \text{ lagging}$$

(c) The power loss in the lines in kW is

$$P_{\text{loss}} = R_a I_a^2 + R_b I_b^2 + R_n I_c^2$$
$$= 0.05(I_a^2 + I_b^2 + I_n^2)/1000$$

The total energy loss for a year is

$$W_{\text{loss}} = (24)(365)P_{\text{loss}} = 8760\,P_{\text{loss}}$$

and the annual cost is

$$\text{Cost} = \$0.08\,W_{\text{loss}}$$

A comparison of the unbalanced and balanced cases is shown in the following table.

	Unbalanced Case	Balanced Case
$P_{\text{loss}}(\text{kW})$	1.147	0.625
$W_{\text{loss}}(\text{kW-hr})$	10,034	5,475
Cost($)	804	438

Therefore, the annual savings obtained using the balanced configuration is

$$\text{Annual energy savings} = 10,048 - 5,475 = 4,573 \text{ kWh}$$
$$\text{Annual cost savings} = 804 - 438 = \$366$$

SUMMARY

- **Instantaneous power** If the current and voltage are sinusoidal functions of time, the instantaneous power is equal to a time-independent average value plus a sinusoidal term that has a frequency twice that of the voltage or current.

- **Average power** $P = 1/2\ VI \cos(\theta_v - \theta_i) = 1/2\ VI \cos\theta$, where θ is the phase of the impedance.

- **Resistive load** $P = 1/2\ I^2R = 1/2\ VI$ since $\mathbf{V}$ and $\mathbf{I}$ are in phase.

- **Reactive load** $P = 1/2\ VI \cos(\pm 90°) = 0$

- **Maximum average power transfer** To obtain the maximum average power transfer to a load, the load impedance should be chosen equal to the complex conjugate of the Thévenin equivalent impedance representing the remainder of the network.

- **rms or effective value of a periodic waveform** The effective, or rms, value of a periodic waveform was introduced as a means of measuring the effectiveness of a source in delivering power to a resistive load. The effective value of a periodic waveform is found by determining the root-mean-square value of the waveform. The rms value of a sinusoidal function is equal to the maximum value of the sinusoid divided by $\sqrt{2}$.

- **Power factor** Apparent power is defined as the product $V_{rms}I_{rms}$. The power factor is defined as the ratio of the average power to the apparent power and is said to be leading when the phase of the current leads the voltage, and lagging when the phase of the current lags the voltage. The power factor of a load with a lagging power factor can be corrected by placing a capacitor in parallel with the load.

- **Complex power** The complex power, $\mathbf{S}$, is defined as the product $\mathbf{V}_{rms}\mathbf{I}^*_{rms}$. The complex power $\mathbf{S}$ can be written as $\mathbf{S} = P + jQ$, where P is the real or average power and Q is the imaginary or quadrature power.

$$\mathbf{S} = I^2\mathbf{Z} = I^2R + jI^2X$$

- **The single-phase three-wire circuit** The single-phase three-wire circuit is the one commonly used in households. Large appliances are connected line to line and small appliances and lights are connected line to neutral.

- **Safety** Safety must be a primary concern in the design and use of any electrical circuit. The National Electric Code is the primary guide for the construction and operation of electrical systems.

PROBLEMS

PSV **CS** both available on the web at: http://www.justask4u.com/irwin

SECTION 9.1

9.1 Determine the equations for the current and the instantaneous power in the network in Fig. P9.1. **CS**

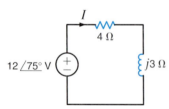

Figure P9.1

9.2 Determine the equations for the voltage and the instantaneous power in the network in Fig. P9.2.

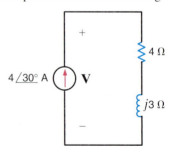

Figure P9.2

SECTION 9.2

9.3 The voltage and current at the input of a circuit are given by the expressions

$$v(t) = 170 \cos(\omega t + 30°)\ \text{V}$$
$$i(t) = 5 \cos(\omega t + 45°)\ \text{A}$$

Determine the average power absorbed by the circuit.

9.4 The voltage and current at the input of a network are given by the expressions

$$v(t) = 6 \cos \omega t\ \text{V}$$
$$i(t) = 4 \sin \omega t\ \text{A}$$

Determine the average power absorbed by the network.

9.5 Find the average power absorbed by the resistor in the circuit shown in Fig. P9.5 if $v_1(t) = 10 \cos(377t + 60°)$ V and $v_2(t) = 20 \cos(377t + 120°)$ V.

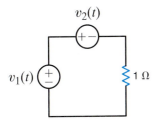

Figure P9.5

9.6 Find the average power absorbed by the resistor in the circuit shown in Fig. P9.6. Let $i_1(t) = 4 \cos(377t + 60°)$ A, $i_2(t) = 6 \cos(754t + 10°)$ A, and $i_3(t) = 4 \cos(377t - 30°)$ A.

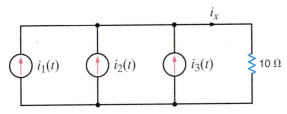

Figure P9.6

9.7 Compute the average power absorbed by each of the elements to the right of the dashed line in the circuit shown in Fig. P9.7.

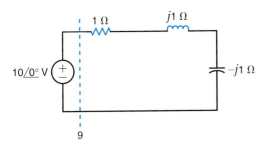

Figure P9.7

9.8 Determine the average power supplied by each source in the network shown in Fig. P9.8. **CS**

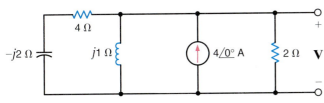

Figure P9.8

9.9 Find the average power absorbed by the network shown in Fig. P9.9.

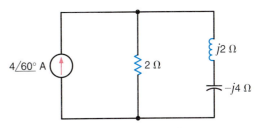

Figure P9.9

9.10 Given the network in Fig. P9.10, find the power supplied and the average power absorbed by each element. **PSV**

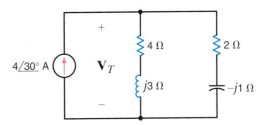

Figure P9.10

9.11 Given the network in Fig. P9.11, determine which elements are supplying power, which ones are absorbing power, and how much power is being supplied and absorbed.

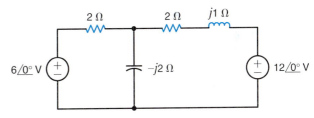

Figure P9.11

9.12 Given the network in Fig. P9.12, show that the power supplied by the sources is equal to the power absorbed by the passive elements.

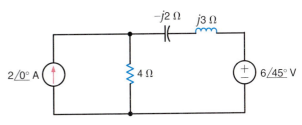

Figure P9.12

9.13 Calculate the average power absorbed by the 1-Ω resistor in the network shown in Fig. P9.13.

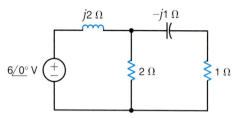

Figure P9.13

9.14 Given the network in Fig. P9.14, find the average power supplied to the circuit. **CS**

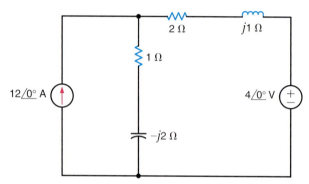

Figure P9.14

9.15 Given $v_S(t) = 100 \cos 100t$ volts, find the average power supplied by the source and the current $i_2(t)$ in the network in Fig. P9.15.

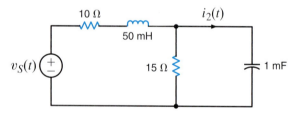

Figure P9.15

9.16 If $i_g(t) = 0.5 \cos 2000t$ A, find the average power absorbed by each element in the circuit in Fig. P9.16.

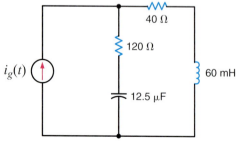

Figure P9.16

9.17 Calculate the average power absorbed by the 1-Ω resistor in the network shown in Fig. P9.17.

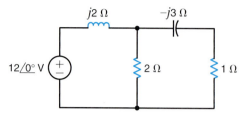

Figure P9.17

9.18 Find the average power supplied and/or absorbed by each element in Fig. P9.18.

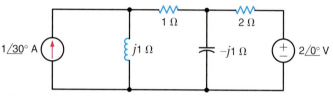

Figure P9.18

9.19 Determine the average power absorbed by the 4-Ω in the network shown in Fig. P9.19. **PSV**

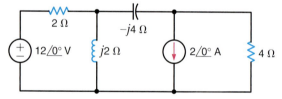

Figure P9.19

9.20 Given the network in Fig. P9.20, find the total average power supplied and the average power absorbed in the 4-Ω resistor.

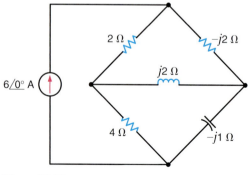

Figure P9.20

9.21 Determine the average power supplied to the network in Fig. P9.21.

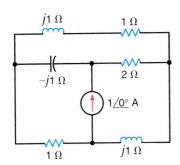

Figure P9.21

9.22 Find the average power absorbed by the 2-Ω resistor in the circuit shown in Fig. P9.22.

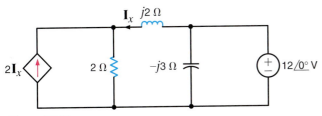

Figure P9.22

9.23 Determine the average power absorbed by a 2-Ω resistor connected at the output terminals of the network shown in Fig. P9.23.

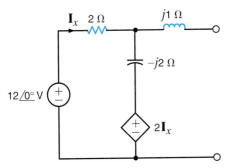

Figure P9.23

9.24 Determine the average power absorbed by the 2-kΩ output resistor in Fig. P9.24.

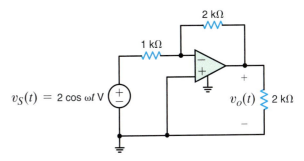

Figure P9.24

9.25 Determine the average power absorbed by the 4-kΩ resistor in Fig. P9.25.

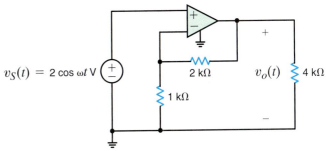

Figure P9.25

<hr>

SECTION 9.3

9.26 Determine the impedance $\mathbf{Z}_L$ for maximum average power transfer and the value of the maximum power transferred to $\mathbf{Z}_L$ for the circuit shown in Fig. P9.26.

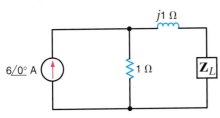

Figure P9.26

9.27 Determine the impedance $\mathbf{Z}_L$ for maximum average power transfer and the value of the maximum average power transferred to $\mathbf{Z}_L$ for the circuit shown in Fig. P9.27. **CS**

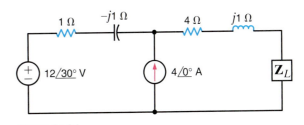

Figure P9.27

9.28 Determine the impedance $\mathbf{Z}_L$ for maximum average power transfer and the value of the maximum average power absorbed by the load in the network shown in Fig. P9.28.

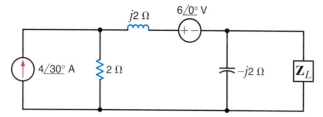

Figure P9.28

9.29 Determine the impedance $\mathbf{Z}_L$ for maximum average power transfer and the value of the maximum average power absorbed by the load in the network shown in Fig. P9.29.

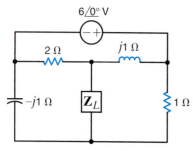

Figure P9.29

9.30 Repeat Problem 9.29 for the network in Fig. P9.30.

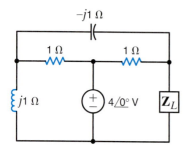

Figure P9.30

9.31 Determine the impedance $\mathbf{Z}_L$ for maximum average power transfer and the value of the maximum average power transferred to $\mathbf{Z}_L$ for the circuit shown in Fig. P9.31. **PSV**

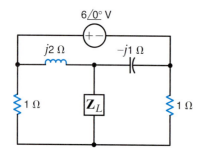

Figure P9.31

9.32 In the network in Fig. P9.32, find $\mathbf{Z}_L$ for maximum average power transfer and the maximum average power transferred. **CS**

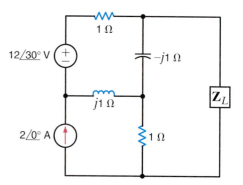

Figure P9.32

9.33 Repeat Problem 9.32 for the network in Fig. P9.33.

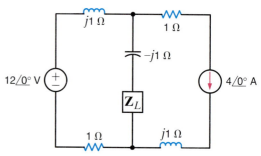

Figure P9.33

9.34 Repeat Problem 9.32 for the network in Fig. P9.34. **CS**

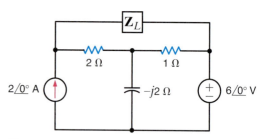

Figure P9.34

9.35 Repeat Problem 9.32 for the network in Fig. P9.35.

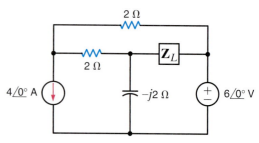

Figure P9.35

9.36 Repeat Problem 9.32 for the network in Fig. P9.36.

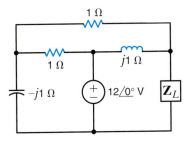

Figure P9.36

9.37 Determine the impedance $\mathbf{Z}_L$ for maximum average power transfer and the value of the maximum average power absorbed by the load in the network shown in Fig. P9.37. **CS**

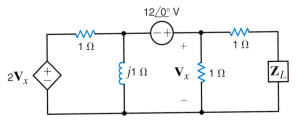

Figure P9.37

9.38 Find the impedance $\mathbf{Z}_L$ for maximum average power transfer and the value of the maximum average power transferred to $\mathbf{Z}_L$ for the circuit shown in Fig. P9.38.

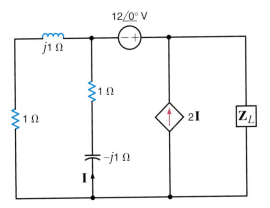

Figure P9.38

SECTION 9.4

9.41 Compute the rms value of the voltage given by the expression $v(t) = 10 + 20 \cos(377t + 30°)$ V.

9.42 Compute the rms value of the voltage given by the waveform shown in Fig. P9.42.

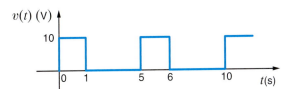

Figure P9.42

9.39 Repeat Problem 9.38 for the network in Fig. P9.39.
PSV

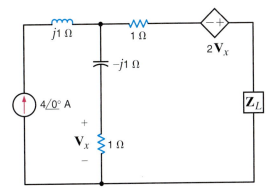

Figure P9.39

9.40 Find the value of $\mathbf{Z}_L$ in the circuit in Fig. P9.40 for maximum average power transfer.

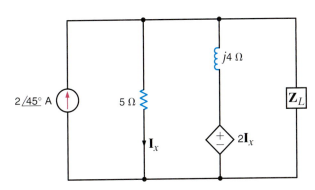

Figure P9.40

9.43 Calculate the rms value of the waveform shown in Fig. P9.43.

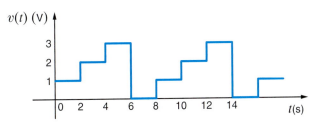

Figure P9.43

9.44 Calculate the rms value of the waveform in Fig. P9.44.

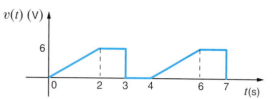

Figure P9.44

9.45 Calculate the rms value of the waveform in Fig. P9.45.

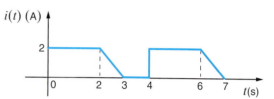

Figure P9.45

9.46 Calculate the rms value of the waveform in Fig. P9.46.

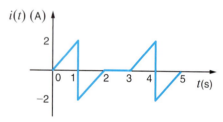

Figure P9.46

9.47 Calculate the rms value of the waveform shown in Fig. P9.47.

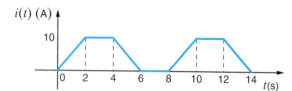

Figure P9.47

9.48 Calculate the rms value of the waveform shown in Fig. P9.48. **CS**

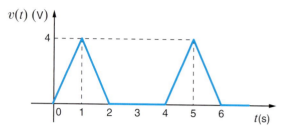

Figure P9.48

9.49 The current waveform in Fig. P9.49 is flowing through a 5-Ω resistor. Find the average power absorbed by the resistor. **PSV**

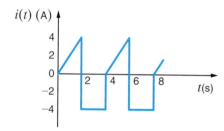

Figure P9.49

9.50 Calculate the rms value of the waveform shown in Fig. P9.50. **CS**

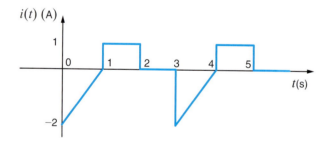

Figure P9.50

SECTION 9.5

9.51 A plant consumes 20 kW of power from a 240-V rms line. If the load power factor is 0.9, what is the angle by which the load voltage leads the load current? What is the load current phasor if the line voltage has a phasor of 240 $\underline{/0°}$ V rms?

9.52 A plant consumes 100 kW of power at 0.9 pf lagging. If the load current is 200 A rms, find the load voltage.

9.53 A plant draws 250 A rms from a 240-V rms line to supply a load with 50 kW. What is the power factor of the load?

9.54 The power company supplies 80 kW to an industrial load. The load draws 220 A rms from the transmission line. If the load voltage is 440 V rms and the load power factor is 0.8 lagging, find the losses in the transmission line. **CS**

9.55 An industrial load that consumes 40 kW is supplied by the power company through a transmission line with 0.1 Ω resistance, with 44 kW. If the voltage at the load is 240 V rms, find the power factor at the load.

9.56 The power company supplies 40 kW to an industrial load. The load draws 200 A rms from the transmission line. If the load voltage is 240 V rms and the load power factor is 0.8 lagging, find the losses in the transmission line.

9.57 A transmission line with impedance $0.08 + j0.25\ \Omega$ is used to deliver power to a load. The load is inductive and the load voltage is $220\ \underline{/0°}$ V rms at 60 Hz. If the load requires 12 kW and the real power loss in the line is 560 W, determine the power factor angle of the load. **CS**

SECTION 9.6

9.58 Determine the real power, the reactive power, the complex power, and the power factor for a load having the following characteristics.

(a) $\mathbf{I} = 2\ \underline{/40°}$ A rms, $\mathbf{V} = 450\ \underline{/70°}$ V rms.

(b) $\mathbf{I} = 1.5\ \underline{/-20°}$ A rms, $\mathbf{Z} = 5000\ \underline{/15°}\ \Omega$.

(c) $\mathbf{V} = 200\ \underline{/+35°}$ V rms, $\mathbf{Z} = 1500\ \underline{/-15°}\ \Omega$.

9.59 An industrial load operates at 30 kW, 0.8 pf lagging. The load voltage is $240\ \underline{/0°}$ V rms. The real and reactive power losses in the transmission-line feeder are 1.8 kW and 2.4 kvar, respectively. Find the impedance of the trasmission line and the input voltage to the line. **PSV**

9.60 A transmission line with impedance $0.1 + j0.2\ \Omega$ is used to deliver power to a load. The load is capacitive and the load voltage is $240\ \underline{/0°}$ V rms at 60 Hz. If the load requires 15 kW and the real power loss in the line is 660 W, determine the input voltage to the line.

9.61 Find the real and reactive power absorbed by each element in the circuit in Fig. P9.61.

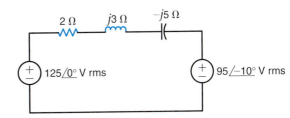

Figure P9.61

9.62 Given the network in Fig. P9.62, determine the input voltage $\mathbf{V}_S$.

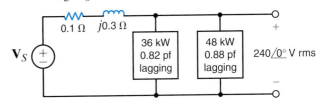

Figure P9.62

9.63 Given the network in Fig. P9.63, compute the input source voltage and the input power factor.

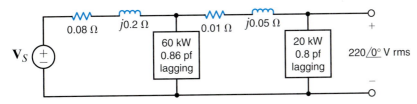

Figure P9.63

9.64 Use Kirchhoff's laws to compute the source voltage of the network shown in Fig. P9.64.

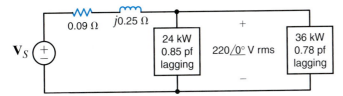

Figure P9.64

9.65 Given the network in Fig. P9.65, determine the input voltage $\mathbf{V}_S$. **PSV**

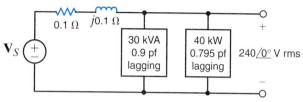

Figure P9.65

9.66 Find the input source voltage and the power factor of the source for the network shown in Fig. P9.66. **CS**

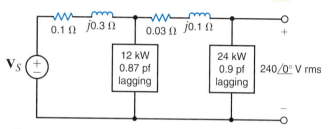

Figure P9.66

9.67 Given the network in Fig. P9.67, find the complex power supplied by the source, the power factor of the source, and the voltage $v_S(t)$. The frequency is 60 Hz.

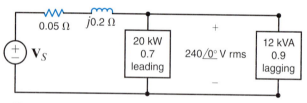

Figure P9.67

9.68 Given the circuit in Fig. P9.68, find the complex power supplied by the source and the source power factor. If $f = 60$ Hz, find $v_s(t)$.

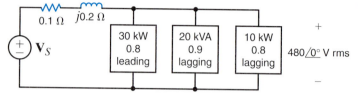

Figure P9.68

9.69 In the circuit in Fig. P9.69, the complex power supplied by source $\mathbf{V}_1$ is $2000\,\underline{/-30°}$ VA. If $\mathbf{V}_1 = 200\,\underline{/10°}$ V rms, find $\mathbf{V}_2$.

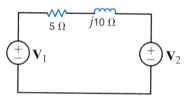

Figure P9.69

9.70 For the network in Fig. P9.70, the complex power absorbed by the source on the right is $0 + j1582.5$ VA. Find the value of R and the unknown element and its value if $f = 60$ Hz. (If the element is a capacitor, give its capacitance; if the element is an inductor, give its inductance.)

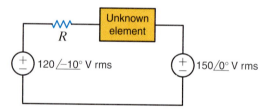

Figure P9.70

SECTION 9.7

9.71 A plant consumes 60 kW at a power factor of 0.75 lagging from a 240-V rms 60-Hz line. Determine the value of the capacitor that when placed in parallel with the load will change the load power factor to 0.9 lagging.

9.72 A particular load has a pf of 0.8 lagging. The power delivered to the load is 40 kW from a 270-V rms 60-Hz line. What value of capacitance placed in parallel with the load will raise the pf to 0.9 lagging? **PSV**

9.73 An industrial load is supplied through a transmission line that has a line impedance of $0.1 + j0.2$ Ω. The 60-Hz line voltage at the load is $480\,\underline{/0°}$ V rms. The load consumes 124 kW at 0.75 pf lagging. What value of capacitance when placed in parallel with the load will change the power factor to 0.9 lagging? **CS**

9.74 The 60-Hz line voltage for a 60-kW, 0.76-pf lagging industrial load is $240\,\underline{/0°}$ V rms. Find the value of capacitance that when placed in parallel with the load will raise the power factor to 0.9 lagging.

9.75 An industrial load consumes 44 kW at 0.82 pf lagging from a $240\,\underline{/0°}$-V-rms 60-Hz line. A bank of capacitors totaling 600 µF is available. If these capacitors are placed in parallel with the load, what is the new power factor of the total load?

9.76 A particular load has a pf of 0.8 lagging. The power delivered to the load is 40 kW from a 220-V rms 60-Hz line. What value of capacitance placed in parallel with the load will raise the pf to 0.9 lagging? **CS**

SECTION 9.8

9.77 A single-phase three-wire 60-Hz circuit serves three loads, as shown in Fig. P9.77. Determine $\mathbf{I}_{aA}$, $\mathbf{I}_{nN}$, $\mathbf{I}_c$, and the energy use over a 24-hour period in kilowatt-hours.

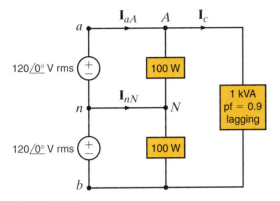

Figure P9.77

9.78 A 5.1-kW household range is designed to operate on a 240-V rms sinusoidal voltage, as shown in Fig. P9.78a. However, the electrician has mistakenly connected the range to 120 V rms, as shown in Fig. P9.78b. What is the effect of this error?

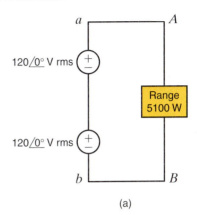

(a)

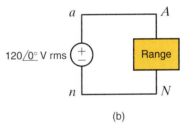

(b)

Figure P9.78

SECTION 9.9

9.79 A man and his son are flying a kite. The kite becomes entangled in a 7200-V rms power line close to a power pole. The man crawls up the pole to remove the kite. While trying to remove the kite, the man accidentally touches the 7200-V rms line. Assuming the power pole is well grounded, what is the potential current through the man's body? **cs**

9.80 A number of 120-V rms household fixtures are to be used to provide lighting for a large room. The total lighting load is 8 kW. The National Electric Code requires that no circuit breaker be larger than 20 A rms with a 25% safety margin. Determine the number of identical branch circuits needed for this requirement.

9.81 To test a light socket, a woman, while standing on cushions that insulate her from the ground, sticks her finger into the socket, as shown in Fig. P9.81. The tip of her finger makes contact with one side of the line, and the side of her finger makes contact with the other side of the line. Assuming that any portion of a limb has a resistance of 95 Ω, is there any current in the body? Is there any current in the vicinity of the heart?

Figure P9.81

9.82 An inexperienced mechanic is installing a 12-V battery in a car. The negative terminal has been connected. He is currently tightening the bolts on the positive terminal. With a tight grip on the wrench, he turns it so that the gold ring on his finger makes contact with the frame of the car. This situation is modeled in Fig. P9.82, where we assume that the resistance of the wrench is negligible and the resistance of the contact is as follows:

$$R_1 = R_{\text{bolt to wrench}} = 0.012 \ \Omega$$
$$R_2 = R_{\text{wrench to ring}} = 0.012 \ \Omega$$
$$R_3 = R_{\text{ring}} = 0.012 \ \Omega$$
$$R_4 = R_{\text{ring to frame}} = 0.012 \ \Omega$$

What power is quickly dissipated in the gold ring, and what is the impact of this power dissipation? **CS**

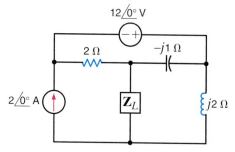

Figure P9.82

9.83 A 5-kW load operates at 60 Hz, 240 V rms and has a power factor of 0.866 lagging. We wish to create a power factor of at least 0.975 lagging using a single capacitor. Can this requirement be met using a single capacitor from Table 6.1?

9.84 Use an *RC* combination to design a circuit that will reduce a 120-V rms line voltage to a voltage between 75 and 80 V rms while dissipating less than 30 W.

TYPICAL PROBLEMS FOUND ON THE FE EXAM

9FE-1 An industrial load consumes 120 kW at 0.707 pf lagging and is connected to a $480 \underline{/0^\circ}$ V rms 60-Hz line. Determine the value of the capacitor that, when connected in parallel with the load, will raise the power factor to 0.95 lagging. **CS**

9FE-2 Determine the average and rms values of the following waveform.

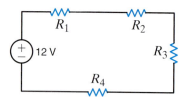

Figure 9PFE-2

9FE-3 Find the impedance $\mathbf{Z}_L$ in the network in Fig. 9PFE-3 for maximum average power transfer. **CS**

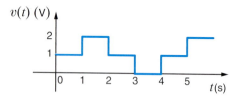

Figure 9PFE-3

9FE-4 An rms-reading voltmeter is connected to the output of the op-amp shown in Fig. 9PFE-4. Determine the meter reading.

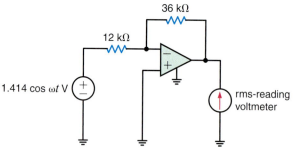

Figure 9PFE-4

9FE-5 Determine the average power delivered to the resistor in Fig. 9PFE-5a if the current waveform is shown in Fig. 9PFE-5b. **CS**

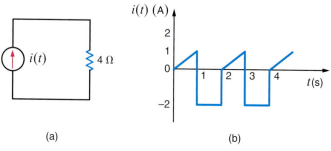

(a)   (b)

Figure 9PFE-5

Magnetically Coupled Networks

Access Problem-Solving Videos **P S V** *and Circuit Solutions* **C S** *at:*
http://www.justask4u.com/irwin
using the registration code on the inside cover and see a website with answers and more!

High-voltage power lines crisscross the country to supply electric power to us in every area of our lives. Many of the electric utilities are tied together to form a power grid. As indicated earlier, in order to deliver power in an efficient manner, it is transmitted at high voltage. Five-hundred-kilovolt lines are typical transmission facilities. However, the power that is supplied to our homes, for example, is typically 240/120 V rms. Thus, it is natural for us to question how the high voltage supplied through the transmission lines is stepped down to this level. Hence, one of the topics that we will address in this chapter is the transformer, which is the primary component in this application.

The recent advances in integrated circuits have caused designers to avoid the use of transformers (and inductors too) in many situations. For example, to avoid the use of a transformer, we may use a large number of other components such as resistors, capacitors, op-amps, and the like. However, although the tradeoff in number of components may seem unreasonable, it is not. It is very easy to fabricate such components as resistors and capacitors, and they require much less space than a single inductor.

Nevertheless, transformers remain an important electrical component. In addition to power systems, where the transformers play a fundamental role, there are other applications. For example, transformers are used to reject high-frequency noise in audio and industrial control systems, and they are built into special wall plugs that are used to step down the voltage for the purpose of recharging batteries for calculators and hand tools. ●

10.1 Mutual Inductance

As we introduce this subject, we feel compelled to remind the reader, once again, that in our analyses we assume that we are dealing with "ideal" elements. For example, we ignore the resistance of the coil used to make an inductor and any stray capacitance that might exist. This approach is especially important in our discussion of mutual inductance because an exact analysis of this topic is quite involved. As is our practice, we will treat the subject in a straightforward manner and ignore issues, beyond the scope of this book, that only serve to complicate the presentation.

To begin our discussion of mutual inductance, we will recall two important laws: Ampère's law and Faraday's law. Ampère's law predicts that the flow of electric current will create a magnetic field. If the field links an electric circuit, and that field is time-varying, Faraday's law predicts the creation of a voltage within the linked circuit. Although this occurs to some extent in all circuits, the effect is magnified in coils because the circuit geometry amplifies the linkage effect. With these ideas in mind, consider the ideal situation in Fig. 10.1 in which a current i flows in an N-turn coil and produces a magnetic field, represented by magnetic flux ϕ. The flux linkage for this coil is

$$\lambda = N\phi \tag{10.1}$$

Figure 10.1
Magnetic flux ϕ linking an N-turn coil.

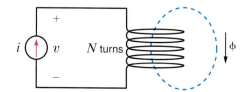

For the linear systems that we are studying in this textbook, the flux linkage and current are related by

$$\lambda = Li \tag{10.2}$$

The constant of proportionality between the flux linkage and current is the inductance, which we studied in Chapter 6. Equations (10.1) and (10.2) can be utilized to express the magnetic flux in terms of the current.

$$\phi = \frac{L}{N}i \tag{10.3}$$

According to Faraday's law, the voltage induced in the coil is related to the time rate of change of the flux linkage λ.

$$v = \frac{d\lambda}{dt} \tag{10.4}$$

Let's substitute Eq. (10.2) into Eq. (10.4) and use the chain rule to take the derivative.

$$v = \frac{d\lambda}{dt} = \frac{d}{dt}(Li) = L\frac{di}{dt} + i\frac{dL}{dt} \tag{10.5}$$

We will not allow our inductances to vary with time, so Eq. (10.5) reduces to the defining equation for the ideal inductor, as shown in Fig. 10.2.

$$v = L\frac{di}{dt} \tag{10.6}$$

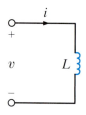

Figure 10.2
An ideal inductor.

Note that the voltage and current in this figure satisfy the passive sign convention. Equation (10.6) tells us that a current i flowing through a coil produces a voltage v across that coil.

Now let's suppose that a second coil with N_2 turns is moved close enough to an N_1-turn coil such that the magnetic flux produced by current i_1 links the second coil. No current flows in the second coil as shown in Fig. 10.3. By Faraday's law, a voltage v_2 will be induced because the magnetic flux ϕ links the second coil. The flux linkage for coil 1 is

$$\lambda_1 = N_1\phi = L_1 i_1 \qquad \textbf{10.7}$$

Current flowing in coil 1 produces a voltage $v_1 = \dfrac{d\lambda_1}{dt} = L_1\dfrac{di_1}{dt}$. We have been referring to L_1 as the inductance. In multiple coil systems, we will refer to L_1 as the self-inductance of coil 1. The flux linkage for coil 2 is $\lambda_2 = N_2\phi$, and from Faraday's law, the voltage v_2 is given as

$$v_2 = \frac{d\lambda_2}{dt} = \frac{d}{dt}(N_2\phi) = \frac{d}{dt}\left(N_2\left(\frac{L_1}{N_1}i_1\right)\right) = \frac{N_2}{N_1}L_1\frac{di_1}{dt} = L_{21}\frac{di_1}{dt} \qquad \textbf{10.8}$$

Note that the voltage v_2 is directly proportional to the time rate of change of i_1. The constant of proportionality, L_{21}, is defined as the mutual inductance and is given in units of henrys. We will say that the coils in Fig. 10.3 are magnetically coupled.

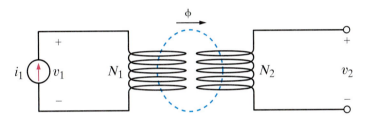

Figure 10.3
Two coils magnetically coupled.

Let's connect a current source to the terminals of coil 2 as shown in Fig. 10.4. Both currents contribute to the magnetic flux ϕ. For the coil configuration and current directions shown in this figure, the flux linkages for each coil are

$$\lambda_1 = L_1 i_1 + L_{12} i_2 \qquad \textbf{10.9}$$

$$\lambda_2 = L_{21} i_1 + L_2 i_2 \qquad \textbf{10.10}$$

Applying Faraday's law,

$$v_1 = \frac{d\lambda_1}{dt} = L_1\frac{di_1}{dt} + L_{12}\frac{di_2}{dt} \qquad \textbf{10.11}$$

$$v_2 = \frac{d\lambda_2}{dt} = L_{21}\frac{di_1}{dt} + L_2\frac{di_2}{dt} \qquad \textbf{10.12}$$

Since we have limited our study to linear systems, $L_{12} = L_{21} = M$, where M is the symbol for mutual inductance. From Eqs. (10.11) and (10.12), we can see that the voltage across each coil is composed of two terms: a "self term" due to current flowing in that coil and a "mutual term" due to current flowing in the other coil.

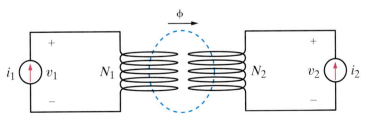

Figure 10.4
Two magnetically coupled coils driven by current sources.

If the direction of i_2 in Fig. 10.4 is reversed, Eqs. (10.9) through (10.12) become

$$\lambda_1 = L_1 i_1 - M i_2 \qquad \textbf{10.13}$$

$$\lambda_2 = -M i_1 + L_2 i_2 \qquad \textbf{10.14}$$

$$v_1 = \frac{d\lambda_1}{dt} = L_1 \frac{di_1}{dt} - M \frac{di_2}{dt} \qquad \textbf{10.15}$$

$$v_2 = \frac{d\lambda_2}{dt} = -M \frac{di_1}{dt} + L_2 \frac{di_2}{dt} \qquad \textbf{10.16}$$

Equations (10.13)–(10.16) can also be obtained from the circuit in Fig. 10.5. Note that coil 2 in this figure has a different winding arrangement as compared to coil 2 in Fig. 10.4.

Figure 10.5

Magnetically coupled coils with different winding configuration.

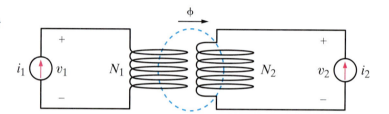

Our circuit diagrams will become quite complex if we have to include details of the winding configuration. The use of the dot convention permits us to maintain these details while simplifying our circuit diagrams. Figure 10.6a is the circuit diagram for the magnetically coupled coils of Fig. 10.4. The coils are represented by two coupled inductors with self-inductances L_1 and L_2 and mutual inductance M. Recall that the voltage across each coil consists of two terms: a self term due to current flowing in that coil and a mutual term due to current flowing in the other coil. The self term is the same voltage that we discussed in an earlier chapter. The mutual term results from current flowing in the other coupled coil.

Figure 10.6

Circuit diagrams for magnetically coupled coils.

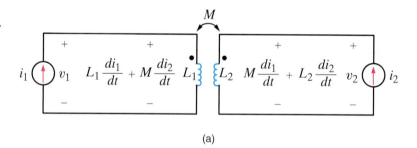

(a)

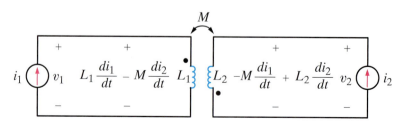

(b)

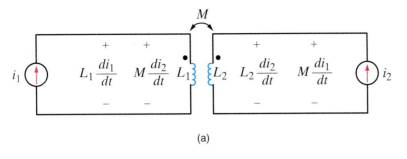

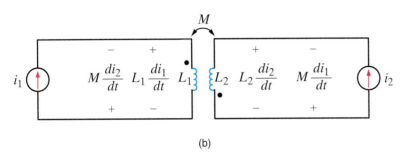

Figure 10.7
Circuit diagrams for magnetically coupled coils showing self and mutual voltage terms.

In Fig. 10.6a, the mutual terms are positive when both currents enter the dots. The opposite is true when one current enters a dot and the other current leaves a dot, as seen in Fig. 10.6b. Let's use this observation to develop a general procedure for writing circuit equations for magnetically coupled inductors. Figure 10.7a is the same diagram as Fig. 10.6a except that the voltage across the inductors is broken into the self term and the mutual term. The polarity of the self terms—$L_1 \, di_1/dt$ and $L_2 \, di_2/dt$—are given by the passive sign convention used extensively throughout this text. These terms would be present even if the coils were not magnetically coupled. The mutual terms in Fig. 10.7a have the same polarity as the self terms. Note that both currents are entering the dots in Fig. 10.7a. The opposite is true in Fig. 10.7b. The self terms have the same polarity as before; however, the polarities for the mutual terms are different from those in Fig. 10.7a. We can now make a general statement:

> When a current is defined to enter the dotted terminal of a coil, it produces a voltage in the coupled coil which is positive at the dotted terminal. Similarly, when a current is defined to enter the undotted terminal of a coil, it produces a voltage in the coupled coil which is positive at the undotted terminal.

Let's illustrate the use of this statement through some examples.

PROBLEM-SOLVING STRATEGY

Magnetically-Coupled Inductors

Step 1. Assign mesh currents. It is usually much easier to write mesh equations for a circuit containing magnetically coupled inductors than to write nodal equations.

Step 2. Write mesh equations by applying KVL. If a defined current enters the dotted terminal on one coil, it produces a voltage in the other coil that is positive at the dotted terminal. If a defined current enters the undotted terminal on one coil, it produces a voltage in the other coil that is positive at the undotted terminal.

Step 3. Solve for the mesh currents.

Example 10.1

Determine the equations for $v_1(t)$ and $v_2(t)$ in the circuit shown in Fig. 10.8a.

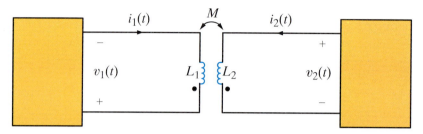

Figure 10.8a *Circuit used in Example 10.1.*

SOLUTION   The different voltage terms for the circuit are shown on the circuit diagram in Fig. 10.8b. The polarity of the self terms is given by the passive sign convention. For both coils, the defined currents are entering the undotted terminals on both coils. As a result, the polarity of the voltages produced by these currents is positive at the undotted terminal of the other coils. The equations for $v_1(t)$ and $v_2(t)$ are

$$v_1(t) = -L_1 \frac{di_1}{dt} - M \frac{di_2}{dt}$$

$$v_2(t) = L_2 \frac{di_2}{dt} + M \frac{di_1}{dt}$$

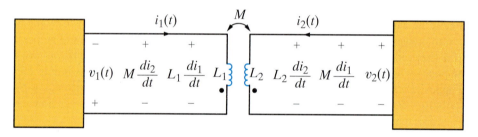

Figure 10.8b *Circuit showing self and mutual voltage terms.*

Example 10.2

Write mesh equations for the circuit of Fig. 10.9a using the assigned mesh currents.

Figure 10.9a
*Circuit used in
Example 10.2.*

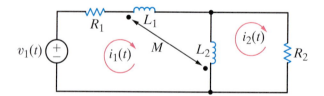

SOLUTION   The circuit in Fig. 10.9b shows the voltage terms for mesh 1. The polarity of the self terms for L_1 and L_2 is determined by the passive sign convention. The current $(i_2 - i_1)$ enters the dotted terminal of inductor L_2. This current produces the mutual term shown across inductor L_1. Current i_1 enters the dotted terminal of L_1 and produces a voltage across L_2 that is positive at its dotted terminal. The equation for this mesh is

$$v_1(t) = R_1 i_1(t) + L_1 \frac{di_1}{dt} + M \frac{d}{dt}(i_2 - i_1) + L_2 \frac{d}{dt}(i_1 - i_2) - M \frac{di_1}{dt}$$

The voltage terms for the second mesh are shown in Fig. 10.9c. The equation for mesh 2 is

$$R_2 i_2(t) + L_2 \frac{d}{dt}(i_2 - i_1) + M \frac{di_1}{dt} = 0$$

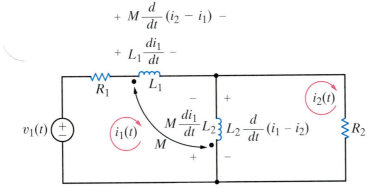

$$+ M \frac{d}{dt}(i_2 - i_1) -$$

$$+ L_1 \frac{di_1}{dt} -$$

Figure 10.9b
Circuit showing voltage terms for mesh 1.

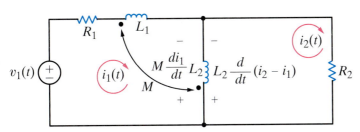

Figure 10.9c
Circuit showing voltage terms for mesh 2.

LEARNING EXTENSION

E10.1 Write the equations for $v_1(t)$ and $v_2(t)$ in the circuit in Fig. E10.1.

ANSWER:

$$v_1(t) = L_1 \frac{di_1(t)}{dt} + M \frac{di_2(t)}{dt};$$

$$v_2(t) = -L_2 \frac{di_2(t)}{dt} - M \frac{di_1(t)}{dt}.$$

Figure E10.1

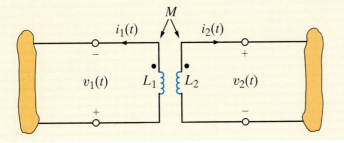

Assume that the coupled circuit in Fig. 10.10 is excited with a sinusoidal source. The voltages will be of the form $\mathbf{V}_1 e^{j\omega t}$ and $\mathbf{V}_2 e^{j\omega t}$, and the currents will be of the form $\mathbf{I}_1 e^{j\omega t}$ and $\mathbf{I}_2 e^{j\omega t}$, where $\mathbf{V}_1$, $\mathbf{V}_2$, $\mathbf{I}_1$, and $\mathbf{I}_2$ are phasors. Substituting these voltages and currents into Eqs. (10.11) and (10.12), and using the fact that $L_{12} = L_{21} = M$, we obtain

$$\mathbf{V}_1 = j\omega L_1 \mathbf{I}_1 + j\omega M \mathbf{I}_2$$

10.17

$$\mathbf{V}_2 = j\omega L_2 \mathbf{I}_2 + j\omega M \mathbf{I}_1$$

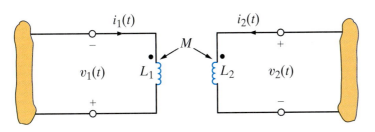

Figure 10.10
Mutually coupled coils.

The model of the coupled circuit in the frequency domain is identical to that in the time domain except for the way the elements and variables are labeled. The sign on the mutual terms is handled in the same manner as is done in the time domain.

Example 10.3

The two mutually coupled coils in Fig. 10.11a can be interconnected in four possible ways. We wish to determine the equivalent inductance of each of the four possible interconnections.

SOLUTION Case 1 is shown in Fig. 10.11b. In this case

$$\mathbf{V} = j\omega L_1\mathbf{I} + j\omega M\mathbf{I} + j\omega L_2\mathbf{I} + j\omega M\mathbf{I}$$
$$= j\omega L_{eq}\mathbf{I}$$

where $L_{eq} = L_1 + L_2 + 2M$.

Case 2 is shown in Fig. 10.11c. Using KVL, we obtain

$$\mathbf{V} = j\omega L_1\mathbf{I} - j\omega M\mathbf{I} + j\omega L_2\mathbf{I} - j\omega M\mathbf{I}$$
$$= j\omega L_{eq}\mathbf{I}$$

where $L_{eq} = L_1 + L_2 - 2M$.

Case 3 is shown in Fig. 10.11d and redrawn in Fig. 10.11e. The two KVL equations are

$$\mathbf{V} = j\omega L_1\mathbf{I}_1 + j\omega M\mathbf{I}_2$$
$$\mathbf{V} = j\omega M\mathbf{I}_1 + j\omega L_2\mathbf{I}_2$$

Solving these equations for $\mathbf{I}_1$ and $\mathbf{I}_2$ yields

$$\mathbf{I}_1 = \frac{\mathbf{V}(L_2 - M)}{j\omega(L_1L_2 - M^2)}$$

$$\mathbf{I}_2 = \frac{\mathbf{V}(L_1 - M)}{j\omega(L_1L_2 - M^2)}$$

Using KCL gives us

$$\mathbf{I} = \mathbf{I}_1 + \mathbf{I}_2 = \frac{\mathbf{V}(L_1 + L_2 - 2M)}{j\omega(L_1L_2 - M^2)} = \frac{\mathbf{V}}{j\omega L_{eq}}$$

where

$$L_{eq} = \frac{L_1L_2 - M^2}{L_1 + L_2 - 2M}$$

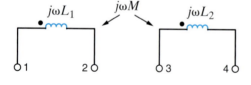

(a)   (b)

Figure 10.11 Circuits used in Example 10.3.

Case 4 is shown in Fig. 10.11f. The voltage equations in this case will be the same as those in case 3 except that the signs of the mutual terms will be negative. Therefore,

$$L_{eq} = \frac{L_1 L_2 - M^2}{L_1 + L_2 + 2M}$$

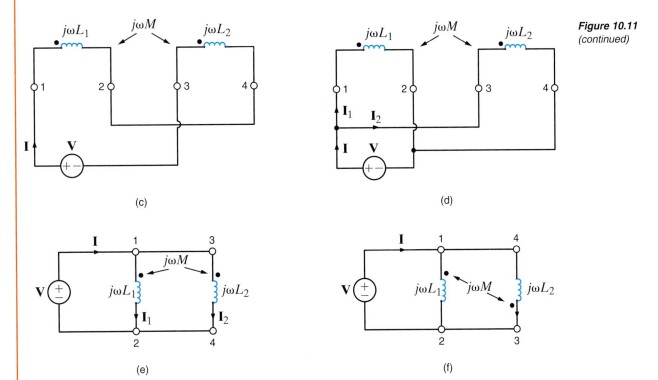

(c)

(d)

(e)

(f)

Figure 10.11
(continued)

Example 10.4

We wish to determine the output voltage $\mathbf{V}_o$ in the circuit in Fig. 10.12.

SOLUTION   The two KVL equations for the network are

$$(2 + j4)\mathbf{I}_1 - j2\mathbf{I}_2 = 24\underline{/30°}$$

$$-j2\mathbf{I}_1 + (2 + j6 - j2)\mathbf{I}_2 = 0$$

Solving the equations yields

$$\mathbf{I}_2 = 2.68\underline{/3.43°}\ \text{A}$$

Therefore,

$$\mathbf{V}_o = 2\mathbf{I}_2$$

$$= 5.36\underline{/3.43°}\ \text{V}$$

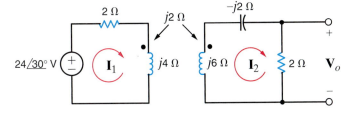

Figure 10.12
Example of a magnetically coupled circuit.

Let us now consider a more complicated example involving mutual inductance.

Example 10.5

Consider the circuit in Fig. 10.13. We wish to write the mesh equations for this network.

SOLUTION Because of the multiple currents that are present in the coupled inductors, we must be very careful in writing the circuit equations.

The mesh equations for the phasor network are

$$\mathbf{I}_1 R_1 + j\omega L_1(\mathbf{I}_1 - \mathbf{I}_2) + j\omega M(\mathbf{I}_2 - \mathbf{I}_3) + \frac{1}{j\omega C_1}(\mathbf{I}_1 - \mathbf{I}_2) = \mathbf{V}$$

$$\frac{1}{j\omega C_1}(\mathbf{I}_2 - \mathbf{I}_1) + j\omega L_1(\mathbf{I}_2 - \mathbf{I}_1) + j\omega M(\mathbf{I}_3 - \mathbf{I}_2) + R_2 \mathbf{I}_2$$

$$+ j\omega L_2(\mathbf{I}_2 - \mathbf{I}_3) + j\omega M(\mathbf{I}_1 - \mathbf{I}_2) + R_3(\mathbf{I}_2 - \mathbf{I}_3) = 0$$

$$R_3(\mathbf{I}_3 - \mathbf{I}_2) + j\omega L_2(\mathbf{I}_3 - \mathbf{I}_2) + j\omega M(\mathbf{I}_2 - \mathbf{I}_1)$$

$$+ \frac{1}{j\omega C_2}\mathbf{I}_3 + R_4 \mathbf{I}_3 = 0$$

which can be rewritten in the form

$$\left(R_1 + j\omega L_1 + \frac{1}{j\omega C_1}\right)\mathbf{I}_1 - \left(j\omega L_1 + \frac{1}{j\omega C_1} - j\omega M\right)\mathbf{I}_2$$

$$- j\omega M\mathbf{I}_3 = \mathbf{V}$$

$$-\left(j\omega L_1 + \frac{1}{j\omega C_1} - j\omega M\right)\mathbf{I}_1$$

$$+\left(\frac{1}{j\omega C_1} + j\omega L_1 + R_2 + j\omega L_2 + R_3 - j2\omega M\right)\mathbf{I}_2$$

$$-\left(j\omega L_2 + R_3 - j\omega M\right)\mathbf{I}_3 = 0$$

$$-j\omega M\mathbf{I}_1 - \left(R_3 + j\omega L_2 - j\omega M\right)\mathbf{I}_2$$

$$+\left(R_3 + j\omega L_2 + \frac{1}{j\omega C_2} + R_4\right)\mathbf{I}_3 = 0$$

Note the symmetrical form of these equations.

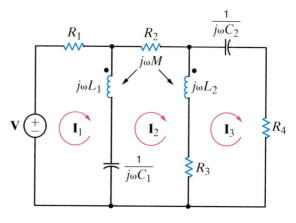

Figure 10.13 *Example of a magnetically coupled circuit.*

E10.2 Find the currents $\mathbf{I}_1$ and $\mathbf{I}_2$ and the output voltage $\mathbf{V}_o$ in the network in Fig. E10.2.   **PSV**

Figure E10.2

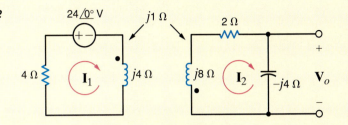

ANSWER:
$\mathbf{I}_1 = +4.29\underline{/137.2°}$ A;
$\mathbf{I}_2 = 0.96\underline{/-16.26°}$ A;
$\mathbf{V}_o = 3.84\underline{/-106.26°}$ V.

E10.3 Write the KVL equations in standard form for the network in Fig. E10.3.

Figure E10.3

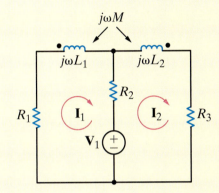

ANSWER:
$(R_1 + j\omega L_1 + R_2)\mathbf{I}_1$
$\qquad - (R_2 + j\omega M)\mathbf{I}_2 = -\mathbf{V}_1;$
$-(R_2 + j\omega M)\mathbf{I}_1$
$\qquad + (R_2 + j\omega L_2 + R_3)\mathbf{I}_2$
$\qquad\qquad = \mathbf{V}_1.$

┌─ **Example 10.6**

Given the network in Fig. 10.14 with the parameters $\mathbf{Z}_S = 3 + j1\ \Omega$, $j\omega L_1 = j2\ \Omega$, $j\omega L_2 = j2\ \Omega$, $j\omega M = j1\ \Omega$, and $\mathbf{Z}_L = 1 - j1\ \Omega$, determine the impedance seen by the source $\mathbf{V}_S$.

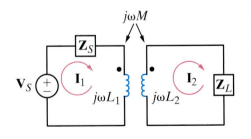

Figure 10.14
Circuit employed in Example 10.6.

SOLUTION   The mesh equations for the network are

$$\mathbf{V}_S = (\mathbf{Z}_S + j\omega L_1)\mathbf{I}_1 - j\omega M \mathbf{I}_2$$

$$0 = -j\omega M \mathbf{I}_1 + (j\omega L_2 + \mathbf{Z}_L)\mathbf{I}_2$$

If we now define $\mathbf{Z}_{11} = \mathbf{Z}_S + j\omega L_1$ and $\mathbf{Z}_{22} = j\omega L_2 + \mathbf{Z}_L$, then the second equation yields

$$\mathbf{I}_2 = \frac{j\omega M}{\mathbf{Z}_{22}}\mathbf{I}_1$$

If this secondary mesh equation is substituted into the primary mesh equation, we obtain

$$\mathbf{V}_S = \mathbf{Z}_{11}\mathbf{I}_1 + \frac{\omega^2 M^2}{\mathbf{Z}_{22}}\mathbf{I}_1$$

and therefore

$$\frac{\mathbf{V}_S}{\mathbf{I}_1} = \mathbf{Z}_{11} + \frac{\omega^2 M^2}{\mathbf{Z}_{22}}$$

which is the impedance seen by $\mathbf{V}_S$. Note that the mutual term is squared, and therefore the impedance is independent of the location of the dots.

Using the values of the circuit parameters, we find that

$$\frac{\mathbf{V}_S}{\mathbf{I}_1} = (3 + j1 + j2) + \frac{1}{j2 + 1 - j1}$$

$$= 3 + j3 + 0.5 - j0.5$$

$$= 3.5 + j2.5 \ \Omega$$

LEARNING EXTENSION

E10.4 Find the impedance seen by the source in the circuit in Fig. E10.4. **PSV**

ANSWER:
$\mathbf{Z}_S = 2.25 \underline{/20.9°} \ \Omega$.

Figure E10.4

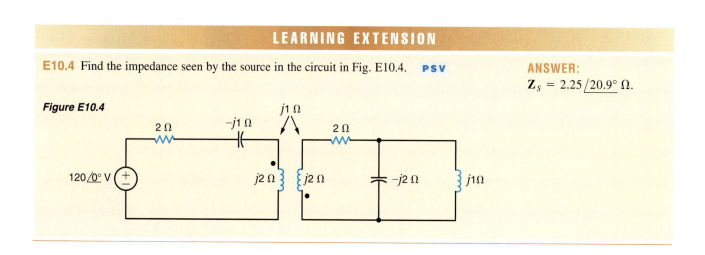

10.2 Energy Analysis

We now perform an energy analysis on a pair of mutually coupled inductors, which will yield some interesting relationships for the circuit elements. Our analysis will involve the performance of an experiment on the network shown in Fig. 10.15. Before beginning the experiment, we set all voltages and currents in the circuit equal to zero. Once the circuit is quiescent, we begin by letting the current $i_1(t)$ increase from zero to some value I_1 with the right-side terminals open circuited. Since the right-side terminals are open, $i_2(t) = 0$, and therefore the power entering these terminals is zero. The instantaneous power entering the left-side terminals is

$$p(t) = v_1(t)i_1(t) = \left[L_1 \frac{di_1(t)}{dt} \right] i_1(t)$$

The energy stored within the coupled circuit at t_1 when $i_1(t) = I_1$ is then

$$\int_0^{t_1} v_1(t)i_1(t) \ dt = \int_0^{I_1} L_1 i_1(t) \ di_1(t) = \frac{1}{2} L_1 I_1^2$$

Continuing our experiment, starting at time t_1, we let the current $i_2(t)$ increase from zero to some value I_2 at time t_2 while holding $i_1(t)$ constant at I_1. The energy delivered through the right-side terminals is

$$\int_{t_1}^{t_2} v_2(t)i_2(t) \ dt = \int_0^{I_2} L_2 i_2(t) \ di_2(t) = \frac{1}{2} L_2 I_2^2$$

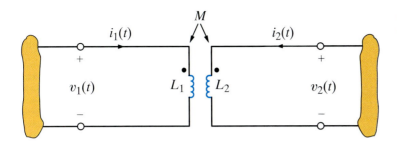

Figure 10.15
Magnetically coupled circuit.

However, during the interval t_1 to t_2 the voltage $v_1(t)$ is

$$v_1(t) = L_1 \frac{di_1(t)}{dt} + M \frac{di_2(t)}{dt}$$

Since $i_1(t)$ is a constant I_1, the energy delivered through the left-side terminals is

$$\int_{t_1}^{t_2} v_1(t) i_1(t)\, dt = \int_{t_1}^{t_2} M \frac{di_2(t)}{dt} I_1\, dt = MI_1 \int_0^{I_2} di_2(t)$$

$$= MI_1 I_2$$

Therefore, the total energy stored in the network for $t > t_2$ is

$$w = \frac{1}{2} L_1 I_1^2 + \frac{1}{2} L_2 I_2^2 + MI_1 I_2 \qquad \textbf{10.18}$$

We could, of course, repeat our entire experiment with either the dot on L_1 or L_2, but not both, reversed, and in this case the sign on the mutual inductance term would be negative, producing

$$w = \frac{1}{2} L_1 I_1^2 + \frac{1}{2} L_2 I_2^2 - MI_1 I_2$$

It is very important for the reader to realize that in our derivation of the preceding equation, by means of the experiment, the values I_1 and I_2 could have been any values at *any time*; therefore, the energy stored in the magnetically coupled inductors at any instant of time is given by the expression

$$w(t) = \frac{1}{2} L_1 [i_1(t)]^2 + \frac{1}{2} L_2 [i_2(t)]^2 \pm M i_1(t) i_2(t) \qquad \textbf{10.19}$$

The two coupled inductors represent a passive network, and therefore, the energy stored within this network must be nonnegative for any values of the inductances and currents.

The equation for the instantaneous energy stored in the magnetic circuit can be written as

$$w(t) = \frac{1}{2} L_1 i_1^2 + \frac{1}{2} L_2 i_2^2 \pm M i_1 i_2$$

Adding and subtracting the term $1/2 (M^2/L_2) i_1^2$ and rearranging the equation yields

$$w(t) = \frac{1}{2} \left(L_1 - \frac{M^2}{L_2} \right) i_1^2 + \frac{1}{2} L_2 \left(i_2 + \frac{M}{L_2} i_1 \right)^2$$

From this expression we recognize that the instantaneous energy stored will be nonnegative if

$$M \leq \sqrt{L_1 L_2} \qquad \textbf{10.20}$$

Note that this equation specifies an upper limit on the value of the mutual inductance.

We define the coefficient of coupling between the two inductors L_1 and L_2 as

$$k = \frac{M}{\sqrt{L_1 L_2}} \qquad \textbf{10.21}$$

and we note from Eq. (10.20) that its range of values is

$$0 \leq k \leq 1 \qquad \textbf{10.22}$$

This coefficient is an indication of how much flux in one coil is linked with the other coil; that is, if all the flux in one coil reaches the other coil, then we have 100% coupling and $k = 1$. For large values of k (i.e., $k > 0.5$), the inductors are said to be tightly coupled, and for small values of k (i.e., $k \leq 0.5$), the coils are said to be loosely coupled. If there is no coupling, $k = 0$. The previous equations indicate that the value for the mutual inductance is confined to the range

$$0 \leq M \leq \sqrt{L_1 L_2}$$ **10.23**

and that the upper limit is the geometric mean of the inductances L_1 and L_2.

Example 10.7

The coupled circuit in Fig. 10.16a has a coefficient of coupling of 1 (i.e., $k = 1$). We wish to determine the energy stored in the mutually coupled inductors at time $t = 5$ ms. $L_1 = 2.653$ mH and $L_2 = 10.61$ mH.

SOLUTION From the data the mutual inductance is

$$M = \sqrt{L_1 L_2} = 5.31 \text{ mH}$$

The frequency-domain equivalent circuit is shown in Fig. 10.16b, where the impedance values for X_{L_1}, X_{L_2}, and X_M are 1, 4, and 2, respectively. The mesh equations for the network are then

$$(2 + j1)\mathbf{I}_1 - j2\mathbf{I}_2 = 24 \underline{/0°}$$

$$-j2\mathbf{I}_1 + (4 + j4)\mathbf{I}_2 = 0$$

Solving these equations for the two mesh currents yields

$$\mathbf{I}_1 = 9.41 \underline{/-11.31°} \text{ A} \quad \text{and} \quad \mathbf{I}_2 = 3.33 \underline{/+33.69°} \text{ A}$$

and therefore,

$$i_1(t) = 9.41 \cos(377t - 11.31°) \text{ A}$$

$$i_2(t) = 3.33 \cos(377t + 33.69°) \text{ A}$$

At $t = 5$ ms, $377t = 1.885$ radians or $108°$, and therefore,

$$i_1(t = 5 \text{ ms}) = 9.41 \cos(108° - 11.31°) = -1.10 \text{ A}$$

$$i_2(t = 5 \text{ ms}) = 3.33 \cos(108° + 33.69°) = -2.61 \text{ A}$$

Therefore, the energy stored in the coupled inductors at $t = 5$ ms is

$$w(t)|_{t=0.005 \text{ s}} = \frac{1}{2}(2.653)(10^{-3})(-1.10)^2 + \frac{1}{2}(10.61)(10^{-3})(-2.61)^2$$

$$-(5.31)(10^{-3})(-1.10)(-2.61)$$

$$= (1.61)(10^{-3}) + (36.14)(10^{-3}) - (15.25)(10^{-3})$$

$$= 22.5 \text{ mJ}$$

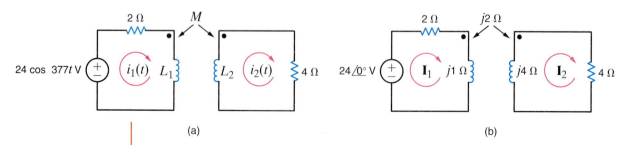

(a)   (b)

Figure 10.16 *Example of a magnetically coupled circuit drawn in the time and frequency domains.*

LEARNING EXTENSION

E10.5   The network in Fig. E10.5 operates at 60 Hz. Compute the energy stored in the mutually coupled inductors at time $t = 10$ ms.

ANSWER:
$w(10 \text{ ms}) = 39$ mJ.

Figure E10.5

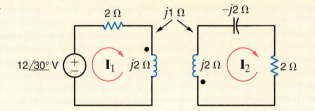

10.3   The Ideal Transformer

Consider the situation illustrated in Fig. 10.17, showing two coils of wire wound around a single closed magnetic core. Assume a core flux ϕ, which links all the turns of both coils. In the ideal case we also neglect wire resistance. Let us now examine the coupling equations under the condition that the same flux goes through each winding and so,

$$v_1(t) = N_1 \frac{d\phi}{dt}$$

and

$$v_2(t) = N_2 \frac{d\phi}{dt}$$

and therefore,

$$\frac{v_1}{v_2} = \frac{N_1 \dfrac{d\phi}{dt}}{N_2 \dfrac{d\phi}{dt}} = \frac{N_1}{N_2} \qquad \textbf{10.24}$$

Ampère's law requires that

$$\oint H \cdot dl = i_{\text{enclosed}} = N_1 i_1 + N_2 i_2 \qquad \textbf{10.25}$$

where H is the magnetic field intensity and the integral is over the closed path traveled by the flux around the transformer core. If $H = 0$, which is the case for an ideal magnetic core with infinite permeability, then

$$N_1 i_1 + N_2 i_2 = 0 \qquad \textbf{10.26}$$

or

$$\frac{i_1}{i_2} = -\frac{N_2}{N_1} \qquad \textbf{10.27}$$

Figure 10.17
Transformer employing a magnetic core.

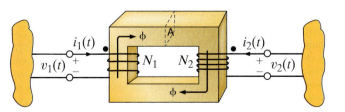

Note that if we divide Eq. (10.26) by N_1 and multiply it by v_1, we obtain

$$v_1 i_1 + \frac{N_2}{N_1} v_1 i_2 = 0$$

However, since $v_1 = (N_1/N_2)v_2$,

$$v_1 i_1 + v_2 i_2 = 0$$

and hence the total power into the device is zero, which means that an ideal transformer is lossless.

The symbol we employ for the ideal transformer is shown in Fig. 10.18a, and the corresponding equations are

$$\frac{v_1}{v_2} = \frac{N_1}{N_2}$$

$$N_1 i_1 + N_2 i_2 = 0$$

10.28

The normal power flow through a transformer occurs from an input current (i_1) on the primary to an output current (i_2) on the secondary. This situation is shown in Fig. 10.18b, and the corresponding equations are

$$\frac{v_1}{v_2} = \frac{N_1}{N_2}$$

$$N_1 i_1 = N_2 i_2$$

10.29

Figure 10.18
Symbol for an ideal transformer: (a) primary and secondary currents into the dots; (b) primary current into, and secondary current out of, the dots.

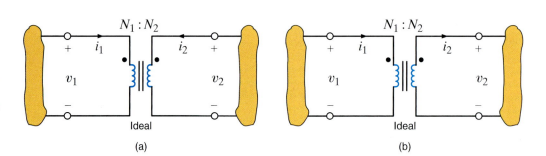

(a)   (b)

Note that although the voltage, current, and impedance levels change through a transformer, the power levels do not. The vertical lines between the coils, shown in the figures, represent the magnetic core. Although practical transformers do not use dots per se, they use markings specified by the National Electrical Manufacturers Association (NEMA) that are conceptually equivalent to the dots.

Thus, our model for the ideal transformer is specified by the circuit in Fig. 10.18a and the corresponding Eq. (10.28), or alternatively by the circuit in Fig. 10.18b, together with Eq. (10.29). Therefore, it is important to note carefully that our model specifies the equations as well as the relationship among the voltages, currents, and the position of the dots. In other words, the equations are valid only for the corresponding circuit diagram. Thus, in a direct analogy to our discussion of the mutual inductance equations and their corresponding circuit, if we change the direction of the current or voltage or the position of the dots, we must make a corresponding change in the equations. The following material will clarify this critical issue.

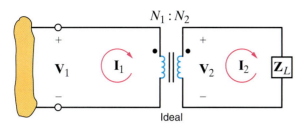

Figure 10.19
Ideal transformer circuit used to illustrate input impedance.

Consider now the circuit shown in Fig. 10.19. If we compare this circuit to that shown in Fig. 10.18b, we find that the direction of both the currents and voltages are the same. Hence the equations for the network are

$$\frac{\mathbf{V}_1}{\mathbf{V}_2} = \frac{N_1}{N_2}$$

and

$$\frac{\mathbf{I}_1}{\mathbf{I}_2} = \frac{N_2}{N_1}$$

These equations can be written as

$$\mathbf{V}_1 = \frac{N_1}{N_2} \mathbf{V}_2 \qquad \text{10.30}$$

$$\mathbf{I}_1 = \frac{N_2}{N_1} \mathbf{I}_2$$

Also note that

$$\mathbf{Z}_L = \frac{\mathbf{V}_2}{\mathbf{I}_2}$$

and therefore the input impedance is

$$\mathbf{Z}_1 = \frac{\mathbf{V}_1}{\mathbf{I}_1} = \left(\frac{N_1}{N_2}\right)^2 \mathbf{Z}_L \qquad \text{10.31}$$

where $\mathbf{Z}_L$ is reflected into the primary side by the turns ratio.

If we now define the turns ratio as

$$n = \frac{N_2}{N_1} \qquad \text{10.32}$$

then the defining equations for the *ideal transformer* in this configuration are

$$\mathbf{V}_1 = \frac{\mathbf{V}_2}{n}$$

$$\mathbf{I}_1 = n\mathbf{I}_2 \qquad \text{10.33}$$

$$\mathbf{Z}_1 = \frac{\mathbf{Z}_L}{n^2}$$

Care must be exercised in using these equations because the signs on the voltages and currents are dependent on the assigned references and their relationship to the dots.

Example 10.8

Given the circuit shown in Fig. 10.20, we wish to determine all indicated voltages and currents.

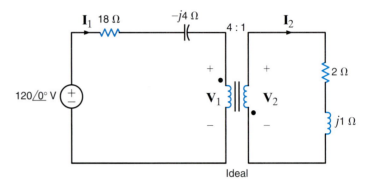

Figure 10.20 *Ideal transformer circuit.*

SOLUTION Because of the relationships between the dots and the currents and voltages, the transformer equations are

$$\mathbf{V}_1 = -\frac{\mathbf{V}_2}{n} \quad \text{and} \quad \mathbf{I}_1 = -n\mathbf{I}_2$$

where $n = 1/4$. The reflected impedance at the input to the transformer is

$$\mathbf{Z}_1 = 4^2\mathbf{Z}_L = 16(2 + j1) = 32 + j16 \ \Omega$$

Therefore, the current in the source is

$$\mathbf{I}_1 = \frac{120\,\underline{/0^\circ}}{18 - j4 + 32 + j16} = 2.33\,\underline{/-13.5^\circ}\ \text{A}$$

The voltage across the input to the transformer is then

$$\mathbf{V}_1 = \mathbf{I}_1\mathbf{Z}_1$$
$$= (2.33\,\underline{/-13.5^\circ})(32 + j16)$$
$$= 83.49\,\underline{/13.07^\circ}\ \text{V}$$

Hence, $\mathbf{V}_2$ is

$$\mathbf{V}_2 = -n\mathbf{V}_1$$
$$= -\frac{1}{4}(83.49\,\underline{/13.07^\circ})$$
$$= 20.87\,\underline{/193.07^\circ}\ \text{V}$$

The current $\mathbf{I}_2$ is

$$\mathbf{I}_2 = -\frac{\mathbf{I}_1}{n}$$
$$= -4(2.33\,\underline{/-13.5^\circ})$$
$$= 9.33\,\underline{/166.50^\circ}\ \text{A}$$

Another technique for simplifying the analysis of circuits containing an ideal transformer involves the use of either Thévenin's or Norton's theorem to obtain an equivalent circuit that replaces the transformer and either the primary or secondary circuit. This technique usually requires more effort, however, than the approach presented thus far. Let us demonstrate this approach by employing Thévenin's theorem to derive an equivalent circuit for the transformer

E10.6 Compute the current $\mathbf{I}_1$ in the network in Fig. E10.6.

ANSWER:
$\mathbf{I}_1 = 3.07\,\underline{/39.81°}\ \text{A}$.

Figure E10.6

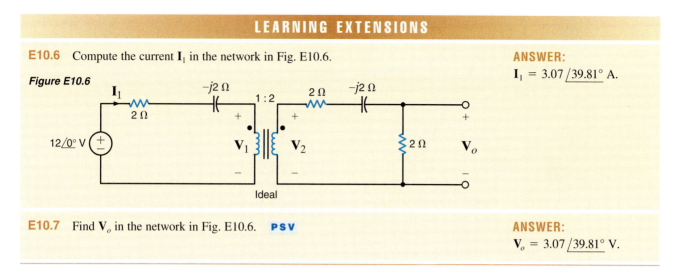

E10.7 Find $\mathbf{V}_o$ in the network in Fig. E10.6. **PSV**

ANSWER:
$\mathbf{V}_o = 3.07\,\underline{/39.81°}\ \text{V}$.

and primary circuit of the network shown in Fig. 10.21a. The equations for the transformer in view of the direction of the currents and voltages and the position of the dots are

$$\mathbf{I}_1 = n\mathbf{I}_2$$

$$\mathbf{V}_1 = \frac{\mathbf{V}_2}{n}$$

Forming a Thévenin equivalent at the secondary terminals $2 = 2'$, as shown in Fig. 10.21b, we note that $\mathbf{I}_2 = 0$ and therefore $\mathbf{I}_1 = 0$. Hence

$$\mathbf{V}_{oc} = \mathbf{V}_2 = n\mathbf{V}_1 = n\mathbf{V}_{S_1}$$

The Thévenin equivalent impedance obtained by looking into the open-circuit terminals with $\mathbf{V}_{S_1}$ replaced by a short circuit is $\mathbf{Z}_1$, which when reflected into the secondary by the turns ratio is

$$\mathbf{Z}_{\text{Th}} = n^2\mathbf{Z}_1$$

Therefore, one of the resulting equivalent circuits for the network in Fig. 10.21a is as shown in Fig. 10.21c. In a similar manner, we can show that replacing the transformer and its secondary circuit by an equivalent circuit results in the network shown in Fig. 10.21d.

It can be shown in general that when developing an equivalent circuit for the transformer and its primary circuit, each primary voltage is multiplied by n, each primary current is divided by n,

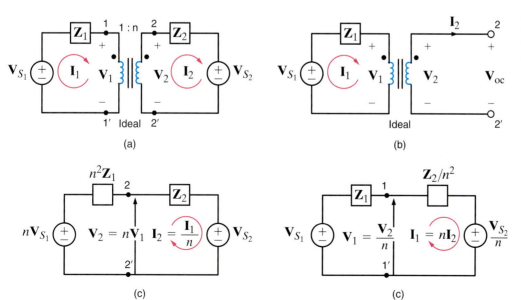

(a)

(b)

(c)

(c)

Figure 10.21
Circuit containing an ideal transformer and some of its equivalent networks.

and each primary impedance is multiplied by n^2. Similarly, when developing an equivalent circuit for the transformer and its secondary circuit, each secondary voltage is divided by n, each secondary current is multiplied by n, and each secondary impedance is divided by n^2. Powers are the same, whether calculated on the primary or secondary side.

Recall from our previous analysis that if either dot on the transformer is reversed, then n is replaced by $-n$ in the equivalent circuits. In addition, note that the development of these equivalent circuits is predicated on the assumption that removing the transformer will divide the network into two parts; that is, there are no connections between the primary and secondary other than through the transformer. If any external connections exist, the equivalent circuit technique cannot in general be used. Finally, we point out that if the primary or secondary circuits are more complicated than those shown in Fig. 10.21a, Thévenin's theorem may be applied to reduce the network to that shown in Fig. 10.21a. Also, we can simply reflect the complicated circuit component by component from one side of the transformer to the other.

Example 10.9

Given the circuit in Fig. 10.22a, we wish to draw the two networks obtained by replacing the transformer and the primary, and the transformer and the secondary, with equivalent circuits.

SOLUTION Due to the relationship between the assigned currents and voltages and the location of the dots, the network containing an equivalent circuit for the primary and the network containing an equivalent circuit for the secondary are shown in Figs. 10.22b and c, respectively. The reader should note carefully the polarity of the voltage sources in the equivalent networks.

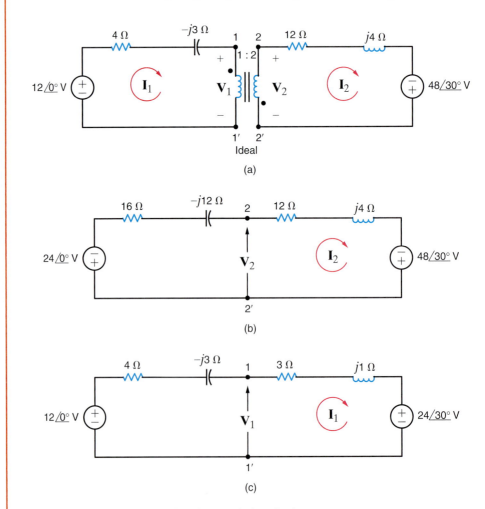

Figure 10.22 Example circuit and two equivalent circuits.

PROBLEM-SOLVING STRATEGY

Circuits Containing Ideal Transformers

Step 1. Carefully examine the circuit diagram to determine the assigned voltage polarities and current directions in relation to the transformer dots.

- If both voltages are referenced positive at the dotted terminals or undotted terminals, then $v_1/v_2 = N_1/N_2$. If this is not true, then $v_1/v_2 = -N_1/N_2$.

- If one current is defined as entering a dotted terminal and the other current is defined as leaving a dotted terminal, then $N_1i_1 = N_2i_2$. If this condition is not satisfied, then $N_1i_1 = -N_2i_2$.

Step 2. If there are no electrical connections between two transformer windings, reflect all circuit elements on one side of the transformer through to the other side, thus eliminating the ideal transformer. Be careful to apply the statements above when reflecting elements through the transformer. Remember that impedances are scaled in magnitude only. Apply circuit analysis techniques to the circuit that results from eliminating all ideal transformers. After this circuit has been analyzed, reflect voltages and currents back through the appropriate ideal transformers to find the answer.

Step 3. As an alternative approach, use Thévenin's or Norton's theorem to simplify the circuit. Typically, calculation of the equivalent circuit eliminates the ideal transformer. Solve the simplified circuit.

Step 4. If there are electrical connections between two transformer windings, use nodal analysis or mesh analysis to write equations for the circuits. Solve the equations using the proper relationships between the voltages and currents for the ideal transformer.

Example 10.10

Let us determine the output voltage $\mathbf{V}_o$ in the circuit in Fig. 10.23a.

SOLUTION We begin our attack by forming a Thévenin equivalent for the primary circuit. From Fig. 10.23b we can show that the open-circuit voltage is

$$\mathbf{V}_{oc} = \frac{24\,\underline{/0°}}{4 - j4}(-j4) - 4\,\underline{/-90°}$$

$$= 12 - j8 = 14.42\,\underline{/-33.69°}\text{ V}$$

The Thévenin equivalent impedance looking into the open-circuit terminals with the voltage sources replaced by short circuits is

$$\mathbf{Z}_{Th} = \frac{(4)(-j4)}{4 - j4} + 2$$

$$= 4 - j2\ \Omega$$

The circuit in Fig. 10.23a thus reduces to that shown in Fig. 10.23c. Forming an equivalent circuit for the transformer and primary results in the network shown in Fig. 10.23d. Therefore, the voltage $\mathbf{V}_o$ is

$$\mathbf{V}_o = \frac{-28.84\,\underline{/-33.69°}}{20 - j5}\quad(2)$$

$$= 2.80\,\underline{/160.35°}\text{ V}$$

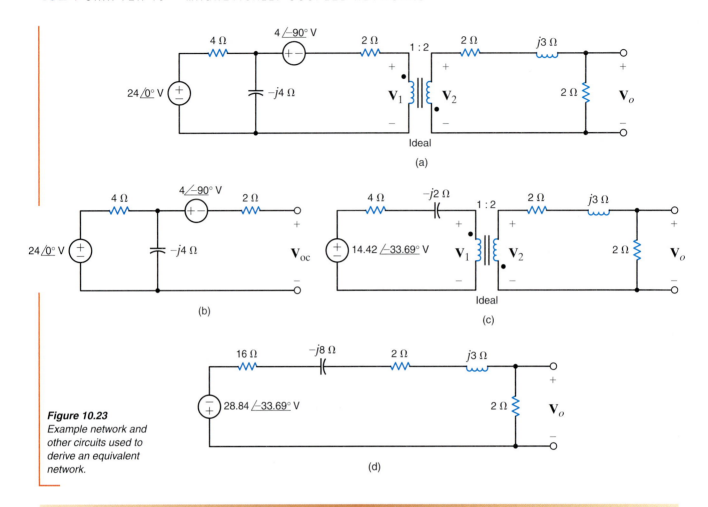

Figure 10.23
Example network and other circuits used to derive an equivalent network.

E10.8 Given the network in Fig. E10.8, form an equivalent circuit for the transformer and secondary, and use the resultant network to compute $\mathbf{I}_1$.

ANSWER:
$\mathbf{I}_1 = 13.12\,\underline{/38.66°}$ A.

Figure E10.8

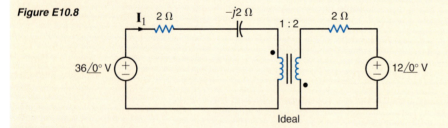

E10.9 Given the network in Fig. E10.9, form an equivalent circuit for the transformer and primary, and use the resultant network to find $\mathbf{V}_o$.

ANSWER:
$\mathbf{V}_o = 3.12\,\underline{/38.66°}$ V.

Figure E10.9

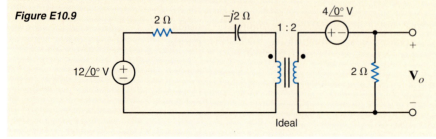

Example 10.11

Determine $\mathbf{I}_1$, $\mathbf{I}_2$, $\mathbf{V}_1$, and $\mathbf{V}_2$ in the network in Fig. 10.24.

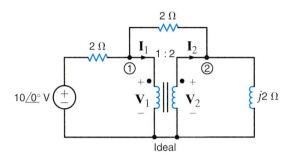

Figure 10.24
Circuit used in Example 10.11.

SOLUTION   The nodal equations at nodes 1 and 2 are

$$\frac{10 - \mathbf{V}_1}{2} = \frac{\mathbf{V}_1 - \mathbf{V}_2}{2} + \mathbf{I}_1$$

$$\mathbf{I}_2 + \frac{\mathbf{V}_1 - \mathbf{V}_2}{2} = \frac{\mathbf{V}_2}{2j}$$

The transformer relationships are $\mathbf{V}_2 = 2\mathbf{V}_1$ and $\mathbf{I}_1 = 2\mathbf{I}_2$. The first nodal equation yields $\mathbf{I}_1 = 5$ A and therefore $\mathbf{I}_2 = 2.5$ A. The second nodal equation, together with the constraint equations specified by the transformer, yields $\mathbf{V}_1 = \sqrt{5} \;\underline{/63°}$ V and $\mathbf{V}_2 = 2\sqrt{5} \;\underline{/63°}$ V.

LEARNING EXTENSION

E10.10 Determine $\mathbf{I}_1$, $\mathbf{I}_2$, $\mathbf{V}_1$, and $\mathbf{V}_2$ in the network in Fig. E10.10.   **PSV**

Figure E10.10

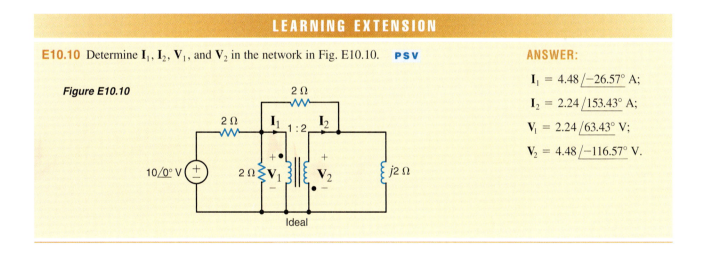

Ideal

ANSWER:

$\mathbf{I}_1 = 4.48\,\underline{/-26.57°}$ A;

$\mathbf{I}_2 = 2.24\,\underline{/153.43°}$ A;

$\mathbf{V}_1 = 2.24\,\underline{/63.43°}$ V;

$\mathbf{V}_2 = 4.48\,\underline{/-116.57°}$ V.

Before we move on to the next topic, let's return to Faraday's law. For the ideal transformer, Faraday's law tells us that $v_1(t) = N_1 \dfrac{d\phi}{dt}$ and $v_2(t) = N_2 \dfrac{d\phi}{dt}$. What if a dc voltage is applied to our transformer? In that case, the magnetic flux ϕ is a constant, $v_1 = v_2 = 0$, and our transformer is not very useful. What if an ac voltage is applied to our transformer? The magnetic flux is sinusoidal and time-varying. Transformers allow the ac voltage value to be stepped up or down easily and efficiently; it is much more difficult to efficiently step up or down the dc voltage value. The ease with which transformers allow us to change the voltage level is one of the main reasons that ac voltages and currents are utilized to transmit the bulk of the world's electrical power.

10.4 Safety Considerations

Transistors are used extensively in modern electronic equipment to provide a low-voltage power supply. As examples, a common voltage level in computer systems is 5 V dc, portable radios use 9 V dc, and military and airplane equipment operates at 28 V dc. When transformers are used to connect these low-voltage transistor circuits to the power line, there is generally less danger of shock within the system because the transformer provides electrical isolation from the line voltage. However, from a safety standpoint, a transformer, although helpful in many situations, is not an absolute solution. When working with any electrical equipment, we must always be vigilant to minimize the dangers of electrical shock.

In power electronics equipment or power systems, the danger is severe. The problem in these cases is that of high voltage from a low-impedance source, and we must constantly remember that the line voltage in our homes can be lethal.

Consider now the following example, which illustrates a hidden danger that could surprise even the experienced professional, with devastating consequences.

Example 10.12

Two adjacent homes, *A* and *B*, are fed from different transformers, as shown in Fig. 10.25a. A surge on the line feeding house *B* has caused the circuit breaker *X-Y* to open. House *B* is now left without power. In an attempt to help his neighbor, the resident of house *A* volunteers to connect a long extension cord between a wall plug in house *A* and a wall plug in house *B*, as shown in Fig. 10.25b. Later, the line technician from the utility company comes to reconnect the circuit breaker. Is the line technician in any danger in this situation?

SOLUTION   Unaware of the extension cord connection, the line technician believes that there is no voltage between points *X* and *Z*. However, because of the electrical connection between the two homes, 7200 V rms exists between the two points, and the line technician could be seriously injured or even killed if he comes in contact with this high voltage.

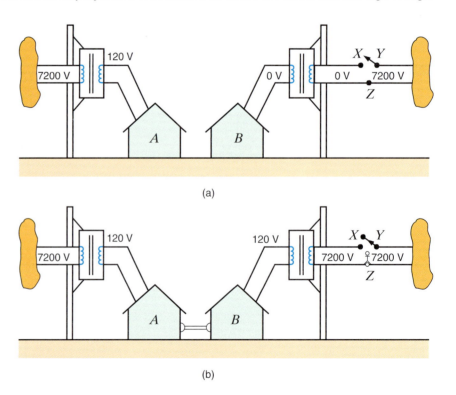

(a)

(b)

Figure 10.25 *Diagrams used in Example 10.12 (voltages in rms).*

10.5 Application Examples

The following examples demonstrate several applications for transformers.

Application Example 10.13

Consider the problem of transporting 24 MW over a distance of 100 miles (160.9 km) using a two-conductor line. Determine the requisite conductor radius to achieve a transmission efficiency of 95%, considering only the line resistance, if the line operates at (a) 240 V rms or (b) 240 kV rms. Assume that conductor resistivity is $\rho = 8 \times 10^{-8}$ Ω-m.

SOLUTION

a. At 240 V:

$$I = \frac{P}{V} = \frac{24\,M}{240} = 100 \text{ kA rms}$$

If

$$\eta = 95\%,$$

$$P_{\text{loss}} = 0.05(24\,M) = 1.2 \text{ MW} = I^2 R$$

Therefore,

$$R = \frac{P_{\text{loss}}}{I^2} = \frac{1.2\,M}{(100k)^2} = 1.2 \times 10^{-4}\Omega$$

Since

$$R = \frac{\rho l}{A} = \frac{(8 \times 10^{-8})(2 \times 160.9 \times 10^{3})}{A}$$

$$A = \frac{0.25744}{1.2 \times 10^{-4}} = 214.5\,\text{m}^2 = \pi r^2$$

Therefore,

$$r = 8.624 \text{ m}$$

(a huge conductor and totally impractical!)

b. At 240 kV rms:

$$I = 100 \text{ A rms}$$

and

$$R = \frac{1.2 \times 10^{6}}{(100)^2} = 120 \ \Omega$$

$$A = \frac{0.25744}{120} = 2.145 \times 10^{-4} \text{ m}$$

and

$$r = 0.8264 \text{ cm}$$

(which is a very practical value!)

The point of this example is that practical transmission of bulk electrical energy requires operation at high voltage. What is needed is an economical device that can efficiently convert one voltage level to another. Such a device is, as we have shown, the power transformer.

Application Example 10.14

The local transformer in Fig. 10.26 provides the last voltage stepdown in a power distribution system. A common sight on utility poles in residential areas, it is a single-phase transformer that typically has a 13.8-kV rms line to neutral on its primary coil, and a center tap secondary coil provides both 120 V rms and 240 V rms to service several residences. Let us find the turns ratio necessary to produce the 240-V rms secondary voltage. Assuming that the transformer provides 200-A rms service to each of 10 houses, let us determine the minimum power rating for the transformer and the maximum current in the primary.

SOLUTION   The turns ratio is given by

$$n = \frac{V_2}{V_1} = \frac{240}{13,800} = \frac{1}{57.5}$$

If I_H is the maximum current per household, then the maximum primary current is

$$I_1 = nI_2 = n(10I_H) = 34.78 \text{ A rms}$$

The maximum power delivered to the primary is then

$$S_1 = V_1 I_1 = (13,800)(34.78) = 480 \text{ kVA}.$$

Therefore, the transformer must have a power rating of at least 480 kVA.

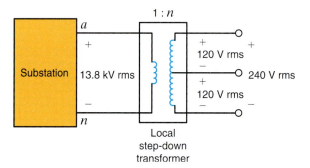

Figure 10.26 *Local transformer subcircuit with center tap.*

Application Example 10.15

Your electric toothbrush sits innocuously in its cradle overnight. Even though there are no direct electrical connections between the cradle and the toothbrush, the internal batteries are being recharged. How can this be?

SOLUTION   Mutually coupled inductors is the answer! One coil is in the cradle and energized by an ac source. The second coil is in the bottom of the toothbrush itself. When the toothbrush is mounted in the cradle, the two coils are physically close and thus mutually coupled, as shown in Fig. 10.27.

Let's investigate a reasonable design for these coils. First, we assume that the coils are poorly coupled with a coupling coefficient of $k = 0.25$. Second, the coil in the cradle is driven at $120 \underline{/0^\circ}$ V rms Third, the coil in the toothbrush should generate $6 \underline{/0^\circ}$ V at 100 mA rms in order to charge the battery. To keep the power "relatively" low, we will limit the primary current to only 0.5 A rms. Finally, to simplify our analysis, we will assume that $\mathbf{I}_1$ and $\mathbf{I}_2$ are in phase.

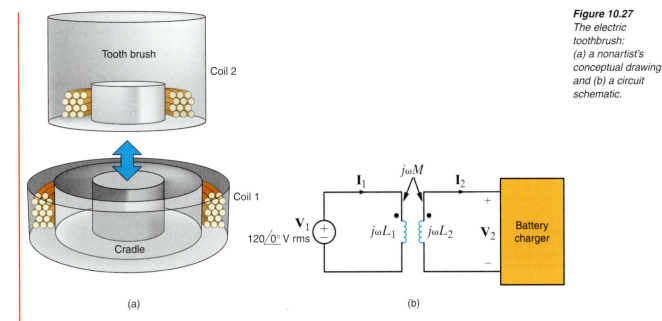

Figure 10.27
*The electric
toothbrush:
(a) a nonartist's
conceptual drawing
and (b) a circuit
schematic.*

First we develop loop equations for our circuit, which are

$$\mathbf{V}_1 = j\omega L_1 \mathbf{I}_1 - j\omega M \mathbf{I}_2$$

$$\mathbf{V}_2 = j\omega M \mathbf{I}_1 - j\omega L_2 \mathbf{I}_2$$

10.34

where $\mathbf{V}_1 = 120\underline{/0°}$ V rms and $\mathbf{V}_2 = 6\underline{/0°}$ V rms By defining a new variable α such that

$$L_2 = \alpha^2 L_1$$

we can eliminate L_2 in Eqs. (10.34). Hence,

$$\mathbf{V}_1 = j\omega L_1 \mathbf{I}_1 - j\omega k\alpha L_1 \mathbf{I}_2$$

$$\mathbf{V}_2 = j\omega k\alpha L_1 \mathbf{I}_1 - j\omega\alpha^2 L_1 \mathbf{I}_2$$

10.35

Taking the ratio of each side of Eqs. (10.35), we can eliminate ω and L_1.

$$\frac{\mathbf{V}_1}{\mathbf{V}_2} = \frac{120}{6} = 20 = \frac{\mathbf{I}_1 - k\alpha \mathbf{I}_2}{k\alpha \mathbf{I}_1 - \alpha^2 \mathbf{I}_2}$$

If we now substitute the design parameter values listed above for $\mathbf{I}_1$, $\mathbf{I}_2$ and k and then solve for α, we find

$$20 I_2 \alpha^2 - (20k I_1 + k I_2)\alpha + I_1 = 0$$

which yields

$$\alpha = \begin{cases} 0.246 \\ 1.02 \end{cases}$$

(We have used our restriction that $\mathbf{I}_1$ and $\mathbf{I}_2$ are in phase to convert current phasors to magnitudes.) Choosing the smaller value for α is the same as choosing a smaller L_2. Hence, this is the result we select since the resulting coil will require fewer turns of wire, reducing cost, weight, and size. Next, using Eqs. (10.35), we can solve for the product ωL_1.

$$\mathbf{V}_1 = \omega L_1(0.5) - \omega(0.25)(0.246)L_1(0.1) = 0.494\omega L_1$$

To investigate the effect of ω on L_1, we use the value for V_1 and the relationship between the two inductors, for the given α. In Table 10.1, L_1 and L_2 have been calculated for a collection of ω values. Note that a 60-Hz excitation requires huge values for the inductances that are completely impractical. Therefore, the table skips past the entire audible range (there's no reason to have to listen to your toothbrush recharging) to 20 kHz. Here the inductance values are much more reasonable but still considerable. However, at 100 kHz, the total inductance is just a few hundred microhenrys. These are very reasonable values and ones which we will use.

Table 10.1 A listing of frequency choices and the resulting inductances.

Frequency (Hz)	Frequency (rad/s)	L_1	L_2
60	377	693 mH	39.0 mH
20k	126k	2.01 mH	117 μH
100k	628k	416 μH	23.4 μH

The final question is this: if we have a 60-Hz sinusoid at the wall outlet, how do we obtain 100 kHz? We add a voltage-controlled switch as shown in Fig. 10.28 that is turned on and off at a rate of 100 kHz. The result is a pulsing voltage applied to the inductor at 100 kHz. Although the result is not exactly a 100-kHz sine wave, it is effective.

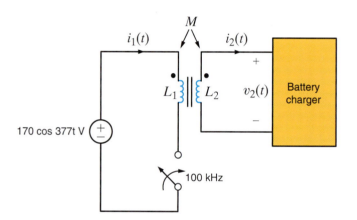

Figure 10.28 *A switch, turned on and off at a 100-kHz rate, can emulate a high-frequency ac input for our toothbrush application.*

Application Example 10.16

As shown in Fig. 10.29, two circuits are placed in close proximity: a high-current ac circuit and a low-current dc circuit. Since each circuit constitutes a loop, we should expect a little inductance in each circuit. Because of their proximity, we could also anticipate some coupling. In this particular situation the inductance in each loop is 10 nH, and the coupling coefficient is $k = 0.1$. Let us consider two scenarios. In the first case, the ac circuit contains an ac motor operating at 60 Hz. In the second case, the ac circuit models a FM radio transmitter operating at 100 MHz. We wish to determine the induced noise in the dc circuit for both cases. Which scenario produces the worst inductively coupled noise? Why?

SOLUTION The voltage induced into the dc circuit is noise and is known to be

$$\mathbf{V}_{\text{noise}} = j\omega M \mathbf{I}_{\text{AC}} = j\omega k \sqrt{L_{\text{AC}} L_{\text{DC}}} \mathbf{I}_{\text{AC}}$$

We are concerned only with the magnitude of the noise. Given the model parameters listed above, the noise voltage magnitude is

$$V_{\text{noise}} = 2\pi f(0.1)(10^{-8})(5) = 3.14 \times 10^{-8} f \text{ V}$$

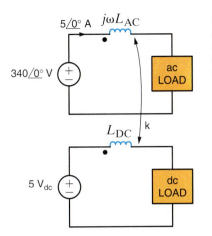

For the ac motor scenario, $f = 60$ Hz and the noise voltage is 1.88 μV—essentially zero when compared against the 5-V dc input. However, when modeling a FM radio transmitter operating at 100 MHz the noise voltage is 3.14 V. That's more than 60% of the 5-V dc level!

Thus, we find that magnetically-induced noise is much worse for high-frequency situations. It should be no surprise then that great care is taken to magnetically "shield" high-frequency–high-current circuitry.

10.6 Design Examples

Design Example 10.17

A linear variable differential transformer, or LVDT, is commonly used to measure linear movement. LVDTs are useful in a wide range of applications such as measuring the thickness of thin material sheets and measuring the physical deformation of objects under mechanical load. (A web search on LVDT will yield a multitude of other example applications with explanations and photographs.) As shown in Figs. 10.30a and b, the LVDT is just a coupled inductor apparatus with one primary winding and two secondaries that are wound and connected such that their induced voltages subtract.

All three windings are contained in a hollow cylinder that receives a rod, usually made of steel or iron, that is physically attached to whatever it is that's moving. The presence of the rod drastically increases the coupling coefficient between the windings. Let us investigate how the LVDT output voltage is related to displacement and how the LVDT is driven. Then, we will design our own LVDT, driven at 10 V rms, 2 kHz, such that at 100% travel, the output voltage magnitude equals that of the input voltage.

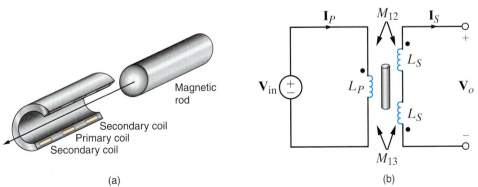

(a)   (b)

Figure 10.30
Two representations of the standard LVDT: (a) the cutaway view and (b) the circuit diagram.

SOLUTION   Typically, the primary winding of the LVDT is excited by an ac sinusoid in the range of 3 to 30 V rms at frequencies between 400 and 5000 Hz. Since we only need to measure the output voltage directly with a voltmeter, no external load is necessary.

The null position for the rod is dead center between the secondary windings. In that position the coupling between the primary and each secondary is identical, and the output voltage is zero. Should the rod move in either direction, the coupling will change linearly, as will the output voltage magnitude. The direction of travel is indicated by the relative phase of the output.

Our LVDT design begins with the circuit in Fig. 10.30b where the mutual coupling coefficient for each secondary winding varies as shown in Fig. 10.31. To create a linear relationship between displacement and output voltage, we restrict the nominal travel to that portion of Fig. 10.31 where the coupling coefficient is linear with displacement. Therefore, in this design, 100% travel will correspond to a coupling coefficient of 0.8.

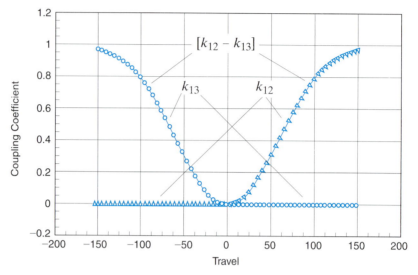

Figure 10.31 *Coupling coefficients for each secondary winding and the coupling difference. It is the difference that will determine the output voltage magnitude.*

Applying KVL to the primary loop yields

$$\mathbf{V}_{in} = j\omega L_P \mathbf{I}_P + j\omega M_{13}\mathbf{I}_S - j\omega M_{12}\mathbf{I}_S \qquad \textbf{10.36}$$

At the secondary, the KVL equation is

$$2(j\omega L_S)\mathbf{I}_S + j\omega M_{13}\mathbf{I}_P - j\omega M_{12}\mathbf{I}_P + \mathbf{V}_o = 0 \qquad \textbf{10.37}$$

With no load at the output, $\mathbf{I}_S = 0$ and Eqs. (10.36) and (10.37) reduce to

$$\mathbf{V}_{in} = j\omega L_P \mathbf{I}_P \quad \text{and} \quad \mathbf{V}_o = \mathbf{I}_P j\omega[M_{12} - M_{13}] \qquad \textbf{10.38}$$

Solving these for the output voltage and recognizing that $M_{1X} = k_{1X}[L_P L_S]^{0.5}$, we obtain

$$\mathbf{V}_o = \mathbf{V}_{in}\sqrt{\frac{L_S}{L_P}}[k_{12} - k_{13}] \qquad \textbf{10.39}$$

We can express the coupling coefficients for each secondary in terms of the percent of travel.

$$k_{12} = \begin{cases} 0.008x & \text{for } 0 < x < 100 \\ 0 & \text{for } x < 0 \end{cases} \qquad k_{13} = \begin{cases} 0.008x & \text{for } -100 < x < 0 \\ 0 & \text{for } x > 0 \end{cases}$$

And, finally, assuming that the input voltage has zero phase angle, the output voltage can be expressed as

$$\mathbf{V}_o = \mathbf{V}_{in}\sqrt{\frac{L_S}{L_P}}[0.008x] = \mathbf{V}_{in}\sqrt{\frac{L_S}{L_P}}[0.008x]\underline{/0°} \qquad 0 < x < 100$$

$$\textbf{10.40}$$

$$\mathbf{V}_o = \mathbf{V}_{in}\sqrt{\frac{L_S}{L_P}}[-0.008x] = \mathbf{V}_{in}\sqrt{\frac{L_S}{L_P}}[+0.008x]\underline{/-180°} \qquad -100 < x < 0$$

Note the phase angle difference for positive versus negative travel.

To complete our analysis, we must determine a value for the secondary to primary inductance ratio. At 100% travel, the magnitudes of the input and output voltages are equal and $k = 0.8$. Using this information in Eq. (10.40), we find that the inductor ratio must be $L_S/L_P = 1.25^2 = 1.5625$.

To determine actual values for the inductances, we must consider the input current we will tolerate at the primary. We would prefer a relatively small current, because a large current would require large-diameter wire in the primary winding. Let us choose a primary current of 25 mA rms with an excitation of 10 V rms at 2000 Hz. From Eq. (10.38), the primary inductance will be

$$L_P = \frac{V_{in}}{\omega I_P} = \frac{10}{2\pi(2000)(0.025)} = 31.8 \text{ mH}$$

which yields a secondary inductance of

$$L_S = 1.5625 L_P = 49.7 \text{ mH}$$

The selection of the two inductances completes this design.

The next example illustrates a technique for employing a transformer in a configuration that will extend the life of a set of Christmas tree lights.

Design Example 10.18

The bulbs in a set of Christmas tree lights normally operate at 120 V rms. However, they last much longer if they are instead connected to 108 V rms. Using a 120 V–12 V transformer, let us design an autotransformer that will provide 108 V rms to the bulbs.

SOLUTION   The two-winding transformers we have presented thus far provide electrical isolation between primary and secondary windings, as shown in Fig. 10.32a. It is possible, however, to interconnect primary and secondary windings serially, creating a three-terminal device, known as an autotransformer, as shown in Fig. 10.32b and represented in Fig. 10.32c.

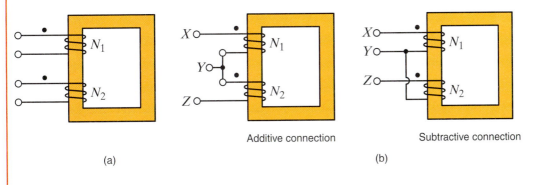

Additive connection   Subtractive connection

(a)   (b)

Figure 10.32
Autotransformer:
(a) normal two-winding transformer with adjacent windings;
(b) two-winding transformer interconnected to create a single-winding, three-terminal autotransformer;
(c) symbolic representation of (b).

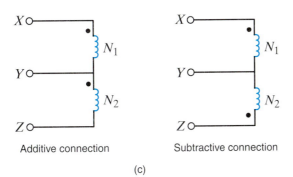

Additive connection   Subtractive connection

(c)

As we shall see, this arrangement offers certain practical advantages over the isolated case. Note that the three-terminal arrangement is essentially one continuous winding with an internal tap.

To reduce the voltage from 120 V rms to 108 V rms, the two coils must be connected such that their voltages are in opposition to each other, corresponding to a subtractive connection (in Fig. 10.32b), as shown in Fig. 10.33. In this arrangement, the voltage across both coils is

$$V_o = V_1 - V_2 = 120 - 12 = 108 \text{ V rms}$$

and the lights are simply connected across both coils.

Figure 10.33
Autotransformer for low-voltage Christmas tree lights.

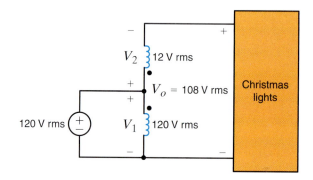

Design Example 10.19

Many electronics products today are powered by low-power ac to dc converters. (These units simply convert an ac signal at the input to a constant dc signal at the output.) They are normally called wall transformers and plug directly into a 120 V rms utility outlet. They typically have dc output voltages in the range of 5 to 18 V. As shown in Fig. 10.34, there are three basic components in a wall transformer: a simple transformer, an ac to dc converter, and a controller. A particular wall transformer is required that has a dc output of 9 V and a maximum power output of 2 W at an efficiency of only 60%. In addition, the ac to dc converter requires a peak ac voltage input of 12 V for proper operation. We wish to design the transformer by selecting its turns ratio and current rating.

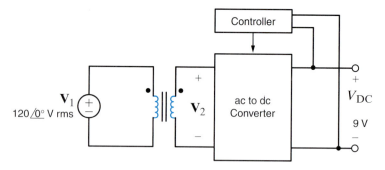

Figure 10.34 A *A block diagram for a simple wall transformer. These devices convert ac voltages (typically 120 V rms) to a dc voltage at a fairly low power level.*

SOLUTION First we consider the necessary turns ratio for the transformer. We must determine the voltage ratio, V_2/V_1. From the specifications, V_2 must have a peak value of at least 12 V. We will include some safety margin and design for V_2 around 13.5 V. Since V_1 is 120 V rms, its peak value is 169.7 V. Therefore,

$$n = \frac{V_2}{V_1} = \frac{169.7}{13.5} = 12.6$$

Thus, the V_2/V_1 ratio is 12.6. We will use a turns ratio of 12.5:1, or 25:2. Next we consider the power requirement. The maximum load is 2 W. At an efficiency of 60%, the maximum input power to the unit is

$$P_{in} = \frac{P_{out}}{\eta} = \frac{2}{0.6} = 3.33 \text{ W}$$

At 120 V rms, the input current is only

$$I_{in} = \frac{P_{in}}{V_{in}} = \frac{3.33}{120} = 27.8 \text{ mA rms}$$

Therefore, specifying a transformer with a turns ratio of 25:2 and a current rating of 100 mA rms should provide an excellent safety margin.

SUMMARY

- **Mutual inductance**   Mutual inductance occurs when inductors are placed in close proximity to one another and share a common magnetic flux.

- **The dot convention for mutual inductance**
  The dot convention governs the sign of the induced voltage in one coil based on the current direction in another.

- **The relationship between the mutual inductance and self-inductance of two coils**
  An energy analysis indicates that $M = k\sqrt{L_1 L_2}$, where k, the coefficient of coupling, has a value between 0 and 1.

- **The ideal transformer**   An ideal transformer has infinite core permeability and winding conductance. The voltage and current can be transformed between the primary

and secondary ends based on the ratio of the number of winding turns between the primary and secondary.

- **The dot convention for an ideal transformer**
  The dot convention for ideal transformers, like that for mutual inductance, specifies the manner in which a current in one winding induces a voltage in another winding.

- **Equivalent circuits involving ideal transformers**   Based on the location of the circuits' unknowns, either the primary or secondary can be reflected to the other side of the transformer to form a single circuit containing the desired unknown. The reflected voltages, currents, and impedances are a function of the dot convention and turns ratio.

PROBLEMS

PSV **CS** both available on the web at: http://www.justask4u.com/irwin

SECTION 10.1

10.1 Given the network in Fig. P10.1,

    (a) find the equations for $v_a(t)$ and $v_b(t)$.

    (b) find the equations for $v_c(t)$ and $v_d(t)$.

10.2 Given the network in Fig. P10.2,

    (a) write the equations for $v_a(t)$ and $v_b(t)$.

    (b) write the equations for $v_c(t)$ and $v_d(t)$.

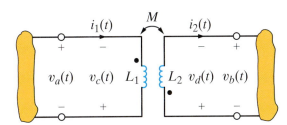

Figure P10.1

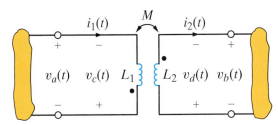

Figure P10.2

10.3 Given the network in Fig. P10.3, **CS**

    (a) find the equations for $v_a(t)$ and $v_b(t)$.

    (b) find the equations for $v_c(t)$ and $v_d(t)$.

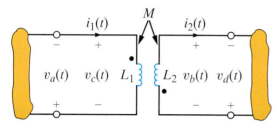

Figure P10.3

10.4 Given the network in Fig. P10.4,

    (a) write the equations for $v_a(t)$ and $v_b(t)$.

    (b) write the equations for $v_c(t)$ and $v_d(t)$.

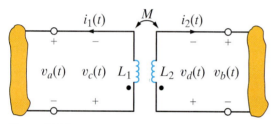

Figure P10.4

10.5 Find the voltage gain V_o/V_S of the network shown in Fig. P10.5. **CS**

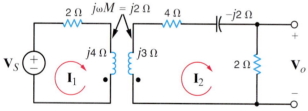

Figure P10.5

10.6 Find V_o in the network in Fig. P10.6.

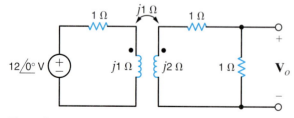

Figure P10.6

10.7 Given the network in Fig. P10.7, find V_o. **PSV**

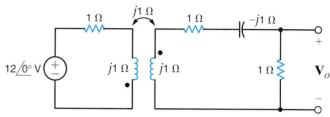

Figure P10.7

10.8 Find V_o in the circuit in Fig. P10.8. **CS**

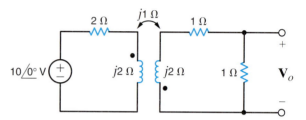

Figure P10.8

10.9 Find the voltage gain V_o/V_S of the network shown in Fig. P10.9.

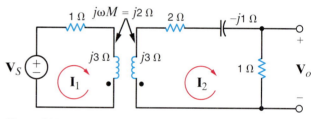

Figure P10.9

10.10 Find V_o in the network in Fig. P10.10.

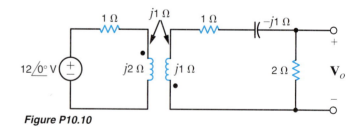

Figure P10.10

10.11 Find V_o in the network in Fig. P10.11.

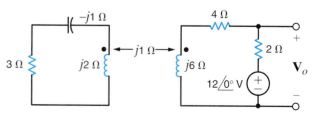

Figure P10.11

10.12 Find $\mathbf{V}_o$ in the circuit in Fig. P10.12.

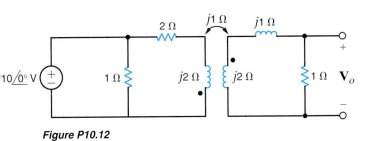

Figure P10.12

10.13 Find $\mathbf{V}_o$ in the network in Fig. P10.13.

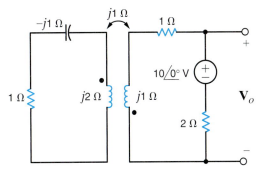

Figure P10.13

10.14 Find $\mathbf{V}_o$ in the network in Fig. P10.14.

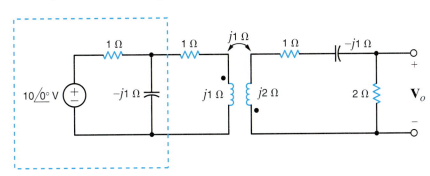

Figure P10.14

10.15 Find $\mathbf{V}_o$ in the network in Fig. P10.15.

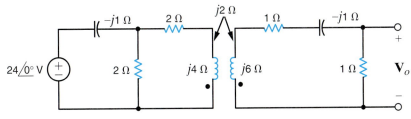

Figure P10.15

10.16 Find $\mathbf{V}_o$ in the circuit in Fig. P10.16. **cs**

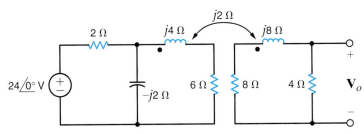

Figure P10.16

10.17 Find $\mathbf{V}_o$ in the network in Fig. P10.17.

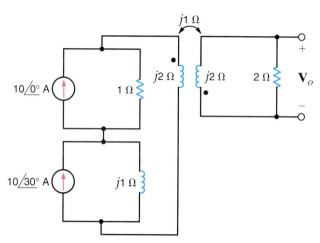

Figure P10.17

10.18 Find $\mathbf{I}_o$ in the circuit in Fig. P10.18.

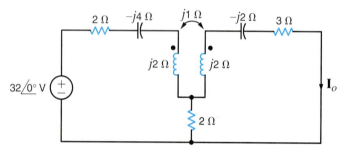

Figure P10.18

10.19 Write the mesh equations for the network in Fig. P10.19.

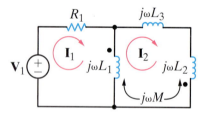

Figure P10.19

10.20 Write the mesh equations for the network in Fig. P10.20.

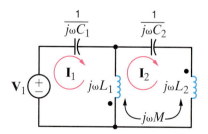

Figure P10.20

10.21 Write the mesh equations for the network shown in Fig. P10.21. **cs**

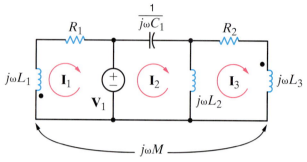

Figure P10.21

10.22 Write the mesh equations for the network shown in Fig. P10.22.

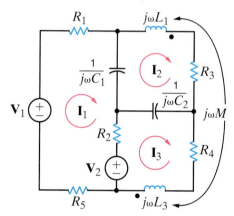

Figure P10.22

10.23 Write the mesh equations for the network in Fig. P10.23.

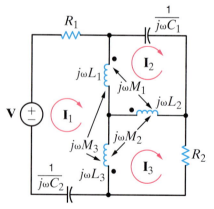

Figure P10.23

10.24 Find $\mathbf{V}_o$ in the network in Fig. P10.24.

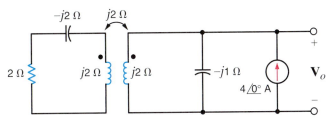

Figure P10.24

10.25 Find $\mathbf{V}_o$ in the network in Fig. P10.25. **CS**

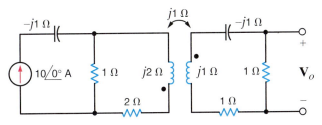

Figure P10.25

10.26 Find $\mathbf{V}_o$ in the network in Fig. P10.26.

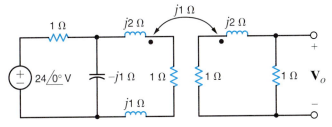

Figure P10.26

10.27 Find $\mathbf{V}_o$ in the network in Fig. P10.27. **PSV**

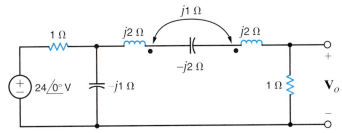

Figure P10.27

10.28 Find $\mathbf{V}_o$ in the network in Fig. P10.28. **CS**

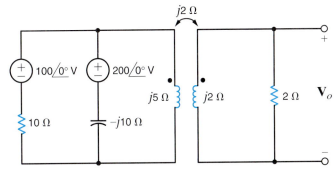

Figure P10.28

10.29 Find $\mathbf{V}_o$ in the network in Fig. P10.29.

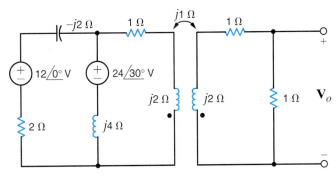

Figure P10.29

10.30 Find $\mathbf{V}_o$ in the network in Fig. P10.30.

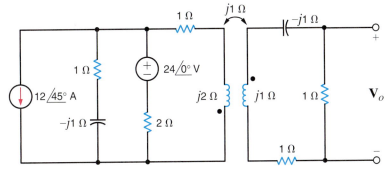

Figure P10.30

10.31 Find $\mathbf{V}_o$ in the network in Fig. P10.31. **CS**

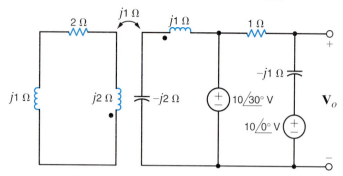

Figure P10.31

10.32 Find $\mathbf{I}_o$ in the circuit in Fig. P10.32.

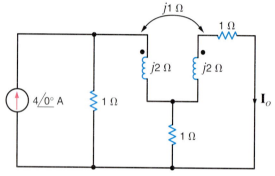

Figure P10.32

10.33 Find $\mathbf{I}_o$ in the circuit in Fig. P10.33.

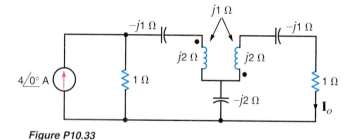

Figure P10.33

10.34 Find $\mathbf{V}_o$ in the network in Fig. P10.34.

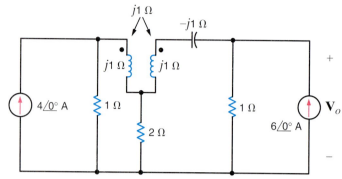

Figure P10.34

10.35 Find $\mathbf{V}_o$ in the circuit in Fig. P10.35. **PSV**

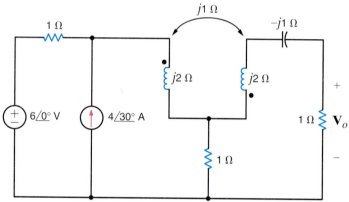

Figure P10.35

10.36 Find $\mathbf{V}_o$ in the network in Fig. P10.36. **CS**

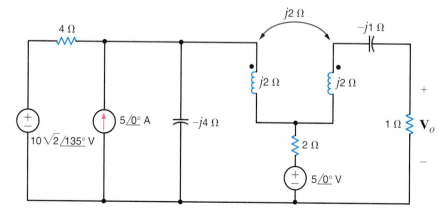

Figure P10.36

10.37 Determine the impedance seen by the source in the network shown in Fig. P10.37.

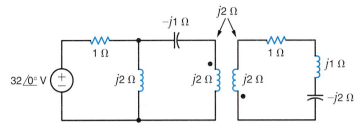

Figure P10.37

10.38 Determine the impedance seen by the source in the network shown in Fig. P10.38.

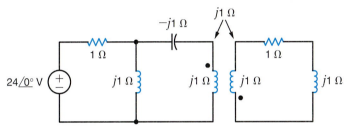

Figure P10.38

10.39 Determine the input impedance $\mathbf{Z}_{in}$ in the network in Fig. P10.39.

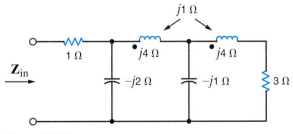

Figure P10.39

10.40 Determine the input impedance $\mathbf{Z}_{in}$ of the circuit in Fig. P10.40. **CS**

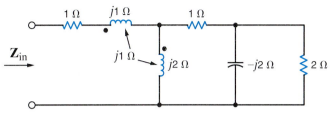

Figure P10.40

10.41 Given the network shown in Fig. P10.41, determine the value of the capacitor C that will cause the impedance seen by the $24\,\underline{/0°}$ V voltage source to be purely resistive. $f = 60$ Hz.

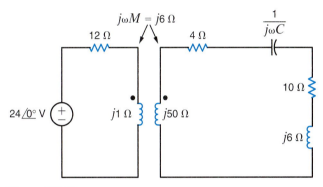

Figure P10.41

10.42 Analyze the network in Fig. P10.42 and determine whether a value of X_C can be found such that the output voltage is equal to twice the input voltage.

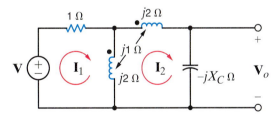

Figure P10.42

SECTION 10.2

10.43 Two coils in a network are positioned such that there is 100% coupling between them. If the inductance of one coil is 10 mH and the mutual inductance is 6 mH, compute the inductance of the other coil. **CS**

10.44 The currents in the network in Fig. P10.44 are known to be $i_1(t) = 10\cos(377t - 30°)$ mA and $i_2(t) = 20\cos(377t - 45°)$ mA. The inductances are $L_1 = 2$ H, $L_2 = 2$ H, and $k = 0.8$. Determine $v_1(t)$ and $v_2(t)$.

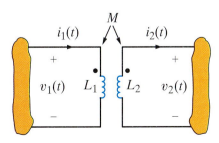

Figure P10.44

10.45 Determine the energy stored in the coupled inductors in the circuit in P10.44 at $t = 1$ ms.

10.46 The currents in the magnetically-coupled inductors shown in Fig. P10.46 are known to be $i_1(t) = 8\cos(377t - 20°)$ mA and $i_2(t) = 4\cos(377t - 50°)$ mA. The inductor values are $L_1 = 2$ H, $L_2 = 1$ H, and $k = 0.6$. Determine $v_1(t)$ and $v_2(t)$.

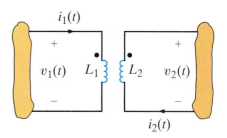

Figure P10.46

10.47 Determine the energy stored in the coupled inductors in Problem 10.46 at $t = 1$ ms. **CS**

SECTION 10.3

10.48 Determine $\mathbf{I}_1$, $\mathbf{I}_2$, $\mathbf{V}_1$, and $\mathbf{V}_2$ in the network in Fig. P10.48.

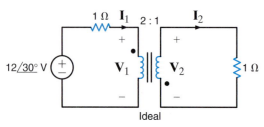

Figure P10.48

10.49 Find all currents and voltages in the network in Fig. P10.49.

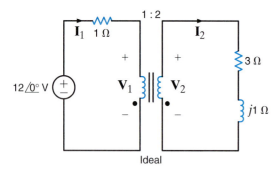

Figure P10.49

10.50 Determine $\mathbf{V}_o$ in the circuit in Fig. P10.50. **PSV**

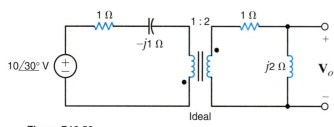

Figure P10.50

10.51 Determine $\mathbf{V}_o$ in the circuit in Fig. P10.51.

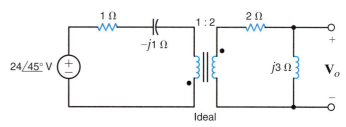

Figure P10.51

10.52 Determine $\mathbf{I}_1$, $\mathbf{I}_2$, $\mathbf{V}_1$, and $\mathbf{V}_2$ in the network in Fig. P10.52. **CS**

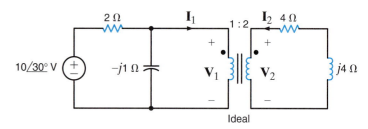

Figure P10.52

10.53 Determine $\mathbf{I}_1$, $\mathbf{I}_2$, $\mathbf{V}_1$, and $\mathbf{V}_2$ in the network in Fig. P10.53.

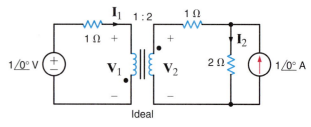

Figure P10.53

10.54 Find **I** in the network in Fig. P10.54.

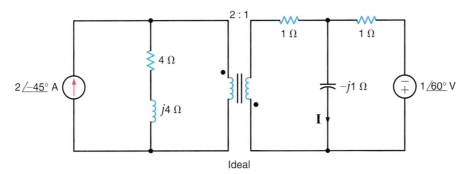

Figure P10.54

10.55 Determine $\mathbf{I}_1$, $\mathbf{I}_2$, $\mathbf{V}_1$, and $\mathbf{V}_2$ in the network in Fig. P10.55.

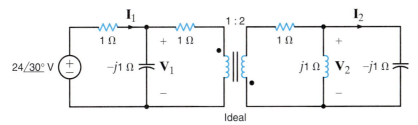

Figure P10.55

10.56 Find the current **I** in the network in Fig. P10.56.

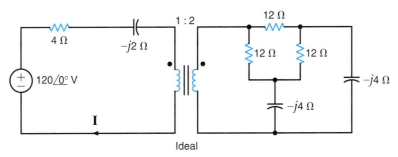

Figure P10.56

10.57 Determine $\mathbf{I}_1$, $\mathbf{I}_2$, $\mathbf{V}_1$, and $\mathbf{V}_2$ in the network in Fig. P10.57.

10.58 Find $\mathbf{V}_o$ in the network in Fig. P10.58.

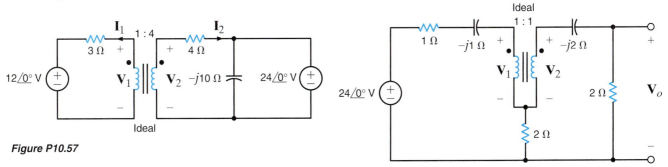

Figure P10.57

Figure P10.58

10.59 Find the current **I** in the network in Fig. P10.59. **CS**

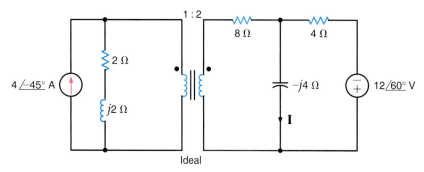

Figure P10.59

10.60 Find the voltage **V**$_o$ in the network in Fig. P10.60. **PSV**

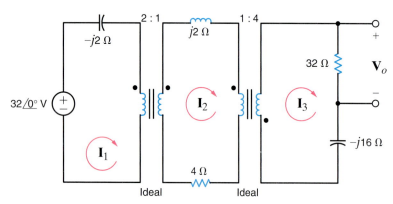

Figure P10.60

10.61 Find **V**$_o$ in the circuit in Fig. P10.61.

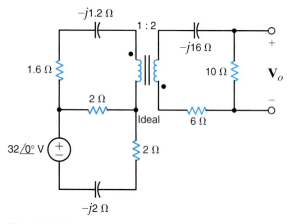

Figure P10.61

10.62 Find **V**$_o$ in the network in Fig. P10.62.

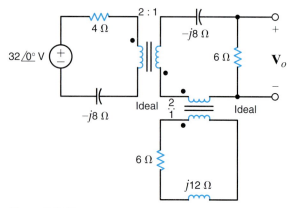

Figure P10.62

10.63 Find $\mathbf{V}_o$ in the circuit in Fig. P10.63. **cs**

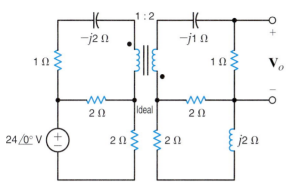

Figure P10.63

10.64 Determine the input impedance seen by the source in the circuit in Fig. P10.64.

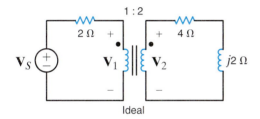

Figure P10.64

10.65 Determine the input impedance seen by the source in the circuit in Fig. P10.65. **cs**

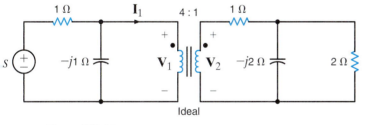

Figure P10.65

10.66 Determine the input impedance seen by the source in the network shown in Fig. P10.66.

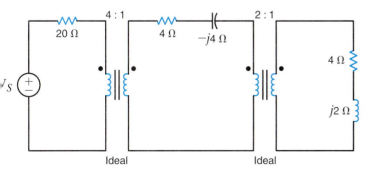

Figure P10.66

10.67 Determine the input impedance seen by the source in the network shown in Fig. P10.67.

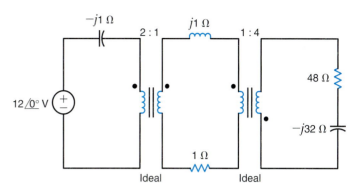

Figure P10.67

10.68 The output stage of an amplifier in an old radio is to be matched to the impedance of a speaker, as shown in Fig. P10.68. If the impedance of the speaker is 8 Ω and the amplifier requires a load impedance of 3.2 kΩ, determine the turns ratio of the ideal transformer.

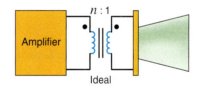

Figure P10.68

10.69 Determine $\mathbf{V}_S$ in the circuit in Fig. P10.69.

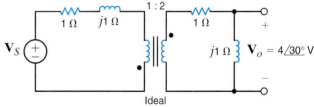

Figure P10.69

10.70 Determine $\mathbf{I}_S$ in the circuit in Fig. P10.70. **PSV**

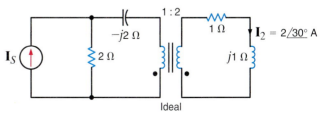

Figure P10.70

10.71 Given that $\mathbf{V}_o = 48\underline{/30°}$ V in the circuit shown in Fig. P10.71, determine $\mathbf{V}_S$. **CS**

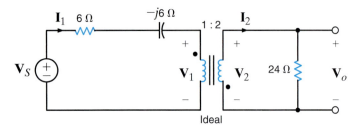

Figure P10.71

10.72 In the circuit in Fig. P10.72, if $\mathbf{I}_x = 4\underline{/30°}$ A, find $\mathbf{V}_o$.

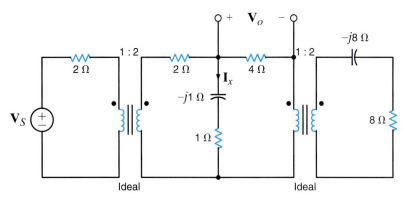

Figure P10.72

10.73 In the network in Fig. P10.73, if $\mathbf{I}_1 = 6\underline{/0°}$ A, find $\mathbf{V}_S$.

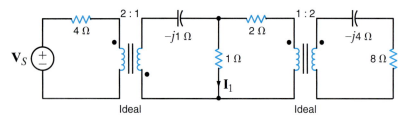

Figure P10.73

10.74 For maximum power transfer, we desire to match the impedance of the inverting amplifier stage in Fig. P10.74 to the 50-Ω equivalent resistance of the ac input source. However, standard op-amps perform best when the resistances around them are at least a few hundred ohms. The gain of the op-amp circuit should be −10. Design the complete circuit by selecting resistors no smaller than 1 kΩ and specifying the turns ratio of the ideal transformer to satisfy both the gain and impedance matching requirements.

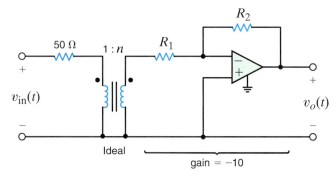

Figure P10.74

10.75 Digital clocks often divide a 60-Hz frequency signal to obtain a 1-second, 1-minute, or 1-hour signal. A convenient source of this 60-Hz signal is the power line. However, 120 volts is too high to be used by the low-power electronics. Instead, a 3-V, 60-Hz signal is needed. If a resistive voltage divider is used to drop the voltage from 120 V to 3 V, the heat generated will be unacceptable. In addition, it is costly to use a trans-

former in this application. Digital clocks are consumer items and must be very inexpensive to be a competitive product. The problem then is to design a circuit that will produce between 2.5 V and 3 V at 60 Hz from the 120-V ac power line without dissipating any heat or the use of a transformer. The design will interface with a circuit that has an input resistance of 1200 ohms.

TYPICAL PROBLEMS FOUND ON THE FE EXAM

10FE-1 In the network in Fig. 10PFE-1, find the impedance seen by the source. **CS**

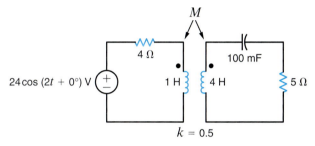

Figure 10PFE-1

10FE-2 In the circuit in Fig. 10PFE-2, select the value of the transformer's turns ratio $n = N_2/N_1$ to achieve impedance matching for maximum power transfer. Using this value of n, calculate the power absorbed by the 3-Ω resistor.

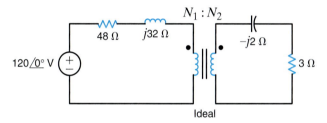

Figure 10PFE-2

10FE-3 In the circuit in Fig. 10FE-3, select the turns ratio of the ideal transformer that will match the output of the transistor amplifier to the speaker represented by the 16-Ω load. **CS**

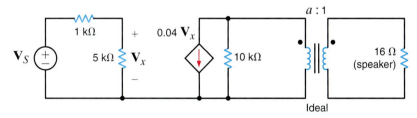

Figure 10PFE-3

Polyphase Circuits

Access Problem-Solving Videos **PSV** *and Circuit Solutions* **CS** *at:* **http://www.justask4u.com/irwin** using the registration code on the inside cover and see a website with answers and more!

While traveling throughout our city or across the countryside, we often encounter power transmission lines overhead. If we look carefully, we find that there are typically three large conductors. Depending on the configuration, there may be one or two smaller conductors that are placed above the power conductors and used for lightning protection. This three-phase power distribution line is employed to transmit power from the generating station to the user. This three-phase circuit is important because we can show that (1) power distributed in this manner is steady, rather than pulsating, as is the case in single-phase distribution and, therefore, there is less wear and tear on mechanical conversion equipment; and (2) by using a three-phase set of voltages we can deliver the same amount of power as that delivered in the single-phase case with fewer conductors and therefore much less material. ●

11.1 Three-Phase Circuits

In this chapter we add a new dimension to our study of ac steady-state circuits. Up to this point we have dealt with what we refer to as single-phase circuits. Now we extend our analysis techniques to polyphase circuits or, more specifically, three-phase circuits (that is, circuits containing three voltage sources that are one-third of a cycle apart in time).

We study three-phase circuits for a number of important reasons. It is more advantageous and economical to generate and transmit electric power in the polyphase mode than with single-phase systems. As a result, most electric power is transmitted in polyphase circuits. In the United States the power system frequency is 60 Hz, whereas in other parts of the world 50 Hz is common.

Power transmission is most efficiently accomplished at very high voltage. Since this voltage can be extremely high in comparison to the level at which it is normally used (e.g., in the household), there is a need to raise and lower the voltage. This can be accomplished in ac systems using transformers, which we studied in Chapter 10.

As the name implies, three-phase circuits are those in which the forcing function is a three-phase system of voltages. If the three sinusoidal voltages have the same magnitude and frequency and each voltage is 120° out of phase with the other two, the voltages are said to be *balanced*. If the loads are such that the currents produced by the voltages are also balanced, the entire circuit is referred to as a *balanced three-phase circuit*.

A balanced set of three-phase voltages can be represented in the frequency domain as shown in Fig. 11.1a, where we have assumed that their magnitudes are 120 V rms. From the figure we note that

$$\mathbf{V}_{an} = 120\,\underline{/0°}\ \text{V rms}$$

$$\mathbf{V}_{bn} = 120\,\underline{/-120°}\ \text{V rms}$$

$$\mathbf{V}_{cn} = 120\,\underline{/-240°}\ \text{V rms}$$

$$= 120\,\underline{/120°}\ \text{V rms}$$

11.1

Note that our double-subscript notation is exactly the same as that employed in the earlier chapters; that is, $\mathbf{V}_{an}$ means the voltage at point a with respect to the point n. We will also employ the double-subscript notation for currents; that is, $\mathbf{I}_{an}$ is used to represent the current from a to n. However, we must be very careful in this case to describe the precise path, since in a circuit there will be more than one path between the two points. For example, in the case of a single loop the two possible currents in the two paths will be 180° out of phase with each other.

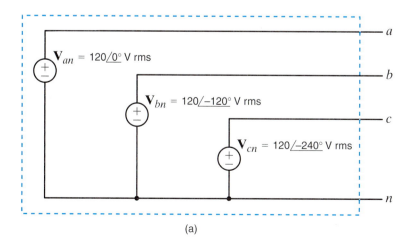

(a)

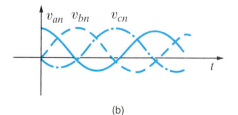

(b)

Figure 11.1
Balanced three-phase voltages.

The preceding phasor voltages can be expressed in the time domain as

$$v_{an}(t) = 120\sqrt{2}\cos\omega t \text{ V}$$

$$v_{bn}(t) = 120\sqrt{2}\cos(\omega t - 120°) \text{ V}$$

$$v_{cn}(t) = 120\sqrt{2}\cos(\omega t - 240°) \text{ V}$$

11.2

These time functions are shown in Fig. 11.1b.

Finally, let us examine the instantaneous power generated by a three-phase system. Assume that the voltages in Fig. 11.1 are

$$v_{an}(t) = V_m\cos\omega t \text{ V}$$

$$v_{bn}(t) = V_m\cos(\omega t - 120°) \text{ V}$$

$$v_{cn}(t) = V_m\cos(\omega t - 240°) \text{ V}$$

11.3

If the load is balanced, the currents produced by the sources are

$$i_a(t) = I_m\cos(\omega t - \theta) \text{ A}$$

$$i_b(t) = I_m\cos(\omega t - \theta - 120°) \text{ A}$$

$$i_c(t) = I_m\cos(\omega t - \theta - 240°) \text{ A}$$

11.4

The instantaneous power produced by the system is

$$
\begin{aligned}
p(t) &= p_a(t) + p_b(t) + p_c(t) \\
&= V_m I_m[\cos\omega t\cos(\omega t - \theta) + \cos(\omega t - 120°)\cos(\omega t - \theta - 120°) \\
&\quad + \cos(\omega t - 240°)\cos(\omega t - \theta - 240°)]
\end{aligned}
$$

11.5

Using the trigonometric identity

$$\cos\alpha\cos\beta = \frac{1}{2}\big[\cos(\alpha - \beta) + \cos(\alpha + \beta)\big]$$

11.6

Eq. (11.5) becomes

$$
\begin{aligned}
p(t) = \frac{V_m I_m}{2}\big[&\cos\theta + \cos(2\omega t - \theta) + \cos\theta \\
&+ \cos(2\omega t - \theta - 240°) + \cos\theta + \cos(2\omega t - \theta - 480°)\big]
\end{aligned}
$$

11.7

which can be written as

$$
\begin{aligned}
p(t) = \frac{V_m I_m}{2}\big[&3\cos\theta + \cos(2\omega t - \theta) \\
&+ \cos(2\omega t - \theta - 120°) + \cos(2\omega t - \theta + 120°)\big]
\end{aligned}
$$

11.8

There exists a trigonometric identity that allows us to simplify the preceding expression. The identity, which we will prove later using phasors, is

$$\cos\phi + \cos(\phi - 120°) + \cos(\phi + 120°) = 0$$

11.9

If we employ this identity, the expression for the power becomes

$$p(t) = 3\frac{V_m I_m}{2}\cos\theta \text{ W}$$

11.10

Note that this equation indicates that the instantaneous power is always constant in time, rather than pulsating as in the single-phase case. Therefore, power delivery from a three-phase voltage source is very smooth, which is another important reason why power is generated in three-phase form.

11.2 Three-Phase Connections

By far the most important polyphase voltage source is the balanced three-phase source. This source, as illustrated by Fig. 11.2, has the following properties. The phase voltages—that is, the voltage from each line a, b, and c to the neutral n—are given by

$$\mathbf{V}_{an} = V_p \underline{/0°}$$

$$\mathbf{V}_{bn} = V_p \underline{/-120°} \qquad\qquad \textbf{11.11}$$

$$\mathbf{V}_{cn} = V_p \underline{/+120°}$$

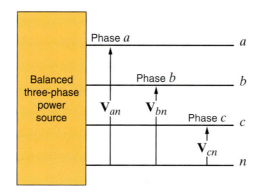

Figure 11.2
Balanced three-phase voltage source.

The phasor diagram for these voltages is shown in Fig. 11.3. The phase sequence of this set is said to be *abc* (called positive phase sequence), meaning that $\mathbf{V}_{bn}$ *lags* $\mathbf{V}_{an}$ by 120°.

We will standardize our notation so that we always label the voltages $\mathbf{V}_{an}$, $\mathbf{V}_{bn}$, and $\mathbf{V}_{cn}$ and observe them in the order *abc*. Furthermore, we will normally assume with no loss of generality that $\underline{/\mathbf{V}_{an}} = 0°$.

An important property of the balanced voltage set is that

$$\mathbf{V}_{an} + \mathbf{V}_{bn} + \mathbf{V}_{cn} = 0 \qquad\qquad \textbf{11.12}$$

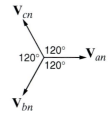

Figure 11.3
Phasor diagram for a balanced three-phase voltage source.

This property can easily be seen by resolving the voltage phasors into components along the real and imaginary axes. It can also be demonstrated via Eq. (11.9).

From the standpoint of the user who connects a load to the balanced three-phase voltage source, it is not important how the voltages are generated. It is important to note, however, that if the load currents generated by connecting a load to the power source shown in Fig. 11.2 are also *balanced*, there are two possible equivalent configurations for the load. The equivalent load can be considered as being connected in either a *wye* (Y) or a *delta* (Δ) configuration. The balanced wye configuration is shown in Fig. 11.4a and equivalently in Fig. 11.4b. The delta configuration is shown in Fig. 11.5a and equivalently in Fig. 11.5b. Note that in the case of the delta connection, there is no neutral line. The actual function of the neutral line in the wye connection will be examined, and it will be shown that in a balanced system the neutral line carries no current and, for purposes of analysis, may be omitted.

The wye and delta connections each have their advantages. In the wye case, we have access to two voltages, the line-to-line and line-to-neutral, and it provides a convenient place to connect to ground for system protection. That is, it limits the magnitude of surge voltages. In the delta case, this configuration stays in balance better when serving unbalanced loads, and it is capable of trapping the third harmonic.

Figure 11.4
Wye (Y)-connected loads.

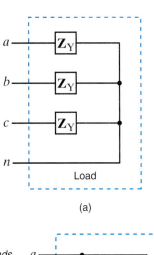

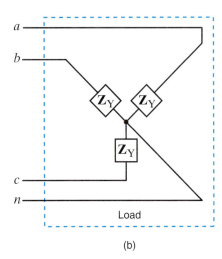

(a)

(b)

Figure 11.5
Delta (Δ)-connected loads.

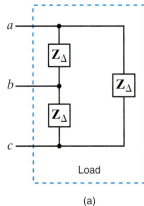

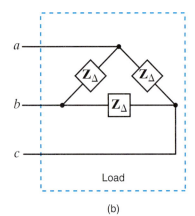

(a)

(b)

11.3 Source/Load Connections

Since the source and the load can each be connected in either Y or Δ, three-phase balanced circuits can be connected Y–Y, Y–Δ, Δ–Y, or Δ–Δ. Our approach to the analysis of all of these circuits will be "Think Y", therefore, we will analyze the Y–Y connection first.

BALANCED WYE–WYE CONNECTION Suppose now that the source and load are both connected in a wye, as shown in Fig. 11.6. The phase voltages with positive phase sequence are

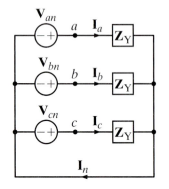

Figure 11.6
Balanced three-phase wye–wye connection.

$$\mathbf{V}_{an} = V_p \underline{/0°}$$
$$\mathbf{V}_{bn} = V_p \underline{/-120°}$$
$$\mathbf{V}_{cn} = V_p \underline{/+120°}$$

11.13

where V_p, the phase voltage, is the magnitude of the phasor voltage from the neutral to any line. The *line-to-line* or, simply, *line voltages* can be calculated using KVL; for example,

$$\mathbf{V}_{ab} = \mathbf{V}_{an} - \mathbf{V}_{bn}$$
$$= V_p \underline{/0°} - V_p \underline{/-120°}$$
$$= V_p - V_p \left[-\frac{1}{2} - j\frac{\sqrt{3}}{2} \right]$$
$$= V_p \left[\frac{3}{2} + j\frac{\sqrt{3}}{2} \right]$$
$$= \sqrt{3}\, V_p \underline{/30°}$$

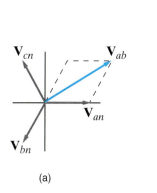

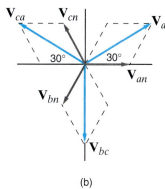

(a)   (b)

Figure 11.7
Phasor representation of phase and line voltages in a balanced wye–wye system.

The phasor addition is shown in Fig. 11.7a. In a similar manner, we obtain the set of line-to-line voltages as

HINT
Conversion rules
$\underline{/\mathbf{V}_{ab}} = \underline{/\mathbf{V}_{an}} + 30°$
$V_{ab} = \sqrt{3}\, V_{an}$

$$\mathbf{V}_{ab} = \sqrt{3}\, V_p \underline{/30°}$$

$$\mathbf{V}_{bc} = \sqrt{3}\, V_p \underline{/-90°} \qquad \textbf{11.14}$$

$$\mathbf{V}_{ca} = \sqrt{3}\, V_p \underline{/-210°}$$

All the line voltages together with the phase voltages are shown in Fig. 11.7b. We will denote the magnitude of the line voltages as V_L, and therefore, for a balanced system,

$$V_L = \sqrt{3}\, V_p \qquad \textbf{11.15}$$

Hence, in a wye-connected system, the line voltage is equal to $\sqrt{3}$ times the phase voltage.
As shown in Fig. 11.6, the line current for the a phase is

$$\mathbf{I}_a = \frac{\mathbf{V}_{an}}{\mathbf{Z}_Y} = \frac{V_p \underline{/0°}}{\mathbf{Z}_Y} \qquad \textbf{11.16}$$

where $\mathbf{I}_b$ and $\mathbf{I}_c$ have the same magnitude but lag $\mathbf{I}_a$ by 120° and 240°, respectively.
The neutral current $\mathbf{I}_n$ is then

$$\mathbf{I}_n = \left(\mathbf{I}_a + \mathbf{I}_b + \mathbf{I}_c\right) = 0 \qquad \textbf{11.17}$$

Since there is no current in the neutral, this conductor could contain any impedance or it could be an open or a short circuit, without changing the results found previously.
As illustrated by the wye–wye connection in Fig. 11.6, the current in the line connecting the source to the load is the same as the phase current flowing through the impedance $\mathbf{Z}_Y$. Therefore, in a *wye–wye connection,*

$$I_L = I_Y \qquad \textbf{11.18}$$

where I_L is the magnitude of the line current and I_Y is the magnitude of the current in a wye-connected load.
Although we have a three-phase system composed of three sources and three loads, we can analyze a single phase and use the phase sequence to obtain the voltages and currents in the other phases. This is, of course, a direct result of the balanced condition. We may even have impedances present in the lines; however, as long as the system remains balanced, we need analyze only one phase. If the line impedances in lines a, b, and c are equal, the system will be balanced. Recall that the balance of the system is unaffected by whatever appears in the neutral line, and since the neutral line impedance is arbitrary, we assume that it is zero (i.e., a short circuit).

Example 11.1

An *abc*-sequence three-phase voltage source connected in a balanced wye has a line voltage of $\mathbf{V}_{ab} = 208\underline{/-30°}$ V rms. We wish to determine the phase voltages.

SOLUTION The magnitude of the phase voltage is given by the expression

$$V_p = \frac{208}{\sqrt{3}}$$

$$= 120 \text{ V rms}$$

The phase relationships between the line and phase voltages are shown in Fig. 11.7. From this figure we note that

$$\mathbf{V}_{an} = 120\underline{/-60°} \text{ V rms}$$

$$\mathbf{V}_{bn} = 120\underline{/-180°} \text{ V rms}$$

$$\mathbf{V}_{cn} = 120\underline{/+60°} \text{ V rms}$$

The magnitudes of these voltages are quite common, and one often hears that the electric service in a building, for example, is three-phase 208/120 V rms.

Example 11.2

A three-phase wye-connected load is supplied by an *abc*-sequence balanced three-phase wye-connected source with a phase voltage of 120 V rms. If the line impedance and load impedance per phase are $1 + j1\ \Omega$ and $20 + j10\ \Omega$, respectively, we wish to determine the value of the line currents and the load voltages.

SOLUTION The phase voltages are

$$\mathbf{V}_{an} = 120\underline{/0°} \text{ V rms}$$

$$\mathbf{V}_{bn} = 120\underline{/-120°} \text{ V rms}$$

$$\mathbf{V}_{cn} = 120\underline{/+120°} \text{ V rms}$$

The per-phase circuit diagram is shown in Fig. 11.8. The line current for the *a* phase is

$$\mathbf{I}_{aA} = \frac{120\underline{/0°}}{21 + j11}$$

$$= 5.06\underline{/-27.65°} \text{ A rms}$$

The load voltage for the *a* phase, which we call $\mathbf{V}_{AN}$, is

$$\mathbf{V}_{AN} = \left(5.06\underline{/-27.65°}\right)(20 + j10)$$

$$= 113.15\underline{/-1.08°} \text{ V rms}$$

Figure 11.8
Per-phase circuit diagram for the problem in Example 11.2.

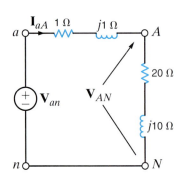

The corresponding line currents and load voltages for the b and c phases are

$$\mathbf{I}_{bB} = 5.06\,\underline{/-147.65°}\,\text{A rms} \qquad \mathbf{V}_{BN} = 113.15\,\underline{/-121.08°}\,\text{V rms}$$

$$\mathbf{I}_{cC} = 5.06\,\underline{/-267.65°}\,\text{A rms} \qquad \mathbf{V}_{CN} = 113.15\,\underline{/-241.08°}\,\text{V rms}$$

To reemphasize and clarify our terminology, phase voltage, V_p, is the magnitude of the phasor voltage from the neutral to any line, while line voltage, V_L, is the magnitude of the phasor voltage between any two lines. Thus, the values of V_L and V_p will depend on the point at which they are calculated in the system.

LEARNING EXTENSIONS

E11.1 The voltage for the a phase of an abc-phase-sequence balanced wye-connected source is $\mathbf{V}_{an} = 120\,\underline{/90°}$ V rms. Determine the line voltages for this source.

ANSWER:
$\mathbf{V}_{ab} = 208\,\underline{/120°}$ V rms;
$\mathbf{V}_{bc} = 208\,\underline{/0°}$ V rms;
$\mathbf{V}_{ca} = 208\,\underline{/-120°}$ V rms.

E11.2 An abc-phase-sequence three-phase voltage source connected in a balanced wye has a line voltage of $\mathbf{V}_{ab} = 208\,\underline{/0°}$ V rms. Determine the phase voltages of the source.

ANSWER:
$\mathbf{V}_{an} = 120\,\underline{/-30°}$ V rms;
$\mathbf{V}_{bn} = 120\,\underline{/-150°}$ V rms;
$\mathbf{V}_{cn} = 120\,\underline{/-270°}$ V rms.

E11.3 A three-phase wye-connected load is supplied by an abc-sequence balanced three-phase wye-connected source through a transmission line with an impedance of $1 + j1\ \Omega$ per phase. The load impedance is $8 + j3\ \Omega$ per phase. If the load voltage for the a phase is $104.02\,\underline{/26.6°}$ V rms (i.e., $V_p = 104.02$ V rms at the load end), determine the phase voltages of the source.

ANSWER:
$\mathbf{V}_{an} = 120\,\underline{/30°}$ V rms;
$\mathbf{V}_{bn} = 120\,\underline{/-90°}$ V rms;
$\mathbf{V}_{cn} = 120\,\underline{/-210°}$ V rms.

The previous analysis indicates that we can simply treat a three-phase balanced circuit on a per phase basis and use the phase relationship to determine all voltages and currents. Let us now examine the situations in which either the source or the load is connected in Δ.

DELTA-CONNECTED SOURCE   Consider the delta-connected source shown in Fig. 11.9a. Note that the sources are connected line to line. We found earlier that the relationship between line-to-line and line-to-neutral voltages was given by Eq. (11.14) and illustrated in Fig. 11.7 for an abc-phase sequence of voltages. Therefore, if the delta sources are

$$\mathbf{V}_{ab} = V_L\,\underline{/0°}$$
$$\mathbf{V}_{bc} = V_L\,\underline{/-120°}$$
$$\mathbf{V}_{ca} = V_L\,\underline{/+120°}$$

11.19

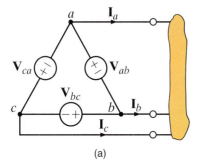

(a)

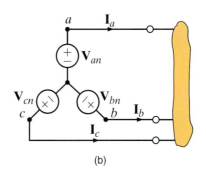

(b)

Figure 11.9
Sources connected in delta and wye.

where V_L is the magnitude of the phase voltage. The equivalent wye sources shown in Fig. 11.9b are

$$\mathbf{V}_{an} = \frac{V_L}{\sqrt{3}} \underline{/-30°} = V_p \underline{/-30°}$$

$$\mathbf{V}_{bn} = \frac{V_L}{\sqrt{3}} \underline{/-150°} = V_p \underline{/-150°} \qquad \textbf{11.20}$$

$$\mathbf{V}_{cn} = \frac{V_L}{\sqrt{3}} \underline{/-270°} = V_p \underline{/+90°}$$

where V_p is the magnitude of the phase voltage of an equivalent wye-connected source. Therefore, if we encounter a network containing a delta-connected source, we can easily convert the source from delta to wye so that all the techniques we have discussed previously can be applied in an analysis.

PROBLEM-SOLVING STRATEGY

Three-Phase Balanced AC Power Circuits

Step 1. Convert the source/load connection to a wye–wye connection if either the source, load, or both are connected in delta since the wye–wye connection can be easily used to obtain the unknown phasors.

Step 2. Only the unknown phasors for the *a*-phase of the circuit need be determined since the three-phase system is balanced.

Step 3. Finally, convert the now known phasors to the corresponding phasors in the original system.

— Example 11.3

Consider the network shown in Fig. 11.10a. We wish to determine the line currents and the magnitude of the line voltage at the load.

SOLUTION The single-phase diagram for the network is shown in Fig. 11.10b. The line current $\mathbf{I}_{aA}$ is

$$\mathbf{I}_{aA} = \frac{(208/\sqrt{3})\underline{/-30°}}{12.1 + j4.2}$$

$$= 9.38\underline{/-49.14°} \text{ A rms}$$

and thus $\mathbf{I}_{bB} = 9.38\underline{/-169.14°}$ V rms and $\mathbf{I}_{cC} = 9.38\underline{/70.86°}$ V rms. The voltage $\mathbf{V}_{AN}$ is then

$$\mathbf{V}_{AN} = \left(9.38\underline{/-49.14°}\right)(12 + j4)$$

$$= 118.65\underline{/-30.71°} \text{ V rms}$$

Therefore, the magnitude of the line voltage at the load is

$$V_L = \sqrt{3}\,(118.65)$$

$$= 205.51 \text{ V rms}$$

The phase voltage at the source is $V_p = 208/\sqrt{3} = 120$ V rms, while the phase voltage at the load is $V_p = 205.51/\sqrt{3} = 118.65$ V rms. Clearly, we must be careful with our notation and specify where the phase or line voltage is taken.

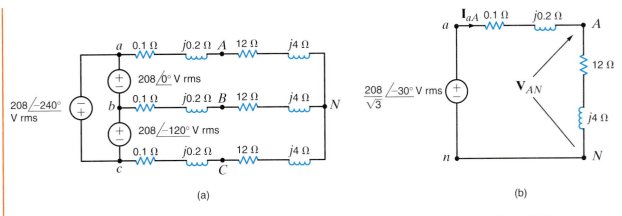

(a)

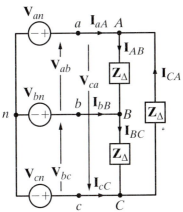

(b)

Figure 11.10
Delta–wye network and an equivalent single-phase (a-phase) diagram.

LEARNING EXTENSION

E11.4 Consider the network shown in Fig. E11.4. Compute the magnitude of the line voltages at the load. **PSV**

ANSWER:
$V_L = 205.2$ V rms.

Figure E11.4

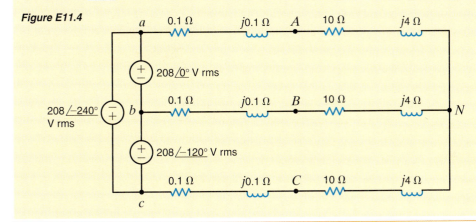

DELTA-CONNECTED LOAD Consider now the Δ-connected load shown in Fig. 11.11. Note that in this connection the line-to-line voltage is the voltage across each load impedance.

If the phase voltages of the source are

$$\mathbf{V}_{an} = V_p\,\underline{/0°}$$
$$\mathbf{V}_{bn} = V_p\,\underline{/-120°} \qquad \text{11.21}$$
$$\mathbf{V}_{cn} = V_p\,\underline{/+120°}$$

then the line voltages are

$$\mathbf{V}_{ab} = \sqrt{3}\,V_p\,\underline{/30°} = V_L\,\underline{/30°} = \mathbf{V}_{AB}$$
$$\mathbf{V}_{bc} = \sqrt{3}\,V_p\,\underline{/-90°} = V_L\,\underline{/-90°} = \mathbf{V}_{BC} \qquad \text{11.22}$$
$$\mathbf{V}_{ca} = \sqrt{3}\,V_p\,\underline{/-210°} = V_L\,\underline{/-210°} = \mathbf{V}_{CA}$$

where V_L is the magnitude of the line voltage at both the delta-connected load and at the source since there is no line impedance present in the network.

Figure 11.11
Balanced three-phase wye–delta system.

From Fig. 11.11 we note that if $\mathbf{Z}_\Delta = Z_\Delta \underline{/\theta}$, the phase currents at the load are

$$\mathbf{I}_{AB} = \frac{\mathbf{V}_{AB}}{\mathbf{Z}_\Delta} \qquad \textbf{11.23}$$

where $\mathbf{I}_{BC}$ and $\mathbf{I}_{CA}$ have the same magnitude but lag $\mathbf{I}_{AB}$ by 120° and 240°, respectively. KCL can now be employed in conjunction with the phase currents to determine the line currents. For example,

$$\mathbf{I}_{aA} = \mathbf{I}_{AB} + \mathbf{I}_{AC}$$
$$= \mathbf{I}_{AB} - \mathbf{I}_{CA}$$

However, it is perhaps easier to simply convert the balanced Δ-connected load to a balanced Y-connected load using the Δ–Y transformation. This conversion is possible since the wye–delta and delta–wye transformations outlined in Chapter 2 are also valid for impedance in the frequency domain. In the balanced case, the transformation equations reduce to

$$\mathbf{Z}_Y = \frac{1}{3}\mathbf{Z}_\Delta$$

and then the line current $\mathbf{I}_{aA}$ is simply

$$\mathbf{I}_{aA} = \frac{\mathbf{V}_{an}}{\mathbf{Z}_Y}$$

Finally, using the same approach as that employed earlier to determine the relationship between the line voltages and phase voltages in a Y–Y connection, we can show that the relationship between the *magnitudes* of the phase currents in the Δ-connected load and the line currents is

$$I_L = \sqrt{3}\, I_\Delta \qquad \textbf{11.24}$$

Example 11.4

A balanced delta-connected load contains a 10-Ω resistor in series with a 20-mH inductor in each phase. The voltage source is an *abc*-sequence three-phase 60-Hz, balanced wye with a voltage $\mathbf{V}_{an} = 120 \underline{/30°}$ V rms. We wish to determine all Δ currents and line currents.

SOLUTION The impedance per phase in the delta load is $\mathbf{Z}_\Delta = 10 + j7.54$ Ω. The line voltage $\mathbf{V}_{ab} = 120\sqrt{3}\underline{/60°}$ V rms. Since there is no line impedance, $\mathbf{V}_{AB} = \mathbf{V}_{ab} = 120\sqrt{3}\underline{/60°}$ V rms. Hence,

$$\mathbf{I}_{AB} = \frac{120\sqrt{3}\underline{/60°}}{10 + j7.54}$$
$$= 16.60 \underline{/+22.98°} \text{ A rms}$$

If $\mathbf{Z}_\Delta = 10 + j7.54$ Ω, then

$$\mathbf{Z}_Y = \frac{1}{3}\mathbf{Z}_\Delta$$
$$= 3.33 + j2.51 \text{ Ω}$$

Then the line current

$$\mathbf{I}_{aA} = \frac{\mathbf{V}_{an}}{\mathbf{Z}_Y} = \frac{120\underline{/30°}}{3.33 + j2.51}$$
$$= \frac{120\underline{/30°}}{4.17\underline{/37.01°}}$$
$$= 28.78\underline{/-7.01°} \text{ A rms}$$

Therefore, the remaining phase and line currents are

$$\mathbf{I}_{BC} = 16.60\underline{/-97.02°} \text{ A rms} \quad \mathbf{I}_{bB} = 28.78\underline{/-127.01°} \text{ A rms}$$

$$\mathbf{I}_{CA} = 16.60\underline{/+142.98°} \text{ A rms} \quad \mathbf{I}_{cC} = 28.78\underline{/+112.99°} \text{ A rms}$$

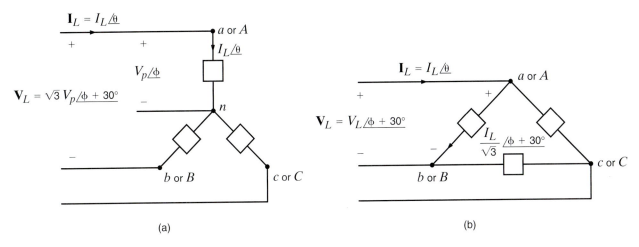

(a)

(b)

Figure 11.12
Voltage and current relationships for Y and Δ configurations.

In summary, the relationship between the line voltage and phase voltage and the line current and phase current for both the Y and Δ configurations are shown in Fig. 11.12. The currents and voltages are shown for one phase. The two remaining phases have the same magnitude but lag by 120° and 240°, respectively.

Careful observation of Table 11.1 indicates that the following rules apply when solving problems in balanced three-phase systems:

- The phase of the voltages and currents in a Δ connection is 30° ahead of those in a Y connection.
- The magnitude of the line voltage or, equivalently, the Δ-connection phase voltage, is $\sqrt{3}$ times that of the Y-connection phase voltage.
- The magnitude of the line current or, equivalently, the Y-connection phase current, is $\sqrt{3}$ times that of the Δ-connection phase current.
- The load impedance in the Y connection is one-third of that in the Δ-connection, and the phase is identical.

Table 11.1   The voltage, current, and impedance relationships for Y and Δ configurations

	Y	Δ
Line voltage	$\sqrt{3}\,V_p\underline{/\phi + 30°}$	$V_L\underline{/\phi + 30°}$
$\left(\mathbf{V}_{ab} \text{ or } \mathbf{V}_{AB}\right)$	$= V_L\underline{/\phi + 30°}$	
Line current $\mathbf{I}_{aA}$	$I_L\underline{/\theta}$	$I_L\underline{/\theta}$
Phase voltage	$V_p\underline{/\phi}\ (\mathbf{V}_{an} \text{ or } \mathbf{V}_{AN})$	$\sqrt{3}\,V_p\underline{/\phi + 30°}$
Phase current	$I_L\underline{/\theta}$	$\dfrac{I_L}{\sqrt{3}}\underline{/\theta + 30°}$
Load impedance	$\mathbf{Z}_Y\underline{/\phi - \theta}$	$3\mathbf{Z}_Y\underline{/\phi - \theta}$

LEARNING EXTENSION

E11.5 An *abc*-sequence three-phase voltage source connected in a balanced wye supplies power to a balanced delta-connected load. The line current for the *a* phase is $\mathbf{I}_{aA} = 12\underline{/40°}$ A rms. Find the phase currents in the delta-connected load.

ANSWER:
$\mathbf{I}_{AB} = 6.93\underline{/70°}$ A rms;
$\mathbf{I}_{BC} = 6.93\underline{/-50°}$ A rms;
$\mathbf{I}_{CA} = 6.93\underline{/-170°}$ A rms.

11.4 Power Relationships

Whether the load is connected in a wye or a delta, the real and reactive power per phase is

$$P_p = V_p I_p \cos \theta$$

$$Q_p = V_p I_p \sin \theta$$

11.25

where θ is the angle between the phase voltage and the line current. For a Y-connected system, $I_p = I_L$ and $V_p = V_L / \sqrt{3}$, and for a Δ-connected system, $I_p = I_L / \sqrt{3}$ and $V_p = V_L$. Therefore,

$$P_p = \frac{V_L I_L}{\sqrt{3}} \cos \theta$$

$$Q_p = \frac{V_L I_L}{\sqrt{3}} \sin \theta$$

11.26

The total real and reactive power for all three phases is then

$$P_T = 3 P_p = \sqrt{3} V_L I_L \cos \theta$$

$$Q_T = 3 Q_P = \sqrt{3} V_L I_L \sin \theta$$

11.27

and, therefore, the magnitude of the complex power (apparent power) is

$$S_T = \sqrt{P_T^2 + Q_T^2}$$

$$= \sqrt{3} V_L I_L$$

and

$$\underline{/S_T} = \theta$$

Example 11.5

A three-phase balanced wye–delta system has a line voltage of 208 V rms. The total real power absorbed by the load is 1200 W. If the power factor angle of the load is 20° lagging, we wish to determine the magnitude of the line current and the value of the load impedance per phase in the delta.

SOLUTION The line current can be obtained from Eq. (11.26). Since the real power per phase is 400 W,

$$400 = \frac{208 I_L}{\sqrt{3}} \cos 20°$$

$$I_L = 3.54 \text{ A rms}$$

The magnitude of the current in each leg of the delta-connected load is

$$I_\Delta = \frac{I_L}{\sqrt{3}}$$

$$= 2.05 \text{ A rms}$$

Therefore, the magnitude of the delta impedance in each phase of the load is

$$|\mathbf{Z}_\Delta| = \frac{V_L}{I_\Delta}$$

$$= \frac{208}{2.05}$$

$$= 101.46 \ \Omega$$

Since the power factor angle is 20° lagging, the load impedance is

$$\mathbf{Z}_\Delta = 101.46 \underline{/20°}$$

$$= 95.34 + j34.70 \ \Omega$$

Example 11.6

For the circuit in Example 11.2 we wish to determine the real and reactive power per phase at the load and the total real power, reactive power, and complex power at the source.

SOLUTION From the data in Example 11.2 the complex power per phase at the load is

$$\mathbf{S}_{\text{load}} = \mathbf{V}\mathbf{I}^*$$

$$= \left(113.15\,\underline{/-1.08°}\right)\left(5.06\,\underline{/27.65°}\right)$$

$$= 572.54\,\underline{/26.57°}$$

$$= 512.07 + j256.09 \text{ VA}$$

Therefore, the real and reactive power per phase at the load are 512.07 W and 256.09 var, respectively.

The complex power per phase at the source is

$$\mathbf{S}_{\text{source}} = \mathbf{V}\mathbf{I}^*$$

$$= \left(120\,\underline{/0°}\right)\left(5.06\,\underline{/27.65°}\right)$$

$$= 607.2\,\underline{/27.65°}$$

$$= 537.86 + j281.78 \text{ VA}$$

Therefore, total real power, reactive power, and apparent power at the source are 1613.6 W, 845.2 var, and 1821.6 VA, respectively.

Example 11.7

A balanced three-phase source serves three loads as follows:

Load 1: 24 kW at 0.6 lagging power factor

Load 2: 10 kW at unity power factor

Load 3: 12 kVA at 0.8 leading power factor

If the line voltage at the loads is 208 V rms at 60 Hz, we wish to determine the line current and the combined power factor of the loads.

HINT

The sum of three complex powers

$\mathbf{S}_{\text{load}} = \mathbf{S}_1 + \mathbf{S}_2 + \mathbf{S}_3$

SOLUTION From the data we find that

$$\mathbf{S}_1 = 24{,}000 + j32{,}000$$

$$\mathbf{S}_2 = 10{,}000 + j0$$

$$\mathbf{S}_3 = 12{,}000\,\underline{/-36.9°} = 9600 - j7200$$

Therefore,

$$\mathbf{S}_{\text{load}} = 43{,}600 + j24{,}800$$

$$= 50{,}160\,\underline{/29.63°} \text{ VA}$$

$$I_L = \frac{|\mathbf{S}_{\text{load}}|}{\sqrt{3}\,V_L}$$

$$= \frac{50{,}160}{208\sqrt{3}}$$

$$I_L = 139.23 \text{ A rms}$$

and the combined power factor is

$$\text{pf}_{\text{load}} = \cos 29.63°$$

$$= 0.869 \text{ lagging}$$

Example 11.8

HINT

Recall that the complex power for all three lines is

$\mathbf{S}_{\text{line}} = 3I_L^2 \mathbf{Z}_{\text{line}}$

Given the three-phase system in Example 11.7, let us determine the line voltage and power factor at the source if the line impedance is $\mathbf{Z}_{\text{line}} = 0.05 + j0.02\ \Omega$.

SOLUTION The complex power absorbed by the line impedances is

$$\mathbf{S}_{\text{line}} = 3\left(R_{\text{line}}I_L^2 + jX_{\text{line}}I_L^2\right)$$
$$= 2908 + j1163\ \text{VA}$$

The complex power delivered by the source is then

$$\mathbf{S}_S = \mathbf{S}_{\text{load}} + \mathbf{S}_{\text{line}}$$
$$= 43{,}600 + j24{,}800 + 2908 + j1163$$
$$= 53{,}264\ \underline{/29.17°}\ \text{VA}$$

The line voltage at the source is then

$$V_{L_S} = \frac{S_S}{\sqrt{3}\,I_L}$$
$$= 220.87\ \text{V rms}$$

and the power factor at the source is

$$\text{pf}_S = \cos 29.17°$$
$$= 0.873\ \text{lagging}$$

LEARNING EXTENSIONS

E11.6 A three-phase balanced wye–wye system has a line voltage of 208 V rms. The total real power absorbed by the load is 12 kW at 0.8 pf lagging. Determine the per-phase impedance of the load.

ANSWER: $\mathbf{Z} = 2.88\ \underline{/36.87°}\ \Omega$.

E11.7 For the balanced wye–wye system described in Extension E11.3, determine the real and reactive power and the complex power at both the source and the load.

ANSWER: $\mathbf{S}_{\text{load}} = 1186.77 + j444.66\ \text{VA}$; $\mathbf{S}_{\text{source}} = 1335.65 + j593.55\ \text{VA}$.

E11.8 A 480-V rms line feeds two balanced three-phase loads. If the two loads are rated as follows, **PSV**

Load 1: 5 kVA at 0.8 pf lagging

Load 2: 10 kVA at 0.9 pf lagging

determine the magnitude of the line current from the 480-V rms source.

ANSWER: $I_L = 17.97\ \text{A rms}$.

11.5 Power Factor Correction

HINT

Major precautions for three-phase power factor correction:

Must distinguish P_T and P_P. Must use appropriate V rms for Y- and Δ-connections.

In Section 9.7 we illustrated a simple technique for raising the power factor of a load. The method involved judiciously selecting a capacitor and placing it in parallel with the load. In a balanced three-phase system, power factor correction is performed in exactly the same manner. It is important to note, however, that the $\mathbf{S}_{\text{cap}}$ specified in Eq. (9.37) is provided by three capacitors, and in addition, V_{rms} in the equation is the voltage across each capacitor. The following example illustrates the technique.

Example 11.9

In the balanced three-phase system shown in Fig. 11.13, the line voltage is 34.5 kV rms at 60 Hz. We wish to find the values of the capacitors C such that the total load has a power factor of 0.94 leading.

HINT

The reactive power to be supplied by C is derived from the expression

$$jQ_{cap} = -j\omega CV^2_{rms}$$

The phase voltage for the Y connection is

$$V_Y = \frac{34.5k}{\sqrt{3}}$$

SOLUTION Following the development outlined in Section 9.7 for single-phase power factor correction, we obtain

$$\mathbf{S}_{old} = 24 \,\underline{/\cos^{-1} 0.78}\, \text{MVA}$$

$$= 18.72 + j15.02 \,\text{MVA}$$

and

$$\theta_{new} = -\cos^{-1} 0.94$$

$$= -19.95°$$

Therefore,

$$\mathbf{S}_{new} = 18.72 + j18.72 \tan(-19.95°)$$

$$= 18.72 - j6.80 \,\text{MVA}$$

and

$$\mathbf{S}_{cap} = \mathbf{S}_{new} - \mathbf{S}_{old}$$

$$= -j21.82 \,\text{MVA}$$

However,

$$-j\omega C \, V^2_{rms} = -j21.82 \,\text{MVA}$$

and since the line voltage is 34.5 kV rms, then

$$(377)\left(\frac{34.5k}{\sqrt{3}}\right)^2 C = \frac{21.82}{3} \,\text{M}$$

Hence,

$$C = 48.6 \,\mu\text{F}$$

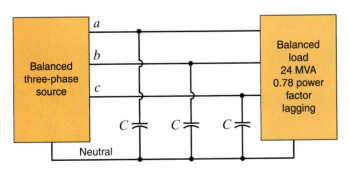

Figure 11.13
Network used in Example 11.9.

LEARNING EXTENSION

E11.9 Find C in Example 11.9 such that the load has a power factor of 0.90 lagging.

ANSWER: $C = 13.26 \,\mu\text{F}$.

Finally, recall that our entire discussion in this chapter has focused on balanced systems. It is extremely important, however, to point out that in the unbalanced three-phase system the problem is much more complicated because of the mutual inductive coupling between phases in power apparatus.

11.6 Application Examples

The first of the following three examples illustrates the manner in which power flow is measured when utilities are interconnected, answering the question of who is supplying power to whom. The last example demonstrates the actual method in which capacitors are specified by the manufacturer for power factor correction.

Application Example 11.10

Two balanced three-phase systems, X and Y, are interconnected with lines having impedance $\mathbf{Z}_{line} = 1 + j2\ \Omega$. The line voltages are $\mathbf{V}_{ab} = 12\ \underline{/0°}$ kV rms and $\mathbf{V}_{AB} = 12\ \underline{/5°}$ kV rms, as shown in Fig. 11.14a. We wish to determine which system is the source, which is the load, and the average power supplied by the source and absorbed by the load.

SOLUTION When we draw the per phase circuit for the system as shown in Fig. 11.14b, the analysis will be essentially the same as that of Example 9.12.

The network in Fig. 11.14b indicates that

$$\mathbf{I}_{aA} = \frac{\mathbf{V}_{an} - \mathbf{V}_{AN}}{\mathbf{Z}_{line}}$$

$$= \frac{\dfrac{12{,}000}{\sqrt{3}}\ \underline{/-30°} - \dfrac{12{,}000}{\sqrt{3}}\ \underline{/-25°}}{\sqrt{5}\ \underline{/63.43°}}$$

$$= 270.30\ \underline{/-180.93°}\ \text{A rms}$$

The average power absorbed by system Y is

$$P_Y = \sqrt{3}\, V_{AB} I_{aA} \cos(\theta_{\mathbf{V}_{an}} - \theta_{\mathbf{I}_{aA}})$$

$$= \sqrt{3}\,(12{,}000)(270.30)\cos(-25° + 180.93°)$$

$$= -5.130\ \text{MW}$$

Note that system Y is not the load, but rather the source and supplies 5.130 MW.

System X absorbs the following average power:

$$P_X = \sqrt{3}\, V_{ab} I_{Aa} \cos(\theta_{\mathbf{V}_{an}} - \theta_{\mathbf{I}_{aA}})$$

where

$$\mathbf{I}_{Aa} = -\mathbf{I}_{aA} = 270.30\ \underline{/-0.93°}\ \text{A rms}$$

Therefore,

$$P_X = \sqrt{3}\,(12{,}000)(270.30)\cos(-30° + 0.93°)$$

$$= 4.910\ \text{MW}$$

and hence system X is the load.

The difference in the power supplied by system Y and that absorbed by system X is, of course, the power absorbed by the resistance of the three lines.

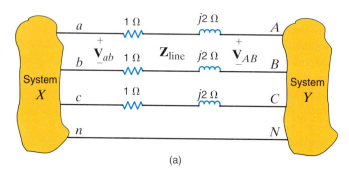

(a)

Figure 11.14 Circuits used in Example 11.10: (a) original three-phase system, (b) per phase circuit.

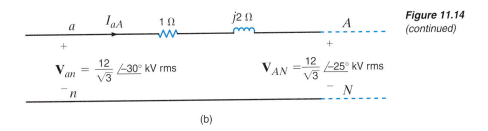

Figure 11.14
(continued)

(b)

The preceding example illustrates an interesting point. Note that the phase difference between the two ends of the power line determines the direction of the power flow. Since the numerous power companies throughout the United States are tied together to form the U.S. power grid, the phase difference across the interconnecting transmission lines reflects the manner in which power is transferred between power companies.

Capacitors for power factor correction are usually specified by the manufacturer in vars rather than in farads. Of course, the supplier must also specify the voltage at which the capacitor is designed to operate, and a frequency of 60 Hz is assumed. The relationship between capacitance and the var rating is

$$Q_R = \frac{V^2}{Z_C}$$

where Q_R is the var rating, V is the voltage rating, and Z_C is the capacitor's impedance at 60 Hz. Thus, a 500-V, 600-var capacitor has a capacitance of

$$C = \frac{Q_R}{\omega V^2} = \frac{600}{(377)(500)^2}$$

or

$$C = 6.37 \ \mu F$$

and can be used in any application where the voltage across the capacitor does not exceed the rated value of 500 V.

Application Example 11.11

Let us examine, in a general sense, the incremental cost of power factor correction; specifically, how much capacitance is required to improve the power factor by a fixed amount, say 0.01?

SOLUTION   The answer to this question depends primarily on two factors: the apparent power and the original power factor before correction. This dependence can be illustrated by developing equations for the old and new power factors, and their corresponding power factor angles. We know that

$$pf_{old} = \cos(\theta_{old}) \qquad \tan(\theta_{old}) = \frac{Q_{old}}{P}$$

11.28

$$pf_{new} = \cos(\theta_{new}) \qquad \tan(\theta_{new}) = \frac{Q_{old} - Q_C}{P}$$

If the difference in the power factors is 0.01, then

$$pf_{new} - pf_{old} = 0.01$$

11.29

Solving for the ratio Q_C/P, since reactive power and capacitance are proportional to one another will yield the reactive power per watt required to improve the power factor by 0.01. Using Eq. (11.28), we can write

$$\frac{Q_C}{P} = \frac{Q_{old}}{P} - \tan(\theta_{new}) = \tan(\theta_{old}) - \tan(\theta_{new}) = \tan\left[acos(pf_{old})\right] - \tan\left[acos(pf_{old} + 0.01)\right]$$

11.30

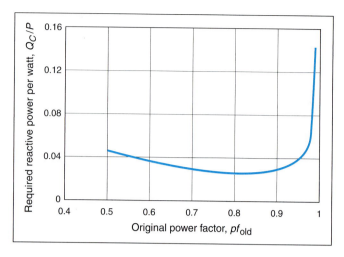

Figure 11.15 *A plot of required reactive power per watt needed to improve the original power factor by 0.01.*

A plot of Eq. (11.30), shown in Fig. 11.15, has some rather interesting implications. First, the improvement required for a power factor change of 0.01 is at a minimum when the original power factor is about 0.81. Thus, an incremental improvement at that point is least expensive. Second, as the original power factor approaches unity, changes in power factor are more expensive to implement.

Application Example 11.12

Table 11.2 lists the voltage and power ratings for three power factor correction capacitors. Let us determine which of them, if any, can be employed in Example 11.9.

Table 11.2 Rated voltage and vars for power factor correction capacitors

Capacitor	Rated Voltage (kV)	Rated Q (Mvars)
1	10.0	4.0
2	50.0	25.0
3	20.0	7.5

SOLUTION From Fig. 11.13 we see that the voltage across the capacitors is the line-to-neutral voltage, which is

$$V_{an} = \frac{V_{ab}}{\sqrt{3}} = \frac{34{,}500}{\sqrt{3}}$$

or

$$V_{an} = 19.9 \text{ kV}$$

Therefore, only those capacitors with rated voltages greater than or equal to 19.9 kV can be used in this application, which eliminates capacitor 1. Let us now determine the capacitance of capacitors 2 and 3. For capacitor 2,

$$C_2 = \frac{Q}{\omega V^2} = \frac{25 \times 10^6}{(377)(50{,}000)^2}$$

or

$$C_2 = 26.53 \text{ μF}$$

which is much smaller than the required 48.6 μF. The capacitance of capacitor 3 is

$$C_3 = \frac{Q}{\omega V^2} = \frac{7.5 \times 10^6}{(377)(20,000)^2}$$

or

$$C_3 = 49.7 \ \mu F$$

which is within 2.5% of the required value. Obviously, capacitor 3 is the best choice.

11.7   Design Examples

In the first example in this section, we examine the selection of both the conductor and the capacitor in a practical power factor correction problem.

Design Example 11.13

Two stores, as shown in Fig. 11.16, are located at a busy intersection. The stores are fed from a balanced three-phase 60-Hz source with a line voltage of 13.8 kV rms. The power line is constructed of a #4ACSR (aluminum cable steel reinforced) conductor that is rated at 170 A rms.

A third store, shown in Fig. 11.16, wishes to locate at the intersection. Let us determine (1) if the #4ACSR conductor will permit the addition of this store, and (2) the value of the capacitors connected in wye that are required to change the overall power factor for all three stores to 0.92 lagging.

Figure 11.16
Circuit used in Example 11.13.

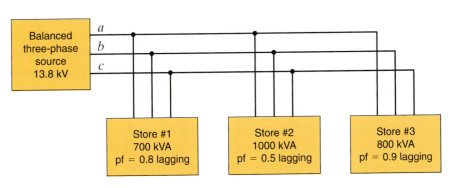

SOLUTION   (1) The complex power for each of the three loads is

$$S_1 = 700 \underline{/36.9°} = 560 + j420 \ kVA$$

$$S_2 = 1000 \underline{/60°} = 500 + j866 \ kVA$$

$$S_3 = 800 \underline{/25.8°} = 720 + j349 \ kVA$$

Therefore, the total complex power is

$$S_T = S_1 + S_2 + S_3$$

$$= 1780 + j1635$$

$$= 2417 \underline{/42.57°} \ kVA$$

Since

$$S_T = \sqrt{3} \ V_L I_L$$

the line current is

$$I_L = \frac{(2417)(10^3)}{\sqrt{3}(13.8)(10^3)}$$

$$= 101.1 \text{ A rms}$$

Since this value is well below the rated value of 170 A rms, the conductor is sized properly and we can safely add the third store.

(2) The combined power factor for the three loads is found from the expression

$$\cos\theta = \text{pf} = \frac{1780}{2417} = 0.7365 \text{ lagging}$$

By adding capacitors we wish to change this power factor to 0.92 lagging. This new power factor corresponds to a θ_{new} of 23.07°. Therefore, the new complex power is

$$\mathbf{S}_{\text{new}} = 1780 + j1780 \tan(23.07°)$$

$$= 1780 + j758.28 \text{ kVA}$$

As illustrated in Fig. 9.17, the difference between $\mathbf{S}_{\text{new}}$ and $\mathbf{S}_T$ is that supplied by the purely reactive capacitor and, therefore,

$$\mathbf{S}_{\text{cap}} = jQ_C = \mathbf{S}_{\text{new}} - \mathbf{S}_T$$

or

$$jQ_C = j(758.28 - 1635)$$

$$= -j876.72 \text{ kVA}$$

Thus,

$$-j\omega C\, V_{\text{rms}}^2 = \frac{-j876.72\text{k}}{3}$$

and

$$377\left(\frac{13.8 \times 10^3}{\sqrt{3}}\right)^2 C = \frac{876.72}{3} \times 10^3$$

Therefore,

$$C = 12.2 \text{ μF}$$

Hence, three capacitors of this value connected in wye at the load will yield a total power factor of 0.92 lagging.

Design Example 11.14

Control circuitry for high-voltage, three-phase equipment usually operates at much lower voltages. For example, a 10-kW power supply might operate at a line voltage of 480 V rms, while its control circuit is powered by internal dc power supplies at ±5 V. Lower voltages are not only safer to operate but also permit engineers to easily incorporate op-amps and digital electronics into the control system. It is a great convenience to test the control circuit without having to connect it directly to a 480 V rms three-phase source. Therefore, let us design a low-power, three-phase emulator that simulates a three-phase system at low voltage and low cost and provides a test bed for the control circuitry. The emulator should generate proper phasing but with a magnitude that is adjustable between 1 and 4 volts peak.

SOLUTION   Our design, shown in Fig. 11.17, will consist of three parts: a magnitude adjustment, a phase angle generator, and a phase B generator. The ac input is a 60-Hz sine wave with a peak of about 5 V. This voltage can be generated from a standard 120 V rms wall outlet using a step-down transformer with a turns ratio of

$$n = \frac{120\sqrt{2}}{5} = 34 : 1$$

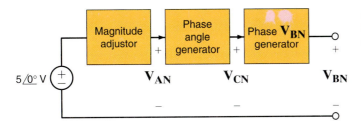

Figure 11.17

A block diagram for a three-phase emulator.

The potentiometer circuit shown in Fig. 11.18a can be used to provide magnitude adjustment. Resistors R_1 and R_2 provide the voltage limits of 1 and 4 V. We can use simple voltage division to determine the relationships between R_1, R_2, and R_p. When the pot's wiper arm is at the bottom of the pot in Fig. 11.18a, we have

$$V_1 = 1 = 5\left[\frac{R_2}{R_1 + R_2 + R_p}\right] \quad \Rightarrow \quad R_1 + R_p = 4R_2 \qquad \textbf{11.31}$$

and, when the wiper is at the top,

$$V_1 = 4 = 5\left[\frac{R_2 + R_p}{R_1 + R_2 + R_p}\right] \quad \Rightarrow \quad R_2 + R_p = 4R_1 \qquad \textbf{11.32}$$

Solving Eqs. (11.31) and (11.32) yields the requirements that $R_1 = R_2 = R_p/3$. To obtain values for these resistors we must simply choose one of them. We know that resistors are available in a wide variety of values in small increments. Potentiometers on the other hand are not. Typical potentiometer values are 10, 20, 50, 100, 200, 500, ...10k ... 100k, 200k, 500k, ... up to about 10 MΩ. Since the potentiometer offers fewer options, we will choose its value to be 10 kΩ, which yields $R_1 = R_2 = 3.3$ kΩ—a standard resistor value. We will set V_1 as the phase voltage $\mathbf{V_{AN}}$.

Next we consider the phase angle generator. Since capacitors are generally smaller physically than inductors, we will use the simple RC network in Fig. 11.18b to shift the phase of $\mathbf{V_1}$. Assigning a phase angle of $0°$ to $\mathbf{V_1}$, we know that the phase of $\mathbf{V_2}$ must be between 0 and -90 degrees. Unfortunately, in order to generate $\mathbf{V_{CN}}$ we need a phase angle of $+120°$. If we create a phase angle of $-60°$ and invert the resulting sine wave, we will produce an equivalent phase angle of $+120°$! The inversion can be performed by an inverting op-amp configuration. To produce a $-60°$ phase angle at $\mathbf{V_2}$ requires

$$\omega C R_3 = \tan(60°) = 1.732 \quad \Rightarrow \quad R_3 C = 4.59 \times 10^{-3}$$

We will choose a standard value of 120 nF for C, which yields $R_3 = 38.3$ kΩ. This is a standard value at 1% tolerance. Using these values, $\mathbf{V_2}$ will be

$$\mathbf{V_2} = \mathbf{V_1}\left[\frac{1}{1 + j\omega C R_3}\right] = \frac{\mathbf{V_1}}{2.0}\underline{/-60°} \qquad \textbf{11.33}$$

From Eq. (11.33) we see that our inverter should also have a gain of 2 to restore the magnitude of $\mathbf{V_2}$. The complete phase angle generator circuit is shown in Fig. 11.18c, where $R_4 = 10$ kΩ and $R_5 = 20$ kΩ have been chosen to produce the required gain. Now $\mathbf{V_3}$ is used to represent $\mathbf{V_{CN}}$. The additional unity gain buffer stage isolates the resistances associated with the inverter from the R-C phase generator. That way the inverter will not alter the phase angle.

Finally, we must create the phase voltage $\mathbf{V_{BN}}$. Since the sum of the three-phase voltages is zero, we can write

$$\mathbf{V_{BN}} = -\mathbf{V_{AN}} - \mathbf{V_{CN}}$$

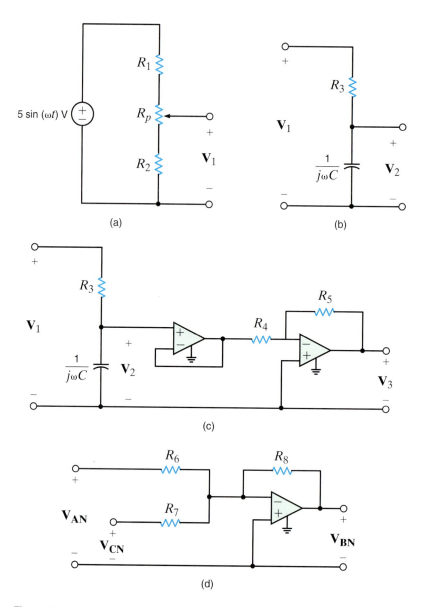

Figure 11.18 *Subcircuits within the three-phase emulator: (a) the magnitude adjustor, (b) the R-C portion of the phase angle generator, (c) the complete phase angle generator, and (d) the generator for phase* $\mathbf{V}_{BN}$.

The simple op-amp summer in Fig. 11.18d will perform this mathematical operation. For the summer

$$\mathbf{V}_{BN} = -\left[\frac{R_8}{R_6}\right]\mathbf{V}_{AN} - \left[\frac{R_8}{R_7}\right]\mathbf{V}_{CN}$$

We require $R_6 = R_7 = R_8$. Since we are already using some 10-kΩ resistors anyway, we just use three more here. The complete circuit is shown in Fig. 11.19 where one more unity gain buffer has been added at the potentiometer. This one isolates the *R-C* phase angle generator from the magnitude adjustment resistors.

It may seem that we have used op-amps too liberally, requiring a total of four. However, most op-amp manufacturers package their op-amps in single (one op-amp), dual (two op-amps), and quad (four op-amps) packages. Using a quad op-amp, we see that our circuit will require just one integrated circuit. As a final note, the op-amp power supply voltages must exceed the maximum input or output voltages at the op-amp terminals, which is 4 V. Therefore, we will specify +10 V supplies.

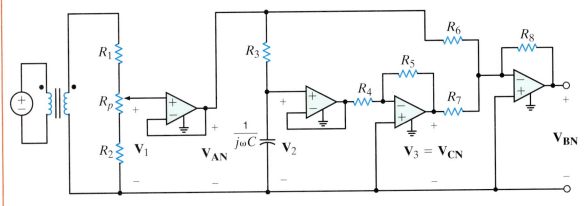

Figure 11.19
The complete three-phase emulator with variable voltage magnitude.

SUMMARY

- An important advantage of the balanced three-phase system is that it provides very smooth power delivery.

- Because of the balanced condition, it is possible to analyze a circuit on a per-phase basis, thereby providing a significant computational shortcut to a solution.

- A balanced three-phase source and load system: A balanced three-phase voltage source has three sinusoidal voltages of the same magnitude and frequency, and each voltage is 120° out of phase with the others. A positive-phase-sequence balanced voltage source is one in which $\mathbf{V}_{bn}$ lags $\mathbf{V}_{an}$ by 120° and $\mathbf{V}_{cn}$ lags $\mathbf{V}_{bn}$ by 120°.

- The relationships between wye- and delta-connected sources are shown in Table 11.1.

- The three-phase terminology is shown in Table 11.3.

Table 11.3 Three-phase terminology

Quantity	Wye	Delta
$\mathbf{I}_a, \mathbf{I}_b, \mathbf{I}_c$	Line current (I_L)	
	Phase current (I_p)	
$\mathbf{V}_{an}, \mathbf{V}_{bn}, \mathbf{V}_{cn}$	Line-to-neutral voltage (V_p)	
	Phase voltage (V_p)	
$\mathbf{V}_{ab}, \mathbf{V}_{bc}, \mathbf{V}_{ca}$	Line-to-line, phase-to-phase, line voltage (V_L)	
		Phase voltage (V_p)
$\mathbf{I}_{ab}, \mathbf{I}_{bc}, \mathbf{I}_{ca}$		Phase current (I_p)

- In a balanced system the voltages and currents sum to zero.

$$\mathbf{V}_{an} + \mathbf{V}_{bn} + \mathbf{V}_{cn} = 0$$

$$\mathbf{I}_a + \mathbf{I}_b + \mathbf{I}_c = 0 \quad \text{(no current in the neutral line)}$$

and

$$\mathbf{V}_{ab} + \mathbf{V}_{bc} + \mathbf{V}_{ca} = 0$$

$$\mathbf{I}_{ab} + \mathbf{I}_{bc} + \mathbf{I}_{ca} = 0$$

- The steps recommended for solving balanced three-phase ac circuits are as follows:

1. If the source/load connection is not wye–wye, then transform the system to a wye–wye connection.
2. Determine the unknown phasors in the wye–wye connection and deal only with the phase a.
3. Convert the now-known phasors back to the corresponding phasors in the original connection.

- Power factor correction in a balanced three-phase environment is performed in the same manner as in the single-phase case. Three capacitors are put in parallel with the load to reduce the lagging phase caused by the three-phase load.

PROBLEMS

PSV CS both available on the web at: http://www.justask4u.com/irwin

SECTION 11.2

11.1 Sketch a phasor representation of an *abc*-sequence balanced three-phase Y-connected source, including $\mathbf{V}_{an}$, $\mathbf{V}_{bn}$, and $\mathbf{V}_{cn}$ if $\mathbf{V}_{an} = 120\,\underline{/15°}$ V rms.

11.2 Sketch a phasor representation of a balanced three-phase system containing both phase voltages and line voltages if $\mathbf{V}_{an} = 100\,\underline{/45°}$ V rms. Label all magnitudes and assume an *abc*-phase sequence. **CS**

11.3 Sketch a phasor representation of a balanced three-phase system containing both phase voltages and line voltages if $\mathbf{V}_{an} = 120\,\underline{/90°}$ V rms. Label all magnitudes and assume an *abc*-phase sequence.

11.4 A positive-sequence three-phase balanced wye voltage source has a phase voltage of $\mathbf{V}_{an} = 240\,\underline{/90°}$ V rms. Determine the line voltages of the source. **CS**

11.5 Sketch a phasor representation of a balanced three-phase system containing both phase voltages and line voltages if $\mathbf{V}_{ab} = 208\,\underline{/60°}$ V rms. Label all phasors and assume an *abc*-phase sequence.

SECTION 11.3

11.6 A positive-sequence balanced three-phase wye-connected source with a phase voltage of 120 V rms supplies power to a balanced wye-connected load. The per phase load impedance is $40 + j10\ \Omega$. Determine the line currents in the circuit if $\underline{/\mathbf{V}_{an}} = 0°$. **PSV**

11.7 A positive-sequence balanced three-phase wye-connected source supplies power to a balanced wye-connected load. The magnitude of the line voltages is 208 V rms. If the load impedance per phase is $36 + j12\ \Omega$, determine the line currents if $\underline{/\mathbf{V}_{an}} = 0°$.

11.8 An *abc*-sequence balanced three-phase wye-connected source supplies power to a balanced wye-connected load. The line impedance per phase is $1 + j5\ \Omega$, and the load impedance per phase is $25 + j25\ \Omega$. If the source line voltage $\mathbf{V}_{ab}$ is $208\,\underline{/0°}$ V rms, find the line currents.

11.9 An *abc*-sequence balanced three-phase wye-connected source supplies power to a balanced wye-connected load. The line impedance per phase is $1 + j0\ \Omega$, and the load impedance per phase is $20 + j20\ \Omega$. If the source line voltage $\mathbf{V}_{ab}$ is $100\,\underline{/0°}$ V rms, find the line currents. **CS**

11.10 An *abc*-sequence set of voltages feeds a balanced three-phase wye–wye system. The line and load impedances are $1 + j1\ \Omega$ and $10 + j10\ \Omega$, respectively. If the load voltage on the *a* phase is $\mathbf{V}_{AN} = 110\,\underline{/30°}$ V rms, determine the line voltages of the input. **PSV**

11.11 In a balanced three-phase wye–wye system, the source is an *abc*-sequence set of voltages. The load voltage on the *a* phase is $\mathbf{V}_{AN} = 110\,\underline{/80°}$ V rms, $\mathbf{Z}_{line} = 1 + j1.4\ \Omega$, and $\mathbf{Z}_{load} = 10 + j13\ \Omega$. Determine the input sequence of the line-to-neutral voltages.

11.12 Find the equivalent impedances $\mathbf{Z}_{ab}$, $\mathbf{Z}_{bc}$, and $\mathbf{Z}_{ca}$ in the network in Fig. P11.12.

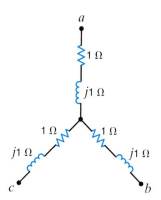

Figure P11.12

11.13 Find the equivalent $\mathbf{Z}$ of the network in Fig. P11.13.

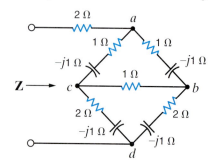

Figure P11.13

11.14 Find the equivalent $\mathbf{Z}$ of the network in Fig. P11.14.

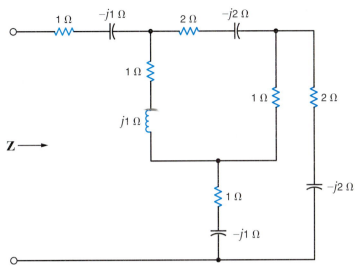

Figure P11.14

11.15 In a balanced three-phase wye–wye system, the source is an *abc*-sequence set of voltages. The load voltage on the *a* phase is $\mathbf{V}_{AN} = 120\,\underline{/60°}$ V rms, $\mathbf{Z}_{line} = 2 + j1.4\ \Omega$, and $\mathbf{Z}_{load} = 10 + j10\ \Omega$. Determine the input voltages.

11.16 A balance *abc*-sequence of voltages feeds a balanced three-phase wye–wye system. The line and load impedances are $0.6 + j0.9\ \Omega$ and $8 + j12\ \Omega$, respectively. The load voltage on the *a* phase is $\mathbf{V}_{AN} = 116.63\,\underline{/10°}$ V rms.

Find the line voltage $\mathbf{V}_{ab}$. **CS**

11.17 In a balanced three-phase wye–wye system, the source is an *abc*-sequence set of voltages. $\mathbf{Z}_{line} = 1 + j1\ \Omega$, $\mathbf{Z}_{load} = 14 + j12\ \Omega$, and the load voltage on the *a* phase is $\mathbf{V}_{AN} = 440\,\underline{/30°}$ V rms. Find the line voltage $\mathbf{V}_{ab}$.

11.18 An *abc*-phase sequence balanced three-phase source feeds a balanced load. The system is connected wye–wye and $\underline{/\mathbf{V}_{an}} = 0°$. The line impedance is $0.5 + j0.2\ \Omega$, the load impedance is $16 + j10\ \Omega$, and the total power absorbed by the load is 2000 W. Determine the magnitude of the source voltage $\mathbf{V}_{an}$.

11.19 In a balanced three-phase wye–wye system, the total power loss in the lines is 400 W. $\mathbf{V}_{AN} = 105.28\,\underline{/31.65°}$ V rms and the power factor of the load is 0.77 lagging. If the line impedance is $2 + j1\ \Omega$, determine the load impedance. **PSV**

11.20 In a balanced three-phase wye–wye system, the load impedance is $8 + j4\ \Omega$. The source has phase sequence *abc* and $\mathbf{V}_{an} = 120\,\underline{/0°}$ V rms. If the load voltage is $\mathbf{V}_{AN} = 111.62\,\underline{/-1.33°}$ V rms, determine the line impedance. **CS**

11.21 In a balanced three-phase wye–wye system, the load impedance is $10 + j1\Omega$. The source has phase sequence *abc* and the line voltage $\mathbf{V}_{ab} = 220\,\underline{/30°}$ V rms. If the load voltage $\mathbf{V}_{AN} = 120\,\underline{/0°}$ V rms, determine the line impedance.

11.22 In a balanced three-phase wye–wye system, the load impedance is $20 + j12\ \Omega$. The source has an *abc*-phase sequence and $\mathbf{V}_{an} = 120\,\underline{/0°}$ V rms. If the load voltage is $\mathbf{V}_{AN} = 111.49\,\underline{/-0.2°}$ V rms, determine the magnitude of the line current if the load is suddenly short-circuited.

11.23 In a balanced three-phase wye–wye system, the source is an *abc*-sequence set of voltages and $\mathbf{V}_{an} = 120\,\underline{/40°}$ V rms. If the *a*-phase line current and line impedance are known to be $7.10\,\underline{/-10.28°}$ A rms and $0.8 + j1\ \Omega$, respectively, find the load impedance. **CS**

11.24 In a three-phase balanced system, a delta-connected source supplies power to a wye-connected load. If the line impedance is $0.2 + j0.4\ \Omega$, the load impedance $3 + j2\ \Omega$, and the source phase voltage $\mathbf{V}_{ab} = 208\,\underline{/10°}$ V rms, find the magnitude of the line voltage at the load.

11.25 An *abc*-phase-sequence three-phase balanced wye-connected 60-Hz source supplies a balanced delta-connected load. The phase impedance in the load consists of a 20-Ω resistor in series with a 20-mH inductor, and the phase voltage at the source is $\mathbf{V}_{an} = 120\,\underline{/30°}$ V rms. If the line impedance is zero, find the line currents in the system.

11.26 In a balanced three-phase wye–wye system, the source is an *abc*-sequence set of voltages and $\mathbf{V}_{an} = 120\,\underline{/50°}$ V rms. The load voltage on the *a* phase is $110\,\underline{/50°}$ V rms, and the load impedance is $16 + j20\ \Omega$. Find the line impedance.

11.27 In a balanced three-phase wye–wye system, the source is an *abc*-sequence set of voltages and $\mathbf{V}_{an} = 120\,\underline{/40°}$ V rms. If the *a*-phase line current and line impedance are known to be $6\,\underline{/15°}$ A rms and $1 + j1\ \Omega$, respectively, find the load impedance.

11.28 An *abc*-sequence set of voltages feeds a balanced three-phase wye-wye system. If $\mathbf{V}_{an} = 440\,\underline{/30°}$ V rms, $\mathbf{V}_{AN} = 413.28\,\underline{/29.78°}$ V rms, and $\mathbf{Z}_{line} = 2 + j1.5\,\Omega$, find the load impedance.

11.29 In a three-phase balanced system, a delta-connected source supplies power to a wye-connected load. If the line impedance is $0.2 + j0.4\ \Omega$, the load impedance $6 + j4\ \Omega$, and the source phase voltage $\mathbf{V}_{ab} = 210\,\underline{/40°}$ V rms, find the magnitude of the line voltage at the load. **PSV**

11.30 In a balanced three-phase delta–wye system, the source has an *abc*-phase sequence. The line and load impedances are $0.6 + j0.3\ \Omega$ and $12 + j7\ \Omega$, respectively. If the line current $\mathbf{I}_{aA} = 9.6\,\underline{/-20°}$ A rms, determine the phase voltages of the source. **cs**

11.31 An *abc*-phase-sequence three-phase balanced wye-connected source supplies a balanced delta-connected load. The impedance per phase in the delta load is $12 + j6\ \Omega$. The line voltage at the source is $\mathbf{V}_{ab} = 120\sqrt{3}\,\underline{/40°}$ V rms. If the line impedance is zero, find the line currents in the balanced wye-delta system.

11.32 An *abc*-phase-sequence three-phase balanced wye-connected source supplies power to a balanced delta-connected load. The impedance per phase in the load is $14 + j7\ \Omega$. If the source voltage for the *a* phase is $\mathbf{V}_{an} = 120\,\underline{/80°}$ V rms and the line impedance is zero, find the phase currents in the wye-connected source.

11.33 An *abc*-phase-sequence three-phase balanced wye-connected source supplies a balanced delta-connected load. The impedance per phase of the delta load is $20 + j4\ \Omega$. If $\mathbf{V}_{AB} = 115\,\underline{/35°}$ V rms, find the line current.

11.34 An *abc*-phase-sequence three-phase balanced wye-connected source supplies a balanced delta-connected load. The impedance per phase of the delta load is $10 + j8\ \Omega$. If the line impedance is zero and the line current in the *a* phase is known to be $\mathbf{I}_{aA} = 28.10\,\underline{/-28.66°}$ A rms, find the load voltage $\mathbf{V}_{AB}$. **cs**

11.35 In a balanced three-phase wye–delta system, the source has an *abc*-phase sequence and $\mathbf{V}_{an} = 120\,\underline{/0°}$ V rms. If the line impedance is zero and the line current $\mathbf{I}_{aA} = 5\,\underline{/20°}$ A rms, find the load impedance per phase in the delta.

11.36 In a balanced three-phase wye–delta system, the source has an *abc*-phase sequence and $\mathbf{V}_{an} = 120\,\underline{/40°}$ V rms. The line and load impedance are $0.5 + j0.4\ \Omega$ and $36 + j18\ \Omega$, respectively. Find the delta currents in the load.

11.37 In a three-phase balanced delta–delta system, the source has an *abc*-phase sequence. The line and load impedances are $0.3 + j0.2\ \Omega$ and $9 + j6\ \Omega$, respectively. If the load current in the delta is $\mathbf{I}_{AB} = 15\,\underline{/40°}$ A rms, find the phase voltages of the source. **cs**

11.38 In a balanced three-phase delta–delta system, the source has an *abc*-phase sequence. The phase angle for the source voltage is $\underline{/\mathbf{V}_{ab}} = 40°$ and $\mathbf{I}_{ab} = 4\,\underline{/15°}$ A rms. If the total power absorbed by the load is 1400 W, find the load impedance.

11.39 A three-phase load impedance consists of a balanced wye in parallel with a balanced delta. What is the equivalent wye load and what is the equivalent delta load if the phase impedances of the wye and delta are $6 + j3\ \Omega$ and $15 + j10\ \Omega$, respectively?

11.40 In a balanced three-phase system, the *abc*-phase-sequence source is wye connected and $\mathbf{V}_{an} = 120\,\underline{/20°}$ V rms. The load consists of two balanced wyes with phase impedances of $8 + j2\ \Omega$ and $12 + j3\ \Omega$. If the line impedance is zero, find the line currents and the phase current in each load.

11.41 In a balanced three-phase system, the source is a balanced wye with an *abc*-phase sequence and $\mathbf{V}_{ab} = 208\,\underline{/60°}$ V rms. The load consists of a balanced wye with a phase impedance of $8 + j5\ \Omega$ in parallel with a balanced delta with a phase impedance of $21 + j12\ \Omega$. If the line impedance is $1.2 + j1\ \Omega$, find the phase currents in the balanced wye load. **cs**

11.42 In a balanced three-phase system, the source is a balanced wye with an *abc*-phase sequence and $\mathbf{V}_{ab} = 215\,\underline{/50°}$ V rms. The load is a balanced wye in parallel with a balanced delta. The phase impedance of the wye is $5 + j3\ \Omega$, and the phase impedance of the delta is $18 + j12\ \Omega$. If the line impedance is $1 + j0.8\ \Omega$, find the line currents and the phase currents in the loads.

11.43 In a balanced three-phase system, the source has an *abc*-phase sequence and is connected in delta. There are two parallel wye-connected loads. The phase impedance of load 1 and load 2 is $4 + j4\ \Omega$ and $10 + j4\ \Omega$, respectively. The line impedance connecting the source to the loads is $0.3 + j0.2\ \Omega$. If the current in the *a* phase of load 1 is $\mathbf{I}_{AN_1} = 10\,\underline{/20°}$ A rms, find the delta currents in the source.

11.44 In a balanced three-phase system, the source has an *abc*-phase sequence and is connected in delta. There are two loads connected in parallel. The line connecting the source to the loads has an impedance of $0.2 + j0.1\ \Omega$. Load 1 is connected in wye, and the phase impedance is $4 + j2\ \Omega$. Load 2 is connected in delta, and the phase impedance is $12 + j9\ \Omega$. The current $\mathbf{I}_{AB}$ in the delta load is $16\,\underline{/45°}$ A rms. Find the phase voltages of the source. **cs**

11.45 A balanced three-phase delta-connected source supplies power to a load consisting of a balanced delta in parallel with a balanced wye. The phase impedance of the delta is $24 + j12\ \Omega$, and the phase impedance of the wye is $12 + j8\ \Omega$. The *abc*-phase-sequence source voltages are $\mathbf{V}_{ab} = 440\,\underline{/60°}$ V rms, $\mathbf{V}_{bc} = 440\,\underline{/-60°}$ V rms, and $\mathbf{V}_{ca} = 440\,\underline{/-180°}$ V rms, and the line impedance per phase is $1 + j0.8\ \Omega$. Find the line currents and the power absorbed by the wye-connected load. **PSV**

SECTION 11.4

11.46 An *abc*-sequence wye-connected source having a phase-*a* voltage of $120 \underline{/0°}$ V rms is attached to a wye-connected load having a per-phase impedance of $100 \underline{/70°}$ Ω. If the line impedance is $1 \underline{/20°}$ Ω, determine the total complex power produced by the voltage sources and the real and reactive power dissipated by the load.

11.47 The magnitude of the complex power (apparent power) supplied by a three-phase balanced wye–wye system is 3600 VA. The line voltage is 208 V rms. If the line impedance is negligible and the power factor angle of the load is 25°, determine the load impedance. **CS**

11.48 A three-phase *abc*-sequence wye-connected source supplies 14 kVA with a power factor of 0.75 lagging to a delta load. If the delta load consumes 12 kVA at a power factor of 0.7 lagging and has a phase current of $10 \underline{/-30°}$ A rms, determine the per-phase impedance of the load and the line.

11.49 A three-phase balanced wye–wye system has a line voltage of 208 V rms. The line current is 6 A rms and the total real power absorbed by the load is 1800 W. Determine the load impedance per-phase, if the line impedance is negligible.

11.50 A balanced three-phase source serves two loads:

Load 1: 36 kVA at 0.8 pf lagging

Load 2: 18 kVA at 0.6 pf lagging

The line voltage at the load is 208 V rms at 60 Hz. Find the line current and the combined power factor at the load. **CS**

11.51 A balanced three-phase source serves the following loads:

Load 1: 60 kVA at 0.8 pf lagging

Load 2: 30 kVA at 0.75 pf lagging

The line voltage at the load is 208 V rms at 60 Hz. Determine the line current and the combined power factor at the load.

11.52 A small shopping center contains three stores that represent three balanced three-phase loads. The power lines to the shopping center represent a three-phase source with a line voltage of 13.8 kV rms. The three loads are

Load 1: 400 kVA at 0.9 pf lagging

Load 2: 200 kVA at 0.85 pf lagging

Load 3: 100 kVA at 0.90 pf lagging

Find the power line current.

11.53 The following loads are served by a balanced three-phase source:

Load 1: 20 kVA at 0.8 pf lagging

Load 2: 4 kVA at 0.8 pf leading

Load 3: 10 kVA at 0.75 pf lagging

The load voltage is 208 V rms at 60 Hz. If the line impedance is negligible, find the power factor at the source.

11.54 A balanced three-phase source supplies power to three loads. The loads are

Load 1: 30 kVA at 0.8 pf lagging

Load 2: 24 kW at 0.6 pf leading

Load 3: unknown

If the line voltage and total complex power at the load are 208 V rms and $60 \underline{/0°}$ kVA, respectively, find the unknown load. **CS**

11.55 A balanced three-phase source serves the following loads:

Load 1: 20 kVA at 0.8 pf lagging

Load 2: 10 kVA at 0.7 pf leading

Load 3: 10 kW at unity pf

Load 4: 16 kVA at 0.6 pf lagging

The line voltage at the load is 208 V rms at 60 Hz, and the line impedance is $0.02 + j0.04$ Ω. Find the line voltage and power factor at the source. **PSV**

11.56 A balanced three-phase source supplies power to three loads. The loads are

Load 1: 24 kW at 0.8 pf lagging

Load 2: 10 kVA at 0.7 pf leading

Load 3: unknown

If the line voltage at the load is 208 V rms, the magnitude of the total complex power is 41.93 kVA, and the combined power factor at the load is 0.86 lagging, find the unknown load.

11.57 A balanced three-phase source supplies power to three loads. The loads are

Load 1: 24 kVA at 0.6 pf lagging

Load 2: 10 kW at 0.75 pf lagging

Load 3: unknown

If the line voltage at the load is 208 V rms, the magnitude of the total complex power is 35.52 kVA, and the combined power factor at the load is 0.88 lagging, find the unknown load.

11.58 A standard practice for utility companies is to divide its customers into single-phase users and three-phase users. The utility must provide three-phase users, typically industries, with all three phases. However, single-phase users, residential, and light commercial are connected to only one phase. To reduce cable costs, all single-phase users in a neighborhood are connected together. This means that even if the three-phase users present perfectly balanced loads to the power grid, the single-phase loads will never be in balance, resulting in current flow in the neutral connection. Consider the 60-Hz, *abc*-sequence network in Fig. P11.58. With a line voltage of $416\,\underline{/30°}$ V rms, phase *a* supplies the single-phase users on A Street, phase *b* supplies B Street, and phase *c* supplies C Street. Furthermore, the three-phase industrial load, which is connected in delta, is balanced. Find the neutral current.   **CS**

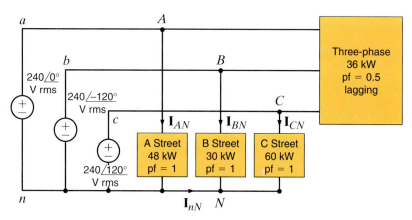

Figure P11.58

11.59 A three-phase *abc*-sequence wye-connected source with $\mathbf{V}_{an} = 220\,\underline{/0°}$ V rms supplies power to a wye-connected load that consumes 50 kW of power in each phase at a pf of 0.8 lagging. Three capacitors are found that each have an impedance of $-j2.0\ \Omega$, and they are connected in parallel with the previous load in a wye configuration. Determine the power factor of the combined load as seen by the source.

11.60 If the three capacitors in the network in Problem 11.59 are connected in a delta configuration, determine the power factor of the combined load as seen by the source.

11.61 Find *C* in the network in Fig. P11.61 such that the total load has a power factor of 0.9 lagging.

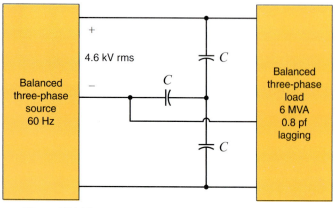

Figure P11.61

11.62 Find *C* in the network in Fig. P11.61 so that the total load has a power factor of 0.9 leading.   **CS**

11.63 Find *C* in the network in Fig. P11.63 such that the total load has a power factor of 0.87 leading.   **PSV**

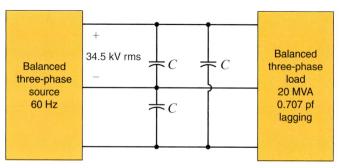

Figure P11.63

11.64 Find the value of *C* in Problem 11.63 such that the total load has a power factor of 0.87 lagging.

11.65 The U.S. Department of Energy estimates that in 2001 the electrical energy consumption in the United States was roughly 3.6 TW. (Check the URL *http://www.eia.doe.gov/emeu/aer/txt/ptb0805.html* for a complete breakdown.) Of that total, 1 TW was consumed in the industrial sector where large electric motor loads represent 60% of the usage. Suppose that the average uncorrected power factor for these motor loads is 0.7 lagging and that the power factor for the remaining loads is 0.9 lagging. Determine the total capacitance required nationwide to correct everything to unity power factor. Assume that all capacitors are connected to 480 V rms line voltages.

11.66 The power utilities typically transmit and distribute power as shown in Fig. P11.66 with high-voltage transmission for long distances and lower voltage distribution within a city or town. Also, the largest loads are, if possible, located "upstream" of lesser loads. List at least two advantages of this arrangement.

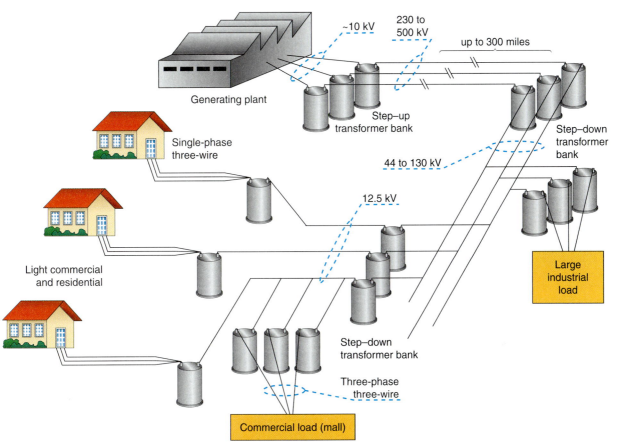

Figure P11.66

TYPICAL PROBLEMS FOUND ON THE FE EXAM

11FE-1 A wye-connected load consists of a series *RL* impedance. Measurements indicate that the rms voltage across each element is 84.85 V. If the rms line current is 6 A, find the total complex power for the three-phase load configuration. **cs**

11FE-2 A balanced three-phase delta-connected load consists of an impedance of $12 + j12\ \Omega$. If the line voltage at the load is measured to be 230 V rms, find the magnitude of the line current and the total real power absorbed by the three-phase configuration.

11FE-3 Two balanced three-phase loads are connected in parallel. One load with a phase impedance of $24 + j18\ \Omega$ is connected in delta, and the other load has a phase impedance of $6 + j4\ \Omega$ and is connected in wye. If the line-to-line voltage is 208 V rms, determine the line current. **cs**

11FE-4 The total complex power at the load of a three-phase balanced system is $24\ \underline{/30°}$ kVA. Find the real power per phase.

12

Variable-Frequency Network Performance

Access Problem-Solving Videos **PSV** **and Circuit Solutions** **CS** **at:** http://www.justask4u.com/irwin using the registration code on the inside cover and see a website with answers and more!

In Chapter 8 we demonstrated that a network containing a capacitor and an inductor operated differently if the frequency was changed from the U.S. power grid frequency of 60 Hz to the aircraft frequency of 400 Hz. This phenomenon, though not surprising since the impedance of both these circuit elements is frequency dependent, indicates that if the frequency of the network sources is varied over some range, we can also expect the network to undergo variations in response to these changes in frequency.

Consider for a moment your stereo amplifier. The input signal contains sound waves of frequencies that range from soup to nuts; and yet, the amplifier must amplify each frequency component exactly the same amount in order to achieve perfect sound reproduction. Achieving perfect sound reproduction is a nontrivial task; and when you buy a very good amplifier, part of the price reflects the design necessary to achieve constant amplification over a wide range of frequencies.

In this chapter we examine the performance of electrical networks when excited from variable-frequency sources. Such effects are important in the analysis and design of networks such as filters, tuners, and amplifiers that have wide application in communication and control systems. Terminology and techniques for frequency-response analysis are introduced, including standard plots (Bode plots) for describing network performance graphically. In particular, Bode plot construction and interpretation are discussed in detail.

The concept of resonance is introduced in reference to frequency selectivity and tuning. The various parameters used to define selectivity such as bandwidth, cutoff frequency, and quality factor are defined and discussed. Network scaling for both magnitude and phase is also presented.

Networks with special filtering properties are examined. Specifically, low-pass, high-pass, band-pass, and band-elimination filters are discussed. Techniques for designing active filters (containing op-amps) are presented. ●

12.1   Variable Frequency-Response Analysis

In previous chapters we investigated the response of *RLC* networks to sinusoidal inputs. In particular, we considered 60-Hz sinusoidal inputs. In this chapter we allow the frequency of excitation to become a variable and evaluate network performance as a function of frequency. To begin, let us consider the effect of varying frequency on elements with which we are already quite familiar—the resistor, inductor, and capacitor. The frequency-domain impedance of the resistor shown in Fig. 12.1a is

$$\mathbf{Z}_R = R = R\,\underline{/0^\circ}$$

The magnitude and phase are constant and independent of frequency. Sketches of the magnitude and phase of $\mathbf{Z}_R$ are shown in Figs. 12.1b and c. Obviously, this is a very simple situation.

For the inductor in Fig. 12.2a, the frequency-domain impedance $\mathbf{Z}_L$ is

$$\mathbf{Z}_L = j\omega L = \omega L\,\underline{/90^\circ}$$

The phase is constant at 90°, but the magnitude of $\mathbf{Z}_L$ is directly proportional to frequency. Figures 12.2b and c show sketches of the magnitude and phase of $\mathbf{Z}_L$ versus frequency. Note that at low frequencies the inductor's impedance is quite small. In fact, at dc, $\mathbf{Z}_L$ is zero, and the inductor appears as a short circuit. Conversely, as frequency increases, the impedance also increases.

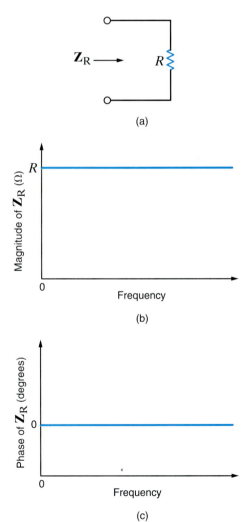

(a)

(b)

(c)

Figure 12.1
Frequency-dependent impedance of a resistor.

Figure 12.2

Frequency-dependent impedance of an inductor.

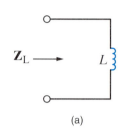

(a)

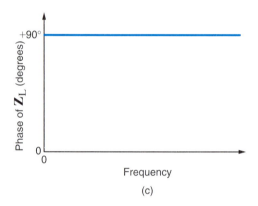

(b)

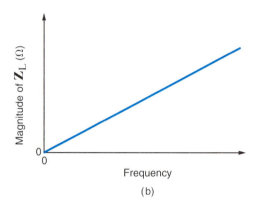

(c)

Next consider the capacitor of Fig. 12.3a. The impedance is

$$\mathbf{Z}_C = \frac{1}{j\omega C} = \frac{1}{\omega C} \angle{-90°}$$

Once again the phase of the impedance is constant, but now the magnitude is inversely proportional to frequency, as shown in Figs. 12.3b and c. Note that the impedance approaches infinity, or an open circuit, as ω approaches zero and $\mathbf{Z}_C$ approaches zero as ω approaches infinity.

Now let us investigate a more complex circuit: the *RLC* series network in Fig. 12.4a. The equivalent impedance is

$$\mathbf{Z}_{eq} = R + j\omega L + \frac{1}{j\omega C}$$

or

$$\mathbf{Z}_{eq} = \frac{(j\omega)^2 LC + j\omega RC + 1}{j\omega C}$$

Sketches of the magnitude and phase of this function are shown in Figs. 12.4b and c.

Note that at very low frequencies, the capacitor appears as an open circuit, and, therefore, the impedance is very large in this range. At high frequencies, the capacitor has very little effect and the impedance is dominated by the inductor, whose impedance keeps rising with frequency.

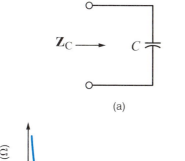

Figure 12.3
Frequency-dependent impedance of a capacitor.

(a)

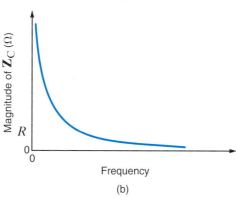

(b)

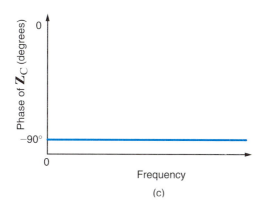

(c)

As the circuits become more complicated, the equations become more cumbersome. In an attempt to simplify them, let us make the substitution $j\omega = s$. (This substitution has a more important meaning, which we will describe in later chapters.) With this substitution, the expression for $\mathbf{Z}_{eq}$ becomes

$$\mathbf{Z}_{eq} = \frac{s^2 LC + sRC + 1}{sC}$$

If we review the four circuits we investigated thus far, we will find that in every case the impedance is the ratio of two polynomials in s and is of the general form

$$\mathbf{Z}(s) = \frac{N(s)}{D(s)} = \frac{a_m s^m + a_{m-1} s^{m-1} + \cdots + a_1 s + a_0}{b_n s^n + b_{n-1} s^{n-1} + \cdots + b_1 s + b_0} \qquad \textbf{12.1}$$

where $N(s)$ and $D(s)$ are polynomials of order m and n, respectively. An extremely important aspect of Eq. (12.1) is that it holds not only for impedances but also for all voltages, currents, admittances, and gains in the network. The only restriction is that the values of all circuit elements (resistors, capacitors, inductors, and dependent sources) must be real numbers.

Let us now demonstrate the manner in which the voltage across an element in a series RLC network varies with frequency.

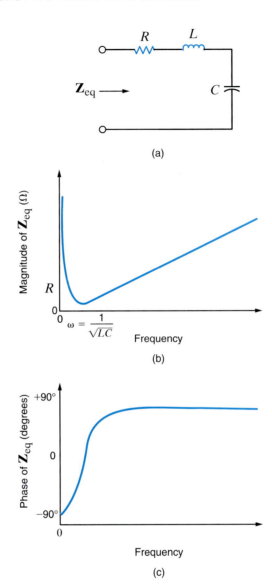

Figure 12.4

Frequency-dependent impedance of an RLC series network.

(a)

(b)

(c)

Example 12.1

Consider the network in Fig. 12.5a. We wish to determine the variation of the output voltage as a function of frequency over the range from 0 to 1 kHz.

SOLUTION Using voltage division, the output can be expressed as

$$\mathbf{V}_o = \left(\frac{R}{R + j\omega L + \dfrac{1}{j\omega C}} \right) \mathbf{V}_S$$

or, equivalently,

$$\mathbf{V}_o = \left(\frac{j\omega CR}{(j\omega)^2 LC + j\omega CR + 1} \right) \mathbf{V}_S$$

Using the element values, the equation becomes

$$\mathbf{V}_o = \left(\frac{(j\omega)(37.95 \times 10^{-3})}{(j\omega)^2 (2.53 \times 10^{-4}) + j\omega(37.95 \times 10^{-3}) + 1} \right) 10\underline{/0^\circ}$$

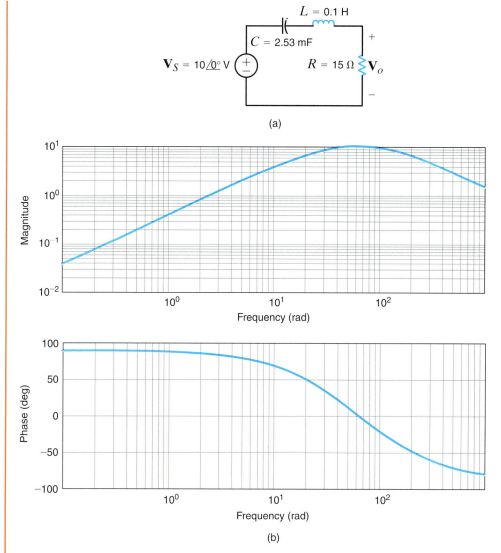

Figure 12.5 (a) Network and (b) its frequency-response simulation.

As we will demonstrate later in this chapter, MATLAB can be effectively employed to determine the magnitude and phase of this voltage as a function of frequency. The resultant magnitude and phase characteristics are semilog plots in which the frequency is displayed on the log axis. The plots, obtained using MATLAB, for the function $\mathbf{V}_o$ are shown in Fig. 12.5b.

We will illustrate in subsequent sections that the use of a semilog plot is a very useful tool in deriving frequency-response information.

As an introductory application of variable frequency-response analysis and characterization, let us consider a stereo amplifier. In particular, we should consider first the frequency range over which the amplifier must perform and then exactly what kind of performance we desire. The frequency range of the amplifier must exceed that of the human ear, which is roughly 50 Hz to 15,000 Hz. Accordingly, typical stereo amplifiers are designed to operate in the frequency range from 50 Hz to 20,000 Hz. Furthermore, we want to preserve the fidelity of the signal as it passes through the amplifier. Thus, the output signal should be an exact duplicate of the input signal times a gain factor. This requires that the gain be independent of frequency over the specified frequency range of 50 Hz to 20,000 Hz. An ideal sketch of this requirement for a gain of 1000 is shown in Fig. 12.6, where the midband region is defined as that portion of the plot where the gain is constant and is bounded by two points, which we will refer to as f_{LO} and f_{HI}. Notice once again that the frequency axis is a log axis and, thus, the frequency response is displayed on a semilog plot.

Figure 12.6
Amplifier frequency-response requirements.

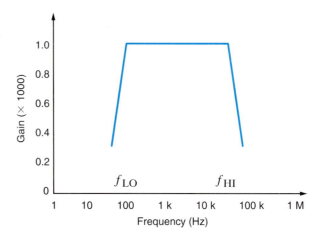

A model for the amplifier described graphically in Fig. 12.6 is shown in Fig. 12.7a with the frequency-domain equivalent circuit in Fig. 12.7b.

If the input is a steady-state sinusoid, we can use frequency-domain analysis to find the gain

$$\mathbf{G}_v(j\omega) = \frac{\mathbf{V}_o(j\omega)}{\mathbf{V}_S(j\omega)}$$

which with the substitution $s = j\omega$ can be expressed as

$$\mathbf{G}_v(s) = \frac{\mathbf{V}_o(s)}{\mathbf{V}_S(s)}$$

Using voltage division, we find that the gain is

$$\mathbf{G}_v(s) = \frac{\mathbf{V}_o(s)}{\mathbf{V}_S(s)} = \frac{\mathbf{V}_{in}(s)}{\mathbf{V}_S(s)}\frac{\mathbf{V}_o(s)}{\mathbf{V}_{in}(s)} = \left[\frac{R_{in}}{R_{in} + 1/sC_{in}}\right](1000)\left[\frac{1/sC_o}{R_o + 1/sC_o}\right]$$

or

$$\mathbf{G}_v(s) = \left[\frac{sC_{in}R_{in}}{1 + sC_{in}R_{in}}\right](1000)\left[\frac{1}{1 + sC_oR_o}\right]$$

Figure 12.7
Amplifier equivalent network.

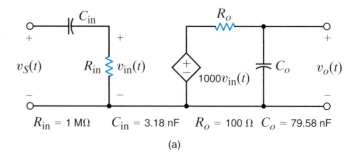

$R_{in} = 1\ M\Omega \qquad C_{in} = 3.18\ nF \qquad R_o = 100\ \Omega \quad C_o = 79.58\ nF$

(a)

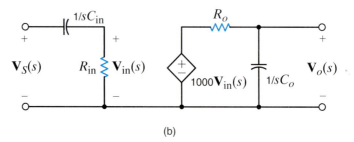

(b)

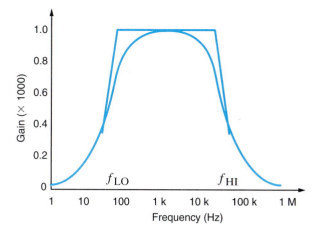

Figure 12.8
Exact and approximate amplifier gain versus frequency plots.

Using the element values in Fig. 12.7a,

$$\mathbf{G}_v(s) = \left[\frac{s}{s + 100\pi}\right](1000)\left[\frac{40,000\pi}{s + 40,000\pi}\right]$$

where 100π and $40,000\pi$ are the radian equivalents of 50 Hz and 20,000 Hz, respectively. Since $s = j\omega$, the network function is indeed complex. An exact plot of $\mathbf{G}_v(s)$ is shown in Fig. 12.8 superimposed over the sketch of Fig. 12.6. The exact plot exhibits smooth transitions at f_{LO} and f_{HI}; otherwise the plots match fairly well.

Let us examine our expression for $\mathbf{G}_v(s)$ more closely with respect to the plot in Fig. 12.8. Assume that f is well within the midband frequency range; that is,

$$f_{\mathrm{LO}} \ll f \ll f_{\mathrm{HI}}$$

or

$$100\pi \ll |s| \ll 40,000\pi$$

Under these conditions, the network function becomes

$$\mathbf{G}_v(s) \approx \left[\frac{s}{s}\right](1000)\left[\frac{1}{1 + 0}\right]$$

or

$$\mathbf{G}_v(s) = 1000$$

Thus, well within midband, the gain is constant. However, if the frequency of excitation decreases toward f_{LO}, then $|s|$ is comparable to 100π and

$$\mathbf{G}_v(s) \approx \left[\frac{s}{s + 100\pi}\right](1000)$$

Since $R_{\mathrm{in}}C_{\mathrm{in}} = 1/100\pi$, we see that C_{in} causes the rolloff in gain at low frequencies. Similarly, when the frequency approaches f_{HI}, the gain rolloff is due to C_o.

Through this amplifier example, we have introduced the concept of frequency-dependent networks and have demonstrated that frequency-dependent network performance is caused by the reactive elements in a network.

NETWORK FUNCTIONS In the previous section, we introduced the term *voltage gain*, $\mathbf{G}_v(s)$. This term is actually only one of several network functions, designated generally as $\mathbf{H}(s)$, which define the ratio of response to input. Since the function describes a reaction due to an excitation at some other point in the circuit, network functions are also called *transfer functions*. Furthermore, transfer functions are not limited to voltage ratios. Since in electrical networks inputs and outputs can be either voltages or currents, there are four possible network functions, as listed in Table 12.1.

There are also *driving point functions*, which are impedances or admittances defined at a single pair of terminals. For example, the input impedance of a network is a driving point function.

Table 12.1 Network transfer functions

Input	Output	Transfer Function	Symbol
Voltage	Voltage	Voltage gain	$\mathbf{G}_v(s)$
Current	Voltage	Transimpedance	$\mathbf{Z}(s)$
Current	Current	Current gain	$\mathbf{G}_i(s)$
Voltage	Current	Transadmittance	$\mathbf{Y}(s)$

Example 12.2

We wish to determine the transfer admittance $\left[\mathbf{I}_2(s)/\mathbf{V}_1(s)\right]$ and the voltage gain of the network shown in Fig. 12.9.

SOLUTION   The mesh equations for the network are

$$(R_1 + sL)\mathbf{I}_1(s) - sL\mathbf{I}_2(s) = \mathbf{V}_1(s)$$

$$-sL\mathbf{I}_1(s) + \left(R_2 + sL + \frac{1}{sC}\right)\mathbf{I}_2(s) = 0$$

$$\mathbf{V}_2(s) = \mathbf{I}_2(s)R_2$$

Solving the equations for $\mathbf{I}_2(s)$ yields

$$\mathbf{I}_2(s) = \frac{sL\mathbf{V}_1(s)}{(R_1 + sL)(R_2 + sL + 1/sC) - s^2L^2}$$

Therefore, the transfer admittance $\left[\mathbf{I}_2(s)/\mathbf{V}_1(s)\right]$ is

$$\mathbf{Y}_T(s) = \frac{\mathbf{I}_2(s)}{\mathbf{V}_1(s)} = \frac{LCs^2}{(R_1 + R_2)LCs^2 + (L + R_1R_2C)s + R_1}$$

and the voltage gain is

$$\mathbf{G}_v(s) = \frac{\mathbf{V}_2(s)}{\mathbf{V}_1(s)} = \frac{LCR_2s^2}{(R_1 + R_2)LCs^2 + (L + R_1R_2C)s + R_1}$$

Figure 12.9
Circuit employed in Example 12.2.

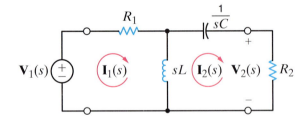

POLES AND ZEROS   As we have indicated, the network function can be expressed as the ratio of the two polynomials in s. In addition, we note that since the values of our circuit elements, or controlled sources, are real numbers, the coefficients of the two polynomials will be real. Therefore, we will express a network function in the form

$$\mathbf{H}(s) = \frac{N(s)}{D(s)} = \frac{a_m s^m + a_{m-1}s^{m-1} + \cdots + a_1 s + a_0}{b_n s^n + b_{n-1}s^{n-1} + \cdots + b_1 s + b_0} \qquad \textbf{12.2}$$

where $N(s)$ is the numerator polynomial of degree m and $D(s)$ is the denominator polynomial of degree n. Equation (12.2) can also be written in the form

$$\mathbf{H}(s) = \frac{K_0(s - z_1)(s - z_2)\cdots(s - z_m)}{(s - p_1)(s - p_2)\cdots(s - p_n)} \qquad \textbf{12.3}$$

where K_0 is a constant, $z_1, \ldots, z_m$ are the roots of $N(s)$, and $p_1, \ldots, p_n$ are the roots of $D(s)$. Note that if $s = z_1$, or $z_2, \ldots, z_m$, then $\mathbf{H}(s)$ becomes zero and hence $z_1, \ldots, z_m$ are called zeros of the transfer function. Similarly, if $s = p_1$, or $p_2, \ldots, p_n$, then $\mathbf{H}(s)$ becomes infinite and, therefore, $p_1, \ldots, p_n$ are called poles of the function. The zeros or poles may actually be complex. However, if they are complex, they must occur in conjugate pairs since the coefficients of the polynomial are real. The representation of the network function specified in Eq. (12.3) is extremely important and is generally employed to represent any linear time-invariant system. The importance of this form lies in the fact that the dynamic properties of a system can be gleaned from an examination of the system poles.

LEARNING EXTENSIONS

E12.1 Find the driving point impedance at $\mathbf{V}_S(s)$ in the amplifier shown in Fig. 12.7b.

ANSWER:
$$\mathbf{Z}(s) = R_{\text{in}} + \frac{1}{sC_{\text{in}}}$$
$$= \left[1 + \left(\frac{100\pi}{s} \right) \right] \text{M}\Omega.$$

E12.2 Find the pole and zero locations in hertz and the value of K_0 for the amplifier network in Fig. 12.7. **PSV**

ANSWER: $z_1 = 0$ Hz (dc);
$p_1 = -50$ Hz;
$p_2 = -20{,}000$ Hz;
$K_0 = (4 \times 10^7)\,\pi.$

12.2 Sinusoidal Frequency Analysis

Although in specific cases a network operates at only one frequency (e.g., a power system network), in general we are interested in the behavior of a network as a function of frequency. In a sinusoidal steady-state analysis, the network function can be expressed as

$$\mathbf{H}(j\omega) = M(\omega)e^{j\phi(\omega)} \qquad \textbf{12.4}$$

where $M(\omega) = |\mathbf{H}(j\omega)|$ and $\phi(\omega)$ is the phase. A plot of these two functions, which are commonly called the *magnitude* and *phase characteristics*, displays the manner in which the response varies with the input frequency ω. We will now illustrate the manner in which to perform a frequency-domain analysis by simply evaluating the function at various frequencies within the range of interest.

FREQUENCY RESPONSE USING A BODE PLOT If the network characteristics are plotted on a semilog scale (that is, a linear scale for the ordinate and a logarithmic scale for the abscissa), they are known as *Bode plots* (named after Hendrik W. Bode). This graph is a powerful tool in both the analysis and design of frequency-dependent systems and networks such as filters, tuners, and amplifiers. In using the graph, we plot $20 \log_{10} M(\omega)$ versus $\log_{10}(\omega)$ instead of $M(\omega)$ versus ω. The advantage of this technique is that rather than plotting the characteristic point by point, we can employ straight-line approximations to obtain the characteristic very efficiently. The ordinate for the magnitude plot is the decibel (dB). This unit was originally employed to measure the ratio of powers; that is,

$$\text{number of dB} = 10 \log_{10} \frac{P_2}{P_1} \qquad \textbf{12.5}$$

If the powers are absorbed by two equal resistors, then

$$\text{number of dB} = 10 \log_{10} \frac{|\mathbf{V}_2|^2/R}{|\mathbf{V}_1|^2/R} = 10 \log_{10} \frac{|\mathbf{I}_2|^2 R}{|\mathbf{I}_1|^2 R}$$

$$= 20 \log_{10} \frac{|\mathbf{V}_2|}{|\mathbf{V}_1|} = 20 \log_{10} \frac{|\mathbf{I}_2|}{|\mathbf{I}_1|}$$

12.6

The term dB has become so popular that it is now used for voltage and current ratios, as illustrated in Eq. (12.6), without regard to the impedance employed in each case.

In the sinusoidal steady-state case, $\mathbf{H}(j\omega)$ in Eq. (12.3) can be expressed in general as

$$\mathbf{H}(j\omega) = \frac{K_0(j\omega)^{\pm N}\left(1 + j\omega\tau_1\right)\left[1 + 2\zeta_3(j\omega\tau_3) + \left(j\omega\tau_3\right)^2\right]\cdots}{\left(1 + j\omega\tau_a\right)\left[1 + 2\zeta_b(j\omega\tau_b) + \left(j\omega\tau_b\right)^2\right]\cdots}$$

12.7

Note that this equation contains the following typical factors:

1. A frequency-independent factor $K_0 > 0$

2. Poles or zeros at the origin of the form $j\omega$; that is, $(j\omega)^{+N}$ for zeros and $(j\omega)^{-N}$ for poles

3. Poles or zeros of the form $(1 + j\omega\tau)$

4. Quadratic poles or zeros of the form $1 + 2\zeta(j\omega\tau) + (j\omega\tau)^2$

Taking the logarithm of the magnitude of the function $\mathbf{H}(j\omega)$ in Eq. (12.7) yields

$$20 \log_{10}|\mathbf{H}(j\omega)| = 20 \log_{10}K_0 \pm 20N \log_{10}|j\omega|$$

$$+ 20 \log_{10}|1 + j\omega\tau_1|$$

$$+ 20 \log_{10}\left|1 + 2\zeta_3(j\omega\tau_3) + \left(j\omega\tau_3\right)^2\right|$$

$$+ \cdots - 20 \log_{10}|1 + j\omega\tau_a|$$

$$- 20 \log_{10}\left|1 + 2\zeta_b(j\omega\tau_b) + \left(j\omega\tau_b\right)^2\right|\cdots$$

12.8

Note that we have used the fact that the log of the product of two or more terms is equal to the sum of the logs of the individual terms, the log of the quotient of two terms is equal to the difference of the logs of the individual terms, and $\log_{10}A^n = n \log_{10}A$.

The phase angle for $\mathbf{H}(j\omega)$ is

$$\underline{/\mathbf{H}(j\omega)} = 0 \pm N(90°) + \tan^{-1}\omega\tau_1 + \tan^{-1}\left(\frac{2\zeta_3\omega\tau_3}{1 - \omega^2\tau_3^2}\right)$$

$$+ \cdots - \tan^{-1}\omega\tau_a - \tan^{-1}\left(\frac{2\zeta_b\omega\tau_b}{1 - \omega^2\tau_b^2}\right)\cdots \quad \textbf{12.9}$$

As Eqs. (12.8) and (12.9) indicate, we will simply plot each factor individually on a common graph and then sum them algebraically to obtain the total characteristic. Let us examine some of the individual terms and illustrate an efficient manner in which to plot them on the Bode diagram.

Constant Term The term $20 \log_{10}K_0$ represents a constant magnitude with zero phase shift, as shown in Fig. 12.10a.

Poles or Zeros at the Origin Poles or zeros at the origin are of the form $(j\omega)^{\pm N}$, where $+$ is used for a zero and $-$ is used for a pole. The magnitude of this function is $\pm 20N \log_{10}\omega$, which is a straight line on semilog paper with a slope of $\pm 20N$ dB/decade; that

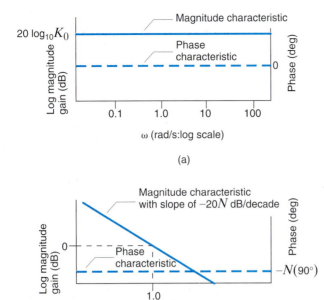

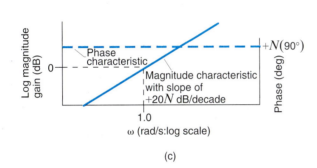

is, the value will change by $20N$ each time the frequency is multiplied by 10, and the phase of this function is a constant $\pm N(90°)$. The magnitude and phase characteristics for poles and zeros at the origin are shown in Figs. 12.10b and c, respectively.

Simple Pole or Zero   Linear approximations can be employed when a simple pole or zero of the form $(1 + j\omega\tau)$ is present in the network function. For $\omega\tau \ll 1$, $(1 + j\omega\tau) \approx 1$, and therefore, $20 \log_{10}|(1 + j\omega\tau)| = 20 \log_{10}1 = 0$ dB. Similarly, if $\omega\tau \gg 1$, then $(1 + j\omega\tau) \approx j\omega\tau$, and hence $20 \log_{10}|(1 + j\omega\tau)| \approx 20 \log_{10}\omega\tau$. Therefore, for $\omega\tau \ll 1$ the response is 0 dB and for $\omega\tau \gg 1$ the response has a slope that is the same as that of a simple pole or zero at the origin. The intersection of these two asymptotes, one for $\omega\tau \ll 1$ and one for $\omega\tau \gg 1$, is the point where $\omega\tau = 1$ or $\omega = 1/\tau$, which is called the *break frequency*. At this break frequency, where $\omega = 1/\tau$, $20 \log_{10}|(1 + j1)| = 20 \log_{10}(2)^{1/2} = 3$ dB. Therefore, the actual curve deviates from the asymptotes by 3 dB at the break frequency. It can be shown that at one-half and twice the break frequency, the deviations are 1 dB. The phase angle associated with a simple pole or zero is $\phi = \tan^{-1}\omega\tau$, which is a simple arctangent curve. Therefore, the phase shift is 45° at the break frequency and 26.6° and 63.4° at one-half and twice the break frequency, respectively. The actual magnitude curve for a pole of this form is shown in Fig. 12.11a. For a zero the magnitude curve and the asymptote for $\omega\tau \gg 1$ have a positive slope, and the phase curve extends from 0° to +90°, as shown in Fig. 12.11b. If multiple poles or zeros of the form $(1 + j\omega\tau)^N$ are present, then the slope of the high-frequency asymptote is multiplied by N, the deviation between the actual curve and the asymptote at the break frequency is $3N$ dB, and the phase curve extends from 0 to $N(90°)$ and is $N(45°)$ at the break frequency.

Figure 12.11

Magnitude and phase plot (a) for a simple pole, and (b) for a simple zero.

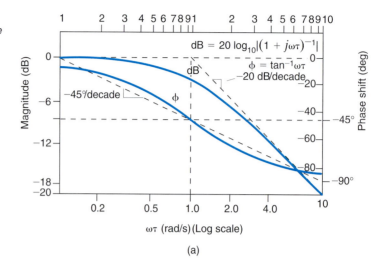

(a)

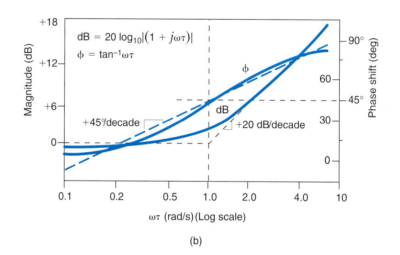

(b)

Quadratic Poles or Zeros Quadratic poles or zeros are of the form $1 + 2\zeta(j\omega\tau) + (j\omega\tau)^2$. This term is a function not only of ω, but also of the dimensionless term ζ, which is called the *damping ratio*. If $\zeta > 1$ or $\zeta = 1$, the roots are real and unequal or real and equal, respectively, and these two cases have already been addressed. If $\zeta < 1$, the roots are complex conjugates, and it is this case that we will examine now. Following the preceding argument for a simple pole or zero, the log magnitude of the quadratic factor is 0 dB for $\omega\tau \ll 1$. For $\omega\tau \gg 1$.

$$20 \log_{10}|1 - (\omega\tau)^2 + 2j\zeta(\omega\tau)| \approx 20 \log_{10}|(\omega\tau)^2| = 40 \log_{10}|\omega\tau|$$

and therefore, for $\omega\tau \gg 1$, the slope of the log magnitude curve is $+40$ dB/decade for a quadratic zero and -40 dB/decade for a quadratic pole. Between the two extremes, $\omega\tau \ll 1$ and $\omega\tau \gg 1$, the behavior of the function is dependent on the damping ratio ζ. Figure 12.12a illustrates the manner in which the log magnitude curve for a quadratic *pole* changes as a function of the damping ratio. The phase shift for the quadratic factor is $\tan^{-1} 2\zeta\omega\tau/[1 - (\omega\tau)^2]$. The phase plot for quadratic *poles* is shown in Fig. 12.12b. Note that in this case the phase changes from $0°$ at frequencies for which $\omega\tau \ll 1$ to $-180°$ at frequencies for which $\omega\tau \gg 1$. For quadratic zeros the magnitude and phase curves are inverted; that is, the log magnitude curve has a slope of $+40$ dB/decade for $\omega\tau \gg 1$, and the phase curve is $0°$ for $\omega\tau \ll 1$ and $+180°$ for $\omega\tau \gg 1$.

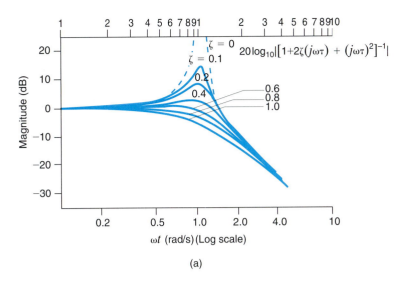

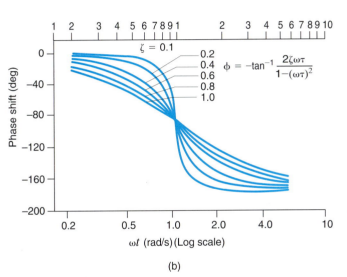

Figure 12.12
Magnitude and phase characteristics for quadratic poles.

Example 12.3

We want to generate the magnitude and phase plots for the transfer function

$$\mathbf{G}_v(j\omega) = \frac{10(0.1j\omega + 1)}{(j\omega + 1)(0.02j\omega + 1)}$$

SOLUTION   Note that this function is in standard form, since every term is of the form $(j\omega\tau + 1)$. To determine the composite magnitude and phase characteristics, we will plot the individual asymptotic terms and then add them as specified in Eqs. (12.8) and (12.9). Let us consider the magnitude plot first. Since $K_0 = 10$, $20 \log_{10} 10 = 20$ dB, which is a constant independent of frequency, as shown in Fig. 12.13a. The zero of the transfer function contributes a term of the form $+20 \log_{10}|1 + 0.1j\omega|$, which is 0 dB for $0.1\omega \ll 1$, has a slope of $+20$ dB/decade for $0.1\omega \gg 1$, and has a break frequency at $\omega = 10$ rad/s. The poles have break frequencies at $\omega = 1$ and $\omega = 50$ rad/s. The pole with break frequency at $\omega = 1$ rad/s contributes a term of the form $-20 \log_{10}|1 + j\omega|$, which is 0 dB for $\omega \ll 1$ and has a slope of -20 dB/decade for $\omega \gg 1$. A similar argument can be made for the pole that has a break frequency at $\omega = 50$ rad/s. These factors are all plotted individually in Fig. 12.13a.

Consider now the individual phase curves. The term K_0 is not a function of ω and does not contribute to the phase of the transfer function. The phase curve for the zero is $+\tan^{-1} 0.1\omega$, which is an arctangent curve that extends from $0°$ for $0.1\omega \ll 1$ to $+90°$ for $0.1\omega \gg 1$ and has a phase of $+45°$ at the break frequency. The phase curves for the two poles are $-\tan^{-1} \omega$ and $-\tan^{-1} 0.02\omega$. The term $-\tan^{-1} \omega$ is $0°$ for $\omega \ll 1$, $-90°$ for $\omega \gg 1$, and $-45°$ at the break frequency $\omega = 1$ rad/s. The phase curve for the remaining pole is plotted in a similar fashion. All the individual phase curves are shown in Fig. 12.13a.

As specified in Eqs. (12.8) and (12.9), the composite magnitude and phase of the transfer function are obtained simply by adding the individual terms. The composite curves are plotted in Fig. 12.13b. Note that the actual magnitude curve (solid line) differs from the straight-line approximation (dashed line) by 3 dB at the break frequencies and 1 dB at one-half and twice the break frequencies.

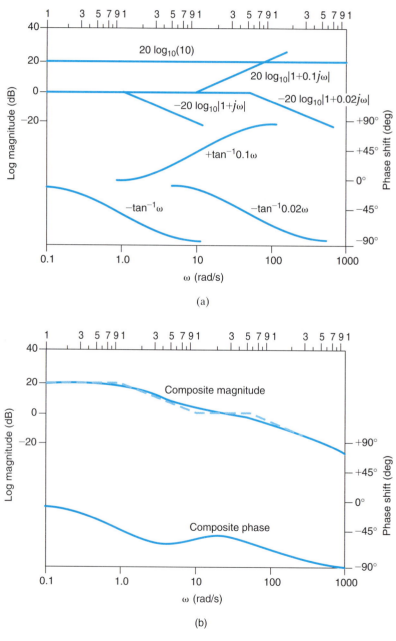

Figure 12.13 (a) Magnitude and phase components for the poles and zeros of the transfer function in Example 12.3; (b) Bode plot for the transfer function in Example 12.3.

Example 12.4

Let us draw the Bode plot for the following transfer function:

$$\mathbf{G}_v(j\omega) = \frac{25(j\omega + 1)}{(j\omega)^2(0.1j\omega + 1)}$$

SOLUTION Once again all the individual terms for both magnitude and phase are plotted in Fig. 12.14a. The straight line with a slope of -40 dB/decade is generated by the double pole at the origin. This line is a plot of $-40\log_{10}\omega$ versus ω and therefore passes through 0 dB at $\omega = 1$ rad/s. The phase for the double pole is a constant $-180°$ for all frequencies. The remainder of the terms are plotted as illustrated in Example 12.3.

The composite plots are shown in Fig. 12.14b. Once again they are obtained simply by adding the individual terms in Fig. 12.14a. Note that for frequencies for which $\omega \ll 1$, the slope of the magnitude curve is -40 dB/decade. At $\omega = 1$ rad/s, which is the break frequency of the zero,

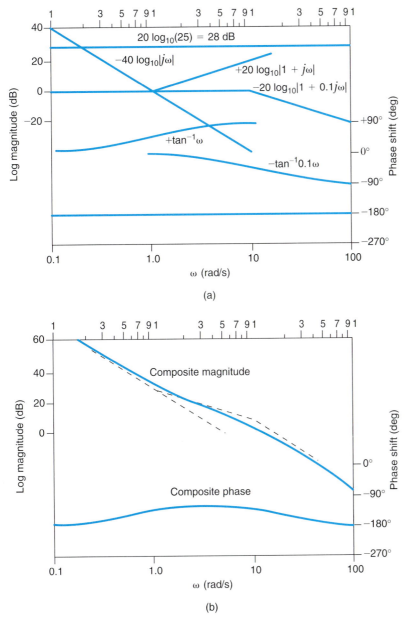

Figure 12.14 (a) Magnitude and phase components for the poles and zeros of the transfer function in Example 12.4; (b) Bode plot for the transfer function in Example 12.4.

the magnitude curve changes slope to -20 dB/decade. At $\omega = 10$ rad/s, which is the break frequency of the pole, the slope of the magnitude curve changes back to -40 dB/decade.

The composite phase curve starts at $-180°$ due to the double pole at the origin. Since the first break frequency encountered is a zero, the phase curve shifts toward $-90°$. However, before the composite phase reaches $-90°$, the pole with break frequency $\omega = 10$ rad/s begins to shift the composite curve back toward $-180°$.

Example 12.4 illustrates the manner in which to plot directly terms of the form $K_0/(j\omega)^N$. For terms of this form, the initial slope of $-20N$ dB/decade will intersect the 0-dB axis at a frequency of $(K_0)^{1/N}$ rad/s; that is, $-20\log_{10}\left|K_0/(j\omega)^N\right| = 0$ dB implies that $K_0/(j\omega)^N = 1$, and therefore, $\omega = (K_0)^{1/N}$ rad/s. Note that the projected slope of the magnitude curve in Example 12.4 intersects the 0-dB axis at $\omega = (25)^{1/2} = 5$ rad/s.

Similarly, it can be shown that for terms of the form $K_0(j\omega)^N$, the initial slope of $+20N$ dB/decade will intersect the 0-dB axis at a frequency of $\omega = (1/K_0)^{1/N}$ rad/s; that is, $+20\log_{10}\left|K_0/(j\omega)^N\right| = 0$ dB implies that $K_0/(j\omega)^N = 1$, and therefore $\omega = (1/K_0)^{1/N}$ rad/s.

By applying the concepts we have just demonstrated, we can normally plot the log magnitude characteristic of a transfer function directly in one step.

LEARNING EXTENSIONS

E12.3 Sketch the magnitude characteristic of the Bode plot, labeling all critical slopes and points for the function

$$\mathbf{G}(j\omega) = \frac{10^4(j\omega + 2)}{(j\omega + 10)(j\omega + 100)}$$

ANSWER:

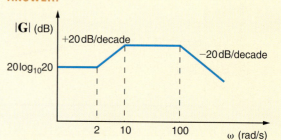

Figure E12.3

E12.4 Sketch the magnitude characteristic of the Bode plot, labeling all critical slopes and points for the function

$$\mathbf{G}(j\omega) = \frac{100(0.02j\omega + 1)}{(j\omega)^2}$$

ANSWER:

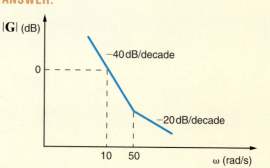

Figure E12.4

E12.5 Sketch the magnitude characteristic of the Bode plot, labeling all critical slopes and points for the function **PSV**

$$\mathbf{G}(j\omega) = \frac{10j\omega}{(j\omega + 1)(j\omega + 10)}$$

ANSWER:

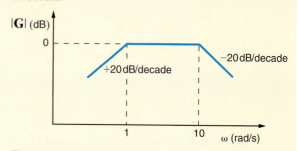

Figure E12.5

Example 12.5

We wish to generate the Bode plot for the following transfer function:

$$\mathbf{G}_v(j\omega) = \frac{25j\omega}{(j\omega + 0.5)\big[(j\omega)^2 + 4j\omega + 100\big]}$$

SOLUTION   Expressing this function in standard form, we obtain

$$\mathbf{G}_v(j\omega) = \frac{0.5j\omega}{(2j\omega + 1)\big[(j\omega/10)^2 + j\omega/25 + 1\big]}$$

The Bode plot is shown in Fig. 12.15. The initial low-frequency slope due to the zero at the origin is $+20$ dB/decade, and this slope intersects the 0-dB line at $\omega = 1/K_0 = 2$ rad/s. At $\omega = 0.5$ rad/s the slope changes from $+20$ dB/decade to 0 dB/decade due to the presence of the pole with a break frequency at $\omega = 0.5$ rad/s. The quadratic term has a center frequency of $\omega = 10$ rad/s (i.e., $\tau = 1/10$). Since

$$2\zeta\tau = \frac{1}{25}$$

and

$$\tau = 0.1$$

then

$$\zeta = 0.2$$

Plotting the curve in Fig. 12.12a with a damping ratio of $\zeta = 0.2$ at the center frequency $\omega = 10$ rad/s completes the composite magnitude curve for the transfer function.

The initial low-frequency phase curve is $+90°$, due to the zero at the origin. This curve and the phase curve for the simple pole and the phase curve for the quadratic term, as defined in Fig. 12.12b, are combined to yield the composite phase curve.

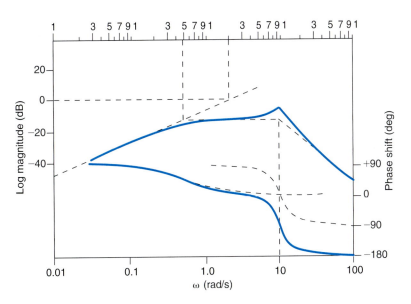

Figure 12.15
Bode plot for the transfer function in Example 12.5.

LEARNING EXTENSION

E12.6 Given the following function $\mathbf{G}(j\omega)$, sketch the magnitude characteristic of the Bode plot, labeling all critical slopes and points.

$$\mathbf{G}(j\omega) = \frac{0.2(j\omega + 1)}{j\omega\left[(j\omega/12)^2 + j\omega/36 + 1\right]}$$

ANSWER:

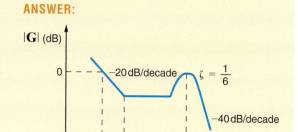

Figure E12.6

Creating Bode Plots using MATLAB

Our approach to the use of MATLAB in the creation of Bode plots will be to explain each step involved in the process using a fairly complicated example. Since only a few steps are involved, one can then follow the procedure to produce a Bode plot for any other transfer function. Suppose that we wish to obtain the Bode plot for the transfer function

$$\mathbf{H}(j\omega) = \frac{2500(10 + j\omega)}{j\omega(2 + j\omega)(2500 + 30j\omega + (j\omega)^2)}$$

over the range of frequencies from 0.1 rad/s to 1000 rad/s using 50 points per decade. The steps involved in obtaining this Bode plot are outlined as follows. In this step-by-step procedure, each statement must end with a semicolon.

Step 1. Clear MATLAB memory and close all open figures.

```
>> clear all;
>> close all;
```

Step 2. Open figure number i, where i is 1 for a single plot, etc.

```
>> figure(1);
```

Step 3. Create a vector of **x** logarithmically spaced, that is, the same number of points per decade, frequencies from y rad/s to z rad/s. The variable w represents omega. In this case, we use 200 points (4 decades and 50 points/decade) and plot from 0.1 rad/s, that is, 10^{-1} to 1000 rad/s (i.e., 10^3).

```
>> w=logspace(-1,3,200);
```

Step 4. Define the transfer function, $\mathbf{H}(j\omega)$, to be plotted in two figures: magnitude and phase. Note that the **.** operators perform array math, element-by-element.

```
>> H=2500*(10+j*w)./(j*w.*(2+j*w).*(2500+j*w*30+(j*w).^2));
```

Step 5. Divide the figure window into two parts: the first part is the magnitude plot, and the second part is the phase plot, as shown on the next page.

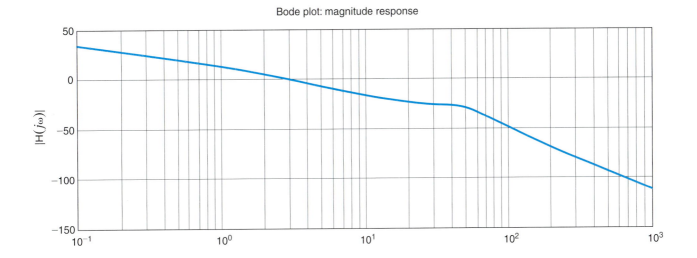

Bode plot: magnitude response

Bode plot: phase response

ω(rad/s)

Figure 12.16
MATLAB-generated Bode plots.

The first subplot statement specifies that there are two plots, one column, and the statements that follow this subplot statement apply to row 1, that is, the magnitude plot.

```
>> subplot(2,1,1);
```

The next statement specifies that the plot is semilog, w is the variable, and $20 \log 10|\mathbf{H}|$ is the ordinate.

```
>> semilogx(w,20*log10(abs(H)));
```

The following statements turn on the grid, specify that the ordinate should be labeled as $|\mathbf{H}(j\omega)|$ and indicate that the title for this plot is "Bode Plot: Magnitude Response." Note that \omega means use the lower case omega symbol here, and anything contained within the single quotes is printed.

```
>> grid;
>> ylabel('|H(j\omega)|');
>> title('Bode Plot: Magnitude Response');
```

The second subplot statement specifies that there are two plots, one column, and the statements that follow this subplot statement apply to row 2, that is, the phase plot.

```
>> subplot(2,1,2);
```

The next statement specifies that the plot is semilog, *w* is the variable, and the ordinate is the angle of **H**, which though normally in rad/s is converted to degrees. The use of the "unwrap" feature eliminates phase jumps from +180 to −180 degrees.

```
>> semilogx(w,unwrap(angle(H))*180/pi);
```

The following statements turn on the grid; they specify that the variable is omega in rad/s and the ordinate is the angle of $\mathbf{H}(j\omega)$ in degrees (circ is the little circle used to represent degrees) and that the title of the plot is "Bode Plot: Phase Response."

```
>> grid;
>> xlabel('\omega(rad/s)');
>> ylabel('\angleH(j\omega)(\circ)');
>> title('Bode Plot: Phase Response');
```

The MATLAB output is shown in Fig. 12.16.

DERIVING THE TRANSFER FUNCTION FROM THE BODE PLOT

The following example will serve to demonstrate the derivation process.

Example 12.6

Given the asymptotic magnitude characteristic shown in Fig. 12.17, we wish to determine the transfer function $\mathbf{G}_v(j\omega)$.

SOLUTION Since the initial slope is 0 dB/decade, and the level of the characteristic is 20 dB, the factor K_0 can be obtained from the expression

$$20 \text{ dB} = 20 \log_{10} K_0$$

and hence

$$K_0 = 10$$

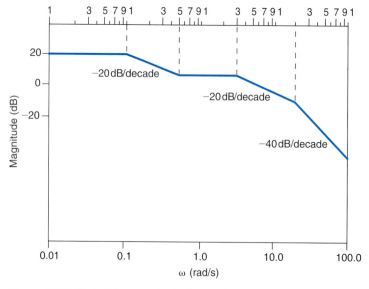

Figure 12.17 *Straight-line magnitude plot employed in Example 12.6.*

The -20-dB/decade slope starting at $\omega = 0.1$ rad/s indicates that the first pole has a break frequency at $\omega = 0.1$ rad/s, and therefore one of the factors in the denominator is $(10j\omega + 1)$. The slope changes by $+20$ dB/decade at $\omega = 0.5$ rad/s, indicating that there is a zero present with a break frequency at $\omega = 0.5$ rad/s, and therefore the numerator has a factor of $(2j\omega + 1)$. Two additional poles are present with break frequencies at $\omega = 2$ rad/s and $\omega = 20$ rad/s. Therefore, the composite transfer function is

$$\mathbf{G}_v(j\omega) = \frac{10(2j\omega + 1)}{(10j\omega + 1)(0.5j\omega + 1)(0.05j\omega + 1)}$$

Note carefully the ramifications of this example with regard to network design.

LEARNING EXTENSION

E12.7 Determine the transfer function $\mathbf{G}(j\omega)$ if the straight-line magnitude characteristic approximation for this function is as shown in Fig. E12.7.

ANSWER:

$$G(j\omega) = \frac{5\left(\dfrac{j\omega}{5} + 1\right)\left(\dfrac{j\omega}{50} + 1\right)}{j\omega\left(\dfrac{j\omega}{20} + 1\right)\left(\dfrac{j\omega}{100} + 1\right)}$$

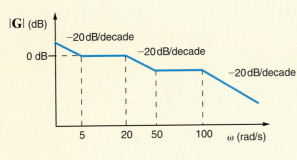

Figure E12.7

12.3   Resonant Circuits

SERIES RESONANCE   A circuit with extremely important frequency characteristics is shown in Fig. 12.18. The input impedance for the series RLC circuit is

$$\mathbf{Z}(j\omega) = R + j\omega L + \frac{1}{j\omega C} = R + j\left(\omega L - \frac{1}{\omega C}\right) \qquad \textbf{12.10}$$

The imaginary term will be zero if

$$\omega L = \frac{1}{\omega C}$$

The value of ω that satisfies this equation is

$$\omega_0 = \frac{1}{\sqrt{LC}} \qquad \textbf{12.11}$$

and at this value of ω the impedance becomes

$$\mathbf{Z}(j\omega_0) = R \qquad \textbf{12.12}$$

This frequency ω_0, at which the impedance of the circuit is purely real, is also called the *resonant frequency,* and the circuit itself at this frequency is said to be *in resonance.* Resonance is a very important consideration in engineering design. For example, engineers designing the

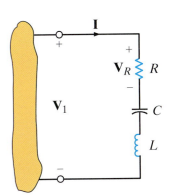

Figure 12.18
Series RLC circuit.

attitude control system for the Saturn vehicles had to ensure that the control system frequency did not excite the body bending (resonant) frequencies of the vehicle. Excitation of the bending frequencies would cause oscillations that, if continued unchecked, would result in a buildup of stress until the vehicle would finally break apart.

Resonance is also a benefit, providing string and wind musical instruments with volume and rich tones.

At resonance the voltage and current are in phase and, therefore, the phase angle is zero and the power factor is unity. At resonance the impedance is a minimum and, therefore, the current is maximum for a given voltage. Figure 12.19 illustrates the frequency response of the series *RLC* circuit. Note that at low frequencies the impedance of the series circuit is dominated by the capacitive term, and at high frequencies the impedance is dominated by the inductive term.

Resonance can be viewed from another perspective—that of the phasor diagram. In the series circuit the current is common to every element. Therefore, the current is employed as reference. The phasor diagram is shown in Fig. 12.20 for the three frequency values $\omega < \omega_0$, $\omega = \omega_0$, $\omega > \omega_0$.

When $\omega < \omega_0$, $\mathbf{V}_C > \mathbf{V}_L$, θ_Z is negative and the voltage $\mathbf{V}_1$ lags the current. If $\omega = \omega_0$, $\mathbf{V}_L = \mathbf{V}_C$, θ_Z is zero, and the voltage $\mathbf{V}_1$ is in phase with the current. If $\omega > \omega_0$, $\mathbf{V}_L > \mathbf{V}_C$, θ_Z is positive, and the voltage $\mathbf{V}_1$ leads the current.

For the series circuit we define what is commonly called the *quality factor Q* as

$$Q = \frac{\omega_0 L}{R} = \frac{1}{\omega_0 C R} = \frac{1}{R}\sqrt{\frac{L}{C}} \qquad \textbf{12.13}$$

Q is a very important factor in resonant circuits, and its ramifications will be illustrated throughout the remainder of this section.

Figure 12.19

Frequency response of a series RLC *circuit.*

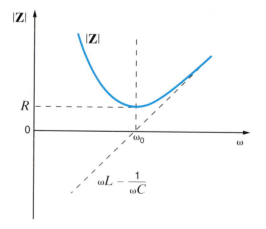

Figure 12.20

Phasor diagrams for the series RLC *circuit.*

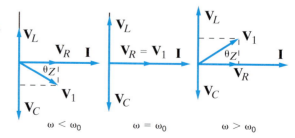

Example 12.7

Consider the network shown in Fig. 12.21. Let us determine the resonant frequency, the voltage across each element at resonance, and the value of the quality factor.

SOLUTION   The resonant frequency is obtained from the expression

$$\omega_0 = \frac{1}{\sqrt{LC}}$$

$$= \frac{1}{\sqrt{(25)(10^{-3})(10)(10^{-6})}}$$

$$= 2000 \text{ rad/s}$$

At this resonant frequency

$$\mathbf{I} = \frac{\mathbf{V}}{\mathbf{Z}} = \frac{\mathbf{V}}{R} = 5\underline{/0^\circ} \text{ A}$$

Therefore,

$$\mathbf{V}_R = (5\underline{/0^\circ})(2) = 10\underline{/0^\circ} \text{ V}$$

$$\mathbf{V}_L = j\omega_0 L\mathbf{I} = 250\underline{/90^\circ} \text{ V}$$

$$\mathbf{V}_C = \frac{\mathbf{I}}{j\omega_0 C} = 250\underline{/-90^\circ} \text{ V}$$

Note the magnitude of the voltages across the inductor and capacitor with respect to the input voltage. Note also that these voltages are equal and are 180° out of phase with one another. Therefore, the phasor diagram for this condition is shown in Fig. 12.20 for $\omega = \omega_0$. The quality factor Q derived from Eq. (12.13) is

$$Q = \frac{\omega_0 L}{R} = \frac{(2)(10^3)(25)(10^{-3})}{2} = 25$$

The voltages across the inductor and capacitor can be written in terms of Q as

$$|\mathbf{V}_L| = \omega_0 L|\mathbf{I}| = \frac{\omega_0 L}{R}|\mathbf{V}_S| = Q|\mathbf{V}_S|$$

and

$$|\mathbf{V}_C| = \frac{|\mathbf{I}|}{\omega_0 C} = \frac{1}{\omega_0 CR}|\mathbf{V}_S| = Q|\mathbf{V}_S|$$

This analysis indicates that for a given current there is a resonant voltage rise across the inductor and capacitor that is equal to the product of Q and the applied voltage.

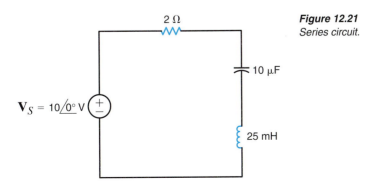

Figure 12.21
Series circuit.

Example 12.8

In an undergraduate circuits laboratory, students are asked to construct an *RLC* network that will demonstrate resonance at $f = 1000$ Hz given a 0.02 H inductor that has a Q of 200. One student produced the circuit shown in Fig. 12.22, where the inductor's internal resistance is represented by R.

If the capacitor chosen to demonstrate resonance was an oil-impregnated paper capacitor rated at 300 V, let us determine the network parameters and the effect of this choice of capacitor.

Figure 12.22
RLC *series resonant network.*

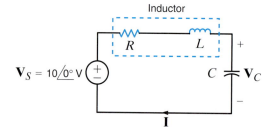

SOLUTION For resonance at 1000 Hz, the student found the required capacitor value using the expression

$$\omega_0 = 2\pi f_0 = \frac{1}{\sqrt{LC}}$$

which yields

$$C = 1.27 \ \mu\text{F}$$

The student selected an oil-impregnated paper capacitor rated at 300 V. The resistor value was found using the expression for Q

$$Q = \frac{\omega_0 L}{R} = 200$$

or

$$R = 1.59 \ \Omega$$

At resonance, the current would be

$$\mathbf{I} = \frac{\mathbf{V}_S}{R}$$

or

$$\mathbf{I} = 6.28 \ \underline{/0°} \ \text{A}$$

When constructed, the current was measured to be only

$$\mathbf{I} \sim 1 \ \underline{/0°} \ \text{mA}$$

This measurement clearly indicated that the impedance seen by the source was about 10 kΩ of resistance instead of 1.59 Ω—quite a drastic difference. Suspecting that the capacitor that was selected was the source of trouble, the student calculated what the capacitor voltage should be. If operated as designed, then at resonance,

$$\mathbf{V}_C = \frac{\mathbf{V}_S}{R}\left(\frac{1}{j\omega C}\right) = Q\mathbf{V}_S$$

or

$$\mathbf{V}_C = 2000 \ \underline{/-90°} \ \text{V}$$

which is more than six times the capacitor's rated voltage! This overvoltage had damaged the capacitor so that it did not function properly. When a new capacitor was selected and the source voltage reduced by a factor of 10, the network performed properly as a high Q circuit.

E12.8 Given the network in Fig. E12.8, find the value C that will place the circuit in resonance at 1800 rad/s.   **PSV**   **ANSWER:** $C = 3.09\ \mu\text{F}$.

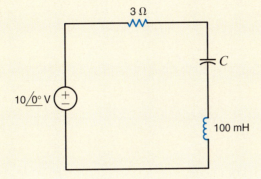

Figure E12.8

3 Ω

C

$10\underline{/0°}$ V

100 mH

E12.9 Given the network in E12.8, determine the Q of the network and the magnitude of the voltage across the capacitor.   **PSV**   **ANSWER:** $Q = 60$, $|\mathbf{V}_C| = 600$ V.

Let us develop a general expression for the ratio of $\mathbf{V}_R/\mathbf{V}_1$ for the network in Fig. 12.18 in terms of Q, ω, and ω_0. The impedance of the circuit, given by Eq. (12.10), can be used to determine the admittance, which can be expressed as

$$\mathbf{Y}(j\omega) = \frac{1}{R\big[1 + j(1/R)(\omega L - 1/\omega C)\big]}$$

$$= \frac{1}{R\big[1 + j(\omega L/R - 1/\omega CR)\big]}$$

$$= \frac{1}{R\big[1 + jQ(\omega L/RQ - 1/\omega CRQ)\big]} \qquad \textbf{12.14}$$

Using the fact that $Q = \omega_0 L/R = 1/\omega_0 CR$, Eq. (12.14) becomes

$$\mathbf{Y}(j\omega) = \frac{1}{R\big[1 + jQ(\omega/\omega_0 - \omega_0/\omega)\big]} \qquad \textbf{12.15}$$

Since $\mathbf{I} = \mathbf{Y}\mathbf{V}_1$ and the voltage across the resistor is $\mathbf{V}_R = \mathbf{I}R$, then

$$\frac{\mathbf{V}_R}{\mathbf{V}_1} = \mathbf{G}_v(j\omega) = \frac{1}{1 + jQ(\omega/\omega_0 - \omega_0/\omega)} \qquad \textbf{12.16}$$

and the magnitude and phase are

$$M(\omega) = \frac{1}{\big[1 + Q^2(\omega/\omega_0 - \omega_0/\omega)^2\big]^{1/2}} \qquad \textbf{12.17}$$

and

$$\phi(\omega) = -\tan^{-1}Q\left(\frac{\omega}{\omega_0} - \frac{\omega_0}{\omega}\right) \qquad \textbf{12.18}$$

The sketches for these functions are shown in Fig. 12.23. Note that the circuit has the form of a band-pass filter. The bandwidth is defined as the difference between the two half-power frequencies. Since power is proportional to the square of the magnitude, these two frequencies may be derived by setting the magnitude $M(\omega) = 1/\sqrt{2}$; that is,

$$\left|\frac{1}{1 + jQ(\omega/\omega_0 - \omega_0/\omega)}\right| = \frac{1}{\sqrt{2}}$$

Figure 12.23

Magnitude and phase curves for Eqs. (12.17) and (12.18).

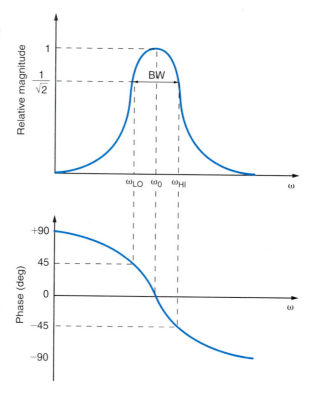

Therefore,

$$Q\left(\frac{\omega}{\omega_0} - \frac{\omega_0}{\omega}\right) = \pm 1 \qquad \textbf{12.19}$$

Solving this equation, we obtain four frequencies,

$$\omega = \pm \frac{\omega_0}{2Q} \pm \omega_0 \sqrt{\left(\frac{1}{2Q}\right)^2 + 1} \qquad \textbf{12.20}$$

Taking only the positive values, we obtain

HINT

Half-power frequencies and their dependance on ω_0 and Q are outlined in these equations.

$$\omega_{LO} = \omega_0 \left[-\frac{1}{2Q} + \sqrt{\left(\frac{1}{2Q}\right)^2 + 1} \right]$$

$$\omega_{HI} = \omega_0 \left[\frac{1}{2Q} + \sqrt{\left(\frac{1}{2Q}\right)^2 + 1} \right] \qquad \textbf{12.21}$$

Subtracting these two equations yields the bandwidth as shown in Fig. 12.23:

HINT

The bandwidth is the difference between the half-power frequencies and a function of ω_0 and Q.

$$BW = \omega_{HI} - \omega_{LO} = \frac{\omega_0}{Q} \qquad \textbf{12.22}$$

and multiplying the two equations yields

$$\omega_0^2 = \omega_{LO}\omega_{HI} \qquad \textbf{12.23}$$

which illustrates that the resonant frequency is the geometric mean of the two half-power frequencies. Recall that the half-power frequencies are the points at which the log-magnitude curve is down 3 dB from its maximum value. Therefore, the difference between the 3-dB frequencies, which is, of course, the bandwidth, is often called the 3-dB bandwidth.

LEARNING EXTENSION

E12.10 For the network in Fig. E12.8, compute the two half-power frequencies and the bandwidth of the network.

ANSWER:
$\omega_{HI} = 1815$ rad/s;
$\omega_{LO} = 1785$ rad/s;
BW $= 30$ rad/s.

Equation (12.13) indicates the dependence of Q on R. A high-Q series circuit has a small value of R.

Equation (12.22) illustrates that the bandwidth is inversely proportional to Q. Therefore, the frequency selectivity of the circuit is determined by the value of Q. A high-Q circuit has a small bandwidth, and, therefore, the circuit is very selective. The manner in which Q affects the frequency selectivity of the network is graphically illustrated in Fig. 12.24. Hence, if we pass a signal with a wide frequency range through a high-Q circuit, only the frequency components within the bandwidth of the network will not be attenuated; that is, the network acts like a band-pass filter.

Q has a more general meaning that we can explore via an energy analysis of the series resonant circuit. Let's excite a series RLC circuit at its resonant frequency as shown in Fig. 12.25. Recall that the impedance of the RLC circuit at resonance is just R. Therefore, the current $i(t) = (V_m/R)\cos \omega_0 t$ A. The capacitor voltage is

$$\mathbf{V}_C = \frac{1}{j\omega_0 C} \mathbf{I} = \frac{1}{j\omega_0 C} \frac{V_m}{R} \underline{/0°} = \frac{V_m}{\omega_0 RC} \underline{/-90°} \qquad \textbf{12.24}$$

and $v_C(t) = \dfrac{V_m}{\omega_0 RC} \cos(\omega_0 t - 90°) = \dfrac{V_m}{\omega_0 RC} \sin \omega_0 t$ volts. Recall from Chapter 6 that the energy stored in an inductor is $(1/2)Li^2$ and the energy stored in a capacitor is $(1/2)Cv^2$. For the inductor,

$$w_L(t) = \frac{1}{2}Li^2(t) = \frac{1}{2}L\left(\frac{V_m}{R}\cos \omega_0 t\right)^2 = \frac{V_m^2 L}{2R^2}\cos^2 \omega_0 t \text{ joules} \qquad \textbf{12.25}$$

and for the capacitor.

$$w_C(t) = \frac{1}{2}Cv_C^2(t) = \frac{1}{2}C\left(\frac{V_m}{\omega_0 RC}\sin \omega_0 t\right)^2 = \frac{V_m^2}{2\omega_0^2 R^2 C}\sin^2 \omega_0 t \text{ joules} \qquad \textbf{12.26}$$

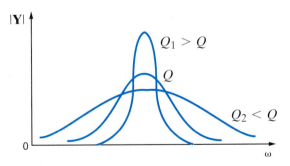

Figure 12.24
Network frequency responce as a function of Q.

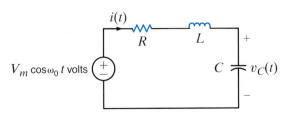

Figure 12.25
Series RLC circuit excited at its resonant frequency.

At resonance, $\omega_0^2 = 1/LC$, so the energy stored in the capacitor can be rewritten as

$$w_C(t) = \frac{V_m^2}{2\left(\dfrac{1}{LC}\right)R^2 C}\sin^2 \omega_0 t = \frac{V_m^2 L}{2R^2}\sin^2 \omega_0 t \text{ joules} \qquad \textbf{12.27}$$

The total energy stored in the circuit is $w_L(t) + w_C(t) = \dfrac{V_m^2 L}{2R^2}\left(\cos^2 \omega_0 t + \sin^2 \omega_0 t\right)$. From trigonometry, we know that $\cos^2 \omega_0 t + \sin^2 \omega_0 t = 1$, so the total energy stored is a constant: $\dfrac{V_m^2 L}{2R^2}$ joules.

Now that we have determined that the total energy stored in the resonant circuit is a constant, let's examine the energy stored in the inductor and capacitor. Figure 12.26 is a plot of the normalized energy stored in each element over two periods. Equations (12.25) and (12.27) have been divided by $\dfrac{V_m^2 L}{2R^2}$ to yield the normalized energy. When a circuit is in resonance, there is a continuous exchange of energy between the magnetic field of the inductor and the electric field of the capacitor. This energy exchange is like the motion of a pendulum. The energy stored in the inductor starts at a maximum value, falls to zero, and then returns to a maximum; the energy stored in the capacitor starts at zero, increases to a maximum, and then returns to zero. Note that when the energy stored in the inductor is a maximum, the energy stored in the capacitor is zero and vice versa. In the first half-cycle, the capacitor absorbs energy as fast as the inductor gives it up; the opposite happens in the next half-cycle. Even though the energy stored in each element is continuously varying, the total energy stored in the resonant circuit is constant and therefore not changing with time.

The maximum energy stored in the RLC circuit at resonance is $W_S = \dfrac{V_m^2 L}{2R^2}$. Let's calculate the energy dissipated per cycle in this series resonant circuit, which is

$$W_D = \int_0^T p_R \, dt = \int_0^T i^2(t)R \, dt = \int_0^T \left(\frac{V_m}{R}\cos^2 \omega_0 t\right)^2 R \, dt = \frac{V_m^2 T}{2R} \qquad \textbf{12.28}$$

The ratio of W_S to W_D is

$$\frac{W_S}{W_D} = \frac{\dfrac{V_m^2 L}{2R^2}}{\dfrac{V_m^2 T}{2R}} = \frac{L}{RT} = \frac{L}{R\dfrac{2\pi}{\omega_0}} = \frac{\omega_0 L}{R(2\pi)} \qquad \textbf{12.29}$$

Figure 12.26
Energy transfer in a resonant circuit.

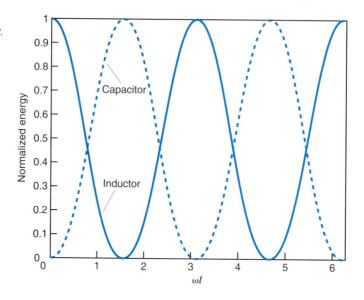

Earlier in this chapter, we defined Q to be $\omega_0 L/R$, so the equation above can be rewritten as

$$Q = 2\pi \frac{W_S}{W_D} \qquad \textbf{12.30}$$

The importance of this expression for Q stems from the fact that this expression is applicable to acoustic, electrical, and mechanical systems and therefore is generally considered to be the basic definition of Q.

Example 12.9

Given a series circuit with $R = 2\ \Omega$, $L = 2$ mH, and $C = 5\ \mu$F, we wish to determine the resonant frequency, the quality factor, and the bandwidth for the circuit. Then we will determine the change in Q and the BW if R is changed from 2 to 0.2 Ω.

SOLUTION Using Eq. (12.11), we have

$$\omega_0 = \frac{1}{\sqrt{LC}} = \frac{1}{\left[(2)(10^{-3})(5)(10^{-6})\right]^{1/2}}$$

$$= 10^4\ \text{rad/s}$$

and therefore, the resonant frequency is $10^4/2\pi = 1592$ Hz.

The quality factor is

$$Q = \frac{\omega_0 L}{R} = \frac{(10^4)(2)(10^{-3})}{2}$$

$$= 10$$

and the bandwidth is

$$\text{BW} = \frac{\omega_0}{Q} = \frac{10^4}{10}$$

$$= 10^3\ \text{rad/s}$$

If R is changed to $R = 0.2\ \Omega$, the new value of Q is 100, and therefore the new BW is 10^2 rad/s.

LEARNING EXTENSIONS

E12.11 A series circuit is composed of $R = 2\ \Omega$, $L = 40$ mH, and $C = 100\ \mu$F. Determine the bandwidth of this circuit and its resonant frequency.

ANSWER: BW = 50 rad/s; $\omega_0 = 500$ rad/s.

E12.12 A series RLC circuit has the following properties: $R = 4\ \Omega$, $\omega_0 = 4000$ rad/s, and the BW = 100 rad/s. Determine the values of L and C.

ANSWER: $L = 40$ mH; $C = 1.56\ \mu$F.

Example 12.10

We wish to determine the parameters R, L, and C so that the circuit shown in Fig. 12.27 operates as a band-pass filter with an ω_0 of 1000 rad/s and a bandwidth of 100 rad/s.

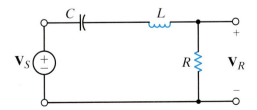

Figure 12.27
Series RLC circuit.

SOLUTION The voltage gain for the network is

$$\mathbf{G}_v(j\omega) = \frac{(R/L)j\omega}{(j\omega)^2 + (R/L)j\omega + 1/LC}$$

Hence,

$$\omega_0 = \frac{1}{\sqrt{LC}}$$

and since $\omega_0 = 10^3$,

$$\frac{1}{LC} = 10^6$$

The bandwidth is

$$\text{BW} = \frac{\omega_0}{Q}$$

Then

$$Q = \frac{\omega_0}{\text{BW}} = \frac{1000}{100}$$

$$= 10$$

However,

$$Q = \frac{\omega_0 L}{R}$$

Therefore,

$$\frac{1000L}{R} = 10$$

Note that we have two equations in the three unknown circuit parameters R, L, and C. Hence, if we select $C = 1\ \mu\text{F}$, then

$$L = \frac{1}{10^6 C} = 1\ \text{H}$$

and

$$\frac{1000(1)}{R} = 10$$

yields

$$R = 100\ \Omega$$

Therefore, the parameters $R = 100\ \Omega$, $L = 1\ \text{H}$, and $C = 1\ \mu\text{F}$ will produce the proper filter characteristics.

In Examples 12.7 and 12.8 we found that the voltage across the capacitor or inductor in the series resonant circuit could be quite high. In fact, it was equal to Q times the magnitude of the source voltage. With this in mind, let us reexamine this network as shown in Fig. 12.28. The output voltage for the network is

$$\mathbf{V}_o = \left(\frac{1/j\omega C}{R + j\omega L + 1/j\omega C} \right) \mathbf{V}_S$$

which can be written as

$$\mathbf{V}_o = \frac{\mathbf{V}_S}{1 - \omega^2 LC + j\omega CR}$$

The magnitude of this voltage can be expressed as

$$|\mathbf{V}_o| = \frac{|\mathbf{V}_S|}{\sqrt{(1 - \omega^2 LC)^2 + (\omega CR)^2}}$$

12.31

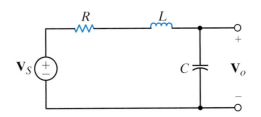

Figure 12.28
Series resonant circuit.

In view of the previous discussion, we might assume that the maximum value of the output voltage would occur at the resonant frequency ω_0. Let us see whether this assumption is correct. The frequency at which $|\mathbf{V}_o|$ is maximum is the nonzero value of ω, which satisfies the equation

$$\frac{d|\mathbf{V}_o|}{d\omega} = 0 \qquad\qquad \textbf{12.32}$$

If we perform the indicated operation and solve for the nonzero ω_{max}, we obtain

$$\omega_{max} = \sqrt{\frac{1}{LC} - \frac{1}{2}\left(\frac{R}{L}\right)^2} \qquad\qquad \textbf{12.33}$$

By employing the relationships $\omega_0^2 = 1/LC$ and $Q = \omega_0 L/R$, the expression for ω_{max} can be written as

$$\omega_{max} = \sqrt{\omega_0^2 - \frac{1}{2}\left(\frac{\omega_0}{Q}\right)^2}$$

$$= \omega_0\sqrt{1 - \frac{1}{2Q^2}} \qquad\qquad \textbf{12.34}$$

Clearly, $\omega_{max} \neq \omega_0$; however, ω_0 closely approximates ω_{max} if the Q is high. In addition, if we substitute Eq. (12.34) into Eq. (12.31) and use the relationships $\omega_0^2 = 1/LC$ and $\omega_0^2 C^2 R^2 = 1/Q^2$, we find that

$$|\mathbf{V}_o|_{max} = \frac{Q|\mathbf{V}_S|}{\sqrt{1 - 1/4Q^2}} \qquad\qquad \textbf{12.35}$$

Again, we see that $|\mathbf{V}_o|_{max} \approx Q|\mathbf{V}_S|$ if the network has a high Q.

Example 12.11

Given the network in Fig. 12.28, we wish to determine ω_0 and ω_{max} for $R = 50\ \Omega$ and $R = 1\ \Omega$ if $L = 50$ mH and $C = 5\ \mu$F.

SOLUTION The network parameters yield

$$\omega_0 = \frac{1}{\sqrt{LC}}$$

$$= \frac{1}{\sqrt{(5)(10^{-2})(5)(10^{-6})}}$$

$$= 2000\ \text{rad/s}$$

If $R = 50\ \Omega$, then

$$Q = \frac{\omega_0 L}{R}$$

$$= \frac{(2000)(0.05)}{50}$$

$$= 2$$

and

$$\omega_{max} = \omega_0 \sqrt{1 - \frac{1}{2Q^2}}$$

$$= 2000 \sqrt{1 - \frac{1}{8}}$$

$$= 1871 \text{ rad/s}$$

If $R = 1\ \Omega$, then $Q = 100$ and $\omega_{max} = 2000$ rad/s.

We can plot the frequency response of the network transfer function for $R = 50\ \Omega$ and $R = 1\ \Omega$. The transfer function is

$$\frac{\mathbf{V}_o}{\mathbf{V}_S} = \frac{1}{2.5 \times 10^{-7}(j\omega)^2 + 2.5 \times 10^{-4}(j\omega) + 1}$$

for $R = 50\ \Omega$ and

$$\frac{\mathbf{V}_o}{\mathbf{V}_S} = \frac{1}{2.5 \times 10^{-7}(j\omega)^2 + 5 \times 10^{-6}(j\omega) + 1}$$

for $R = 1\ \Omega$. The magnitude and phase characteristics for the network with $R = 50\ \Omega$ and $R = 1\ \Omega$ are shown in Figs. 12.29a and b, respectively.

Note that when the Q of the network is small, the frequency response is not selective and $\omega_0 \neq \omega_{max}$. However, if the Q is large, the frequency response is very selective and $\omega_0 \cong \omega_{max}$.

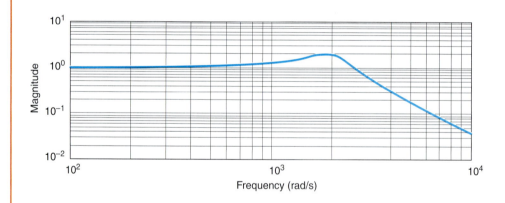

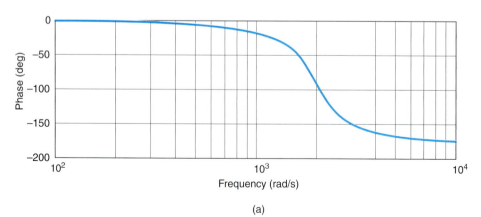

(a)

Figure 12.29 *Frequency response plots for the network in Fig. 12.28 with (a) $R = 50\ \Omega$ and (b) $R = 1\ \Omega$.*

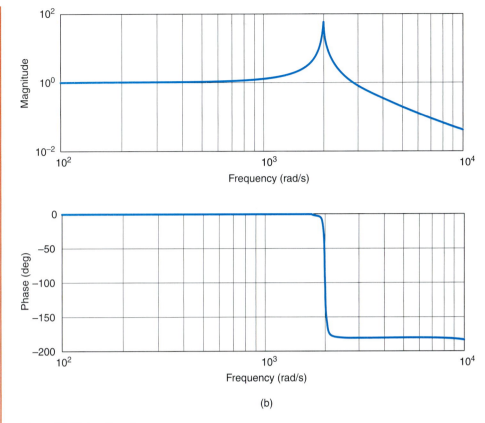

Figure 12.29 (continued)

(b)

Example 12.12

On July 1, 1940, the third longest bridge in the nation, the Tacoma Narrows Bridge, was opened to traffic across Puget Sound in Washington. On November 7, 1940, the structure collapsed in what has become the most celebrated structural failure of that century. A photograph of the bridge, taken as it swayed back and forth just before breaking apart, is shown in Fig. 12.30. Explaining the disaster in quantitative terms is a feat for civil engineers and structures experts, and several theories have been presented. However, the one common denominator in each explanation is that wind blowing across the bridge caused the entire structure to resonate to such an extent that the bridge tore itself apart. One can theorize that the wind, fluctuating at a frequency near the natural frequency of the bridge (0.2 Hz), drove the structure into resonance. Thus, the bridge can be roughly modeled as a second-order system. Let us design an *RLC* resonance network to demonstrate the bridge's vertical movement and investigate the effect of the wind's frequency.

SOLUTION   The *RLC* network shown in Fig. 12.31 is a second-order system in which $v_{in}(t)$ is analogous to vertical deflection of the bridge's roadway (1 volt = 1 foot). The values of C, L, R_A, and R_B can be derived from the data taken at the site and from scale models, as follows:

$$\text{vertical deflection at failure} \approx 4 \text{ feet}$$
$$\text{wind speed at failure} \approx 42 \text{ mph}$$
$$\text{resonant frequency} = f_0 \approx 0.2 \text{ Hz}$$

The output voltage can be expressed as

$$\mathbf{V}_o(j\omega) = \frac{j\omega\left(\dfrac{R_B}{L}\right)\mathbf{V}_{in}(j\omega)}{-\omega^2 + j\omega\left(\dfrac{R_A + R_B}{L}\right) + \dfrac{1}{LC}}$$

Figure 12.30 *Tacoma Narrows Bridge on the verge of collapse. (Used with permission from Special Collection Division, University of Washington Libraries. Photo by Farguharson, negative number 12.)*

Figure 12.31
RLC *resonance
network for a simple
Tacoma Narrows
Bridge simulation.*

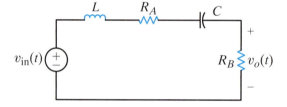

from which we can easily extract the following expressions:

$$\omega_0 = \frac{1}{\sqrt{LC}} = 2\pi(0.2) \text{ rad/s}$$

$$2\zeta\omega_0 = \frac{R_A + R_B}{L}$$

and

$$\frac{\mathbf{V}_o(j\omega_0)}{\mathbf{V}_{in}(j\omega_0)} = \frac{R_B}{R_A + R_B} \approx \frac{4 \text{ feet}}{42 \text{ mph}}$$

Let us choose $R_B = 1 \ \Omega$ and $R_A = 9.5 \ \Omega$. Having no data for the damping ratio, ζ, we will select $L = 20$ H, which yields $\zeta = 0.209$ and $Q = 2.39$, which seem reasonable for such a large structure. Given the aforementioned choices, the required capacitor value is $C = 31.66$ mF. Using these circuit values, we now simulate the effect of 42 mph winds fluctuating at 0.05 Hz, 0.1 Hz, and 0.2 Hz using an ac analysis at the three frequencies of interest.

The results are shown in Fig. 12.32. Note that at 0.05 Hz the vertical deflection (1 ft/V) is only 0.44 feet, whereas at 0.1 Hz the bridge undulates about 1.07 feet. Finally, at the bridge's resonant frequency of 0.2 Hz, the bridge is oscillating 3.77 feet—catastrophic failure.

Clearly, we have used an extremely simplistic approach to modeling something as complicated as the Tacoma Narrows Bridge. However, we will revisit this event in Chapter 14 and examine it more closely with a more accurate model (K. Y. Billah and R. H. Scalan, "Resonance, Tacoma Narrows Bridge Failure, and Undergraduate Physics Textbooks," *American Journal of Physics*, (1991, vol. 59, no. 2, pp. 118–124).

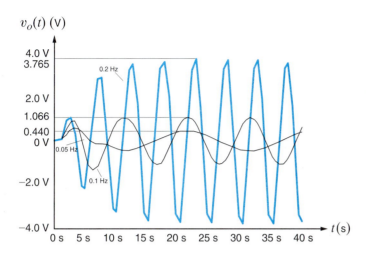

Figure 12.32
Simulated vertical deflection (1 volt = 1 foot) for the Tacoma Narrows Bridge for wind shift frequencies of 0.05, 0.1, and 0.2 Hz.

PARALLEL RESONANCE   In our presentation of resonance thus far, we have focused our discussion on the series resonant circuit. Of course, resonance and all its ramifications still apply if the *RLC* elements are arranged in parallel. In fact, the series and parallel resonant circuits possess many similarities and a few differences.

Consider the network shown in Fig. 12.33. The source current $\mathbf{I}_S$ can be expressed as

$$\mathbf{I}_S = \mathbf{I}_G + \mathbf{I}_C + \mathbf{I}_L$$

$$= \mathbf{V}_S G + j\omega C \mathbf{V}_S + \frac{\mathbf{V}_S}{j\omega L}$$

$$= \mathbf{V}_S \left[G + j\left(\omega C - \frac{1}{\omega L} \right) \right]$$

When the network is in resonance,

$$\mathbf{I}_S = G\mathbf{V}_S \qquad \textbf{12.36}$$

The input admittance for the parallel *RLC* circuit is

$$\mathbf{Y}(j\omega) = G + j\omega C + \frac{1}{j\omega L} \qquad \textbf{12.37}$$

and the admittance of the parallel circuit, at resonance, is

$$\mathbf{Y}(j\omega_0) = G \qquad \textbf{12.38}$$

that is, all the source current flows through the conductance G. Does this mean that there is no current in L or C? Definitely not! $\mathbf{I}_C$ and $\mathbf{I}_L$ are equal in magnitude but 180° out of phase with one another. Therefore, $\mathbf{I}_x$, as shown in Fig. 12.33, is zero. In addition, if $G = 0$, the source current is zero. What is actually taking place, however, is an energy exchange between the electric field of the capacitor and the magnetic field of the inductor. As one increases, the other decreases, and vice versa.

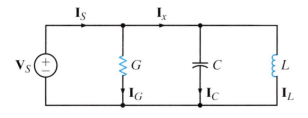

Figure 12.33
Parallel RLC circuit.

Figure 12.34
(a) The frequency plot of the admittance and
(b) the phasor diagram for the parallel resonant circuit.

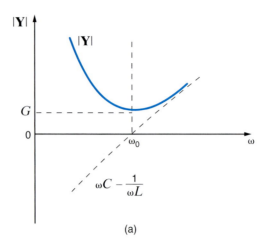

(a)

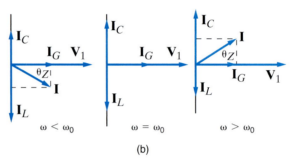

(b)

Analogous to the series resonant case, the frequency response, shown in Fig. 12.34a, for the parallel resonant circuit reveals that the admittance is dominated by the inductive term at low frequencies and by the capacitive term at high frequencies. Similarly, the phasor diagram for the parallel resonant circuit, shown in Figure 12.34b, again has much in common with that of the series circuit. For $\omega < \omega_0$, the impedance phase angle, θ_Z, is positive, again indicating that inductance dominates in the parallel circuit at low frequencies. For $\omega > \omega_0$, θ_Z is negative, and the capacitance dominates.

Applying the general definition of resonance in Fig. 12.23 to the parallel resonant circuit yields an interesting result

$$Q = \frac{R}{\omega_0 L} = \frac{1}{G\omega_0 L} = R\omega_0 C = \frac{\omega_0 C}{G} \qquad \textbf{12.39}$$

This result appears to be the reciprocal of Q for the series case. However, the *RLC* currents in the parallel case mimic the voltages in the series case.

$$|\mathbf{I}_C| = Q|\mathbf{I}_S|$$

and

$$|\mathbf{I}_L| = Q|\mathbf{I}_S|$$

12.40

Example 12.13

The network in Fig. 12.33 has the following parameters:

$$\mathbf{V}_S = 120 \underline{/0°} \text{ V}, \qquad G = 0.01 \text{ S},$$
$$C = 600 \text{ μF}, \quad \text{and} \quad L = 120 \text{ mH}$$

If the source operates at the resonant frequency of the network, compute all the branch currents.

SOLUTION   The resonant frequency for the network is

$$\omega_0 = \frac{1}{\sqrt{LC}}$$

$$= \frac{1}{\sqrt{(120)(10^{-3})(600)(10^{-6})}}$$

$$= 117.85 \text{ rad/s}$$

At this frequency

$$\mathbf{Y}_C = j\omega_0 C = j7.07 \times 10^{-2} \text{ S}$$

and

$$\mathbf{Y}_L = -j\left(\frac{1}{\omega_0 L}\right) = -j7.07 \times 10^{-2} \text{ S}$$

The branch currents are then

$$\mathbf{I}_G = G\mathbf{V}_S = 1.2 \underline{/0°} \text{ A}$$

$$\mathbf{I}_C = \mathbf{Y}_C \mathbf{V}_S = 8.49 \underline{/90°} \text{ A}$$

$$\mathbf{I}_L = \mathbf{Y}_L \mathbf{V}_S = 8.49 \underline{/-90°} \text{ A}$$

and

$$\mathbf{I}_S = \mathbf{I}_G + \mathbf{I}_C + \mathbf{I}_L$$

$$= \mathbf{I}_G = 1.2 \underline{/0°} \text{ A}$$

As the analysis indicates, the source supplies only the losses in the resistive element. In addition, the source voltage and current are in phase and, therefore, the power factor is unity.

Example 12.14

Given the parallel *RLC* circuit in Fig. 12.35,

a. Derive the expression for the resonant frequency, the half-power frequencies, the bandwidth, and the quality factor for the transfer characteristic $\mathbf{V}_{\text{out}}/\mathbf{I}_{\text{in}}$ in terms of the circuit parameters R, L, and C.

b. Compute the quantities in part (a) if $R = 1 \text{ k}\Omega$, $L = 10 \text{ mH}$, and $C = 100 \text{ μF}$.

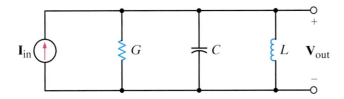

Figure 12.35 *Circuit used in Example 12.14.*

SOLUTION   (a) The output voltage can be written as

$$\mathbf{V}_{\text{out}} = \frac{\mathbf{I}_{\text{in}}}{\mathbf{Y}_T}$$

and, therefore, the magnitude of the transfer characteristic can be expressed as

$$\left|\frac{\mathbf{V}_{\text{out}}}{\mathbf{I}_{\text{in}}}\right| = \frac{1}{\sqrt{(1/R^2) + (\omega C - 1/\omega L)^2}}$$

The transfer characteristic is a maximum at the resonant frequency

$$\omega_0 = \frac{1}{\sqrt{LC}}$$

12.41

and at this frequency

$$\left|\frac{\mathbf{V}_{out}}{\mathbf{I}_{in}}\right|_{max} = R$$

12.42

As demonstrated earlier, at the half-power frequencies the magnitude is equal to $1/\sqrt{2}$ of its maximum value, and hence the half-power frequencies can be obtained from the expression

$$\frac{1}{\sqrt{(1/R^2) + (\omega C - 1/\omega L)^2}} = \frac{R}{\sqrt{2}}$$

Solving this equation and taking only the positive values of ω yields

$$\omega_{LO} = -\frac{1}{2RC} + \sqrt{\frac{1}{(2RC)^2} + \frac{1}{LC}}$$

12.43

and

$$\omega_{HI} = \frac{1}{2RC} + \sqrt{\frac{1}{(2RC)^2} + \frac{1}{LC}}$$

12.44

Subtracting these two half-power frequencies yields the bandwidth

$$BW = \omega_{HI} - \omega_{LO}$$

$$= \frac{1}{RC}$$

12.45

Therefore, the quality factor is

$$Q = \frac{\omega_0}{BW}$$

$$= \frac{RC}{\sqrt{LC}}$$

$$= R\sqrt{\frac{C}{L}}$$

12.46

Using Eqs. (12.41), (12.45), and (12.46), we can write Eqs. (12.43) and (12.44) as

$$\omega_{LO} = \omega_0 \left[\frac{-1}{2Q} + \sqrt{\frac{1}{(2Q)^2} + 1}\right]$$

12.47

$$\omega_{HI} = \omega_0 \left[\frac{1}{2Q} + \sqrt{\frac{1}{(2Q)^2} + 1}\right]$$

12.48

(b) Using the values given for the circuit components, we find that

$$\omega_0 = \frac{1}{\sqrt{(10^{-2})(10^{-4})}} = 10^3 \text{ rad/s}$$

The half-power frequencies are

$$\omega_{LO} = \frac{-1}{(2)(10^3)(10^{-4})} + \sqrt{\frac{1}{[(2)(10^{-1})]^2} + 10^6}$$

$$= 995 \text{ rad/s}$$

and

$$\omega_{HI} = 1005 \text{ rad/s}$$

Therefore, the bandwidth is

$$BW = \omega_{HI} = \omega_{LO} = 10 \text{ rad/s}$$

and

$$Q = 10^3 \sqrt{\frac{10^{-4}}{10^{-2}}}$$

$$= 100$$

Example 12.15

Two radio stations, WHEW and WHAT, broadcast in the same listening area: WHEW broadcasts at 100 MHz and WHAT at 98 MHz. A single-stage tuned amplifier, such as that shown in Fig. 12.36, can be used as a tuner to filter out one of the stations. However, single-stage tuned amplifiers have poor selectivity due to their wide bandwidths. To reduce the bandwidth (increase the quality factor) of single-stage tuned amplifiers, designers employ a technique called synchronous tuning. In this process, identical tuned amplifiers are cascaded. To demonstrate this phenomenon, let us generate a Bode plot for the amplifier shown in Fig. 12.36 when it is tuned to WHEW (100 MHz), using one, two, three, and four stages of amplification.

SOLUTION   Using the circuit for a single-stage amplifier shown in Fig. 12.36, we can cascade the stages to form a four-stage synchronously tuned amplifier. If we now plot the frequency response over the range from 90 MHz to 110 MHz, which is easily done using PSPICE, we obtain the Bode plot shown in Fig. 12.37.

From the Bode plot in Fig. 12.37 we see that increasing the number of stages does indeed decrease the bandwidth without altering the center frequency. As a result, the quality factor and selectivity increase. Accordingly, as we add stages, the gain at 98 MHz (WHAT's frequency) decreases, and that station is "tuned out."

Figure 12.36
Single-stage tuned amplifier.

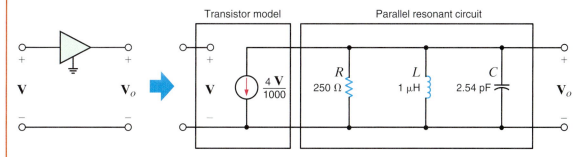

Figure 12.37
Bode plots for one-, two-, three-, and four-stage tuned amplifiers.

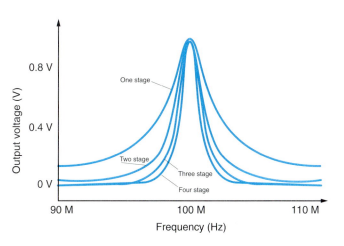

LEARNING EXTENSIONS

E12.13 A parallel *RLC* circuit has the following parameters: $R = 2$ kΩ, $L = 20$ mH, and $C = 150$ μF. Determine the resonant frequency, the Q, and the bandwidth of the circuit. **PSV**

ANSWER:
$\omega_0 = 577$ rad/s; $Q = 173$; and BW $= 3.33$ rad/s.

E12.14 A parallel *RLC* circuit has the following parameters: $R = 6$ kΩ, BW $= 1000$ rad/s, and $Q = 120$. Determine the values of L, C, and ω_0.

ANSWER:
$L = 417.5$ μH; $C = 0.167$ μF; and $\omega_0 = 119{,}760$ rad/s.

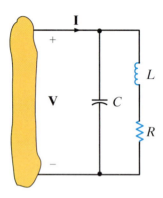

Figure 12.38
Practical parallel resonant circuit.

In general, the resistance of the winding of an inductor cannot be neglected, and hence a more practical parallel resonant circuit is the one shown in Fig. 12.38. The input admittance of this circuit is

$$\mathbf{Y}(j\omega) = j\omega C + \frac{1}{R + j\omega L}$$

$$= j\omega C + \frac{R - j\omega L}{R^2 + \omega^2 L^2}$$

$$= \frac{R}{R^2 + \omega^2 L^2} + j\left(\omega C - \frac{\omega L}{R^2 + \omega^2 L^2}\right)$$

The frequency at which the admittance is purely real is

$$\omega_r C - \frac{\omega_r L}{R^2 + \omega_r^2 L^2} = 0$$

$$\omega_r = \sqrt{\frac{1}{LC} - \frac{R^2}{L^2}} \qquad \textbf{12.49}$$

Example 12.16

Given the tank circuit in Fig. 12.39, let us determine ω_0 and ω_r for $R = 50\ \Omega$ and $R = 5\ \Omega$.

Figure 12.39
Tank circuit used in Example 12.16.

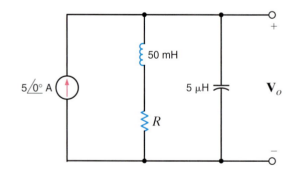

SOLUTION Using the network parameter values, we obtain

$$\omega_0 = \frac{1}{\sqrt{LC}}$$

$$= \frac{1}{\sqrt{(0.05)(5)(10^{-6})}}$$

$$= 2000 \text{ rad/s}$$

$$f_0 = 318.3 \text{ Hz}$$

If $R = 50\ \Omega$, then

$$\omega_r = \sqrt{\frac{1}{LC} - \frac{R^2}{L^2}}$$

$$= \sqrt{\frac{1}{(0.05)(5)(10^{-6})} - \left(\frac{50}{0.05}\right)^2}$$

$$= 1732\ \text{rad/s}$$

$$f_r = 275.7\ \text{Hz}$$

If $R = 5\ \Omega$, then

$$\omega_r = \sqrt{\frac{1}{(0.05)(5)(10^{-6})} - \left(\frac{5}{0.05}\right)^2}$$

$$= 1997\ \text{rad/s}$$

$$f_r = 317.9\ \text{Hz}$$

Note that as $R \rightarrow 0$, $\omega_r \rightarrow \omega_0$. This fact is also illustrated in the frequency-response curves in Figs. 12.40a and b, where we have plotted $|\mathbf{V}_o|$ versus frequency for $R = 50\ \Omega$ and $R = 5\ \Omega$, respectively.

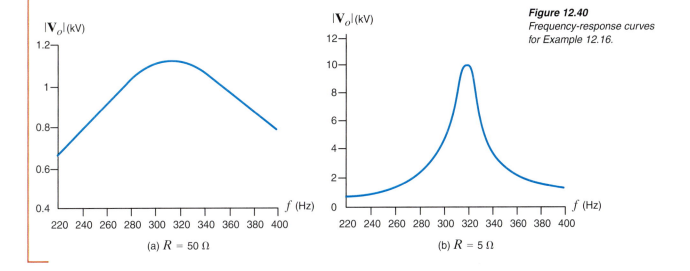

Figure 12.40
Frequency-response curves for Example 12.16.

Let us now try to relate some of the things we have learned about resonance to the Bode plots we presented earlier. The admittance for the series resonant circuit is

$$\mathbf{Y}(j\omega) = \frac{1}{R + j\omega L + 1/j\omega C}$$

$$= \frac{j\omega C}{(j\omega)^2 LC + j\omega CR + 1} \qquad \textbf{12.50}$$

The standard form for the quadratic factor is

$$(j\omega\tau)^2 + 2\zeta\omega\tau j + 1$$

where $\tau = 1/\omega_0$, and hence in general the quadratic factor can be written as

$$\frac{(j\omega)^2}{\omega_0^2} + \frac{2\zeta\omega}{\omega_0}j + 1 \qquad \textbf{12.51}$$

If we now compare this form of the quadratic factor with the denominator of $\mathbf{Y}(j\omega)$, we find that

$$\omega_0^2 = \frac{1}{LC}$$

$$\frac{2\zeta}{\omega_0} = CR$$

and therefore,

$$\zeta = \frac{R}{2}\sqrt{\frac{C}{L}}$$

However, from Eq. (12.13),

$$Q = \frac{1}{R}\sqrt{\frac{L}{C}}$$

and hence,

$$Q = \frac{1}{2\zeta} \qquad \textbf{12.52}$$

To illustrate the significance of this equation, consider the Bode plot for the function $\mathbf{Y}(j\omega)$. The plot has an initial slope of $+20$ dB/decade due to the zero at the origin. If $\zeta > 1$, the poles represented by the quadratic factor in the denominator will simply roll off the frequency response, as illustrated in Fig. 12.12a, and at high frequencies the slope of the composite characteristic will be -20 dB/decade. If $0 < \zeta < 1$, the frequency response will peak as shown in Fig. 12.12a, and the sharpness of the peak will be controlled by ζ. If ζ is very small, the peak of the frequency response is very narrow, the Q of the network is very large, and the circuit is very selective in filtering the input signal. Equation (12.52) and Fig. 12.24 illustrate the connections among the frequency response, the Q, and the ζ of a network.

12.4 Scaling

Throughout this book we have employed a host of examples to illustrate the concepts being discussed. In many cases the actual values of the parameters were unrealistic in a practical sense, even though they may have simplified the presentation. In this section we illustrate how to *scale* the circuits to make them more realistic.

There are two ways to scale a circuit: *magnitude* or *impedance scaling* and *frequency scaling*. To magnitude scale a circuit, we simply multiply the impedance of each element by a scale factor K_M. Therefore, a resistor R becomes $K_M R$. Multiplying the impedance of an inductor $j\omega L$ by K_M yields a new inductor $K_M L$, and multiplying the impedance of a capacitor $1/j\omega C$ by K_M yields a new capacitor C/K_M. Therefore, in magnitude scaling,

HINT

Magnitude or impedance scaling.

$$R' \rightarrow K_M R$$

$$L' \rightarrow K_M L \qquad \textbf{12.53}$$

$$C' \rightarrow \frac{C}{K_M}$$

since

$$\omega_0' = \frac{1}{\sqrt{L'C'}} = \frac{1}{\sqrt{K_M LC/K_M}} = \omega_0$$

and Q' is

$$Q' = \frac{\omega_0 L'}{R'} = \frac{\omega_0 K_M L}{K_M R} = Q$$

The resonant frequency, the quality factor, and therefore the bandwidth are unaffected by magnitude scaling.

In frequency scaling the scale factor is denoted as K_F. The resistor is frequency independent and, therefore, unaffected by this scaling. The new inductor L', which has the same impedance at the scaled frequency ω_1', must satisfy the equation

$$j\omega_1 L = j\omega_1' L'$$

where $\omega_1' = k_F \omega_1$. Therefore,

$$j\omega_1 L = jK_F \omega_1 L'$$

Hence, the new inductor value is

$$L' = \frac{L}{K_F}$$

Using a similar argument, we find that

$$C' = \frac{C}{K_F}$$

Therefore, to frequency scale by a factor K_F,

HINT
Frequency scaling.

$$R' \rightarrow R$$

$$L' \rightarrow \frac{L}{K_F}$$

$$C' \rightarrow \frac{C}{K_F}$$

12.54

Note that

$$\omega_0' = \frac{1}{\sqrt{(L/K_F)(C/K_F)}} = K_F \omega_0$$

and

$$Q' = \frac{K_F \omega_0 L}{R K_F} = Q$$

and therefore,

$$\text{BW}' = K_F(\text{BW})$$

Hence, the resonant frequency and bandwidth of the circuit are affected by frequency scaling.

Example 12.17

If the values of the circuit parameters in Fig 12.38 are $R = 2\ \Omega$, $L = 1$ H, and $C = 1/2$ F, let us determine the values of the elements if the circuit is magnitude scaled by a factor $K_M = 10^2$ and frequency scaled by a factor $K_F = 10^2$.

SOLUTION   The magnitude scaling yields

$$R' = 2K_M = 200\ \Omega$$

$$L' = (1)K_M = 100\ \text{H}$$

$$C' = \frac{1}{2}\frac{1}{K_M} = \frac{1}{200}\ \text{F}$$

Applying frequency scaling to these values yields the final results:

$$R'' = 200\ \Omega$$

$$L'' = \frac{100}{K_F} = 100\ \mu\text{H}$$

$$C'' = \frac{1}{200}\frac{1}{K_F} = 0.005\ \mu\text{F}$$

LEARNING EXTENSION

E12.15 An *RLC* network has the following parameter values: $R = 10\ \Omega$, $L = 1$ H, and $C = 2$ F. Determine the values of the circuit elements if the circuit is magnitude scaled by a factor of 100 and frequency scaled by a factor of 10,000.

ANSWER: $R = 1\ \text{k}\Omega$; $L = 10$ mH; $C = 2\ \mu\text{F}$.

12.5 Filter Networks

PASSIVE FILTERS A filter network is generally designed to pass signals with a specific frequency range and reject or attenuate signals whose frequency spectrum is outside this passband. The most common filters are *low-pass* filters, which pass low frequencies and reject high frequencies; *high-pass* filters, which pass high frequencies and block low frequencies; *band-pass* filters, which pass some particular band of frequencies and reject all frequencies outside the range; and *band-rejection* filters, which are specifically designed to reject a particular band of frequencies and pass all other frequencies.

The ideal frequency characteristic for a low-pass filter is shown in Fig. 12.41a. Also shown is a typical or physically realizable characteristic. Ideally, we would like the low-pass filter to pass all frequencies to some frequency ω_0 and pass no frequency above that value; however, it is not possible to design such a filter with linear circuit elements. Hence, we must be content to employ filters that we can actually build in the laboratory, and these filters have frequency characteristics that are simply not ideal.

A simple low-pass filter network is shown in Fig. 12.41b. The voltage gain for the network is

$$\mathbf{G}_v(j\omega) = \frac{1}{1 + j\omega RC} \qquad\qquad \textbf{12.55}$$

which can be written as

$$\mathbf{G}_v(j\omega) = \frac{1}{1 + j\omega\tau} \qquad\qquad \textbf{12.56}$$

where $\tau = RC$, the time constant. The amplitude characteristic is

$$M(\omega) = \frac{1}{\left[1 + (\omega\tau)^2\right]^{1/2}} \qquad\qquad \textbf{12.57}$$

and the phase characteristic is

$$\phi(\omega) = -\tan^{-1}\omega\tau \qquad\qquad \textbf{12.58}$$

Note that at the break frequency, $\omega = \dfrac{1}{\tau}$, the amplitude is

$$M\left(\omega = \frac{1}{\tau}\right) = \frac{1}{\sqrt{2}} \qquad\qquad \textbf{12.59}$$

The break frequency is also commonly called the *half-power frequency*. This name is derived from the fact that if the voltage or current is $1/\sqrt{2}$ of its maximum value, then the power, which is proportional to the square of the voltage or current, is one-half its maximum value.

The magnitude, in decibels, and phase curves for this simple low-pass circuit are shown in Fig. 12.41c. Note that the magnitude curve is flat for low frequencies and rolls off at high frequencies. The phase shifts from 0° at low frequencies to −90° at high frequencies.

The ideal frequency characteristic for a high-pass filter is shown in Fig. 12.42a, together with a typical characteristic that we could achieve with linear circuit components. Ideally, the high-pass filter passes all frequencies above some frequency ω_0 and no frequencies below that value.

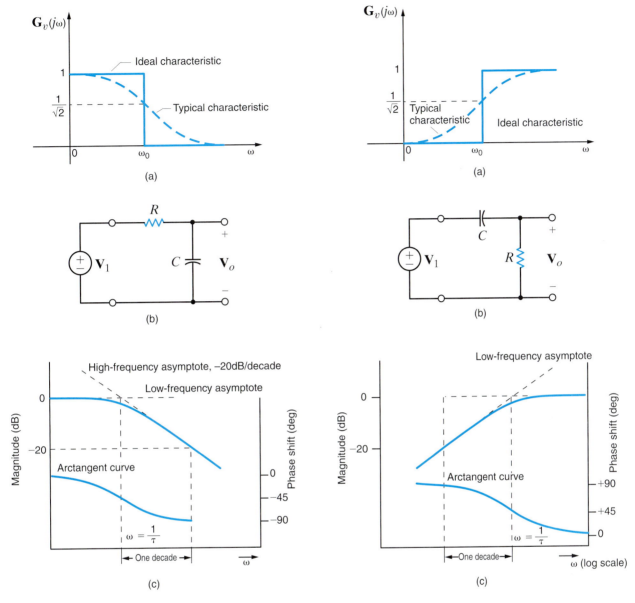

Figure 12.41
Low-pass filter circuit and its frequency characteristics.

Figure 12.42
High-pass filter circuit and frequency characteristics.

A simple high-pass filter network is shown in Fig. 12.42b. This is the same network as shown in Fig. 12.41b, except that the output voltage is taken across the resistor. The voltage gain for this network is

$$\mathbf{G}_v(j\omega) = \frac{j\omega\tau}{1 + j\omega\tau}$$ **12.60**

where once again $\tau = RC$. The magnitude of this function is

$$M(\omega) = \frac{\omega\tau}{\left[1 + (\omega\tau)^2\right]^{1/2}}$$ **12.61**

and the phase is

$$\phi(\omega) = \frac{\pi}{2} - \tan^{-1}\omega\tau$$ **12.62**

The half-power frequency is $\omega = 1/\tau$, and the phase at this frequency is $45°$.

Figure 12.43

Band-pass and band-rejection filters and characteristics.

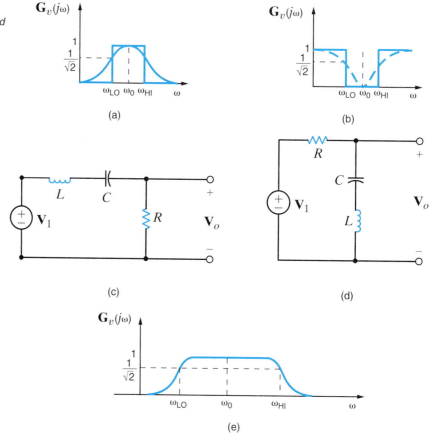

The magnitude and phase curves for this high-pass filter are shown in Fig. 12.42c. At low frequencies the magnitude curve has a slope of +20 dB/decade due to the term $\omega\tau$ in the numerator of Eq. (12.61). Then at the break frequency the curve begins to flatten out. The phase curve is derived from Eq. (12.62).

Ideal and typical amplitude characteristics for simple band-pass and band-rejection filters are shown in Figs. 12.43a and b, respectively. Simple networks that are capable of realizing the typical characteristics of each filter are shown below the characteristics in Figs. 12.43c and d. ω_0 is the center frequency of the pass or rejection band and the frequency at which the maximum or minimum amplitude occurs. ω_{LO} and ω_{HI} are the lower and upper break frequencies or *cutoff frequencies*, where the amplitude is $1/\sqrt{2}$ of the maximum value. The width of the pass or rejection band is called *bandwidth*, and hence

$$\text{BW} = \omega_{HI} - \omega_{LO}$$

12.63

To illustrate these points, let us consider the band-pass filter. The voltage transfer function is

$$\mathbf{G}_v(j\omega) = \frac{R}{R + j(\omega L - 1/\omega C)}$$

and, therefore, the amplitude characteristic is

$$M(\omega) = \frac{RC\omega}{\sqrt{(RC\omega)^2 + (\omega^2 LC - 1)^2}}$$

At low frequencies

$$M(\omega) \approx \frac{RC\omega}{1} \approx 0$$

At high frequencies

$$M(\omega) \simeq \frac{RC\omega}{\omega^2 LC} \approx \frac{R}{\omega L} \approx 0$$

In the midfrequency range $(RC\omega)^2 \gg (\omega^2 LC - 1)^2$, and thus $M(\omega) \approx 1$. Therefore, the frequency characteristic for this filter is shown in Fig. 12.43e. The center frequency is $\omega_0 = 1/\sqrt{LC}$. At the lower cutoff frequency

$$\omega^2 LC - 1 = -RC\omega$$

or

$$\omega^2 + \frac{R\omega}{L} - \omega_0^2 = 0$$

Solving this expression for ω_{LO}, we obtain

$$\omega_{LO} = \frac{-(R/L) + \sqrt{(R/L)^2 + 4\omega_0^2}}{2}$$

At the upper cutoff frequency

$$\omega^2 LC - 1 = +RC\omega$$

or

$$\omega^2 - \frac{R}{L}\omega - \omega_0^2 = 0$$

Solving this expression for ω_{HI}, we obtain

$$\omega_{HI} = \frac{+(R/L) + \sqrt{(R/L)^2 + 4\omega_0^2}}{2}$$

Therefore, the bandwidth of the filter is

$$BW = \omega_{HI} - \omega_{LO} = \frac{R}{L}$$

Example 12.18

Consider the frequency-dependent network in Fig. 12.44. Given the following circuit parameter values: $L = 159\ \mu H$, $C = 159\ \mu F$, and $R = 10\ \Omega$, let us demonstrate that this one network can be used to produce a low-pass, high-pass, or band-pass filter.

SOLUTION The voltage gain $\mathbf{V}_R/\mathbf{V}_S$ is found by voltage division to be

$$\frac{\mathbf{V}_R}{\mathbf{V}_S} = \frac{R}{j\omega L + R + 1/(j\omega C)} = \frac{j\omega\left(\dfrac{R}{L}\right)}{(j\omega)^2 + j\omega\left(\dfrac{R}{L}\right) + \dfrac{1}{LC}}$$

$$= \frac{(62.9 \times 10^3)j\omega}{-\omega^2 + (62.9 \times 10^3)j\omega + 39.6 \times 10^6}$$

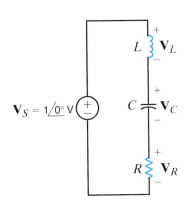

Figure 12.44
Circuit used in Example 12.18.

which is the transfer function for a band-pass filter. At resonance, $\omega^2 = 1/LC$, and hence

$$\frac{\mathbf{V}_R}{\mathbf{V}_S} = 1$$

Now consider the gain $\mathbf{V}_L/\mathbf{V}_S$:

$$\frac{\mathbf{V}_L}{\mathbf{V}_S} = \frac{j\omega L}{j\omega L + R + 1/(j\omega C)} = \frac{-\omega^2}{(j\omega)^2 + j\omega\left(\dfrac{R}{L}\right) + \dfrac{1}{LC}}$$

$$= \frac{-\omega^2}{-\omega^2 + (62.9 \times 10^3)j\omega + 39.6 \times 10^6}$$

which is a second-order high-pass filter transfer function. Again, at resonance,

$$\frac{\mathbf{V}_L}{\mathbf{V}_S} = \frac{j\omega L}{R} = jQ = j0.1$$

Similarly, the gain $\mathbf{V}_C/\mathbf{V}_S$ is

$$\frac{\mathbf{V}_C}{\mathbf{V}_S} = \frac{1/(j\omega C)}{j\omega L + R + 1/(j\omega C)} = \frac{\dfrac{1}{LC}}{(j\omega)^2 + j\omega\left(\dfrac{R}{L}\right) + \dfrac{1}{LC}}$$

$$= \frac{39.6 \times 10^6}{-\omega^2 + (62.9 \times 10^3)j\omega + 39.6 \times 10^6}$$

which is a second-order low-pass filter transfer function. At the resonant frequency,

$$\frac{\mathbf{V}_C}{\mathbf{V}_S} = \frac{1}{j\omega CR} = -jQ = -j0.1$$

Thus, one circuit produces three different filters depending on where the output is taken. This can be seen in the Bode plot for each of the three voltages in Fig. 12.45, where $\mathbf{V}_S$ is set to $1\,\underline{/0°}$ V.

We know that Kirchhoff's voltage law must be satisfied at all times. Note from the Bode plot that the $\mathbf{V}_R + \mathbf{V}_C + \mathbf{V}_L$ also equals $\mathbf{V}_S$ at all frequencies! Finally, let us demonstrate KVL by adding $\mathbf{V}_R$, $\mathbf{V}_L$, and $\mathbf{V}_C$.

$$\mathbf{V}_L + \mathbf{V}_R + \mathbf{V}_C = \frac{\left((j\omega)^2 + j\omega\left(\dfrac{R}{L}\right) + \dfrac{1}{\sqrt{LC}}\right)\mathbf{V}_S}{(j\omega)^2 + j\omega\left(\dfrac{R}{L}\right) + \dfrac{1}{\sqrt{LC}}} = \mathbf{V}_S$$

Thus, even though $\mathbf{V}_S$ is distributed between the resistor, capacitor, and inductor based on frequency, the sum of the three voltages completely reconstructs $\mathbf{V}_S$.

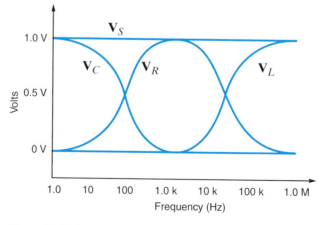

Figure 12.45 Bode plots for network in Fig. 12.44.

Example 12.19

A telephone transmission system suffers from 60-Hz interference caused by nearby power utility lines. Let us use the network in Fig. 12.46 to design a simple notch filter to eliminate the 60-Hz interference.

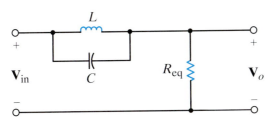

Figure 12.46
Circuit used in Example 12.19.

SOLUTION The resistor R_{eq} represents the equivalent resistance of the telephone system to the right of the LC combination. The LC parallel combination has an equivalent impedance of

$$\mathbf{Z} = (j\omega L)//(1/j\omega C) = \frac{(L/C)}{j\omega L + 1/(j\omega C)}$$

Now the voltage transfer function is

$$\frac{\mathbf{V}_o}{\mathbf{V}_{in}} = \frac{R_{eq}}{R_{eq} + \mathbf{Z}} = \frac{R_{eq}}{R_{eq} + \dfrac{(L/C)}{j\omega L + (1/j\omega C)}}$$

which can be written

$$\frac{\mathbf{V}_o}{\mathbf{V}_{in}} = \frac{(j\omega)^2 + \dfrac{1}{LC}}{(j\omega)^2 + \left(\dfrac{j\omega}{R_{eq}C}\right) + \dfrac{1}{LC}}$$

Note that at resonance, the numerator and thus $\mathbf{V}_o$ go to zero. We want resonance to occur at 60 Hz. Thus,

$$\omega_0 = \frac{1}{\sqrt{LC}} = 2\pi(60) = 120\pi$$

If we select $C = 100\ \mu\text{F}$, then the required value for L is 70.3 mH—both are reasonable values. To demonstrate the effectiveness of the filter, let the input voltage consist of a 60-Hz sinusoid and a 1000-Hz sinusoid of the form

$$v_{in}(t) = 1 \sin\big[(2\pi)60t\big] + 0.2 \sin\big[(2\pi)1000t\big]$$

The input and output waveforms are both shown in Fig. 12.47. Note that the output voltage, as desired, contains none of the 60-Hz interference.

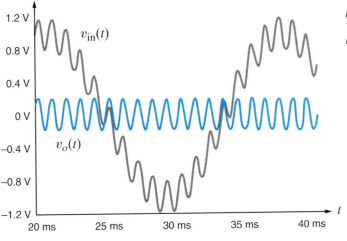

Figure 12.47
Transient analysis of the network in Fig. 12.46.

E12.16 Given the filter network shown in Fig. E12.16, sketch the magnitude characteristic of the Bode plot for $\mathbf{G}_v(j\omega)$.

ANSWER:

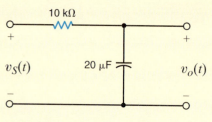

Figure E12.16

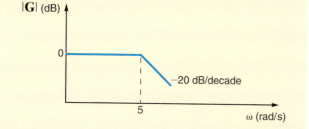

E12.17 Given the filter network in Fig. E12.17, sketch the magnitude characteristic of the Bode plot for $\mathbf{G}_v(j\omega)$.

ANSWER:

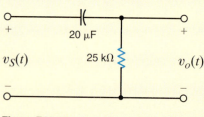

Figure E12.17

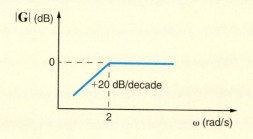

E12.18 A band-pass filter network is shown in Fig. E12.18. Sketch the magnitude characteristic of the Bode plot for $\mathbf{G}_v(j\omega)$.

ANSWER:

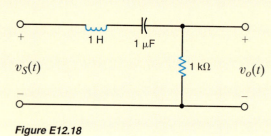

Figure E12.18

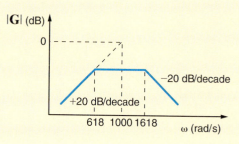

ACTIVE FILTERS   In the preceding section we saw that the four major classes of filters (*low pass, high pass, band pass,* and *band rejection*) are realizable with simple, passive element circuits. However, passive filters do have some serious drawbacks. One obvious problem is the inability to generate a network with a gain > 1 since a purely passive network cannot add energy to a signal. Another serious drawback of passive filters is the need in many topologies for inductive elements. Inductors are generally expensive and are not usually available in precise values. In addition, inductors usually come in odd shapes (toroids, bobbins, E-cores, etc.) and are not easily handled by existing automated printed circuit board assembly machines. By applying operational amplifiers in linear feedback circuits, one can generate all of the primary filter types using only resistors, capacitors, and the op-amp integrated circuits themselves.

The equivalent circuits for the operational amplifiers derived in Chapter 4 are also valid in the sinusoidal steady-state case, when we replace the attendant resistors with impedances.

The equivalent circuits for the basic inverting and noninverting op-amp circuits are shown in Figs. 12.48a and b, respectively. Particular filter characteristics are obtained by judiciously selecting the impedances $\mathbf{Z}_1$ and $\mathbf{Z}_2$.

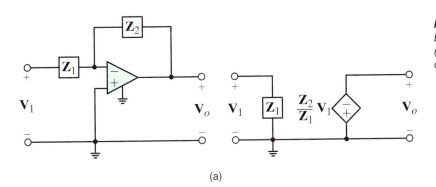

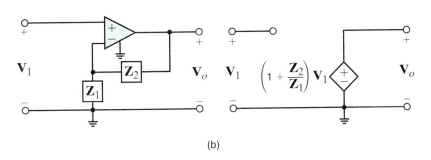

(a)

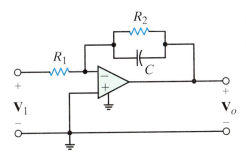

(b)

Figure 12.48

Equivalent circuits for the (a) inverting and (b) noninverting operational amplifier circuits.

Example 12.20

Let us determine the filter characteristics of the network shown in Fig. 12.49.

Figure 12.49

Operational amplifier filter circuit.

SOLUTION   The impedances as illustrated in Fig. 12.48a are

$$\mathbf{Z}_1 = R_1$$

and

$$\mathbf{Z}_2 = \frac{R_2/j\omega C}{R_2 + 1/j\omega C} = \frac{R_2}{j\omega R_2 C + 1}$$

Therefore, the voltage gain of the network is

$$\mathbf{G}_v(j\omega) = \frac{\mathbf{V}_0(j\omega)}{\mathbf{V}_1(j\omega)} = \frac{-R_2/R_1}{j\omega R_2 C + 1}$$

Note that the transfer function is that of a low-pass filter.

Example 12.21

We will show that the amplitude characteristic for the filter network in Fig. 12.50a is as shown in Fig. 12.50b.

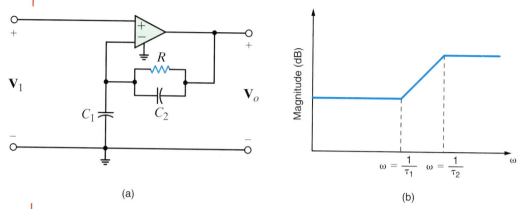

(a)   (b)

Figure 12.50 *Operational amplifier circuit and its amplitude characteristic.*

SOLUTION   Comparing this network with that in Fig. 12.48b, we see that

and

$$\mathbf{Z}_1 = \frac{1}{j\omega C_1}$$

$$\mathbf{Z}_2 = \frac{R}{j\omega R C_2 + 1}$$

Therefore, the voltage gain for the network as a function of frequency is

$$G_v(j\omega) = \frac{\mathbf{V}_0(j\omega)}{\mathbf{V}_1(j\omega)} = 1 + \frac{R/(j\omega R C_2 + 1)}{1/j\omega C_1}$$

$$= \frac{j\omega(RC_1 + RC_2) + 1}{j\omega R C_2 + 1}$$

$$= \frac{j\omega \tau_1 + 1}{j\omega \tau_2 + 1}$$

where $\tau_1 = R(C_1 + C_2)$ and $\tau_2 = RC_2$. Since $\tau_1 > \tau_2$, the amplitude characteristic is of the form shown in Fig. 12.50b. Note that the low frequencies have a gain of 1; however, the high frequencies are amplified. The exact amount of amplification is determined through selection of the circuit parameters.

Example 12.22

A low-pass filter is shown in Fig. 12.51, together with the op-amp subcircuit. We wish to plot the frequency response of the filter over the range from 1 to 10,000 Hz.

Figure 12.51
*Circuit used in Example 12.22:
(a) low-pass filter;
(b) op-amp subcircuit.*

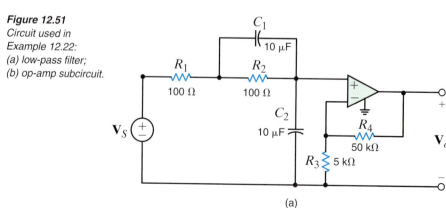

(a)

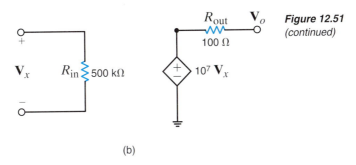

Figure 12.51
(continued)

(b)

SOLUTION   The frequency-response plot, which can be determined by any convenient method, such as a PSPICE simulation, is shown in Fig. 12.52.

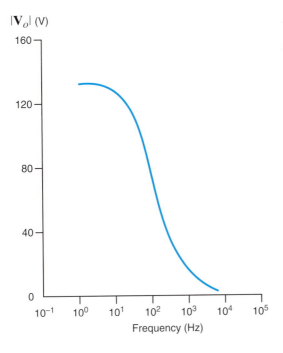

Figure 12.52
Frequency-response plot for the network in Example 12.22.

Example 12.23

A high-pass filter network is shown in Fig. 12.53 together with the op-amp subcircuit. We wish to plot the frequency response of the filter over the range from 1 to 100 kHz.

Figure 12.53
Circuits used in Example 12.23:
(a) high-pass filter; (b) op-amp subcircuit.

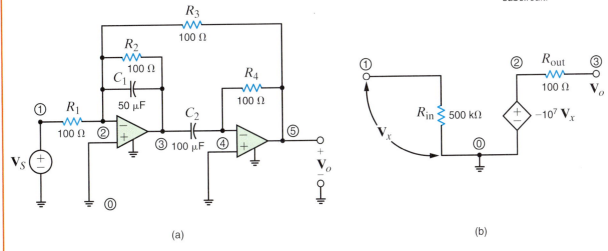

(a)

(b)

SOLUTION The frequency-response plot for this high-pass filter is shown in Fig. 12.54.

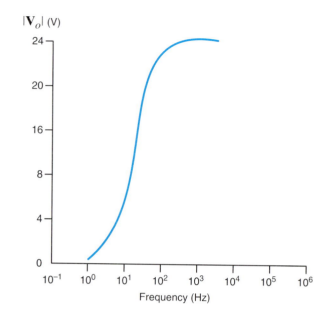

Figure 12.54 Frequency-response plot for the network in Example 12.23.

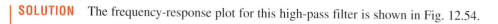
LEARNING EXTENSION

E12.19 Given the filter network shown in Fig. E12.19, determine the transfer function $\mathbf{G}_v(j\omega)$, sketch the magnitude characteristic of the Bode plot for $\mathbf{G}_v(j\omega)$, and identify the filter characteristics of the network.

ANSWER: $\mathbf{G}_v(j\omega) = \dfrac{-j\omega C R_2}{1 + j\omega C R_1}$; this is a high-pass filter.

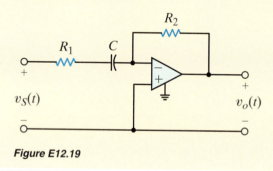

Figure E12.19

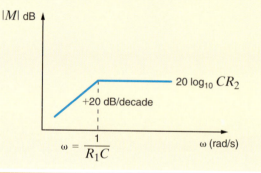

All the circuits considered so far in this section have been first-order filters. In other words, they all had no more than one pole and/or one zero. In many applications, it is desired to generate a circuit with a frequency selectivity greater than that afforded by first-order circuits. The next logical step is to consider the class of second-order filters. For most active-filter applications, if an order greater than two is desired, one usually takes two or more active filter circuits and places them in series so that the total response is the desired higher-order response. This is done because first- and second-order filters are well understood and easily obtained with single op-amp circuits.

In general, second-order filters will have a transfer function with a denominator containing quadratic poles of the form $s^2 + As + B$. For high-pass and low-pass circuits, $B = \omega_c^2$ and $A = 2\zeta\omega_c$. For these circuits, ω_c is the cutoff frequency, and ζ is the damping ratio discussed earlier.

For band-pass circuits, $B = \omega_0^2$ and $A = \omega_0/Q$, where ω_0 is the center frequency and Q is the quality factor for the circuit. Notice that $Q = 1/2\zeta$. Q is a measure of the selectivity of these circuits. The bandwidth is ω_0/Q, as discussed previously.

The transfer function of the second-order low-pass active filter can generally be written as

$$\mathbf{H}(s) = \frac{H_0\omega_c^2}{s^2 + 2\zeta\omega_c s + \omega_c^2} \qquad \textbf{12.64}$$

where H_0 is the dc gain. A circuit that exhibits this transfer function is illustrated in Fig. 12.55 and has the following transfer function:

$$\mathbf{H}(s) = \frac{\mathbf{V}_o(s)}{\mathbf{V}_{in(s)}} = \frac{-\left(\dfrac{R_3}{R_1}\right)\left(\dfrac{1}{R_3 R_2 C_1 C_2}\right)}{s^2 + s\left(\dfrac{1}{R_1 C_1} + \dfrac{1}{R_2 C_1} + \dfrac{1}{R_3 C_1}\right) + \dfrac{1}{R_3 R_2 C_1 C_2}} \qquad \textbf{12.65}$$

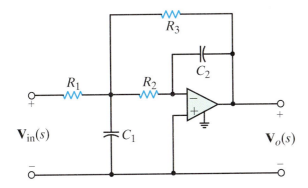

Figure 12.55
Second-order low-pass filter.

Example 12.24

We wish to determine the damping ratio, cutoff frequency, and dc gain H_o for the network in Fig. 12.55 if $R_1 = R_2 = R_3 = 5\,k\Omega$ and $C_1 = C_2 = 0.1\,\mu F$.

SOLUTION   Comparing Eqs. (12.64) and (12.65), we find that

$$\omega_c = \frac{1}{\sqrt{R_3 R_2 C_1 C_2}}$$

$$2\zeta\omega_c = \frac{1}{C_1}\left(\frac{1}{R_1} + \frac{1}{R_2} + \frac{1}{R_3}\right)$$

and therefore,

$$\zeta = \frac{1}{2}\sqrt{\frac{C_2}{C_1}}\left(\frac{1}{R_1} + \frac{1}{R_2} + \frac{1}{R_3}\right)\sqrt{R_2 R_3}$$

In addition, we note that

$$H_o = -\frac{R_3}{R_1}$$

Substituting the given parameter values into the preceding equation yields

$$\omega_0 = 2000 \text{ rad/s}$$

$$\zeta = 1.5$$

and

$$H_o = -1$$

Example 12.25

We wish to vary the capacitors C_1 and C_2 in Example 12.24 to achieve damping ratios of 1.0, 0.75, 0.50, and 0.25 while maintaining ω_c constant at 2000 rad/s.

SOLUTION As shown in the cutoff-frequency equation in Example 12.24, if ω_c is to remain constant at 2000 rad/s, the product of C_1 and C_2 must remain constant. Using the capacitance values in Example 12.24, we have

$$C_1 C_2 = (10)^{-4}$$

or

$$C_2 = \frac{(10)^{-14}}{C_1}$$

Substituting this expression into the equation for the damping ratio yields

$$\zeta = \frac{\sqrt{10^{-14}}}{\sqrt{C_1}\sqrt{C_1}} \left[\frac{1}{2}\left(\frac{1}{R_1} + \frac{1}{R_2} + \frac{1}{R_3} \right) \right] \sqrt{R_2 R_3}$$

$$= \frac{(0.15)(10^{-6})}{C_1}$$

or

$$C_1 = \frac{(0.15)(10^{-6})}{\zeta}$$

Therefore, for $\zeta = 1.0, 0.75, 0.50$, and 0.25, the corresponding values for C_1 are 0.15, 0.20, 0.30, and 0.6 μF, respectively. The values of C_2 that correspond to these values of C_1 are 67, 50, 33, and 17 nF, respectively.

This example illustrates that we can adjust the network parameters to achieve a specified transient response while maintaining the cutoff frequency of the filter constant. In fact, as a general rule we design filters with specific characteristics through proper manipulation of the network parameters.

Example 12.26

We will now demonstrate that the transient response of the circuits generated in Example 12.25 will exhibit increasing overshoot and ringing as ζ is decreased. We will apply a −1-V step function to the input of the network and employ the op-amp model with $R_i = \infty\ \Omega$, $R_o = 0\ \Omega$, and $A = 10^5$.

SOLUTION The transient response for the four cases of damping, including the associated values for the capacitors, can be computed using any convenient method (e.g., PSPICE).

The results are shown in Fig. 12.56. The curves indicate that a $\zeta = 0.75$ might be a good design compromise between rapid step response and minimum overshoot.

Figure 12.56
Transient analysis of Example 12.25.

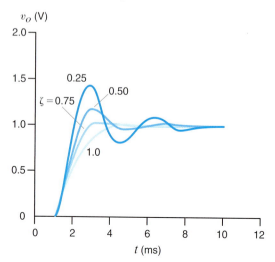

E12.20 Verify that Eq. (12.65) is the transfer function for the network in Fig. 12.55.

The general transfer function for the second-order band-pass filter is

$$\frac{\mathbf{V}_o(s)}{\mathbf{V}_S(s)} = \frac{sH_o}{s^2 + \dfrac{\omega_0}{Q}s + \omega_0^2} \qquad \textbf{12.66}$$

As discussed earlier, ω_0 is the center frequency of the band-pass characteristic and Q is the quality factor. Recall that for low-pass filters, H_o was the passband or dc gain. For a band-pass filter, the gain is maximum at the center frequency, ω_0. To find this maximum gain we substitute $s = j\omega_0$ in the preceding expression to obtain

$$\frac{\mathbf{V}_o(j\omega_0)}{\mathbf{V}_S(j\omega_0)} = \frac{j\omega_0 H_o}{-\omega_0^2 + j\omega_0(\omega_0/Q) + \omega_0^2}$$

$$= \frac{QH_o}{\omega_0} \qquad \textbf{12.67}$$

In addition, the difference between the high and low half-power frequencies (i.e., $\omega_{HI} - \omega_{LO}$) is, of course, the bandwidth

$$\omega_{HI} - \omega_{LO} = \text{BW} = \frac{\omega_0}{Q} \qquad \textbf{12.68}$$

Q is a measure of the selectivity of the band-pass filter, and as the equation indicates, as Q is increased, the bandwidth is decreased.

An op-amp implementation of a band-pass filter is shown in Fig. 12.57. The transfer function for this network is

$$\frac{\mathbf{V}_o(s)}{\mathbf{V}_S(s)} = \frac{-\left(\dfrac{1}{R_1C_1}\right)s}{s^2 + \left(\dfrac{1}{R_2C_1} + \dfrac{1}{R_2C_2}\right)s + \dfrac{1 + R_1/R_3}{R_1R_2C_1C_2}} \qquad \textbf{12.69}$$

Comparing this expression to the more general expression for the band-pass filter yields the following definitions:

$$\omega_0 = \left(\frac{1 + R_1/R_3}{R_1R_2C_1C_2}\right)^{1/2}$$

$$\frac{Q}{\omega_0} = \frac{R_2C_1C_2}{C_1 + C_2} \qquad \textbf{12.70}$$

$$\frac{Q}{\omega_0}\omega_0 = \frac{R_2C_1C_2}{C_1 + C_2}\left(\frac{1 + R_1/R_3}{R_1R_2C_1C_2}\right)^{1/2}$$

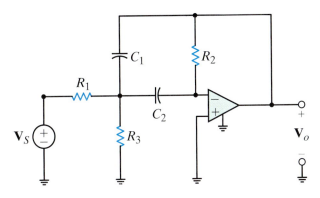

Figure 12.57
Second-order band-pass filter.

These expressions can be simplified to yield

$$Q = \frac{(1 + R_1/R_3)^{1/2}}{1 + C_1/C_2} \left(\frac{R_2 C_1}{R_1 C_2}\right)^{1/2} \qquad \textbf{12.71}$$

$$\text{BW} = \frac{\omega_0}{Q} = \frac{1}{R_2}\left(\frac{1}{C_1} + \frac{1}{C_2}\right) \qquad \textbf{12.72}$$

and

$$\left.\frac{\mathbf{V}_o}{\mathbf{V}_S}\right|_{\omega = \omega_0} = \frac{QH_o}{\omega_0} = -\frac{R_2}{R_1}\left(\frac{1}{1 + C_1/C_2}\right) \qquad \textbf{12.73}$$

Example 12.27

We wish to find a new expression for Eqs. (12.70) to (12.73) under the condition that $C_1 = C_2 = C$.

SOLUTION Using the condition the equations reduce to

$$\omega_0 = \frac{1}{C}\sqrt{\frac{1 + R_1/R_3}{R_1 R_2}}$$

$$Q = \frac{1}{2}\sqrt{\frac{R_2}{R_1}}\sqrt{1 + \frac{R_1}{R_3}}$$

$$\text{BW} = \frac{2}{R_2 C}$$

and

$$\left.\frac{\mathbf{V}_o}{\mathbf{V}_S}\right|_{\omega = \omega_0} = -\frac{R_2}{2R_1}$$

Example 12.28

Let us use the equations in Example 12.27 to design a band-pass filter of the form shown in Fig. 12.57 with a BW = 2000 rad/s, $(\mathbf{V}_o/\mathbf{V}_S)(\omega_0) = -5$, and $Q = 3$. Use $C = 0.1\ \mu\text{F}$, and determine the center frequency of the filter.

SOLUTION Using the filter equations, we find that

$$\text{BW} = \frac{2}{R_2 C}$$

$$2000 = \frac{2}{R_2(10)^{-7}}$$

$$R_2 = 10\ \text{k}\Omega$$

$$\frac{\mathbf{V}_o}{\mathbf{V}_S}(\omega_0) = -\frac{R_2}{2R_1}$$

$$-5 = -\frac{10,000}{2R_1}$$

$$R_1 = 1\ \text{k}\Omega$$

and

$$Q = \frac{1}{2}\sqrt{\frac{R_2}{R_1}}\sqrt{1 + \frac{R_1}{R_3}}$$

$$3 = \frac{1}{2}\sqrt{\frac{10,000}{1000}}\sqrt{1 + \frac{1000}{R_3}}$$

or

$$R_3 = 385 \ \Omega$$

Therefore, $R = 1 \text{ k}\Omega$, $R_2 = 10 \text{ k}\Omega$, $R_3 = 385 \ \Omega$, and $C = 0.1 \ \mu\text{F}$ completely define the band-pass filter shown in Fig. 12.57. The center frequency of the filter is

$$\omega_0 = \frac{1}{C} \sqrt{\frac{1 + R_1/R_3}{R_1 R_2}}$$

$$= \frac{1}{10^{-7}} \sqrt{\frac{1 + (1000/385)}{(1000)(10,000)}}$$

$$= 6000 \text{ rad/s}$$

Example 12.29

We wish to obtain the Bode plot for the filter designed in Example 12.28. We will employ the op-amp model, in which $R_i = \infty$, $R_o = 0$, and $A = 10^5$, and plot over the frequency range from 600 to 60 kHz.

SOLUTION The equivalent circuit for the filter is shown in Fig. 12.58a. The Bode plot is shown in Fig. 12.58b. As can be seen from the plot, the center frequency is 6 krad/s and BW = 2 krad/s.

Figure 12.58

Figures employed in Example 12.29: (a) band-pass filter equivalent circuit; (b) Bode plot.

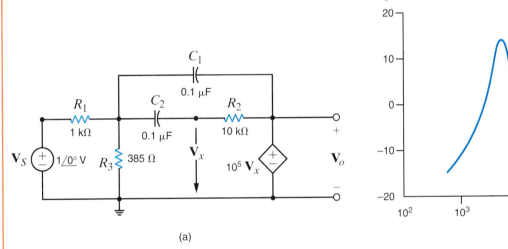

(a)

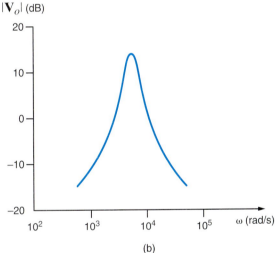

(b)

LEARNING EXTENSION

E12.21 Verify that Eq. (12.69) is the transfer function for the band-pass filter in Fig. 12.57.

Although op-amps are very popular and extremely useful in a wide variety of filter applications, they are not always the best choices as a result of limitations associated with their internal circuitry. Two examples are high-frequency active filters and low-voltage ($< 3 \text{ V}$) circuitry. Given the evolution of the wireless market (cell phones, pagers, etc.), these applications will only grow in prominence. There is, however, an op-amp variant called the *operational transconductance amplifier*, or OTA, that performs excellently in these scenarios, allowing, for example, very advanced filters to be implemented on a single chip. In this text, we will introduce OTA fundamentals and applications, including analog multipliers, automatic gain control amplifiers, and the aforementioned filters.

Figure 12.59

Block diagrams depicting the physical construction of the op-amp (a) and the OTA (b).

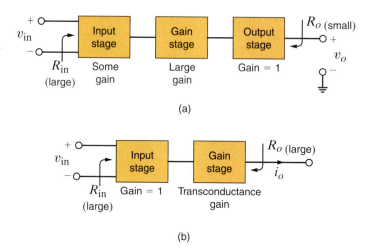

(a)

(b)

Advantages of the OTA over the op-amp can be deduced from the diagrams in Fig. 12.59. In the three-stage op-amp model, the input stage provides the large-input resistance, converts the differential input voltage $v_{in}(t)$ to a single-ended (referenced to ground) voltage, and produces some voltage gain. The gain stage provides the bulk of the op-amp's voltage gain. Finally, the output stage has little if any voltage gain but produces a low-output resistance. This three-stage model accurately depicts the physical design of most op-amps.

Now consider the two-stage OTA model. As in the op-amp, the input stage provides a large-input resistance, but its voltage gain is minimal. Unlike the op-amp, the gain stage produces a current output rather than a voltage. Since the output signal is a current, the gain is amps per volt, or transconductance, in A/V or siemen. With no output stage, the OTA is more compact and consumes less power than the op-amp and has an overall output resistance of R_o—a large value. Having all of the OTAs gain in a single stage further simplifies the internal design, resulting in a simple, fast, compact amplifier that can be efficiently replicated many times on a single silicon chip. The schematic symbol for the OTA and a simpler model are shown in Figs. 12.60a and b, respectively.

To compare the performance of the op-amp and OTA, consider the circuits in Fig. 12.61. For the op-amp, the overall voltage gain is

$$A = \frac{v_o}{v_{in}} = \left[\frac{R_{in}}{R_S + R_{in}} \right] A_V \left[\frac{R_L}{R_L + R_o} \right]$$

12.74

Figure 12.60

The OTA schematic symbol (a) and simple model (b).

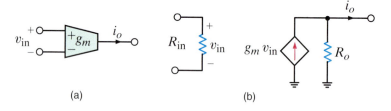

(a)

(b)

Figure 12.61

Simple circuits that demonstrate the relative strengths of the op-amp (a) and OTA (b).

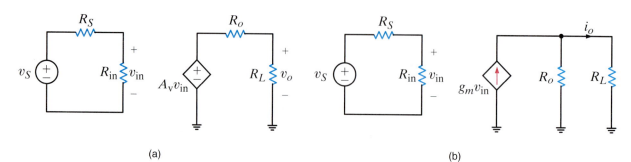

(a)

(b)

Ideally, $R_{in} \to \infty$, $R_o \to 0$, and the output voltage is independent of external components R_S and R_L. The overall gain of the OTA is

$$G_m = \frac{i_o}{v_{in}} = \left[\frac{R_{in}}{R_S + R_{in}} \right] g_m \left[\frac{R_o}{R_L + R_o} \right] \qquad \textbf{12.75}$$

For an ideal OTA, both R_{in} and $R_o \to \infty$, yielding a transconductance that is independent of R_S and R_L. Similarities and differences between ideal OTAs and ideal op-amps are listed in Table 12.2.

Table 12.2 A comparison of ideal op-amps and OTA features.

Amplifier Type	Ideal R_{in}	Ideal R_o	Ideal Gain	Input Currents	Input Voltage
Op-amp	∞	0	∞	0	0
OTA	∞	∞	g_m	0	nonzero

As with op-amps, OTAs can be used to create mathematical circuits. We will focus on three OTA circuits used extensively in active filters: the integrator, the simulated resistor, and the summer. To simplify our analyses, we assume the OTA is ideal with infinite input and output resistances. The integrator in Fig. 12.62, which forms the heart of our OTA active filters, can be analyzed as follows:

$$i_o = g_m v_1 \qquad v_o = \frac{1}{C} \int i_o \, dt \qquad v_o = \frac{g_m}{C} \int v_1 \, dt \qquad \textbf{12.76}$$

Or, in the frequency domain,

$$\mathbf{I_O} = g_m \mathbf{V_1} \qquad \mathbf{V_O} = \frac{\mathbf{I_O}}{j\omega C} \qquad \mathbf{V_O} = \frac{g_m}{j\omega C} \mathbf{V_1} \qquad \textbf{12.77}$$

An interesting aspect of IC fabrication is that resistors (especially large-valued resistors, that is, $> 10\,\text{k}\Omega$) are physically very large compared to other devices such as transistors. In addition, producing accurate values is quite difficult. This has motivated designers to use OTAs to simulate resistors. One such circuit is the grounded resistor, shown in Fig. 12.63. Applying the ideal OTA equations in Table 12.2,

$$i_o = g_m(0 - v_{in}) = -g_m v_{in} \qquad i_{in} = -i_o \qquad R_{eq} = \frac{v_{in}}{i_{in}} = \frac{1}{g_m} \qquad \textbf{12.78}$$

A simple summer circuit is shown in Fig. 12.64a, where OTA 3 is a simulated resistor. Based on Eq. (12.78), we produce the equivalent circuit in Fig. 12.64b. The analysis is straightforward.

$$i_{o1} = g_{m1} v_1 \quad i_{o2} = g_{m2} v_2 \quad i_o = i_{o1} + i_{o2} \quad v_o = \frac{i_o}{g_{m3}} = \frac{g_{m1}}{g_{m3}} v_1 + \frac{g_{m2}}{g_{m3}} v_2 \qquad \textbf{12.79}$$

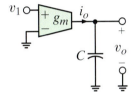

Figure 12.62
The OTA integrator.

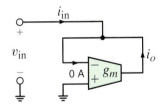

Figure 12.63
The OTA simulated resistor.

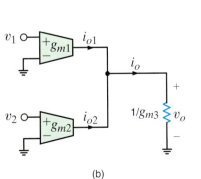

Figure 12.64
An OTA voltage summer.

(a)   (b)

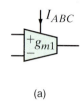

(a)

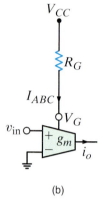

(b)

Figure 12.65

A modified OTA schematic symbol showing (a) the amplifier input bias current and (b) setting I_{ABC} with a single resistor.

At this point, we introduce our last important feature of the OTA—programmability. The transconductance, g_m, is linearly controlled by a current called the amplifier bias current, or I_{ABC}, as seen in Fig. 12.65a. Unfortunately, the I_{ABC} input is not part of the schematic symbol. The sensitivity of g_m to I_{ABC} is typically 20 S/A, but the range of g_m and its maximum value depend on the OTA design. Typical values are 10 mS for the maximum g_m and 3 to 7 powers of ten, or decades, for the transconductance range. For example, if the maximum g_m were 10 mS and the range were 4 decades, then the minimum g_m would be 1 μS and the usable range of I_{ABC} would be 0.05 μA to 0.5 mA.

Figure 12.65b shows a simple means for setting I_{ABC}. The gain set resistor, R_G, limits I_{ABC}.

$$I_{ABC} = \frac{V_{CC} - V_G}{R_G} \qquad \textbf{12.80}$$

where V_{CC} is the positive power supply. If the voltage at the pin labeled V_G is known, then I_{ABC} can be set by R_G. Unfortunately, different manufacturers design their OTAs with different V_G values, which are listed in the amplifier's data sheet. For our work, we will assume that V_G is zero volts.

$$I_{ABC} = \frac{V_{CC}}{R_G} \qquad \textbf{12.81}$$

Example 12.30

An ideal OTA has a $g_m - I_{ABC}$ sensitivity of 20, a maximum g_m of 4 mS, and a g_m range of 4 decades. Using the circuit in Fig. 12.63, produce an equivalent resistance of 25 kΩ, giving both g_m and I_{ABC}.

SOLUTION From Eq. (12.78), the equivalent resistance is $R_{eq} = 1/g_m = 25$ kΩ, yielding $g_m = 40$ μS. Since $g_m = 20I_{ABC}$, the required amplifier bias current is $I_{ABC} = 2$ μA.

Example 12.31

The circuit in Fig. 12.66 is a floating simulated resistor. For an ideal OTA, find an expression for $R_{eq} = v_1/i_1$. Using the OTA described in Example 12.30, produce an 80-kΩ resistance. Repeat for a 10-MΩ resistor.

Figure 12.66

The floating simulated resistor.

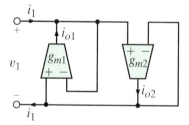

SOLUTION For OTA 1, we have $i_{o1} = g_{m1}(-v_1)$ and $i_1 = -i_{o1}$. Thus, $R_{eq} = v_1/i_1 = 1/g_{m1}$. We must also consider the return current that is contributed by OTA 2, where $i_{o2} = g_{m2}(v_1)$ and $i_{o2} = i_1$. Now $R_{eq} = v_1/i_1 = 1/g_{m2}$. For proper operation, we must ensure that $g_{m1} = g_{m2}$.

For $R_{eq} = 1/g_m = 80$ kΩ, we have $g_{m1} = g_{m2} = g_m = 12.5$ μS. Since $g_m = 20I_{ABC}$, the required bias current for both OTAs is $I_{ABC} = 0.625$ μA. Changing to $R_{eq} = 1/g_m = 10$ MΩ, the transconductance becomes $g_m = 0.1$ μS. However, the minimum g_m for these OTAs is specified at 0.4 μS. We must find either suitable OTAs or a better circuit.

Example 12.32

Using the summer in Fig. 12.64, and the OTAs specified in Example 12.30, produce the following function

$$v_o = 10v_1 + 2v_2$$

Repeat for the function

$$v_o = 10v_1 - 2v_2$$

SOLUTION   Comparing Eq. (12.79) with the desired function, we see that $g_{m1}/g_{m3} = 10$ and $g_{m2}/g_{m3} = 2$. With only two equations and three unknowns, we must choose one g_m value. Arbitrarily selecting $g_{m3} = 0.1$ mS yields $g_{m1} = 1$ mS and $g_{m2} = 0.2$ mS. The corresponding bias currents are $I_{ABC1} = 50$ μA, $I_{ABC2} = 10$ μA, and $I_{ABC3} = 5$ μA.

For the second case, we simply invert the sign of v_2 as shown in Fig. 12.67. This is yet another advantage of OTA versus op-amps. Again choosing $g_{m3} = 0.1$ mS yields the same bias current as the first case.

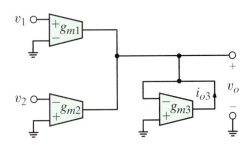

Figure 12.67
A slight modification of the adder in Fig. 12.64 yields this subtracting circuit.

Since the gain of the OTA is controlled by I_{ABC}, is it possible to design an analog multiplier whose output is the product of two voltages? This is shown in Fig. 12.68 where the output current can be written as

$$i_o = v_1 g_m = v_1(20I_{ABC}) = 20v_1\left[\frac{v_2}{R_G}\right]$$ **12.82**

and the output voltage as

$$v_o = i_o R_L = 20\left[\frac{R_L}{R_G}\right]v_1 v_2$$ **12.83**

The resistor ratio is used to set the scale factor for the output voltage. Using ± supply voltages, V_{CC} and V_{EE}, we see that the multiplier can support positive and negative voltages at v_1 and at v_o. However, v_2 must supply positive I_{ABC} into the bias current pin. Thus, v_2 must be positive. This kind of multiplier, where only one input can be positive or negative, is called a two-quadrant multiplier. When both inputs can be of either sign, the classification is a four-quadrant multiplier.

Consider a Sunday drive through the city out to the countryside and back again. You happen to pass the antenna for the very same FM radio station you're listening to on your car radio. You know that your car antenna receives a larger signal when you are nearer the antenna, but the radio's volume is the same whether you are near to or far from it. How does the car know?

Of course, the car has no idea where the station antenna is. Instead, the amplification between the car's antenna and its speakers is controlled based on the strength of the received signal—a technique called automatic gain control, or AGC. The circuit in Fig. 12.69 shows how this can be implemented with OTAs. There are two critical features here. First, the gain of OTA 1 is dependent on its own output voltage such that an increase in v_o causes a decrease in gain. This is called automatic gain control. Second, g_{m1} should be a function of the magnitude of v_o rather than its instantaneous value. A subcircuit called a peak detector performs this function. Although its internal workings are beyond our scope, we should understand the necessity of it.

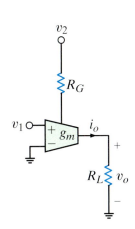

Figure 12.68
A two-quadrant analog multiplier.

Figure 12.69

An amplifier with automatic gain control implemented using two OTAs. A third OTA could be used to realize the load resistor if desired.

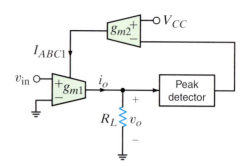

While OTA1 provides variable gain, OTA2 adjusts the gain to yield an output voltage dependent on itself.

$$v_o = i_o R_L = v_{in} g_{m1} R_L = v_{in} R_L (20 I_{ABC1}) = 20 v_{in} R_L g_{m2} (V_{CC} - |v_o|) = A v_{in} - B v_{in} |v_o| \quad \textbf{12.84}$$

We see that the output voltage has two terms, both of which are proportional to v_{in}. It is in the second term, where the proportionality constant depends on v_o itself, that automatic gain control occurs. Solving Eq. (12.84) for v_o shows the impact of AGC more clearly. (To facilitate the point we are making, we have dropped the absolute value operator for now. The peak detector is, of course, still required.)

$$v_o = \frac{A v_{in}}{1 + B v_{in}} \quad \textbf{12.85}$$

When the received signal v_{in} is small (we are far from the station's antenna), the denominator approaches unity and the output is approximately $A v_{in}$. However, as we get nearer to the antenna, v_{in} increases and the denominator grows until $B v_{in} \gg 1$. Now v_o approaches the ratio A/B, essentially independent of the received signal, and the radio volume is less sensitive to our distance from the antenna!

Using the subcircuits in Figs. 12.62 and 12.63, we can create active filters called OTA-C filters, which contain only OTAs and capacitors. The lack of resistors makes OTA-C filters ideal for single-chip, or monolithic, implementations. As an introduction, consider the circuit in Fig. 12.70. For ideal OTAs, the transfer function can be determined as follows.

$$\mathbf{I_{o1}} = g_{m1} \mathbf{V_{i1}} \quad \mathbf{I_{o2}} = -g_{m2} \mathbf{V_o} \quad \mathbf{I_C} = \mathbf{V_o}(j\omega C) = \mathbf{I_{o1}} + \mathbf{I_{o2}} \quad \mathbf{V_o} = \frac{\mathbf{I_{o1}} + \mathbf{I_{o2}}}{j\omega C} = \frac{g_{m1}}{j\omega C} \mathbf{V_{i1}} - \frac{g_{m2}}{j\omega C} \mathbf{V_o}$$

Solving the transfer function yields the low-pass function

$$\frac{\mathbf{V_o}}{\mathbf{V_{i1}}} = \frac{g_{m1}/g_{m2}}{\dfrac{j\omega C}{g_{m2}} + 1} \quad \textbf{12.86}$$

From Eq. (12.86), the circuit is a first-order low-pass filter with the asymptotic Bode plot shown in Fig. 12.71. Both the corner frequency, $f_C = g_{m2}/(2\pi C)$, and dc gain, $A_{DC} = g_{m1}/g_{m2}$, are programmable.

Figure 12.70

A simple first-order low-pass OTA-C filter.

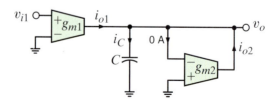

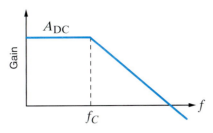

Figure 12.71
Asymptotic Bode plot for a first-order low-pass filter.

In monolithic OTA-C filters, the capacitors and OTAs are fabricated on a single chip. Typical OTA capacitor values range from about 1 pF up to 50 pF.

Example 12.33

The low-pass filter in Fig. 12.70 is implemented using a 25-pF capacitor and OTAs with a $g_m - I_{ABC}$ sensitivity of 20, a maximum g_m of 1 mS, and 3 decades of range. Find the required bias currents for the filter transfer function:

$$\frac{\mathbf{V_O}}{\mathbf{V_{in}}} = \frac{4}{\dfrac{j\omega}{2\pi(10^5)} + 1}$$

12.87

SOLUTION Comparing Eq. (12.86) to the desired function, $g_{m2}/C = (2\pi)10^5$. For $C = 25$ pF, $g_{m2} = 15.7$ μS. Since $g_m = 20I_{ABC}$, the bias current for OTA 2 is $I_{ABC2} = 0.785$ μA. Finally, $g_{m1}/g_{m2} = 4$ yields $I_{ABC1} = 3.14$ μA.

Of the dozens of OTA filter topologies, a very popular one is the two-integrator biquad filter. The term *biquad* is short for biquadratic, which, in filter terminology, means the filter gain is a ratio of two quadratic functions such as

$$\frac{\mathbf{V_O}}{\mathbf{V_{in}}} = \frac{A(j\omega)^2 + B(j\omega) + C}{(j\omega)^2 + \dfrac{\omega_0}{Q}(j\omega) + \omega_0^2}$$

12.88

By selecting appropriate values for A, B, and C, low-pass, band-pass, and high-pass functions can be created, as listed in Table 12.3. Figure 12.72 shows the most popular two-integrator biquad used in practice—the Tow-Thomas filter. Assuming ideal OTAs, we can derive the filter's transfer function. For OTA 1, an integrator,

$$\frac{\mathbf{V_{O1}}}{\mathbf{V_{i1}} - \mathbf{V_{O2}}} = \frac{g_{m1}}{j\omega C_1}$$

Table 12.3 Various Tow-Thomas biquad filter possibilities.

Filter Type	A	B	C
Low-pass	0	0	nonzero
Band-pass	0	nonzero	0
High-pass	nonzero	0	0

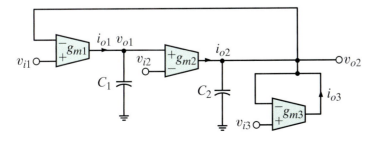

Figure 12.72
The Tow-Thomas OTA-C biquad filter.

The output current of OTA 2 is

$$\mathbf{I_{O2}} = g_{m2}\big[\mathbf{V_{O1}} - \mathbf{V_{i2}}\big]$$

Applying KCL at the second output node, we find

$$\mathbf{I_{O3}} + \mathbf{I_{O2}} = \big(j\omega C_2\big)\mathbf{V_{O2}}$$

where

$$\mathbf{I_{O3}} = \big[\mathbf{V_{i3}} - \mathbf{V_{O2}}\big]g_{m3}$$

Solving for $\mathbf{V_{O1}}$ and $\mathbf{V_{O2}}$ yields

$$\mathbf{V_{O1}} = \frac{\left[\dfrac{j\omega C_2}{g_{m2}} + \dfrac{g_{m3}}{g_{m2}}\right]\mathbf{V_{i1}} + \mathbf{V_{i2}} - \left[\dfrac{g_{m3}}{g_{m2}}\right]\mathbf{V_{i3}}}{\left[\dfrac{C_1 C_2}{g_{m1} g_{m2}}\right](j\omega)^2 + \left[\dfrac{g_{m3} C_1}{g_{m2} g_{m1}}\right](j\omega) + 1}$$

and

$$\mathbf{V_{O2}} = \frac{\mathbf{V_{i1}} - \left[\dfrac{j\omega C_1}{g_{m1}}\right]\mathbf{V_{i2}} + \left[\dfrac{j\omega C_1 g_{m3}}{g_{m1} g_{m2}}\right]\mathbf{V_{i3}}}{\left[\dfrac{C_1 C_2}{g_{m1} g_{m2}}\right](j\omega)^2 + \left[\dfrac{g_{m3} C_1}{g_{m2} g_{m1}}\right](j\omega) + 1} \qquad \textbf{12.89}$$

Note that this single circuit can implement both low-pass and band-pass filters depending on where the input is applied! Table 12.4 lists the possibilities. Comparing Eqs. (12.88) and (12.89), design equations for ω_0, Q, and the bandwidth can be written as

$$\omega_0 = \sqrt{\frac{g_{m1} g_{m2}}{C_1 C_2}} \qquad \frac{\omega_0}{Q} = BW = \frac{g_{m3}}{C_2} \qquad Q = \sqrt{\frac{g_{m1} g_{m2}}{g_{m3}^2}}\sqrt{\frac{C_2}{C_1}} \qquad \textbf{12.90}$$

Table 12.4 Low-pass and band-pass combinations for the Tow-Thomas biquad filter in Figure 12.72

Filter Type	Input	Output	Sign
Low-pass	V_{i2}	V_{o1}	positive
	V_{i3}	V_{o1}	negative
Band-pass	V_{i1}	V_{o2}	positive
	V_{i2}	V_{o2}	negative
	V_{i3}	V_{o2}	positive

Consider a Tow-Thomas band-pass filter. From Eq. (12.90), if $g_{m1} = g_{m2} = g_m$ and $C_1 = C_2 = C$, the following relationships are easily derived.

$$\omega_0 = \frac{g_m}{C} = \frac{k}{C} I_{ABC} \qquad \frac{\omega_0}{Q} = BW = \frac{g_{m3}}{C} = \frac{k}{C} I_{ABC3} \qquad Q = \frac{g_m}{g_{m3}} = \frac{I_{ABC}}{I_{ABC3}} \qquad \textbf{12.91}$$

where k is the $g_m - I_{ABC}$ sensitivity. Based on Eq. (12.91), we have efficient control over the filter characteristics. In particular, tuning I_{ABC} with I_{ABC3} fixed scales both the center frequency and Q directly without affecting bandwidth. Tuning I_{ABC3} only changes the bandwidth but not the center frequency. Finally, tuning all three bias currents scales both the center frequency and bandwidth proportionally, producing a constant Q factor.

Example 12.34

Using the OTAs specifications from Example 12.30 and 5-pF capacitors, design a Tow-Thomas low-pass filter with a corner frequency of 6 MHz, $Q = 5$, and dc gain of 1.

SOLUTION   Using the $v_{i1} - v_{o2}$ input–output pair with $g_{m1} = g_{m2} = g_m$ and $C_1 = C_2$, allows us to use Eq. (12.91).

$$g_m = \omega_0 C = (2\pi)(6 \times 10^6)(5 \times 10^{-12}) = 188.5 \ \mu S \qquad g_{m3} = \frac{g_m}{Q} = 37.7 \ \mu S$$

The required bias currents are

$$I_{ABC1} = I_{ABC2} = \frac{g_m}{20} = 9.425 \ \mu A \qquad I_{ABC3} = \frac{g_{m3}}{20} = 1.885 \ \mu A$$

From the Bode plot, shown in Fig. 12.73, we see that the corner frequency is indeed 6 MHz.

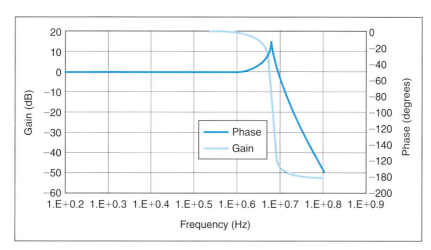

Figure 12.73

Bode plot of the Tow-Thomas low-pass filter of Example 12.34.

12.6   Application Examples

Application Example 12.35

The ac-dc converter in Fig. 12.74a is designed for use with a hand-held calculator. Ideally, the circuit should convert a 120-V rms sinusoidal voltage to a 9-V dc output. In actuality the output is

$$v_o(t) = 9 + 0.5 \sin 377t \ V$$

Let us use a low-pass filter to reduce the 60-Hz component of $v_o(t)$.

SOLUTION   The Thévenin equivalent circuit for the converter is shown in Fig. 12.74b. By placing a capacitor across the output terminals, as shown in Fig. 12.74c, we create a low-pass filter at the output. The transfer function of the filtered converter is

$$\frac{\mathbf{V}_{OF}}{\mathbf{V}_{Th}} = \frac{1}{1 + sR_{Th}C}$$

which has a pole at a frequency of $f = 1/2\pi R_{Th}C$. To obtain significant attenuation at 60 Hz, we choose to place the pole at 6 Hz, yielding the equation

$$\frac{1}{2\pi R_{Th}C} = 6$$

or

$$C = 53.05 \ \mu F$$

A transient simulation of the converter is used to verify performance.

Figure 12.74d shows the output without filtering, $v_o(t)$, and with filtering, $v_{OF}(t)$. The filter has successfully reduced the unwanted 60-Hz component by a factor of roughly six.

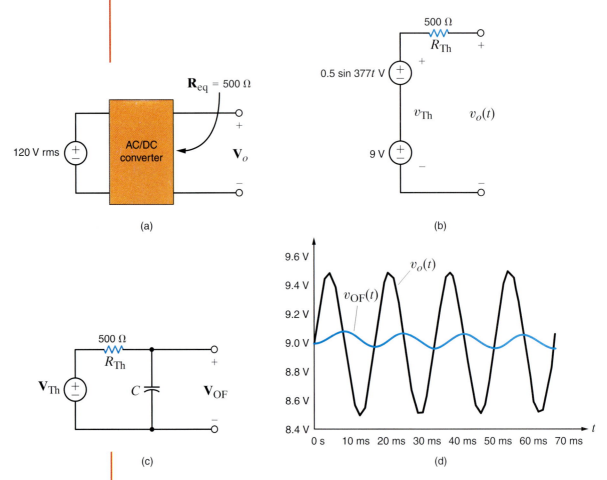

Figure 12.74 Circuits and output plots for ac/dc converter.

Application Example 12.36

The antenna of an FM radio picks up stations across the entire FM frequency range—approximately 87.5 MHz to 108 MHz. The radio's circuitry must have the capability to first reject all of the stations except the one that the listener wants to hear and then to boost the minute antenna signal. A tuned amplifier incorporating parallel resonance can perform both tasks simultaneously.

The network in Fig. 12.75a is a circuit model for a single-stage tuned transistor amplifier where the resistor, capacitor, and inductor are discrete elements. Let us find the transfer function $\mathbf{V}_o(s)/\mathbf{V}_A(s)$, where $\mathbf{V}_A(s)$ is the antenna voltage and the value of C for maximum gain at 91.1 MHz. Finally, we will simulate the results.

SOLUTION Since $\mathbf{V}(s) = \mathbf{V}_A(s)$, the transfer function is

$$\frac{\mathbf{V}_o(s)}{\mathbf{V}_A(s)} = -\frac{4}{1000}\left[R//sL//\frac{1}{sC}\right]$$

$$\frac{\mathbf{V}_o(s)}{\mathbf{V}_A(s)} = -\frac{4}{1000}\left[\frac{s/C}{s^2 + \dfrac{s}{RC} + \dfrac{1}{LC}}\right]$$

The parallel resonant network is actually a band-pass filter. Maximum gain occurs at the center frequency, f_0. This condition corresponds to a minimum value in the denominator. Isolating the denominator polynomial, $D(s)$, and letting $s = j\omega$, we have

$$\mathbf{D}(j\omega) = \frac{R}{LC} - \omega^2 + \frac{j\omega}{C}$$

which has a minimum value when the real part goes to zero, or

$$\frac{1}{LC} - \omega_0^2 = 0$$

yielding a center frequency of

$$\omega_0 = \frac{1}{\sqrt{LC}}$$

Thus, for a center frequency of 91.1 MHz, we have

$$2\pi(91.1 \times 10^6) = \frac{1}{\sqrt{LC}}$$

and the required capacitor value is

$$C = 3.05 \text{ pF}$$

The Bode plot for the tuned amplifier, as shown in Fig. 12.75b, confirms the design, since the center frequency is 91.1 MHz, as specified.

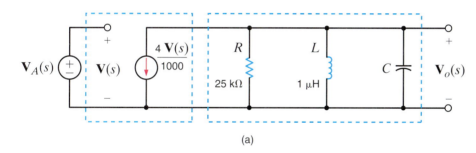

(a)

Figure 12.75
Circuit and Bode plot for the parallel resonant tuned amplifier.

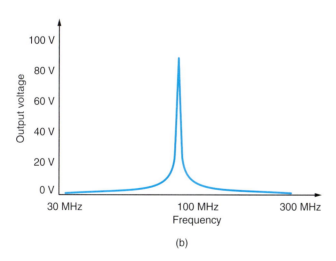

(b)

12.7 Design Examples

Throughout this chapter we have presented a number of design examples. In this section we consider some additional ones that also have practical ramifications.

Design Example 12.37

Compact discs (CDs) have become a very popular medium for recording and playing music. CDs store information in a digital manner; that is, the music is sampled at a very high rate, and the samples are recorded on the disc. The trick is to sample so quickly that the reproduction sounds continuous. The industry standard sampling rate is 44.1 kHz—one sample every 22.7 μs.

One interesting aspect regarding the analog-to-digital conversion that takes place inside the unit recording a CD is called the Nyquist criterion. This criterion states that in the analog conversion, any signal components at frequencies above half the sampling rate (22.05 kHz in this case) cannot be faithfully reproduced. Therefore, recording technicians filter these frequencies out before any sampling occurs, yielding higher fidelity to the listener.

Let us design a series of low-pass filters to perform this task.

SOLUTION Suppose, for example, that our specification for the filter is unity gain at dc and 20 dB of attenuation at 22.05 kHz. Let us consider first the simple *RC* filter in Fig. 12.76.

The transfer function is easily found to be

$$\mathbf{G}_{v1}(s) = \frac{\mathbf{V}_{o1}}{\mathbf{V}_{in}} = \frac{1}{1 + sRC}$$

Figure 12.76
Single-pole low-pass filter.

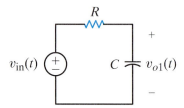

Since a single-pole transfer function attenuates at 20 dB/decade, we should place the pole frequency one decade before the −20 dB point of 22.05 kHz.

Thus,

$$f_P = \frac{1}{2\pi RC} = 2.205 \text{ kHz}$$

If we arbitrarily choose $C = 1$ nF, the resulting value for R is 72.18 kΩ, which is reasonable. A Bode plot of the magnitude of $\mathbf{G}_{v1}(s)$ is shown in Fig. 12.77. All specifications are met but at the cost of severe attenuation in the audible frequency range. This is undesirable.

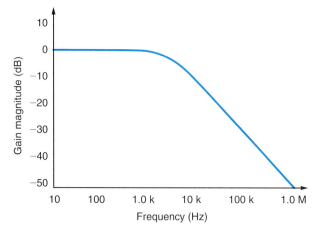

Figure 12.77 *Bode plot for single-pole filter.*

An improved filter is shown in Fig. 12.78. It is a two-stage low-pass filter with identical filter stages separated by a unity-gain buffer.

The presence of the op-amp permits us to consider the stages independently. Thus, the transfer function becomes

$$\mathbf{G}_{v2}(s) = \frac{\mathbf{V}_{o2}}{\mathbf{V}_{\text{in}}} = \frac{1}{[1 + sRC]^2}$$

To find the required pole frequencies, let us employ the equation for $\mathbf{G}_{v2}(s)$ at 22.05 kHz, since we know that the gain must be 0.1 (attenuated 20 dB) at that frequency. Using the substitution $s = j\omega$, we can express the magnitude of $\mathbf{G}_{v2}(s)$ as

$$|\mathbf{G}_{v2}| = \left\{ \frac{1}{1 + (22{,}050/f_p)^2} \right\} = 0.1$$

and the pole frequency is found to be 7.35 kHz. The corresponding resistor value is 21.65 kΩ. Bode plots for $\mathbf{G}_{v1}$ and $\mathbf{G}_{v2}$ are shown in Fig. 12.79. Note that the two-stage filter has a wider bandwidth, which improves the fidelity of the recording.

Let us try one more improvement—expanding the two-stage filter to a four-stage filter. Again, the gain magnitude is 0.1 at 22.05 kHz and can be written

$$|\mathbf{G}_{v3}| = \left\{ \frac{1}{[1 + (22{,}050/f_p)^2]^2} \right\} = 0.1$$

The resulting pole frequencies are at 15 kHz, and the required resistor value is 10.61 kΩ. Figure 12.80 shows all three Bode plots. Obviously, the four-stage filter, having the widest bandwidth, is the best option (discounting any extra cost associated with the additional active and passive circuit elements).

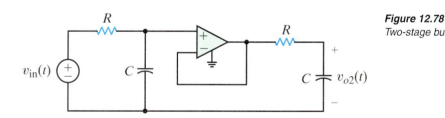

Figure 12.78
Two-stage buffered filter.

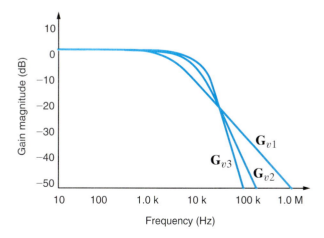

Figure 12.79
Bode plot for single- and two-stage filters.

Figure 12.80
Bode plots for single-, two-, and four-stage filters.

Design Example 12.38

The circuit in Fig. 12.81a is called a notch filter. From a sketch of its Bode plot in Fig. 12.81b, we see that at the notch frequency, f_n, the transfer function gain is zero, while at frequencies above and below f_n the gain is unity. Let us design a notch filter to remove an annoying 60-Hz hum from the output voltage of a cassette tape player and generate its Bode plot.

SOLUTION Figure 12.81c shows a block diagram for the filter implementation. The tape output contains both the desired music and the undesired hum. After filtering, the voltage $\mathbf{V}_{amp}$ will have no 60-Hz component as well as some attenuation at frequencies around 60 Hz. An equivalent circuit for the block diagram including a Thévenin equivalent for the tape deck and an equivalent resistance for the power amp is shown in Fig. 12.81d. Applying voltage division, we find the transfer function to be

$$\frac{\mathbf{V}_{amp}}{\mathbf{V}_{tape}} = \frac{R_{amp}}{R_{amp} + R_{tape} + \left(sL//\dfrac{1}{Cs}\right)}$$

After some manipulation, the transfer function can be written as

$$\frac{\mathbf{V}_{amp}}{\mathbf{V}_{tape}} = \frac{R_{amp}}{R_{amp} + R_{tape}} \left[\frac{s^2LC + 1}{s^2LC + s\left(\dfrac{L}{R_{tape} + R_{amp}}\right) + 1} \right]$$

We see that the transfer function contains two zeros and two poles. Letting $s = j\omega$, the zero frequencies, ω_z, are found to be at

$$\omega_z = \pm \frac{1}{\sqrt{LC}}$$

Obviously, we would like the zero frequencies to be at 60 Hz. If we arbitrarily choose $C = 10\ \mu\text{F}$, then $L = 0.704\ \text{mH}$.

The Bode plot, shown in Fig. 12.81e, confirms that there is indeed zero transmission at 60 Hz.

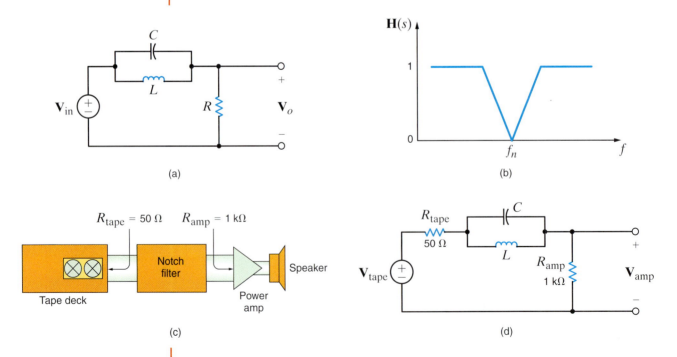

(a)   (b)

(c)   (d)

Figure 12.81

Circuits and Bode plots for 60-Hz notch filter.

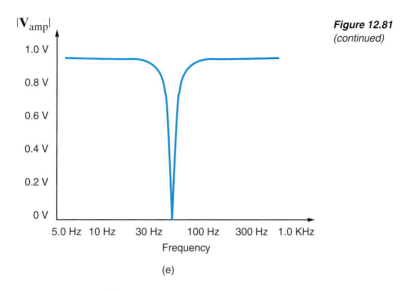

Figure 12.81
(continued)

(e)

Design Example 12.39

A fast growing field within electrical engineering is mixed-mode circuitry, which combines digital and analog networks to create a larger system. A key component in these systems is the analog-to-digital converter, or ADC. It "measures" an analog voltage and converts it to a digital representation. If these conversions are done quickly enough, the result is a sequence of data points, as shown in Fig. 12.82a. Connecting the dots reveals the original analog signal, $v_A(t)$. Unfortunately, as seen in Fig. 12.82b, undesired signals such as $v_B(t)$ at higher frequencies can also have the same set of data points. This phenomenon is called aliasing and can be avoided by employing a low-pass filter, called an anti-aliasing filter, before the ADC as shown in Fig. 12.82c. In general, the half-power frequency of the filter should be greater than the frequency of the signals you wish to convert but less than those you want to reject.

We wish to design an anti-aliasing filter, with a half-power frequency at 100 Hz, that will permit us to acquire a 60-Hz signal. In this design we will assume the ADC has infinite input resistance.

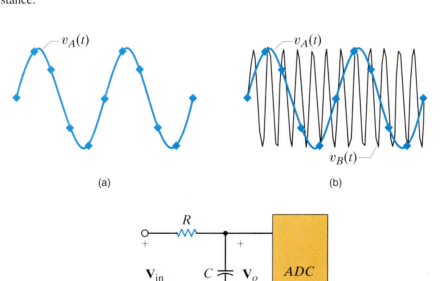

(a)

(b)

(c)

Figure 12.82
A brief explanation of ADC basics. (a) The ADC samples are like data points on the acquired waveform. (b) Higher frequency signals can have the same data points. After acquisition, it appears that $v_B(t)$ has been shifted to a lower frequency, an effect called aliasing. (c) The solution, an anti-aliasing low-pass filter.

SOLUTION Assuming the ADC has infinite input resistance, we find that the transfer function for the filter is quite simple.

$$\frac{\mathbf{V_o}}{\mathbf{V_{in}}} = \frac{\dfrac{1}{j\omega C}}{R + \dfrac{1}{j\omega C}} = \frac{1}{1 + j\omega RC}$$

The half-power frequency is

$$f_P = \frac{1}{2\pi RC} = 100 \text{ Hz}$$

If we somewhat arbitrarily choose C at 100 nF, a little larger than the resistor but smaller than the ADC integrated circuit in size, the resulting resistor value is 15.9 kΩ.

Design Example 12.40

The circuit in Fig. 12.83a is an inexpensive "bass boost" amplifier that amplifies only low-frequency audio signals, as illustrated by the Bode plot sketch in Fig. 12.83b. We wish to derive the transfer function $\mathbf{V_o}/\mathbf{V_{in}}$ when the switch is open. Then, from this transfer function and Fig. 12.83b, select appropriate values for R_1 and R_2. What is the resulting value of f_P?

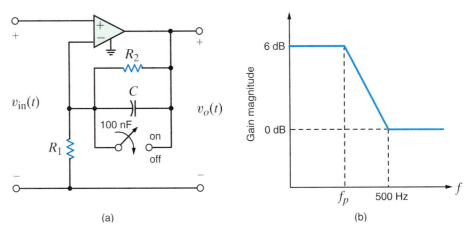

(a)   (b)

Figure 12.83 *A "bass boost" circuit (a) and a sketch of its magnitude Bode plot (b).*

SOLUTION With the switch open, we can use the classic noninverting op-amp configuration expression to write the transfer function as

$$\frac{\mathbf{V_o}}{\mathbf{V_{in}}} = 1 + \frac{\mathbf{Z_2}}{\mathbf{Z_1}}$$

where $\mathbf{Z_1} = R_1$ and $\mathbf{Z_2}$ is the parallel combination of the R_2 and $1/j\omega C$. $\mathbf{Z_2}$ can be written as

$$\mathbf{Z_2} = R_2 // \frac{1}{j\omega C} = \frac{R_2}{1 + j\omega R_2 C}$$

and the transfer function as

$$\frac{\mathbf{V_o}}{\mathbf{V_{in}}} = 1 + \frac{\dfrac{R_2}{1 + j\omega R_2 C}}{R_1} = \frac{R_1 + R_2 + j\omega R_1 R_2 C}{R_1 (1 + j\omega R_2 C)} = \left[\frac{R_1 + R_2}{R_1} \right] \left[\frac{1 + j\omega R_P C}{1 + j\omega R_2 C} \right] \qquad \textbf{12.92}$$

where $R_P = R_1 // R_2$. The network has one zero at $1/R_P C$ and one pole at $1/R_2 C$. Since R_P must be less than R_2, the zero frequency must be greater than the pole frequency, and the sketch in Fig. 12.83b is appropriate.

Now let us determine the component values. At dc ($\omega = 0$) the gain must be 6 dB or a factor of 2. From Eq. (12.92).

$$\frac{\mathbf{V_o}}{\mathbf{V_{in}}}(j0) = \frac{R_1 + R_2}{R_1} = 2$$

Thus, $R_1 = R_2$ and $R_P = R_1/2$. From Fig. 12.83b, the zero frequency is 500 Hz, and given this information we can determine R_P as

$$\frac{1}{R_P C} = 2\pi(500) \Rightarrow R_P = \frac{1}{1000\pi C} = 3.18\ \text{k}\Omega$$

Of course, $R_1 = R_2 = 2R_P = 6.37\ \text{k}\Omega$. Finally, the pole frequency is

$$\frac{1}{2\pi R_2 C} = 250\ \text{Hz}$$

Design Example 12.41

An audiophile has discovered that his tape player has the limited high-frequency response shown in Fig. 12.84a. Anxious to the point of sleeplessness, he decides to insert a "treble boost" circuit between the tape deck and the main amplifier that has the transfer function shown in Fig. 12.84b. Passing the tape audio through the boost should produce a "flat" response out to about 20 kHz. The circuit in Fig. 12.84c is his design. Show that the circuit's transfer function has the correct form and select R_1 and R_2 for proper operation.

SOLUTION Recognizing the circuit as a noninverting gain configuration, the transfer is

$$\frac{\mathbf{V_o}}{\mathbf{V_{in}}} = 1 + \frac{\mathbf{Z_2}}{\mathbf{Z_1}} \qquad \textbf{12.93}$$

where $\mathbf{Z_2}$ is R_2 and $\mathbf{Z_1}$ is the series combination of R_1 and $1/j\omega C$. Substituting these values into Eq. 12.93 yields

$$\frac{\mathbf{V_o}}{\mathbf{V_{in}}} = 1 + \frac{R_2}{R_1 + \dfrac{1}{j\omega C}} = \frac{1 + j\omega C(R_1 + R_2)}{1 + j\omega C R_1}$$

Since the pole frequency should be 20 kHz,

$$f_P = 2 \times 10^4 = \frac{1}{2\pi C R_1} \Rightarrow R_1 = 7.96\ \text{k}\Omega$$

The zero frequency is at 8 kHz, and thus

$$f_Z = 8 \times 10^3 = \frac{1}{2\pi C(R_1 + R_2)} \Rightarrow R_1 + R_2 = 19.9\ \text{k}\Omega$$

Therefore, R_2 is then 12.0 kΩ.

Figure 12.84
Correcting a deficient audio response. (a) The original response, (b) the corrective transfer function, and (c) the circuit implementation.

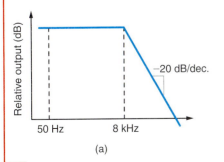

(a)

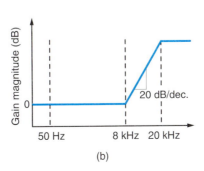

(b)

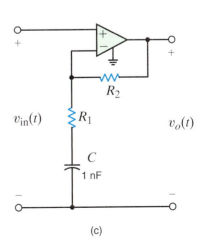

(c)

SUMMARY

- There are four types of network or transfer functions:

  1. $\mathbf{Z}(j\omega)$: the ratio of the output voltage to the input current

  2. $\mathbf{Y}(j\omega)$: the ratio of the output current to the input voltage

  3. $\mathbf{G}_v(j\omega)$: the ratio of the output voltage to the input voltage

  4. $\mathbf{G}_i(j\omega)$: the ratio of the output current to the input current

- Driving point functions are impedances or admittances defined at a single pair of terminals, such as the input impedance of a network.

- When the network function is expressed in the form

  $$\mathbf{H}(s) = \frac{N(s)}{D(s)}$$

  the roots of $N(s)$ cause $\mathbf{H}(s)$ to become zero and are called zeros of the function, and the roots of $D(s)$ cause $\mathbf{H}(s)$ to become infinite and are called poles of the function.

- Bode plots are semilog plots of the magnitude and phase of a transfer function as a function of frequency. Straight-line approximations can be used to sketch quickly the magnitude characteristic. The error between the actual characteristic and the straight-line approximation can be calculated when necessary.

- The resonant frequency, given by the expression

  $$\omega_0 = \frac{1}{\sqrt{LC}}$$

  is the frequency at which the impedance of a series RLC circuit or the admittance of a parallel RLC circuit is purely real.

- The quality factor is a measure of the sharpness of the resonant peak. A higher Q yields a sharper peak.
  For series RLC circuits, $Q = (1/R)\sqrt{L/C}$. For parallel RLC circuits, $Q = R\sqrt{C/L}$.

- The half-power, cutoff, or break frequencies are the frequencies at which the magnitude characteristic of the Bode plot is $1/\sqrt{2}$ of its maximum value.

- The parameter values for passive circuit elements can be both magnitude and frequency scaled.

- The four common types of filters are low-pass, high-pass, band-pass, and band rejection.

- The bandwidth of a band-pass or band-rejection filter is the difference in frequency between the half-power points; that is,

  $$BW = \omega_{HI} - \omega_{LO}$$

  For a series RLC circuit, $BW = R/L$. For a parallel RLC circuit, $BW = 1/RC$.

PROBLEMS

PSV **CS** both available on the web at: **http://www.justask4u.com/irwin**

SECTION 12.1

12.1 Determine the driving point impedance at the input terminals of the network shown in Fig. P12.1 as a function of s. **CS**

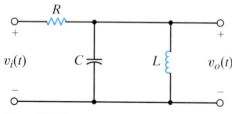

Figure P12.1

12.2 Determine the voltage transfer function $\mathbf{V}_o(s)/\mathbf{V}_i(s)$ as a function of s for the network shown in Fig. P12.2. **PSV**

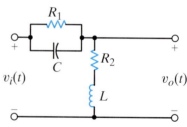

Figure P12.2

12.3 Determine the driving point impedance at the input terminals of the network shown in Fig. P12.3 as a function of s.

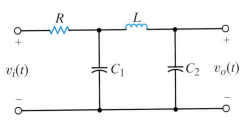

Figure P12.3

12.4 Find the transfer impedance $\mathbf{V}_o(s)/\mathbf{I}_S(s)$ for the network shown in Fig. P12.4.

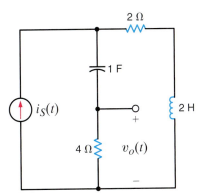

Figure P12.4

12.5 Find the driving point impedance at the input terminals of the circuit in Fig. P12.5 as a function of s. **CS**

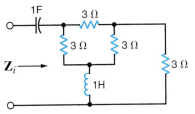

Figure P12.5

SECTION 12.2

12.6 Draw the Bode plot for the network function

$$\mathbf{H}(j\omega) = \frac{j\omega 4 + 1}{j\omega 20 + 1}$$

12.7 Draw the Bode plot for the network function

$$\mathbf{H}(j\omega) = \frac{j\omega 5 + 1}{j\omega 10 + 1}$$

12.8 Draw the Bode plot for the network function

$$\mathbf{H}(j\omega) = \frac{10(10 j\omega + 1)}{(100 j\omega + 1)(0.01 j\omega + 1)}$$

12.9 Draw the Bode plot for the network function **CS**

$$\mathbf{H}(j\omega) = \frac{j\omega}{(j\omega + 1)(0.1 j\omega + 1)}$$

12.10 Draw the Bode plot for the network function

$$\mathbf{H}(j\omega) = \frac{10 j\omega + 1}{j\omega(0.01 j\omega + 1)}$$

12.11 Sketch the magnitude characteristic of the Bode plot for the transfer function **PSV**

$$\mathbf{H}(j\omega) = \frac{100}{j\omega(0.1 j\omega + 1)}$$

12.12 Sketch the magnitude characteristic of the Bode plot for the transfer function

$$\mathbf{H}(j\omega) = \frac{20(0.1 j\omega + 1)}{j\omega(j\omega + 1)(0.01 j\omega + 1)}$$

12.13 Sketch the magnitude characteristic of the Bode plot for the transfer function **CS**

$$\mathbf{H}(j\omega) = \frac{100(j\omega)}{(j\omega + 1)(j\omega + 10)(j\omega + 50)}$$

12.14 Draw the Bode plot for the network function

$$\mathbf{H}(j\omega) = \frac{100}{(j\omega)^2(j\omega 2 + 1)}$$

12.15 Sketch the magnitude characteristic of the Bode plot for the transfer function

$$\mathbf{H}(j\omega) = \frac{36 \times 10^5(5j\omega + 1)^2}{(j\omega)^2(j\omega + 10)(j\omega + 100)^2}$$

12.16 Sketch the magnitude characteristic of the Bode plot for the transfer function

$$\mathbf{G}(j\omega) = \frac{400(j\omega + 2)(j\omega + 50)}{-\omega^2(j\omega + 100)^2}$$

12.17 Sketch the magnitude characteristic of the Bode plot for the transfer function **CS**

$$\mathbf{G}(j\omega) = \frac{10j\omega}{(j\omega + 1)(j\omega + 10)^2}$$

12.18 Sketch the magnitude characteristic of the Bode plot for the transfer function

$$\mathbf{G}(j\omega) = \frac{-\omega^2 10^4}{(j\omega + 1)^2(j\omega + 10)(j\omega + 100)^2}$$

12.19 Sketch the magnitude characteristic of the Bode plot for the transfer function

$$\mathbf{G}(j\omega) = \frac{8(j\omega + 1)^2}{-j\omega^3(0.1j\omega + 1)}$$

12.20 Sketch the magnitude characteristic of the Bode plot for the transfer function **CS**

$$\mathbf{G}(j\omega) = \frac{-\omega^2}{(j\omega + 1)^3}$$

12.21 Sketch the magnitude characteristic of the Bode plot for the transfer function

$$\mathbf{G}(j\omega) = \frac{10^4(j\omega + 1)(-\omega^2 + 6j\omega + 225)}{j\omega(j\omega + 50)^2(j\omega + 450)}$$

12.22 Sketch the magnitude characteristic of the Bode plot for the transfer function

$$\mathbf{H}(j\omega) = \frac{+6.4}{(j\omega + 1)(-\omega^2 + 8j\omega + 16)}$$

12.23 Sketch the magnitude characteristic of the Bode plot for the transfer function

$$\mathbf{G}(j\omega) = \frac{10(j\omega + 2)(j\omega + 100)}{j\omega(-\omega^2 + 4j\omega + 100)}$$

12.24 Sketch the magnitude characteristic of the Bode plot for the transfer function

$$\mathbf{G}(j\omega) = \frac{j\omega(j\omega + 100)}{(j\omega + 1)(-\omega^2 + 6j\omega + 400)}$$

12.25 Sketch the magnitude characteristic of the Bode plot for the transfer function **CS**

$$\mathbf{H}(j\omega) = \frac{+6.4(j\omega)}{(j\omega + 1)(-\omega^2 + 8j\omega + 64)}$$

12.26 Use MATLAB to generate the Bode plot for the following transfer function over the frequency range from $\omega = 0.01$ to 1000 rad/s.

$$\mathbf{G}(j\omega) = \frac{10(j\omega + 1)}{(j\omega)(j\omega + 10)}$$

12.27 Use MATLAB to generate the Bode plot for the following transfer function over the frequency range from $\omega = 0.1$ to 10,000 rad/s.

$$\mathbf{G}(j\omega) = \frac{20(j\omega + 10)}{(j\omega)(j\omega + 1)(j\omega + 100)}$$

12.28 The magnitude characteristic of a band-elimination filter is shown in Fig. P12.28. Determine $\mathbf{H}(j\omega)$.

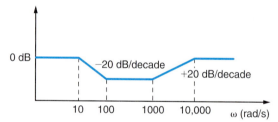

Figure P12.28

12.29 Find $\mathbf{H}(j\omega)$ if its magnitude characteristic is shown in Fig. P12.29. **PSV**

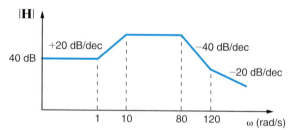

Figure P12.29

12.30 Find $\mathbf{H}(j\omega)$ if its magnitude characteristic is shown in Fig. P12.30. **CS**

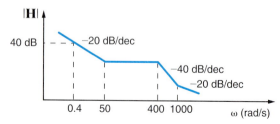

Figure P12.30

12.31 Find $\mathbf{H}(j\omega)$ if its amplitude characteristic is shown in Fig. P12.31.

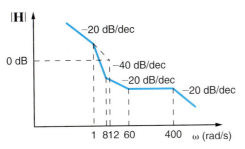

Figure P12.31

12.32 Given the magnitude characteristic in Fig. P12.32, find $\mathbf{H}(j\omega)$.

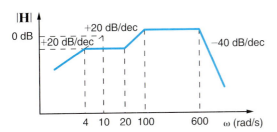

Figure P12.32

12.33 Determine $\mathbf{H}(j\omega)$ if its magnitude cha shown in Fig. P12.33.

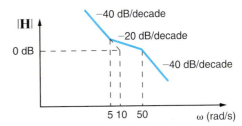

Figure P12.33

12.34 Find $\mathbf{G}(j\omega)$ for the magnitude characteristic shown in Fig. P12.34. **CS**

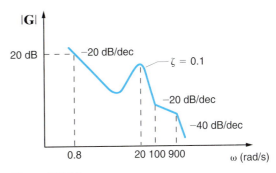

Figure P12.34

SECTION 12.3

12.35 A series *RLC* circuit resonates at 1000 rad/s. If $C = 20\ \mu\text{F}$, and it is known that the impedance at resonance is 2.4 Ω, compute the value of *L*, the *Q* of the circuit, and the bandwidth.

12.36 A series resonant circuit has a *Q* of 120 and a resonant frequency of 10,000 rad/s. Determine the half-power frequencies and the bandwidth of the circuit.

12.37 The series *RLC* circuit in Fig. P12.37 is driven by a variable-frequency source. If the resonant frequency of the network is selected as $\omega_0 = 1600$ rad/s, find the value of *C*. In addition, compute the current at resonance and at $\omega_0/4$ and $4\ \omega_0$. **PSV**

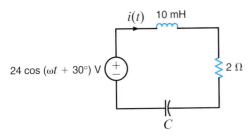

Figure P12.37

12.38 Given the series *RLC* circuit in Fig. P12.38, (a) derive the expression for the half-power frequencies, the resonant frequency, the bandwidth, and the quality factor for the transfer characteristic $\mathbf{I}/\mathbf{V}_{\text{in}}$ in terms of *R*, *L*, and *C*, (b) compute the quantities in part (a) if $R = 10\ \Omega$, $L = 50$ mH, and $C = 10\ \mu\text{F}$.

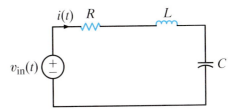

Figure P12.38

12.39 Given the network in Fig. P12.39, find $\omega_0, Q, \omega_{\text{max}}$, and $|\mathbf{V}_o|_{\text{max}}$. **CS**

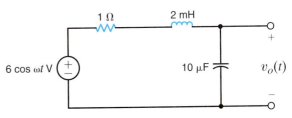

Figure P12.39

12.40 A series RLC circuit is driven by a signal generator. The resonant frequency of the network is known to be 1600 rad/s, and at that frequency the impedance seen by the signal generator is 5 Ω. If $C = 20$ μF, find L, Q, and the bandwidth.

12.41 A variable-frequency voltage source drives the network in Fig. P12.41. Determine the resonant frequency, Q, BW, and the average power dissipated by the network at resonance.

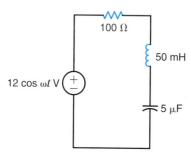

Figure P12.41

12.42 A parallel RLC resonant circuit with a resonant frequency of 20,000 rad/s has an admittance at resonance of 1 mS. If the capacitance of the network is 2 μF, find the values of R and L.   **PSV**

12.43 A parallel RLC resonant circuit has a resistance of 200 Ω. If it is known that the bandwidth is 80 rad/s and the lower half-power frequency is 800 rad/s, find the values of the parameters L and C.

12.44 In the network in Fig. P12.44, the inductor value is 10 mH, and the circuit is driven by a variable-frequency source. If the magnitude of the current at resonance is 12 A, $\omega_0 = 1000$ rad/s, and $L = 10$ mH, find C, Q, and the bandwidth of the circuit.   **CS**

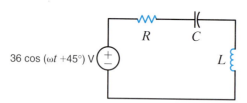

Figure P12.44

12.45 A parallel RLC circuit, which is driven by a variable-frequency 2-A current source, has the following values: $R = 1$ kΩ, $L = 400$ mH, and $C = 10$ μF. Find the bandwidth of the network, the half-power frequencies, and the voltage across the network at the half-power frequencies.   **CS**

12.46 A parallel RLC circuit, which is driven by a variable-frequency 2-A current source, has the following values: $R = 1$ kΩ, $L = 100$ mH, and $C = 10$ μF. Find the bandwidth of the network, the half-power frequencies, and the voltage across the network at the half-power frequencies.   **CS**

12.47 Consider the network in Fig. P12.47. If $R = 1$ kΩ, $L = 20$ mH, $C = 50$ μF, and $R_S = \infty$, determine the resonant frequency ω_0, the Q of the network, and the bandwidth of the network. What impact does an R_S of 10 kΩ have on the quantities determined?

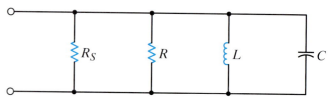

Figure P12.47

12.48 The source in the network in Fig. P12.48 is $i_S(t) = \cos 1000t + \cos 1500t$ A. $R = 200$ Ω and $C = 500$ μF. If $\omega_0 = 1000$ rad/s, find L, Q, and the BW. Compute the output voltage $v_o(t)$ and discuss the magnitude of the output voltage at the two input frequencies.

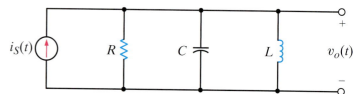

Figure P12.48

12.49 Determine the parameters of a parallel resonant circuit that has the following properties: $\omega_0 = 2$ Mrad/s, BW = 20 rad/s, and an impedance at resonance of 2000 Ω.   **CS**

12.50 Determine the value of C in the network shown in Fig. P12.50 for the circuit to be in resonance.

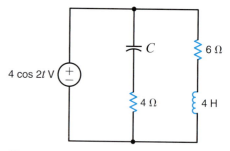

Figure P12.50

12.51 Determine the equation for the nonzero resonant frequency of the impedance shown in Fig. P12.51.

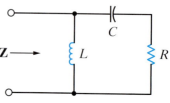

Figure P12.51

SECTION 12.4

12.52 Determine the new parameters of the network shown in Fig. P12.52 if $\mathbf{Z}_{new} = 10^4 \mathbf{Z}_{old}$.

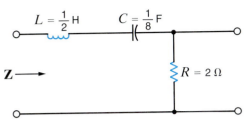

Figure P12.52

12.53 Determine the new parameters of the network in Problem 12.52 if $\omega_{new} = 10^4 \omega_{old}$. **CS**

SECTION 12.5

12.54 Given the network in Fig. P12.54, sketch the magnitude characteristic of the transfer function

$$\mathbf{G}_v(j\omega) = \frac{\mathbf{V}_o}{\mathbf{V}_1}(j\omega)$$

Identify the type of filter.

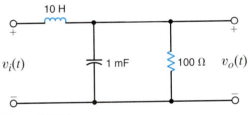

Figure P12.54

12.55 Given the network in Fig. P12.55, sketch the magnitude characteristic of the transfer function

$$\mathbf{G}_v(j\omega) = \frac{\mathbf{V}_o}{\mathbf{V}_1}(j\omega)$$

Identify the type of filter.

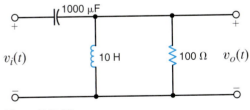

Figure P12.55

12.56 Determine what type of filter the network shown in Fig. P12.56 represents by determining the voltage transfer function. **CS**

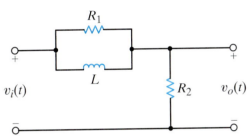

Figure P12.56

12.57 Determine what type of filter the network shown in Fig. P12.57 represents by determining the voltage transfer function. **PSV**

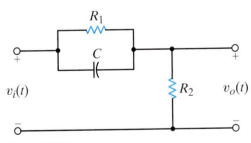

Figure P12.57

12.58 Given the lattice network shown in Fig. P12.58, determine what type of filter this network represents by determining the voltage transfer function.

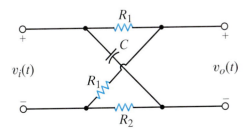

Figure P12.58

12.59 Given the network in Fig. P12.59, and employing the voltage follower analyzed in Chapter 4, determine the voltage transfer function and its magnitude characteristic. What type of filter does the network represent?

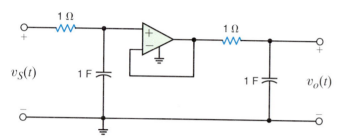

Figure P12.59

12.60 Given the network in Fig. P12.60, find the transfer function

$$\frac{\mathbf{V}_o}{\mathbf{V}_1}(j\omega)$$

and determine what type of filter the network represents.

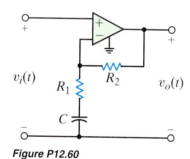

Figure P12.60

12.61 Repeat Problem 12.55 for the network shown in Fig. P12.61.

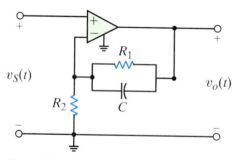

Figure P12.61

12.62 Determine the voltage transfer function and its magnitude characteristic for the network shown in Fig. P12.62 and identify the filter properties.

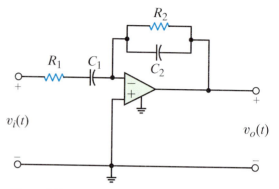

Figure P12.62

12.63 An OTA with a transconductance of 1 mS is required. A 5-V supply is available, and the sensitivity of g_m to I_{ABC} is 20.

(a) What values of I_{ABC} and R_G do you recommend?

(b) If R_G has a tolerance of +5%, what is the possible range of g_m in the final circuit?

12.64 A particular OTA has a maximum transconductance of 5 mS with a range of 6 decades.

(a) What is the minimum possible transconductance?

(b) What is the range of I_{ABC}?

(c) Using a 5-V power supply and resistor to set I_{ABC}, what is the range of values for the resistor and the power it consumes?

12.65 The OTA and 5-V source described in Problem 12.64 are used to create a transconductance of 2.5 mS.

(a) What resistor value is required?

(b) If the input voltage to the amplifier is $v_{in}(t) = 1.5\cos(\omega t)$ V, what is the output current function?

12.66 A fluid level sensor, used to measure water level in a reservoir, outputs a voltage directly proportional to fluid level. Unfortunately, the sensitivity of the sensor drifts about 10% over time. Some means for tuning the sensitivity is required. Your engineering team produces the simple OTA circuit in Fig. P12.66.

(a) Show that either V_G or R_G can be used to vary the sensitivity.

(b) List all pros and cons you can think of for these two options.

(c) What's your recommendation?

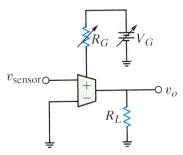

Figure P12.66

12.67 A circuit is required that can double the frequency of a sinusoidal voltage.

(a) If $v_{in}(t) = 1 \sin(\omega t)$ V, show that the multiplier circuit in Fig. P12.67 can produce an output that contains a sinusoid at frequency 2ω.

(b) We want the magnitude of the double-frequency sinusoid to be 1 V. Determine values for R_G and R_L if the transconductance range is limited between 10 μS and 10 mS.

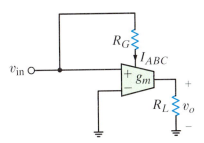

Figure P12.67

12.68 The frequency doubler in Problem 12.67 uses a two-quadrant multiplier.

(a) What effect does this have on the output signal?

(b) The circuit in Fig. P12.68 is one solution. Show that v_o has a double-frequency term.

(c) How would you propose to eliminate the other terms?

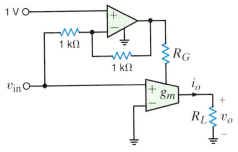

Figure P12.68

12.69 In Fig. P12.69, V_x is a dc voltage. The circuit is intended to be a dc wattmeter where the output voltage value equals the power consumed by R_L in watts.

(a) The $g_m - I_{ABC}$ sensitivity is 20 S/A. Find R_G such that $I_x/I_1 = 10^4$.

(b) Choose R_L such that 1 V at V_o corresponds to 1 W dissipated in R_L.

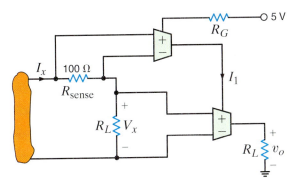

Figure P12.69

12.70 The automatic gain control circuit in Fig. P12.70 is used to limit the transconductance, i_o/v_{in}.

(a) Find an expression for v_o in terms of v_{in}, R_G, and R_L.

(b) Express the asymptotic transconductance, i_o/v_{in}, in terms of R_G and R_L at $v_{in} = 0$ and as v_{in} approaches infinity. Given R_L and R_G values in the circuit diagram, what are the values of the asymptotic transconductance?

(c) What are the consequences of your results in (b)?

(d) If v_{in} must be no more than V_{CC} for proper operation, what is the minimum transconductance for the functional circuit?

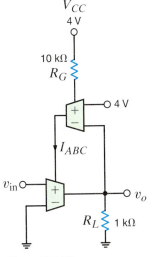

Figure P12.70

12.71 Design a low-pass filter using one resistor and one capacitor that will produce a 4.24-volt output at 159 Hz when 6 volts at 159 Hz is applied at the input.

12.72 Design a low-pass filter with a cutoff frequency between 15 and 16 kHz.

12.73 Design a high-pass filter with a half-power frequency between 159 and 161 Hz.

12.74 Design a band-pass filter with a low cutoff frequency of approximately 4535 Hz and a high cutoff frequency of approximately 5535 Hz.

12.75 An engineer has proposed the circuit shown in Fig. P12.75 to filter out high-frequency noise. Determine the values of the capacitor and resistor to achieve a 3-dB voltage drop at 23.16 kHz.

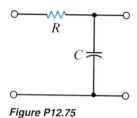

Figure P12.75

12.76 For the low-pass active filter in Fig. P12.76, choose R_2 and C such that $H_o = -7$ and $f_c = 10$ kHz.

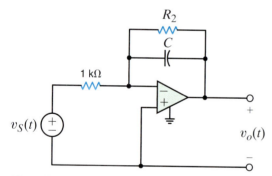

Figure P12.76

12.77 For the high-pass active filter in Fig. P12.77, choose C, R_1, and R_2 such that $H_o = 5$ and $f_c = 3$ kHz.

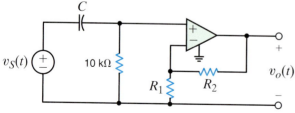

Figure P12.77

12.78 Given the second-order low-pass filter in Fig. P12.78, design a filter that has $H_o = 100$ and $f_c = 5$ kHz. Set $R_1 = R_3 = 1$ kΩ, and let $R_2 = R_4$ and $C_1 = C_2$. Use an op-amp model with $R_i = \infty$, $R_o = 0$, and $A = (2)10^5$.

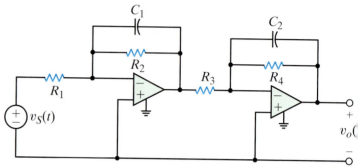

Figure P12.78

12.79 The second-order low-pass filter shown in Fig. P12.79 has the transfer function

$$\frac{\mathbf{V}_o}{\mathbf{V}_1}(s) = \frac{\dfrac{-R_3}{R_1}\left(\dfrac{1}{R_2 R_3 C_1 C_2}\right)}{s^2 + \dfrac{s}{C_1}\left(\dfrac{1}{R_1} + \dfrac{1}{R_2} + \dfrac{1}{R_3}\right) + \dfrac{1}{R_2 R_3 C_1 C_2}}$$

Design a filter with $H_o = -10$ and $f_c = 500$ Hz, assuming that $C_1 = C_2 = 10$ nF and $R_1 = 1$ kΩ.

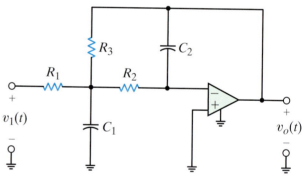

Figure P12.79

12.80 Given the circuit in Figure 12.57, design a second-order bandpass filter with a center frequency gain of -5, $\omega_0 = 50$ krad/s, and a BW $= 10$ krad/s. Let $C_1 = C_2 = C$ and $R_1 = 1$ kΩ. What is the Q of this filter? Sketch the Bode plot for the filter. Use the ideal op-amp model.

12.81 Referring to Example 12.38, design a notch filter for the tape deck for use in Europe, where power utilities generate at 50 Hz.

TYPICAL PROBLEMS FOUND ON THE FE EXAM

12FE-1 Determine the resonant frequency of the circuit in Fig. 12PFE-1, and find the voltage $\mathbf{V}_o$ at resonance. **CS**

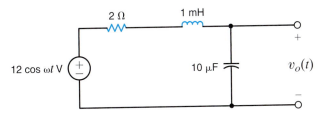

Figure 12PFE-1

12FE-2 Given the series circuit in Fig. 12PFE-2, determine the resonant frequency, and find the value of R so that the BW of the network about the resonant frequency is 200 rad/s.

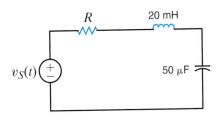

Figure 12PFE-2

12FE-3 Given the low-pass filter circuit shown in Fig. 12PFE-3, find the frequency in Hz at which the output is down 3 dB from the dc, or very low frequency, output. **CS**

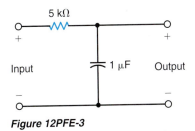

Figure 12PFE-3

12FE-4 Given the band-pass filter shown in Fig. 12PFE-4, find the components L and R necessary to provide a resonant frequency of 1000 rad/s and a BW of 100 rad/s.

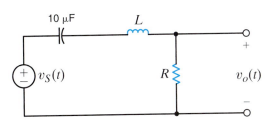

Figure 12PFE-4

12FE-5 Given the low-pass filter shown in Fig. 12PFE-5, find the half-power frequency and the gain of this circuit, if the source frequency is 8 Hz. **CS**

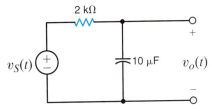

Figure 12PFE-5

The Laplace Transform

Access Problem-Solving Videos **PSV** **and Circuit Solutions** **CS** **at:**
http://www.justask4u.com/irwin
using the registration code on the inside cover and see a website with answers and more!

Recall in Chapter 7 that we studied how to solve first- and second-order transient circuits. The solution to these circuits consisted of a natural response, determined by initial conditions and the nature of the circuit, and a forced response, which has the same form as the forcing function or source for the circuit. Since most circuits in Chapter 7 contained dc sources, the forced response was always a constant. For a first-order circuit, the natural response was an exponential term, while the natural response for the second-order circuit has three forms depending on the roots of the characteristic equation. The initial conditions were utilized to determine numerical values in the natural response. What if we have a transient circuit that is driven by an exponential source or a sinusoidal source? What if we have a circuit that is third order? We will now introduce the Laplace transform as a general technique for solving first-, second-, or higher-order transient circuits. This is an extremely important technique in that, for a given set of initial conditions, it will yield both the natural and forced responses in one operation.

Our use of the Laplace transform to solve circuit problems is analogous to the use of phasors in sinusoidal steady-state analysis. Using the Laplace transform, we transform the circuit problem from the time domain to the complex frequency domain, solve the problem using algebra in the complex frequency domain, and then convert the solution in the complex frequency domain back to the time domain. Therefore, as we shall see, the Laplace transform is an integral transform that converts a set of linear simultaneous integrodifferential equations to a set of simultaneous algebraic equations.

Our approach is to define the Laplace transform, derive some of the transform pairs, consider some of the important properties of the transform, illustrate the inverse transform operation, introduce the convolution integral, and finally apply the transform to circuit analysis. ●

13.1   **Definition**

The Laplace transform of a function $f(t)$ is defined by the equation

$$\mathcal{L}[f(t)] = \mathbf{F}(s) = \int_0^\infty f(t)e^{-st}\, dt \qquad \textbf{13.1}$$

where s is the complex frequency

$$s = \sigma + j\omega \qquad \textbf{13.2}$$

and the function $f(t)$ is assumed to possess the property that

$$f(t) = 0 \qquad \text{for } t < 0$$

Note that the Laplace transform is unilateral $(0 \le t < \infty)$, in contrast to the Fourier transform (see Chapter 15), which is bilateral $(-\infty < t < \infty)$. In our analysis of circuits using the Laplace transform, we will focus our attention on the time interval $t \ge 0$. It is important to note that it is the initial conditions that account for the operation of the circuit prior to $t = 0$; therefore our analyses will describe the circuit operation for $t \ge 0$.

For a function $f(t)$ to possess a Laplace transform, it must satisfy the condition

$$\int_0^\infty e^{-\sigma t}|f(t)|\, dt < \infty \qquad \textbf{13.3}$$

for some real value of σ. Because of the convergence factor $e^{-\sigma t}$, a number of important functions have Laplace transforms, even though Fourier transforms for these functions do not exist. All of the inputs we will apply to circuits possess Laplace transforms. Functions that do not have Laplace transforms $\left(\text{e.g., } e^{t^2}\right)$ are of no interest to us in circuit analysis.

The inverse Laplace transform, which is analogous to the inverse Fourier transform, is defined by the relationship

$$\mathcal{L}^{-1}[\mathbf{F}(s)] = f(t) = \frac{1}{2\pi j} \int_{\sigma_1 - j\infty}^{\sigma_1 + j\infty} \mathbf{F}(s)e^{st}\, ds \qquad \textbf{13.4}$$

where σ_1 is real and $\sigma_1 > \sigma$ in Eq. (13.3).

Since evaluation of this integral is based on complex variable theory, we will avoid its use. How then will we be able to convert our solution in the complex frequency domain back to the time domain? The Laplace transform has a uniqueness property: for a given $f(t)$, there is a unique $F(s)$. In other words, two different functions $f_1(t)$ and $f_2(t)$ cannot have the same $F(s)$. Our procedure then will be to use Eq. (13.1) to determine the Laplace transform for a number of functions common to electric circuits and store them in a table of transform pairs. We will use a partial fraction expansion to break our complex frequency-domain solution into a group of terms for which we can utilize our table of transform pairs to identify a time function corresponding to each term.

13.2   **Two Important Singularity Functions**

Two singularity functions are very important in circuit analysis: (1) the unit step function, $u(t)$, discussed in Chapter 7, and (2) the unit impulse or delta function, $\delta(t)$. They are called *singularity functions* because they are either not finite or they do not possess finite derivatives everywhere. They are mathematical models for signals that we employ in circuit analysis.

The *unit step function* $u(t)$ shown in Fig. 13.1a was defined in Section 7.2 as

$$u(t) = \begin{cases} 0 & t < 0 \\ 1 & t > 0 \end{cases}$$

Figure 13.1
Representations of the unit step function.

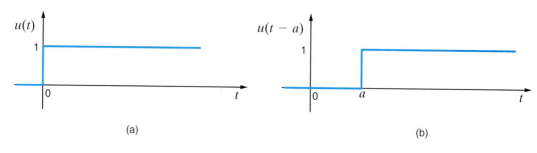

(a)

(b)

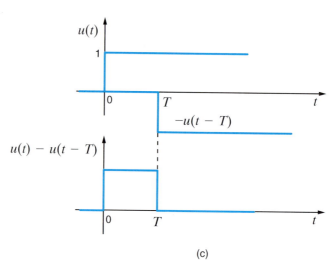

(c)

Recall that the physical analogy of this function, as illustrated earlier, corresponds to closing a switch at $t = 0$ and connecting a voltage source of 1 V or a current source of 1 A to a given circuit. The following example illustrates the calculation of the Laplace transform for unit step functions.

Example 13.1

Let us determine the Laplace transform for the waveforms in Fig. 13.1.

SOLUTION The Laplace transform for the unit step function in Fig. 13.1a is

$$\mathbf{F}(s) = \int_0^\infty u(t)e^{-st}\,dt$$

$$= \int_0^\infty 1e^{-st}\,dt$$

$$= -\frac{1}{s}e^{-st}\Big|_0^\infty$$

$$= \frac{1}{s} \qquad \sigma > 0$$

Therefore,

$$\mathcal{L}[u(t)] = \mathbf{F}(s) = \frac{1}{s}$$

The Laplace transform of the time-shifted unit step function shown in Fig. 13.1b is

$$\mathbf{F}(s) = \int_0^\infty u(t - a)e^{-st}\,dt$$

Note that

$$u(t - a) = \begin{cases} 1 & a < t < \infty \\ 0 & t < a \end{cases}$$

Therefore,

$$\mathbf{F}(s) = \int_a^\infty e^{-st}\, dt$$

$$= \frac{e^{-as}}{s} \qquad \sigma > 0$$

Finally, the Laplace transform of the pulse shown in Fig. 13.1c is

$$\mathbf{F}(s) = \int_0^\infty \left[u(t) - u(t - T)\right]e^{-st}\, dt$$

$$= \frac{1 - e^{-Ts}}{s} \qquad \sigma > 0$$

The unit impulse function can be represented in the limit by the rectangular pulse shown in Fig. 13.2a as $a \to 0$. The function is defined by the following:

$$\delta(t - t_0) = 0 \qquad t \neq t_0$$

$$\int_{t_0-\varepsilon}^{t_0+\varepsilon} \delta(t - t_0)\, dt = 1 \qquad \varepsilon > 0$$

The unit impulse is zero except at $t = t_0$, where it is undefined, but it has unit area (sometimes referred to as *strength*). We represent the unit impulse function on a plot as shown in Fig. 13.2b.

An important property of the unit impulse function is what is often called the *sampling property*, which is exhibited by the following integral:

$$\int_{t_1}^{t_2} f(t)\delta(t - t_0)\, dt = \begin{cases} f(t_0) & t_1 < t_0 < t_2 \\ 0 & t_0 < t_1, t_0 > t_2 \end{cases}$$

for a finite t_0 and any $f(t)$ continuous at t_0. Note that the unit impulse function simply samples the value of $f(t)$ at $t = t_0$.

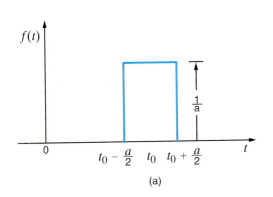

(a)

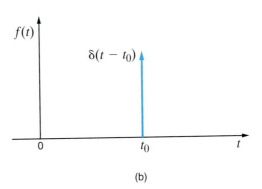

(b)

Figure 13.2
Representations of the unit impulse.

Now that we have defined the unit impulse function, let's consider the following question: why introduce the unit impulse function? We certainly cannot produce a voltage or current signal with zero width and infinite height in a physical system. For engineers, the unit impulse function is a convenient mathematical function that can be utilized to model a physical process. For example, a lightning stroke is a short-duration event. If we were analyzing a system that was struck by lightning, we might consider modeling the lightning stroke as a unit impulse function. Another example is the process of sampling where an analog-to-digital converter (ADC) is utilized to convert a time signal into values that can be used in a computer. The ADC captures the value of the time signal at certain instants of time. The sampling property of the unit impulse function described above is very useful in modeling the sampling process.

Example 13.2

Let us determine the Laplace transform of an impulse function.

SOLUTION   The Laplace transform of the impulse function is

$$F(s) = \int_0^\infty \delta(t - t_0)e^{-st}\, dt$$

Using the sampling property of the delta function, we obtain

$$\mathcal{L}[\delta(t - t_0)] = e^{-t_0 s}$$

In the limit as $t_0 \to 0$, $e^{-t_0 s} \to 1$, and therefore

$$\mathcal{L}[\delta(t)] = F(s) = 1$$

13.3   Transform Pairs

We will now illustrate the development of a number of basic transform pairs that are very useful in circuit analysis.

Example 13.3

Let us find the Laplace transform of $f(t) = t$.

SOLUTION   The Laplace transform of the function $f(t) = t$ is

$$F(s) = \int_0^\infty te^{-st}\, dt$$

Integrating the function by parts, we let

$$u = t \quad \text{and} \quad dv = e^{-st}\, dt$$

Then

$$du = dt \quad \text{and} \quad v = \int e^{-st}\, dt = -\frac{1}{s}e^{-st}$$

Therefore,

$$F(s) = \frac{-t}{s}e^{-st}\Big|_0^\infty + \int_0^\infty \frac{e^{-st}}{s}\, dt$$

$$= \frac{1}{s^2} \quad \sigma > 0$$

HINT

$$t \leftrightarrow \frac{1}{s^2}$$

Example 13.4

Let us determine the Laplace transform of the cosine function.

SOLUTION The Laplace transform for the cosine function is

$$\mathbf{F}(s) = \int_0^\infty \cos \omega t \, e^{-st} \, dt$$

$$= \int_0^\infty \frac{e^{+j\omega t} + e^{-j\omega t}}{2} e^{-st} \, dt$$

$$= \int_0^\infty \frac{e^{-(s-j\omega)t} + e^{-(s+j\omega)t}}{2} \, dt$$

$$= \frac{1}{2}\left(\frac{1}{s - j\omega} + \frac{1}{s + j\omega}\right) \qquad \sigma > 0$$

$$= \frac{s}{s^2 + \omega^2}$$

HINT

$$\cos \omega t \leftrightarrow \frac{s}{s^2 + \omega^2}$$

A short table of useful Laplace transform pairs is shown in Table 13.1.

Once the transform pairs are known, we can easily move back and forth between the time domain and the complex frequency domain without having to use Eqs. (13.1) and (13.4).

Table 13.1 Short table of Laplace transform pairs

$f(t)$	$F(s)$
$\delta(t)$	1
$u(t)$	$\dfrac{1}{s}$
e^{-at}	$\dfrac{1}{s + a}$
t	$\dfrac{1}{s^2}$
$\dfrac{t^n}{n!}$	$\dfrac{1}{s^{n+1}}$
te^{-at}	$\dfrac{1}{(s + a)^2}$
$\dfrac{t^n e^{-at}}{n!}$	$\dfrac{1}{(s + a)^{n+1}}$
$\sin bt$	$\dfrac{b}{s^2 + b^2}$
$\cos bt$	$\dfrac{s}{s^2 + b^2}$
$e^{-at} \sin bt$	$\dfrac{b}{(s + a)^2 + b^2}$
$e^{-at} \cos bt$	$\dfrac{s + a}{(s + a)^2 + b^2}$

LEARNING EXTENSIONS

E13.1 If $f(t) = e^{-at}$, show that $\mathbf{F}(s) = 1/(s + a)$.

E13.2 If $f(t) = \sin \omega t$, show that $\mathbf{F}(s) = \omega/(s^2 + \omega^2)$. **PSV**

13.4 Properties of the Transform

A number of useful theorems describe important properties of the Laplace transform. We will first demonstrate a couple of these theorems, provide a concise listing of a number of them, and, finally, illustrate their usefulness via several examples.

The *time-scaling theorem* states that

$$\mathcal{L}[f(at)] = \frac{1}{a}\mathbf{F}\left(\frac{s}{a}\right) \qquad a > 0 \qquad\qquad \textbf{13.5}$$

The *Laplace transform* of $f(at)$ is

$$\mathcal{L}[f(at)] = \int_0^\infty f(at)e^{-st}\,dt$$

Now let $\lambda = at$ and $d\lambda = a\,dt$. Then

$$\mathcal{L}[f(at)] = \int_0^\infty f(\lambda)e^{-(\lambda/a)s}\frac{d\lambda}{a}$$

$$= \frac{1}{a}\int_0^\infty f(\lambda)e^{-(s/a)\lambda}\,d\lambda$$

$$= \frac{1}{a}\mathbf{F}\left(\frac{s}{a}\right) \qquad a > 0$$

The *time-shifting theorem* states that

$$\mathcal{L}[f(t - t_0)u(t - t_0)] = e^{-t_0 s}\mathbf{F}(s) \qquad t_0 \geq 0 \qquad\qquad \textbf{13.6}$$

This theorem is illustrated as follows:

$$\mathcal{L}[f(t - t_0)u(t - t_0)] = \int_0^\infty f(t - t_0)u(t - t_0)e^{-st}\,dt$$

$$= \int_{t_0}^\infty f(t - t_0)e^{-st}\,dt$$

If we now let $\lambda = t - t_0$ and $d\lambda = dt$, then

$$\mathcal{L}[f(t - t_0)u(t - t_0)] = \int_0^\infty f(\lambda)e^{-s(\lambda + t_0)}\,d\lambda$$

$$= e^{-t_0 s}\int_0^\infty f(\lambda)e^{-s\lambda}\,d\lambda$$

$$= e^{-t_0 s}\mathbf{F}(s) \qquad t_0 \geq 0$$

The *frequency-shifting* or *modulation theorem* states that

$$\mathcal{L}[e^{-at}f(t)] = \mathbf{F}(s + a) \qquad\qquad \textbf{13.7}$$

By definition,

$$\mathcal{L}[e^{-at}f(t)] = \int_0^\infty e^{-at}f(t)e^{-st}\,dt$$

$$= \int_0^\infty f(t)e^{-(s+a)t}\,dt$$

$$= \mathbf{F}(s + a)$$

The three theorems we have demonstrated, together with a number of other important properties, are listed in a concise manner in Table 13.2. Let us now provide several simple examples that illustrate how these properties can be used.

Table 13.2   Some useful properties of the Laplace transform

Property Number	$f(t)$	$F(s)$
1. Magnitude scaling	$Af(t)$	$A\mathbf{F}(s)$
2. Addition/subtraction	$f_1(t) \pm f_2(t)$	$\mathbf{F}_1(s) \pm \mathbf{F}_2(s)$
3. Time scaling	$f(at)$	$\dfrac{1}{a}\mathbf{F}\left(\dfrac{s}{a}\right), a > 0$
4. Time shifting	$f(t - t_0)u(t - t_0), t_0 \geq 0$	$e^{-t_0 s}\mathbf{F}(s)$
	$f(t)u(t - t_0)$	$e^{-t_0 s}\mathcal{L}[f(t + t_0)]$
5. Frequency shifting	$e^{-at}f(t)$	$\mathbf{F}(s + a)$
6. Differentiation	$\dfrac{d^n f(t)}{dt^n}$	$s^n\mathbf{F}(s) - s^{n-1}f(0) - s^{n-2}f^1(0) \cdots - s^0 f^{n-1}(0)$
7. Multiplication by t	$tf(t)$	$-\dfrac{d\mathbf{F}(s)}{ds}$
	$t^n f(t)$	$(-1)^n \dfrac{d^n \mathbf{F}(s)}{ds^n}$
8. Division by t	$\dfrac{f(t)}{t}$	$\displaystyle\int_s^\infty \mathbf{F}(\lambda)\, d\lambda$
9. Integration	$\displaystyle\int_0^t f(\lambda)\, d\lambda$	$\dfrac{1}{s}\mathbf{F}(s)$
10. Convolution	$\displaystyle\int_0^t f_1(\lambda)f_2(t - \lambda)\, d\lambda$	$\mathbf{F}_1(s)\mathbf{F}_2(s)$

Example 13.5

Use the Laplace transform of $\cos \omega t$ to find the Laplace transform of $e^{-\alpha t}\cos \omega t$.

SOLUTION   Since the Laplace transform of $\cos \omega t$ is known to be

$$\mathcal{L}[\cos \omega t] = \frac{s}{s^2 + \omega^2}$$

then using property number 5

$$\mathcal{L}[e^{-at}\cos \omega t] = \frac{s + a}{(s + a)^2 + \omega^2}$$

Example 13.6

Let us demonstrate property number 8.

SOLUTION   If $f(t) = te^{-at}$, then

$$\mathbf{F}(\lambda) = \frac{1}{(\lambda + a)^2}$$

Therefore,

$$\int_s^{\infty} \mathbf{F}(\lambda)\, d\lambda = \int_s^{\infty} \frac{1}{(\lambda + a)^2}\, d\lambda = \frac{-1}{\lambda + a}\Big|_s^{\infty} = \frac{1}{s + a}.$$

Hence,

$$f_1(t) = \frac{f(t)}{t} = \frac{te^{-at}}{t} = e^{-at} \quad \text{and} \quad \mathbf{F}_1(s) = \frac{1}{s + a}$$

Example 13.7

Let us employ the Laplace transform to solve the equation

$$\frac{dy(t)}{dt} + 2y(t) + \int_0^t y(\lambda)e^{-2(t-\lambda)}\, d\lambda = 10u(t) \qquad y(0) = 0$$

SOLUTION Applying property numbers 6 and 10, we obtain

$$s\mathbf{Y}(s) + 2\mathbf{Y}(s) + \frac{\mathbf{Y}(s)}{s + 2} = \frac{10}{s}$$

$$\mathbf{Y}(s)\left(s + 2 + \frac{1}{s + 2}\right) = \frac{10}{s}$$

$$\mathbf{Y}(s) = \frac{10(s + 2)}{s(s^2 + 4s + 5)}$$

This is the solution of the linear constant-coefficient integrodifferential equation in the s-domain. However, we want the solution $y(t)$ in the time domain. $y(t)$ is obtained by performing the inverse transform, which is the topic of the next section, and the solution $y(t)$ is derived in Example 13.9.

LEARNING EXTENSIONS

E13.3 Find $\mathbf{F}(s)$ if $f(t) = \frac{1}{2}(t - 4e^{-2t})$.

ANSWER:

$$\mathbf{F}(s) = \frac{1}{2s^2} - \frac{2}{s + 2}.$$

E13.4 If $f(t) = te^{-(t-1)}u(t - 1) - e^{-(t-1)}u(t - 1)$, determine $\mathbf{F}(s)$ using the time-shifting theorem.

ANSWER:

$$\mathbf{F}(s) = \frac{e^{-s}}{(s + 1)^2}.$$

E13.5 Find $\mathbf{F}(s)$ if $f(t) = e^{-4t}(t - e^{-t})$. Use property number 2. **PSV**

ANSWER:

$$\mathbf{F}(s) = \frac{1}{(s + 4)^2} - \frac{1}{s + 5}.$$

13.5 Performing the Inverse Transform

As we begin our discussion of this topic, let us outline the procedure we will use in applying the Laplace transform to circuit analysis. First, we will transform the problem from the time domain to the complex frequency domain (that is, s-domain). Next we will solve the circuit equations algebraically in the complex frequency domain. Finally, we will transform the solution from the s-domain back to the time domain. It is this latter operation that we discuss now.

The algebraic solution of the circuit equations in the complex frequency domain results in a rational function of s of the form

$$\mathbf{F}(s) = \frac{\mathbf{P}(s)}{\mathbf{Q}(s)} = \frac{a_m s^m + a_{m-1} s^{m-1} + \cdots + a_1 s + a_0}{b_n s^n + b_{n-1} s^{n-1} + \cdots + b_1 s + b_0} \qquad \textbf{13.8}$$

The roots of the polynomial $\mathbf{P}(s)$ (i.e., $-z_1, -z_2 \cdots -z_m$) are called the *zeros* of the function $\mathbf{F}(s)$ because at these values of s, $\mathbf{F}(s) = 0$. Similarly, the roots of the polynomial $\mathbf{Q}(s)$ (i.e., $-p_1, -p_2 \cdots -p_n$) are called *poles* of $\mathbf{F}(s)$, since at these values of s, $\mathbf{F}(s)$ becomes infinite.

If $\mathbf{F}(s)$ is a proper rational function of s, then $n > m$. However, if this is not the case, we simply divide $\mathbf{P}(s)$ by $\mathbf{Q}(s)$ to obtain a quotient and a remainder; that is,

$$\frac{\mathbf{P}(s)}{\mathbf{Q}(s)} = C_{m-n} s^{m-n} + \cdots + C_2 s^2 + C_1 s + C_0 + \frac{\mathbf{P}_1(s)}{\mathbf{Q}(s)} \qquad \textbf{13.9}$$

Now $\mathbf{P}_1(s)/\mathbf{Q}(s)$ is a proper rational function of s. Let us examine the possible forms of the roots of $\mathbf{Q}(s)$.

1. If the roots are simple, $\mathbf{P}_1(s)/\mathbf{Q}(s)$ can be expressed in partial fraction form as

$$\frac{\mathbf{P}_1(s)}{\mathbf{Q}(s)} = \frac{K_1}{s + p_1} + \frac{K_2}{s + p_2} + \cdots + \frac{K_n}{s + p_n} \qquad \textbf{13.10}$$

2. If $\mathbf{Q}(s)$ has simple complex roots, they will appear in complex-conjugate pairs, and the partial fraction expansion of $\mathbf{P}_1(s)/\mathbf{Q}(s)$ for each pair of complex-conjugate roots will be of the form

$$\frac{\mathbf{P}_1(s)}{\mathbf{Q}_1(s)(s + \alpha - j\beta)(s + \alpha + j\beta)} = \frac{K_1}{s + \alpha - j\beta} + \frac{K_1^*}{s + \alpha + j\beta} + \cdots \qquad \textbf{13.11}$$

where $\mathbf{Q}(s) = \mathbf{Q}_1(s)(s + a - j\beta)(s + \alpha + j\beta)$ and K_1^* in the complex conjugate of K_1.

3. If $\mathbf{Q}(s)$ has a root of multiplicity r, the partial fraction expansion for each such root will be of the form

$$\frac{\mathbf{P}_1(s)}{\mathbf{Q}_1(s)(s + p_1)^r} = \frac{K_{11}}{(s + p_1)} + \frac{K_{12}}{(s + p_1)^2} + \cdots + \frac{K_{1r}}{(s + p_1)^r} + \cdots \qquad \textbf{13.12}$$

The importance of these partial fraction expansions stems from the fact that once the function $\mathbf{F}(s)$ is expressed in this form, the individual inverse Laplace transforms can be obtained from known and tabulated transform pairs. The sum of these inverse Laplace transforms then yields the desired time function, $f(t) = \mathcal{L}^{-1}[\mathbf{F}(s)]$.

SIMPLE POLES   Let us assume that all the poles of $\mathbf{F}(s)$ are simple, so that the partial fraction expansion of $\mathbf{F}(s)$ is of the form

$$\mathbf{F}(s) = \frac{\mathbf{P}(s)}{\mathbf{Q}(s)} = \frac{K_1}{s + p_1} + \frac{K_2}{s + p_2} + \cdots + \frac{K_n}{s + p_n} \qquad \textbf{13.13}$$

Then the constant K_i can be computed by multiplying both sides of this equation by $(s + p_i)$ and evaluating the equation at $s = -p_i$; that is,

$$\left. \frac{(s + p_i)\mathbf{P}(s)}{\mathbf{Q}(s)} \right|_{s=-p_i} = 0 + \cdots + 0 + K_i + 0 + \cdots + 0 \qquad i = 1, 2, \ldots, n \qquad \textbf{13.14}$$

Once all of the K_i terms are known, the time function $f(t) = \mathcal{L}^{-1}[\mathbf{F}(s)]$ can be obtained using the Laplace transform pair.

$$\mathcal{L}^{-1}\left[\frac{1}{s + a}\right] = e^{-at} \qquad \textbf{13.15}$$

Example 13.8

Given that

$$\mathbf{F}(s) = \frac{12(s+1)(s+3)}{s(s+2)(s+4)(s+5)}$$

let us find the function $f(t) = \mathcal{L}^{-1}[\mathbf{F}(s)]$.

SOLUTION Expressing $\mathbf{F}(s)$ in a partial fraction expansion, we obtain

$$\frac{12(s+1)(s+3)}{s(s+2)(s+4)(s+5)} = \frac{K_0}{s} + \frac{K_1}{s+2} + \frac{K_2}{s+4} + \frac{K_3}{s+5}$$

To determine K_0, we multiply both sides of the equation by s to obtain the equation

$$\frac{12(s+1)(s+3)}{(s+2)(s+4)(s+5)} = K_0 + \frac{K_1 s}{s+2} + \frac{K_2 s}{s+4} + \frac{K_3 s}{s+5}$$

Evaluating the equation at $s = 0$ yields

$$\frac{(12)(1)(3)}{(2)(4)(5)} = K_0 + 0 + 0 + 0$$

or

$$K_0 = \frac{36}{40}$$

Similarly,

$$(s+2)\mathbf{F}(s)\Big|_{s=-2} = \frac{12(s+1)(s+3)}{s(s+4)(s+5)}\Big|_{s=-2} = K_1$$

or

$$K_1 = 1$$

Using the same approach, we find that $K_2 = \frac{36}{8}$ and $K_3 = -\frac{32}{5}$. Hence $\mathbf{F}(s)$ can be written as

$$\mathbf{F}(s) = \frac{36/40}{s} + \frac{1}{s+2} + \frac{36/8}{s+4} - \frac{32/5}{s+5}$$

Then $f(t) = \mathcal{L}^{-1}[\mathbf{F}(s)]$ is

$$f(t) = \left(\frac{36}{40} + 1e^{-2t} + \frac{36}{8}e^{-4t} - \frac{32}{5}e^{-5t}\right)u(t)$$

MATLAB can also be used to obtain the inverse Laplace transform. Two statements are required: the first states that the symbolic objects used in the workspace are s and t, and the second statement is the inverse Laplace (ilaplace) command. In this case, the two statements, together with the MATLAB results, are listed as follows.

```
>> syms s t
>> ilaplace
(12*(s+1)*(s+3)/(s*(s+2)*(s+4)*(s+5)))
ans =
9/10+exp (-2*t)+9/2*exp(-4*t)-32/5*exp(-5*t)
```

$$f(t) = 0.9 + e^{-2t} + 4.5e^{-4t} - 6.4e^{-5t}$$

LEARNING EXTENSIONS

E13.6 Find $f(t)$ if $\mathbf{F}(s) = 10(s+6)/(s+1)(s+3)$.

ANSWER:
$f(t) = (25e^{-t} - 15e^{-3t})u(t)$.

E13.7 If $\mathbf{F}(s) = 12(s+2)/s(s+1)$, find $f(t)$.

ANSWER:
$f(t) = (24 - 12e^{-t})u(t)$.

COMPLEX-CONJUGATE POLES Let us assume that $\mathbf{F}(s)$ has one pair of complex-conjugate poles. The partial fraction expansion of $\mathbf{F}(s)$ can then be written as

$$\mathbf{F}(s) = \frac{\mathbf{P}_1(s)}{\mathbf{Q}_1(s)(s + \alpha - j\beta)(s + \alpha + j\beta)} = \frac{K_1}{s + \alpha - j\beta} + \frac{K_1^*}{s + \alpha + j\beta} + \cdots \qquad \textbf{13.16}$$

The constant K_1 can then be determined using the procedure employed for simple poles; that is,

$$(s + \alpha - j\beta)\mathbf{F}(s)\Big|_{s=-\alpha+j\beta} = K_1 \qquad \textbf{13.17}$$

In this case K_1 is in general a complex number that can be expressed as $|K_1|\underline{/\theta}$. Then $K_1^* = |K_1|\underline{/-\theta}$. Hence, the partial fraction expansion can be expressed in the form

$$\mathbf{F}(s) = \frac{|K_1|\underline{/\theta}}{s + \alpha - j\beta} + \frac{|K_1|\underline{/-\theta}}{s + \alpha - j\beta} + \cdots$$

$$= \frac{|K_1|e^{j\theta}}{s + \alpha - j\beta} + \frac{|K_1|e^{-j\theta}}{s + \alpha + j\beta} + \cdots \qquad \textbf{13.18}$$

The corresponding time function is then of the form

$$f(t) = \mathcal{L}^{-1}[\mathbf{F}(s)] = |K_1|e^{j\theta}e^{-(\alpha - j\beta)t} + |K_1|e^{-j\theta}e^{-(\alpha + j\beta)t} + \cdots$$

$$= |K_1|e^{-\alpha t}\left[e^{j(\beta t + \theta)} + e^{-j(\beta t + \theta)}\right] + \cdots \qquad \textbf{13.19}$$

$$= 2|K_1|e^{-\alpha t}\cos{(\beta t + \theta)} + \cdots$$

> **HINT**
> Recall that
> $$\cos x = \frac{e^{jx} + e^{-jx}}{2}$$

Example 13.9

Let us determine the time function $y(t)$ for the function

$$\mathbf{Y}(s) = \frac{10(s + 2)}{s(s^2 + 4s + 5)}$$

SOLUTION Expressing the function in a partial fraction expansion, we obtain

$$\frac{10(s + 2)}{s(s + 2 - j1)(s + 2 + j1)}$$

$$= \frac{K_0}{s} + \frac{K_1}{s + 2 - j1} + \frac{K_1^*}{s + 2 - j1}$$

$$\frac{10(s + 2)}{s^2 + 4s + 5}\Big|_{s=0} = K_0$$

$$4 = K_0$$

In a similar manner,

$$\frac{10(s + 2)}{s(s + 2 + j1)}\Big|_{s=-2+j1} = K_1$$

$$2.236\underline{/-153.43°} = K_1$$

Therefore,

$$2.236\underline{/153.43°} = K_1^*$$

The partial fraction expansion of $\mathbf{Y}(s)$ is then

$$\mathbf{Y}(s) = \frac{4}{s} + \frac{2.236\underline{/-153.43°}}{s + 2 - j1} + \frac{2.236\underline{/153.43°}}{s + 2 + j1}$$

and therefore,

$$y(t) = \left[4 + 4.472e^{-2t}\cos{(t - 153.43°)}\right]u(t)$$

The MATLAB statements and resulting solution in this case are

```
>> syms s t
>>
ilaplace(10*(s+2)/(s*(s^2+4*s+5)))
ans =
4-4*exp(-2*t)*cos(t)+2*exp(-2*t)*sin(t)
```

$$y(t) = 4 - 4e^{-2t}\cos(t) + 2e^{-2t}\sin(t)$$

Using Eq. (8.11) the reader can easily verify that this MATLAB result is the same as that obtained using the partial fraction expansion technique.

LEARNING EXTENSION

E13.8 Determine $f(t)$ if $\mathbf{F}(s) = s/(s^2 + 4s + 8)$. **PSV**

ANSWER:
$$f(t) = 1.41e^{-2t}\cos(2t + 45°)u(t).$$

MULTIPLE POLES Let us suppose that $\mathbf{F}(s)$ has a pole of multiplicity r. Then $\mathbf{F}(s)$ can be written in a partial fraction expansion of the form

$$\mathbf{F}(s) = \frac{\mathbf{P}_1(s)}{\mathbf{Q}_1(s)(s + p_1)^r}$$

$$= \frac{K_{11}}{s + p_1} + \frac{K_{12}}{(s + p_1)^2} + \cdots + \frac{K_{1r}}{(s + p_1)^r} + \cdots \qquad \textbf{13.20}$$

Employing the approach for a simple pole, we can evaluate K_{1r} as

$$(s + p_1)^r \mathbf{F}(s)\Big|_{s=-p_1} = K_{1r} \qquad \textbf{13.21}$$

To evaluate K_{1r-1} we multiply $\mathbf{F}(s)$ by $(s + p_1)^r$ as we did to determine K_{1r}; however, prior to evaluating the equation at $s = -p_1$, we take the derivative with respect to s. The proof that this will yield K_{1r-1} can be obtained by multiplying both sides of Eq. (13.20) by $(s + p_1)^r$ and then taking the derivative with respect to s. Now when we evaluate the equation at $s = -p_1$, the only term remaining on the right side of the equation is K_{1r-1}, and therefore,

$$\frac{d}{ds}\left[(s + p_1)^r \mathbf{F}(s)\right]\Big|_{s=-p_1} = K_{1r-1} \qquad \textbf{13.22}$$

K_{1r-2} can be computed in a similar fashion, and in that case the equation is

$$\frac{d^2}{ds^2}\left[(s + p_1)^r \mathbf{F}(s)\right]\Big|_{s=-p_1} = (2!)K_{1r-2} \qquad \textbf{13.23}$$

The general expression for this case is

$$K_{1j} = \frac{1}{(r - j)!}\frac{d^{r-j}}{ds^{r-j}}\left[(s + p_1)^r \mathbf{F}(s)\right]\Big|_{s=-p_1} \qquad \textbf{13.24}$$

Let us illustrate this procedure with an example.

Example 13.10

Given the following function $\mathbf{F}(s)$, let us determine the corresponding time function $f(t) = \mathcal{L}^{-1}[\mathbf{F}(s)]$.

$$\mathbf{F}(s) = \frac{10(s + 3)}{(s + 1)^3(s + 2)}$$

SOLUTION Expressing $\mathbf{F}(s)$ as a partial fraction expansion, we obtain

$$\mathbf{F}(s) = \frac{10(s + 3)}{(s + 1)^3(s + 2)}$$

$$= \frac{K_{11}}{s + 1} + \frac{K_{12}}{(s + 1)^2} + \frac{K_{13}}{(s + 1)^3} + \frac{K_2}{s + 2}$$

Then

$$(s + 1)^3\mathbf{F}(s)\bigg|_{s=-1} = K_{13}$$

$$20 = K_{13}$$

K_{12} is now determined by the equation

$$\frac{d}{ds}\left[(s + 1)^3\mathbf{F}(s)\right]\bigg|_{s=-1} = K_{12}$$

$$\frac{-10}{(s + 2)^2}\bigg|_{s=-1} = -10 = K_{12}$$

In a similar fashion K_{11} is computed from the equation

$$\frac{d^2}{ds^2}\left[(s + 1)^3\mathbf{F}(s)\right]\bigg|_{s=-1} = 2K_{11}$$

$$\frac{20}{(s + 2)^3}\bigg|_{s=-1} = 20 = 2K_{11}$$

Therefore,

$$10 = K_{11}$$

In addition,

$$(s + 2)\mathbf{F}(s)\bigg|_{s=-2} = K_2$$

$$-10 = K_2$$

Hence, $\mathbf{F}(s)$ can be expressed as

$$\mathbf{F}(s) = \frac{10}{s + 1} - \frac{10}{(s + 1)^2} + \frac{20}{(s + 1)^3} - \frac{10}{s + 2}$$

Now we employ the transform pair

$$\mathcal{L}^{-1}\left[\frac{1}{(s + a)^{n+1}}\right] = \frac{t^n}{n!}e^{-at}$$

and hence,

$$f(t) = \left(10e^{-t} - 10te^{-t} + 10t^2e^{-t} - 10e^{-2t}\right)u(t)$$

Once again MATLAB can be used to obtain a solution as outlined next.

```
>> syms s t
>> ilaplace(10*(s+3)/((s+1)^3*(s+2)))

ans =

10*t^2*exp(-t)-10*t*exp(-t)+10*exp(-t)-10*exp(-2*t)
```

$$f(t) = 10t^2e^{-t} - 10te^{-t} + 10e^{-t} - 10e^{-2t}$$

LEARNING EXTENSIONS

E13.9 Determine $f(t)$ if $\mathbf{F}(s) = s/(s + 1)^2$.

ANSWER:
$f(t) = \left(e^{-t} - te^{-t}\right)u(t)$.

E13.10 If $\mathbf{F}(s) = (s + 2)/s^2(s + 1)$, find $f(t)$. **PSV**

ANSWER:
$f(t) = \left(-1 + 2t + e^{-t}\right)u(t)$.

Back in Chapter 7 we discussed the characteristic equation for a second-order transient circuit. The polynomial $\mathbf{Q}(s) = 0$ is the characteristic equation for our circuit. The roots of the characteristic equation, also called the poles of $\mathbf{F}(s)$, determine the time response for our circuit. If $\mathbf{Q}(s) = 0$ has simple roots, then the time response will be characterized by decaying exponential functions. Multiple roots produce a time response that contains decaying exponential terms such as e^{-at}, te^{-at}, and $t^2 e^{-at}$. The time response for simple complex-conjugate roots is a sinusoidal function whose amplitude decays exponentially. Note that all of these time responses decay to zero with time. Suppose our circuit response contained a term such as $3e^{2t}$. A quick plot of this function reveals that it increases without bound for $t > 0$. Certainly, if our circuit was characterized by this type of response, we would need eye protection as our circuit destructed before us!

Earlier, in Eq. (13.8), we defined $\mathbf{F}(s)$ as the ratio of two polynomials. Let's suppose that $m = n$ in this equation. In this case, only C_0 is nonzero in Eq. (13.9). Recall that we perform a partial fraction expansion on $\mathbf{P}_1(s)/\mathbf{Q}(s)$ and use our table of Laplace transform pairs to determine the corresponding time function for each term in the expansion. What do we do with this constant C_0? Looking at our table of transform pairs in Table 13.1, we note that the Laplace transform of the unit impulse function is a constant. As a result, our circuit response would contain a unit impulse function. Earlier we noted that unit impulse functions don't exist in physical systems; therefore, $m < n$ for physical systems.

13.6 Convolution Integral

Convolution is a very important concept and has wide application in circuit and systems analysis. We first illustrate the connection that exists between the convolution integral and the Laplace transform. We then indicate the manner in which the convolution integral is applied in circuit analysis.

Property number 10 in Table 13.2 states the following.
If

$$f(t) = f_1(t) \otimes f_2(t) = \int_0^t f_1(t - \lambda)f_2(\lambda)\,d\lambda = \int_0^t f_1(\lambda)f_2(t - \lambda)\,d\lambda \qquad \textbf{13.25}$$

and

$$\mathcal{L}\big[f(t)\big] = \mathbf{F}(s), \mathcal{L}\big[f_1(t)\big] = \mathbf{F}_1(s) \qquad \text{and} \qquad \mathcal{L}\big[f_2(t)\big] = \mathbf{F}_2(s)$$

then

$$\mathbf{F}(s) = \mathbf{F}_1(s)\mathbf{F}_2(s) \qquad \textbf{13.26}$$

Our demonstration begins with the definition

$$\mathcal{L}[f(t)] = \int_0^\infty \left[\int_0^t f_1(t - \lambda)f_2(\lambda)\,d\lambda \right] e^{-st}\,dt$$

We now force the function into the proper format by introducing into the integral within the brackets the unit step function $u(t - \lambda)$. We can do this because

$$u(t - \lambda) = \begin{cases} 1 & \text{for } \lambda < t \\ 0 & \text{for } \lambda > t \end{cases} \qquad \textbf{13.27}$$

The first condition in Eq. (13.27) ensures that the insertion of the unit step function has no impact within the limits of integration. The second condition in Eq. (13.27) allows us to change the upper limit of integration from t to ∞. Therefore,

$$\mathcal{L}[f(t)] = \int_0^\infty \left[\int_0^\infty f_1(t-\lambda)u(t-\lambda)f_2(\lambda)d\lambda \right] e^{-st} dt$$

which can be written as

$$\mathcal{L}[f(t)] = \int_0^\infty f_2(\lambda) \left[\int_0^\infty f_1(t-\lambda)u(t-\lambda)e^{-st} dt \right] d\lambda$$

Note that the integral within the brackets is the time-shifting theorem illustrated in Eq. (13.6). Hence, the equation can be written as

$$\mathcal{L}[f(t)] = \int_0^\infty f_2(\lambda)\mathbf{F}_1(s)e^{-s\lambda} d\lambda$$

$$= \mathbf{F}_1(s) \int_0^\infty f_2(\lambda)e^{-s\lambda} d\lambda$$

$$= \mathbf{F}_1(s)\mathbf{F}_2(s)$$

Note that convolution in the time domain corresponds to multiplication in the frequency domain.

Let us now illustrate the use of this property in the evaluation of an inverse Laplace transform.

Example 13.11

The transfer function for a network is given by the expression

$$\mathbf{H}(s) = \frac{\mathbf{V}_o(s)}{\mathbf{V}_S(s)} = \frac{10}{s+5}$$

The input is a unit step function $\mathbf{V}_S(s) = \dfrac{1}{s}$. Let us use convolution to determine the output voltage $v_o(t)$.

SOLUTION   Since $\mathbf{H}(s) = \dfrac{10}{(s+5)}$, $h(t) = 10e^{-5t}$ and therefore

$$v_o(t) = \int_0^t 10u(\lambda)e^{-5(t-\lambda)} d\lambda$$

$$= 10e^{-5t} \int_0^t e^{5\lambda} d\lambda$$

$$= \frac{10e^{-5t}}{5}\left[e^{5t} - 1 \right]$$

$$= 2\left[1 - e^{-5t} \right]u(t) \text{ V}$$

For comparison, let us determine $v_o(t)$ from $\mathbf{H}(s)$ and $\mathbf{V}_S(s)$ using the partial fraction expansion method. $\mathbf{V}_o(s)$ can be written as

$$\mathbf{V}_o(s) = \mathbf{H}(s)\mathbf{V}_S(s)$$

$$= \frac{10}{s(s+5)} = \frac{K_0}{s} + \frac{K_1}{s+5}$$

Evaluating the constants, we obtain $K_0 = 2$ and $K_1 = -2$. Therefore,

$$\mathbf{V}_o(s) = \frac{2}{s} - \frac{2}{s+5}$$

and hence

$$v_o(t) = 2\left[1 - e^{-5t} \right]u(t) \text{ V}$$

Although we can employ convolution to derive an inverse Laplace transform, the example, though quite simple, illustrates that this is a very poor approach. If the function $\mathbf{F}(s)$ is very complicated, the mathematics can become unwieldy. Convolution is, however, a very powerful and useful tool. For example, if we know the impulse response of a network, we can use convolution to determine the network's response to an input that may be available only as an experimental curve obtained in the laboratory. Thus, convolution permits us to obtain the network response to inputs that cannot be written as analytical functions but can be simulated on a digital computer. In addition, we can use convolution to model a circuit, which is completely unknown to us, and use this model to determine the circuit's response to some input signal.

Example 13.12

To demonstrate the power of convolution, we will create a model for a "black-box" linear band-pass filter, shown as a block in Fig. 13.3. We have no details about the filter circuitry at all—no circuit diagram, no component list, no component values. As a result, our filter model must be based solely on measurements. Using our knowledge of convolution and the Laplace transform, let us discuss appropriate measurement techniques, the resulting model, and how to employ the model in subsequent simulations.

Figure 13.3
Conceptual diagram for a band-pass filter.

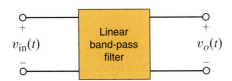

SOLUTION Because the filter is linear, $v_o(t)$ can be written

$$v_o(t) = h(t) \otimes v_{in}(t) \qquad \textbf{13.28}$$

Thus, the function $h(t)$ will be our model for the filter. To determine $h(t)$, we must input some $v_{in}(t)$, measure the response, $v_o(t)$, and perform the appropriate mathematics. One obvious option for $v_{in}(t)$ is the impulse function, $\delta(t)$, then $\mathbf{V}_{in}(s)$ is 1, and the output is the desired model, $h(t)$.

$$v_o(t) = h(t)$$

Unfortunately, creating an adequate impulse, infinite amplitude, and zero width in the laboratory is nontrivial. It is much easier, and more common, to apply a step function such as $10\,u(t)$. Then $\mathbf{V}_{in}(s)$ is $10/s$, and the output can be expressed in the s-domain as

$$\mathbf{V}_o(s) = \mathbf{H}(s)\left[\frac{10}{s}\right]$$

or

$$\mathbf{H}(s) = \left[\frac{s}{10}\right]\mathbf{V}_o(s)$$

Since multiplication by s is equivalent to the time derivative, we have for $h(t)$

$$h(t) = \left[\frac{1}{10}\right]\frac{dv_o(t)}{dt} \qquad \textbf{13.29}$$

Thus, $h(t)$ can be obtained from the derivative of the filter response to a step input!

In the laboratory, the input $10\ u(t)$ was applied to the filter and the output voltage was measured using a digital oscilloscope. Data points for time and $v_o(t)$ were acquired every 50 μs over the interval 0 to 50 ms, that is, 1,000 data samples. The digital oscilloscope formats the data as a text file, which can be transferred to a personal computer where the data can be processed. (In other words, we can find our derivative in Eq. (13.29), $dv_o(t)/dt$.) The results are shown in Table 13.3. The second and third columns in the table show the elapsed time and the output voltage for the first few data samples. To produce $h(t)$, the derivative was approximated in software using the simple algorithm,

$$\frac{dv_o(t)}{dt} \approx \frac{\Delta V_o}{\Delta t} = \frac{V_o\big[(n+1)T_S\big] - V_o\big[nT_S\big]}{T_S}$$

Table 13.3   The first five data samples of the step response and the evaluation of $h(t)$

n	Time (s)	Step Response (V)	h(t)
0	0.00E+00	0.00E+00	3.02E+02
1	5.00E−05	1.51E−01	8.98E+02
2	1.00E−04	6.00E−01	9.72E+02
3	1.50E−04	1.09E+00	9.56E+02
4	2.00E−04	1.56E+00	9.38E+02

where T_S is the sample time, 50 μs, and n is the sample number. Results for $h(t)$ are shown in the fourth column of the table. At this point, $h(t)$ exists as a table of data points and the filter is now modeled.

To test our model, $h(t)$, we let the function $v_{in}(t)$ contain a combination of dc and sinusoid components such as

$$v_{in}(t) = \begin{cases} 1\sin\big[(2\pi)100t\big] + 1\sin\big[(2\pi)1234t\big] + 4 & 0 \le t < 25\ \text{ms} \\ 0 & t \ge 25\ \text{ms} \end{cases} \qquad \textbf{13.30}$$

How will the filter perform? What will the output voltage look like? To find out, we must convolve $h(t)$ and $v_{in}(t)$. A data file for $v_{in}(t)$ can be created by simply evaluating the function in Eq. (13.30) every 50 μs. We will perform the convolution using MATLAB's `conv` function and simple file input/output functions. First, let us agree that text files for $h(t)$ and $v_{in}(t)$ have been saved as `ht.txt` and `vin.txt` in MATLAB's Work subdirectory. A set of suitable commands is

```
h=textread('ht.txt','%f');
v=textread('vin.txt','%f');
out=conv(h,v);
dlmwrite('vout.txt',out,'\t')
```

The first two commands read files `ht.txt` and `vin.txt` as MATLAB variables h and v, respectively. Entry `'%f'` specifies that each file's data are read as floating point numbers. The third command puts the convolution results into the MATLAB variable `out`. Finally, the last command puts the convolution results into a tab-delimited text file named `vout.txt`, in the Work subdirectory. The entry `'\t'` sets the delimiter to TAB.

Plots of the resulting $v_o(t)$ and $v_{in}(t)$ are shown in Fig. 13.4. An examination of the output waveform indicates that the 100-Hz component of $v_{in}(t)$ is amplified, whereas the dc and 1234-Hz components are attenuated. That is, $v_o(t)$ has an amplitude of approximately 3 V and an average value of near zero. Indeed, the circuit performs as a band-pass filter. Remember that these waveforms are not measured; they are simulation results obtained from our model, $h(t)$.

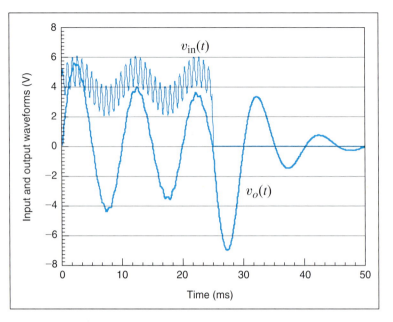

Figure 13.4 *Plots of input and output waveforms reveal the nature of the band-pass filter—particularly, attenuation of dc and higher-frequency components.*

13.7 Initial-Value and Final-Value Theorems

Suppose that we wish to determine the initial or final value of a circuit response in the time domain from the Laplace transform of the function in the s-domain without performing the inverse transform. If we determine the function $f(t) = \mathcal{L}^{-1}[\mathbf{F}(s)]$, we can find the initial value by evaluating $f(t)$ as $t \to 0$ and the final value by evaluating $f(t)$ as $t \to \infty$. It would be very convenient, however, if we could simply determine the initial and final values from $\mathbf{F}(s)$ without having to perform the inverse transform. The initial- and final-value theorems allow us to do just that.

The *initial-value theorem* states that

$$\lim_{t \to 0} f(t) = \lim_{s \to \infty} s\mathbf{F}(s) \qquad\qquad \textbf{13.31}$$

provided that $f(t)$ and its first derivative are transformable.

The proof of this theorem employs the Laplace transform of the function $df(t)/dt$.

$$\int_0^\infty \frac{df(t)}{dt} e^{-st}\, dt = s\mathbf{F}(s) - f(0)$$

Taking the limit of both sides as $s \to \infty$, we find that

$$\lim_{s \to \infty} \int_0^\infty \frac{df(t)}{dt} e^{-st}\, dt = \lim_{s \to \infty}\left[s\mathbf{F}(s) - f(0) \right]$$

and since

$$\int_0^\infty \frac{df(t)}{dt} \lim_{s \to \infty} e^{-st}\, dt = 0$$

then

$$f(0) = \lim_{s \to \infty} s\mathbf{F}(s)$$

which is, of course,

$$\lim_{t \to 0} f(t) = \lim_{s \to \infty} s\mathbf{F}(s)$$

The *final-value theorem* states that

$$\lim_{t \to \infty} f(t) = \lim_{s \to 0} s\mathbf{F}(s) \qquad \textbf{13.32}$$

provided that $f(t)$ and its first derivative are transformable and that $f(\infty)$ exists. This latter requirement means that the poles of $\mathbf{F}(s)$ must have negative real parts with the exception that there can be a simple pole at $s = 0$.

The proof of this theorem also involves the Laplace transform of the function $df(t)/dt$.

$$\int_0^\infty \frac{df(t)}{dt} e^{-st} \, dt = s\mathbf{F}(s) - f(0)$$

Taking the limit of both sides as $s \to 0$ gives us

$$\lim_{s \to 0} \int_0^\infty \frac{df(t)}{dt} e^{-st} \, dt = \lim_{s \to 0} \big[s\mathbf{F}(s) - f(0) \big]$$

Therefore,

$$\int_0^\infty \frac{df(t)}{dt} \, dt = \lim_{s \to 0} \big[s\mathbf{F}(s) - f(0) \big]$$

and

$$f(\infty) - f(0) = \lim_{s \to 0} s\mathbf{F}(s) - f(0)$$

and hence,

$$f(\infty) = \lim_{t \to \infty} f(t) = \lim_{s \to 0} s\mathbf{F}(s)$$

Example 13.13

Let us determine the initial and final values for the function

$$\mathbf{F}(s) = \frac{10(s + 1)}{s(s^2 + 2s + 2)}$$

and corresponding time function

$$f(t) = 5 + 5\sqrt{2} \, e^{-t} \cos(t - 135°)u(t)$$

SOLUTION   Applying the initial-value theorem, we have

$$f(0) = \lim_{s \to \infty} s\mathbf{F}(s)$$

$$= \lim_{s \to \infty} \frac{10(s + 1)}{s^2 + 2s + 2}$$

$$= 0$$

The poles of $\mathbf{F}(s)$ are $s = 0$ and $s = -1 \pm j1$, so the final-value theorem is applicable. Thus,

$$f(\infty) = \lim_{s \to 0} s\mathbf{F}(s)$$

$$= \lim_{s \to 0} \frac{10(s + 1)}{s^2 + 2s + 2}$$

$$= 5$$

Note that these values could be obtained directly from the time function $f(t)$.

13.8 Application Example

As a prelude to Chapter 14 in which we will employ the power and versatility of the Laplace transform in a wide variety of circuit analysis problems, we will now demonstrate how the techniques outlined in this chapter can be used in the solution of a circuit problem via the differential equation that describes the network.

Application Example 13.14

Consider the network shown in Fig. 13.5a. Assume that the network is in steady state prior to $t = 0$. Let us find the current $i(t)$ for $t > 0$.

SOLUTION In steady state prior to $t = 0$, the network is as shown in Fig. 13.5b, since the inductor acts like a short circuit to dc and the capacitor acts like an open circuit to dc. From Fig. 13.5b we note that $i(0) = 4$ A and $v_C(0) = 4$ V. For $t > 0$, the KVL equation for the network is

$$12u(t) = 2i(t) + 1\frac{di(t)}{dt} + \frac{1}{0.1}\int_0^t i(x)\,dx + v_C(0)$$

Using the results of Example 13.1 and properties 7 and 10, the transformed expression becomes

$$\frac{12}{s} = 2\mathbf{I}(s) + s\mathbf{I}(s) - i(0) + \frac{10}{s}\mathbf{I}(s) + \frac{v_C(0)}{s}$$

Using the initial conditions, we find that the equation becomes

$$\frac{12}{s} = \mathbf{I}(s)\left(2 + s + \frac{10}{s}\right) - 4 + \frac{4}{s}$$

or

$$\mathbf{I}(s) = \frac{4(s + 2)}{s^2 + 2s + 10} = \frac{4(s + 2)}{(s + 1 - j3)(s + 1 + j3)}$$

and then

$$K_1 = \left.\frac{4(s + 2)}{s + 1 + j3}\right|_{s = -1 + j3}$$

$$= 2.11\underline{/-18.4°}$$

Therefore,

$$i(t) = 2(2.11)e^{-t}\cos(3t - 18.4°)u(t)\ \text{A}$$

Note that this expression satisfies the initial condition $i(0) = 4$ A.

In the introduction to this chapter, we stated that the Laplace transform would yield both the natural and forced responses for a circuit. Our solution to this problem contains only one term. Is it the forced response or the natural response? Remember that the forced response always has the same form as the forcing function or source. The source for this problem is a dc voltage source, so the forced response should be a constant. In fact, the forced response is zero for our circuit, and the natural response is the damped cosine function. Does a zero forced response make sense? Yes! If we look at our circuit, the capacitor is going to charge up to the source voltage. Once the capacitor voltage reaches the source voltage, the current will become zero.

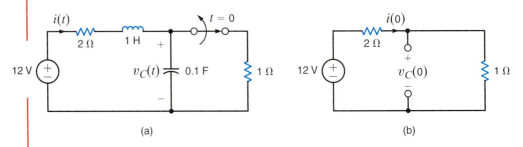

Figure 13.5 *Circuits used in Example 13.14.*

LEARNING EXTENSION

E13.11 Find the initial and final values of the function $f(t)$ if $\mathbf{F}(s) = \mathcal{L}[f(t)]$ is given by the expression

$$\mathbf{F}(s) = \frac{(s+1)^2}{s(s+2)(s^2+2s+2)} \quad \textbf{PSV}$$

ANSWER:

$f(0) = 0$ and $f(\infty) = \dfrac{1}{4}$.

PROBLEM-SOLVING STRATEGY

The Laplace Transform and Transient Circuits

Step 1. Assume that the circuit has reached steady state before a switch is moved. Draw the circuit valid for $t = 0^-$ replacing capacitors with open circuits and inductors with short circuits. Solve for the initial conditions: voltages across capacitors and currents flowing through inductors. Remember $v_C(0-) = v_C(0+) = v_C(0)$ and $i_L(0-) = i_L(0+) = i_L(0)$.

Step 2. Draw the circuit valid for $t > 0$. Use circuit analysis techniques to determine the differential or integrodifferential equation that describes the behavior of the circuit.

Step 3. Convert this differential/integrodifferential equation to an algebraic equation using the Laplace transform.

Step 4. Solve this algebraic equation for the variable of interest. Your result will be a ratio of polynomials in the complex variable s.

Step 5. Perform an inverse Laplace transform to solve for the circuit response in the time domain.

LEARNING EXTENSION

E13.12 Assuming the network in Fig. E13.12 is in steady state prior to $t = 0$, find $i(t)$ for $t > 0$. **PSV**

ANSWER:
$i(t) = (3 - e^{-2t})u(t)$ A.

Figure E13.12

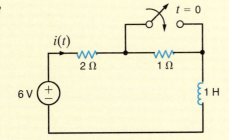

SUMMARY

- In applying the Laplace transform, we convert an integrodifferential equation in the time domain to an algebraic equation, which includes initial conditions, in the s-domain. We solve for the unknowns in the s-domain and convert the results back to the time domain.

- The Laplace transform is defined by the expression

$$\mathcal{L}[f(t)] = \mathbf{F}(s) = \int_0^\infty f(t)e^{-st}\,dt$$

- Laplace transform pairs, as listed in Table 13.1, can be used to convert back and forth between the time and frequency domains.

- The Laplace transform properties, as listed in Table 13.2, are useful in performing the Laplace transform and its inverse.

- The partial fraction expansion of a function in the *s*-domain permits the use of the transform pairs in Table 13.1 and the properties in Table 13.2 to convert the function to the time domain.

- The convolution of two functions in the time domain corresponds to a simple multiplication of the two functions in the *s*-domain.

- The initial and final values of a time-domain function can be obtained from its Laplace transform in the frequency domain.

PROBLEMS

PSV **CS** both available on the web at: **http://www.justask4u.com/irwin**

SECTION 13.2

13.1 Find the Laplace transform of the function

$$f(t) = te^{-at}\delta(t - 1) \quad \text{CS}$$

13.2 Find the Laplace transform of the function

$$f(t) = te^{-at} \sin(\omega t)\delta(t - 4).$$

SECTION 13.3

13.3 If $f(t) = e^{-at}$, show that $\mathbf{F}(s) = \dfrac{1}{s + a}$.

13.4 If $f(t) = e^{-at} \sin \omega t$, show that $\mathbf{F}(s) = \dfrac{\omega}{(s + a)^2 + \omega^2}$.

13.5 If $f(t) = t \cos(\omega t)u(t - 1)$, find $\mathbf{F}(s)$. **CS**

SECTION 13.4

13.6 Use the time-shifting theorem to determine $\mathcal{L}[f(t)]$, where $f(t) = \left[e^{-(t-2)} - e^{-2(t-2)}\right]u(t - 2)$.

13.7 Use the time-shifting theorem to determine $\mathcal{L}[f(t)]$, where $f(t) = \left[t - 1 + e^{-(t-1)}\right]u(t - 1)$. **PSV**

13.8 Use property number 5 to find $\mathcal{L}[f(t)]$ if $f(t) = e^{-at}u(t - 1)$. **CS**

13.9 Use property number 7 to find $\mathcal{L}[f(t)]$ if $f(t) = te^{-at}u(t - 1)$.

SECTION 13.5

13.10 Given the following functions $\mathbf{F}(s)$, find $f(t)$.

   (a) $\mathbf{F}(s) = \dfrac{4}{(s + 3)(s + 4)}$

   (b) $\mathbf{F}(s) = \dfrac{10s}{(s + 1)(s + 6)}$

13.11 Given the following functions $\mathbf{F}(s)$, find $f(t)$. **PSV**

   (a) $\mathbf{F}(s) = \dfrac{s + 1}{(s + 2)(s + 6)}$

   (b) $\mathbf{F}(s) = \dfrac{24}{(s + 2)(s + 3)}$

13.12 Given the following functions $\mathbf{F}(s)$, find $f(t)$. **CS**

   (a) $\mathbf{F}(s) = \dfrac{s + 1}{s(s + 2)(s + 3)}$

   (b) $\mathbf{F}(s) = \dfrac{s^2 + s + 1}{s(s + 1)(s + 2)}$

13.13 Given the following functions $\mathbf{F}(s)$, find $f(t)$.

   (a) $\mathbf{F}(s) = \dfrac{s^2 + 5s + 4}{(s + 2)(s + 4)(s + 6)}$

   (b) $\mathbf{F}(s) = \dfrac{(s + 3)(s + 6)}{s(s^2 + 8s + 12)}$

13.14 Given the following functions $\mathbf{F}(s)$, find $f(t)$.

   (a) $\mathbf{F}(s) = \dfrac{s^2 + 7s + 12}{(s + 2)(s + 4)(s + 6)}$

   (b) $\mathbf{F}(s) = \dfrac{(s + 3)(s + 6)}{s(s^2 + 10s + 24)}$

13.15 Use MATLAB to solve Problem 13.14.

13.16 Given the following functions $\mathbf{F}(s)$, find $f(t)$. **CS**

   (a) $\mathbf{F}(s) = \dfrac{10}{s^2 + 2s + 2}$

   (b) $\mathbf{F}(s) = \dfrac{10(s + 2)}{s^2 + 4s + 5}$

13.17 Given the following functions $\mathbf{F}(s)$, find $f(t)$.

(a) $\mathbf{F}(s) = \dfrac{s(s + 6)}{(s + 3)(s^2 + 6s + 18)}$

(b) $\mathbf{F}(s) = \dfrac{(s + 4)(s + 8)}{s(s^2 + 4s + 8)}$

13.18 Given the following functions $\mathbf{F}(s)$, find the inverse Laplace transform of each function.

(a) $\mathbf{F}(s) = \dfrac{10(s + 1)}{s^2 + 2s + 2}$

(b) $\mathbf{F}(s) = \dfrac{s + 1}{s(s^2 + 4s + 5)}$

13.19 Given the following functions $\mathbf{F}(s)$, find $f(t)$.

(a) $\mathbf{F}(s) = \dfrac{s(s + 6)}{(s + 3)(s^2 + 6s + 18)}$

(b) $\mathbf{F}(s) = \dfrac{(s + 4)(s + 8)}{s(s^2 + 8s + 32)}$

13.20 Use MATLAB to solve Problem 13.19. **CS**

13.21 Given the following functions $\mathbf{F}(s)$, find $f(t)$. **PSV**

(a) $\mathbf{F}(s) = \dfrac{(s + 1)(s + 3)}{(s + 2)(s^2 + 2s + 2)}$

(b) $\mathbf{F}(s) = \dfrac{(s + 2)^2}{s^2 + 4s + 5}$

13.22 Given the following functions $\mathbf{F}(s)$, find $f(t)$.

(a) $\mathbf{F}(s) = \dfrac{s^2 + 4s + 8}{(s + 1)(s + 4)}$

(b) $\mathbf{F}(s) = \dfrac{s + 4}{s^2}$

13.23 Given the following functions $\mathbf{F}(s)$, find the inverse Laplace transform of each function. **PSV**

(a) $\mathbf{F}(s) = \dfrac{s + 6}{s^2(s + 2)}$

(b) $\mathbf{F}(s) = \dfrac{s + 3}{(s + 1)^2(s + 3)}$

13.24 Given the following functions $\mathbf{F}(s)$, find $f(t)$.

(a) $\mathbf{F}(s) = \dfrac{s + 4}{(s + 2)^2}$

(b) $\mathbf{F}(s) = \dfrac{s + 6}{s(s + 1)^2}$

13.25 Given the following functions $\mathbf{F}(s)$, find $f(t)$.

(a) $\mathbf{F}(s) = \dfrac{s + 8}{s^2(s + 4)}$

(b) $\mathbf{F}(s) = \dfrac{1}{s^2(s + 1)^2}$

13.26 Given the following functions $\mathbf{F}(s)$, find $f(t)$.

(a) $\mathbf{F}(s) = \dfrac{s + 3}{(s + 2)^2}$

(b) $\mathbf{F}(s) = \dfrac{s + 6}{s(s + 2)^2}$

13.27 Given the following functions $\mathbf{F}(s)$, find $f(t)$. **CS**

(a) $\mathbf{F}(s) = \dfrac{s^2}{(s + 1)^2(s + 2)}$

(b) $\mathbf{F}(s) = \dfrac{s^2 + 9s + 20}{s(s + 4)^3(s + 5)}$

13.28 Find $f(t)$ if $\mathbf{F}(s)$ is given by the expression

$$\mathbf{F}(s) = \dfrac{s(s + 1)}{(s + 2)^3(s + 3)}$$

13.29 Use MATLAB to solve Problem 13.27.

13.30 Find the inverse Laplace transform of the following functions. **CS**

(a) $\mathbf{F}(s) = \dfrac{e^{-s}}{s + 1}$

(b) $\mathbf{F}(s) = \dfrac{1 - e^{-2s}}{s}$

(c) $\mathbf{F}(s) = \dfrac{1 - e^{-s}}{s + 2}$

13.31 Find the inverse Laplace transform of the following functions.

(a) $\mathbf{F}(s) = \dfrac{(s + 2)e^{-s}}{s(s + 2)}$

(b) $\mathbf{F}(s) = \dfrac{e^{-10s}}{(s + 2)(s + 3)}$

13.32 Find the inverse Laplace transform $f(t)$ if $\mathbf{F}(s)$ is

$$\mathbf{F}(s) = \dfrac{se^{-s}}{(s + 4)(s + 8)}$$

13.33 Find $f(t)$ if $\mathbf{F}(s)$ is given by the following functions:

(a) $\mathbf{F}(s) = \dfrac{(s^2 + 2s + 1)e^{-2s}}{s(s + 1)(s + 2)}$

(b) $\mathbf{F}(s) = \dfrac{(s + 1)e^{-4s}}{s^2(s + 2)}$

13.34 Find $f(t)$ if $\mathbf{F}(s)$ is given by the following functions: **CS**

(a) $\mathbf{F}(s) = \dfrac{2(s + 1)e^{-s}}{(s + 2)(s + 4)}$

(b) $\mathbf{F}(s) = \dfrac{10(s + 2)e^{-2s}}{(s + 1)(s + 4)}$

13.35 Solve the following differential equations using Laplace transforms.

(a) $\dfrac{dx(t)}{dt} + 4x(t) = e^{-2t}, \quad x(0) = 1$

(b) $\dfrac{dx(t)}{dt} + 6x(t) = 4u(t), \quad x(0) = 2$

13.36 Solve the following differential equations using Laplace transforms.

(a) $\dfrac{d^2y(t)}{dt^2} + \dfrac{2dy(t)}{dt} + y(t) = e^{-2t},$

$$y(0) = y'(0) = 0$$

(b) $\dfrac{d^2y(t)}{dt^2} + \dfrac{4dy(t)}{dt} + 4y(t) = u(t), \quad y(0) = 0,$

$$y'(0) = 1$$

13.37 Use Laplace transforms to find $y(t)$ if

$$\dfrac{dy(t)}{dt} + 3y(t) + 2\int_0^t y(x)\,dx = u(t), \quad y(0) = 0, \quad t > 0$$

PSV

13.38 Solve the following integrodifferential equation using Laplace transforms. **CS**

$$\dfrac{dy(t)}{dt} + 2y(t) + \int_0^t y(\lambda)\,d\lambda = 1 - e^{-2t}, \quad y(0) = 0, \quad t > 0$$

SECTION 13.6

13.39 Find $f(t)$ using convolution if $\mathbf{F}(s)$ is

$$\mathbf{F}(s) = \dfrac{1}{(s+1)(s+4)}$$

13.40 Use convolution to find $f(t)$ if

$$\mathbf{F}(s) = \dfrac{10}{(s+1)(s+3)^2}$$

SECTION 13.7

13.41 Determine the initial and final values of $f(t)$ if $\mathbf{F}(s)$ is given by the expressions

(a) $\mathbf{F}(s) = \dfrac{2(s+2)}{s(s+1)}$

(b) $\mathbf{F}(s) = \dfrac{2(s^2+2s+6)}{(s+1)(s+2)(s+3)}$

(c) $\mathbf{F}(s) = \dfrac{2s^2}{(s+1)(s^2+2s+2)}$ **CS**

13.42 Find the initial and final values of the time function $f(t)$ if $\mathbf{F}(s)$ is given as

(a) $\mathbf{F}(s) = \dfrac{10(s+2)}{(s+1)(s+4)}$

(b) $\mathbf{F}(s) = \dfrac{s^2+2s+2}{(s+6)(s^3+4s^2+8s+4)}$

(c) $\mathbf{F}(s) = \dfrac{2s}{s^2+2s+3}$ **PSV**

13.43 Find the final values of the time function $f(t)$ given that

(a) $\mathbf{F}(s) = \dfrac{10(s+6)}{(s+2)(s+3)}$

(b) $\mathbf{F}(s) = \dfrac{2}{s^2+4s+8}$

SECTION 13.8

13.44 In the network in Fig. P13.44, the switch opens at $t = 0$. Use Laplace transforms to find $i(t)$ for $t > 0$.

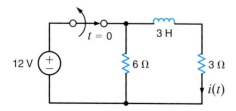

Figure P13.44

13.45 The switch in the circuit in Fig. P13.45 opens at $t = 0$. Find $i(t)$ for $t > 0$ using Laplace transforms. **CS**

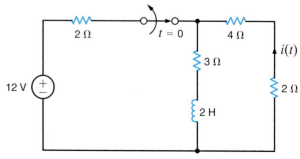

Figure P13.45

13.46 In the circuit in Fig. P13.46, the switch moves from position 1 to position 2 at $t = 0$. Use Laplace transforms to find $v(t)$ for $t > 0$.   **PSV**

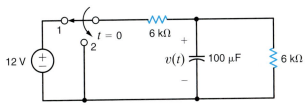

Figure P13.46

13.47 In the network in Fig. P13.47, the switch opens at $t = 0$. Use Laplace transforms to find $i_L(t)$ for $t > 0$.
PSV

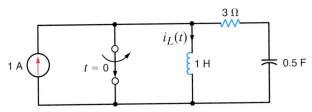

Figure P13.47

13.48 In the network in Fig. P13.48, the switch opens at $t = 0$. Use Laplace transforms to find $v_o(t)$ for $t > 0$.   **CS**

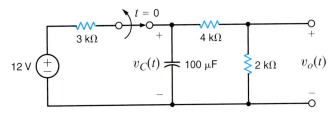

Figure P13.48

TYPICAL PROBLEMS FOUND ON THE FE EXAM

13FE-1 The output function of a network is expressed using Laplace transforms, in the following form.   **CS**

$$\mathbf{V}_o(s) = \frac{12}{s(s^2 + 3s + 2)}$$

Find the output $v_o(t)$ as a function of time.

13FE-2 The Laplace transform function representing the output voltage of a network is expressed as

$$\mathbf{V}_o(s) = \frac{120}{s(s + 10)(s + 20)}$$

Determine the time-domain function and the value of $v_o(t)$ at $t = 100$ ms.

13FE-3 The Laplace transform function for the output voltage of a network is expressed in the following form.

$$\mathbf{V}_o(s) = \frac{12(s + 2)}{s(s + 1)(s + 3)(s + 4)}$$

Determine the final value of this voltage, that is, $v_o(t)$ as $t \to \infty$.   **CS**

Application of the Laplace Transform to Circuit Analysis

Access Problem-Solving Videos **PSV**
and Circuit Solutions **CS** **at:**
http://www.justask4u.com/irwin
using the registration code on the inside cover
and see a website with answers and more!

We demonstrated in Chapter 13 how the Laplace transform can be used to solve linear constant-coefficient differential equations. Since all of the linear networks with which we are dealing can be described by linear constant-coefficient differential equations, use of the Laplace transform for circuit analysis would appear to be a feasible approach. The terminal characteristics for each circuit element can be described in the *s*-domain by transforming the appropriate time domain equations. Kirchhoff's laws when applied to a circuit produce a set of integrodifferential equations in terms of the terminal characteristics of the network elements, which when transformed yield a set of algebraic equations in the *s*-domain. Therefore, a complex frequency-domain analysis, in which the passive network elements are represented by their transform impedance or admittance and sources (independent or dependent) are represented in terms of their transform variables, can be performed. Such an analysis in the *s*-domain is algebraic, and all of the techniques derived in dc analysis will apply. Therefore, the analysis is similar to the dc analysis of resistive networks, and all of the network analysis techniques and network theorems that were applied in dc analysis are valid in the *s*-domain (e.g., node analysis, loop analysis, superposition, source transformation, Thévenin's theorem, Norton's theorem, and combinations of impedance or admittance). Therefore, our approach will be to transform each of the circuit elements, draw an *s*-domain equivalent circuit, and then, using the transformed network, solve the circuit equations algebraically in the *s*-domain. Finally, a table of transform pairs can be employed to obtain a complete (transient plus steady-state) solution. Our approach in this chapter is to employ the circuit element models and demonstrate, using a variety of examples, many of the concepts and techniques presented earlier. ●

14.1   **Laplace Circuit Solutions**

To introduce the utility of the Laplace transform in circuit analysis, let us consider the *RL* series circuit shown in Fig. 14.1. In particular, let us find the current, $i(t)$.

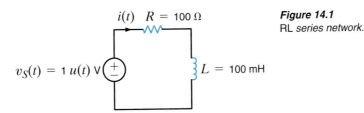

$i(t)$ $R = 100\ \Omega$

$v_S(t) = 1\ u(t)\ \text{V}$

$L = 100\ \text{mH}$

Figure 14.1
RL *series network.*

Using Kirchhoff's voltage law, we can write the time domain differential equation,

$$v_S(t) = L\left(\frac{di(t)}{dt}\right) + Ri(t)$$

The complementary differential equation is

$$L\left(\frac{di(t)}{dt}\right) + Ri(t) = 0 \qquad\qquad \textbf{14.1}$$

and has the solution

$$i_C(t) = K_C e^{-\alpha t}$$

Substituting $i_C(t)$ into the complementary equation yields the relationship

$$R - \alpha L = 0$$

or

$$\alpha = \frac{R}{L} = 1000$$

The particular solution is of the same form as the forcing function, $v_S(t)$.

$$i_p(t) = K_p$$

Substituting $i_p(t)$ into the original differential equation yields the expression

$$1 = RK_p$$

or

$$K_p = 1/R = 1/100$$

The final solution is the sum of $i_p(t)$ and $i_C(t)$,

$$i(t) = K_p + K_C e^{-\alpha t} = \frac{1}{100} + K_C e^{-1000t}$$

To find K_C, we must use the value of the current at some particular instant of time. For $t < 0$, the unit step function is zero and so is the current. At $t = 0$, the unit step goes to one; however, the inductor forces the current to instantaneously remain at zero. Therefore, at $t = 0$, we can write

$$i(0) = 0 = K_p + K_C$$

or

$$K_C = -K_p = -\frac{1}{100}$$

Thus, the current is

$$i(t) = 10\left(1 - e^{-1000t}\right)u(t)\ \text{mA}$$

Let us now try a different approach to the same problem. Making use of Table 13.2, let us take the Laplace transform of both sides of Eq. (14.1).

$$\mathcal{L}[v_S(t)] = \mathbf{V}_S(s) = L[s\mathbf{I}(s) - i(0)] + R\mathbf{I}(s)$$

Since the initial value for the inductor $i(0) = 0$, this equation becomes

$$\mathcal{L}[v_S(t)] = \mathbf{V}_S(s) = L[s\mathbf{I}(s)] + R\mathbf{I}(s)$$

Now the circuit is represented not by a time-domain differential equation, but rather by an algebraic expression in the s-domain. Solving for $\mathbf{I}(s)$, we can write

$$\mathbf{I}(s) = \frac{\mathbf{V}_S(s)}{sL + R} = \frac{1}{s[sL + R]}$$

We find $i(t)$ using the inverse Laplace transform. First, let us express $\mathbf{I}(s)$ as a sum of partial products

$$\mathbf{I}(s) = \frac{1/L}{s\left[s + \dfrac{R}{L}\right]} = \frac{1}{sR} - \frac{1}{R\left[s + \dfrac{R}{L}\right]}$$

The inverse transform is simply

$$i(t) = \frac{1}{R}\left(1 - e^{-Rt/L}\right)$$

Given the circuit element values in Fig. 14.1, the current is

$$i(t) = 10\left(1 - e^{-1000t}\right)u(t) \text{ mA}$$

which is exactly the same as that obtained using the differential equation approach. Note carefully that the solution using the Laplace transform approach yields the entire solution in one step.

We have shown that the Laplace transform can be used to transform a differential equation into an algebraic equation. Since the voltage–current relationships for resistors, capacitors, and inductors involve only constants, derivatives, and integrals, we can represent and solve any circuit in the s-domain.

14.2 Circuit Element Models

The Laplace transform technique employed earlier implies that the terminal characteristics of circuit elements can be expressed as algebraic expressions in the s-domain. Let us examine these characteristics for the resistor, capacitor, and inductor.

The voltage–current relationship for a resistor in the time domain using the passive sign convention is

$$v(t) = Ri(t) \qquad \textbf{14.2}$$

Using the Laplace transform, we find that this relationship in the s-domain is

$$\mathbf{V}(s) = R\mathbf{I}(s) \qquad \textbf{14.3}$$

Therefore, the time domain and complex frequency-domain representations of this element are as shown in Fig. 14.2a.

The time domain relationships for a capacitor using the passive sign convention are

$$v(t) = \frac{1}{C}\int_0^t i(x)\,dx + v(0) \qquad \textbf{14.4}$$

$$i(t) = C\frac{dv(t)}{dt} \qquad \textbf{14.5}$$

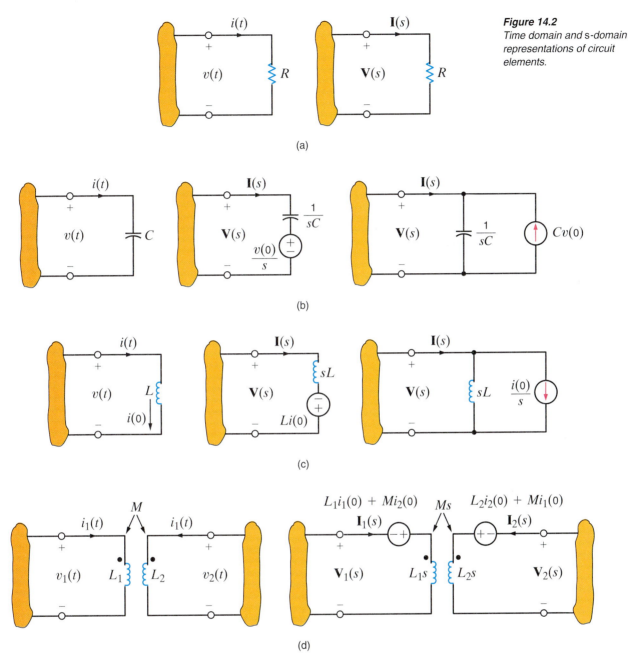

Figure 14.2
Time domain and s-domain representations of circuit elements.

The *s*-domain equations for the capacitor are then

$$\mathbf{V}(s) = \frac{\mathbf{I}(s)}{sC} + \frac{v(0)}{s} \qquad \textbf{14.6}$$

$$\mathbf{I}(s) = sC\mathbf{V}(s) - Cv(0) \qquad \textbf{14.7}$$

and hence the *s*-domain representation of this element is as shown in Fig. 14.2b.

For the inductor, the voltage–current relationships using the passive sign convention are

$$v(t) = L\frac{di(t)}{dt} \qquad \textbf{14.8}$$

$$i(t) = \frac{1}{L}\int_0^t v(x)\,dx + i(0) \qquad \textbf{14.9}$$

The relationships in the s-domain are then

$$\mathbf{V}(s) = sL\mathbf{I}(s) - Li(0) \qquad \textbf{14.10}$$

$$\mathbf{I}(s) = \frac{\mathbf{V}(s)}{sL} + \frac{i(0)}{s} \qquad \textbf{14.11}$$

The s-domain representation of this element is shown in Fig. 14.2c.

Using the passive sign convention, we find that the voltage–current relationships for the coupled inductors shown in Fig. 14.2d are

$$v_1(t) = L_1 \frac{di_1(t)}{dt} + M \frac{di_2(t)}{dt} \qquad \textbf{14.12}$$

$$v_2(t) = L_2 \frac{di_2(t)}{dt} + M \frac{di_1(t)}{dt}$$

The relationships in the s-domain are then

$$\mathbf{V}_1(s) = L_1 s\mathbf{I}_1(s) - L_1 i_1(0) + Ms\mathbf{I}_2(s) - Mi_2(0) \qquad \textbf{14.13}$$

$$\mathbf{V}_2(s) = L_2 s\mathbf{I}_2(s) - L_2 i_2(0) + Ms\mathbf{I}_1(s) - Mi_1(0)$$

Independent and dependent voltage and current sources can also be represented by their transforms; that is,

$$\mathbf{V}_1(s) = \mathcal{L}[v_1(t)] \qquad \textbf{14.14}$$

$$\mathbf{I}_2(s) = \mathcal{L}[i_2(t)]$$

and if $v_1(t) = Ai_2(t)$, which represents a current-controlled voltage source, then

$$\mathbf{V}_1(s) = A\mathbf{I}_2(s) \qquad \textbf{14.15}$$

Note carefully the direction of the current sources and the polarity of the voltage sources in the transformed network that result from the initial conditions. If the polarity of the initial voltage or direction of the initial current is reversed, the sources in the transformed circuit that results from the initial condition are also reversed.

PROBLEM-SOLVING STRATEGY

s-domain Circuits

Step 1. Solve for initial capacitor voltages and inductor currents. This may require the analysis of a circuit valid for $t < 0$ drawn with all capacitors replaced by open circuits and all inductors replaced by short circuits.

Step 2. Draw an s-domain circuit by substituting an s-domain representation for all circuit elements. Be sure to include initial conditions for capacitors and inductors if nonzero.

Step 3. Use the circuit analysis techniques presented in this textbook to solve for the appropriate voltages and/or currents. The voltages and/or currents will be described by a ratio of polynomials in s.

Step 4. Perform an inverse Laplace transform to convert the voltages and/or currents back to the time domain.

14.3 Analysis Techniques

Now that we have the s-domain representation for the circuit elements, we are in a position to analyze networks using a transformed circuit.

Example 14.1

Given the network in Fig. 14.3a, let us draw the s-domain equivalent circuit and find the output voltage in both the s and time domains.

SOLUTION   The s-domain network is shown in Fig. 14.3b. We can write the output voltage as

$$\mathbf{V}_o(s) = \left[R // \frac{1}{sC} \right] \mathbf{I}_S(s)$$

or

$$\mathbf{V}_o(s) = \left[\frac{1/C}{s + (1/RC)} \right] \mathbf{I}_S(s)$$

Given the element values, $\mathbf{V}_o(s)$ becomes

$$\mathbf{V}_o(s) = \left(\frac{40{,}000}{s + 4} \right) \left(\frac{0.003}{s + 1} \right) = \frac{120}{(s + 4)(s + 1)}$$

Expanding $\mathbf{V}_o(s)$ into partial fractions yields

$$\mathbf{V}_o(s) = \frac{120}{(s + 4)(s + 1)} = \frac{40}{s + 1} - \frac{40}{s + 4}$$

Performing the inverse Laplace transform yields the time domain representation

$$v_o(t) = 40\left[e^{-t} - e^{-4t} \right] u(t) \text{ V}$$

Figure 14.3
Time domain and s-domain representations of an RC parallel network.

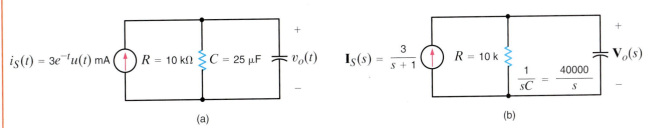

(a)   (b)

Now that we have demonstrated the use of the Laplace transform in the solution of a simple circuit, let us consider the more general case. Note that in Fig. 14.2 we have shown two models for the capacitor and inductor when initial conditions are present. Let us now consider an example in which we will illustrate the use of these models in deriving both the node and loop equations for the circuit.

Example 14.2

Given the circuits in Figs. 14.4a and b, we wish to write the mesh equations in the s-domain for the network in Fig. 14.4a and the node equations in the s-domain for the network in Fig. 14.4b.

SOLUTION   The transformed circuit for the network in Fig. 14.4a is shown in Fig. 14.4c. The mesh equations for this network are

$$\left(R_1 + \frac{1}{sC_1} + \frac{1}{sC_2} + sL_1 \right) \mathbf{I}_1(s) - \left(\frac{1}{sC_2} + sL_1 \right) \mathbf{I}_2(s)$$

$$= \mathbf{V}_A(s) - \frac{v_1(0)}{s} + \frac{v_2(0)}{s} - L_1 i_1(0)$$

$$-\left(\frac{1}{sC_2} + sL_1 \right) \mathbf{I}_1(s) + \left(\frac{1}{sC_2} + sL_1 + sL_2 + R_2 \right) \mathbf{I}_2(s)$$

$$= L_1 i_1(0) - \frac{v_2(0)}{s} - L_2 i_2(0) + \mathbf{V}_B(s)$$

HINT

Note that the equations employ the same convention as that used in dc analysis.

The transformed circuit for the network in Fig. 14.4b is shown in Fig. 14.4d. The node equations for this network are

$$\left(G_1 + \frac{1}{sL_1} + sC_1 + \frac{1}{sL_2}\right)\mathbf{V}_1(s) - \left(\frac{1}{sL_2} + sC_1\right)\mathbf{V}_2(s)$$

$$= \mathbf{I}_A(s) - \frac{i_1(0)}{s} + \frac{i_2(0)}{s} - C_1v_1(0)$$

$$-\left(\frac{1}{sL_2} + sC_1\right)\mathbf{V}_1(s) + \left(\frac{1}{sL_2} + sC_1 + G_2 + sC_2\right)\mathbf{V}_2(s)$$

$$= C_1v_1(0) - \frac{i_2(0)}{s} - C_2v_2(0) + \mathbf{I}_B(s)$$

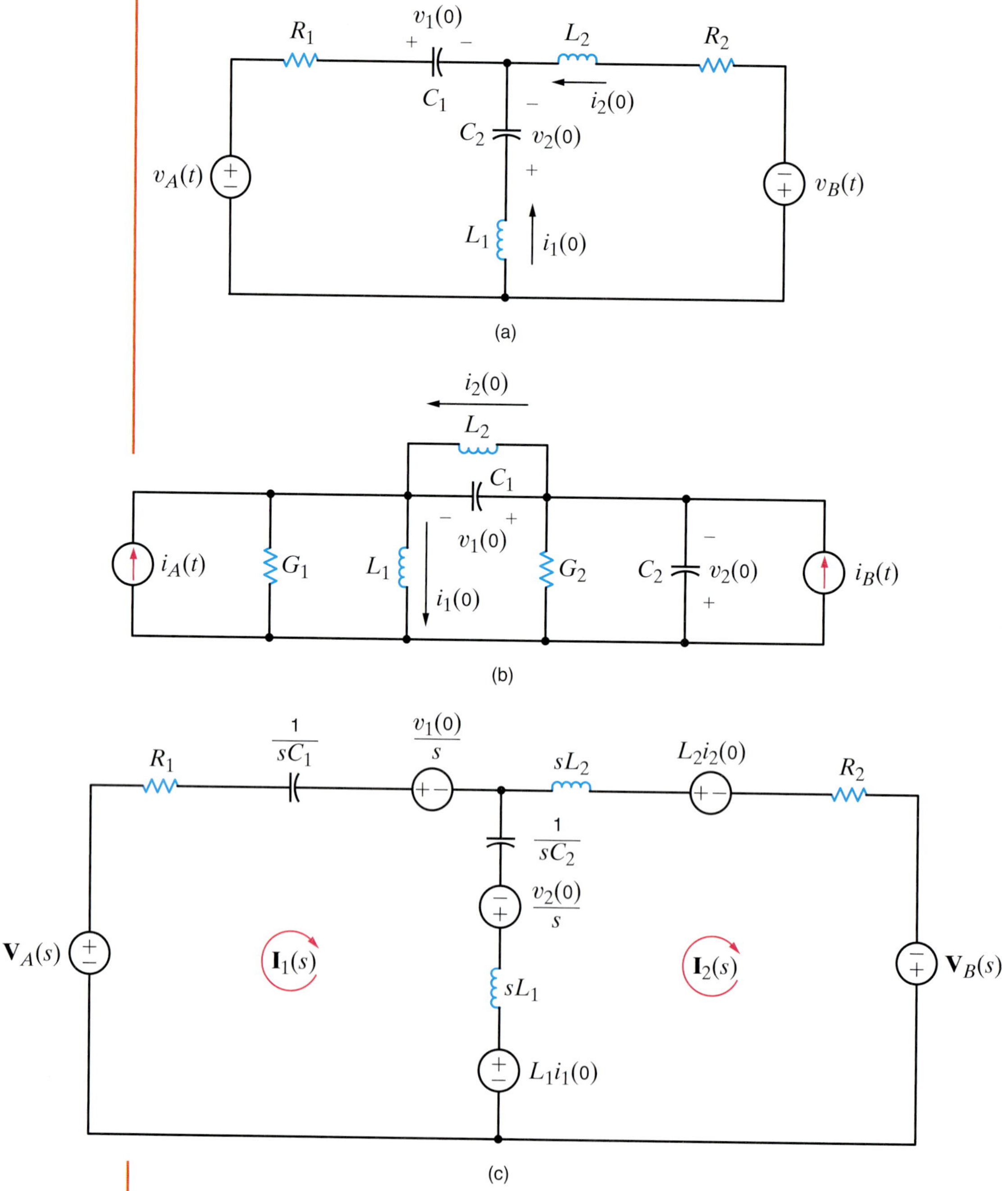

(a)

(b)

(c)

Figure 14.4 Circuits used in Example 14.2.

Figure 14.4
(continued)

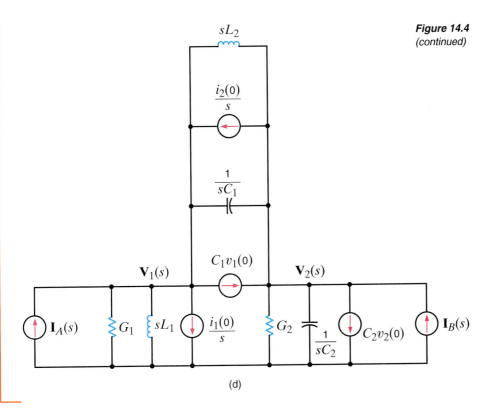

(d)

Example 14.2 attempts to illustrate the manner in which to employ the two s-domain representations of the inductor and capacitor circuit elements when initial conditions are present. In the following examples, we illustrate the use of a number of analysis techniques in obtaining the complete response of a transformed network. The circuits analyzed have been specifically chosen to demonstrate the application of the Laplace transform to circuits with a variety of passive and active elements.

Example 14.3

Let us examine the network in Fig. 14.5a. We wish to determine the output voltage $v_o(t)$.

SOLUTION   As a review of the analysis techniques presented earlier in this text, we will solve this problem using nodal analysis, mesh analysis, superposition, source exchange, Thévenin's theorem, and Norton's theorem.

The transformed network is shown in Fig. 14.5b. In our employment of nodal analysis, rather than writing KCL equations at the nodes labeled $V_1(s)$ and $V_o(s)$, we will use only the former node and use voltage division to find the latter.

KCL at the node labeled $V_1(s)$ is

$$ -\frac{4}{s} + \frac{V_1(s) - \dfrac{12}{s}}{s} + \frac{V_1(s)}{\dfrac{1}{s} + 2} = 0 $$

Solving for $V_1(s)$ we obtain

$$ V_1(s) = \frac{4(s + 3)(2s + 1)}{s(s^2 + 2s + 1)} $$

Figure 14.5
Circuits used in
Example 14.3.

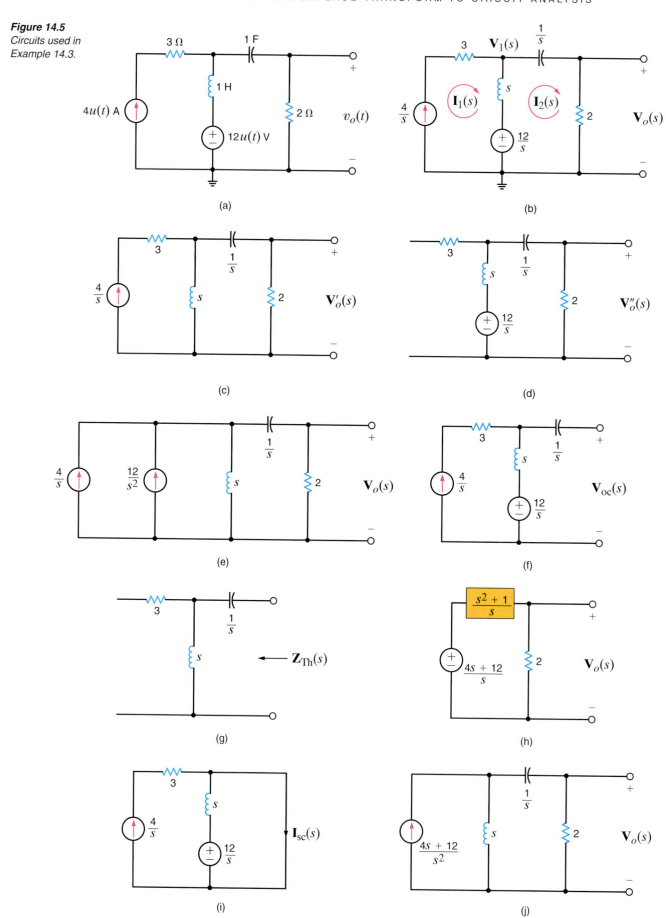

(a)

(b)

(c)

(d)

(e)

(f)

(g)

(h)

(i)

(j)

Now employing voltage division

$$\mathbf{V}_o(s) = \mathbf{V}_1(s)\left[\dfrac{2}{\dfrac{1}{s} + 2}\right] = \mathbf{V}_1(s)\left(\dfrac{2s}{2s + 1}\right)$$

$$= \dfrac{8(s + 3)}{(s + 1)^2}$$

In our mesh analysis we note that the current $\mathbf{I}_1(s)$ goes through the current source, and therefore KVL for the right-hand loop is

$$\dfrac{12}{s} - \left[\mathbf{I}_2(s) - \mathbf{I}_1(s)\right]s - \dfrac{\mathbf{I}_2(s)}{s} - 2\mathbf{I}_2(s) = 0$$

However, $\mathbf{I}_1(s) = 4/s$, and hence

$$\mathbf{I}_2(s) = \dfrac{4(s + 3)}{(s + 1)^2}$$

Therefore,

$$\mathbf{V}_o(s) = \dfrac{8(s + 3)}{(s + 1)^2}$$

The 3-Ω resistor never enters our equations. Furthermore, it will not enter our other analyses either. Why?

In using superposition, we first consider the current source acting alone as shown in Fig. 14.5c. Applying current division, we obtain

$$\mathbf{V}_o'(s) = \left[\dfrac{\dfrac{4}{s}(s)}{s + \dfrac{1}{s} + 2}\right](2) \quad (2)$$

$$= \dfrac{8s}{s^2 + 2s + 1}$$

With the voltage source acting alone, as shown in Fig. 14.5d, we obtain

$$\mathbf{V}_o''(s) = \left[\dfrac{\dfrac{12}{s}}{s + \dfrac{1}{s} + 2}\right](2) \quad (2)$$

$$= \dfrac{24}{s^2 + 2s + 1}$$

Hence,

$$\mathbf{V}_o(s) = \mathbf{V}_o'(s) + \mathbf{V}_o''(s)$$

$$= \dfrac{8(s + 3)}{(s + 1)^2}$$

In applying source exchange, we transform the voltage source and series inductor into a current source with the inductor in parallel as shown in Fig. 14.5e. Adding the current sources and applying current division yields

$$\mathbf{V}_o(s) = \left(\dfrac{12}{s^2} + \dfrac{4}{s}\right)\left[\dfrac{s}{s + \dfrac{1}{s} + 2}\right](2)$$

$$= \dfrac{\left(\dfrac{12}{s} + 4\right)(2)}{s + \dfrac{1}{s} + 2}$$

$$\mathbf{V}_o(s) = \frac{8(s + 3)}{(s + 1)^2}$$

To apply Thévenin's theorem, we first find the open-circuit voltage shown in Fig. 14.5f. $\mathbf{V}_{oc}(s)$ is then

$$\mathbf{V}_{oc}(s) = \left(\frac{4}{s}\right)(s) + \frac{12}{s}$$

$$= \frac{4s + 12}{s}$$

The Thévenin equivalent impedance derived from Fig. 14.5g is

$$\mathbf{Z}_{Th}(s) = \frac{1}{s} + s$$

$$= \frac{s^2 + 1}{s}$$

Now, connecting the Thévenin equivalent circuit to the load produces the circuit shown in Fig. 14.5h. Then, applying voltage division, we obtain

$$\mathbf{V}_o(s) = \frac{4s + 12}{s}\left[\frac{2}{\frac{s^2 + 1}{s} + 2}\right]$$

$$= \frac{8(s + 3)}{(s + 1)^2}$$

In applying Norton's theorem, for simplicity we break the network to the right of the first mesh. In this case, the short-circuit current is obtained from the circuit in Fig. 14.5i, that is,

$$\mathbf{I}_{sc}(s) = \frac{\frac{12}{s}}{s} + \frac{4}{s}$$

$$= \frac{4s + 12}{s^2}$$

The Thévenin equivalent impedance in this application of Norton's theorem is $\mathbf{Z}_{Th}(s) = s$. Connecting the Norton equivalent circuit to the remainder of the original network yields the circuit in Fig. 14.5j. Then

$$\mathbf{V}_o(s) = \frac{4s + 12}{s^2}\left[\frac{s}{s + \frac{1}{s} + 2}\right](2)$$

$$= \frac{8(s + 3)}{(s + 1)^2}$$

Finally, $\mathbf{V}_o(s)$ can now be transformed to $v_o(t)$. $\mathbf{V}_o(s)$ can be written as

$$\mathbf{V}_o(s) = \frac{8(s + 3)}{(s + 1)^2} = \frac{K_{11}}{(s + 1)^2} + \frac{K_{12}}{s + 1}$$

Evaluating the constants, we obtain

$$8(s + 3)|_{s=-1} = K_{11}$$

$$16 = K_{11}$$

and

$$\frac{d}{ds}\big[8(s+3)\big]\bigg|_{s=-1} = K_{12}$$

$$8 = K_{12}$$

Therefore,

$$v_o(t) = \big(16te^{-t} + 8e^{-t}\big)u(t)\ \text{V}$$

Example 14.4

Consider the network shown in Fig. 14.6a. We wish to determine the output voltage $v_o(t)$.

SOLUTION   As we begin to attack the problem, we note two things. First, because the source $12u(t)$ is connected between $v_1(t)$ and $v_2(t)$, we have a supernode. Second, if $v_2(t)$ is known, $v_o(t)$ can be easily obtained by voltage division. Hence, we will use nodal analysis in conjunction with voltage division to obtain a solution. Then for purposes of comparison, we will find $v_o(t)$ using Thévenin's theorem.

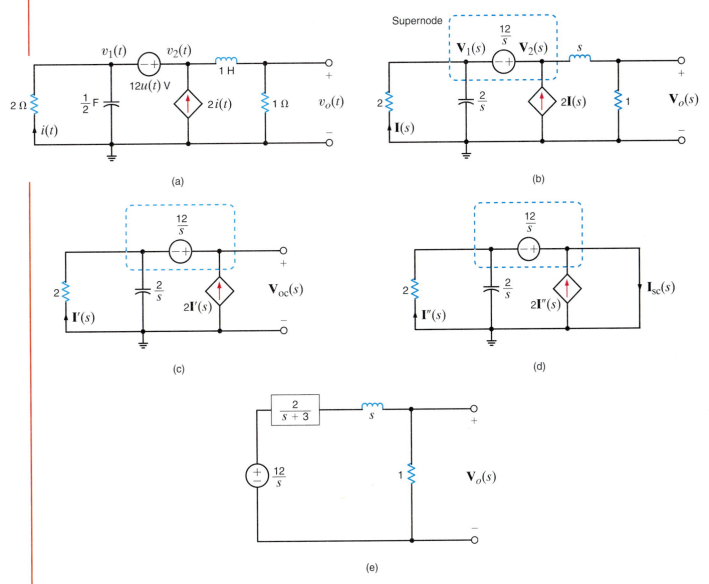

Figure 14.6 *Circuits used in Example 14.4.*

HINT

Summing the currents
leaving the supernode.

The transformed network is shown in Fig. 14.6b. KCL for the supernode is

$$\frac{\mathbf{V}_1(s)}{2} + \mathbf{V}_1(s)\frac{s}{2} - 2\mathbf{I}(s) + \frac{\mathbf{V}_2(s)}{s+1} = 0$$

However,

$$\mathbf{I}(s) = -\frac{\mathbf{V}_1(s)}{2}$$

and

$$\mathbf{V}_1(s) = \mathbf{V}_2(s) - \frac{12}{s}$$

Substituting the last two equations into the first equation yields

$$\left[\mathbf{V}_2(s) - \frac{12}{s}\right]\frac{s+3}{2} + \frac{\mathbf{V}_2(s)}{s+1} = 0$$

or

$$\mathbf{V}_2(s) = \frac{12(s+1)(s+3)}{s(s^2+4s+5)}$$

Employing a voltage divider, we obtain

$$\mathbf{V}_o(s) = \mathbf{V}_2(s)\frac{1}{s+1}$$

$$= \frac{12(s+3)}{s(s^2+4s+5)}$$

To apply Thévenin's theorem, we break the network to the right of the dependent current source as shown in Fig. 14.6c. KCL for the supernode is

$$\frac{\mathbf{V}_{oc}(s) - \frac{12}{s}}{2} + \frac{\mathbf{V}_{oc}(s) - \frac{12}{s}}{\frac{2}{s}} - 2\mathbf{I}'(s) = 0$$

where

$$\mathbf{I}'(s) = -\left(\frac{\mathbf{V}_{oc}(s) - \frac{12}{s}}{2}\right)$$

Solving these equations for $\mathbf{V}_{oc}(s)$ yields

$$\mathbf{V}_{oc}(s) = \frac{12}{s}$$

The short-circuit current is derived from the network in Fig. 14.6d as

$$\mathbf{I}_{sc}(s) = 2\mathbf{I}''(s) + \frac{\frac{12}{s}}{\frac{(2)\left(\frac{2}{s}\right)}{2+\frac{2}{s}}}$$

where

$$\mathbf{I}''(s) = \frac{\frac{12}{s}}{2}$$

Solving these equations for $\mathbf{I}_{sc}(s)$ yields

$$\mathbf{I}_{sc}(s) = \frac{6(s+3)}{s}$$

The Thévenin equivalent impedance is then

$$\mathbf{Z}_{\text{Th}}(s) = \frac{\mathbf{V}_{\text{oc}}(s)}{\mathbf{I}_{\text{sc}}(s)}$$

$$= \frac{\dfrac{12}{s}}{\dfrac{6(s + 3)}{s}}$$

$$= \frac{2}{s + 3}$$

If we now connect the Thévenin equivalent circuit to the remainder of the original network, we obtain the circuit shown in Fig. 14.6e. Using voltage division,

$$\mathbf{V}_{\text{oc}}(s) = \frac{1}{\dfrac{2}{s + 3} + s + 1}\left(\frac{12}{s}\right)$$

$$= \frac{12(s + 3)}{s(s^2 + 4s + 5)}$$

or

$$\mathbf{V}_{\text{oc}}(s) = \frac{12(s + 3)}{s(s + 2 - j1)(s + 2 + j1)}$$

To obtain the inverse transform, the function is written as

$$\frac{12(s + 3)}{s(s + 2 - j1)(s + 2 + j1)} = \frac{K_0}{s} + \frac{K_1}{s + 2 - j1} + \frac{K_1^*}{s + 2 + j1}$$

Evaluating the constants, we obtain

$$\left.\frac{12(s + 3)}{s^2 + 4s + 5}\right|_{S=0} = K_0$$

$$\frac{36}{5} = K_0$$

and

$$\left.\frac{12(s + 3)}{s(s + 2 + j1)}\right|_{S=-2+j1} = K_1$$

$$3.79\,\underline{/161.57^\circ} = K_1$$

Therefore,

$$v_o(t) = \left[7.2 + 7.58e^{-2t}\cos(t + 161.57^\circ)\right]u(t)\ \text{V}$$

LEARNING EXTENSIONS

E14.1 Find $i_o(t)$ in the network in Fig. E14.1 using node equations.   **ANSWER:**
PSV $i_o(t) = 6.53e^{-t/4}\cos\left[(\sqrt{15}/4)t - 156.72^\circ\right]u(t)\ \text{A}.$

Figure E14.1

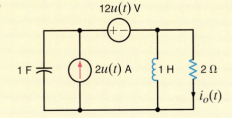

(continues on the next page)

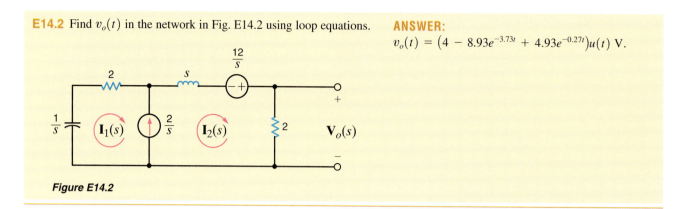

E14.2 Find $v_o(t)$ in the network in Fig. E14.2 using loop equations.

ANSWER:
$$v_o(t) = \left(4 - 8.93e^{-3.73t} + 4.93e^{-0.27t}\right)u(t) \text{ V.}$$

Figure E14.2

We will now illustrate the use of the Laplace transform in the transient analysis of circuits. We will analyze networks such as those considered in Chapter 7. Our approach will first be to determine the initial conditions for the capacitors and inductors in the network, and then we will employ the element models that were specified at the beginning of this chapter together with the circuit analysis techniques to obtain a solution. The following example demonstrates the approach.

Example 14.5

Let us determine the output voltage of the network shown in Fig. 14.7a for $t > 0$.

SOLUTION At $t = 0$ the initial voltage across the capacitor is 1 V, and the initial current drawn through the inductor is 1 A. The circuit for $t > 0$ is shown in Fig. 14.7b with the initial conditions. The transformed network is shown in Fig. 14.7c.

The mesh equations for the transformed network are

$$(s + 1)\mathbf{I}_1(s) - s\mathbf{I}_2(s) = \frac{4}{s} + 1$$

$$-s\mathbf{I}_1(s) + \left(s + \frac{2}{s} + 1\right)\mathbf{I}_2(s) = \frac{-1}{s} - 1$$

which can be written in matrix form as

$$\begin{bmatrix} s + 1 & -s \\ -s & \dfrac{s^2 + s + 2}{s} \end{bmatrix} \begin{bmatrix} \mathbf{I}_1(s) \\ \mathbf{I}_2(s) \end{bmatrix} = \begin{bmatrix} \dfrac{s + 4}{s} \\ \dfrac{-(s + 1)}{s} \end{bmatrix}$$

Solving for the currents, we obtain

$$\begin{bmatrix} \mathbf{I}_1(s) \\ \mathbf{I}_2(s) \end{bmatrix} = \begin{bmatrix} s + 1 & -s \\ -s & \dfrac{s^2 + s + 2}{s} \end{bmatrix}^{-1} \begin{bmatrix} \dfrac{s + 4}{s} \\ \dfrac{-(s + 1)}{s} \end{bmatrix}$$

$$= \frac{s}{2s^2 + 3s + 2} \begin{bmatrix} \dfrac{s^2 + s + 2}{s} & s \\ s & s + 1 \end{bmatrix} \begin{bmatrix} \dfrac{s + 4}{s} \\ \dfrac{-(s + 1)}{s} \end{bmatrix}$$

$$= \begin{bmatrix} \dfrac{4s^2 + 6s + 8}{s(2s^2 + 3s + 2)} \\ \dfrac{2s - 1}{2s^2 + 3s + 2} \end{bmatrix}$$

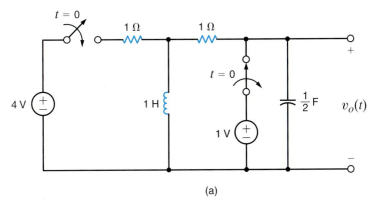

(a)

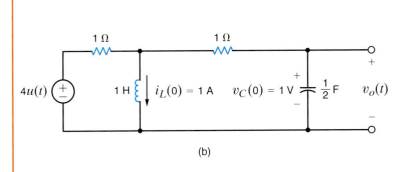

(b)

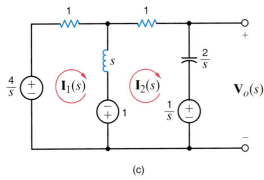

(c)

Figure 14.7
Circuits employed in Example 14.5.

The output voltage is then

$$\mathbf{V}_o(s) = \frac{2}{s}\mathbf{I}_2(s) + \frac{1}{s}$$

$$= \frac{2}{s}\left(\frac{2s - 1}{2s^2 + 3s + 2}\right) + \frac{1}{s}$$

$$= \frac{s + \dfrac{7}{2}}{s^2 + \dfrac{3}{2}s + 1} =$$

This function can be written in a partial fraction expansion as

$$\frac{s + \dfrac{7}{2}}{s^2 + \dfrac{3}{2}s + 1} = \frac{K_1}{s + \dfrac{3}{4} - j(\sqrt{7}/4)} + \frac{K_1^*}{s + \dfrac{3}{4} + j(\sqrt{7}/4)}$$

Evaluating the constants, we obtain

$$\left.\frac{s + \dfrac{7}{2}}{s + \dfrac{3}{4} + j(\sqrt{7}/4)}\right|_{s=-(3/4)+j(\sqrt{7}/4)} = K_1$$

$$2.14\,\underline{/-76.5^\circ} = K_1$$

Therefore,

$$v_o(t) = \left[4.29e^{-(3/4)t}\cos\left(\frac{\sqrt{7}}{4}t - 76.5^\circ\right)\right]u(t)\ \text{V}$$

LEARNING EXTENSIONS

E14.3 Solve Learning Extension E7.2 using Laplace transforms.

ANSWER:
$i_1(t) = (1e^{-9t})u(t)$ A.

E14.4 Solve Learning Extension E7.4 using Laplace transforms. **PSV**

ANSWER:
$$v_o(t) = \left(6 - \frac{10}{3}e^{-2t}\right)u(t) \text{ V.}$$

14.4 Transfer Function

In Chapter 12 we introduced the concept of network or transfer function. It is essentially nothing more than the ratio of some output variable to some input variable. If both variables are voltages, the transfer function is a voltage gain. If both variables are currents, the transfer function is a current gain. If one variable is a voltage and the other is a current, the transfer function becomes a transfer admittance or impedance.

In deriving a transfer function, all initial conditions are set equal to zero. In addition, if the output is generated by more than one input source in a network, superposition can be employed in conjunction with the transfer function for each source.

To present this concept in a more formal manner, let us assume that the input/output relationship for a linear circuit is

$$b_n \frac{d^n y_o(t)}{dt^n} + b_{n-1} \frac{d^{n-1} y_o(t)}{dt^{n-1}} + \cdots + b_1 \frac{dy_o(t)}{dt} + b_0 y_o(t)$$
$$= a_m \frac{d^m x_i(t)}{dt^m} + a_{m-1} \frac{d^{m-1} x_i(t)}{dt^{m-1}} + \cdots + a_1 \frac{dx_i(t)}{dt} + a_0 x_i(t)$$

If all the initial conditions are zero, the transform of the equation is

$$\left(b_n s^n + b_{n-1} s^{n-1} + \cdots + b_1 s + b_0\right)\mathbf{Y}_o(s) = \left(a_m s^m + a_{m-1} s^{m-1} + \cdots + a_1 s + a_0\right)\mathbf{X}_i(s)$$

or

$$\frac{\mathbf{Y}_o(s)}{\mathbf{X}_i(s)} = \frac{a_m s^m + a_{m-1} s^{m-1} + \cdots + a_1 s + a_0}{b_n s^n + b_{n-1} s^{n-1} + \cdots + b_1 s + b_0}$$

This ratio of $\mathbf{Y}_o(s)$ to $\mathbf{X}_i(s)$ is called the *transfer* or *network function*, which we denote as $\mathbf{H}(s)$; that is,

$$\frac{\mathbf{Y}_o(s)}{\mathbf{X}_i(s)} = \mathbf{H}(s)$$

or

$$\mathbf{Y}_o(s) = \mathbf{H}(s)\mathbf{X}_i(s) \qquad \textbf{14.16}$$

This equation states that the output response $\mathbf{Y}_o(s)$ is equal to the network function multiplied by the input $\mathbf{X}_i(s)$. Note that if $x_i(t) = \delta(t)$ and therefore $\mathbf{X}_i(s) = 1$, the impulse response is equal to the inverse Laplace transform of the network function. This is an extremely important concept because it illustrates that if we know the impulse response of a network, we can find the response due to some other forcing function using Eq. (14.16).

At this point, it is informative to review briefly the natural response of both first-order and second-order networks. We demonstrated in Chapter 7 that if only a single storage element is present, the natural response of a network to an initial condition is always of the form

$$x(t) = X_0 e^{-t/\tau}$$

where $x(t)$ can be either $v(t)$ or $i(t)$, X_0 is the initial value of $x(t)$, and τ is the time constant of the network. We also found that the natural response of a second-order network is controlled by the roots of the *characteristic equation*, which is of the form

$$s^2 + 2\zeta\omega_0 s + \omega_0^2 = 0$$

where ζ is the *damping ratio* and ω_0 is the *undamped natural frequency*. These two key factors, ζ and ω_0, control the response, and there are basically three cases of interest.

Case 1, $\zeta > 1$: Overdamped Network   The roots of the characteristic equation are $s_1, s_2 = -\zeta\omega_0 \pm \omega_0 \sqrt{\zeta^2 - 1}$ and, therefore, the network response is of the form

$$x(t) = K_1 e^{-(\zeta\omega_0 + \omega_0 \sqrt{\zeta^2-1})t} + K_2 e^{-(\zeta\omega_0 - \omega_0 \sqrt{\zeta^2-1})t}$$

Case 2, $\zeta < 1$: Underdamped Network   The roots of the characteristic equation are $s_1, s_2 = -\zeta\omega_0 \pm j\omega_0 \sqrt{1 - \zeta^2}$ and, therefore, the network response is of the form

$$x(t) = K e^{-\zeta\omega_0 t} \cos\left(\omega_0 \sqrt{1 - \zeta^2}\, t + \phi\right)$$

Case 3, $\zeta = 1$: Critically Damped Network   The roots of the characteristic equation are $s_1, s_2 = -\omega_0$ and, hence, the response is of the form

$$x(t) = K_1 t e^{-\omega_0 t} + K_2 e^{-\omega_0 t}$$

The reader should note that the characteristic equation is the denominator of the transfer function $\mathbf{H}(s)$, and the roots of this equation, which are the poles of the network, determine the form of the network's natural response.

A convenient method for displaying the network's poles and zeros in graphical form is the use of a pole-zero plot. A pole-zero plot of a function can be accomplished using what is commonly called the *complex* or *s-plane*. In the complex plane the abscissa is σ and the ordinate is $j\omega$. Zeros are represented by 0's, and poles are represented by ×'s. Although we are concerned only with the finite poles and zeros specified by the network or response function, we should point out that a rational function must have the same number of poles and zeros. Therefore, if $n > m$, there are $n - m$ zeros at the point at infinity, and if $n < m$, there are $m - n$ poles at the point at infinity. A systems engineer can tell a lot about the operation of a network or system by simply examining its pole-zero plot.

In order to correlate the natural response of a network to an initial condition with the network's pole locations, we have illustrated in Fig. 14.8 the correspondence for all three cases:

Figure 14.8

Natural response of a second-order network together with network pole locations for the three cases: (a) overdamped, (b) underdamped, and (c) critically damped.

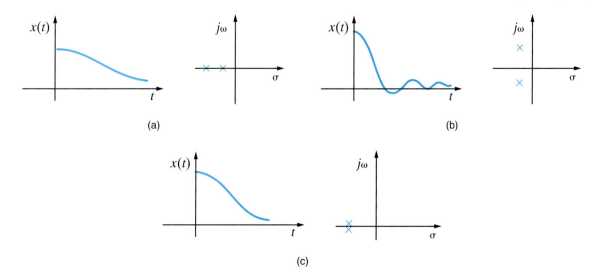

overdamped, underdamped, and critically damped. Note that if the network poles are real and unequal, the response is slow and, therefore, $x(t)$ takes a long time to reach zero. If the network poles are complex conjugates, the response is fast; however, it overshoots and is eventually damped out. The dividing line between the overdamped and underdamped cases is the critically damped case in which the roots are real and equal. In this case the transient response dies out as quickly as possible, with no overshoot.

Example 14.6

If the impulse response of a network is $h(t) = e^{-t}$, let us determine the response $v_o(t)$ to an input $v_i(t) = 10e^{-2t}u(t)$ V.

SOLUTION   The transformed variables are

$$\mathbf{H}(s) = \frac{1}{s + 1}$$

$$\mathbf{V}_i(s) = \frac{10}{s + 2}$$

Therefore,

$$\mathbf{V}_o(s) = \mathbf{H}(s)\mathbf{V}_i(s)$$

$$= \frac{10}{(s + 1)(s + 2)}$$

and hence,

$$v_o(t) = 10(e^{-t} - e^{-2t})u(t) \text{ V}$$

The importance of the transfer function stems from the fact that it provides the systems engineer with a great deal of knowledge about the system's operation, since its dynamic properties are governed by the system poles.

Example 14.7

Let us derive the transfer function $\mathbf{V}_o(s)/\mathbf{V}_i(s)$ for the network in Fig. 14.9a.

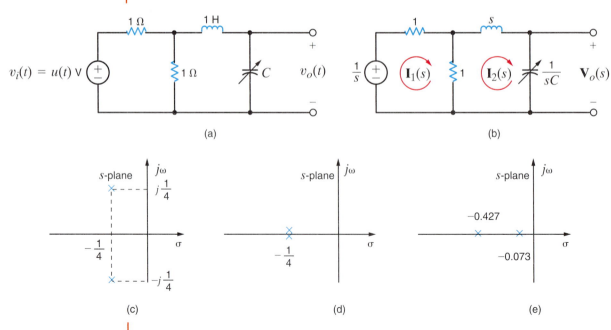

Figure 14.9 *Networks and pole zero plots used in Example 14.7.*

SOLUTION   Our output variable is the voltage across a variable capacitor, and the input voltage is a unit step. The transformed network is shown in Fig. 14.9b. The mesh equations for the network are

$$2\mathbf{I}_1(s) - \mathbf{I}_2(s) = \mathbf{V}_i(s)$$

$$-\mathbf{I}_1(s) + \left(s + \frac{1}{sC} + 1\right)\mathbf{I}_2(s) = 0$$

and the output equation is

$$\mathbf{V}_o(s) = \frac{1}{sC}\mathbf{I}_2(s)$$

From these equations we find that the transfer function is

$$\frac{\mathbf{V}_o(s)}{\mathbf{V}_i(s)} = \frac{1/2C}{s^2 + \frac{1}{2}s + 1/C}$$

Since the transfer function is dependent on the value of the capacitor, let us examine the transfer function and the output response for three values of the capacitor.

a. $C = 8 F$

$$\frac{\mathbf{V}_o(s)}{\mathbf{V}_i(s)} = \frac{\frac{1}{16}}{\left(s^2 + \frac{1}{2}s + \frac{1}{8}\right)} = \frac{\frac{1}{16}}{\left(s + \frac{1}{4} - j\frac{1}{4}\right)\left(s + \frac{1}{4} + j\frac{1}{4}\right)}$$

The output response is

$$\mathbf{V}_o(s) = \frac{\frac{1}{16}}{s\left(s + \frac{1}{4} - j\frac{1}{4}\right)\left(s + \frac{1}{4} + j\frac{1}{4}\right)}$$

As illustrated in Chapter 7, the poles of the transfer function, which are the roots of the characteristic equation, are complex conjugates, as shown in Fig. 14.9c; therefore, the output response will be *underdamped*. The output response as a function of time is

$$v_o(t) = \left[\frac{1}{2} + \frac{1}{\sqrt{2}}e^{-t/4}\cos\left(\frac{t}{4} + 135°\right)\right]u(t)\text{ V}$$

Note that for large values of time the transient oscillations, represented by the second term in the response, become negligible and the output settles out to a value of 1/2 V. This can also be seen directly from the circuit since for large values of time the input looks like a dc source, the inductor acts like a short circuit, the capacitor acts like an open circuit, and the resistors form a voltage divider.

b. $C = 16 F$

$$\frac{\mathbf{V}_o(s)}{\mathbf{V}_i(s)} = \frac{\frac{1}{32}}{s^2 + \frac{1}{2}s + \frac{1}{16}} = \frac{\frac{1}{32}}{\left(s + \frac{1}{4}\right)^2}$$

The output response is

$$\mathbf{V}_o(s) = \frac{\frac{1}{32}}{s\left(s + \frac{1}{4}\right)^2}$$

Since the poles of the transfer function are real and equal as shown in Fig. 14.9d, the output response will be *critically damped*. $v_o(t) = \mathcal{L}^{-1}[\mathbf{V}_o(s)]$ is

$$v_o(t) = \left[\frac{1}{2} - \left(\frac{t}{8} + \frac{1}{2}\right)e^{-t/4}\right]u(t)\text{ V}$$

c. *C = 32 F*

$$\frac{V_o(s)}{V_i(s)} = \frac{\frac{1}{64}}{s^2 + \frac{1}{2}s + \frac{1}{32}} = \frac{\frac{1}{64}}{(s + 0.427)(s + 0.073)}$$

The output response is

$$V_o(s) = \frac{\frac{1}{64}}{s(s + 0.427)(s + 0.073)}$$

The poles of the transfer function are real and unequal, as shown in Fig. 14.9e and, therefore, the output response will be *overdamped*. The response as a function of time is

$$v_o(t) = (0.5 + 0.103e^{-0.427t} - 0.603e^{-0.073t})u(t) \text{ V}$$

Although the values selected for the network parameters are not very practical, remember that both magnitude and frequency scaling, as outlined in Chapter 12, can be applied here also.

Example 14.8

For the network in Fig. 14.10a let us compute (a) the transfer function, (b) the type of damping exhibited by the network, and (c) the unit step response.

SOLUTION Recall that the voltage across the op-amp input terminals is zero and therefore KCL at the node labeled $V_1(s)$ in Fig. 14.10b yields the following equation:

$$\frac{V_S(s) - V_1(s)}{1} = sV_1(s) + \frac{V_1(s) - V_o(s)}{1} + \frac{V_1(s)}{1}$$

Since the current into the negative input terminal of the op-amp is zero, KCL requires that

$$sV_o(s) = -\frac{V_1(s)}{1}$$

Combining the two equations yields the transfer function

$$\frac{V_o(s)}{V_S(s)} = \frac{-1}{s^2 + 3s + 1}$$

which can be expressed in the form

$$\frac{V_o(s)}{V_S(s)} = \frac{-1}{(s + 2.62)(s + 0.38)}$$

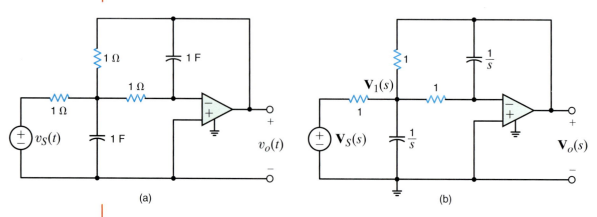

(a)   (b)

Figure 14.10 Circuits used in Example 14.8.

Since the roots are real and unequal, the step response of the network will be overdamped. The step response is

$$\mathbf{V}_o(s) = \frac{-1}{s(s + 2.62)(s + 0.38)}$$

$$= \frac{-1}{s} + \frac{-0.17}{s + 2.62} + \frac{1.17}{s + 0.38}$$

Therefore,

$$v_o(t) = \left(-1 - 0.17e^{-2.62t} + 1.17e^{-0.38t}\right)u(t) \text{ V}$$

LEARNING EXTENSIONS

E14.5 If the unit impulse response of a network is known to be $10/9\left(e^{-t} - e^{-10t}\right)$, determine the unit step response. **PSV**

ANSWER:

$$x(t) = \left(1 - \frac{10}{9}e^{-t} + \frac{1}{9}e^{-10t}\right)u(t).$$

E14.6 The transfer function for a network is

$$\mathbf{H}(s) = \frac{s + 10}{s^2 + 4s + 8}$$

Determine the pole-zero plot of $\mathbf{H}(s)$, the type of damping exhibited by the network, and the unit step response of the network.

ANSWER: The network is underdamped;

$$x(t) = \left[\frac{10}{8} + 1.46e^{-2t}\cos\left(2t - 210.96°\right)\right]u(t).$$

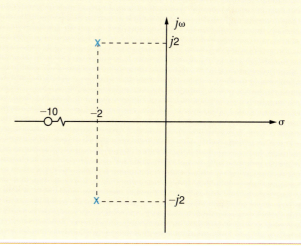

Recall from our previous discussion that if a second-order network is underdamped, the characteristic equation of the network is of the form

$$s^2 + 2\zeta\omega_0 s + \omega_0^2 = 0$$

and the roots of this equation, which are the network poles, are of the form

$$s_1, s_2 = -\zeta\omega_0 \pm j\omega_0\sqrt{1 - \zeta^2}$$

The roots s_1 and s_2, when plotted in the s-plane, generally appear as shown in Fig. 14.11, where

$$\zeta = \text{damping ratio}$$

$$\omega_0 = \text{undamped natural frequency}$$

and as shown in Fig. 14.11,

$$\zeta = \cos\theta$$

The damping ratio and the undamped natural frequency are exactly the same quantities as those employed in Chapter 12 when determining a network's frequency response. We find that it is these same quantities that govern the network's transient response.

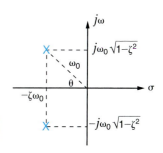

Figure 14.11
Pole locations for second-order underdamped network.

Example 14.9

Let us examine the effect of pole position in the *s*-plane on the transient response of the second-order *RLC* series network shown in Fig. 14.12.

Figure 14.12
RLC *series network.*

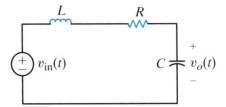

SOLUTION The voltage gain transfer function is

$$\mathbf{G}_v(s) = \frac{\dfrac{1}{LC}}{s^2 + s\left(\dfrac{R}{L}\right) + \dfrac{1}{LC}} = \frac{\omega_0^2}{s^2 + 2\zeta\omega_0 s + \omega_0^2}$$

For this analysis we will let $\omega_0 = 2000$ rad/s for $\zeta = 0.25, 0.50, 0.75$, and 1.0. From the preceding equation we see that

$$LC = \frac{1}{\omega_0^2} = 2.5 \times 10^{-7}$$

and

$$R = 2\zeta\sqrt{\frac{L}{C}}$$

If we arbitrarily let $L = 10$ mH, then $C = 25$ μF. Also, for $\zeta = 0.25, 0.50, 0.75$, and 1.0, $R = 10 \ \Omega, 20 \ \Omega, 30 \ \Omega$, and $40 \ \Omega$, respectively. Over the range of ζ values, the network ranges from underdamped to critically damped. Since poles are complex for underdamped systems, the real and imaginary components and the magnitude of the poles of $\mathbf{G}_v(s)$ are given in Table 14.1 for the ζ values listed previously.

Table 14.1 Pole locations for $\zeta = 0.25$ to 1.0

Damping Ratio	Real	Imaginary	Magnitude
1.00	2000.0	0.0	2000.0
0.75	1500.0	1322.9	2000.0
0.50	1000.0	1732.1	2000.0
0.25	500.0	1936.5	2000.0

Figure 14.13 shows the pole-zero diagrams for each value of ζ. Note first that all the poles lie on a circle; thus, the pole magnitudes are constant, consistent with Table 14.1. Second, as ζ decreases, the real part of the pole decreases while the imaginary part increases. In fact, when ζ goes to zero, the poles become imaginary.

A PSPICE simulation of a unit step transient excitation for all four values of R is shown in Fig. 14.14. We see that as ζ decreases, the overshoot in the output voltage increases. Furthermore, when the network is critically damped ($\zeta = 1$), there is no overshoot at all. In most applications excessive overshoot is not desired. To correct this, the damping ratio, ζ, should be increased, which for this circuit would require an increase in the resistor value.

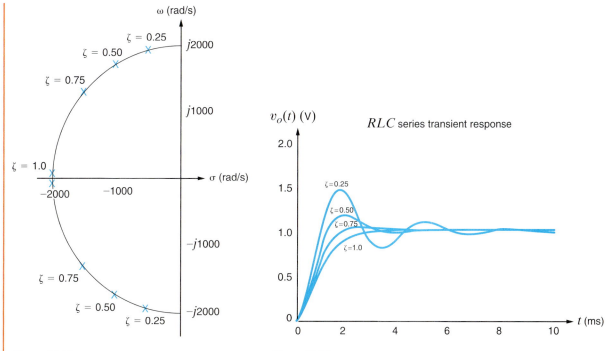

Figure 14.13
Pole-zero diagrams for ζ = 0.25 to 1.0.

Figure 14.14
PSPICE transient response output for ζ = 0.25 to 1.0.

Example 14.10

Let us revisit the Tacoma Narrows Bridge disaster examined in Example 12.12. A photograph of the bridge as it collapsed is shown in Fig. 14.15.

In Chapter 12 we assumed that the bridge's demise was brought on by winds oscillating back and forth at a frequency near that of the bridge (0.2 Hz). We found that we could create an RLC circuit, shown in Fig. 12.31, that resonates at 0.2 Hz and has an output voltage

Figure 14.15
Tacoma Narrows Bridge as it collapsed on November 7, 1940. (Used with permission from Special Collection Division, University of Washington Libraries. Photo by Farguharson, negative number 12.)

consistent with the vertical deflection of the bridge. This kind of forced resonance never happened at Tacoma Narrows. The real culprit was not so much wind fluctuations but the bridge itself. This is thoroughly explained in the paper "Resonance, Tacoma Narrows Bridge Failure, and Undergraduate Physics Textbooks," by K. Y. Billah and R. H. Scalan published in the *American Journal of Physics*, vol. 59, no. 2, pp. 118–124, in which the authors determined that changes in wind speed affected the coefficients of the second-order differential equation that models the resonant behavior. In particular, the damping ratio, ζ, was dependent on the wind speed and is roughly given as

$$\zeta = 0.00460 - 0.00013U \qquad \textbf{14.17}$$

where U is the wind speed in mph. Note, as shown in Fig. 14.16, that ζ becomes negative at wind speeds in excess of 35 mph—a point we will demonstrate later. Furthermore, Billah and Scalan report that the bridge resonated in a twisting mode, which can be easily seen in Fig. 12.30 and is described by the differential equation

$$\frac{d^2\theta(t)}{dt^2} + 2\zeta\omega_0\frac{d\theta(t)}{dt} + \omega_0^2\theta(t) = 0$$

or

$$\ddot{\theta} + 2\zeta\omega_0\dot{\theta} + \omega_0^2\theta = 0 \qquad \textbf{14.18}$$

where $\theta(t)$ is the angle of twist in degrees and wind speed is implicit in ζ through Eq. (14.17). Billah and Scalan list the following data obtained either by direct observation at the bridge site or through scale model experiments afterward.

Wind speed at failure $\approx$ 42 mph

Twist at failure $\approx \pm 12°$

Time to failure $\approx$ 45 minutes

We will start the twisting oscillations using an initial condition on $\theta(0)$ and see whether the bridge oscillations decrease or increase over time. Let us now design a network that will simulate the true Tacoma Narrows disaster.

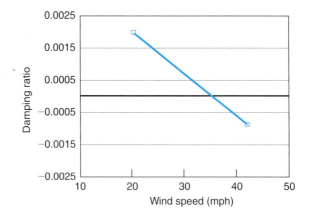

Figure 14.16 Damping ratio versus wind speed for the second-order twisting model of the Tacoma Narrows Bridge.

SOLUTION First, we solve for $\ddot{\theta}(t)$ in Eq. (14.18)

$$\ddot{\theta} = -2\zeta\omega_0\dot{\theta} - \omega_0^2\theta$$

$$\qquad \textbf{14.19}$$

$$\ddot{\theta} = -2(2\pi)(0.2)(0.0046 - 0.00013U)\dot{\theta} - \left[2(2\pi)(0.2)\right]^2\theta$$

or

$$\ddot{\theta} = -(0.01156 - 0.00033U)\dot{\theta} - 1.579\theta$$

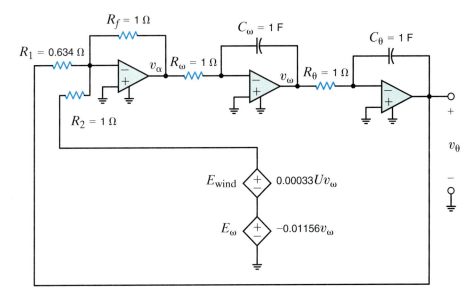

Figure 14.17
Circuit diagram for Tacoma Narrows Bridge simulations.

We now wish to model this equation to produce a voltage proportional to $\ddot{\theta}(t)$. We can accomplish this using the op-amp integrator circuit shown in Fig. 14.17.

The circuit's operation can perhaps be best understood by first assigning the voltage v_α to be proportional to $\ddot{\theta}(t)$, where 1 V represents 1 deg/s^2. Thus, the output of the first integrator, v_ω, must be

$$v_\omega = -\frac{1}{R_\omega C_\omega} \int v_\alpha \, dt$$

or, since $R_\omega = 1\ \Omega$ and $C_\omega = 1\ F$,

$$v_\omega = -\int v_\alpha \, dt$$

So v_ω is proportional to $-\dot{\theta}(t)$ and 1 V equals -1 deg/s. Similarly, the output of the second integrator must be

$$v_\theta = -\int v_\omega \, dt$$

where $v_\theta(t)$ is proportional to $\theta(t)$ and 1 V equals 1 degree. The outputs of the integrators are then fed back as inputs to the summing op-amp. Note that the dependent sources, E_ω and E_{wind}, re-create the coefficient on $\dot{\theta}(t)$ in Eq. (14.17); that is,

$$2\zeta\omega_0 = (2)(0.2)(2\pi)\zeta = 0.01156 - 0.00033U$$

To simulate various wind speeds, we need only change the gain factor of E_{wind}. Finally, we can solve the circuit for $v_\alpha(t)$.

$$v_\alpha(t) = -\left(\frac{R_f}{R_2}\right)(E_\omega - E_{\text{wind}}) - \left(\frac{R_f}{R_1}\right)v_\theta$$

which matches Eq. (14.19) if

$$\frac{R_f}{R_1} = \omega_0^2 = \left[2\pi(0.2)\right]^2 = 1.579$$

and

$$\frac{R_f}{R_2}\left[E_\omega - E_{\text{wind}}\right] = 2\zeta\omega_0$$

or

$$\frac{R_f}{R_2} = 1$$

Thus, if $R_f = R_2 = 1\ \Omega$ and $R_1 = 0.634\ \Omega$, the circuit will simulate the bridge's twisting motion. We will start the twisting oscillations using an initial condition $\theta(0)$ and see whether the bridge oscillations decrease or increase over time.

The first simulation is for a wind speed of 20 mph and one degree of twist. The corresponding output voltage is shown in Fig. 14.18. The bridge twists at a frequency of 0.2 Hz and the oscillations decrease exponentially, indicating a nondestructive situation.

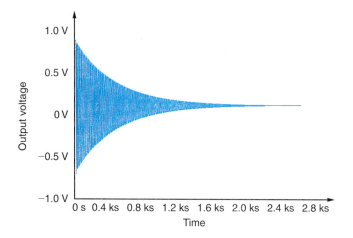

Figure 14.18 *Tacoma Narrows Bridge simulation at 20 mph wind speed and one degree twist initial condition.*

Fig. 14.19 shows the output for 35-mph winds and an initial twist of one degree. Notice that the oscillations neither increase nor decrease. This indicates that the damping ratio is zero.

Finally, the simulation at a wind speed of 42 mph and one degree initial twist is shown in Fig. 14.20. The twisting becomes worse and worse until after 45 minutes, the bridge is twisting ±12.5 degrees, which matches values reported by Billah and Scalan for collapse.

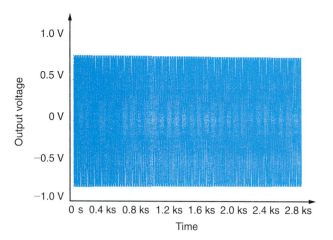

Figure 14.19 *Tacoma Narrows Bridge simulation at 35-mph winds and one degree of initial twist.*

The dependency of the damping ratio on wind speed can also be demonstrated by investigating how the system poles change with the wind. The characteristic equation for the system is

$$s^2 + 2\zeta\omega_0 s + \omega_0^2 = 0$$

or

$$s^2 + (0.01156 - 0.00033U)s + 1.579 = 0$$

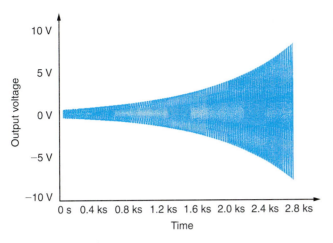

Figure 14.20
*Tacoma Narrows Bridge
simulation at 42 mph wind
speed and one degree of
initial twist.*

The roots of the characteristic equation yield the pole locations. Figure 14.21 shows the system poles at wind speeds of 20, 35, and 42 mph. Note that at 20 mph, the stable situation is shown in Fig. 14.18, and the poles are in the left-half of the *s*-plane. At 35 mph ($\zeta = 0$) the poles are on the $j\omega$ axis and the system is oscillatory, as shown in Fig. 14.19. Finally, at 42 mph, we see that the poles are in the right half of the *s*-plane and, from Fig. 14.20 we know this is an unstable system. This relationship between pole location and transient response is true for all systems—right-half plane poles result in unstable systems.

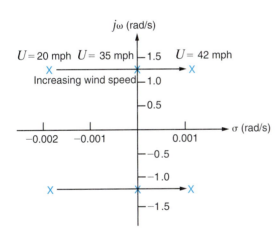

Figure 14.21
*Polo-zero plot for Tacoma
Narrows Bridge second-order
model at wind speeds of 20, 35,
and 42 mph.*

14.5 Pole-Zero Plot/Bode Plot Connection

In Chapter 12 we introduced the Bode plot as an analysis tool for sinusoidal frequency-response studies. Let us now investigate the relationship between the *s*-plane pole-zero plot and the Bode plot. As an example, consider the transfer function of the *RLC* high-pass filter shown in Fig. 14.22.

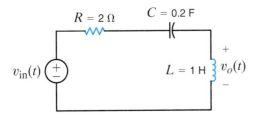

Figure 14.22
RLC high-pass filter.

The transfer function is

$$\mathbf{G}_v(s) = \frac{sL}{sL + R + \dfrac{1}{sC}} = \frac{s^2}{s^2 + s\left(\dfrac{R}{L}\right) + \dfrac{1}{LC}}$$

Using the element values, we find that the transfer function becomes

$$\mathbf{G}_v(s) = \frac{s^2}{s^2 + 2s + 5} = \frac{s^2}{(s + 1 + j2)(s + 1 - j2)}$$

We see that the transfer function has two zeros at the origin ($s = 0$) and two complex-conjugate poles $s = -1 \pm j2$. The standard pole-zero plot for this function is shown in Fig. 14.23a. A three-dimensional s-plane plot of the magnitude of $\mathbf{G}_v(s)$ is shown in Fig. 14.23b. Note carefully that when $s = 0$, $\mathbf{G}_v(s) = 0$ and when $s = -1 \pm j2$, the function is infinite.

Recall that the Bode plot of a transfer function's magnitude is in reality a plot of the magnitude of the gain versus frequency. The frequency domain, where $s = j\omega$, corresponds to the $j\omega$-axis in the s-plane obtained by setting σ, the real part of s, to zero. Thus, the frequency domain corresponds directly to that part of the s-domain where $\sigma = 0$, as illustrated in the three-dimensional plot in Fig. 14.23c.

Let us develop the Bode plot by first rotating Fig. 14.23c such that the real axis is perpendicular to the page as shown in Fig. 14.23d. Note that the transfer function maximum occurs at $\omega = \sqrt{5} = 2.24$ rad/s, which is the magnitude of the complex pole frequencies.

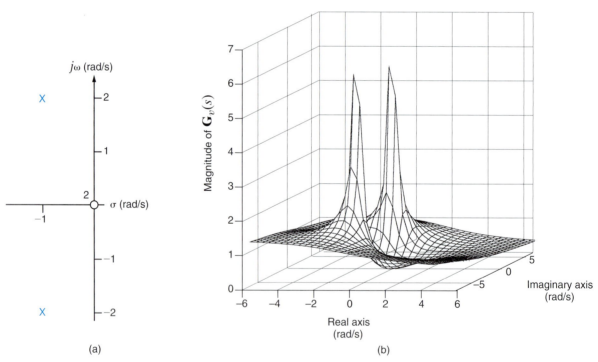

(a)

(b)

Figure 14.23 Figures used to demonstrate pole-zero plot/Bode plot connection.

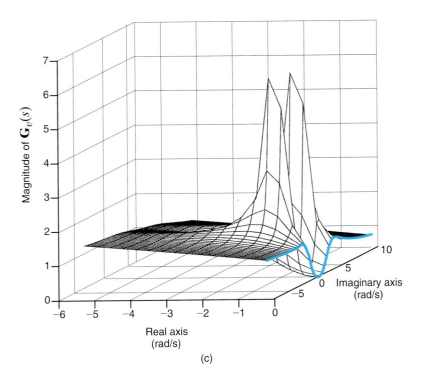

(c)

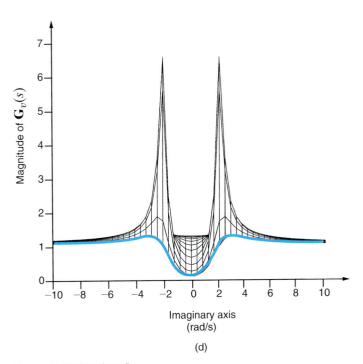

(d)

Figure 14.23 (continued)

Figure 14.23 (continued)

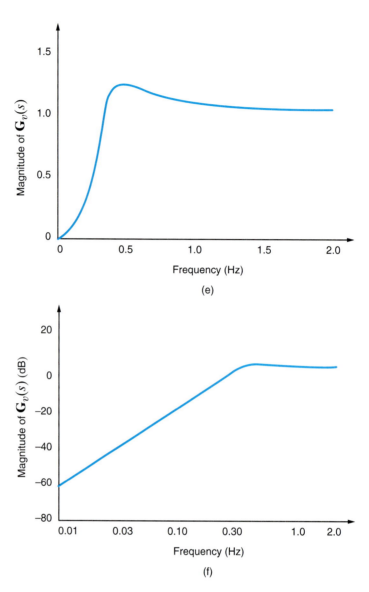

(e)

(f)

In addition, the symmetry of the pole around the real axis becomes readily apparent. As a result of this symmetry, we can restrict our analysis to positive values of $j\omega$, with no loss of information. This plot for $\omega \geq 0$ is shown in Fig. 14.23e where frequency is plotted in Hz rather than rad/s. Finally, converting the transfer function magnitude to dB and using a log axis for frequency, we produce the Bode plot in Fig. 14.23f.

14.6 Steady-State Response

In Section 14.3 we have demonstrated, using a variety of examples, the power of the Laplace transform technique in determining the complete response of a network. This complete response is composed of transient terms, which disappear as $t \rightarrow \infty$, and steady-state terms, which are present at all times. Let us now examine a method by which to determine the steady-state response of a network directly. Recall from previous examples that the network response can be written as

$$\mathbf{Y}(s) = \mathbf{H}(s)\mathbf{X}(s) \qquad \textbf{14.20}$$

where $\mathbf{Y}(s)$ is the output or response, $\mathbf{X}(s)$ is the input or forcing function, and $\mathbf{H}(s)$ is the network function or transfer function defined in Section 12.1. The transient portion of the response $\mathbf{Y}(s)$ results from the poles of $\mathbf{H}(s)$, and the steady-state portion of the response results from the poles of the input or forcing function.

As a direct parallel to the sinusoidal response of a network as outlined in Section 8.2, we assume that the forcing function is of the form

$$x(t) = X_M e^{j\omega_0 t} \qquad\qquad \textbf{14.21}$$

which by Euler's identity can be written as

$$x(t) = X_M \cos\omega_0 t + jX_M \sin\omega_0 t \qquad\qquad \textbf{14.22}$$

The Laplace transform of Eq. (14.21) is

$$\mathbf{X}(s) = \frac{X_M}{s - j\omega_0} \qquad\qquad \textbf{14.23}$$

and therefore,

$$\mathbf{Y}(s) = \mathbf{H}(s)\left(\frac{X_M}{s - j\omega_0}\right) \qquad\qquad \textbf{14.24}$$

At this point we tacitly assume that $\mathbf{H}(s)$ does not have any poles of the form $(s - j\omega_k)$. If, however, this is the case, we simply encounter difficulty in defining the steady-state response.

Performing a partial fraction expansion of Eq. (14.24) yields

$$\mathbf{Y}(s) = \frac{X_M \mathbf{H}(j\omega_0)}{s - j\omega_0} + \text{terms that occur due to the poles of } \mathbf{H}(s) \qquad\qquad \textbf{14.25}$$

The first term to the right of the equal sign can be expressed as

$$\mathbf{Y}(s) = \frac{X_M |\mathbf{H}(j\omega_0)| e^{j\phi(j\omega_0)}}{s - j\omega_0} + \cdots \qquad\qquad \textbf{14.26}$$

since $\mathbf{H}(j\omega_0)$ is a complex quantity with a magnitude and phase that are a function of $j\omega_0$.

Performing the inverse transform of Eq. (14.26), we obtain

$$y(t) = X_M |\mathbf{H}(j\omega_0)| e^{j\omega_0 t} e^{j\phi(j\omega_0)} + \cdots \qquad\qquad \textbf{14.27}$$

$$= X_M |\mathbf{H}(j\omega_0)| e^{(j\omega_0 t + \phi(j\omega_0))} + \cdots$$

and hence the steady-state response is

$$y_{ss}(t) = X_M |\mathbf{H}(j\omega_0)| e^{j(\omega_0 t + \phi(j\omega_0))} \qquad\qquad \textbf{14.28}$$

HINT
The transient terms disappear in steady state.

Since the actual forcing function is $X_M \cos\omega_0(t)$, which is the real part of $X_M e^{j\omega_0 t}$, the steady-state response is the real part of Eq. (14.28).

$$y_{ss}(t) = X_M |\mathbf{H}(j\omega_0)| \cos[\omega_0 t + \phi(j\omega_0)] \qquad\qquad \textbf{14.29}$$

In general, the forcing function may have a phase angle θ. In this case, θ is simply added to $\phi(j\omega_0)$ so that the resultant phase of the response is $\phi(j\omega_0) + \theta$.

Example 14.11

For the circuit shown in Fig. 14.24a, we wish to determine the steady-state voltage $v_{oss}(t)$ for $t > 0$ if the initial conditions are zero.

SOLUTION   As illustrated earlier, this problem could be solved using a variety of techniques, such as node equations, mesh equations, source transformation, and Thévenin's theorem. We will employ node equations to obtain the solution. The transformed network using the impedance values for the parameters is shown in Fig. 14.24b. The node equations for this network are

$$\left(\frac{1}{2} + \frac{1}{s} + \frac{s}{2}\right)\mathbf{V}_1(s) - \left(\frac{s}{2}\right)\mathbf{V}_o(s) = \frac{1}{2}\mathbf{V}_i(s)$$

$$-\left(\frac{s}{2}\right)\mathbf{V}_1(s) + \left(\frac{s}{2} + 1\right)\mathbf{V}_o(s) = 0$$

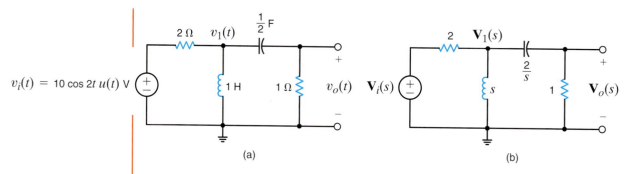

Figure 14.24 Circuits used in Example 14.11.

Solving these equations for $\mathbf{V}_o(s)$, we obtain

$$\mathbf{V}_o(s) = \frac{s^2}{3s^2 + 4s + 4} \mathbf{V}_i(s)$$

Note that this equation is in the form of Eq. (14.20), where $\mathbf{H}(s)$ is

$$\mathbf{H}(s) = \frac{s^2}{3s^2 + 4s + 4}$$

Since the forcing function is $10 \cos 2t \, u(t)$, then $V_M = 10$ and $\omega_0 = 2$. Hence,

$$\mathbf{H}(j2) = \frac{(j2)^2}{3(j2)^2 + 4(j2) + 4}$$

$$= 0.354 \underline{/45°}$$

Therefore,

$$\left|\mathbf{H}(j2)\right| = 0.354$$

$$\phi(j2) = 45°$$

and, hence, the steady-state response is

$$v_{oss}(t) = V_M \left|\mathbf{H}(j2)\right| \cos\left[2t + \phi(j2)\right]$$

$$= 3.54 \cos(2t + 45°) \text{ V}$$

The complete (transient plus steady-state) response can be obtained from the expression

$$\mathbf{V}_o(s) = \frac{s^2}{3s^2 + 4s + 4} \mathbf{V}_i(s)$$

$$= \frac{s^2}{3s^2 + 4s + 4} \left(\frac{10s}{s^2 + 4}\right)$$

$$= \frac{10s^3}{(s^2 + 4)(3s^2 + 4s + 4)}$$

Determining the inverse Laplace transform of this function using the techniques of Chapter 13, we obtain

$$v_o(t) = 3.54 \cos(2t + 45°) + 1.44 e^{-(2/3)t} \cos\left(\frac{2\sqrt{2}}{3} t - 55°\right) \text{ V}$$

Note that as $t \to \infty$ the second term approaches zero, and thus the steady-state response is

$$v_{oss}(t) = 3.54 \cos(2t + 45°) \text{ V}$$

which can easily be checked using a phasor analysis.

E14.7 Determine the steady-state voltage $v_{oss}(t)$ in the network in Fig. E14.7 for $t > 0$ if the initial conditions in the network are zero.   **PSV**

ANSWER:
$v_{oss}(t) = 3.95 \cos(2t - 99.46°)$ V.

Figure E14.7

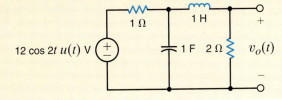

12 cos 2t u(t) V

1 Ω   1 H   1 F   2 Ω $v_o(t)$

14.6   Application Example

Application Example 14.12

The Recording Industry Association of America (RIAA) uses standardized recording and playback filters to improve the quality of phonographic disk recordings. This process is demonstrated in Fig. 14.25. During a recording session, the voice or music signal is passed through the recording filter, which deemphasizes the bass content. This filtered signal is then recorded into the vinyl. On playback, the phonograph needle assembly senses the recorded message and reproduces the filtered signal, which proceeds to the playback filter. The purpose of the playback filter is to emphasize the bass content and reconstruct the original voice/music signal. Next, the reconstructed signal can be amplified and sent on to the speakers.

Let us examine the pole-zero diagrams for the record and playback filters.

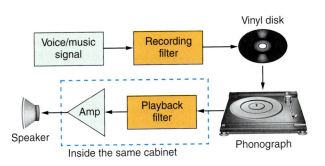

Figure 14.25
Block diagram for phonograph disk recording and playback.

SOLUTION   The transfer function for the recording filter is

$$\mathbf{G}_{vR}(s) = \frac{K(1 + s\tau_{z1})(1 + s\tau_{z2})}{1 + s\tau_p}$$

where the time constants are $\tau_{z1} = 75$ μs, $\tau_{z2} = 3180$ μs; and $\tau_p = 318$ μs; K is a constant chosen such that $\mathbf{G}_{vR}(s)$ has a magnitude of 1 at 1000 Hz. The resulting pole and zero frequencies in radians/second are

$$\omega_{z1} = 1/\tau_{z1} = 13.33 \text{ krad/s}$$
$$\omega_{z2} = 1/\tau_{z2} = 313.46 \text{ rad/s}$$
$$\omega_p = 1/\tau_p = 3.14 \text{ krad/s}$$

Fig. 14.26a shows the pole-zero diagram for the recording filter.

The playback filter transfer function is the reciprocal of the record transfer function.

$$\mathbf{G}_{vp}(s) = \frac{1}{\mathbf{G}_{vR}(s)} = \frac{A_o(1 + s\tau_z)}{(1 + s\tau_{p1})(1 + s\tau_{p2})}$$

where the time constants are now $\tau_{p1} = 75\ \mu s$, $\tau_{p2} = 3180\ \mu s$, $\tau_z = 318\ \mu s$, and A_o is $1/K$. Pole and zero frequencies, in radians/second, are

$$\omega_{p1} = 1/\tau_{z1} = 13.33\ \text{krad/s}$$

$$\omega_{p2} = 1/\tau_{z2} = 313.46\ \text{rad/s}$$

$$\omega_z = 1/\tau_p = 3.14\ \text{krad/s}$$

which yields the pole-zero diagram in Fig. 14.26b. The voice/music signal eventually passes through both filters before proceeding to the amplifier. In the s-domain, this is equivalent to multiplying $\mathbf{V}_s(s)$ by both $\mathbf{G}_{vR}(s)$ and $\mathbf{G}_{vp}(s)$. In the pole-zero diagram, we simply superimpose the pole-zero diagrams of the two filters, as shown in Fig. 14.26c. Note that at each pole frequency there is a zero and vice versa. The pole-zero pairs cancel one another, yielding a pole-zero diagram that contains no poles and no zeros. This effect can be seen mathematically by multiplying the two transfer functions, $\mathbf{G}_{vR}(s)\mathbf{G}_{vp}(s)$, which yields a product independent of s. Thus, the original voice/music signal is reconstructed and fidelity is preserved.

Figure 14.26
Pole-zero diagrams for RIAA phonographic filters.

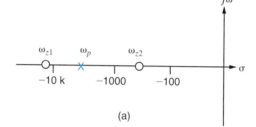

(a)

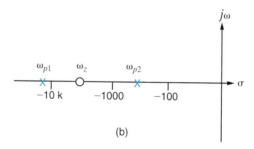

(b)

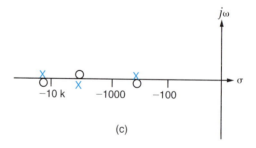

(c)

14.7 Design Examples

Design Example 14.13

In a large computer network, two computers are transferring digital data on a single wire at a rate of 1000 bits/s. The voltage waveform, v_{data}, in Fig. 14.27 shows a possible sequence of bits alternating between "high" and "low" values. Also present in the environment is a source of 100 kHz (628 krad/s) noise, which is corrupting the data.

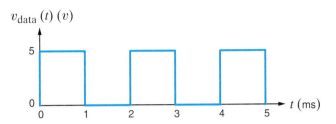

Figure 14.27
1000 bit/s digital data
waveform.

It is necessary to filter out the high-frequency noise without destroying the data waveform. Let us place the second-order low-pass active filter of Fig. 14.28 in the data path so that the data and noise signals will pass through it.

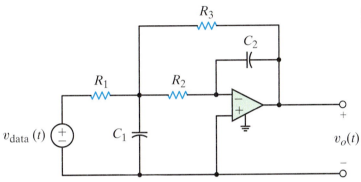

Figure 14.28
Second-order low-pass filter.

SOLUTION   The filter's transfer function is found to be

$$\mathbf{G}_v(s) = \frac{\mathbf{V}_o(s)}{\mathbf{V}_{data}(s)} = \frac{-\left(\dfrac{R_3}{R_1}\right)\left(\dfrac{1}{R_2 R_3 C_1 C_2}\right)}{s^2 + s\left(\dfrac{1}{R_1 C_1} + \dfrac{1}{R_2 C_1} + \dfrac{1}{R_3 C_1}\right) + \dfrac{1}{R_2 R_3 C_1 C_2}}$$

To simplify our work, let $R_1 = R_2 = R_3 = R$. From our work in Chapter 12, we know that the characteristic equation of a second-order system can be expressed

$$s^2 + 2s\zeta\omega_0 + \omega_0^2 = 0$$

Comparing the two preceding equations, we find that

$$\omega_0 = \frac{1}{R\sqrt{C_1 C_2}}$$

$$2\zeta\omega_0 = \frac{3}{RC_1}$$

and therefore,

$$\zeta = \frac{3}{2}\sqrt{\frac{C_2}{C_1}}$$

The poles of the filter are at

$$s_1, s_2 = -\zeta\omega_0 \pm \omega_0\sqrt{\zeta^2 - 1}$$

To eliminate the 100-kHz noise, at least one pole should be well below 100 kHz, as shown in the Bode plot sketched in Fig. 14.29. By placing a pole well below 100 kHz, the gain of the filter will be quite small at 100 kHz, effectively filtering the noise.

If we arbitrarily choose an overdamped system with $\omega_0 = 25$ krad/s and $\zeta = 2$, the resulting filter is overdamped with poles at $s_1 = -6.7$ krad/s and $s_2 = -93.3$ krad/s. The pole-zero diagram for the filter is shown in Fig. 14.30.

Figure 14.29
Bode plot sketch for
a second-order
low-pass filter.

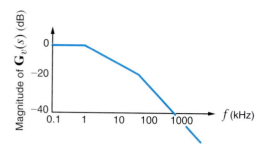

Figure 14.30
Pole-zero diagram
for low-pass filter.

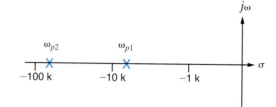

If we let $R = 40 \text{ k}\Omega$, then we may write

$$\omega_0 = 25{,}000 = \frac{1}{40{,}000\sqrt{C_1 C_2}}$$

or

$$C_1 C_2 = 10^{-18}$$

Also,

$$\zeta = 2 = \frac{3}{2}\sqrt{\frac{C_2}{C_1}}$$

which can be expressed as

$$\frac{C_2}{C_1} = \frac{16}{9}$$

Solving for C_1 and C_2 yields

$$C_1 = 0.75 \text{ nF}$$
$$C_2 = 1.33 \text{ nF}$$

The circuit used to simulate the filter is shown in Fig. 14.31. The sinusoidal source has a
frequency of 100 kHz and is used to represent the noise source.

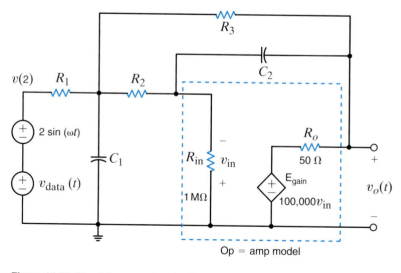

Figure 14.31 Circuit for second-order filter.

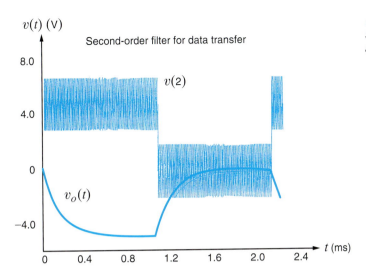

Figure 14.32
*Simulation outputs for node 2
and $v_o(t)$.*

Plots for the input to the filter and the output voltage for 2 ms are shown in Fig. 14.32. Note that output indeed contains much less of the 100-kHz noise. Also, the fast rise and fall times of the data signal are slower in the output voltage. Despite this slower response, the output voltage is fast enough to keep pace with the 1000-bits/s transfer rate.

Let us now increase the data transfer rate from 1000 to 25,000 bits/s, as shown in Fig. 14.33. The total input and output signals are plotted in Fig. 14.34 for 200 μs. Now the output cannot keep pace with the input, and the data information is lost. Let us investigate why this occurs. We know that the filter is second order with poles at s_1 and s_2. If we represent the data input as a 5-V step function, the output voltage is

$$\mathbf{V}_o(s) = \mathbf{G}_v(s)\left(\frac{5}{s}\right) = \frac{K}{(s + s_1)(s + s_2)}\left(\frac{5}{s}\right)$$

where K is a constant. Since the filter is overdamped, s_1 and s_2 are real and positive. A partial fraction expansion of $\mathbf{V}_o(s)$ is of the form

$$\mathbf{V}_o(s) = \frac{K_1}{s} + \frac{K_2}{(s + s_1)} + \frac{K_3}{(s + s_2)}$$

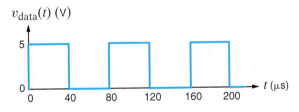

Figure 14.33
*25,000-bit/s digital data
waveform.*

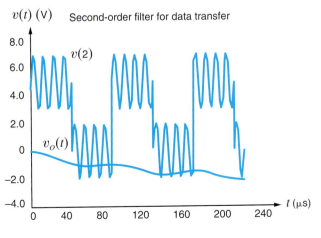

Figure 14.34
*Simulation output for node 2
and $v_o(t)$ with 25,000-bit/s data
transfer rate.*

yielding the time domain expression

$$v_o(t) = \left[K_1 + K_2 e^{-s_1 t} + K_3 e^{-s_2 t}\right]u(t) \text{ V}$$

where K_1, K_2, and K_3 are real constants. The exponential time constants are the reciprocals of the pole frequencies.

$$\tau_1 = \frac{1}{s_1} = \frac{1}{6.7k} = 149 \text{ μs}$$

$$\tau_2 = \frac{1}{s_2} = \frac{1}{93.3k} = 10.7 \text{ μs}$$

Since exponentials reach steady state in roughly 5τ, the exponential associated with τ_2 affects the output for about 50 μs and the τ_1 exponential will reach steady state after about 750 μs. From Fig. 14.33 we see that at a 25,000-bits/s data transfer rate, each bit (a "high" or "low" voltage value) occupies a 40-μs time span. Therefore, the exponential associated with s_1, and thus $v_o(t)$, is still far from its steady-state condition when the next bit is transmitted. In short, s_1 is too small.

Let us remedy this situation by increasing the pole frequencies and changing to a critically damped system, $\zeta = 1$. If we select $\omega_0 = 125$ krad/s, the poles will be at $s_1 = s_2 = -125$ krad/s or 19.9 kHz—both below the 100-kHz noise we wish to filter out. Figure 14.35 shows the new pole positions moved to the left of their earlier positions, which we expect will result in a quicker response to the v_data pulse train.

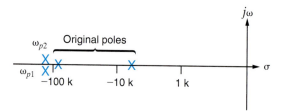

Figure 14.35 Pole-zero diagram for both original and critically damped systems.

Now the expressions for ω_0 and ζ are

$$\omega_0 = 125,000 = \frac{1}{40,000\sqrt{C_1 C_2}}$$

or

$$C_1 C_2 = 4 \times 10^{-20}$$

Also,

$$\zeta = 1 = \frac{3}{2}\sqrt{\frac{C_2}{C_1}}$$

which can be expressed

$$\frac{C_2}{C_1} = \frac{4}{9}$$

Solving for C_1 and C_2 yields

$$C_1 = 300 \text{ pF}$$
$$C_2 = 133.3 \text{ pF}$$

A simulation using these new capacitor values produces the input-output data shown in Fig. 14.36. Now the output voltage just reaches the "high" and "low" levels just before v_data makes its next transition and the 100-kHz noise is still much reduced.

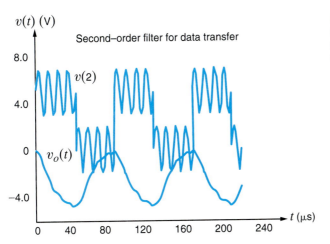

Figure 14.36
Simulation outputs for node 2 and $v_o(t)$ for the critically damped system.

Design Example 14.14

The circuit in Fig. 14.37 is an existing low-pass filter. On installation, we find that its output exhibits too much oscillation when responding to pulses. We wish to alter the filter in order to make it critically damped.

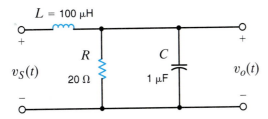

Figure 14.37
A second-order low-pass filter.

SOLUTION   First, we must determine the existing transfer function, $\mathbf{H}(s)$.

$$\mathbf{H}(s) = \frac{\mathbf{V}_O}{\mathbf{V}_S} = \frac{\dfrac{R}{1 + sRC}}{\dfrac{R}{1 + sRC} + sL} = \frac{\dfrac{1}{LC}}{s^2 + \dfrac{s}{RC} + \dfrac{1}{LC}} \qquad \textbf{14.30}$$

where the term $\dfrac{R}{1 + sRC}$ is just the parallel combination of the resistor and capacitor. Given our component values, the transfer function is

$$\mathbf{H}(s) = \frac{10^{10}}{s^2 + (5 \times 10^4)s + 10^{10}} \qquad \textbf{14.31}$$

and the resonant frequency and damping ratio are

$$\omega_0 = \frac{1}{\sqrt{LC}} = 10^5 \text{ rad/s} \quad \text{and} \quad 2\zeta\omega_0 = \frac{1}{RC} \Rightarrow \zeta = \frac{5 \times 10^4}{2\omega_0} = \frac{5 \times 10^4}{2 \times 10^5} = 0.25 \qquad \textbf{14.32}$$

The network is indeed underdamped. From Eq. (14.32), we find that raising the damping ratio by a factor of 4 to 1.0 requires that R be lowered by the same factor of 4 to 5 Ω. This can be done by adding a resistor, R_X, in parallel with R as shown in Fig. 14.38. The required resistor value can be obtained by solving Eq. (14.33) for R_X.

$$R_{eq} = 5 = \frac{RR_X}{R + R_X} = \frac{20R_X}{20 + R_X} \qquad \textbf{14.33}$$

The solution is $R_X = 6.67 \ \Omega$.

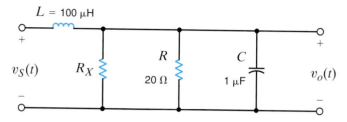

Figure 14.38 The addition of a resistor to change the damping ratio of the network.

Design Example 14.15

The circuit in Fig. 14.39 is called a Wein bridge oscillator. Its output voltage is a sine wave whose frequency can be tuned. Let us design this circuit for an oscillation frequency of 10 kHz.

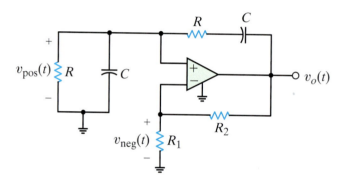

Figure 14.39 The classic Wein bridge oscillator.

SOLUTION This network looks odd for two reasons. First, there is no input signal! Second, we have not seen an op-amp circuit in which the output is connected back to the noninverting input terminal. However, we do know that if the op-amp is working properly, its input currents are zero and the difference in voltage between the two input terminals is zero. We will employ these constraints to write two transfer functions from the op-amp *output* back to each of the op-amp *inputs*. The first is defined as

$$\mathbf{H}_{\text{neg}}(s) = \frac{\mathbf{V}_{\text{neg}}}{\mathbf{V}_O} = \frac{R_1}{R_1 + R_2} \qquad \textbf{14.34}$$

and the second is

$$\mathbf{H}_{\text{pos}}(s) = \frac{\mathbf{V}_{\text{pos}}}{\mathbf{V}_O} = \frac{\mathbf{Z}_1}{\mathbf{Z}_1 + \mathbf{Z}_2} \qquad \textbf{14.35}$$

where $\mathbf{Z}_1$ is the parallel R-C network and $\mathbf{Z}_2$ is the series R-C network. Thus,

$$\mathbf{Z}_1 = \frac{R/sC}{(1/sC) + R} = \frac{R}{1 + sRC} \quad \text{and} \quad \mathbf{Z}_2 = R + \frac{1}{sC} = \frac{1 + sRC}{sC} \qquad \textbf{14.36}$$

Substituting Eq. (14.36) into (14.35) yields

$$\mathbf{H}_{\text{pos}}(s) = \frac{\mathbf{V}_{\text{pos}}}{\mathbf{V}_O} = \frac{\dfrac{R}{1 + sRC}}{\dfrac{R}{1 + sRC} + \dfrac{1 + sRC}{sC}} = \frac{sRC}{s^2(RC)^2 + 3sRC + 1} \qquad \textbf{14.37}$$

Since the voltage across the op-amp inputs is zero, $\mathbf{V}_{\text{neg}} = \mathbf{V}_{\text{pos}}$ and, thus, $\mathbf{H}_{\text{pos}}(s) = \mathbf{H}_{\text{neg}}(s)$!

Note that $\mathbf{H}_{neg}(s)$ in Eq. (14.34) is just a resistor ratio and is therefore real. The op-amp forces the same to be true for $\mathbf{H}_{pos}(s)$ at the frequency of oscillation! Now look at Eq. (14.37). Its numerator is purely imaginary. If $\mathbf{H}_{pos}(s)$ is to be real, then its denominator must also be purely imaginary. The result is

$$(j\omega)^2(RC)^2 + 1 = 0 \Rightarrow \omega = \frac{1}{RC} \Rightarrow f = \frac{1}{2\pi RC}$$

We arbitrarily select $C = 10$ nF and find

$$R = \frac{1}{2\pi Cf} = \frac{1}{2\pi(10^{-8})(10^4)} = 1.59 \text{ k}\Omega$$

We still must determine values for R_1 and R_2. Examine once again the fact that $\mathbf{H}_{pos}(s) = \mathbf{H}_{neg}(s)$. At 10 kHz, $\mathbf{H}_{pos}(s)$ becomes

$$\mathbf{H}_{pos}(s) = \frac{sRC}{s^2(RC)^2 + 3sRC + 1} = \frac{sRC}{3sRC} = \frac{1}{3}$$

The same must be true for $\mathbf{H}_{neg}(s)$.

$$\mathbf{H}_{neg}(s) = \frac{R_1}{R_1 + R_2} = \frac{1}{3}$$

The only possible solution is $R_2 = 2R_1$. Arbitrarily selecting $R_1 = R = 1.59$ kΩ, we find $R_2 = 3.18$ kΩ.

What happens if $\mathbf{H}_{pos}(s)$ does not equal $\mathbf{H}_{neg}(s)$ in the constructed circuit? If $\mathbf{H}_{neg}(s)$ is larger, the oscillations will die out. But if $\mathbf{H}_{pos}(s)$ is larger, the oscillations grow until the op-amp output reaches the power supply limits. At that point, the output is more of a square wave than a sinusoid. Since it is physically impossible to ensure that $\mathbf{H}_{pos}(s)$ and $\mathbf{H}_{neg}(s)$ are exactly the same at 10 kHz, engineers usually replace R_2 with a nonlinear resistor whose resistance decreases with increasing temperature. In this way, if the output oscillations begin to grow, more power is dissipated in the nonlinear resistance, decreasing its value. This decrease in resistance will increase $\mathbf{H}_{neg}(s)$ and bring the oscillator back to a balanced operating point.

SUMMARY

- The use of s-domain models for circuit elements permits us to describe them with algebraic, rather than differential, equations.

- All the dc analysis techniques, including the network theorems, are applicable in the s-domain. Once the s-domain solution is obtained, the inverse transform is used to obtain a time domain solution.

- The roots of the network's characteristic equation (i.e., the poles) determine the type of network response. A plot of these roots in the left half of the s-plane provides an immediate indication of the network's behavior. The relationship between the pole-zero plot and the Bode plot provides further insight.

- The transfer (network) function for a network is expressed as

$$\mathbf{H}(s) = \frac{\mathbf{Y}(s)}{\mathbf{X}(s)}$$

where $\mathbf{Y}(s)$ is the network response and $\mathbf{X}(s)$ is the input forcing function. If the transfer function is known, the output response is simply given by the product $\mathbf{H}(s)\mathbf{X}(s)$. If the input is an impulse function so that $\mathbf{X}(s) = 1$, the impulse response is equal to the inverse Laplace transform of the network function.

- The dc properties of the storage elements, L and C, can be used to obtain initial and final conditions. The initial conditions are required as a part of the s-domain model, and final conditions are often useful in verifying a solution.

- The Laplace transform solution for the network response is composed of transient terms, which disappear as $t \rightarrow \infty$, and steady-state terms, which are present at all times.

- The network response can be expressed as

$$\mathbf{Y}(s) = \mathbf{H}(s)\mathbf{X}(s)$$

- The transient portion of the response $\mathbf{Y}(s)$ results from the poles of $\mathbf{H}(s)$, and the steady-state portion of the response results from the poles of the forcing function $\mathbf{X}(s)$.

PROBLEMS

PSV **CS** both available on the web at: http://www.justask4u.com/irwin

SECTION 14.3

14.1 Find the input impedance $\mathbf{Z}(s)$ of the network in Fig. P14.1. **CS**

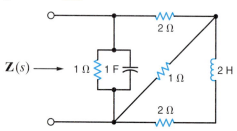

Figure P14.1

14.2 Given the network in Fig. P14.2, determine the value of the output voltage as $t \to \infty$.

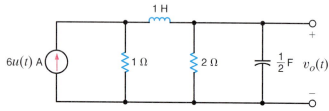

Figure P14.2

14.3 For the network shown in Fig. P14.3, determine the value of the output voltage as $t \to \infty$.

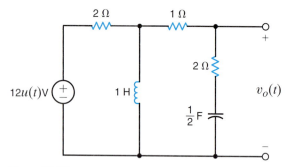

Figure P14.3

14.4 Use Laplace transforms to find $v(t)$ for $t > 0$ in the network shown in Fig. P14.4. Assume zero initial conditions.

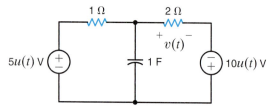

Figure P14.4

14.5 Use Laplace transforms and nodal analysis to find $i_1(t)$ for $t > 0$ in the network shown in Fig. P14.5. Assume zero initial conditions.

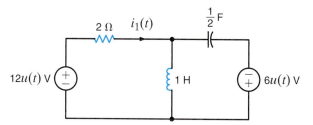

Figure P14.5

14.6 For the network shown in Fig. P14.6, find $v_o(t)$, $t > 0$, using node equations. **PSV**

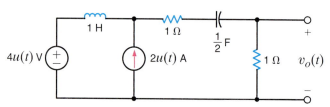

Figure P14.6

14.7 For the network shown in Fig. P14.7, find $i_o(t)$, $t > 0$. **CS**

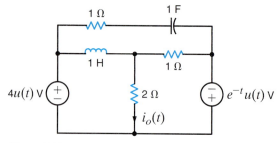

Figure P14.7

14.8 Find $v_o(t)$, $t > 0$, in the network in Fig. P14.8 using node equations.

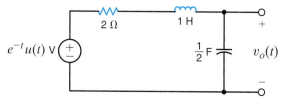

Figure P14.8

14.9 Find $v_o(t)$, $t > 0$, in the network shown in Fig. P14.9 using nodal analysis. **CS**

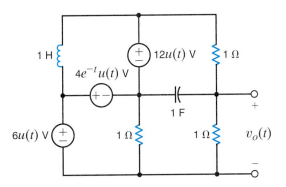

Figure P14.9

14.10 Use nodal analysis to find $v_o(t)$, $t > 0$, in the network in Fig. P14.10. **PSV**

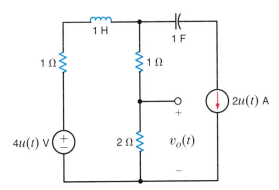

Figure P14.10

14.11 For the network shown in Fig. P14.11, find $v_o(t)$, $t > 0$, using loop equations.

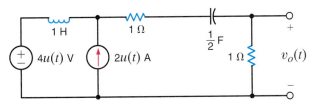

Figure P14.11

14.12 For the network shown in Fig. P14.12, find $v_o(t)$, $t > 0$, using mesh equations.

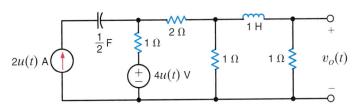

Figure P14.12

14.13 Use mesh equations to find $v_o(t)$, $t > 0$, in the network in Fig. P14.13. **CS**

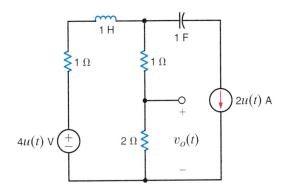

Figure P14.13

14.14 Use loop equations to find $i_o(t)$, $t > 0$, in the network shown in Fig. P14.14.

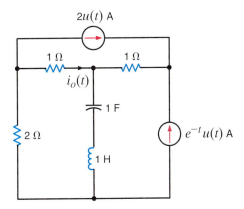

Figure P14.14

14.15 Use loop analysis to find $v_o(t)$ for $t > 0$ in the network in Fig. P14.15.

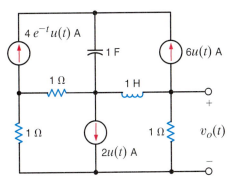

Figure P14.15

14.16 Use mesh analysis to find $v_o(t)$, $t > 0$, in the network in Fig. P14.16. **CS**

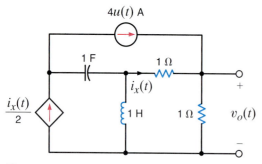

Figure P14.16

14.17 Use superposition to find $v_o(t)$, $t > 0$, in the network shown in Fig. P14.17.

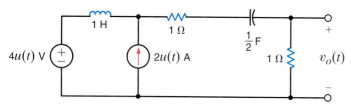

Figure P14.17

14.18 Use source transformation to solve Problem 14.17.

14.19 Use Thévenin's theorem to solve Problem 14.17. **CS**

14.20 Use Thévenin's theorem to solve Problem 14.13.

14.21 Use Thévenin's theorem to find $v_o(t)$, $t > 0$, in the network in Fig. P14.21. **CS**

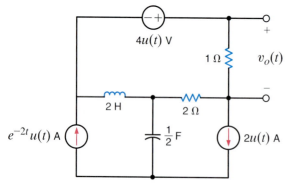

Figure P14.21

14.22 Find $v_o(t)$, for $t > 0$, in the network in Fig. P14.22 using Thévenin's theorem.

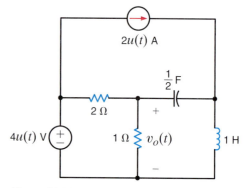

Figure P14.22

14.23 Use Thévenin's theorem to determine $i_o(t)$, $t > 0$, in the circuit shown in Fig. P14.23. **PSV**

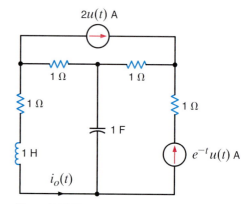

Figure P14.23

14.24 Use Thévenin's theorem to find $v_o(t)$, $t > 0$, in the network in Fig. P14.24.

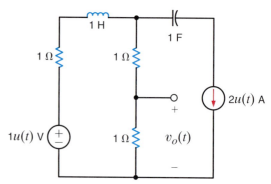

Figure P14.24

14.25 Use Thévenin's theorem to find $v_o(t)$, $t > 0$, in the network shown in Fig. P14.25. **PSV**

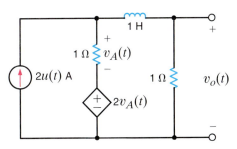

Figure P14.25

14.26 Find $v_o(t)$, $t > 0$, in the network shown in Fig. P14.26 using Laplace transforms. Assume that the circuit has reached steady state at $t = 0-$.

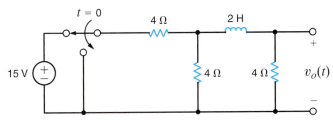

Figure P14.26

14.27 Find $i_o(t)$, $t > 0$, in the network shown in Fig. P14.27.

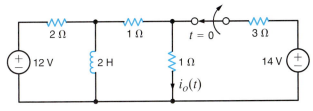

Figure P14.27

14.28 Find $i_o(t)$, $t > 0$, in the network shown in Fig. P14.28. **CS**

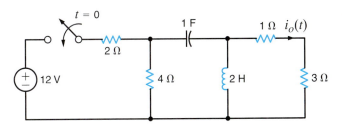

Figure P14.28

14.29 Find $v_o(t)$, $t > 0$, in the circuit shown in Fig. P14.29.

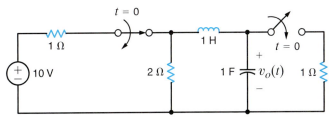

Figure P14.29

14.30 Find $v_o(t)$, $t > 0$, in the circuit in Fig. P14.30. **CS**

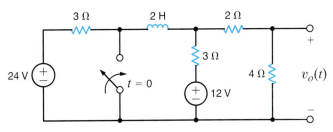

Figure P14.30

14.31 Find $i_o(t)$, $t > 0$, in the network in Fig. P14.31. **PSV**

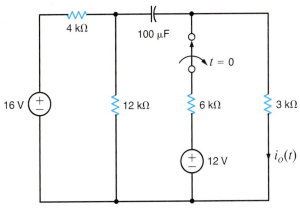

Figure P14.31

14.32 Find $v_o(t)$, $t > 0$, in the network in Fig. P14.32.

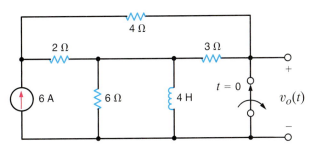

Figure P14.32

14.33 Find $v_o(t)$, for $t > 0$, in the network in Fig. P14.33.

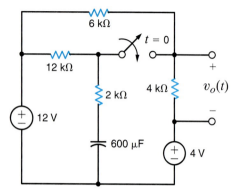

Figure P14.33

14.34 Find $v_o(t)$, for $t > 0$, in the network in Fig. P14.34.

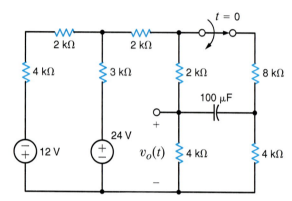

Figure P14.34

14.35 Find $v_o(t)$, for $t > 0$, in the network in Fig. P14.35.
CS

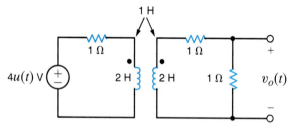

Figure P14.35

14.36 Find $v_o(t)$, for $t > 0$, in the network in Fig. P14.36.
PSV

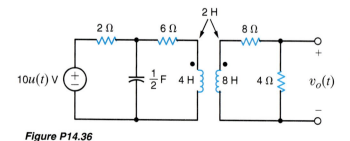

Figure P14.36

14.37 Find $v_o(t)$, for $t > 0$, in the network in Fig. P14.37.

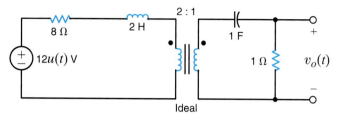

Figure P14.37

14.38 Find $v_o(t)$, for $t > 0$, in the network in Fig. P14.38.
CS

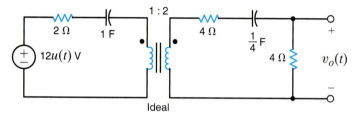

Figure P14.38

14.39 Determine the initial and final values of the voltage $v_o(t)$ in the network in Fig. P14.39.

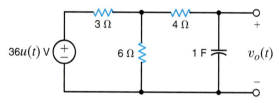

Figure P14.39

14.40 Determine the initial and final values of the voltage $v_o(t)$ in the network in Fig. P14.40.

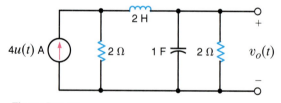

Figure P14.40

14.41 Determine the output voltage $v_o(t)$ in the network in Fig. P14.41a if the input is given by the source in Fig. P14.41b. **PSV**

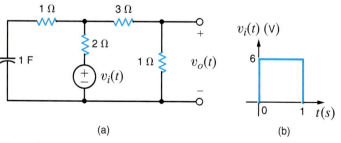

(a)   (b)

Figure P14.41

14.42 Find the output voltage, $v_o(t)$, $t > 0$, in the network in Fig. P14.42a if the input is represented by the waveform shown in Fig. P14.42b.

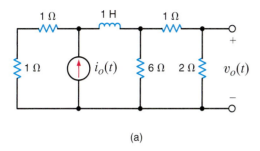

(a)

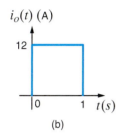

(b)

Figure P14.42

14.43 Determine the output voltage, $v_o(t)$, in the circuit in Fig. P14.43a if the input is given by the source described in Fig. P14.43b.

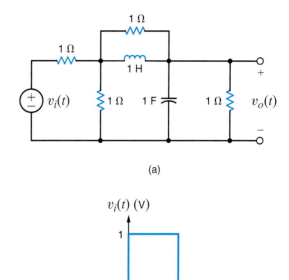

(a)

(b)

Figure P14.43

SECTION 14.4

14.44 Determine the transfer function $\mathbf{I}_o(s)/\mathbf{I}_i(s)$ for the network shown in Fig. P14.44.

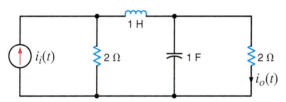

Figure P14.44

14.45 Find the transfer function $\mathbf{V}_o(s)/\mathbf{V}_i(s)$ for the network shown in Fig. P14.45. **CS**

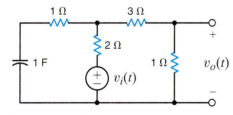

Figure P14.45

14.46 Find the transfer function for the network shown in Fig. P14.46.

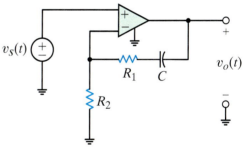

Figure P14.46

14.47 Find the transfer function for the network shown in Fig. P14.47.

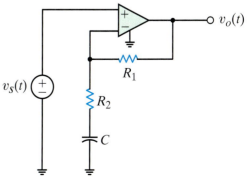

Figure P14.47

14.48 Find the transfer function for the network in Fig. P14.48. **PSV**

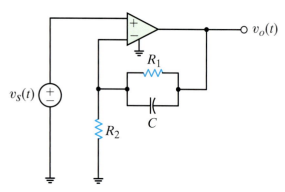

Figure P14.48

14.49 Find the transfer function for the network in Fig. P14.49. If a step function is applied to the network, will the response be overdamped, underdamped, or critically damped?

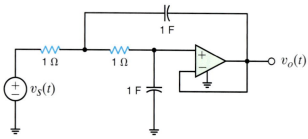

Figure P14.49

14.50 Find the transfer function for the network in Fig. P14.50.

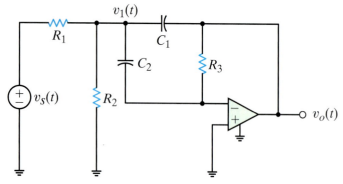

Figure P14.50

14.51 Determine the transfer function for the network shown in Fig. P14.51. If a step function is applied to the network, what type of damping will the network exhibit?

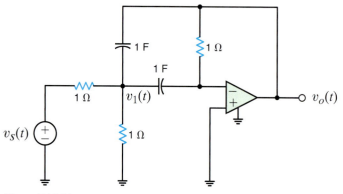

Figure P14.51

14.52 The voltage response of the network to a unit step input is

$$\mathbf{V}_o(s) = \frac{2(s+1)}{s(s^2 + 10s + 25)}$$

Is the response overdamped?

14.53 The transfer function of the network is given by the expression

$$\mathbf{G}(s) = \frac{100s}{s^2 + 13s + 40}$$

Determine the damping ratio, the undamped natural frequency, and the type of response that will be exhibited by the network.

14.54 The transfer function of the network is given by the expression

$$\mathbf{G}(s) = \frac{100s}{s^2 + 22s + 40}$$

Determine the damping ratio, the undamped natural frequency, and the type of response that will be exhibited by the network. **CS**

14.55 The voltage response of a network to a unit step input is

$$\mathbf{V}_o(s) = \frac{10}{s(s^2 + 8s + 18)}$$

Is the response critically damped?

14.56 For the network in Fig. P14.56, choose the value of C for critical damping.

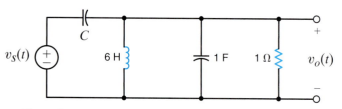

Figure P14.56

14.57 For the filter in Fig. P14.57, choose the values of C_1 and C_2 to place poles at $s = -2$ and $s = -5$ rad/s.

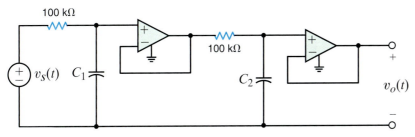

Figure P14.57

SECTION 14.6

14.58 Find the steady-state response $v_o(t)$ for the network in Fig. P14.58.

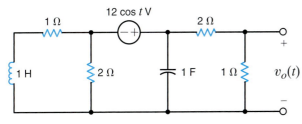

Figure P14.58

14.59 Find the steady-state response $v_o(t)$ for the circuit shown in Fig. P14.59. **PSV**

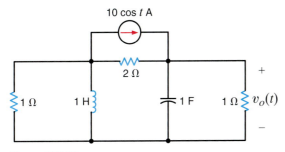

Figure P14.59

14.60 Determine the steady-state response $i_o(t)$ for the network in Fig. P14.60.

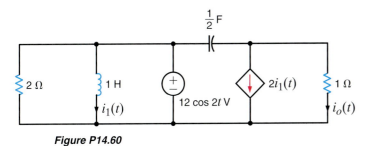

Figure P14.60

14.61 Find the steady-state response $i_o(t)$ for the network shown in Fig. P14.61. **CS**

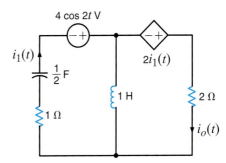

Figure P14.61

14.62 Find the steady-state response $v_o(t)$, for $t > 0$, in the network in Fig. P14.62.

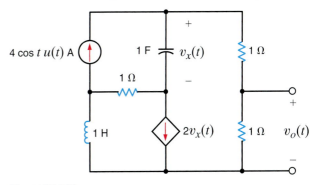

Figure P14.62

14.63 Find the steady-state response $v_o(t)$, for $t > 0$, in the network in Fig. P14.63.

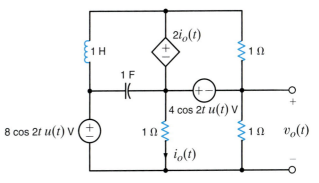

Figure P14.63

TYPICAL PROBLEMS FOUND ON THE FE EXAM

14FE-1 A single loop, second-order circuit is described by the following differential equation.

$$2\frac{dv^2(t)}{dt^2} + 4\frac{dv(t)}{dt} + 4v(t) = 12u(t) \qquad t > 0$$

Which is the correct form of the total (natural plus forced) response? **CS**

(a) $v(t) = K_1 + K_2 e^{-t}$

(b) $v(t) = K_1 \cos t + K_2 \sin t$

(c) $v(t) = K_1 + K_2 t e^{-t}$

(d) $v(t) = K_1 + K_2 e^{-t} \cos t + K_3 e^{-t} \sin t$

14FE-2 If all initial conditions are zero in the network in Fig. 14PFE-2, find the transfer function $\mathbf{V}_o(s)/\mathbf{V}_S(s)$, and determine the type of damping exhibited by the network.

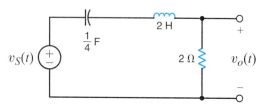

Figure 14PFE-2

14FE-3 The initial conditions in the circuit in Fig. 14PFE-3 are zero. Find the transfer function $\mathbf{I}_o(s)/\mathbf{I}_s(s)$, and determine the type of damping exhibited by the circuit.

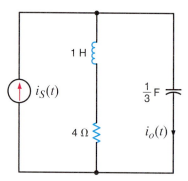

Figure 14PFE-3

Fourier Analysis Techniques

15

In this chapter we examine two very important topics: the Fourier series and the Fourier transform. These two techniques vastly expand our circuit analysis capabilities because they provide a means of effectively dealing with nonsinusoidal periodic signals and aperiodic signals. Using the Fourier series, we show that we can determine the steady-state response of a network to a nonsinusoidal periodic input. The Fourier transform will allow us to analyze circuits with aperiodic inputs by transforming the problem to the frequency domain, solving it algebraically, and then transforming back to the time domain in a manner similar to that used with Laplace transforms. ●

Access Problem-Solving Videos **PSV** *and Circuit Solutions* **CS** *at:*
http://www.justask4u.com/irwin
using the registration code on the inside cover and see a website with answers and more!

15.1 Fourier Series

A periodic function is one that satisfies the relationship

$$f(t) = f(t + nT_0), \qquad n = \pm 1, \pm 2, \pm 3, \ldots$$

for every value of t where T_0 is the period. As we have shown in previous chapters, the sinusoidal function is a very important periodic function. However, there are many other periodic functions that have wide applications. For example, laboratory signal generators produce the pulse-train and square-wave signals shown in Figs. 15.1a and b, respectively, which are used for testing circuits. The oscilloscope is another laboratory instrument, and the sweep of its electron beam across the face of the cathode ray tube is controlled by a triangular signal of the form shown in Fig. 15.1c.

The techniques we will explore are based on the work of Jean Baptiste Joseph Fourier. Although our analyses will be confined to electric circuits, it is important to point out that the techniques are applicable to a wide range of engineering problems. In fact, it was Fourier's work in heat flow that led to the techniques that will be presented here.

In his work, Fourier demonstrated that a periodic function $f(t)$ could be expressed as a sum of sinusoidal functions. Therefore, given this fact and the fact that if a periodic function is expressed as a sum of linearly independent functions, each function in the sum must be periodic with the same period, and the function $f(t)$ can be expressed in the form

$$f(t) = a_0 + \sum_{n=1}^{\infty} D_n \cos(n\omega_0 t + \theta_n) \qquad \textbf{15.1}$$

where $\omega_0 = 2\pi/T_0$ and a_0 is the average value of the waveform. An examination of this expression illustrates that all sinusoidal waveforms that are periodic with period T_0 have been included. For example, for $n = 1$, one cycle covers T_0 seconds and $D_1 \cos(\omega_0 t + \theta_1)$ is called the *fundamental*. For $n = 2$, two cycles fall within T_0 seconds, and the term $D_2 \cos(2\omega_0 t + \theta_2)$ is called the *second harmonic*. In general, for $n = k$, k cycles fall within T_0 seconds and $D_k \cos(k\omega_0 t + \theta_k)$ is the *kth harmonic term*.

Since the function $\cos(n\omega_0 t + \theta_k)$ can be written in exponential form using Euler's identity or as a sum of cosine and sine terms of the form $\cos n\omega_0 t$ and $\sin n\omega_0 t$ as demonstrated in Chapter 8, the series in Eq. (15.1) can be written as

$$f(t) = a_0 + \sum_{\substack{n=-\infty \\ n \neq 0}}^{\infty} \mathbf{c}_n e^{jn\omega_0 t} = \sum_{n=-\infty}^{\infty} \mathbf{c}_n e^{jn\omega_0 t} \qquad \textbf{15.2}$$

Figure 15.1
Some useful periodic signals.

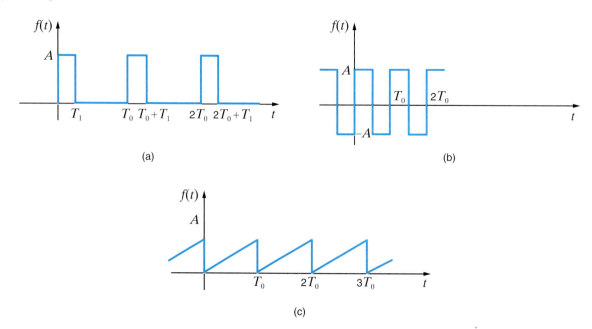

(a)

(b)

(c)

Using the real-part relationship employed as a transformation between the time domain and the frequency domain, we can express $f(t)$ as

$$f(t) = a_0 + \sum_{n=1}^{\infty} \text{Re}\left[\left(D_n \underline{/\theta_n}\right)e^{jn\omega_0 t}\right] \qquad \textbf{15.3}$$

$$= a_0 + \sum_{n=1}^{\infty} \text{Re}\left(2\mathbf{c}_n e^{jn\omega_0 t}\right) \qquad \textbf{15.4}$$

$$= a_0 + \sum_{n=1}^{\infty} \text{Re}\left[\left(a_n - jb_n\right)e^{jn\omega_0 t}\right] \qquad \textbf{15.5}$$

$$= a_0 + \sum_{n=1}^{\infty} \left(a_n \cos n\omega_0 t + b_n \sin n\omega_0 t\right) \qquad \textbf{15.6}$$

These equations allow us to write the Fourier series in a number of equivalent forms. Note that the *phasor* for the *n*th harmonic is

$$D_n \underline{/\theta_n} = 2\mathbf{c}_n = a_n - jb_n \qquad \textbf{15.7}$$

The approach we will take will be to represent a nonsinusoidal periodic input by a sum of complex exponential functions, which because of Euler's identity is equivalent to a sum of sines and cosines. We will then use (1) the superposition property of linear systems and (2) our knowledge that the steady-state response of a time-invariant linear system to a sinusoidal input of frequency ω_0 is a sinusoidal function of the same frequency to determine the response of such a system.

To illustrate the manner in which a nonsinusoidal periodic signal can be represented by a Fourier series, consider the periodic function shown in Fig. 15.2a. In Figs. 15.2b–d we can see the impact of using a specific number of terms in the series to represent the original function. Note that the series more closely represents the original function as we employ more and more terms.

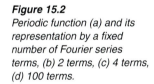

Figure 15.2
Periodic function (a) and its representation by a fixed number of Fourier series terms, (b) 2 terms, (c) 4 terms, (d) 100 terms.

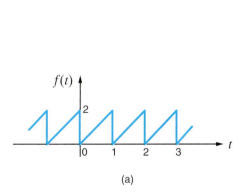

(a)

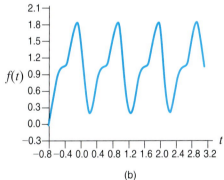

(b)

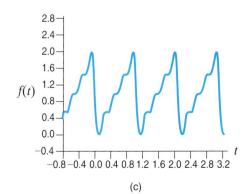

(c)

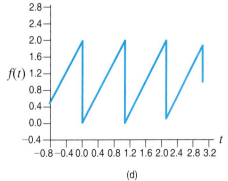

(d)

EXPONENTIAL FOURIER SERIES *Any physically realizable periodic signal may be represented over the interval* $t_1 < t < t_1 + T_0$ *by the exponential Fourier series*

$$f(t) = \sum_{n=-\infty}^{\infty} \mathbf{c}_n e^{jn\omega_0 t} \qquad \textbf{15.8}$$

where the $\mathbf{c}_n$ are the complex (phasor) Fourier coefficients. These coefficients are derived as follows. Multiplying both sides of Eq. (15.8) by $e^{-jk\omega_0 t}$ and integrating over the interval t_1 to $t_1 + T_0$, we obtain

$$\int_{t_1}^{t_1+T_0} f(t)e^{-jk\omega_0 t}\, dt = \int_{t_1}^{t_1+T_0} \left(\sum_{n=-\infty}^{\infty} \mathbf{c}_n e^{jn\omega_0 t} \right) e^{-jk\omega_0 t}\, dt$$

$$= \mathbf{c}_k T_0$$

since

$$\int_{t_1}^{t_1+T_0} e^{j(n-k)\omega_0 t}\, dt = \begin{cases} 0 & \text{for } n \neq k \\ T_0 & \text{for } n = k \end{cases}$$

Therefore, the Fourier coefficients are defined by the equation

$$\mathbf{c}_n = \frac{1}{T_0} \int_{t_1}^{t_1+T_0} f(t)e^{-jn\omega_0 t}\, dt \qquad \textbf{15.9}$$

The following example illustrates the manner in which we can represent a periodic signal by an exponential Fourier series.

Example 15.1

We wish to determine the exponential Fourier series for the periodic voltage waveform shown in Fig. 15.3.

Figure 15.3
Periodic voltage waveform.

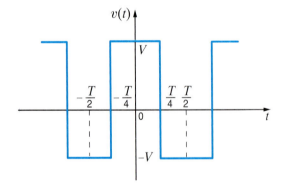

SOLUTION The Fourier coefficients are determined using Eq. (15.9) by integrating over one complete period of the waveform.

$$\mathbf{c}_n = \frac{1}{T} \int_{-T/2}^{T/2} f(t)e^{-jn\omega_0 t}\, dt$$

$$= \frac{1}{T} \int_{-T/2}^{-T/4} -Ve^{-jn\omega_0 t}\, dt$$

$$+ \int_{-T/4}^{T/4} Ve^{-jn\omega_0 t}\, dt + \int_{T/4}^{T/2} -Ve^{-jn\omega_0 t}\, dt$$

$$= \frac{V}{jn\omega_0 T} \left[+e^{-jn\omega_0 t}\Big|_{-T/2}^{-T/4} - e^{-jn\omega_0 t}\Big|_{-T/4}^{T/4} + e^{-jn\omega_0 t}\Big|_{T/4}^{T/2} \right]$$

$$= \frac{V}{jn\omega_0 T}\left(2e^{jn\pi/2} - 2e^{-jn\pi/2} + e^{-jn\pi} - e^{+jn\pi}\right)$$

$$= \frac{V}{n\omega_0 T}\left[4\sin\frac{n\pi}{2} - 2\sin(n\pi)\right]$$

$$= 0 \qquad \text{for } n \text{ even}$$

$$= \frac{2V}{n\pi}\sin\frac{n\pi}{2} \qquad \text{for } n \text{ odd}$$

$\mathbf{c}_0$ corresponds to the average value of the waveform. This term can be evaluated using the original equation for $\mathbf{c}_n$. Therefore,

$$c_0 = \frac{1}{T}\int_{-\frac{T}{2}}^{\frac{T}{2}} v(t)\, dt$$

$$= \frac{1}{T}\left[\int_{-\frac{T}{2}}^{-\frac{T}{4}} -V\, dt + \int_{-\frac{T}{4}}^{\frac{T}{4}} V\, dt + \int_{\frac{T}{4}}^{\frac{T}{2}} -V\, dt\right]$$

$$= \frac{1}{T}\left[-\frac{VT}{4} + \frac{VT}{2} - \frac{VT}{4}\right] = 0$$

Therefore,

$$v(t) = \sum_{\substack{n=-\infty \\ n \neq 0 \\ n \text{ odd}}}^{\infty} \frac{2V}{n\pi}\sin\frac{n\pi}{2}\, e^{jn\omega_0 t}$$

This equation can be written as

$$v(t) = \sum_{\substack{n=1 \\ n \text{ odd}}}^{\infty} \frac{2V}{n\pi}\sin\frac{n\pi}{2}\, e^{jn\omega_0 t} + \sum_{\substack{n=-1 \\ n \text{ odd}}}^{-\infty} \frac{2V}{n\pi}\sin\frac{n\pi}{2}\, e^{jn\omega_0 t}$$

$$= \sum_{\substack{n=1 \\ n \text{ odd}}}^{\infty} \left(\frac{2V}{n\pi}\sin\frac{n\pi}{2}\right)e^{jn\omega_0 t} + \left(\frac{2V}{n\pi}\sin\frac{n\pi}{2}\right)^* e^{-jn\omega_0 t}$$

Since a number plus its complex conjugate is equal to two times the real part of the number, $v(t)$ can be written as

$$v(t) = \sum_{\substack{n=1 \\ n \text{ odd}}}^{\infty} 2\,\text{Re}\left(\frac{2V}{n\pi}\sin\frac{n\pi}{2}\, e^{jn\omega_0 t}\right)$$

or

$$v(t) = \sum_{\substack{n=1 \\ n \text{ odd}}}^{\infty} \frac{4V}{n\pi}\sin\frac{n\pi}{2}\cos n\omega_0 t$$

Note that this same result could have been obtained by integrating over the interval $-T/4$ to $3T/4$.

LEARNING EXTENSIONS

E15.1 Find the Fourier coefficients for the waveform in Fig. E15.1.   **PSV**

Figure E15.1

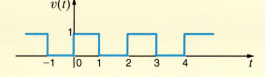

ANSWER:

$$\mathbf{c}_n = \frac{1 - e^{-jn\pi}}{j2\pi n}; \quad \mathbf{c}_0 = \frac{1}{2}.$$

(continues on the next page)

E15.2 Find the Fourier coefficients for the waveform in Fig. E15.2.

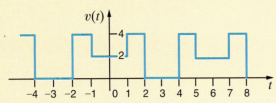

Figure E15.2

ANSWER:

$$\mathbf{c}_n = \frac{2}{n\pi}\left(2\sin\frac{2\pi n}{3} - \sin\frac{n\pi}{3}\right); \mathbf{c}_0 = 2.$$

TRIGONOMETRIC FOURIER SERIES Let us now examine another form of the Fourier series. Since

$$2\mathbf{c}_n = a_n - jb_n \qquad\qquad \textbf{15.10}$$

we will examine this quantity $2\mathbf{c}_n$ and separate it into its real and imaginary parts. Using Eq. (15.9), we find that

$$2\mathbf{c}_n = \frac{2}{T_0}\int_{t_1}^{t_1+T_0} f(t)e^{-jn\omega_0 t}\,dt \qquad\qquad \textbf{15.11}$$

Using Euler's identity, we can write this equation in the form

$$2\mathbf{c}_n = \frac{2}{T_0}\int_{t_1}^{t_1+T_0} f(t)\,(\cos n\omega_0 t - j\sin n\omega_0 t)\,dt$$

$$= \frac{2}{T_0}\int_{t_1}^{t_1+T_0} f(t)\cos n\omega_0 t\,dt - j\frac{2}{T_0}\int_{t_1}^{t_1+T_0} f(t)\sin n\omega_0 t\,dt$$

From Eq. (15.10) we note then that

$$a_n = \frac{2}{T_0}\int_{t_1}^{t_1+T_0} f(t)\cos n\omega_0 t\,dt \qquad\qquad \textbf{15.12}$$

$$b_n = \frac{2}{T_0}\int_{t_1}^{t_1+T_0} f(t)\sin n\omega_0 t\,dt \qquad\qquad \textbf{15.13}$$

These are the coefficients of the Fourier series described by Eq. (15.6), which we call the *trigonometric Fourier series*. These equations are derived directly in most textbooks using the orthogonality properties of the cosine and sine functions. Note that we can now evaluate $\mathbf{c}_n$, a_n, b_n, and since

$$2\mathbf{c}_n = D_n\underline{/\theta_n} \qquad\qquad \textbf{15.14}$$

we can derive the coefficients for the *cosine Fourier series* described by Eq. (15.1). This form of the Fourier series is particularly useful because it allows us to represent each harmonic of the function as a phasor.

From Eq. (15.9) we note that $\mathbf{c}_0$, which is written as a_0, is

$$a_0 = \frac{1}{T}\int_{t_1}^{t_1+T_0} f(t)\,dt \qquad\qquad \textbf{15.15}$$

This is the average value of the signal $f(t)$ and can often be evaluated directly from the waveform.

SYMMETRY AND THE TRIGONOMETRIC FOURIER SERIES If a signal exhibits certain symmetrical properties, we can take advantage of these properties to simplify the calculations of the Fourier coefficients. There are three types of symmetry: (1) even-function symmetry, (2) odd-function symmetry, and (3) half-wave symmetry.

Even-Function Symmetry   A function is said to be even if

$$f(t) = f(-t) \qquad\qquad \textbf{15.16}$$

An even function is symmetrical about the vertical axis, and a notable example is the function $\cos n\omega_0 t$. Note that the waveform in Fig. 15.3 also exhibits even-function symmetry. Let us now determine the expressions for the Fourier coefficients if the function satisfies Eq. (15.16).

If we let $t_1 = -T_0/2$ in Eq. (15.15), we obtain

$$a_0 = \frac{1}{T_0} \int_{-T_0/2}^{T_0/2} f(t)\, dt$$

which can be written as

$$a_0 = \frac{1}{T_0} \int_{-T_0/2}^{0} f(t)\, dt + \frac{1}{T_0} \int_{0}^{T_0/2} f(t)\, dt$$

If we now change the variable on the first integral (i.e., let $t = -x$), then $f(-x) = f(x)$, $dt = -dx$, and the range of integration is from $x = T_0/2$ to 0. Therefore, the preceding equation becomes

$$a_0 = \frac{1}{T_0} \int_{T_0/2}^{0} f(x)(-dx) + \frac{1}{T_0} \int_{0}^{T_0/2} f(t)\, dt$$

$$= \frac{1}{T_0} \int_{0}^{T_0/2} f(x)\, dx + \frac{1}{T_0} \int_{0}^{T_0/2} f(t)\, dt \qquad\qquad \textbf{15.17}$$

$$= \frac{2}{T_0} \int_{0}^{T_0/2} f(t)\, dt$$

The other Fourier coefficients are derived in a similar manner. The a_n coefficient can be written

$$a_n = \frac{2}{T_0} \int_{-T_0/2}^{0} f(t) \cos n\omega_0 t\, dt + \frac{2}{T_0} \int_{0}^{T_0/2} f(t) \cos n\omega_0 t\, dt$$

Employing the change of variable that led to Eq. (15.17), we can express the preceding equation as

$$a_n = \frac{2}{T_0} \int_{T_0/2}^{0} f(x) \cos(-n\omega_0 x)(-dx) + \frac{2}{T_0} \int_{0}^{T_0/2} f(t) \cos n\omega_0 t\, dt$$

$$= \frac{2}{T_0} \int_{0}^{T_0/2} f(x) \cos n\omega_0 x\, dx + \frac{2}{T_0} \int_{0}^{T_0/2} f(t) \cos n\omega_0 t\, dt$$

$$a_n = \frac{4}{T_0} \int_{0}^{T_0/2} f(t) \cos n\omega_0 t\, dt \qquad\qquad \textbf{15.18}$$

Once again, following the preceding development, we can write the equation for the b_n coefficient as

$$b_n = \frac{2}{T_0} \int_{-T_0/2}^{0} f(t) \sin n\omega_0 t\, dt + \int_{0}^{T_0/2} f(t) \sin n\omega_0 t\, dt$$

The variable change employed previously yields

$$b_n = \frac{2}{T_0} \int_{T_0/2}^{0} f(x) \sin(-n\omega_0 x)(-dx) + \frac{2}{T_0} \int_{0}^{T_0/2} f(t) \sin n\omega_0 t\, dt$$

$$= \frac{-2}{T_0} \int_{0}^{T_0/2} f(x) \sin n\omega_0 x\, dx + \frac{2}{T_0} \int_{0}^{T_0/2} f(t) \sin n\omega_0 t\, dt$$

$$b_n = 0 \qquad\qquad \textbf{15.19}$$

The preceding analysis indicates that the Fourier series for an even periodic function consists only of a constant term and cosine terms. Therefore, if $f(t)$ is even, $b_n = 0$ and from Eqs. (15.10) and (15.14), $\mathbf{c}_n$ are real and θ_n are multiples of $180°$.

Odd-Function Symmetry A function is said to be odd if

$$f(t) = -f(-t) \qquad \textbf{15.20}$$

An example of an odd function is $\sin n\omega_0 t$. Another example is the waveform in Fig. 15.4a. Following the mathematical development that led to Eqs. (15.17) to (15.19), we can show that for an odd function the Fourier coefficients are

$$a_0 = 0 \qquad \textbf{15.21}$$

$$a_n = 0 \qquad \text{for all } n > 0 \qquad \textbf{15.22}$$

$$b_n = \frac{4}{T_0} \int_0^{T_0/2} f(t) \sin n\omega_0 t \, dt \qquad \textbf{15.23}$$

Therefore, if $f(t)$ is odd, $a_n = 0$ and, from Eqs. (15.10) and (15.14), $\mathbf{c}_n$ are pure imaginary and θ_n are odd multiples of $90°$.

Figure 15.4
Three waveforms;
(a) and (c) possess
half-wave symmetry.

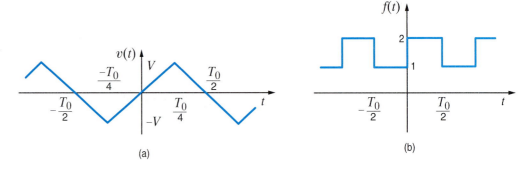

(a)

(b)

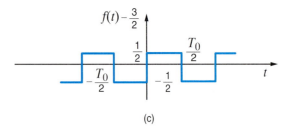

(c)

Half-Wave Symmetry A function is said to possess *half-wave symmetry* if

$$f(t) = -f\left(t - \frac{T_0}{2}\right) \qquad \textbf{15.24}$$

Basically, this equation states that each half-cycle is an inverted version of the adjacent half-cycle; that is, if the waveform from $-T_0/2$ to 0 is inverted, it is identical to the waveform from 0 to $T_0/2$. The waveforms shown in Figs. 15.4a and c possess half-wave symmetry.

   Once again we can derive the expressions for the Fourier coefficients in this case by repeating the mathematical development that led to the equations for even-function symmetry using the change of variable $t = x + T_0/2$ and Eq. (15.24). The results of this development are the following equations:

$$a_0 = 0 \qquad\qquad\qquad\qquad\qquad\qquad\qquad\qquad \textbf{15.25}$$

$$a_n = b_n = 0 \qquad\qquad\qquad \text{for } n \text{ even} \qquad\qquad \textbf{15.26}$$

$$a_n = \frac{4}{T_0} \int_0^{T_0/2} f(t) \cos n\omega_0 t\, dt \qquad \text{for } n \text{ odd} \qquad\qquad \textbf{15.27}$$

$$b_n = \frac{4}{T_0} \int_0^{T_0/2} f(t) \sin n\omega_0 t\, dt \qquad \text{for } n \text{ odd} \qquad\qquad \textbf{15.28}$$

The following equations are often useful in the evaluation of the trigonometric Fourier series coefficients:

$$\int \sin ax\, dx = -\frac{1}{a} \cos ax$$

$$\int \cos ax\, dx = \frac{1}{a} \sin ax$$

$$\textbf{15.29}$$

$$\int x \sin ax\, dx = \frac{1}{a^2} \sin ax - \frac{1}{a} x \cos ax$$

$$\int x \cos ax\, dx = \frac{1}{a^2} \cos ax + \frac{1}{a} x \sin ax$$

Example 15.2

We wish to find the trigonometric Fourier series for the periodic signal in Fig. 15.3.

SOLUTION   The waveform exhibits even-function symmetry and therefore

$$a_0 = 0$$

$$b_n = 0 \qquad \text{for all } n$$

The waveform exhibits half-wave symmetry and therefore

$$a_n = 0 \qquad \text{for } n \text{ even}$$

Hence,

$$a_n = \frac{4}{T_0} \int_0^{T/2} f(t) \cos n\omega_0 t\, dt \qquad \text{for } n \text{ odd}$$

$$= \frac{4}{T} \left(\int_0^{T/4} V \cos n\omega_0 t\, dt - \int_{T/4}^{T/2} V \cos n\omega_0 t\, dt \right)$$

$$= \frac{4V}{n\omega_0 T} \left(\sin n\omega_0 t \Big|_0^{T/4} - \sin n\omega_0 t \Big|_{T/4}^{T/2} \right)$$

$$= \frac{4V}{n\omega_0 T} \left(\sin \frac{n\pi}{2} - \sin n\pi + \sin \frac{n\pi}{2} \right)$$

$$= \frac{8V}{n2\pi} \sin \frac{n\pi}{2} \qquad \text{for } n \text{ odd}$$

$$= \frac{4V}{n\pi} \sin \frac{n\pi}{2} \qquad \text{for } n \text{ odd}$$

The reader should compare this result with that obtained in Example 15.1.

Example 15.3

Let us determine the trigonometric Fourier series expansion for the waveform shown in Fig. 15.4a.

SOLUTION The function not only exhibits odd-function symmetry, but it possesses half-wave symmetry as well. Therefore, it is necessary to determine only the coefficients b_n for n odd. Note that

$$v(t) = \begin{cases} \dfrac{4Vt}{T_0} & 0 \le t \le T_0/4 \\ 2V - \dfrac{4Vt}{T_0} & T_0/4 < t \le T_0/2 \end{cases}$$

The b_n coefficients are then

$$b_n = \frac{4}{T_0} \int_0^{T_0/4} \frac{4Vt}{T_0} \sin n\omega_0 t \, dt + \frac{4}{T_0} \int_{T_0/4}^{T_0/2} \left(2V - \frac{4Vt}{T_0} \right) \sin n\omega_0 t \, dt$$

The evaluation of these integrals is tedious but straightforward and yields

$$b_n = \frac{8V}{n^2\pi^2} \sin \frac{n\pi}{2} \qquad \text{for } n \text{ odd}$$

Hence, the Fourier series expansion is

$$v(t) = \sum_{\substack{n=1 \\ n \text{ odd}}}^{\infty} \frac{8V}{n^2\pi^2} \sin \frac{n\pi}{2} \sin n\omega_0 t$$

Example 15.4

We wish to find the trigonometric Fourier series expansion of the waveform in Fig. 15.4b.

SOLUTION Note that this waveform has an average value of 3/2. Therefore, instead of determining the Fourier series expansion of $f(t)$, we will determine the Fourier series for $f(t) - 3/2$, which is the waveform shown in Fig. 15.4c. The latter waveform possesses half-wave symmetry. The function is also odd and therefore

$$b_n = \frac{4}{T_0} \int_0^{T_0/2} \frac{1}{2} \sin n\omega_0 t \, dt$$

$$= \frac{2}{T_0} \left(\frac{-1}{n\omega_0} \cos n\omega_0 t \Big|_0^{T_0/2} \right)$$

$$= \frac{-2}{n\omega_0 T_0} (\cos n\pi - 1)$$

$$= \frac{2}{n\pi} \qquad \text{for } n \text{ odd}$$

Therefore, the Fourier series expansion for $f(t) - 3/2$ is

$$f(t) - \frac{3}{2} = \sum_{\substack{n=1 \\ n \text{ odd}}}^{\infty} \frac{2}{n\pi} \sin n\omega_0 t$$

or

$$f(t) = \frac{3}{2} + \sum_{\substack{n=1 \\ n \text{ odd}}}^{\infty} \frac{2}{n\pi} \sin n\omega_0 t$$

LEARNING EXTENSIONS

E15.3 Determine the type of symmetry exhibited by the waveform in Figs. E15.2 and E15.3.

ANSWER: Figure E15.2, even symmetry; Fig. E15.3, half-wave symmetry.

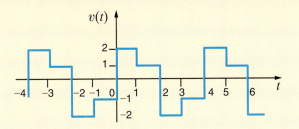

Figure E15.3

E15.4 Find the trigonometric Fourier series for the voltage waveform in Fig. E15.2.   **PSV**

ANSWER:

$$v(t) = 2 + \sum_{n=1}^{\infty} \frac{4}{n\pi} \left(2\sin\frac{2\pi n}{3} - \sin\frac{n\pi}{3} \right) \cos\frac{n\pi}{3} t.$$

E15.5 Find the trigonometric Fourier series for the voltage waveform in Fig. E15.3.

ANSWER:

$$v(t) = \sum_{\substack{n=1 \\ n\ \text{odd}}}^{\infty} \frac{2}{n\pi} \sin\frac{n\pi}{2} \cos\frac{n\pi}{2} t$$

$$+ \frac{2}{n\pi} (2 - \cos n\pi) \sin\frac{n\pi}{2} t.$$

Fourier Series Via PSPICE Simulation

Most of the circuit simulators are capable of performing Fourier series calculations. The following example illustrates the manner in which PSPICE determines the Fourier components for a specified waveform.

Example 15.5

Let us use PSPICE to determine the Fourier series of the waveform in Figure 15.5a.

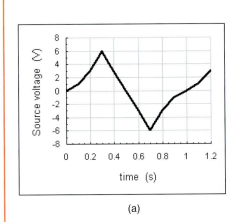

(a)

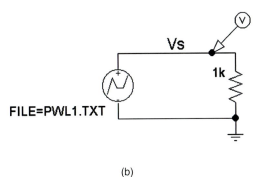

FILE=PWL1.TXT

(b)

Figure 15.5

(a) The waveform for Example 15.5. (b) The PSPICE schematic, (c) the VPWL_FILE source attributes,
(d) the PWL1.TXT file, and (e) the Transient setup requirements for the Fourier analysis simulation.

Figure 15.5
(continued)

V1   PartName: VPWL_FILE   ☒

Name   Value

| DC | = | 0 |

| DC=0 |
| AC=0 |
| TSF=1 |
| VSF=1 |
| FILE=PWL1.TXT |
| REPEAT_VALUE=500 |

Save Attr
Change Display
Delete

☐ Include Non-changeable Attributes
☐ Include System-defined Attributes

OK
Cancel

(c)

```
* Piecewise Linear File for
* BECA 8 Fourier Series Example
*
0,0
0.1,1
0.2,3
0.3,6
0.4,3
0.5,0
0.6,-3
0.7,-6
0.8,-3
0.9,-1
1.0,0
```

(d)

Transient   ☒

┌─ Transient Analysis ──────────────────────┐

Print Step: | 1 |

Final Time: | 1 |

No-Print Delay: | |

Step Ceiling: | 1m |

☐ Detailed Bias Pt.
☐ Skip initial transient solution
└──┘

┌─ Fourier Analysis ────────────────────────┐

☑ Enable Fourier

Center Frequency: | 1 |

Number of harmonics: | 20 |

Output Vars.: | V(Vs) |
└──┘

OK   Cancel

(e)

SOLUTION   Our procedure consists of four steps: (1) create the PSPICE schematic; (2) create the waveform of interest; (3) set up the simulation particulars; and (4) view the results. The schematic in Fig. 15.5b satisfies step 1 where the piecewise linear source VPWL_FILE from the SOURCE library will provide the signal in Fig. 15.5a.

For step 2, we edit the source attributes by double-clicking on the VPWL_FILE symbol to open the dialog box in Fig. 15.5c. Attributes **DC** and **AC** are the voltage values for **DC** and **AC** Sweep analyses, respectively. Since the Fourier series is calculated from a transient analysis, both the **DC** and **AC** attributes can be set to zero. The **FILE** attribute specifies the filename for the piecewise linear waveform. Here, we have chosen the name PWL1.TXT. Attributes **TSF** and **VSF** are scale factors for the time and voltage axes, respectively, of the waveform in PWL1.TXT. For example, if **VSF=2**, then the voltage values in PWL1.TXT are multiplied by 2. Finally, **REPEAT_VALUE** is the number of cycles the waveform repeats. Next, we must enter (time, voltage) data pairs into PWL1.TXT to create the actual waveform. Double-click on the FILE=PWL1.TXT text and, when asked whether you would like to create the file, select **OK**. The NOTEPAD text editor will open as shown in Fig. 15.5d where the required data pairs for the waveform of interest have already been entered. When finished, save and exit the file to return to PSPICE.

In step 3, we must specify a transient analysis with the appropriate Fourier series information. To do this, select SETUP from the **Analysis** menu, then choose TRANSIENT. A dialog box similar to that in Fig. 15.5e will appear. The box has already been edited to (1) enable the Fourier analysis, (2) set the fundamental frequency at 1 Hz (PSPICE calls this the Center Frequency), (3) request 20 harmonics, and (4) specify the voltage Vs for Fourier analysis.

To simulate the circuit, select SIMULATE from the **Analysis** menu. The Fourier results are in the output file, which can be accessed from the *Schematics* menu **Analysis/Examine Output**, or from the PROBE menu **View/Output File**. The results are shown in Table 15.1 for the trigonometric series

$$v_S(t) = a_0 + \sum_{n=1}^{\infty} b_n \sin(n\omega_0 t + \theta_n)$$

Note that the dc component, a_0, is essentially zero, as expected from Fig. 15.5a. The 5th, 10th, 15th, ... harmonics are also zero. In addition, all phases are either zero (very small) or 180 degrees.

Table 15.1 Fourier analysis results for Example 15.5. Fourier components of transient response $V(Vs)$
DC COMPONENT = −1.353267E-08

Harmonic No.	Frequency (Hz)	Fourier Component	Normalized Component	Phase (deg)	Normalized Phase (deg)
1	1.000E+00	4.222E+00	1.000E+00	2.969E−07	0.000E+00
2	2.000E+00	1.283E+00	3.039E−01	1.800E+02	1.800E+02
3	3.000E+00	4.378E−01	1.037E−00	−1.800E+02	−1.800E+02
4	4.000E+00	3.838E−01	9.090E−02	4.620E−07	−7.254E−07
5	5.000E+00	1.079E−04	2.556E−05	1.712E−03	1.711E−03
6	6.000E+00	1.703E−01	4.034E−02	1.800E+02	1.800E+02
7	7.000E+00	8.012E−02	1.898E−02	−9.548E−06	−1.163E−05
8	8.000E+00	8.016E−02	1.899E−02	5.191E−06	2.816E−06
9	9.000E+00	5.144E−02	1.218E−02	−1.800E+02	−1.800E+02
10	1.000E+01	1.397E−04	3.310E−05	1.800E+02	1.800E+02
11	1.100E+01	3.440E−02	8.149E−03	−1.112E−04	−1.145E−04
12	1.200E+01	3.531E−02	8.364E−03	1.800E+02	1.800E+02
13	1.300E+01	2.343E−02	5.549E−03	1.800E+02	1.800E+02
14	1.400E+01	3.068E−02	7.267E−03	−3.545E−05	−3.960E−05
15	1.500E+01	3.379E−04	8.003E−05	−3.208E−03	−3.212E−03
16	1.600E+01	2.355E−02	5.579E−03	−1.800E+02	−1.800E+02
17	1.700E+01	1.309E−02	3.101E−03	2.905E−04	2.854E−04
18	1.800E+01	1.596E−02	3.781E−03	−5.322E−05	−5.856E−05
19	1.900E+01	1.085E−02	2.569E−03	−1.800E+02	−1.800E+02
20	2.000E+01	2.994E−04	7.092E−05	1.800E+02	1.800E+02

Total harmonic distortion = 3.378352E+01 percent

This procedure is very useful in validating calculations done by hand. As an aside, the accuracy of the Fourier analysis is affected by the TRANSIENT setup parameters in Fig. 15.5e. In general, accuracy increases as the number of cycles in the simulation decreases and as the number of data points increases. The **Final Time** determines the number of cycles, and increasing the **Step Ceiling** will increase the number of data points.

TIME-SHIFTING Let us now examine the effect of time-shifting a periodic waveform $f(t)$ defined by the equation

$$f(t) = \sum_{n=-\infty}^{\infty} \mathbf{c}_n e^{jn\omega_0 t}$$

Note that

$$f(t - t_0) = \sum_{n=-\infty}^{\infty} \mathbf{c}_n e^{jn\omega_0(t-t_0)}$$

$$f(t - t_0) = \sum_{n=-\infty}^{\infty} \left(\mathbf{c}_n e^{-jn\omega_0 t_0}\right) e^{jn\omega_0 t} \qquad \textbf{15.30}$$

Since $e^{-jn\omega_0 t_0}$ corresponds to a phase shift, the Fourier coefficients of the time-shifted function are the Fourier coefficients of the original function, with the angle shifted by an amount directly proportional to frequency. Therefore, time shift in the time domain corresponds to phase shift in the frequency domain.

Example 15.6

Let us time delay the waveform in Fig. 15.3 by a quarter period and compute the Fourier series.

SOLUTION The waveform in Fig. 15.3 time delayed by $T_0/4$ is shown in Fig. 15.6. Since the time delay is $T_0/4$,

$$n\omega_0 t_d = n\frac{2\pi}{T_0}\frac{T_0}{4} = n\frac{\pi}{2} = n\,90°$$

Therefore, using Eq. (15.30) and the results of Example 15.1, the Fourier coefficients for the time-shifted waveform are

$$\mathbf{c}_n = \frac{2V}{n\pi} \sin\frac{n\pi}{2} \,\underline{/-n\,90°} \qquad n\text{ odd}$$

and therefore,

$$v(t) = \sum_{\substack{n=1 \\ n\text{ odd}}}^{\infty} \frac{4V}{n\pi} \sin\frac{n\pi}{2} \cos\left(n\omega_0 t - n\,90°\right)$$

If we compute the Fourier coefficients for the time-shifted waveform in Fig. 15.6, we obtain

$$\mathbf{c}_n = \frac{1}{T_0} \int_{-T_0/2}^{T_0/2} f(t) e^{-jn\omega_0 t}\, dt$$

$$= \frac{1}{T_0} \int_{-T_0/2}^{0} -V e^{-jn\omega_0 t}\, dt + \frac{1}{T_0} \int_{0}^{T_0/2} V e^{-jn\omega_0 t}\, dt$$

$$= \frac{2V}{jn\pi} \qquad \text{for } n \text{ odd}$$

Figure 15.6
*Waveform in Fig. 15.3
time-shifted by $T_0/4$.*

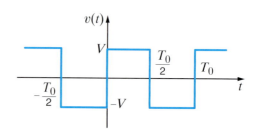

Therefore,

$$\mathbf{c}_n = \frac{2V}{n\pi} \underline{/-90°} \qquad n \text{ odd}$$

Since n is odd, we can show that this expression is equivalent to the one obtained earlier.

In general, we can compute the phase shift in degrees using the expression

$$\text{phase shift(deg)} = \omega_0 t_d = (360°) \frac{t_d}{T_0} \qquad \textbf{15.31}$$

so that a time shift of one-quarter period corresponds to a 90° phase shift.

As another interesting facet of the time shift, consider a function $f_1(t)$ that is nonzero in the interval $0 \le t \le T_0/2$ and is zero in the interval $T_0/2 < t \le T_0$. For purposes of illustration, let us assume that $f_1(t)$ is the triangular waveform shown in Fig. 15.7a. $f_1(t - T_0/2)$ is then shown in Fig. 15.7b. Then the function $f(t)$ defined as

$$f(t) = f_1(t) - f_1\left(t - \frac{T_0}{2}\right) \qquad \textbf{15.32}$$

is shown in Fig. 15.7c. Note that $f(t)$ has half-wave symmetry. In addition, note that if

$$f_1(t) = \sum_{n=-\infty}^{\infty} \mathbf{c}_n e^{-jn\omega_0 t}$$

then

$$f(t) = f_1(t) - f_1\left(t - \frac{T_0}{2}\right) = \sum_{n=-\infty}^{\infty} \mathbf{c}_n\left(1 - e^{-jn\pi}\right)e^{jn\omega_0 t}$$

$$= \begin{cases} \displaystyle\sum_{n=-\infty}^{\infty} 2\mathbf{c}_n e^{jn\omega_0 t} & n \text{ odd} \\[2mm] 0 & n \text{ even} \end{cases} \qquad \textbf{15.33}$$

Figure 15.7
Waveforms that illustrate the generation of half-wave symmetry.

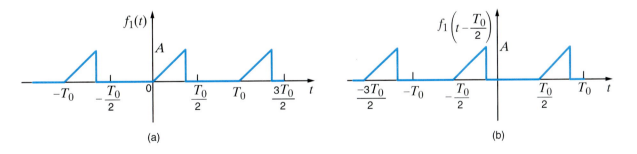

(a)   (b)

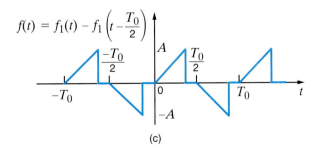

(c)

Therefore, we see that any function with half-wave symmetry can be expressed in the form of Eq. (15.32), where the Fourier series is defined by Eq. (15.33), and $\mathbf{c}_n$ is the Fourier coefficient for $f_1(t)$.

LEARNING EXTENSION

E15.6 If the waveform in Fig. E15.1 is time-delayed 1 s, we obtain the waveform in Fig. E15.6. Compute the exponential Fourier coefficients for the waveform in Fig. E15.6 and show that they differ from the coefficients for the waveform in Fig. E15.1 by an angle $n(180°)$.

ANSWER: $\mathbf{c}_0 = \dfrac{1}{2}$;

$$\mathbf{c}_n = -\left(\frac{1 - e^{-jn\pi}}{j2\pi n}\right).$$

Figure E15.6

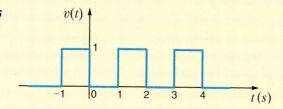

WAVEFORM GENERATION   The magnitude of the harmonics in a Fourier series is independent of the time scale for a given wave shape. Therefore, the equations for a variety of waveforms can be given in tabular form without expressing a specific time scale. Table 15.2 is a set of commonly occurring periodic waves where the advantage of symmetry has been used to simplify the coefficients. These waveforms can be used to generate other waveforms. The level of a wave can be adjusted by changing the average value component; the time can be shifted by adjusting the angle of the harmonics; and two waveforms can be added to produce a third waveform. For example, the waveforms in Figs. 15.8a and b can be added to produce the waveform in Fig. 15.8c.

Figure 15.8
Example of waveform generation.

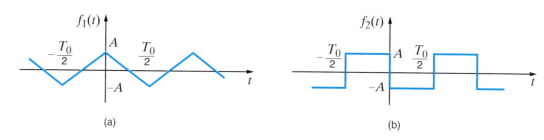

(a)   (b)

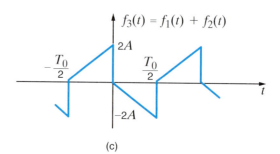

(c)

Table 15.2   Fourier series for some common waveforms

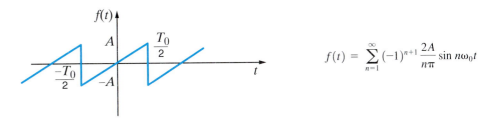

$$f(t) = \sum_{n=1}^{\infty} (-1)^{n+1} \frac{2A}{n\pi} \sin n\omega_0 t$$

$$f(t) = \sum_{\substack{n=1 \\ n\ \text{odd}}}^{\infty} \frac{8A}{n^2\pi^2} \sin \frac{n\pi}{2} \sin n\omega_0 t$$

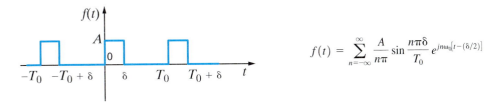

$$f(t) = \sum_{n=-\infty}^{\infty} \frac{A}{n\pi} \sin \frac{n\pi\delta}{T_0} e^{jn\omega_0[t-(\delta/2)]}$$

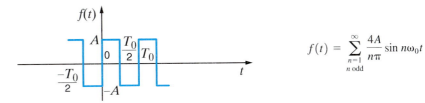

$$f(t) = \sum_{\substack{n=1 \\ n\ \text{odd}}}^{\infty} \frac{4A}{n\pi} \sin n\omega_0 t$$

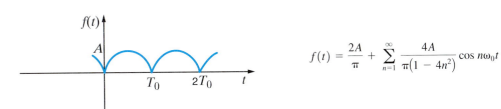

$$f(t) = \frac{2A}{\pi} + \sum_{n=1}^{\infty} \frac{4A}{\pi(1-4n^2)} \cos n\omega_0 t$$

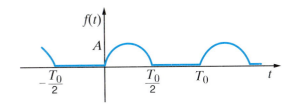

$$f(t) = \frac{A}{\pi} + \frac{A}{2} \sin \omega_0 t + \sum_{\substack{n=2 \\ n\ \text{even}}}^{\infty} \frac{2A}{\pi(1-n^2)} \cos n\omega_0 t$$

(continues on the next page)

Table 15.2 *(continued)*

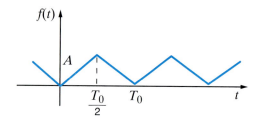

$$f(t) = \frac{A}{2} + \sum_{\substack{n=-\infty \\ n \neq 0 \\ n\ \text{odd}}}^{\infty} \frac{-2A}{n^2\pi^2}\, e^{jn\omega_0 t}$$

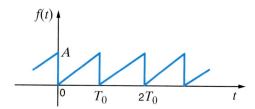

$$f(t) = \frac{A}{2} + \sum_{n=1}^{\infty} \frac{-A}{n\pi}\, \sin n\omega_0 t$$

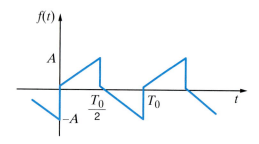

$$f(t) = \sum_{n=1}^{\infty} \frac{-4A}{\pi^2 n^2}\, \cos n\omega_0 t + \frac{2A}{\pi n}\, \sin n\omega_0 t$$

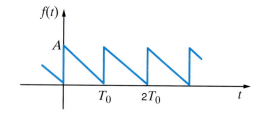

$$f(t) = \frac{A}{2} + \sum_{n=1}^{\infty} \frac{A}{\pi n}\, \sin n\omega_0 t$$

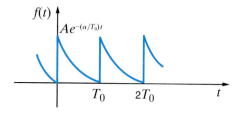

$$f(t) = \sum_{n=-\infty}^{\infty} \frac{A(1 - e^{-\alpha})}{\alpha + j2\pi n}\, e^{jn\omega_0 t}$$

LEARNING EXTENSION

E15.7 Two periodic waveforms are shown in Fig. E15.7. Compute the exponential Fourier series for each waveform, and then add the results to obtain the Fourier series for the waveform in Fig. E15.2.

ANSWER:

$$v_1(t) = \frac{2}{3} + \sum_{\substack{n=-\infty \\ n \neq 0}}^{\infty} \frac{2}{n\pi} \sin \frac{n\pi}{3} e^{jn\omega_0 t};$$

$$v_2(t) = \frac{4}{3} + \sum_{n=-\infty}^{\infty} -\frac{4}{n\pi}$$
$$\left(\sin \frac{n\pi}{3} - \sin \frac{2n\pi}{3} \right) e^{jn\omega_0 t}.$$

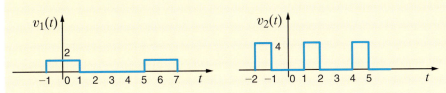

Figure E15.7

FREQUENCY SPECTRUM   The *frequency spectrum* of the function $f(t)$ expressed as a Fourier series consists of a plot of the amplitude of the harmonics versus frequency, which we call the *amplitude spectrum*, and a plot of the phase of the harmonics versus frequency, which we call the *phase spectrum*. Since the frequency components are discrete, the spectra are called *line spectra*. Such spectra illustrate the frequency content of the signal. Plots of the amplitude and phase spectra are based on Eqs. (15.1), (15.3), and (15.7) and represent the amplitude and phase of the signal at specific frequencies.

Example 15.7

The Fourier series for the triangular-type waveform shown in Fig. 15.8c with $A = 5$ is given by the equation

$$v(t) = \sum_{\substack{n=1 \\ n \text{ odd}}}^{\infty} \left(\frac{20}{n\pi} \sin n\omega_0 t - \frac{40}{n^2 \pi^2} \cos n\omega_0 t \right)$$

We wish to plot the first four terms of the amplitude and phase spectra for this signal.

SOLUTION   Since $D_n \underline{/\theta_n} = a_n - jb_n$, the first four terms for this signal are

$$D_1 \underline{/\theta_1} = -\frac{40}{\pi^2} - j\frac{20}{\pi} = 7.5 \underline{/-122°}$$

$$D_3 \underline{/\theta_3} = -\frac{40}{9\pi^2} - j\frac{20}{3\pi} = 2.2 \underline{/-102°}$$

$$D_5 \underline{/\theta_5} = -\frac{40}{25\pi^2} - j\frac{20}{5\pi} = 1.3 \underline{/-97°}$$

$$D_7 \underline{/\theta_7} = -\frac{40}{49\pi^2} - j\frac{20}{7\pi} = 0.91 \underline{/-95°}$$

Therefore, the plots of the amplitude and phase versus ω are as shown in Fig. 15.9.

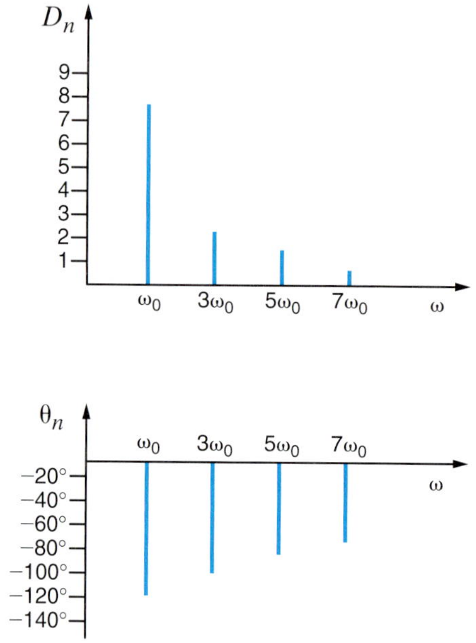

Figure 15.9
Amplitude and phase spectra.

D_n

9—
8—
7—
6—
5—
4—
3—
2—
1—

ω_0 $3\omega_0$ $5\omega_0$ $7\omega_0$ ω

θ_n

ω_0 $3\omega_0$ $5\omega_0$ $7\omega_0$ ω

−20°—
−40°—
−60°—
−80°—
−100°—
−120°—
−140°—

LEARNING EXTENSION

E15.8 Determine the trigonometric Fourier series for the voltage waveform in Fig. E15.8 and plot the first four terms of the amplitude and phase spectra for this signal. **PSV**

ANSWER: $a_0 = 1/2$;
$D_1 = -j(1/\pi)$;
$D_2 = -j(1/2\pi)$;
$D_3 = -j(1/3\pi)$;
$D_4 = -j(1/4\pi)$.

Figure E15.8

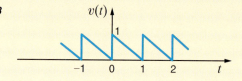

$v(t)$

1

−1 0 1 2 t

STEADY-STATE NETWORK RESPONSE   If a periodic signal is applied to a network, the steady-state voltage or current response at some point in the circuit can be found in the following manner. First, we represent the periodic forcing function by a Fourier series. If the input forcing function for a network is a voltage, the input can be expressed in the form

$$v(t) = v_0 + v_1(t) + v_2(t) + \cdots$$

and therefore represented in the time domain as shown in Fig. 15.10. Each source has its own amplitude and frequency. Next we determine the response due to each component of the input Fourier series; that is, we use phasor analysis in the frequency domain to determine the network response due to each source. The network response due to each source in the frequency domain is then transformed to the time domain. Finally, we add the time domain solutions due to each source using the Principle of Superposition to obtain the Fourier series for the total *steady-state* network response.

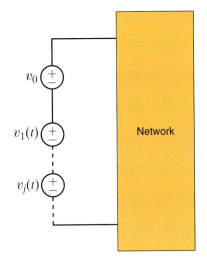

Figure 15.10
Network with a periodic voltage forcing function.

Example 15.8

We wish to determine the steady-state voltage $v_o(t)$ in Fig. 15.11 if the input voltage $v(t)$ is given by the expression

$$v(t) = \sum_{\substack{n=1 \\ n \text{ odd}}}^{\infty} \left(\frac{20}{n\pi} \sin 2nt - \frac{40}{n^2 \pi^2} \cos 2nt \right) \text{V}$$

SOLUTION   Note that this source has no constant term, and therefore its dc value is zero. The amplitude and phase for the first four terms of this signal are given in Example 15.7, and therefore the signal $v(t)$ can be written as

$$v(t) = 7.5 \cos(2t - 122°) + 2.2 \cos(6t - 102°)$$
$$+ 1.3 \cos(10t - 97°) + 0.91 \cos(14t - 95°) + \cdots$$

From the network we find that

$$\mathbf{I} = \frac{\mathbf{V}}{2 + \dfrac{2/j\omega}{2 + 1/j\omega}} = \frac{\mathbf{V}(1 + 2j\omega)}{4 + 4j\omega}$$

$$\mathbf{I}_1 = \frac{\mathbf{I}(1/j\omega)}{2 + 1/j\omega} = \frac{\mathbf{I}}{1 + 2j\omega}$$

$$\mathbf{V}_o = (1)\mathbf{I}_1 = 1 \cdot \frac{\mathbf{V}(1 + 2j\omega)}{4 + 4j\omega} \frac{1}{1 + 2j\omega} = \frac{\mathbf{V}}{4 + 4j\omega}$$

Therefore, since $\omega_0 = 2$,

$$\mathbf{V}_o(n) = \frac{\mathbf{V}(n)}{4 + j8n}$$

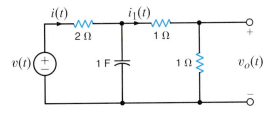

Figure 15.11
RC circuit employed in Example 15.8.

The individual components of the output due to the components of the input source are then

$$\mathbf{V}_o(\omega_0) = \frac{7.5\,\underline{/-122°}}{4 + j8} = 0.84\,\underline{/-185.4°}\ \text{V}$$

$$\mathbf{V}_o(3\omega_0) = \frac{2.2\,\underline{/-102°}}{4 + j24} = 0.09\,\underline{/-182.5°}\ \text{V}$$

$$\mathbf{V}_o(5\omega_0) = \frac{1.3\,\underline{/-97°}}{4 + j40} = 0.03\,\underline{/-181.3°}\ \text{V}$$

$$\mathbf{V}_o(7\omega_0) = \frac{0.91\,\underline{/-95°}}{4 + j56} = 0.016\,\underline{/-181°}\ \text{V}$$

Hence, the steady-state output voltage $v_o(t)$ can be written as

$$v_o(t) = 0.84 \cos(2t - 185.4°) + 0.09 \cos(6t - 182.5°)$$

$$+ \ 0.03 \cos(10t - 181.3°) + 0.016 \cos(14t - 181°) + \cdots \text{V}$$

AVERAGE POWER We have shown that when a linear network is forced with a non-sinusoidal periodic signal, voltages and currents throughout the network are of the form

$$v(t) = V_{\text{DC}} + \sum_{n=1}^{\infty} V_n \cos\left(n\omega_0 t - \theta_{v_n}\right)$$

and

$$i(t) = I_{\text{DC}} + \sum_{n=1}^{\infty} I_n \cos\left(n\omega_0 t - \theta_{i_n}\right)$$

If we employ the passive sign convention and assume that the voltage across an element and the current through it are given by the preceding equations, then from Eq. (9.6),

$$P = \frac{1}{T} \int_{t_0}^{t_0+T} p(t)\, dt$$

$$= \frac{1}{T} \int_{t_0}^{t_0+T} v(t)i(t)\, dt$$

15.34

Note that the integrand involves the product of two infinite series. However, the determination of the average power is actually easier than it appears. First, note that the product $V_{\text{DC}}I_{\text{DC}}$ when integrated over a period and divided by the period is simply $V_{\text{DC}}I_{\text{DC}}$. Second, the product of V_{DC} and any harmonic of the current or I_{DC} and any harmonic of the voltage when integrated over a period yields zero. Third, the product of any two *different* harmonics of the voltage and the current when integrated over a period yields zero. Finally, nonzero terms result only from the products of voltage and current at the *same* frequency. Hence, using the mathematical development that follows Eq. (9.6), we find that

$$P = V_{\text{DC}}I_{\text{DC}} + \sum_{n=1}^{\infty} \frac{V_n I_n}{2} \cos\left(\theta_{v_n} - \theta_{i_n}\right)$$

15.35

Example 15.9

In the network in Fig. 15.12, $v(t) = 42 + 16 \cos(377t + 30°) + 12 \cos(754t - 20°)$ V. We wish to compute the current $i(t)$ and determine the average power absorbed by the network.

SOLUTION   The capacitor acts as an open circuit to dc, and therefore $I_{DC} = 0$. At $\omega = 377$ rad/s.

$$\frac{1}{j\omega C} = \frac{1}{j(377)(100)(10)^{-6}} = -j26.53 \ \Omega$$

$$j\omega L = j(377)(20)10^{-3} = j7.54 \ \Omega$$

Hence,

$$\mathbf{I}_{377} = \frac{16 \underline{/30°}}{16 + j7.54 - j26.53} = 0.64 \underline{/79.88°} \text{ A}$$

At $\omega = 754$ rad/s,

$$\frac{1}{j\omega C} = \frac{1}{j(754)(100)(10)^{-6}} = -j13.26 \ \Omega$$

$$j\omega L = j(754)(20)10^{-3} = j15.08 \ \Omega$$

Hence,

$$\mathbf{I}_{754} = \frac{12 \underline{/20°}}{16 + j15.08 - j13.26} = 0.75 \underline{/-26.49°} \text{ A}$$

Therefore, the current $i(t)$ is

$$i(t) = 0.64 \cos(377t + 79.88°)$$

$$+ 0.75 \cos(754t + 26.49°) \text{ A}$$

and the average power absorbed by the network is

$$P = (42)(0) + \frac{(16)(0.64)}{2} \cos(30° - 79.88°)$$

$$+ \frac{(12)(0.75)}{2} \cos(-20° + 26.49°)$$

$$= 7.77 \text{ W}$$

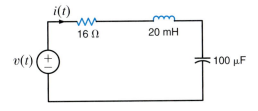

Figure 15.12
Network used in Example 15.9.

PROBLEM-SOLVING STRATEGY

Steady-State Response to Periodic Forcing Functions

Step 1. Determine the Fourier series for the periodic forcing function, which is now expressed as a summation of harmonically-related sinusoidal functions.

Step 2. Use phasor analysis to determine the network response due to each sinusoidal function acting alone.

Step 3. Use the Principle of Superposition to add the time domain solution from each source acting alone to determine the total steady-state network response.

Step 4. If you need to calculate the average power dissipated in a network element, determine the average power dissipated in that element due to each source acting alone and then sum these for the total power dissipation from the periodic forcing function.

LEARNING EXTENSIONS

E15.9 Determine the expression for the steady-state current $i(t)$ in Fig. E15.9 if the input voltage $v_S(t)$ is given by the expression **PSV**

$$v_S(t) = \frac{20}{\pi} + \sum_{n=1}^{\infty} \frac{-40}{\pi(4n^2 - 1)} \cos 2nt \text{ V}$$

ANSWER: $i(t) = 2.12$

$$+ \sum_{n=1}^{\infty} \frac{-40}{\pi(4n^2 - 1)} \frac{1}{A_n} \cos(2nt - \theta_n) \text{ A}.$$

Figure E15.9

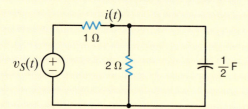

E15.10 At the input terminals of a network, the voltage $v(t)$ and the current $i(t)$ are given by the following expression:

$$v(t) = 64 + 36 \cos(377t + 60°) - 24 \cos(754t + 102°) \text{ V}$$

$$i(t) = 1.8 \cos(377t + 45°) + 1.2 \cos(754t + 100°) \text{ A}$$

Find the average power absorbed by the network. **PSV**

ANSWER:
$P_{\text{ave}} = 16.91$ W.

15.2 Fourier Transform

The preceding sections of this chapter have illustrated that the exponential Fourier series can be used to represent a periodic signal for all time. We will now consider a technique for representing an aperiodic signal for all values of time.

Suppose that an aperiodic signal $f(t)$ is as shown in Fig. 15.13a. We now construct a new signal $f_p(t)$ that is identical to $f(t)$ in the interval $-T/2$ to $T/2$ but is *periodic* with period T, as shown in Fig. 15.13b. Since $f_p(t)$ is periodic, it can be represented in the interval $-\infty$ to ∞ by an exponential Fourier series;

$$f_p(t) = \sum_{n=-\infty}^{\infty} \mathbf{c}_n e^{jn\omega_0 t} \qquad \textbf{15.36}$$

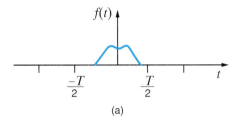

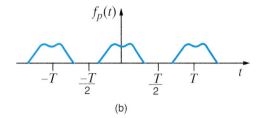

Figure 15.13
Aperiodic and periodic signals.

where

$$\mathbf{c}_n = \frac{1}{T} \int_{-T/2}^{T/2} f_p(t) e^{-jn\omega_0 t} \, dt \qquad\qquad \textbf{15.37}$$

and

$$\omega_0 = \frac{2\pi}{T} \qquad\qquad \textbf{15.38}$$

At this point we note that if we take the limit of the function $f_p(t)$ as $T \to \infty$, the periodic signal in Fig. 15.13b approaches the aperiodic signal in Fig. 15.13a; that is, the repetitious signals centered at $-T$ and $+T$ in Fig. 15.13b are moved to infinity.

The line spectrum for the periodic signal exists at harmonic frequencies $(n\omega_0)$, and the incremental spacing between the harmonics is

$$\Delta\omega = (n + 1)\omega_0 - n\omega_0 = \omega_0 = \frac{2\pi}{T} \qquad\qquad \textbf{15.39}$$

As $T \to \infty$ the lines in the frequency spectrum for $f_p(t)$ come closer and closer together, $\Delta\omega$ approaches the differential $d\omega$, and $n\omega_0$ can take on any value of ω. Under these conditions, the line spectrum becomes a continuous spectrum. Since as $T \to \infty$, $\mathbf{c}_n \to 0$ in Eq. (15.37), we will examine the product $\mathbf{c}_n T$, where

$$\mathbf{c}_n T = \int_{-T/2}^{T/2} f_p(t) e^{-jn\omega_0 t} \, dt$$

In the limit as $T \to \infty$,

$$\lim_{T \to \infty} (\mathbf{c}_n T) = \lim_{T \to \infty} \int_{-T/2}^{T/2} f_p(t) e^{-jn\omega_0 t} \, dt$$

which in view of the previous discussion can be written as

$$\lim_{T \to \infty} (\mathbf{c}_n T) = \int_{-\infty}^{\infty} f(t) e^{-j\omega t} \, dt$$

This integral is the Fourier transform of $f(t)$, which we will denote as $\mathbf{F}(\omega)$, and hence

$$\mathbf{F}(\omega) = \int_{-\infty}^{\infty} f(t) e^{-j\omega t} \, dt \qquad\qquad \textbf{15.40}$$

Similarly, $f_p(t)$ can be expressed as

$$f_p(t) = \sum_{n=-\infty}^{\infty} \mathbf{c}_n e^{jn\omega_0 t}$$

$$= \sum_{n=-\infty}^{\infty} (\mathbf{c}_n T) e^{jn\omega_0 t} \frac{1}{T}$$

$$= \sum_{n=-\infty}^{\infty} (\mathbf{c}_n T) e^{jn\omega_0 t} \frac{\Delta\omega}{2\pi}$$

which in the limit as $T \to \infty$ becomes

$$f(t) = \frac{1}{2\pi} \int_{-\infty}^{\infty} \mathbf{F}(\omega) e^{j\omega t} \, d\omega \qquad\qquad \textbf{15.41}$$

Equations (15.40) and (15.41) constitute what is called the *Fourier transform pair*. Since $\mathbf{F}(\omega)$ is the Fourier transform of $f(t)$ and $f(t)$ is the inverse Fourier transform of $\mathbf{F}(\omega)$, they are normally expressed in the form

$$\mathbf{F}(\omega) = \mathcal{F}\big[f(t)\big] = \int_{-\infty}^{\infty} f(t)e^{-j\omega t}\, dt \qquad \textbf{15.42}$$

$$f(t) = \mathcal{F}^{-1}\big[\mathbf{F}(\omega)\big] = \frac{1}{2\pi}\int_{-\infty}^{\infty} \mathbf{F}(\omega)e^{j\omega t}\, d\omega \qquad \textbf{15.43}$$

SOME IMPORTANT TRANSFORM PAIRS There are a number of important Fourier transform pairs, and in the following material we derive a number of them and then list some of the more common ones in tabular form.

Example 15.10

We wish to derive the Fourier transform for the voltage pulse shown in Fig. 15.14a.

SOLUTION Using Eq. (15.42), the Fourier transform is

$$\mathbf{F}(\omega) = \int_{-\delta/2}^{\delta/2} V e^{-j\omega t}\, dt$$

$$= \frac{V}{-j\omega} e^{-j\omega t}\bigg|_{-\delta/2}^{\delta/2}$$

$$= V\,\frac{e^{-j\omega\delta/2} - e^{+j\omega\delta/2}}{-j\omega}$$

$$= V\delta\,\frac{\sin(\omega\delta/2)}{\omega\delta/2}$$

Therefore, the Fourier transform for the function

$$f(t) = \begin{cases} 0 & -\infty < t \le -\dfrac{\delta}{2} \\[2mm] V & -\dfrac{\delta}{2} < t \le \dfrac{\delta}{2} \\[2mm] 0 & \dfrac{\delta}{2} < t < \infty \end{cases}$$

is

$$\mathbf{F}(\omega) = V\delta\,\frac{\sin(\omega\delta/2)}{\omega\delta/2}$$

A plot of this function is shown in Fig. 15.14c. Let us explore this example even further. Consider now the pulse train shown in Fig. 15.14b. Using the techniques that have been demonstrated earlier, we can show that the Fourier coefficients for this waveform are

$$\mathbf{c}_n = \frac{V\delta}{T_0}\,\frac{\sin(n\omega_0\delta/2)}{n\omega_0\delta/2}$$

The line spectrum for $T_0 = 5\delta$ is shown in Fig. 15.14d.

The equations and figures in this example indicate that as $T_0 \to \infty$ and the periodic function becomes aperiodic, the lines in the discrete spectrum become denser and the amplitude gets smaller, and the amplitude spectrum changes from a line spectrum to a continuous spectrum. Note that the envelope for the discrete spectrum has the same shape as the continuous spectrum. Since the Fourier series represents the amplitude and phase of the signal at specific frequencies, the Fourier transform also specifies the frequency content of a signal.

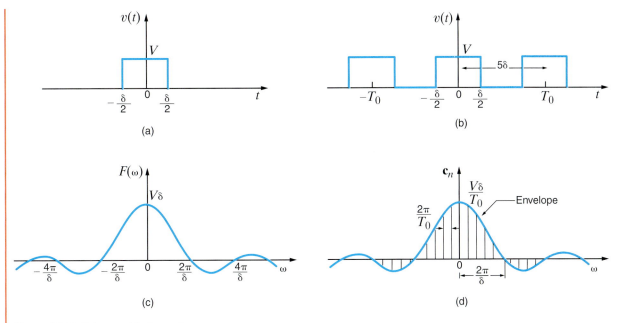

Figure 15.14 *Pulses and their spectra.*

Example 15.11

Find the Fourier transform for the unit impulse function $\delta(t)$.

SOLUTION   The Fourier transform of the unit impulse function $\delta(t - a)$ is

$$\mathbf{F}(\omega) = \int_{-\infty}^{\infty} \delta(t - a)e^{-j\omega t}\, dt$$

Using the sampling property of the unit impulse, we find that

$$\mathbf{F}(\omega) = e^{-j\omega a}$$

and if $a = 0$, then

$$\mathbf{F}(\omega) = 1$$

Note then that the $\mathbf{F}(\omega)$ for $f(t) = \delta(t)$ is *constant for all frequencies*. This is an important property, as we shall see later.

Example 15.12

We wish to determine the Fourier transform of the function $f(t) = e^{j\omega_0 t}$.

SOLUTION   In this case note that if $\mathbf{F}(\omega) = 2\pi\delta(\omega - \omega_0)$, then

$$f(t) = \frac{1}{2\pi} \int_{-\infty}^{\infty} 2\pi\delta(\omega - \omega_0)e^{j\omega t}\, d\omega$$

$$= e^{j\omega_0 t}$$

Therefore, $f(t) = e^{j\omega_0 t}$ and $\mathbf{F}(\omega) = 2\pi\delta(\omega - \omega_0)$ represent a Fourier transform pair.

LEARNING EXTENSION

E15.11 If $f(t) = \sin \omega_0 t$, find $\mathbf{F}(\omega)$.

ANSWER:
$$\mathbf{F}(\omega) = \pi j[\delta(\omega + \omega_0) - \delta(\omega - \omega_0)].$$

A number of useful Fourier transform pairs are shown in Table 15.3.

Table 15.3   Fourier transform pairs

$f(t)$	$\mathbf{F}(\omega)$
$\delta(t - a)$	$e^{-j\omega a}$
A	$2\pi A\delta(\omega)$
$e^{j\omega_0 t}$	$2\pi\delta(\omega - \omega_0)$
$\cos \omega_0 t$	$\pi\delta(\omega - \omega_0) + \pi\delta(\omega + \omega_0)$
$\sin \omega_0 t$	$j\pi\delta(\omega + \omega_0) - j\pi\delta(\omega - \omega_0)$
$e^{-at}u(t), a > 0$	$\dfrac{1}{a + j\omega}$
$e^{-\alpha|t|}, a > 0$	$\dfrac{2a}{a^2 + \omega^2}$
$e^{-at}\cos \omega_0 t u(t), a > 0$	$\dfrac{j\omega + a}{(j\omega + a)^2 + \omega_0^2}$
$e^{-at}\sin \omega_0 t u(t), a > 0$	$\dfrac{\omega_0}{(j\omega + a)^2 + \omega_0^2}$

SOME PROPERTIES OF THE FOURIER TRANSFORM   The Fourier transform defined by the equation

$$\mathbf{F}(\omega) = \int_{-\infty}^{\infty} f(t)e^{-j\omega t}\, dt$$

has a number of important properties. Table 15.4 provides a short list of a number of these properties.

The proofs of these properties are generally straightforward; however, as an example we will demonstrate the time convolution property.

If $\mathcal{F}[f_1(t)] = \mathbf{F}_1(\omega)$ and $\mathcal{F}[f_2(t)] = \mathbf{F}_2(\omega)$, then

$$\mathcal{F}\left[\int_{-\infty}^{\infty} f_1(x)f_2(t - x)\, dx\right] = \int_{t=-\infty}^{\infty}\int_{x=-\infty}^{\infty} f_1(x)f_2(t - x)\, dx\, e^{-j\omega t}\, dt$$

$$= \int_{x=-\infty}^{\infty} f_1(x)\int_{t=-\infty}^{\infty} f_2(t - x)e^{-j\omega t}\, dt\, dx$$

If we now let $u = t - x$, then

$$\mathcal{F}\left[\int_{-\infty}^{\infty} f_1(x)f_2(t - x)\, dx\right] = \int_{x=-\infty}^{\infty} f_1(x)\int_{u=-\infty}^{\infty} f_2(u)e^{-j\omega(u+x)}\, du\, dx$$

$$= \int_{x=-\infty}^{\infty} f_1(x)e^{-j\omega x}\int_{u=-\infty}^{\infty} f_2(u)e^{-j\omega u}\, du\, dx$$

$$= \mathbf{F}_1(\omega)\mathbf{F}_2(\omega) \qquad\qquad\qquad \textbf{15.44}$$

Table 15.4   Properties of the Fourier transform

$f(t)$	$F(\omega)$	Property
$Af(t)$	$A\mathbf{F}(\omega)$	Linearity
$f_1(t) \pm f_2(t)$	$\mathbf{F}_1(\omega) \pm \mathbf{F}_2(\omega)$	
$f(at)$	$\dfrac{1}{a}\mathbf{F}\left(\dfrac{\omega}{a}\right), a > 0$	Time-scaling
$f(t - t_0)$	$e^{-j\omega t_0}\mathbf{F}(\omega)$	Time-shifting
$e^{j\omega t_0}f(t)$	$\mathbf{F}(\omega - \omega_0)$	Modulation
$\dfrac{d^n f(t)}{dt^n}$	$(j\omega)^n\mathbf{F}(\omega)$	
$t^n f(t)$	$(j)^n\dfrac{d^n\mathbf{F}(\omega)}{d\omega^n}$	Differentiation
$\displaystyle\int_{-\infty}^{\infty} f_1(x)f_2(t - x)\, dx$	$\mathbf{F}_1(\omega)\mathbf{F}_2(\omega)$	
$f_1(t)f_2(t)$	$\dfrac{1}{2\pi}\displaystyle\int_{-\infty}^{\infty}\mathbf{F}_1(x)\mathbf{F}_2(\omega - x)\, dx$	Convolution

We should note very carefully the time convolution property of the Fourier transform. With reference to Fig. 15.15, this property states that if $\mathbf{V}_i(\omega) = \mathcal{F}[v_i(t)]$, $\mathbf{H}(\omega) = \mathcal{F}[h(t)]$, and $\mathbf{V}_o(\omega) = \mathcal{F}[v_o(t)]$, then

$$\mathbf{V}_o(\omega) = \mathbf{H}(\omega)\mathbf{V}_i(\omega) \qquad \textbf{15.45}$$

where $\mathbf{V}_i(\omega)$ represents the input signal, $\mathbf{H}(\omega)$ is the network transfer function, and $\mathbf{V}_o(\omega)$ represents the output signal. Equation (15.45) tacitly assumes that the initial conditions of the network are zero.

Figure 15.15
Representation of the time convolution property.

LEARNING EXTENSION

E15.12 Determine the output $v_o(t)$ in Fig. E15.12 if the signal $v_i(t) = e^{-t}u(t)$ V, the network impulse response $h(t) = e^{-2t}u(t)$, and all initial conditions are zero. **PSV**

ANSWER:
$v_o(t) = (e^{-t} - e^{-2t})u(t)$ V.

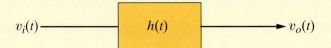

Figure E15.12

PARSEVAL'S THEOREM A mathematical statement of Parseval's theorem is

$$\int_{-\infty}^{\infty} f^2(t)\, dt = \frac{1}{2\pi} \int_{-\infty}^{\infty} |\mathbf{F}(\omega)|^2\, d\omega \qquad \textbf{15.46}$$

This relationship can be easily derived as follows:

$$\int_{-\infty}^{\infty} f^2(t)\, dt = \int_{-\infty}^{\infty} f(t) \frac{1}{2\pi} \int_{-\infty}^{\infty} \mathbf{F}(\omega) e^{j\omega t}\, d\omega\, dt$$

$$= \frac{1}{2\pi} \int_{-\infty}^{\infty} \mathbf{F}(\omega) \int_{-\infty}^{\infty} f(t) e^{-j(-\omega)t}\, dt\, d\omega$$

$$= \int_{-\infty}^{\infty} \frac{1}{2\pi} \mathbf{F}(\omega)\mathbf{F}(-\omega)\, d\omega$$

$$= \int_{-\infty}^{\infty} \frac{1}{2\pi} \mathbf{F}(\omega)\mathbf{F}^*(\omega)\, d\omega$$

$$= \int_{-\infty}^{\infty} \frac{1}{2\pi} |\mathbf{F}(\omega)|^2\, d\omega$$

The importance of Parseval's theorem can be seen if we imagine that $f(t)$ represents the current in a 1-Ω resistor. Since $f^2(t)$ is power and the integral of power over time is energy, Eq. (15.46) shows that we can compute this 1-Ω energy or normalized energy in either the time domain or the frequency domain.

Example 15.13

Using the transform technique, we wish to determine $v_o(t)$ in Fig. 15.16 if (a) $v_i(t) = 5e^{-2t}u(t)$ V and (b) $v_i(t) = 5\cos 2t$ V.

Figure 15.16
Simple RL *circuit.*

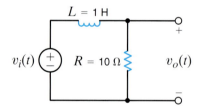

SOLUTION

a. In this case since $v_i(t) = 5e^{-2t}u(t)$ V, then

$$\mathbf{V}_i(\omega) = \frac{5}{2 + j\omega}\ \text{V}$$

$\mathbf{H}(\omega)$ for the network is

$$\mathbf{H}(\omega) = \frac{R}{R + j\omega L}$$

$$= \frac{10}{10 + j\omega}$$

From Eq. (15.45),

$$\mathbf{V}_o(\omega) = \mathbf{H}(\omega)\mathbf{V}_i(\omega)$$

$$= \frac{50}{(2 + j\omega)(10 + j\omega)}$$

$$= \frac{50}{8} \left(\frac{1}{2 + j\omega} - \frac{1}{10 + j\omega} \right)$$

Hence, from Table 15.3, we see that

$$v_o(t) = 6.25\big[e^{-2t}u(t) - e^{-10t}u(t)\big] \text{ V}$$

b. In this case, since $v_i(t) = 5\cos 2t$,

$$\mathbf{V}_i(\omega) = 5\pi\delta(\omega - 2) + 5\pi\delta(\omega + 2) \text{ V}$$

The output voltage in the frequency domain is then

$$\mathbf{V}_o(\omega) = \frac{50\pi\big[\delta(\omega - 2) + \delta(\omega + 2)\big]}{(10 + j\omega)}$$

Using the inverse Fourier transform gives us

$$v_o(t) = \mathcal{F}^{-1}\big[\mathbf{V}_o(\omega)\big] = \frac{1}{2\pi}\int_{-\infty}^{\infty} 50\pi\,\frac{\delta(\omega - 2) + \delta(\omega + 2)}{10 + j\omega}\,e^{j\omega t}\,d\omega$$

Employing the sampling property of the unit impulse function, we obtain

$$\begin{aligned}
v_o(t) &= 25\left(\frac{e^{j2t}}{10 + j2} + \frac{e^{-j2t}}{10 - j2}\right)\\[4pt]
&= 25\left(\frac{e^{j2t}}{10.2e^{j11.31°}} + \frac{e^{-j2t}}{10.2e^{-j1.31°}}\right)\\[4pt]
&= 4.90\cos(2t - 11.31°) \text{ V}
\end{aligned}$$

This result can be easily checked using phasor analysis.

15.3   Application Examples

Application Example 15.14

Consider the network shown in Fig. 15.17a. This network represents a simple low-pass filter as shown in Chapter 12. We wish to illustrate the impact of this network on the input signal by examining the frequency characteristics of the output signal and the relationship between the 1-Ω or normalized energy at the input and output of the network.

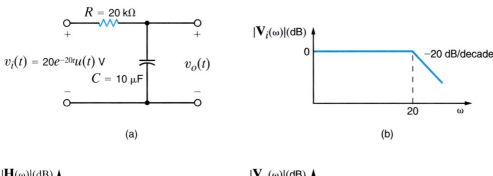

(a)   (b)

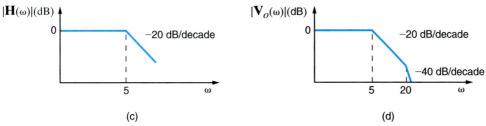

(c)   (d)

Figure 15.17 *Low-pass filter, its frequency characteristic, and its output spectra.*

SOLUTION   The network transfer function is

$$\mathbf{H}(\omega) = \frac{1/RC}{1/RC + j\omega} = \frac{5}{5 + j\omega} = \frac{1}{1 + 0.2j\omega}$$

The Fourier transform of the input signal is

$$\mathbf{V}_i(\omega) = \frac{20}{20 + j\omega} = \frac{1}{1 + 0.05j\omega}$$

Then, using Eq. (15.45), the Fourier transform of the output is

$$\mathbf{V}_o(\omega) = \frac{1}{(1 + 0.2j\omega)(1 + 0.05j\omega)}$$

Using the techniques of Chapter 12, we note that the straight-line log-magnitude plot (frequency characteristic) for these functions is shown in Figs. 15.17b–d. Note that the low-pass filter passes the low frequencies of the input signal but attenuates the high frequencies.

The normalized energy at the filter input is

$$W_i = \int_0^\infty \left(20e^{-20t}\right)^2 dt$$

$$= \frac{400}{-40} e^{-40t} \bigg|_0^\infty$$

$$= 10 \text{ J}$$

The normalized energy at the filter output can be computed using Parseval's theorem. Since

$$\mathbf{V}_o(\omega) = \frac{100}{(5 + j\omega)(20 + j\omega)}$$

and

$$|\mathbf{V}_o(\omega)|^2 = \frac{10^4}{(\omega^2 + 25)(\omega^2 + 400)}$$

$|\mathbf{V}_o(\omega)|^2$ is an even function, and therefore

$$W_o = 2\left(\frac{1}{2\pi}\right) \int_0^\infty \frac{10^4 \, d\omega}{(\omega^2 + 25)(\omega^2 + 400)}$$

However, we can use the fact that

$$\frac{10^4}{(\omega^2 + 25)(\omega^2 + 400)} = \frac{10^4/375}{\omega^2 + 25} - \frac{10^4/375}{\omega^2 + 400}$$

Then

$$W_o = \frac{1}{\pi} \left(\int_0^\infty \frac{10^4/375}{\omega^2 + 25} \, d\omega - \int_0^\infty \frac{10^4/375}{\omega^2 + 400} \, d\omega \right)$$

$$= \frac{10^4}{375}\left(\frac{1}{\pi}\right)\left[\frac{1}{5}\left(\frac{\pi}{2}\right) - \frac{1}{20}\left(\frac{\pi}{2}\right)\right]$$

$$= 2.0 \text{ J}$$

Example 15.14 illustrates the effect that $\mathbf{H}(\omega)$ has on the frequency spectrum of the input signal. In general, $\mathbf{H}(\omega)$ can be selected to shape that spectrum in some prescribed manner. As an illustration of this effect, consider the *ideal* frequency spectrums shown in Fig. 15.18.

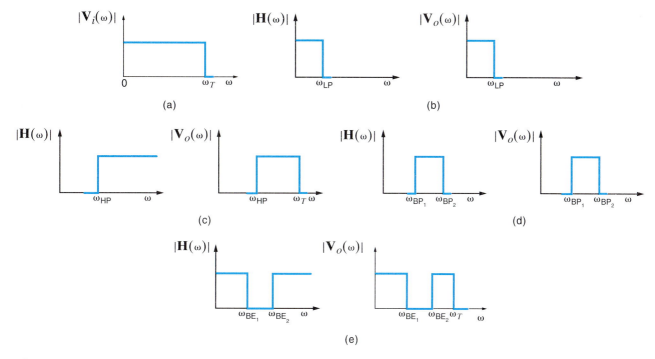

Figure 15.18 *Frequency spectra for the input and output of ideal low-pass, high-pass, band-pass, and band-elimination filters.*

Figure 15.18a shows an ideal input magnitude spectrum $|\mathbf{V}_i(\omega)|$. $|\mathbf{H}(\omega)|$ and the output magnitude spectrum $|\mathbf{V}_o(\omega)|$, which are related by Eq. (15.45), are shown in Figs. 15.18b–e for an *ideal* low-pass, high-pass, band-pass, and band-elimination filter, respectively.

LEARNING EXTENSIONS

E15.13 Compute the total 1-Ω energy content of the signal $v_i(t) = e^{-2t}u(t)$ V using both the time-domain and frequency-domain approaches.

ANSWER: $W_T = 0.25$ J.

E15.14 Compute the 1-Ω energy content of the signal $v_i(t) = e^{-2t}u(t)$ V in the frequency from 0 to 1 rad/s.   **PSV**

ANSWER: $W = 0.07$ J.

We note that by using Parseval's theorem we can compute the total energy content of a signal using either a time-domain or frequency-domain approach. However, the frequency-domain approach is more flexible in that it permits us to determine the energy content of a signal within some specified frequency band.

Application Example 15.15

In AM (amplitude modulation) radio, there are two very important waveforms—the signal, $s(t)$, and the carrier. All of the information we desire to transmit, voice, music, and so on, is contained in the signal waveform, which is in essence transported by the carrier. Therefore, the Fourier transform of $s(t)$ contains frequencies from about 50 Hz to 20,000 Hz. The carrier, $c(t)$, is a sinusoid oscillating at a frequency much greater than those in $s(t)$. For example, the FCC (Federal Communications Commission) rules and regulations have allocated the frequency range 540 kHz to 1.7 MHz for AM radio station carrier frequencies. Even the lowest possible carrier frequency allocation of 540 kHz is much greater than the audio frequencies in $s(t)$. In fact, when a station broadcasts its call letters and frequency, they are telling you the carrier's frequency, which the FCC assigned to that station!

In simple cases, the signal, $s(t)$, is modified to produce a voltage of the form,

$$v(t) = [A + s(t)]\cos(\omega_c t)$$

where A is a constant and ω_c is the carrier frequency in rad/s. This voltage, $v(t)$, with the signal "coded" within, is sent to the antenna and is broadcast to the public whose radios "pick up" a faint replica of the waveform $v(t)$.

Let us plot the magnitude of the Fourier transform of both $s(t)$ and $v(t)$ given that $s(t)$ is

$$s(t) = \cos(2\pi f_a t)$$

where f_a is 1000 Hz, the carrier frequency is 900 kHz, and the constant A is unity.

SOLUTION   The Fourier transform of $s(t)$ is

$$\mathbf{S}(\omega) = \mathcal{F}[\cos(\omega_a t)] = \pi\delta(\omega - \omega_a) + \pi\delta(\omega + \omega_a)$$

and is shown in Fig. 15.19.

Figure 15.19
*Fourier transform
magnitude for s(t)
versus frequency.*

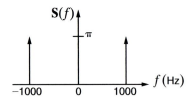

The voltage $v(t)$ can be expressed in the form

$$v(t) = [1 + s(t)]\cos(\omega_c t) = \cos(\omega_c t) + s(t)\cos(\omega_c t)$$

The Fourier transform for the carrier is

$$\mathcal{F}[\cos(\omega_c t)] = \pi\delta(\omega - \omega_c) + \pi\delta(\omega + \omega_c)$$

The term $s(t)\cos(\omega_c t)$ can be written as

$$s(t)\cos(\omega_c t) = s(t)\left\{\frac{e^{j\omega_c t} + e^{-j\omega_c t}}{2}\right\}$$

Using the property of modulation as given in Table 15.4, we can express the Fourier transform of $s(t)\cos(\omega_c t)$ as

$$\mathcal{F}[s(t)\cos(\omega_c t)] = \frac{1}{2}\{\mathbf{S}(\omega - \omega_c) + \mathbf{S}(\omega + \omega_c)\}$$

Employing $\mathbf{S}(\omega)$, we find

$$\begin{aligned}
\mathcal{F}[s(t)\cos(\omega_c t)] &= \mathcal{F}[\cos(\omega_a t)\cos(\omega_c t)] \\
&= \frac{\pi}{2}\{\delta(\omega - \omega_a - \omega_c) + \delta(\omega + \omega_a - \omega_c) \\
&\quad + \delta(\omega - \omega_a + \omega_c) + \delta(\omega + \omega_a + \omega_c)\}
\end{aligned}$$

Finally, the Fourier transform of $v(t)$ is

$$\begin{aligned}
\mathbf{V}(\omega) = \frac{\pi}{2}\{&2\delta(\omega - \omega_c) + 2\delta(\omega + \omega_c) + \delta(\omega - \omega_a - \omega_c) \\
&+ \delta(\omega + \omega_a - \omega_c) + \delta(\omega - \omega_a + \omega_c) + \delta(\omega + \omega_a + \omega_c)\}
\end{aligned}$$

which is shown in Fig. 15.20. Notice that $\mathbf{S}(\omega)$ is centered about the carrier frequency. This is the effect of modulation.

Figure 15.20
*Fourier transform of
the transmitted
waveform v(t) versus
frequency.*

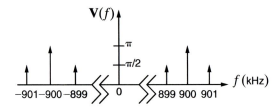

Application Example 15.16

Harmonics can be quite detrimental in a power distribution system. As an example, consider the scenario shown in Fig. 15.21 where a nonlinear three-phase load creates harmonic currents on the lines. Table 15.5 shows the magnitude of the current for each of the harmonics. If the resistance of each line is 0.2 Ω, what is the total power loss in the system? In addition, how much of the power loss is caused by the harmonic content?

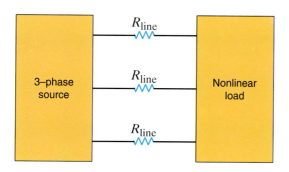

Figure 15.21
A simple model for a three-phase distribution system connecting a nonlinear load.

Table 15.5   Harmonic line current content for a nonlinear load.

Harmonic	Magnitude (A)
Fundamental	100
1st	5
2nd	30
3rd	1
4th	10

SOLUTION   The power loss in a line at any single frequency is

$$P_{line} = \frac{I_{line}^2}{2} R_{line}$$

Using this equation, we can calculate the power loss in a line for the fundamental and each harmonic frequency. The results of this calculation are shown in Table 15.6.

Table 15.6   Line power loss at the fundamental and each harmonic frequency.

Harmonic	P_{line} (W)
Fundamental	1000
1st	0.5
2nd	90
3rd	0.1
4th	10

The power lost in each line is simply the sum of each of the powers shown in Table 15.6. Since this is a three-phase system, the total power that is lost in the lines is simply

$$P_{total} = 3[1000 + 0.5 + 90 + 0.1 + 10] = 3301.8 \text{ W}$$

Note that the harmonics account for 301.8 W, or 9.14% of the total power that is lost.

15.4 Design Examples

Design Example 15.17

Two nearby AM stations are broadcasting at carrier frequencies

$$f_1 = 900 \text{ kHz}$$

and

$$f_2 = 960 \text{ kHz}$$

respectively. To simplify the analysis, we will assume that the information signals, $s_1(t)$ and $s_2(t)$, are identical. It follows that the Fourier transforms $\mathbf{S}_1(\omega)$ and $\mathbf{S}_2(\omega)$ are also identical, and a sketch of what they might look like is shown in Fig. 15.22.

Figure 15.22
Sketch of an arbitrary AM Fourier transform.

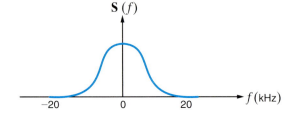

The broadcast waveforms are

$$v_1(t) = \big[1 + s_1(t)\big]\cos(\omega_1 t)$$

and

$$v_2(t) = \big[1 + s_2(t)\big]\cos(\omega_2 t)$$

An antenna in the vicinity will "pick up" both broadcasts. Assuming that $v_1(t)$ and $v_2(t)$ are of equal strength at the antenna, the voltage received is

$$v_r(t) = K\big[v_1(t) + v_2(t)\big]$$

where K is a constant much less than 1. (Typical antenna voltages are in the μV-to-mV range). A sketch of the Fourier transform of $v_r(t)$ is shown in Fig. 15.23.

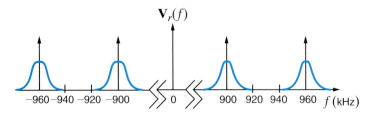

Figure 15.23 *Fourier transform of the antenna waveform, $v_r(t)$*

Before passing $v_r(t)$ on to amplifying and decoding circuitry, we must first employ a tuner to select a particular station. Let us design an *RLC* band-pass filter that contains a variable capacitor to serve as our tuner. Such a circuit is shown in Fig. 15.24.

Figure 15.24
RLC band-pass filter-tuner circuit.

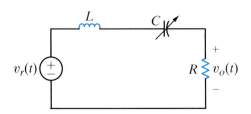

SOLUTION The transfer function is easily found to be

$$\mathbf{G}_v(s) = \frac{\mathbf{V}_o(s)}{\mathbf{V}_r(s)} = \frac{s\left(\dfrac{R}{L}\right)}{s^2 + s\left(\dfrac{R}{L}\right) + \dfrac{1}{LC}}$$

As shown in Chapter 12, the center frequency and bandwidth can be expressed in hertz as

$$f_o = \frac{1}{2\pi\sqrt{LC}}$$

and

$$\mathrm{BW} = \frac{1}{2\pi}\frac{R}{L}$$

Since the two carrier frequencies are separated by only 60 kHz, the filter bandwidth should be less than 60 kHz. Let us arbitrarily choose a bandwidth of 10 kHz and $R = 10\ \Omega$. Based on this selection, our design involves determining the resulting L and C values. From the expression for bandwidth,

$$L = \frac{1}{2\pi}\frac{R}{\mathrm{BW}}$$

or

$$L = 159.2\ \mu\mathrm{H}$$

Placing the center frequency at 900 kHz, we find C to be

$$C = \frac{1}{L[2\pi f_o]^2}$$

or

$$C = 196.4\ \mathrm{pF}$$

To tune to 960 kHz we need only change C to 172.6 pF and the bandwidth is unchanged. A Bode plot of the magnitude of $\mathbf{G}_v(s)$ tuned to 960 kHz is shown in Fig. 15.25. Note that the broadcast at 900 kHz, though attenuated, is not completely eliminated. If this is a problem, we can either narrow the bandwidth through R and/or L or design a more complex tuner filter.

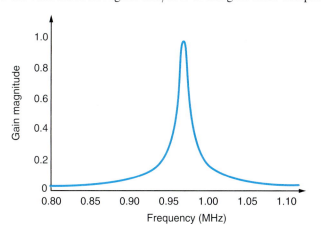

Figure 15.25
Bode plot for RLC *tuner circuit of Fig. 15.24.*

Design Example 15.18

The signal expressed in Eq. (15.47) describes a 10-kHz signal swamped in noise that has two frequency components—1 kHz and 100 kHz. From the equation we see that the signal amplitude is only 1/10 that of the noise components. Let us use the circuit in Fig. 15.26 to design a band-pass filter such that the signal amplitude is 100 times that of the noise components. Assume that the op-amps are ideal.

$$v_S(t) = 0.1\sin\left[(2\pi)10^3 t\right]$$
$$+ 0.01\sin\left[(2\pi)10^4 t\right] + 0.1\sin\left[(2\pi)10^5 t\right]\ \mathrm{V} \qquad \mathbf{15.47}$$

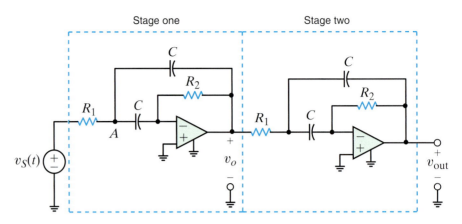

Figure 15.26 *A two-stage, fourth-order band-pass filter.*

SOLUTION Note that the band-pass filter in Fig. 15.26 consists of two identical cascaded stages. We need only determine the gain of a single stage, $\mathbf{A}_1(j\omega)$, since the total gain is

$$\mathbf{A}(j\omega) = \mathbf{A}_1(j\omega)\mathbf{A}_1(j\omega) = \left[\mathbf{A}_1(j\omega)\right]^2$$

Applying KCL at the first op-amp's inverting input, we have

$$\mathbf{V}_\mathbf{A} j\omega C = \frac{-\mathbf{V}_o}{R_2}$$

or

$$\mathbf{V}_\mathbf{A} = \frac{-\mathbf{V}_o}{j\omega C R_2} \qquad\qquad \mathbf{15.48}$$

Using KCL at node A yields

$$\frac{\mathbf{V}_S - \mathbf{V}_\mathbf{A}}{R_1} = \left(\mathbf{V}_\mathbf{A} - \mathbf{V}_o\right)j\omega C + \mathbf{V}_\mathbf{A} j\omega C$$

Multiplying both sides by R_1 and collecting terms produces

$$\mathbf{V}_S = \mathbf{V}_\mathbf{A}\left[2j\omega C R_1 + 1\right] - \mathbf{V}_o j\omega C R_1$$

Substituting from Eq. (15.48) and solving for the gain, $\mathbf{A}_1(j\omega) = \mathbf{V}_o/\mathbf{V}_S$,

$$\mathbf{A}_1(j\omega) = \frac{\mathbf{V}_o}{\mathbf{V}_S} = \frac{-j\omega C R_2}{-\omega^2 C^2 R_1 R_2 + j\omega 2 C R_1 + 1}$$

Finally, we rearrange the gain expression in the form

$$\frac{\mathbf{V}_o}{\mathbf{V}_S} = \frac{-\left[\dfrac{1}{CR_1}\right]j\omega}{-\omega^2 + \left[\dfrac{2}{CR_2}\right]j\omega + \dfrac{1}{C^2 R_1 R_2}} \qquad\qquad \mathbf{15.49}$$

which matches the general form of a band-pass filter given by the expression

$$\frac{\mathbf{V}_o}{\mathbf{V}_S} = \frac{A_o\left[\dfrac{\omega_0}{Q}\right]j\omega}{-\omega^2 + \left[\dfrac{\omega_0}{Q}\right]j\omega + \omega_0^2} \qquad\qquad \mathbf{15.50}$$

Comparing Eqs. (15.49) and (15.50), we find

$$\omega_0 = \frac{1}{C\sqrt{R_1 R_2}} \qquad \frac{\omega_0}{Q} = \frac{2}{CR_2}$$

$$Q = \frac{1}{2}\sqrt{\frac{R_2}{R_1}} \qquad A_o = \frac{-R_2}{2R_1} \qquad\qquad \mathbf{15.51}$$

There are two requirements for the filter performance. First, given the signal and noise amplitudes at the filter input, producing the desired ratio of signal to noise components at the output requires the ratio of the center frequency gain, A_o, to the gains at 1 kHz and 100 kHz to be 1000/1. Since the band-pass gain is symmetric about the center frequency on a log axis, the gains at 1 kHz and 100 kHz will be the same. Thus, we will focus on the gain at 1 kHz only. From Eq. (15.50) the ratio of the single-stage gain at ω_0 and $\omega_0/10$ is

$$\frac{A(\omega_0)}{A(\omega_0/10)} = \frac{-A_o}{\dfrac{-A_o\omega_0^2}{10Q}}$$

$$= \frac{\left| \omega_0^2 - \dfrac{\omega_0^2}{100} + j\,\dfrac{\omega_0^2}{10Q} \right|}{\dfrac{\omega_0^2}{10Q}} = \sqrt{1000}$$

For simplicity sake, we will assume that $\omega_0^2 \gg \omega_0^2/100$. Solving for Q yields

$$\sqrt{100Q^2 + 1} = \sqrt{1000}$$

or, employing Eq. (15.51),

$$Q \approx \sqrt{10} = \frac{1}{2}\sqrt{\frac{R_2}{R_1}}$$

Thus, the gain requirement forces $R_2 = 40R_1$. Arbitrarily choosing $R_1 = 1$ kΩ fixes R_2 at 40 kΩ. The second requirement is that $\omega_0/2\pi$ must equal 10 kHz. From Eq. (15.51) and our resistor values, we have

$$\omega_0 = (2\pi)10^4 = \frac{1}{C\sqrt{R_1 R_2}} = \frac{1}{10^3\sqrt{40}C}$$

which yields $C = 2.5$ nF.

The resulting Bode plot is shown in Fig. 15.27a where the center frequency is at 10 kHz, the gain at 10 kHz is roughly 400, and the gains at 1 kHz and 100 kHz are 0.4—a ratio of 1000/1. Output voltage results from a PSPICE transient analysis are shown in Fig. 15.27b for 10 cycles of the 10 kHz signal. Appropriately, the waveform is a 4-V sinewave at 10 kHz

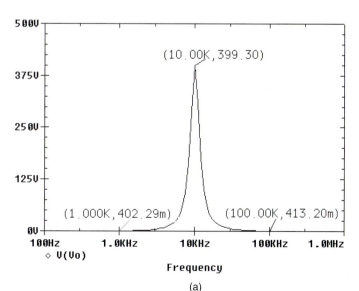

Figure 15.27
(a) Results from a PSPICE AC Sweep analysis showing the amplification of the signal over the noise. (b) PSPICE transient simulation results and (c) the corresponding FFT.

(a)

with little visible distortion. To view the Fourier components of the output, simply click on the Fast Fourier Transform hot button ⬛ at the top of the PROBE window. From the resulting plot, shown in Fig. 15.27c, we confirm that the signal amplitude is 1000 times larger than the noise components. It should be mentioned that the FFT in Fig. 15.27c is the result of a 5-ms transient simulation (i.e., 50 cycles at 10 kHz). In general, the more cycles in the transient analysis, the better is the frequency resolution of the FFT. Op-amps from the ANALOG library with gains of 10^6 and supply rails of ± 15 V were used in the simulation.

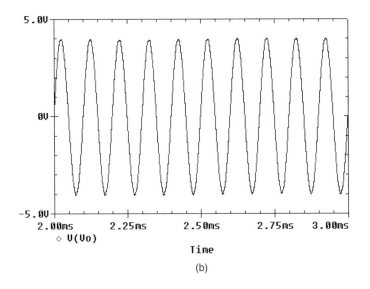

(b)

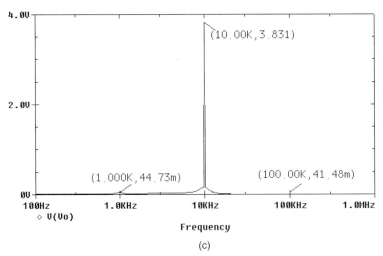

(c)

Figure 15.27 (continued)

Design Example 15.19

The circuit shown in Fig. 15.28 is a notch filter. At its resonant frequency, the *L-C* series circuit has zero effective impedance, and, as a result, any signal at that frequency is short-circuited. For this reason, the filter is often referred to as a trap.

Consider the following scenario. A system operating at 1 kHz has picked up noise at a fundamental frequency of 10 kHz, as well as some second- and third-harmonic junk. Given this information, we wish to design a filter that will eliminate both the noise and its attendant harmonics.

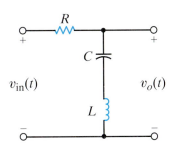

Figure 15.28
A notch filter, or trap, utilizing a series L-C branch.

SOLUTION The key to the trap is setting the resonant frequency of the *L-C* series branch to the frequency we wish to eliminate. Since we have three frequency components to remove, 10 kHz, 20 kHz, and 30 kHz, we will simply use three different *L-C* branches as shown in Fig. 15.29 and set L_1C_1 to trap at 10 kHz, L_2C_2 at 20 kHz, and L_3C_3 at 30 kHz.

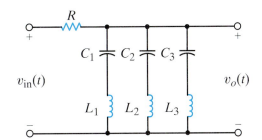

Figure 15.29
The notch filter in Fig. 15.28 expanded to remove three different frequency components.

If we arbitrarily set the value of all inductors to 10 μH and calculate the value of each capacitor, we obtain

$$C_1 = \frac{1}{(2\pi)^2 f^2 L} = \frac{1}{(2\pi)^2 (10^8)(10^{-5})} = 25.3 \ \mu F$$

$$C_2 = \frac{1}{(2\pi)^2 (4 \times 10^8)(10^{-5})} = 6.34 \ \mu F$$

$$C_3 = \frac{1}{(2\pi)^2 (9 \times 10^8)(10^{-5})} = 2.81 \ \mu F$$

The three traps shown in Fig. 15.29 should eliminate the noise and its harmonics.

SUMMARY

A periodic function, its representation using a Fourier series, and some of the useful properties of a Fourier series are outlined here.

- **A periodic function**

$$f(t) = f(t + nT_0), \quad n = 1, 2, 3, \dots \text{ and } T_0 \text{ is the period}$$

- **Exponential Fourier series of a periodic function**

$$f(t) = \sum_{n=-\infty}^{\infty} c_n e^{jn\omega_0 t}, \qquad c_n = \frac{1}{T_0} \int_{t_1}^{t_1+T_0} f(t) e^{-jn\omega_0 t} \, dt$$

- **Trigonometric Fourier series of a periodic function**

$$f(t) = a_0 + \sum_{n=1}^{\infty} \left(a_n \cos n\omega_0 t + b_n \sin n\omega_0 t \right)$$

$$a_n = \frac{2}{T_0} \int_{t_1}^{t_1+T_0} f(t) \cos n\omega_0 t \, dt,$$

$$b_n = \frac{2}{T_0} \int_{t_1}^{t_1+T_0} f(t) \sin n\omega_0 t \, dt,$$

and

$$a_0 = \frac{1}{T_0} \int_{t_1}^{t_1+T_0} f(t) \, dt$$

- ### *Even symmetry of a periodic function*

$$f(t) = f(-t)$$

$$a_n = \frac{4}{T_0} \int_0^{T_0/2} f(t) \cos n\omega_0 t \, dt,$$

$$b_n = 0,$$

and

$$a_0 = \frac{2}{T_0} \int_0^{T_0/2} f(t) \, dt$$

- ### *Odd symmetry of a periodic function*

$$f(t) = -f(-t)$$

$$a_n = 0, \quad b_n = \frac{4}{T_0} \int_0^{T_0/2} f(t) \sin n\omega_0 t \, dt, \quad \text{and} \quad a_0 = 0$$

- ### *Half-wave symmetry of a periodic function*

$$f(t) = -f(t - T_0/2)$$

$$a_n = b_n = 0, \quad \text{for } n \text{ even}$$

$$a_n = \frac{4}{T_0} \int_0^{T_0/2} f(t) \cos n\omega_0 t \, dt \quad \text{for } n \text{ odd}$$

$$b_n = \frac{4}{T_0} \int_0^{T_0/2} f(t) \sin n\omega_0 t \, dt \quad \text{for } n \text{ odd and } a_0 = 0$$

- ### *Time-shifting of a periodic function*

$$f(t - t_0) = \sum_{n=-\infty}^{\infty} \left(c_n e^{-jn\omega_0 t_0} \right) e^{jn\omega_0 t}$$

- ### *Frequency spectrum of a periodic function*

A Fourier series contains discrete frequency components, called line spectra.

- ### *Steady-state response of a periodic function input*
The periodic function input is expressed as a Fourier series, and phasor analysis is used to determine the response of each component of the series. Each component is transformed to the time domain, and superposition is used to determine the total output.

The Fourier transform, its features and properties, as well as its use in circuit analysis, are outlined here.

- ### *Fourier transform for an aperiodic function*

$$\mathbf{F}(\omega) = \int_{-\infty}^{\infty} f(t) e^{-j\omega t} \quad \text{and} \quad f(t) = \frac{1}{2\pi} \int_{-\infty}^{\infty} \mathbf{F}(\omega) e^{j\omega t} \, d\omega$$

- ### *Fourier transform pairs and properties*
The Fourier transform pairs in Table 15.3 and the properties in Table 15.4 can be used together to transform time domain functions to the frequency domain and vice versa.

- ### *Parseval's theorem for determining the energy content of a signal*

$$\int_{-\infty}^{\infty} f^2(t) \, dt = \frac{1}{2\pi} \int_{-\infty}^{\infty} |\mathbf{F}(\omega)|^2 \, d\omega$$

- ### *Network response to an aperiodic input*
An aperiodic input $x(t)$ can be transformed to the frequency domain as $\mathbf{X}(\omega)$. Then using the network transfer function $\mathbf{H}(\omega)$, the output can be computed as $\mathbf{Y}(\omega) = \mathbf{H}(\omega)\mathbf{X}(\omega)$. $y(t)$ can be obtained transforming $\mathbf{Y}(\omega)$ to the time domain.

PROBLEMS

PSV **CS** both available on the web at: http://www.justask4u.com/irwin

SECTION 15.1

15.1 Find the exponential Fourier series for the periodic signal shown in Fig. P15.1.

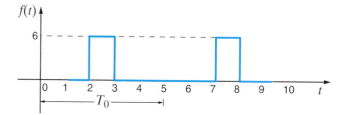

Figure P15.1

15.2 Find the exponential Fourier series for the periodic pulse train shown in Fig. P15.2. **PSV**

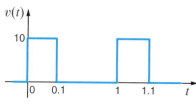

Figure P15.2

15.3 Find the exponential Fourier series for the signal shown in Fig. P15.3. **cs**

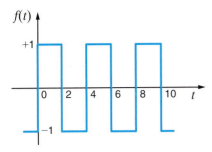

Figure P15.3

15.4 Find the exponential Fourier series for the signal shown in Fig. P15.4. **cs**

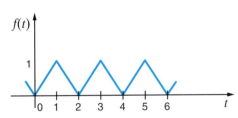

Figure P15.4

15.5 Compute the exponential Fourier series for the waveform that is the sum of the two waveforms in Fig. P15.5 by computing the exponential Fourier series of the two waveforms and adding them.

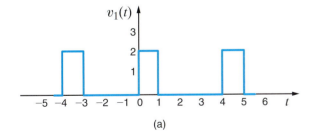

(a)

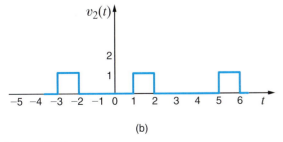

(b)

Figure P15.5

15.6 Given the waveform in Fig. P15.6, determine the type of symmetry that exists if the origin is selected at (a) l_1 and (b) l_2.

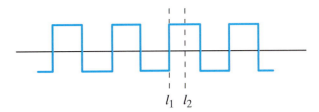

Figure P15.6

15.7 What type of symmetry is exhibited by the two waveforms in Fig. P15.7?

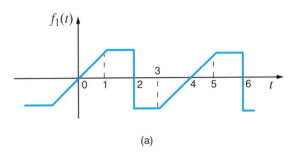

(a)

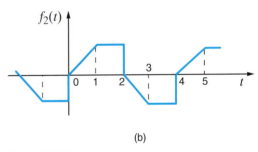

(b)

Figure P15.7

15.8 Find the trigonometric Fourier series for the waveform shown in Fig. P15.8.

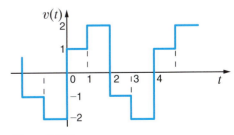

Figure P15.8

15.9 Find the trigonometric Fourier series for the periodic waveform shown in Fig. P15.9.

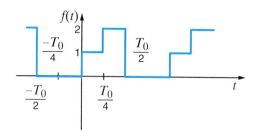

Figure P15.9

15.10 Given the waveform in Fig. P15.10 show that

$$f(t) = \frac{A}{2} + \sum_{n=1}^{\infty} \frac{-A}{n\pi} \sin \frac{2n\pi}{T_0} t \quad \text{PSV}$$

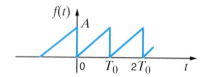

Figure P15.10

15.11 Find the trigonometric Fourier series coefficients for the waveform in Fig. P15.11. **CS**

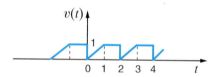

Figure P15.11

15.12 Find the trigonometric Fourier series coefficients for the waveform in Fig. P15.12.

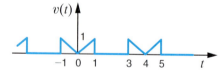

Figure P15.12

15.13 Find the trigonometric Fourier series coefficients for the waveform in Fig. P15.13.

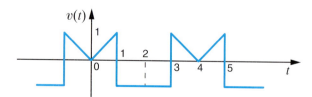

Figure P15.13

15.14 Find the trigonometric Fourier series coefficients for the waveform in Fig. P15.14.

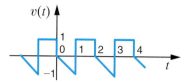

Figure P15.14

15.15 Find the trigonometric Fourier series coefficients for the waveform in Fig. P15.15.

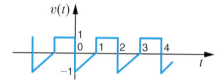

Figure P15.15

15.16 Derive the trigonometric Fourier series for the waveform shown in Fig. P15.16.

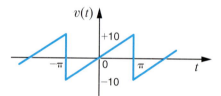

Figure P15.16

15.17 Find the trigonometric Fourier series coefficients for the waveform in Fig. P15.17.

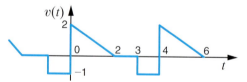

Figure P15.17

15.18 Find the trigonometric Fourier series for the waveform shown in Fig. P15.18. **CS**

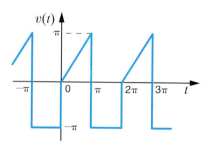

Figure P15.18

15.19 Derive the trigonometric Fourier series for the function shown in Fig. P15.19. **cs**

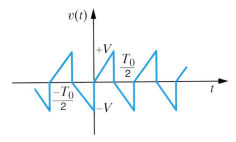

Figure P15.19

15.20 Derive the trigonometric Fourier series for the function $v(t) = A|\sin t|$ as shown in Fig. P15.20.

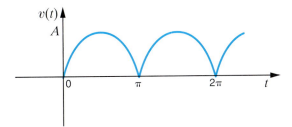

Figure P15.20

15.21 Derive the trigonometric Fourier series for the waveform shown in Fig. P15.21.

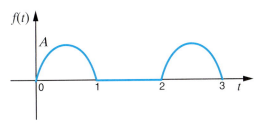

Figure P15.21

15.22 Use PSPICE to determine the Fourier series of the waveform in Fig. P15.22 in the form

$$v_S(t) = a_0 + \sum_{n=1}^{\infty} b_n \sin(n\omega_t + \theta_n)$$

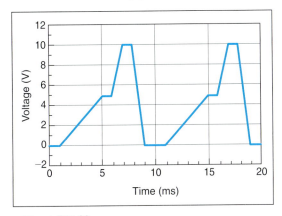

Figure P15.22

15.23 Use PSPICE to determine the Fourier series of the waveform in Fig. P15.23 in the form

$$i_S(t) = a_0 + \sum_{n=1}^{\infty} b_n \sin(n\omega_0 t + \theta_n)$$ **cs**

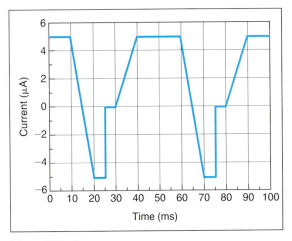

Figure P15.23

15.24 The discrete line spectrum for a periodic function $f(t)$ is shown in Fig. P15.24. Determine the expression for $f(t)$.
PSV

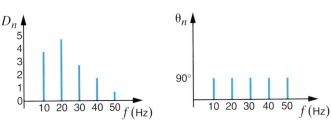

Figure P15.24

15.25 The amplitude and phase spectra for a periodic function $v(t)$ that has only a small number of terms is shown in Fig. P15.25. Determine the expression for $v(t)$ if $T_0 = 0.1$ s.

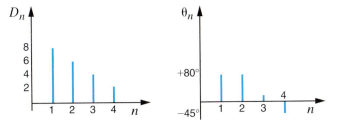

Figure P15.25

15.26 Plot the first four terms of the amplitude and phase spectra for the signal

$$f(t) = \sum_{\substack{n=1 \\ n \text{ odd}}}^{\infty} \frac{-2}{n\pi} \sin\frac{n\pi}{2} \cos n\omega_0 t + \frac{6}{n\pi} \sin n\omega_0 t$$

15.27 Determine the steady-state response of the current $i_o(t)$ in the circuit shown in Fig. P15.27 if the input voltage is described by the waveform shown in Problem 15.16.
CS

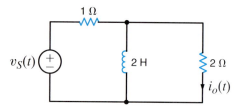

Figure P15.27

15.28 If the input voltage in Problem 15.27 is

$$v_S(t) = 1 - \frac{2}{\pi} \sum_{n=1}^{\infty} \frac{1}{n} \sin 0.2\pi nt \text{ V}$$

find the expression for the steady-state current $i_o(t)$.

15.29 Determine the first three terms of the steady-state voltage $v_o(t)$ in Fig. P15.29 if the input voltage is a periodic signal of the form

$$v(t) = \frac{1}{2} + \sum_{n=1}^{\infty} \frac{1}{n\pi} (\cos n\pi - 1) \sin nt \text{ V} \quad \textbf{PSV}$$

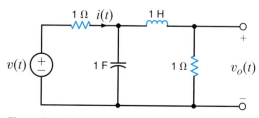

Figure P15.29

15.30 Determine the steady-state voltage $v_o(t)$ in the network in Fig. P15.30a if the input current is given in Fig. P15.30b.

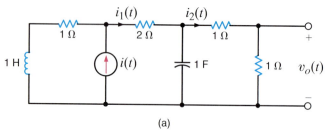

(a)

(b)

Figure P15.30

15.31 Find the average power absorbed by the network in Fig. P15.31 if

$$v(t) = 12 + 6\cos(377t - 10°) + 4\cos(754t - 60°) \text{ V}$$
$$i(t) = 0.2 + 0.4\cos(377t - 150°)$$
$$-0.2\cos(754t - 80°) + 0.1\cos(1131t - 60°) \text{ A}$$

PSV

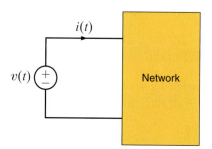

Figure P15.31

15.32 Find the average power absorbed by the network in Fig. P15.32 if $v(t) = 60 + 36\cos(377t + 45°) + 24\cos(754t - 60°)$ V. **CS**

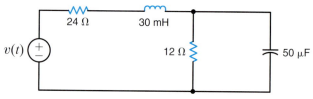

Figure P15.32

SECTION 15.2

15.33 Determine the Fourier transform of the waveform shown in Fig. P15.33.

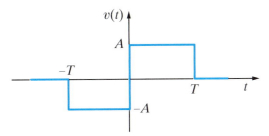

Figure P15.33

15.34 Derive the Fourier transform for the following functions:

(a) $f(t) = e^{-2t} \cos 4tu(t)$

(b) $f(t) = e^{-2t} \sin 4tu(t)$

15.35 Show that

$$\mathcal{F}\left[f_1(t)f_2(t)\right] = \frac{1}{2\pi} \int_{-\infty}^{\infty} \mathbf{F}_1(x)\mathbf{F}_2(\omega - x)dx \quad \boxed{\text{CS}}$$

15.36 Find the Fourier transform of the function $f(t) = 12e^{-2|t|} \cos 4t$.

15.37 Use the transform technique to find $v_o(t)$ in the network in Fig. P15.30a if (a) $i(t) = 4(e^{-t} - e^{-2t})u(t)$ A and (b) $i(t) = 12 \cos 4t$ A. $\boxed{\text{PSV}}$

15.38 The input signal to a network is $v_i(t) = e^{-3t}u(t)$ V. The transfer function of the network is $\mathbf{H}(j\omega) = 1/(j\omega + 4)$. Find the output of the network $v_o(t)$ if the initial conditions are zero. $\boxed{\text{CS}}$

15.39 Determine $v_o(t)$ in the circuit shown in Fig. P15.39 using the Fourier transform if the input signal is $i_S(t) = (e^{-2t} + \cos t)u(t)$ A.

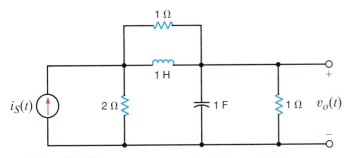

Figure P15.39

15.40 The input signal for the network in Fig. P15.40 is $v_i(t) = 10e^{-5t}u(t)$ V. Determine the total 1-Ω energy content of the output $v_o(t)$. $\boxed{\text{PSV}}$

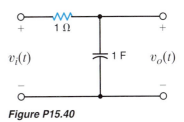

Figure P15.40

15.41 Compute the 1-Ω energy content of the signal $v_o(t)$ in Fig. P15.40 in the frequency range from $\omega = 2$ to $\omega = 4$ rad/s. $\boxed{\text{CS}}$

15.42 Determine the 1-Ω energy content of the signal $v_o(t)$ in Fig. P15.40 in the frequency range from 0 to 1 rad/s.

15.43 Compare the 1-Ω energy at both the input and output of the network in Fig. P15.43 for the given input forcing function $i_i(t) = 2e^{-4t}u(t)$ A.

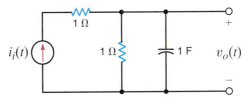

Figure P15.43

15.44 The waveform shown in Fig. P15.44 demonstrates what is called the duty cycle; that is, D illustrates the fraction of the total period that is occupied by the pulse. Determine the average value of this waveform.

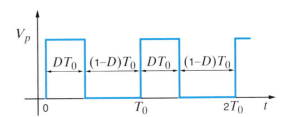

Figure P15.44

TYPICAL PROBLEMS FOUND ON THE FE EXAM

15FE-1 Given the waveform in Fig. 15PFE-1, determine which of the trigonometric Fourier coefficients have zero value, which have nonzero value, and why.

CS

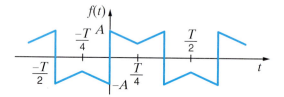

Figure 15PFE-1

15FE-2 Given the waveform in Fig. 15PFE-2, describe the type of symmetry and its impact on the trigonometric coefficients in the Fourier series—that is, a_0, a_n, and b_n.

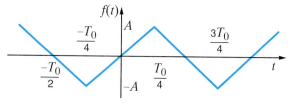

Figure 15PFE-2

Two-Port Networks

16

We study two-ports and the parameters that describe them for a number of reasons. For example, most circuits or systems have at least two ports. We may put an input signal into one port and obtain an output signal from another. The parameters of the two-port completely describe its behavior in terms of the voltage and current at each port. Thus, knowing the parameters of a two-port network permits us to describe its operation when it is connected into a larger network. Two-port networks are also important in modeling electronic devices and system components. For example, in electronics, two-port networks are employed to model such things as transistors and op-amps. Other examples of electrical components modeled by two-ports are transformers and transmission lines.

In general, we describe the two-port as a network consisting of R, L, and C elements, transformers, op-amps, dependent sources, but no independent sources. The network has an input port and an output port, and as is the case with an op-amp, one terminal may be common to both ports.

Four popular types of two-port parameters are examined: admittance, impedance, hybrid, and transmission. We demonstrate the usefulness of each set of parameters, show how they are related to one another, and finally illustrate how two-port networks can be interconnected in parallel, series, or cascade. ●

Access Problem-Solving Videos **PSV** *and Circuit Solutions* **CS** *at:*
http://www.justask4u.com/irwin
using the registration code on the inside cover and see a website with answers and more!

16.1 Admittance Parameters

We say that the linear network in Fig. 16.1a has a single *port*—that is, a single pair of terminals. The pair of terminals A-B that constitute this port could represent a single element (e.g., R, L, or C), or it could be some interconnection of these elements. The linear network in Fig. 16.1b is called a two-port. As a general rule the terminals A-B represent the input port, and the terminals C-D represent the output port.

Figure 16.1
(a) Single-port network;
(b) two-port network.

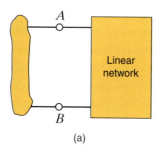 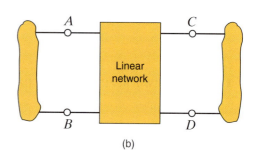

(a)   (b)

In the two-port network shown in Fig. 16.2, it is customary to label the voltages and currents as shown; that is, the upper terminals are positive with respect to the lower terminals, the currents are into the two-port at the upper terminals, and, because KCL must be satisfied at each port, the current is out of the two-port at the lower terminals. Since the network is linear and contains no independent sources, the principle of superposition can be applied to determine the current $\mathbf{I}_1$, which can be written as the sum of two components, one due to $\mathbf{V}_1$ and one due to $\mathbf{V}_2$. Using this principle, we can write

$$\mathbf{I}_1 = \mathbf{y}_{11}\mathbf{V}_1 + \mathbf{y}_{12}\mathbf{V}_2$$

where $\mathbf{y}_{11}$ and $\mathbf{y}_{12}$ are essentially constants of proportionality with units of siemens. In a similar manner $\mathbf{I}_2$ can be written as

$$\mathbf{I}_2 = \mathbf{y}_{21}\mathbf{V}_1 + \mathbf{y}_{22}\mathbf{V}_2$$

Therefore, the two equations that describe the two-port network are

$$\mathbf{I}_1 = \mathbf{y}_{11}\mathbf{V}_1 + \mathbf{y}_{12}\mathbf{V}_2$$

$$\mathbf{I}_2 = \mathbf{y}_{21}\mathbf{V}_1 + \mathbf{y}_{22}\mathbf{V}_2$$

16.1

or in matrix form,

$$\begin{bmatrix} \mathbf{I}_1 \\ \mathbf{I}_2 \end{bmatrix} = \begin{bmatrix} \mathbf{y}_{11} & \mathbf{y}_{12} \\ \mathbf{y}_{21} & \mathbf{y}_{22} \end{bmatrix} \begin{bmatrix} \mathbf{V}_1 \\ \mathbf{V}_2 \end{bmatrix}$$

Note that subscript 1 refers to the input port and subscript 2 refers to the output port, and the equations describe what we will call the *Y parameters* for a network. If these parameters $\mathbf{y}_{11}, \mathbf{y}_{12}, \mathbf{y}_{21}$, and $\mathbf{y}_{22}$ are known, the input/output operation of the two-port is completely defined.

Figure 16.2
Generalized two-port network.

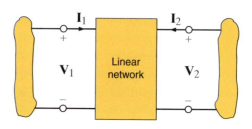

From Eq. (16.1) we can determine the Y parameters in the following manner. Note from the equations that $\mathbf{y}_{11}$ is equal to $\mathbf{I}_1$ divided by $\mathbf{V}_1$ with the output short-circuited (i.e., $\mathbf{V}_2 = 0$).

$$\mathbf{y}_{11} = \left.\frac{\mathbf{I}_1}{\mathbf{V}_1}\right|_{\mathbf{V}_2=0}$$

16.2

Since $\mathbf{y}_{11}$ is an admittance at the input measured in siemens with the output short-circuited, it is called the *short-circuit input admittance*. The equations indicate that the other Y parameters can be determined in a similar manner:

$$\mathbf{y}_{12} = \left.\frac{\mathbf{I}_1}{\mathbf{V}_2}\right|_{\mathbf{V}_1=0}$$

$$\mathbf{y}_{21} = \left.\frac{\mathbf{I}_2}{\mathbf{V}_1}\right|_{\mathbf{V}_2=0}$$

16.3

$$\mathbf{y}_{22} = \left.\frac{\mathbf{I}_2}{\mathbf{V}_2}\right|_{\mathbf{V}_1=0}$$

$\mathbf{y}_{12}$ and $\mathbf{y}_{21}$ are called the *short-circuit transfer admittances*, and $\mathbf{y}_{22}$ is called the *short-circuit output admittance*. As a group, the Y parameters are referred to as the *short-circuit admittance parameters*. Note that by applying the preceding definitions, these parameters could be determined experimentally for a two-port network whose actual configuration is unknown.

Example 16.1

We wish to determine the Y parameters for the two-port network shown in Fig. 16.3a. Once these parameters are known, we will determine the current in a 4-Ω load, which is connected to the output port when a 2-A current source is applied at the input port.

SOLUTION   From Fig. 16.3b, we note that

$$\mathbf{I}_1 = \mathbf{V}_1\left(\frac{1}{1} + \frac{1}{2}\right)$$

Therefore,

$$\mathbf{y}_{11} = \frac{3}{2}\,\mathrm{S}$$

Figure 16.3
Networks employed in Example 16.1.

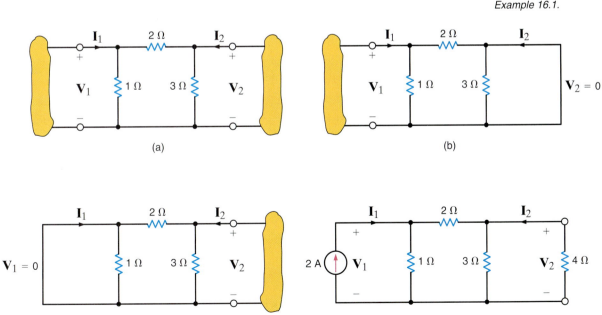

(a)   (b)   (c)   (d)

As shown in Fig. 16.3c,

$$I_1 = -\frac{V_2}{2}$$

and hence,

$$y_{12} = -\frac{1}{2} S$$

Also, y_{21} is computed from Fig. 16.3b using the equation

$$I_2 = -\frac{V_1}{2}$$

and therefore,

$$y_{21} = -\frac{1}{2} S$$

Finally, y_{22} can be derived from Fig. 16.3c using

$$I_2 = V_2 \left(\frac{1}{3} + \frac{1}{2} \right)$$

and

$$y_{22} = \frac{5}{6} S$$

Therefore, the equations that describe the two-port itself are

$$I_1 = \frac{3}{2} V_1 - \frac{1}{2} V_2 \qquad I_2 = -\frac{1}{2} V_1 + \frac{5}{6} V_2$$

These equations can now be employed to determine the operation of the two-port for some given set of terminal conditions. The terminal conditions we will examine are shown in Fig. 16.3d. From this figure we note that

$$I_1 = 2 \text{ A} \qquad \text{and} \qquad V_2 = -4 I_2$$

Combining these with the preceding two-port equations yields

$$2 = \frac{3}{2} V_1 - \frac{1}{2} V_2$$

$$0 = -\frac{1}{2} V_1 + \frac{13}{12} V_2$$

or in matrix form

$$\begin{bmatrix} \frac{3}{2} & -\frac{1}{2} \\ -\frac{1}{2} & \frac{13}{12} \end{bmatrix} \begin{bmatrix} V_1 \\ V_2 \end{bmatrix} = \begin{bmatrix} 2 \\ 0 \end{bmatrix}$$

Note carefully that these equations are simply the nodal equations for the network in Fig. 16.3d. Solving the equations, we obtain $V_2 = 8/11$ V and therefore $I_2 = -2/11$ A.

LEARNING EXTENSIONS

E16.1 Find the Y parameters for the two-port network shown in Fig. E16.1.

Figure E16.1

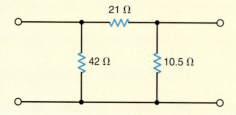

ANSWER: $y_{11} = \frac{1}{14} S;$

$$y_{12} = y_{21} = -\frac{1}{21} S; \ y_{22} = \frac{1}{7} S.$$

E16.2 If a 10-A source is connected to the input of the two-port network in Fig. E16.1, find the current in a 5-Ω resistor connected to the output port. **PSV**

ANSWER: $I_2 = -4.29$ A.

16.2   Impedance Parameters

Once again if we assume that the two-port network is a linear network that contains no independent sources, then by means of superposition we can write the input and output voltages as the sum of two components, one due to $\mathbf{I}_1$ and one due to $\mathbf{I}_2$:

$$\mathbf{V}_1 = \mathbf{z}_{11}\mathbf{I}_1 + \mathbf{z}_{12}\mathbf{I}_2$$
$$\mathbf{V}_2 = \mathbf{z}_{21}\mathbf{I}_1 + \mathbf{z}_{22}\mathbf{I}_2$$

16.4

These equations, which describe the two-port network, can also be written in matrix form as

$$\begin{bmatrix} \mathbf{V}_1 \\ \mathbf{V}_2 \end{bmatrix} = \begin{bmatrix} \mathbf{z}_{11} & \mathbf{z}_{12} \\ \mathbf{z}_{21} & \mathbf{z}_{22} \end{bmatrix}\begin{bmatrix} \mathbf{I}_1 \\ \mathbf{I}_2 \end{bmatrix}$$

16.5

Like the Y parameters, these *Z parameters* can be derived as follows:

$$\mathbf{z}_{11} = \left.\frac{\mathbf{V}_1}{\mathbf{I}_1}\right|_{\mathbf{I}_2=0}$$

$$\mathbf{z}_{12} = \left.\frac{\mathbf{V}_1}{\mathbf{I}_2}\right|_{\mathbf{I}_1=0}$$

16.6

$$\mathbf{z}_{21} = \left.\frac{\mathbf{V}_2}{\mathbf{I}_1}\right|_{\mathbf{I}_2=0}$$

$$\mathbf{z}_{22} = \left.\frac{\mathbf{V}_2}{\mathbf{I}_2}\right|_{\mathbf{I}_1=0}$$

In the preceding equations, setting $\mathbf{I}_1$ or $\mathbf{I}_2 = 0$ is equivalent to open-circuiting the input or output port. Therefore, the Z parameters are called the *open-circuit impedance parameters*. $\mathbf{z}_{11}$ is called the *open-circuit input impedance*, $\mathbf{z}_{22}$ is called the *open-circuit output impedance*, and $\mathbf{z}_{12}$ and $\mathbf{z}_{21}$ are termed *open-circuit transfer impedances*.

Example 16.2

We wish to find the Z parameters for the network in Fig. 16.4a. Once the parameters are known, we will use them to find the current in a 4-Ω resistor that is connected to the output terminals when a $12\,\underline{/0°}$-V source with an internal impedance of $1 + j0\ \Omega$ is connected to the input.

SOLUTION   From Fig. 16.4a we note that

$$\mathbf{z}_{11} = 2 - j4\ \Omega$$

$$\mathbf{z}_{12} = -j4\ \Omega$$

$$\mathbf{z}_{21} = -j4\ \Omega$$

$$\mathbf{z}_{22} = -j4 + j2 = -j2\ \Omega$$

The equations for the two-port network are, therefore,

$$\mathbf{V}_1 = (2 - j4)\mathbf{I}_1 - j4\mathbf{I}_2$$

$$\mathbf{V}_2 = -j4\mathbf{I}_1 - j2\mathbf{I}_2$$

The terminal conditions for the network shown in Fig. 16.4b are

$$\mathbf{V}_1 = 12\,\underline{/0°} - (1)\mathbf{I}_1$$

$$\mathbf{V}_2 = -4\mathbf{I}_2$$

Combining these with the two-port equations yields

$$12\underline{/0°} = (3 - j4)\mathbf{I}_1 - j4\mathbf{I}_2$$

$$0 = -j4\mathbf{I}_1 + (4 - j2)\mathbf{I}_2$$

It is interesting to note that these equations are the mesh equations for the network. If we solve the equations for $\mathbf{I}_2$, we obtain $\mathbf{I}_2 = 1.61\underline{/137.73°}$ A, which is the current in the 4-Ω load.

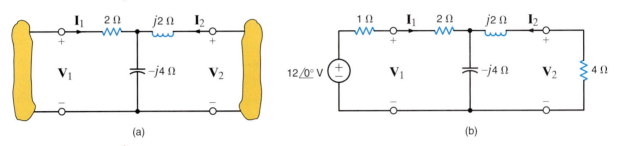

(a)   (b)

Figure 16.4 *Circuits employed in Example 16.2.*

LEARNING EXTENSION

E16.3 Find the Z parameters for the network in Fig. E16.3. Then compute the current in a 4-Ω load if a $24\underline{/0°}$-V source is connected at the input port.

ANSWER:
$\mathbf{I}_2 = -0.73\underline{/0°}$ A.

Figure E16.3

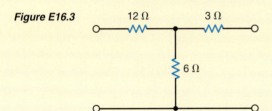

16.3 Hybrid Parameters

Under the assumptions used to develop the Y and Z parameters, we can obtain what are commonly called the *hybrid parameters*. In the pair of equations that define these parameters, $\mathbf{V}_1$ and $\mathbf{I}_2$ are the independent variables. Therefore, the two-port equations in terms of the hybrid parameters are

$$\mathbf{V}_1 = \mathbf{h}_{11}\mathbf{I}_1 + \mathbf{h}_{12}\mathbf{V}_2$$

$$\mathbf{I}_2 = \mathbf{h}_{21}\mathbf{I}_1 + \mathbf{h}_{22}\mathbf{V}_2$$

16.7

or in matrix form,

$$\begin{bmatrix} \mathbf{V}_1 \\ \mathbf{I}_2 \end{bmatrix} = \begin{bmatrix} \mathbf{h}_{11} & \mathbf{h}_{12} \\ \mathbf{h}_{21} & \mathbf{h}_{22} \end{bmatrix} \begin{bmatrix} \mathbf{I}_1 \\ \mathbf{V}_2 \end{bmatrix}$$

16.8

These parameters are especially important in transistor circuit analysis. The parameters are determined via the following equations:

$$\mathbf{h}_{11} = \left.\frac{\mathbf{V}_1}{\mathbf{I}_1}\right|_{\mathbf{V}_2=0}$$

$$\mathbf{h}_{12} = \left.\frac{\mathbf{V}_1}{\mathbf{V}_2}\right|_{\mathbf{I}_1=0}$$

16.9

$$\mathbf{h}_{21} = \left.\frac{\mathbf{I}_2}{\mathbf{I}_1}\right|_{\mathbf{V}_2=0}$$

$$\mathbf{h}_{22} = \left.\frac{\mathbf{I}_2}{\mathbf{V}_2}\right|_{\mathbf{I}_1=0}$$

The parameters $\mathbf{h}_{11}$, $\mathbf{h}_{12}$, $\mathbf{h}_{21}$, and $\mathbf{h}_{22}$ represent the *short-circuit input impedance*, the *open-circuit reverse voltage gain*, the *short-circuit forward current gain*, and the *open-circuit output admittance*, respectively. Because of this mix of parameters, they are called *hybrid parameters*. In transistor circuit analysis, the parameters $\mathbf{h}_{11}$, $\mathbf{h}_{12}$, $\mathbf{h}_{21}$, and $\mathbf{h}_{22}$ are normally labeled $\mathbf{h}_i$, $\mathbf{h}_r$, $\mathbf{h}_f$, and $\mathbf{h}_o$.

— **Example 16.3**

An equivalent circuit for the op-amp in Fig. 16.5a is shown in Fig. 16.5b. We will determine the hybrid parameters for this network.

SOLUTION   Parameter $\mathbf{h}_{11}$ is derived from Fig. 16.5c. With the output shorted, $\mathbf{h}_{11}$ is a function of only R_i, R_1, and R_2 and

$$\mathbf{h}_{11} = R_i + \frac{R_1 R_2}{R_1 + R_2}$$

Figure 16.5
Circuit employed in Example 16.3.

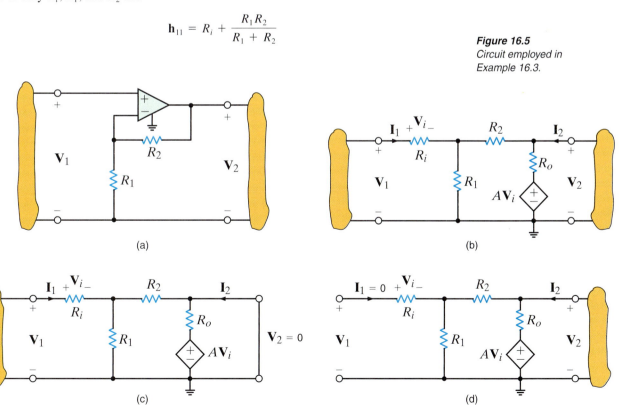

(a)

(b)

(c)

(d)

Figure 16.5d is used to derive $\mathbf{h}_{12}$. Since $\mathbf{I}_1 = 0$, $\mathbf{V}_i = 0$ and the relationship between $\mathbf{V}_1$ and $\mathbf{V}_2$ is a simple voltage divider.

$$\mathbf{V}_1 = \frac{\mathbf{V}_2 R_1}{R_1 + R_2}$$

Therefore,

$$\mathbf{h}_{12} = \frac{R_1}{R_1 + R_2}$$

KVL and KCL can be applied to Fig. 16.5c to determine $\mathbf{h}_{21}$. The two equations that relate $\mathbf{I}_2$ to $\mathbf{I}_1$ are

$$\mathbf{V}_i = \mathbf{I}_1 R_i$$

$$\mathbf{I}_2 = \frac{-A\mathbf{V}_i}{R_o} - \frac{\mathbf{I}_1 R_1}{R_1 + R_2}$$

Therefore,

$$\mathbf{h}_{21} = -\left(\frac{A R_i}{R_o} + \frac{R_1}{R_1 + R_2}\right)$$

Finally, the relationship between $\mathbf{I}_2$ and $\mathbf{V}_2$ in Fig. 16.5d is

$$\frac{\mathbf{V}_2}{\mathbf{I}_2} = \frac{R_o(R_1 + R_2)}{R_o + R_1 + R_2}$$

and therefore,

$$\mathbf{h}_{22} = \frac{R_o + R_1 + R_2}{R_o(R_1 + R_2)}$$

The network equations are, therefore,

$$\mathbf{V}_1 = \left(R_i + \frac{R_1 R_2}{R_1 + R_2}\right)\mathbf{I}_1 + \frac{R_1}{R_1 + R_2}\mathbf{V}_2$$

$$\mathbf{I}_2 = -\left(\frac{A R_i}{R_o} + \frac{R_1}{R_1 + R_2}\right)\mathbf{I}_1 + \frac{R_o + R_1 + R_2}{R_o(R_1 + R_2)}\mathbf{V}_2$$

LEARNING EXTENSIONS

E16.4 Find the hybrid parameters for the network shown in Fig. E16.3. **PSV**

ANSWER: $\mathbf{h}_{11} = 14\ \Omega$; $\mathbf{h}_{12} = \frac{2}{3}$; $\mathbf{h}_{21} = -\frac{2}{3}$; $\mathbf{h}_{22} = \frac{1}{9}\,\text{S}$.

E16.5 If a 4-Ω load is connected to the output port of the network examined in Extension 16.4, determine the input impedance of the two-port with the load connected.

ANSWER: $\mathbf{Z}_i = 15.23\ \Omega$.

16.4   Transmission Parameters

The final parameters we will discuss are called the *transmission parameters*. They are defined by the equations

$$V_1 = AV_2 - BI_2$$

$$I_1 = CV_2 - DI_2$$

16.10

or in matrix form,

$$\begin{bmatrix} V_1 \\ I_1 \end{bmatrix} = \begin{bmatrix} A & B \\ C & D \end{bmatrix} \begin{bmatrix} V_2 \\ -I_2 \end{bmatrix}$$

16.11

These parameters are very useful in the analysis of circuits connected in cascade, as we will demonstrate later. The parameters are determined via the following equations:

$$A = \frac{V_1}{V_2}\bigg|_{I_2=0}$$

$$B = \frac{V_1}{-I_2}\bigg|_{V_2=0}$$

16.12

$$C = \frac{I_1}{V_2}\bigg|_{I_2=0}$$

$$D = \frac{I_1}{-I_2}\bigg|_{V_2=0}$$

A, **B**, **C**, and **D** represent the *open-circuit voltage ratio*, the *negative short-circuit transfer impedance*, the *open-circuit transfer admittance*, and the *negative short-circuit current ratio*, respectively. For obvious reasons the transmission parameters are commonly referred to as the *ABCD parameters*.

Example 16.4

We will now determine the transmission parameters for the network in Fig. 16.6.

SOLUTION   Let us consider the relationship between the variables under the conditions stated in the parameters in Eq. (16.12). For example, with $I_2 = 0$, V_2 can be written as

$$V_2 = \frac{V_1}{1 + 1/j\omega} \left(\frac{1}{j\omega} \right)$$

or

$$A = \frac{V_1}{V_2}\bigg|_{I_2=0} = 1 + j\omega$$

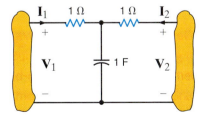

Figure 16.6
Circuit used in Example 16.4

Similarly, with $\mathbf{V}_2 = 0$, the relationship between $\mathbf{I}_2$ and $\mathbf{V}_1$ is

$$-\mathbf{I}_2 = \frac{\mathbf{V}_1}{1 + \dfrac{1/j\omega}{1 + 1/j\omega}}\left(\frac{1/j\omega}{1 + 1/j\omega}\right)$$

or

$$\mathbf{B} = \frac{\mathbf{V}_1}{-\mathbf{I}_2} = 2 + j\omega$$

In a similar manner, we can show that $\mathbf{C} = j\omega$ and $\mathbf{D} = 1 + j\omega$.

LEARNING EXTENSION

E16.6 Compute the transmission parameters for the two-port in Fig. E16.1. **PSV**

ANSWER: $\mathbf{A} = 3$; $\mathbf{B} = 21\ \Omega$; $\mathbf{C} = \dfrac{1}{6}\mathbf{S}$; $\mathbf{D} = \dfrac{3}{2}$.

16.5 Parameter Conversions

If all the two-port parameters for a network exist, it is possible to relate one set of parameters to another since the parameters interrelate the variables $\mathbf{V}_1$, $\mathbf{I}_1$, $\mathbf{V}_2$, and $\mathbf{I}_2$.

Table 16.1 lists all the conversion formulas that relate one set of two-port parameters to another. Note that Δ_Z, Δ_Y, Δ_H, and Δ_T refer to the determinants of the matrices for the Z, Y, hybrid, and ABCD parameters, respectively. Therefore, given one set of parameters for a network, we can use Table 16.1 to find others.

Table 16.1 Two-port parameter conversion formulas

$$
\begin{bmatrix} \mathbf{z}_{11} & \mathbf{z}_{12} \\ \mathbf{z}_{21} & \mathbf{z}_{22} \end{bmatrix}
\quad
\begin{bmatrix} \dfrac{\mathbf{y}_{22}}{\Delta_Y} & \dfrac{-\mathbf{y}_{12}}{\Delta_Y} \\ \dfrac{-\mathbf{y}_{21}}{\Delta_Y} & \dfrac{\mathbf{y}_{11}}{\Delta_Y} \end{bmatrix}
\quad
\begin{bmatrix} \dfrac{\mathbf{A}}{\mathbf{C}} & \dfrac{\Delta_T}{\mathbf{C}} \\ \dfrac{1}{\mathbf{C}} & \dfrac{\mathbf{D}}{\mathbf{C}} \end{bmatrix}
\quad
\begin{bmatrix} \dfrac{\Delta_H}{\mathbf{h}_{22}} & \dfrac{\mathbf{h}_{12}}{\mathbf{h}_{22}} \\ \dfrac{-\mathbf{h}_{21}}{\mathbf{h}_{22}} & \dfrac{1}{\mathbf{h}_{22}} \end{bmatrix}
$$

$$
\begin{bmatrix} \dfrac{\mathbf{z}_{22}}{\Delta_Z} & \dfrac{-\mathbf{z}_{12}}{\Delta_Z} \\ \dfrac{-\mathbf{z}_{21}}{\Delta_Z} & \dfrac{\mathbf{z}_{11}}{\Delta_Z} \end{bmatrix}
\quad
\begin{bmatrix} \mathbf{y}_{11} & \mathbf{y}_{12} \\ \mathbf{y}_{21} & \mathbf{y}_{22} \end{bmatrix}
\quad
\begin{bmatrix} \dfrac{\mathbf{D}}{\mathbf{B}} & \dfrac{-\Delta_T}{\mathbf{B}} \\ -\dfrac{1}{\mathbf{B}} & \dfrac{\mathbf{A}}{\mathbf{B}} \end{bmatrix}
\quad
\begin{bmatrix} \dfrac{1}{\mathbf{h}_{11}} & \dfrac{-\mathbf{h}_{12}}{\mathbf{h}_{11}} \\ \dfrac{\mathbf{h}_{21}}{\mathbf{h}_{11}} & \dfrac{\Delta_H}{\mathbf{h}_{11}} \end{bmatrix}
$$

$$
\begin{bmatrix} \dfrac{\mathbf{z}_{11}}{\mathbf{z}_{21}} & \dfrac{\Delta_Z}{\mathbf{z}_{21}} \\ \dfrac{1}{\mathbf{z}_{21}} & \dfrac{\mathbf{z}_{22}}{\mathbf{z}_{21}} \end{bmatrix}
\quad
\begin{bmatrix} \dfrac{-\mathbf{y}_{22}}{\mathbf{y}_{21}} & \dfrac{-1}{\mathbf{y}_{21}} \\ \dfrac{-\Delta_Y}{\mathbf{y}_{21}} & \dfrac{-\mathbf{y}_{11}}{\mathbf{y}_{21}} \end{bmatrix}
\quad
\begin{bmatrix} \mathbf{A} & \mathbf{B} \\ \mathbf{C} & \mathbf{D} \end{bmatrix}
\quad
\begin{bmatrix} \dfrac{-\Delta_H}{\mathbf{h}_{21}} & \dfrac{-\mathbf{h}_{11}}{\mathbf{h}_{21}} \\ \dfrac{-\mathbf{h}_{22}}{\mathbf{h}_{21}} & \dfrac{-1}{\mathbf{h}_{21}} \end{bmatrix}
$$

$$
\begin{bmatrix} \dfrac{\Delta_Z}{\mathbf{z}_{22}} & \dfrac{\mathbf{z}_{12}}{\mathbf{z}_{22}} \\ \dfrac{-\mathbf{z}_{21}}{\mathbf{z}_{22}} & \dfrac{1}{\mathbf{z}_{22}} \end{bmatrix}
\quad
\begin{bmatrix} \dfrac{1}{\mathbf{y}_{11}} & \dfrac{-\mathbf{y}_{12}}{\mathbf{y}_{11}} \\ \dfrac{\mathbf{y}_{21}}{\mathbf{y}_{11}} & \dfrac{\Delta_Y}{\mathbf{y}_{11}} \end{bmatrix}
\quad
\begin{bmatrix} \dfrac{\mathbf{B}}{\mathbf{D}} & \dfrac{\Delta_T}{\mathbf{D}} \\ -\dfrac{1}{\mathbf{D}} & \dfrac{\mathbf{C}}{\mathbf{D}} \end{bmatrix}
\quad
\begin{bmatrix} \mathbf{h}_{11} & \mathbf{h}_{12} \\ \mathbf{h}_{21} & \mathbf{h}_{22} \end{bmatrix}
$$

LEARNING EXTENSION

E16.7 Determine the Y parameters for a two-port if the Z parameters are

$$Z = \begin{bmatrix} 18 & 6 \\ 6 & 9 \end{bmatrix}$$

ANSWER: $y_{11} = \dfrac{1}{14} S;$

$y_{12} = y_{21} = -\dfrac{1}{21} S;$

$y_{22} = \dfrac{1}{7} S.$

16.6 Interconnection of Two-Ports

Interconnected two-port circuits are important because when designing complex systems it is generally much easier to design a number of simpler subsystems that can then be interconnected to form the complete system. If each subsystem is treated as a two-port network, the interconnection techniques described in this section provide some insight into the manner in which a total system may be analyzed and/or designed. Thus, we will now illustrate techniques for treating a network as a combination of subnetworks. We will, therefore, analyze a two-port network as an interconnection of simpler two-ports. Although two-ports can be interconnected in a variety of ways, we will treat only three types of connections: parallel, series, and cascade.

For the two-port interconnections to be valid, they must satisfy certain specific requirements that are outlined in the book *Network Analysis and Synthesis* by L. Weinberg (McGraw-Hill, 1962). The following examples will serve to illustrate the interconnection techniques.

In the parallel interconnection case, a two-port N is composed of two-ports N_a and N_b connected as shown in Fig. 16.7. *Provided that the terminal characteristics of the two networks N_a and N_b are not altered by the interconnection illustrated in the figure*, then the Y parameters for the total network are

$$\begin{bmatrix} y_{11} & y_{12} \\ y_{21} & y_{22} \end{bmatrix} = \begin{bmatrix} y_{11a} + y_{11b} & y_{12a} + y_{12b} \\ y_{21a} + y_{21b} & y_{22a} + y_{22b} \end{bmatrix} \qquad \textbf{16.13}$$

and hence to determine the Y parameters for the total network, we simply add the Y parameters of the two networks N_a and N_b.

Likewise, if the two-port N is composed of the series connection of N_a and N_b, as shown in Fig. 16.8, then *once again, as long as the terminal characteristics of the two networks N_a and N_b are not altered by the series interconnection*, the Z parameters for the total network are

$$\begin{bmatrix} z_{11} & z_{12} \\ z_{21} & z_{22} \end{bmatrix} = \begin{bmatrix} z_{11a} + z_{11b} & z_{12a} + z_{12b} \\ z_{21a} + z_{21b} & z_{22a} + z_{22b} \end{bmatrix} \qquad \textbf{16.14}$$

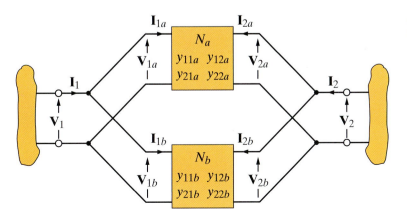

Figure 16.7
Parallel interconnection of two-ports.

Figure 16.8
Series interconnection of two-ports.

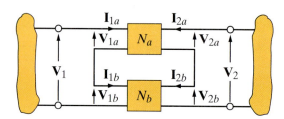

Therefore, the Z parameters for the total network are equal to the sum of the Z parameters for the networks N_a and N_b.

Finally, if a two-port N is composed of a cascade interconnection of N_a and N_b, as shown in Fig. 16.9, the equations for the total network are

$$\begin{bmatrix} \mathbf{V}_1 \\ \mathbf{I}_1 \end{bmatrix} = \begin{bmatrix} \mathbf{A}_a & \mathbf{B}_a \\ \mathbf{C}_a & \mathbf{D}_a \end{bmatrix} \begin{bmatrix} \mathbf{A}_b & \mathbf{B}_b \\ \mathbf{C}_b & \mathbf{D}_b \end{bmatrix} \begin{bmatrix} \mathbf{V}_2 \\ -\mathbf{I}_2 \end{bmatrix}$$

16.15

Figure 16.9
Cascade interconnection of networks.

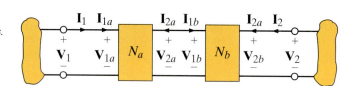

Hence, the transmission parameters for the total network are derived by matrix multiplication as indicated previously. The order of the matrix multiplication is important and is performed in the order in which the networks are interconnected.

The cascade interconnection is very useful. Many large systems can be conveniently modeled as the cascade interconnection of a number of stages. For example, the very weak signal picked up by a radio antenna is passed through a number of successive stages of amplification—each of which can be modeled as a two-port subnetwork. In addition, in contrast to the other interconnection schemes, no restrictions are placed on the parameters of N_a and N_b in obtaining the parameters of the two-port resulting from their interconnection.

Example 16.5

We wish to determine the Y parameters for the network shown in Fig. 16.10a by considering it to be a parallel combination of two networks as shown in Fig. 16.10b. The capacitive network will be referred to as N_a, and the resistive network will be referred to as N_b.

SOLUTION The Y parameters for N_a are

$$\mathbf{y}_{11a} = j\frac{1}{2}\,\text{S} \qquad \mathbf{y}_{12a} = -j\frac{1}{2}\,\text{S}$$

$$\mathbf{y}_{21a} = -j\frac{1}{2}\,\text{S} \qquad \mathbf{y}_{22a} = j\frac{1}{2}\,\text{S}$$

and the Y parameters for N_b are

$$\mathbf{y}_{11b} = \frac{3}{5}\,\text{S} \qquad \mathbf{y}_{12b} = -\frac{1}{5}\,\text{S}$$

$$\mathbf{y}_{21b} = -\frac{1}{5}\,\text{S} \qquad \mathbf{y}_{22b} = \frac{2}{5}\,\text{S}$$

Hence, the Y parameters for the network in Fig. 16.10 are

$$\mathbf{y}_{11} = \frac{3}{5} + j\frac{1}{2}\,\text{S} \qquad \mathbf{y}_{12} = -\left(\frac{1}{5} + j\frac{1}{2}\right)\text{S}$$

$$\mathbf{y}_{21} = -\left(\frac{1}{5} + j\frac{1}{2}\right)\text{S} \qquad \mathbf{y}_{22} = \frac{2}{5} + j\frac{1}{2}\,\text{S}$$

To gain an appreciation for the simplicity of this approach, you need only try to find the Y parameters for the network in Fig. 16.10a directly.

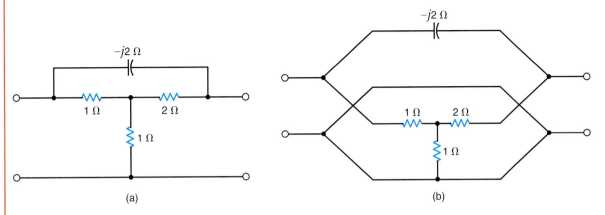

(a)   (b)

Figure 16.10 *Network composed of the parallel combination of two subnetworks.*

Example 16.6

Let us determine the Z parameters for the network shown in Fig. 16.10a. The circuit is redrawn in Fig. 16.11, illustrating a series interconnection. The upper network will be referred to as N_a, and the lower network as N_b.

SOLUTION   The Z parameters for N_a are

$$\mathbf{z}_{11a} = \frac{2 - 2j}{3 - 2j}\,\Omega \qquad \mathbf{z}_{12a} = \frac{2}{3 - 2j}\,\Omega$$

$$\mathbf{z}_{21a} = \frac{2}{3 - 2j}\,\Omega \qquad \mathbf{z}_{22a} = \frac{2 - 4j}{3 - 2j}\,\Omega$$

and the Z parameters for N_b are

$$\mathbf{z}_{11b} = \mathbf{z}_{12b} = \mathbf{z}_{21b} = \mathbf{z}_{22b} = 1\,\Omega$$

Hence the Z parameters for the total network are

$$\mathbf{z}_{11} = \frac{5 - 4j}{3 - 2j}\,\Omega \qquad \mathbf{z}_{12} = \frac{5 - 2j}{3 - 2j}\,\Omega$$

$$\mathbf{z}_{21} = \frac{5 - 2j}{3 - 2j}\,\Omega \qquad \mathbf{z}_{22} = \frac{5 - 6j}{3 - 2j}\,\Omega$$

We could easily check these results against those obtained in Example 16.5 by applying the conversion formulas in Table 16.1.

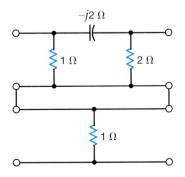

Figure 16.11
Network in Fig. 16.10a redrawn as a series interconnection of two networks.

Example 16.7

Let us derive the two-port parameters of the network in Fig. 16.12 by considering it to be a cascade connection of two networks as shown in Fig. 16.6.

SOLUTION The ABCD parameters for the identical T networks were calculated in Example 16.4 to be

$$\mathbf{A} = 1 + j\omega \qquad \mathbf{B} = 2 + j\omega$$

$$\mathbf{C} = j\omega \qquad \mathbf{D} = 1 + j\omega$$

Therefore, the transmission parameters for the total network are

$$\begin{bmatrix} \mathbf{A} & \mathbf{B} \\ \mathbf{C} & \mathbf{D} \end{bmatrix} = \begin{bmatrix} 1 + j\omega & 2 + j\omega \\ j\omega & 1 + j\omega \end{bmatrix} \begin{bmatrix} 1 + j\omega & 2 + j\omega \\ j\omega & 1 + j\omega \end{bmatrix}$$

Performing the matrix multiplication, we obtain

$$\begin{bmatrix} \mathbf{A} & \mathbf{B} \\ \mathbf{C} & \mathbf{D} \end{bmatrix} = \begin{bmatrix} 1 + 4j\omega - 2\omega^2 & 4 + 6j\omega - 2\omega^2 \\ 2j\omega - 2\omega^2 & 1 + 4j\omega - 2\omega^2 \end{bmatrix}$$

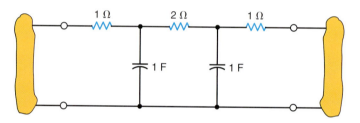

Figure 16.12 Circuit used in Example 16.7.

16.7 Application Examples

Application Example 16.8

Figure 16.13 is a per-phase model used in the analysis of three-phase high-voltage transmission lines. As a general rule in these systems, the voltage and current at the receiving end are known, and it is the conditions at the sending end that we wish to find. The transmission parameters perfectly fit this scenario. Thus, we will find the transmission parameters for a reasonable transmission line model, and, then, given the receiving end voltages, power, and power factor, we will find the receiving end current, sending end voltage and current, and the transmission efficiency. Finally, we will plot the efficiency versus the power factor.

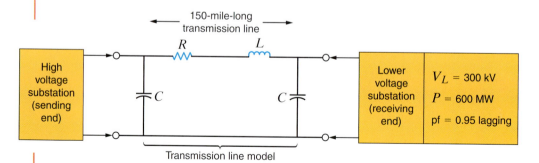

Figure 16.13 A π-circuit model for power transmission lines.

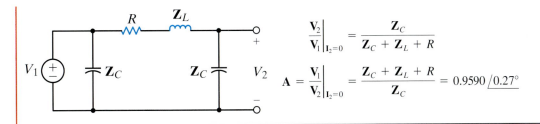

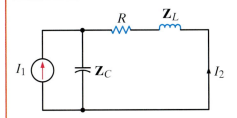

$$\left.\frac{\mathbf{V}_2}{\mathbf{V}_1}\right|_{\mathbf{I}_2=0} = \frac{\mathbf{Z}_C}{\mathbf{Z}_C + \mathbf{Z}_L + R}$$

$$\mathbf{A} = \left.\frac{\mathbf{V}_1}{\mathbf{V}_2}\right|_{\mathbf{I}_2=0} = \frac{\mathbf{Z}_C + \mathbf{Z}_L + R}{\mathbf{Z}_C} = 0.9590\,\underline{/0.27°}$$

$$\left.\frac{-\mathbf{I}_2}{\mathbf{V}_1}\right|_{\mathbf{V}_2=0} = \frac{1}{\mathbf{Z}_L + R}$$

$$\mathbf{B} = \left.\frac{\mathbf{V}_1}{-\mathbf{I}_2}\right|_{\mathbf{V}_2=0} = \mathbf{Z}_L + R = 100.00\,\underline{/84.84°}\ \Omega$$

$$\left.\frac{\mathbf{V}_2}{\mathbf{I}_1}\right|_{\mathbf{I}_2=0} = \frac{\mathbf{Z}_C^2}{2\mathbf{Z}_C + \mathbf{Z}_L + R}$$

$$\mathbf{C} = \left.\frac{\mathbf{V}_2}{\mathbf{I}_1}\right|_{\mathbf{I}_2=0} = \frac{2\mathbf{Z}_C + \mathbf{Z}_L + R}{\mathbf{Z}_C^2} = 975.10\,\underline{/90.13°}\ \mu\text{S}$$

$$\left.\frac{-\mathbf{I}_2}{\mathbf{I}_1}\right|_{\mathbf{V}_2=0} = \frac{\mathbf{Z}_C}{\mathbf{Z}_C + \mathbf{Z}_L + R}$$

$$\mathbf{D} = \left.\frac{\mathbf{I}_1}{-\mathbf{I}_2}\right|_{\mathbf{V}_2=0} = \frac{\mathbf{Z}_C + \mathbf{Z}_L + R}{\mathbf{Z}_C} = 0.950\,\underline{/0.27°}$$

Figure 16.14 *Equivalent circuits used to determine the transmission parameters.*

SOLUTION   Given a 150-mile-long transmission line, reasonable values for the π-circuit elements of the transmission line model are $C = 1.326\ \mu\text{F}$, $R = 9.0\ \Omega$, and $L = 264.18\ \text{mH}$. The transmission parameters can be easily found using the circuits in Fig. 16.14. At 60 Hz, the transmission parameters are

$$\mathbf{A} = 0.9590\,\underline{/0.27°} \qquad \mathbf{C} = 975.10\,\underline{/90.13°}\ \mu\text{S}$$
$$\mathbf{B} = 100.00\,\underline{/84.84°}\ \Omega \qquad \mathbf{D} = 0.9590\,\underline{/0.27°}$$

To use the transmission parameters, we must know the receiving end current, $\mathbf{I}_2$. Using standard three-phase circuit analysis outlined in Chapter 11, we find the line current to be

$$\mathbf{I}_2 = -\frac{600\,\underline{/\cos^{-1}(\text{pf})}}{\sqrt{3}(300)(\text{pf})} = -1.215\,\underline{/-18.19°}\ \text{kA}$$

where the line-to-neutral (i.e., phase) voltage at the receiving end, $\mathbf{V}_2$, is assumed to have zero phase. Now, we can use the transmission parameters to determine the sending end voltage and power. Since the line-to-neutral voltage at the receiving end is $300/\sqrt{3} = 173.21\ \text{kV}$, the results are

$$\mathbf{V}_1 = \mathbf{A}\mathbf{V}_2 - \mathbf{B}\mathbf{I}_2 = (0.9590\,\underline{/0.27°})(173.21\,\underline{/0°})$$

$$+ (100.00\,\underline{/84.84°})(1.215\,\underline{/-18.19°}) = 241.92\,\underline{/27.67°}\ \text{kV}$$

$$\mathbf{I}_1 = \mathbf{C}\mathbf{V}_2 - \mathbf{D}\mathbf{I}_2 = (975.10 \times 10^{-6}\,\underline{/90.13°})(173.21\,\underline{/0°})$$

$$+ (0.9590\,\underline{/0.27°})(1.215\,\underline{/-18.19°}) = 1.12\,\underline{/-9.71°}\ \text{kA}$$

At the sending end, the power factor and power are

$$\text{pf} = \cos(27.67 - (-9.71)) = \cos(37.38) = 0.80 \text{ lagging}$$

$$P_1 = 3V_1I_1(\text{pf}) = (3)(241.92)(1.12)(0.80) = 650.28 \text{ MW}$$

Finally, the transmission efficiency is

$$\eta = \frac{P_2}{P_1} = \frac{600}{650.28} = 92.3\%$$

This entire analysis can be easily programmed into an EXCEL spreadsheet or MATLAB. A plot of the transmission efficiency versus power factor at the receiving end is shown in Fig. 16.15. We see that as the power factor decreases, the transmission efficiency drops, which increases the cost of production for the power utility. This is precisely why utilities encourage industrial customers to operate as close to unity power factor as possible.

Figure 16.15

The results of an EXCEL simulation showing the effect of the receiving-end power factor on the transmission efficiency. Because the EXCEL simulation used more significant digits, slight differences exist between the values in the plot and those in the text.

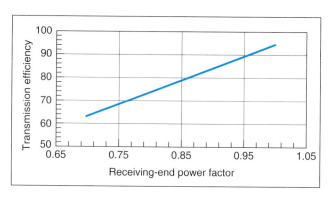

Application Example 16.9

We have available the noninverting op-amp circuit shown in Fig. 16.16 with the following parameters: $A = 20{,}000$, $R_i = 1 \text{ M}\Omega$, $R_o = 500 \ \Omega$, $R_1 = 1 \text{ k}\Omega$, and $R_2 = 49 \text{ k}\Omega$. To determine the possible applications for this network configuration, we will determine the effect of the load R_L on the gain and the gain error (a comparison of the actual gain with the ideal gain).

Figure 16.16

The classic noninverting gain configuration with load.

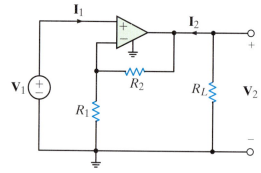

SOLUTION In Example 16.3, the hybrid parameters for the noninverting op-amp were found to be

$$\mathbf{h}_{11} = R_i + \frac{R_1 R_2}{R_1 + R_2} \qquad \mathbf{h}_{12} = \frac{R_1}{R_1 + R_2}$$

$$\mathbf{h}_{21} = -\left[\frac{AR_i}{R_o} + \frac{R_1}{R_1 + R_2} \right] \qquad \mathbf{h}_{22} = \frac{R_o + R_1 + R_2}{R_o(R_1 + R_2)}$$

If we solve the hybrid parameter two-port Eqs. (16.7) for $\mathbf{V}_2$, we obtain

$$\mathbf{V}_2 = \frac{-\mathbf{h}_{21}\mathbf{V}_1 + \mathbf{h}_{11}\mathbf{I}_2}{\mathbf{h}_{11}\mathbf{h}_{22} - \mathbf{h}_{12}\mathbf{h}_{21}}$$

Since the op-amp is connected to a load R_L, then

$$I_2 = -\frac{V_2}{R_L}$$

Combining these expressions, we obtain the equation for the gain,

$$\frac{V_2}{V_1} = \frac{-h_{21}}{h_{11}h_{22} - h_{12}h_{21} + \dfrac{h_{11}}{R_L}}$$

Using the parameter values, the equation becomes

$$\frac{V_2}{V_1} = \frac{4 \times 10^7}{8.02 \times 10^5 + \dfrac{10^6}{R_L}} = \frac{49.88}{1 + \dfrac{1.247}{R_L}} \qquad \textbf{16.16}$$

If the term involving R_L remains small compared to unity, then the gain will be largely independent of R_L.

It is convenient to view the gain of the amplifier with respect to its ideal value of

$$A_{\text{ideal}} = \left.\frac{V_2}{V_1}\right|_{\text{ideal op-amp}} = 1 + \frac{R_2}{R_1} = 1 + \frac{49}{1} = 50$$

From Eq. (16.16), if R_L is infinite, the gain is only 49.88. This deviation from the ideal performance is caused by nonideal values for the op-amp gain, input resistance, and output resistance. We define the gain error as

$$\text{Gain error} = \frac{A_{\text{actual}} - A_{\text{ideal}}}{A_{\text{ideal}}} = \frac{0.998}{1 + \dfrac{1.247}{R_L}} - 1 \qquad \textbf{16.17}$$

A plot of the gain and the gain error versus R_L is shown in Fig. 16.17. Note that as the load resistance decreases, the gain drops and the error increases—consistent with Eq. (16.17). In addition, as R_L increases, the gain asymptotically approaches the ideal value, never quite reaching it.

To identify specific uses for this amplifier, recognize that at a gain of 50, a 0.1-V input will produce a 5-V output. Three possible applications are as follows:

1. Low-budget audio preamplifier—Amplifies low voltages from tape heads and phonograph needle cartridges to levels suitable for power amplification to drive speakers.

2. Sensor amplifier—In many sensors—for example, temperature dependent resistors—changes in the electrical characteristic (resistance) can be much less than changes in the environmental parameter (temperature). The resulting output voltage changes are also small and usually require amplification.

3. Current sensing—Monitoring large currents can be done inexpensively by using low-value sense resistors and a voltmeter. By Ohm's law, the resulting voltage is $I R_{\text{sense}}$ where I is the current of interest. A voltmeter can be used to measure the voltage and, knowing the value of R_{sense}, the current can be determined. The power lost in the sense resistor is $I^2 R_{\text{sense}}$. Thus, low power loss implies low sense voltage values, which most inexpensive voltmeters cannot accurately measure. Our simple amplifier can boost the sense voltage to more reasonable levels.

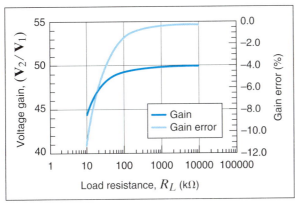

Figure 16.17

The gain and gain error of the noninverting gain configuration described in Example 16.9.

16.8 Design Example

── **Design Example 16.10**

For a particular application, we need an amplifier with a gain of 10,000 when connected to a 1-kΩ load. We have available to us some noninverting op-amps that could be used for this application.

SOLUTION The noninverting op-amp, together with some available components that, based on the results of the previous example (i.e., the ideal gain formula), should yield a gain of 10,000, are shown in Fig. 16.18. Using the hybrid parameter equations for the amplifier, as outlined in the previous example, yields

$$\mathbf{h}_{11} = 1.001 \text{ M}\Omega \qquad \mathbf{h}_{12} = 1.000 \times 10^{-4}$$
$$\mathbf{h}_{21} = -4.000 \times 10^{7} \qquad \mathbf{h}_{22} = 2.000 \text{ mS}$$

Figure 16.18
A single-stage amplifier that should have a gain of 10,000.

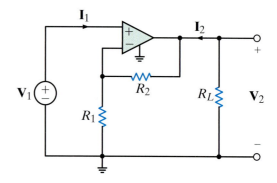

Op-amp Specifications	Components
$A = 20{,}000$	$R_1 = 1 \text{ k}\Omega$
$R_i = 1 \text{ M}\Omega$	$R_2 = 9.999 \text{ M}\Omega$
$R_o = 500 \ \Omega$	

Using Table 16.1, we can convert to the transmission parameters.

$$\mathbf{A} = \frac{-\Delta_H}{\mathbf{h}_{21}} = \frac{\mathbf{h}_{12}\mathbf{h}_{21} - \mathbf{h}_{11}\mathbf{h}_{22}}{\mathbf{h}_{21}} \qquad \mathbf{B} = \frac{-\mathbf{h}_{11}}{\mathbf{h}_{21}}$$

$$\mathbf{C} = \frac{-\mathbf{h}_{22}}{\mathbf{h}_{21}} \qquad \mathbf{D} = \frac{-1}{\mathbf{h}_{21}}$$

Based on the hybrid parameter values above, we find

$$\mathbf{A} = 1.501 \times 10^{-4} \qquad \mathbf{B} = 2.502 \times 10^{-2}$$
$$\mathbf{C} = 5.000 \times 10^{-11} \qquad \mathbf{D} = 2.500 \times 10^{-8}$$

The circuit is now modeled by the two-port equations

$$\begin{bmatrix} \mathbf{V}_1 \\ \mathbf{I}_1 \end{bmatrix} = \begin{bmatrix} \mathbf{A} & \mathbf{B} \\ \mathbf{C} & \mathbf{D} \end{bmatrix} \begin{bmatrix} \mathbf{V}_2 \\ -\mathbf{I}_2 \end{bmatrix}$$

Since $\mathbf{V}_2 = -\mathbf{I}_2 R_L$, we can write the equation for $\mathbf{V}_1$ as

$$\mathbf{V}_1 = \mathbf{A}\mathbf{V}_2 + \frac{\mathbf{B}}{R_L}\mathbf{V}_2$$

and the gain as

$$\frac{V_2}{V_1} = \frac{1}{A + \dfrac{B}{R_L}} = \frac{6667}{1 + \dfrac{166.7}{R_L}}$$

Although the ideal model predicts a gain of 10,000, the actual gain for infinite R_L is only 6667—a rather large discrepancy! A careful analysis of the parameters indicates two problems: (1) the gain of the op-amp is on the same order as the circuit gain, and (2) R_2 is actually larger than R_i. Recall that the ideal op-amp assumptions are that both A and R_i should approach infinity, or, in essence, A should be much larger than the overall gain and R_i should be the largest resistor in the circuit—neither condition exists. We will address these issues by cascading two op-amps. We will design each stage for a gain of 100 by selecting R_2 to be 99 kΩ, thus alleviating the two issues above. Since the stages are cascaded, the ideal overall gain should be $100 \times 100 = 10,000$.

The transmission parameters with the new values of R_2 are

$$A = 1.005 \times 10^{-2} \qquad B = 2.502 \times 10^{-2}$$

$$C = 5.025 \times 10^{-11} \qquad D = 2.500 \times 10^{-8}$$

Since the two stages are cascaded, the transmission parameter equations that describe the overall circuit are

$$\begin{bmatrix} V_1 \\ I_1 \end{bmatrix} = \begin{bmatrix} A_a & B_a \\ C_a & D_a \end{bmatrix} \begin{bmatrix} A_b & B_b \\ C_b & D_b \end{bmatrix} \begin{bmatrix} V_2 \\ -I_2 \end{bmatrix}$$

or

$$\begin{bmatrix} V_1 \\ I_1 \end{bmatrix} = \begin{bmatrix} A_a A_b - B_a C_b & A_a B_b - B_a D_b \\ C_a A_b - D_a C_b & C_a B_b - D_a D_b \end{bmatrix} \begin{bmatrix} V_2 \\ -I_2 \end{bmatrix}$$

where the subscripts a and b indicate the first and second op-amp stages. Since the stages are identical, we can just use A, B, C, and D. Still, $V_2 = -I_2 R_L$, and the gain is

$$\frac{V_2}{V_1} = \frac{1}{A^2 - BC + \dfrac{AB - BD}{R_L}}$$

Using our transmission parameter values, when R_L is infinite, the gain is 9900.75, an error of less than 1%. This is a significant improvement over the single-stage amplifier. Figure 16.19 shows the single- and two-stage gains versus load resistance, R_L. As R_L decreases, the superiority of the two-stage amp is even more marked.

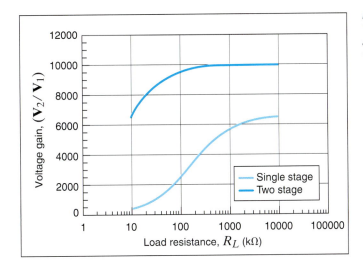

Figure 16.19
The voltage gain of the single- and two-stage op-amp circuits versus load resistance.

SUMMARY

- Four of the most common parameters used to describe a two-port network are the admittance, impedance, hybrid, and transmission parameters.

- If all the two-port parameters for a network exist, a set of conversion formulas can be used to relate one set of two-port parameters to another.

- When interconnecting two-ports, the Y parameters are added for a parallel connection, the Z parameters are added for a series connection, and the transmission parameters in matrix form are multiplied together for a cascade connection.

PROBLEMS

PSV **CS** both available on the web at: http://www.justask4u.com/irwin

SECTIONS 16.1 AND 16.2

16.1 Given the two networks in Fig. P16.1, find the Y parameters for the circuit in (a) and the Z parameters for the circuit in (b). **CS**

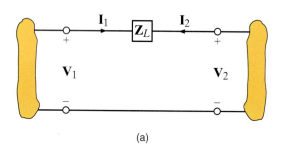

(a)

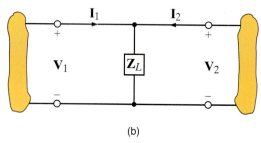

(b)

Figure P16.1

16.2 Find the Y parameters for the two-port network shown in Fig. P16.2.

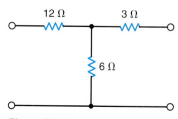

Figure P16.2

16.3 Find the Y parameters for the two-port network shown in Fig. P16.3. **PSV**

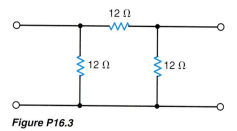

Figure P16.3

16.4 Determine the Y parameters for the network shown in Fig. P16.4.

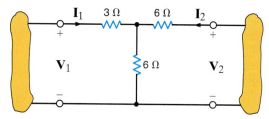

Figure P16.4

16.5 Find the Z parameters for the two-port network in Fig. P16.5.

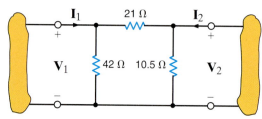

Figure P16.5

16.6 Determine the admittance parameters for the network shown in Fig. P16.6.

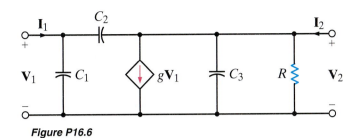

Figure P16.6

16.7 Find the Y parameters for the two-port network in Fig. P16.7. **CS**

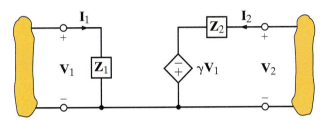

Figure P16.7

16.8 Find the Z parameters for the network in Fig. P16.7. **CS**

16.9 Find the Z parameters for the two-port network shown in Fig. P16.9 and determine the voltage gain of the entire circuit with a 4-kΩ load attached to the output. **PSV**

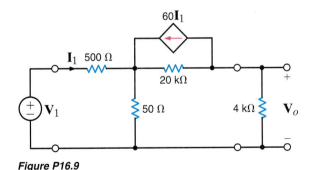

Figure P16.9

16.10 Find the Z parameters for the two-port network shown in Fig. P16.10.

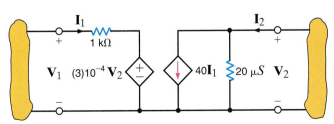

Figure P16.10

16.11 Find the voltage gain of the two-port network in Fig. P16.10 if a 12-kΩ load is connected to the output port. **CS**

16.12 Find the input impedance of the network in Fig. P16.10.

16.13 Find the Z parameters of the two-port network in Fig. P16.13.

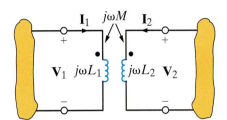

Figure P16.13

16.14 Determine the Z parameters for the two-port network in Fig. P16.14.

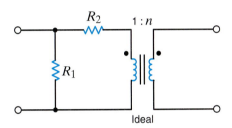

Figure P16.14

16.15 Draw the circuit diagram (with all passive elements in ohms) for a network that has the following Y parameters:

$$[Y] = \begin{bmatrix} \dfrac{3}{2} & -\dfrac{1}{2} \\ -\dfrac{1}{2} & \dfrac{5}{6} \end{bmatrix}$$

16.16 Draw the circuit diagram for a network that has the following Z parameters:

$$[Z] = \begin{bmatrix} 6 - j2 & 4 - j6 \\ 4 - j6 & 7 + j2 \end{bmatrix}$$

16.17 Show that the network in Fig. P16.17 does not have a set of Y parameters unless the source has an internal impedance.

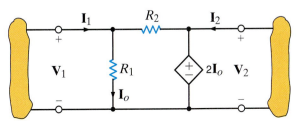

Figure P16.17

SECTION 16.3

16.18 Compute the hybrid parameters for the network in Fig. E16.1.

16.19 Find the hybrid parameters for the network in Fig. P16.3. **PSV**

16.20 Consider the network in Fig. P16.20. The two-port network is a hybrid model for a basic transistor. Determine the voltage gain of the entire network, $\mathbf{V}_2/\mathbf{V}_S$, if a source $\mathbf{V}_S$ with internal resistance R_1 is applied at the input to the two-port network and a load R_L is connected at the output port.

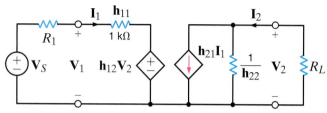

Figure P16.20

16.21 Determine the hybrid parameters for the network shown in Fig. P16.21.

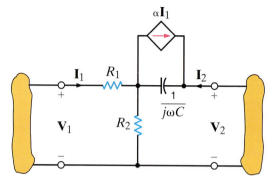

Figure P16.21

SECTION 16.4

16.22 Find the ABCD parameters for the networks in Fig. P16.1. **CS**

16.23 Find the transmission parameters for the network in Fig. P16.23.

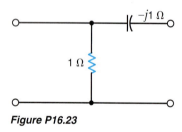

Figure P16.23

16.24 Find the transmission parameters for the network shown in Fig. P16.2. **PSV**

16.25 Find the ABCD parameters for the circuit in Fig. P16.7.

16.26 Determine the transmission parameters for the network in Fig. P16.26. **CS**

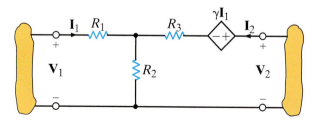

Figure P16.26

16.27 Find the transmission parameters for the circuit in Fig. P16.27.

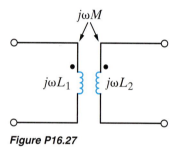

Figure P16.27

16.28 Given the network in Fig. P16.28, find the transmission parameters for the two-port network and then find $\mathbf{I}_o$ using the terminal conditions.

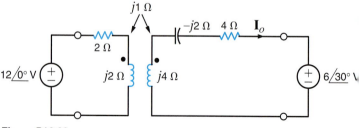

Figure P16.28

SECTION 16.5

16.29 Find the input admittance of the two-port in Fig. P16.29 in terms of the Y parameters and the load $\mathbf{Y}_L$.

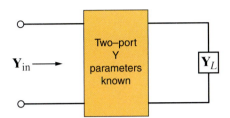

Figure P16.29

16.30 Find the voltage gain $\mathbf{V}_2/\mathbf{V}_1$ for the network in Fig. P16.30 using the Z parameters.

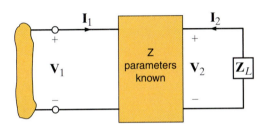

Figure P16.30

16.31 Following are the hybrid parameters for a network.

$$\begin{bmatrix} \mathbf{h}_{11} & \mathbf{h}_{12} \\ \mathbf{h}_{21} & \mathbf{h}_{22} \end{bmatrix} = \begin{bmatrix} \dfrac{11}{5} & \dfrac{2}{5} \\ -\dfrac{2}{5} & \dfrac{1}{5} \end{bmatrix}$$

Determine the Y parameters for the network.

16.32 If the Y parameters for a network are known to be

$$\begin{bmatrix} \mathbf{y}_{11} & \mathbf{y}_{12} \\ \mathbf{y}_{21} & \mathbf{y}_{22} \end{bmatrix} = \begin{bmatrix} \dfrac{5}{11} & -\dfrac{2}{11} \\ -\dfrac{2}{11} & \dfrac{3}{11} \end{bmatrix}$$

find the Z parameters. **CS**

16.33 Find the Z parameters in terms of the ABCD parameters.

16.34 Find the hybrid parameters in terms of the impedance parameters.

SECTION 16.6

16.35 Find the Y parameters for the network in Fig. P16.35.

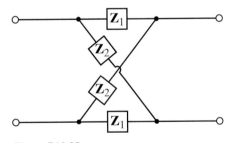

Figure P16.35

16.36 Determine the Y parameters for the network shown in Fig. P16.36.

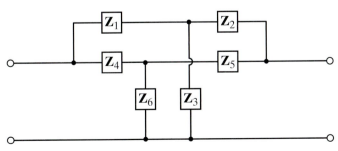

Figure P16.36

16.37 Find the Y parameters of the two-port network in Fig. P16.37. Find the input admittance of the network when the capacitor is connected to the output port. **CS**

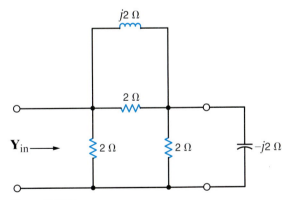

Figure P16.37

16.38 Find the Z parameters of the network in Fig. E16.3 by considering the circuit to be a series interconnection of two two-port networks as shown in Fig. P16.38.

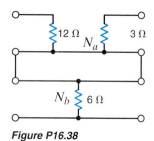

Figure P16.38

16.39 Find the transmission parameters of the network in Fig. E16.3 by considering the circuit to be a cascade interconnection of three two-port networks as shown in Fig. P16.39. **CS**

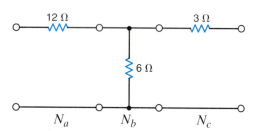

Figure P16.39

16.40 Find the ABCD parameters for the circuit in Fig. P16.40.

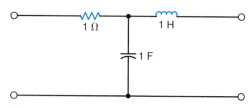

Figure P16.40

16.41 Find the Z parameters for the two-port network in Fig. P16.41 and then determine $\mathbf{I}_o$ for the specified terminal conditions.

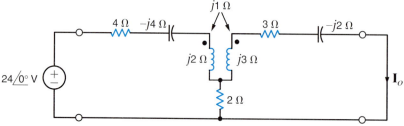

Figure P16.41

16.42 Determine the output voltage $\mathbf{V}_o$ in the network in Fig. P16.42 if the Z parameters for the two-port are

$$\mathbf{Z} = \begin{bmatrix} 3 & 2 \\ 2 & 3 \end{bmatrix} \quad \textbf{PSV}$$

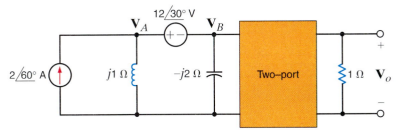

Figure P16.42

TYPICAL PROBLEMS FOUND ON THE FE EXAM

16FE-1 A two-port network is known to have the following parameters:

$$y_{11} = \frac{1}{14}\,\text{S} \qquad y_{12} = y_{21} = -\frac{1}{21}\,\text{S} \qquad y_{22} = \frac{1}{7}\,\text{S}$$

If a 2-A current source is connected to the input terminals as shown in Fig. 16PFE-1, find the voltage across this current source. **CS**

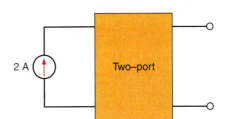

Figure 16PFE-1

16FE-2 Find the Thévenin equivalent circuit at the output terminals of the network in Fig. 16PFE-1.

Appendix
Complex Numbers

Complex numbers are typically represented in three forms: exponential, polar, or rectangular. In the exponential form a complex number $\mathbf{A}$ is written as

$$\mathbf{A} = ze^{j\theta} \qquad\qquad 1$$

The real quantity z is known as the amplitude or magnitude, the real quantity θ is called the *angle* as shown in Fig. 1, and j is the imaginary operator $j = \sqrt{-1}$. θ, which is the angle between the real axis and $\mathbf{A}$, may be expressed in either radians or degrees.

The polar form of a complex number $\mathbf{A}$, which is symbolically equivalent to the exponential form, is written as

$$\mathbf{A} = z\,\underline{/\theta} \qquad\qquad 2$$

and the rectangular representation of a complex number is written as

$$\mathbf{A} = x + jy \qquad\qquad 3$$

where x is the real part of $\mathbf{A}$ and y is the imaginary part of $\mathbf{A}$.

The connection between the various representations of $\mathbf{A}$ can be seen via Euler's identity, which is

$$e^{j\theta} = \cos\theta + j\sin\theta \qquad\qquad 4$$

Figure 2 illustrates that this function in rectangular form is a complex number with a unit amplitude.

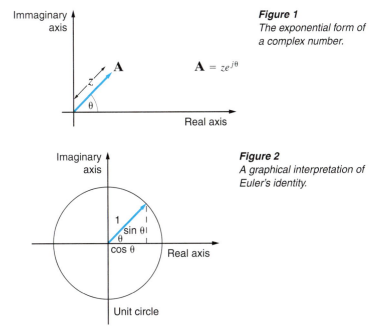

Figure 1
The exponential form of a complex number.

Figure 2
A graphical interpretation of Euler's identity.

Figure 3
The relationship between the exponential and rectangular representation of a complex number.

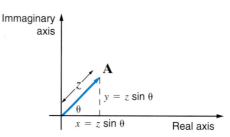

Using this identity, we can write the complex number **A** as

$$\mathbf{A} = ze^{j\theta} = z\cos\theta + jz\sin\theta \qquad \textbf{5}$$

which, as shown in Fig. 3, can be written as

$$\mathbf{A} = x + jy$$

Equating the real and imaginary parts of these two equations yields

$$x = z\cos\theta$$
$$y = z\sin\theta \qquad \textbf{6}$$

From these equations we obtain

$$x^2 + y^2 = z^2\cos^2\theta + z^2\sin^2\theta = z^2 \qquad \textbf{7}$$

Therefore,

$$z = \sqrt{x^2 + y^2} \qquad \textbf{8}$$

In addition,

$$\frac{z\sin\theta}{z\cos\theta} = \tan\theta = \frac{y}{x}$$

and hence

$$\theta = \tan^{-1}\frac{y}{x} \qquad \textbf{9}$$

The interrelationships among the three representations of a complex number are as follows.

Exponential	Polar	Rectangular
$ze^{j\theta}$	$z\underline{/\theta}$	$x + jy$
$\theta = \tan^{-1}y/x$	$\theta = \tan^{-1}y/x$	$x = z\cos\theta$
$z = \sqrt{x^2 + y^2}$	$z = \sqrt{x^2 + y^2}$	$y = z\sin\theta$

We will now show that the operations of addition, subtraction, multiplication, and division apply to complex numbers in the same manner that they apply to real numbers.

The *sum* of two complex numbers $\mathbf{A} = x_1 + jy_1$ and $\mathbf{B} = x_2 + jy_2$ is

$$\mathbf{A} + \mathbf{B} = x_1 + jy_1 + x_2 + jy_2$$
$$= (x_1 + x_2) + j(y_1 + y_2) \qquad \textbf{10}$$

That is, we simply add the individual real parts, and we add the individual imaginary parts to obtain the components of the resultant complex number.

Example 1

Suppose we wish to calculate the sum $\mathbf{A} + \mathbf{B}$ if $\mathbf{A} = 5\,\underline{/36.9°}$ and $\mathbf{B} = 5\,\underline{/53.1°}$.

SOLUTION We must first convert from polar to rectangular form.

$$\mathbf{A} = 5\,\underline{/36.9°} = 4 + j3$$
$$\mathbf{B} = 5\,\underline{/53.1°} = 3 + j4$$

Therefore,

$$\mathbf{A} + \mathbf{B} = 4 + j3 + 3 + j4 = 7 + j7$$
$$= 9.9\,\underline{/45°}$$

HINT

Addition and subtraction of complex numbers are most easily performed when the numbers are in rectangular form.

The *difference* of two complex numbers $\mathbf{A} = x_1 + jy_1$ and $\mathbf{B} = x_2 + jy_2$ is

$$\mathbf{A} - \mathbf{B} = (x_1 + jy_1) - (x_2 + jy_2)$$
$$= (x_1 - x_2) + j(y_1 - y_2)$$

11

That is, we simply subtract the individual real parts, and we subtract the individual imaginary parts to obtain the components of the resultant complex number.

Example 2

Let us calculate the difference $\mathbf{A} - \mathbf{B}$ if $\mathbf{A} = 5\,\underline{/36.9°}$ and $\mathbf{B} = 5\,\underline{/53.1°}$.

SOLUTION Converting both numbers from polar to rectangular form,

$$\mathbf{A} = 5\,\underline{/36.9°} = 4 + j3$$
$$\mathbf{B} = 5\,\underline{/53.1°} = 3 + j4$$

Then

$$\mathbf{A} - \mathbf{B} = (4 + j3) - (3 + j4) = 1 - j1 = \sqrt{2}\,\underline{/-45°}$$

The *product* of two complex numbers $\mathbf{A} = z_1\,\underline{/\theta_1} = x_1 + jy_1$ and $\mathbf{B} = z_2\,\underline{/\theta_2} = x_2 + jy_2$ is

$$\mathbf{AB} = (z_1 e^{j\theta_1})(z_2 e^{j(\theta_2)}) = z_1 z_2\,\underline{/\theta_1 + \theta_2}$$

12

Example 3

Given $\mathbf{A} = 5\,\underline{/36.9°}$ and $\mathbf{B} = 5\,\underline{/53.1°}$, we wish to calculate the product in both polar and rectangular forms.

SOLUTION

$$\mathbf{AB} = (5\,\underline{/36.9°})(5\,\underline{/53.1°}) = 25\,\underline{/90°}$$
$$= (4 + j3)(3 + j4)$$
$$= 12 + j16 + j9 + j^2 12$$
$$= 25j$$
$$= 25\,\underline{/90°}$$

HINT

Multiplication and division of complex numbers are most easily performed when the numbers are in exponential or polar form.

Example 4

Given $\mathbf{A} = 2 + j2$ and $\mathbf{B} = 3 + j4$, we wish to calculate the product $\mathbf{AB}$.

SOLUTION

$$\mathbf{A} = 2 + j2 = 2.828\,\underline{/45°}$$
$$\mathbf{B} = 3 + j4 = 5\,\underline{/53.1°}$$

and

$$\mathbf{AB} = (2.828\,\underline{/45°})(5\,\underline{/53.1°}) = 14.14\,\underline{/98.1°}$$

The *quotient* of two complex numbers $\mathbf{A} = z_1\underline{/\theta_1} = x_1 + jy_1$ and $\mathbf{B} = z_2\underline{/\theta_2} = x_2 + jy_2$ is

$$\frac{\mathbf{A}}{\mathbf{B}} = \frac{z_1 e^{j\theta_1}}{z_2 e^{j\theta_2}} = \frac{z_1}{z_2} e^{j(\theta_1 - \theta_2)} = \frac{z_1}{z_2}\underline{/\theta_1 - \theta_2} \qquad \textbf{13}$$

Example 5

Given $\mathbf{A} = 10\underline{/30°}$ and $\mathbf{B} = 5\underline{/53.1°}$, we wish to determine the quotient $\mathbf{A/B}$ in both polar and rectangular forms.

SOLUTION

$$\frac{\mathbf{A}}{\mathbf{B}} = \frac{10\underline{/30°}}{5\underline{/53.1°}}$$
$$= 2\underline{/-23.1°}$$
$$= 1.84 - j0.79$$

Example 6

Given $\mathbf{A} = 3 + j4$ and $\mathbf{B} = 1 + j2$, we wish to calculate the quotient $\mathbf{A/B}$.

SOLUTION

$$\mathbf{A} = 3 + j4 = 5\underline{/53.1°}$$
$$\mathbf{B} = 1 + j2 = 2.236\underline{/63°}$$

and

$$\frac{\mathbf{A}}{\mathbf{B}} = \frac{5\underline{/53.1°}}{2.236\underline{/63°}} = 2.236\underline{/-9.9°}$$

Example 7

If $\mathbf{A} = 3 + j4$, let us compute $1/\mathbf{A}$.

SOLUTION

$$\mathbf{A} = 3 + j4 = 5\underline{/53.1°}$$

and

$$\frac{1}{\mathbf{A}} = \frac{1\underline{/0°}}{5\underline{/53.1°}} = 0.2\underline{/-53.1°}$$

or

$$\frac{1}{\mathbf{A}} = \frac{1}{3 + j4} = \frac{3 - j4}{(3 + j4)(3 - j4)}$$
$$= \frac{3 - j4}{25} = 0.12 - j0.16$$

Index

Photo Credits

Chapter 2

Fig. 2.1b: Courtesy of Mark Nelms and Jo Ann Loden.

Chapter 4

Fig. 4.1a & Fig. 4.1b: Courtesy of Mark Nelms and Jo Ann Loden.

Chapter 6

Fig. 6.2, Fig. 6.3 & Fig. 6.7: Courtesy of Mark Nelms and Jo Ann Loden.
Fig. 6.24: Tom Way/Ginger Conly/IBM.

Chapter 9

Fig. 9.31: Courtesy of Fluke Corporation.

Chapter 12

Fig. 12.30: Special Collections Division, University of Washington Libraries, UW21413.
Photo by Farguharson.

Chapter 14

Fig. 14.15: Special Collections Division, University of Washington Libraries, UW21422.
Photo by Farguharson.

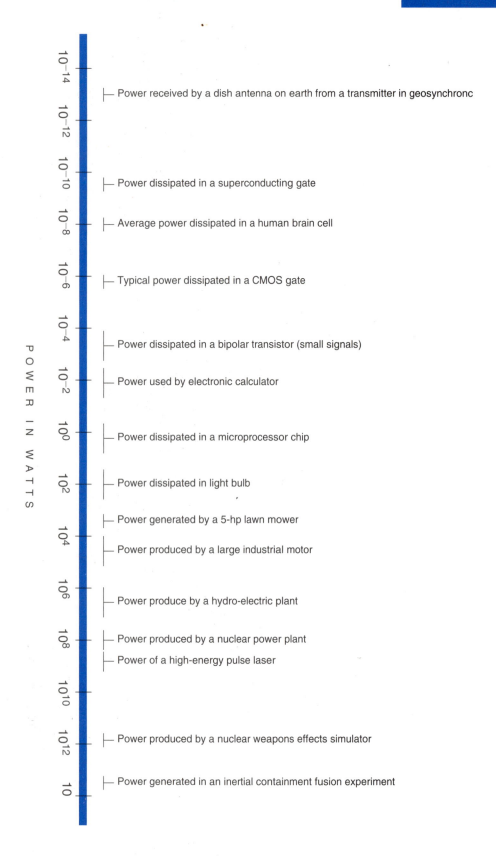

POWER IN WATTS

10^{-14}
— Power received by a dish antenna on earth from a transmitter in geosynchronc

10^{-12}

10^{-10}
— Power dissipated in a superconducting gate

10^{-8}
— Average power dissipated in a human brain cell

10^{-6}
— Typical power dissipated in a CMOS gate

10^{-4}
— Power dissipated in a bipolar transistor (small signals)

10^{-2}
— Power used by electronic calculator

10^{0}
— Power dissipated in a microprocessor chip

10^{2}
— Power dissipated in light bulb

— Power generated by a 5-hp lawn mower

10^{4}
— Power produced by a large industrial motor

10^{6}
— Power produce by a hydro-electric plant

10^{8}
— Power produced by a nuclear power plant
— Power of a high-energy pulse laser

10^{10}

10^{12}
— Power produced by a nuclear weapons effects simulator

10
— Power generated in an inertial containment fusion experiment